Remote Sensing of Regional Soil Moisture

Common and key statements from the editors are

i) More diverse information about local conditions should be utilized to increase the accuracy of soil moisture retrieved by any method.

ii) Thermal infrared-based methods have a better spatial resolution than passive microwave observations, which makes them better suited for regional-scale analyses.

iii) With the current state-of-the-art in remote sensing of soil moisture, the data, methods and software are available to conduct regional-scale soil moisture monitoring. The most limiting factors are the lack of regional knowledge (or its use) and technical resources.

Remote Sensing of Regional Soil Moisture

Editors

Marion Pause
Thomas Wöhling
Karsten Schulz
Thomas Jagdhuber
Martin Schrön

MDPI • Basel • Beijing • Wuhan • Barcelona • Belgrade • Manchester • Tokyo • Cluj • Tianjin

Editors

Marion Pause
Technical University of Dresden
Germany

Thomas Wöhling
Technical University of Dresden
Germany

Karsten Schulz
University of Natural Resources and Life Sciences
Austria

Thomas Jagdhuber
Microwaves and Radar Institute
Germany

Martin Schrön
Monitoring and Exploration Technologies
Germany

Editorial Office
MDPI
St. Alban-Anlage 66
4052 Basel, Switzerland

This is a reprint of articles from the Special Issue published online in the open access journal *Remote Sensing* (ISSN 2072-4292) (available at: https://www.mdpi.com/journal/remotesensing/special_issues/regional_soil_rs).

For citation purposes, cite each article independently as indicated on the article page online and as indicated below:

LastName, A.A.; LastName, B.B.; LastName, C.C. Article Title. *Journal Name* **Year**, *Volume Number*, Page Range.

ISBN 978-3-0365-2956-1 (Hbk)
ISBN 978-3-0365-2957-8 (PDF)

Cover image courtesy of PD Dr. Thomas Wöhling

Contents

About the Editors

Marion Pause (PhD) was born in Grimma (Saxony), Germany, in 1979. She received her degree in Geodesy from the Technical University Dresden in 2006. From 2006–2013, she worked with the Helmholtz Centre for Environmental Research in Germany and completed her PhD at the Ludwig Maximilian University of Munich in 2011. In 2009/2010 she joined Airborne Research South Australia and Flinders University in Adelaide, Australia. Dr. Pause is an expert in multi-source remote sensing, and her key research interests are the development of observation-based frameworks using multi-source remote sensing data to support the climate adaption of landscapes and urban areas. Dr. Pause has a strong background in microwave remote sensing of soil moisture and has experience in airborne remote sensing campaign management, including ground truth data sampling. Dr. Pause serves as lecturer at various academic institutions and provides wide didactic experience in remote sensing, photogrammetry, and sensor technology.

Thomas Wöhling (PD Dr. habil.) leads the Stochastic Modelling of Hydrosystems Group and is the Chair of Hydrology at the Technische Universität Dresden (TUD), Germany. He has a background in integrated hydrosystem modelling with land–surface models, soil–plant–atmosphere models, and coupled surface–subsurface models. After receiving a PhD in hydrology (2005), he joined Lincoln Agritech in Hamilton, New Zealand, and became senior scientist at the Spydia Lysimeter Facility. In 2010, he moved to the University of Tübingen to lead a group on Model-based Optimisation of Monitoring and was the PI in the Research Training Group "Integrated Hydrosystem Modelling" (2012–2021). Dr. Wöhling re-joined TUD in 2015 and habilitated in the field of stochastic hydrology (2021). His research interests are centered around best-practice modelling techniques for hydrosystems and include inverse modelling, uncertainty quantification, data-worth analysis, model diagnosis, surrogate modelling, and Bayesian methods.

Karsten Schulz has been a University Professor of Hydrology and Integrated Water Management at the University of Natural Resources and Life Sciences, Vienna, Austria, since 2013. He studied Geoecology at the University of Bayreuth, Germany, and received his PhD in Soil Physics in 1997. His interest in remote sensing data started during his Postdoc with Prof. Keit Beven at Lancaster University when developing and applying uncertainty estimation techniques to large scale evapotranspiration estimates from thermal remote sensing. After 3 years as an Assistant Professor at the TU Braunschweig, he built up a RS hydrology research group at the Helmholtz Centre for Environmental Research before accepting a Professorship in Physical Geography and Environmental Monitoring at the LMU Munich. The improvement of large-scale predictions of hydrological processes using remote sensing information from all available platforms is still one of his main research foci.

Thomas Jagdhuber (PhD) received his degree in physical geography, physics, remote sensing, and geoinformatics from Ludwig Maximilian University of Munich and the Technical University of Munich (geoinformatics), Germany, in 2006, and his PhD from the Faculty of Science at the University of Potsdam, Germany, in 2012. His main research interests include physics-based multi-sensor data integration with a focus on active and passive microwave interaction theory and on polarimetric SAR techniques for hydrological, cryospheric, plant ecological, and agricultural parameter modeling and estimation. Since 2007 he has been affiliated with the Microwaves and Radar Institute (HR) of the German Aerospace Center (DLR). In 2014, he was honored with the DLR Science Award for his research on polarimetric decomposition techniques. From 2014 through 2019, he was a yearly visiting scientist at the Massachusetts Institute of Technology (MIT), Cambridge, USA. Together with Prof. Entekhabi (MIT), he was awarded with the MIT-MISTI grant for global water cycle and environmental monitoring using active and passive satellite-based microwave instruments. In addition, Dr. Jagdhuber serves as an Associated Lecturer for the University of Jena and the University of Augsburg.

Martin Schrön (PhD) is an expert in regional soil moisture observations with stationary as well as mobile cosmic-ray neutron sensing. He studied Physics at the University of Heidelberg, joined the UFZ Leipzig in 2012, received his PhD from the University of Potsdam, worked as a Research Fellow at the Dep. of Civil Engineering, University of Bristol (2016), and is currently a PostDoc at the UFZ Leipzig. The main research interests focus on fundamental physics of cosmic-ray neutron measurements as well as on new mobile platforms for regional-scale soil moisture monitoring on the ground, on rails, and in the air. Dr. Schrön currently leads a project funded by the BMBF Joint German-Israeli Water Technology Research Program, co-leads two modules of the DFG Research Unit "CosmicSense" Phase I, and leads one module in Phase II. He is also a member of the scientific steering committee of the Terrestrial Environmental Observatories in Germany (TERENO).

remote sensing

MDPI

Article

Improving Estimation of Soil Moisture Content Using a Modified Soil Thermal Inertia Model

Zhenhua Liu [1], Li Zhao [1], Yiping Peng [1], Guangxing Wang [1,2] and Yueming Hu [1,3,4,5,6,*]

1 College of Natural Resources and Environment, South China Agricultural University, Guangzhou 510642, China; zhenhua@scau.edu.cn (Z.L.); zhaoli@stu.scau.edu.cn (L.Z.); pyppyp@stu.scau.edu.cn (Y.P.); gxwang@siu.edu (G.W.)
2 Department of Geography and Environmental Resources, College of Liberal Arts, Southern Illinois University Carbondale (SIUC), Carbondale, IL 62901, USA
3 Guangdong Provincial Key Laboratory of Land Use and Consolidation, South China Agricultural University, Guangzhou 510642, China
4 Guangdong Province Engineering Research Center for Land Information Technology, South China Agricultural University, Guangzhou 510642, China
5 Key Laboratory of Construction Land Transformation, Ministry of Land and Resources, South China Agricultural University, Guangzhou 510642, China
6 College of Agriculture and Animal Husbandry, Qinghai University, Xining 810016, China
* Correspondence: ymhu@scau.edu.cn; Tel.: +86-186-8888-2020

Received: 18 April 2020; Accepted: 25 May 2020; Published: 27 May 2020

Abstract: There has been substantial research for estimating and mapping soil moisture content (SMC) of large areas using remotely sensed images by developing models of soil thermal inertia (STI). However, it is still a great challenge to accurately estimate SMC because of the impact of vegetation canopies and vegetation-induced shadows in mixed pixels on the estimates. In this study, a new method was developed to increase the estimation accuracy of SMC for an irrigated area located in YingKe of Heihe, China, using ASTER data. In the method, an original model of estimating bare STI was modified by decomposing a mixed pixel into three components, bare soil, vegetated soil, and shaded soil, as well as extracting their fractions using a spectral unmixing analysis and then deriving their fluxes. Moreover, the 90 m spatial resolution thermal images were scaled down to the 15 m spatial resolution by data fusion of a discrete wavelet transform (DWT) and re-sampling using the nearest neighbor method (NNM). The modified model was compared with the original model based on the mean absolute error (MAE) and relative root mean square error (RRMSE) between the SMC estimates and observations from 30 validation soil samples. The results indicated that compared to the original model based on the parallel dual layer, the modified STI model based on the serial dual layer statistically significantly decreased the MAE and RRMSE of the SMC estimates by 63.0–63.2% and 63.0–63.5%, respectively. The 15 m spatial resolution thermal bands obtained by the DWT data fusion provided more detailed information of SMC but did not significantly improve its estimation accuracy than the 15 m spatial resolution thermal bands by re-sampling using NNM. This implied that the novel method offered insights on how to increase the accuracy of retrieving SMC estimates in vegetated areas.

Keywords: ASTER imagery; soil moisture content; thermal inertia model; serial dual-source model; surface component temperature; shadow impact

1. Introduction

Variation of soil moisture content (SMC) affects energy balance of land surfaces, soil erosion, and vegetation growth [1]. Thus, it is very important to retrieve the information of SMC. Traditionally, however, the information is often obtained using soil samples from sampled locations, which is time-consuming and costly. The traditional method also ignores the spatial distributions and patterns of SMC. On the other hand, remotely sensed images with spatiotemporal coverages of a study area provide the possibility for rapidly and cost-efficiently mapping SMC and monitoring its dynamics at a regional, national, and global scale [2,3]. For this purpose, substantial research has been conducted, and the used remote sensing data include optical and thermal infrared images and microwave observations to estimate spatial distributions of SMC over large areas [4–6].

Because the dielectric property of soils is closely related to SMC, various methods have been developed for estimating SMC using microwave images [7,8]. Microwave remote sensing is suitable for estimating large-scale SMC due to its capability of penetrating surface soil and being less disturbed by clouds and sparse vegetation cover [9,10]. However, microwave images are easily affected by ground roughness and dense vegetation canopy [11,12].

Moreover, optical and thermal image-based methods for mapping SMC became popular because the methods can provide maps of SMC with a wide range of spatial resolutions and a long-term history. Previous studies have shown that SMC has a linear or non-linear relationship with spectral reflectance of surfaces [13,14]. Thus, spectral-vegetation-index-based models are developed to estimate SMC. Gao et al. [13] used red-near infrared reflectance of soil to estimate SMC over a vegetated area. However, surface reflectance has a weak relationship with SMC when there is a dense vegetation cover. The methods are also limited by weather conditions such as clouds and night time.

There are also the methods that combine optical and thermal infrared (TIR) images to estimate SMC [15–18]. Thermal inertia is affected by the characteristics of surface layers, including thermal conductivity and capacity of heat storage. The heat energy from the surface layers is propagated to and stored in sub-surfaces during days and then returned to the surfaces during nights. Thus, the capacity of the sub-surfaces is characterized by thermal inertia. The surface heat capacity and thermal conductivity vary depending on SMC. The TIR methods are developed based on the sensitivity of land surface temperature (LST) to surface SMC [2]. These kinds of methods consist of (1) estimating LST using TIR images; (2) deriving soil latent and sensible fluxes using LST; (3) estimating soil thermal inertia (STI) using soil latent and sensible fluxes; and (4) developing an empirical model that accounts for the relationship of SMC with STI.

Based on soil thermal characteristics and infrared images, Watson et al. [19] proposed the first thermal inertia model. Based on a generalized theory in the study of Price [20], thermal inertia can be calculated using remote sensing data at a regional and global scale. The thermal properties of soils can also be estimated using the model proposed by Pratt [21] and the results are dependent on the changes in composition, porosity, and moisture content of the soils. However, this model is determined by many physical variables, and among the variables, the estimation of wind speed and humidity using remote sensing data is limited. Therefore, the model of Pratt [21] is hard to use. Moreover, by making an assumption that the coefficients of variation in energy fluxes into the atmosphere and the ground are constant, Price [22] proposed an Apparent Thermal Inertia model that can be only applied to bare soil dominant areas [23]. Based on the phase angle of diurnal temperature change, in addition, Xue and Cracknell [24] advanced the estimation of thermal inertia by presenting a real-time model.

The air–soil interface energy exchange implies a method of improving the estimation of surface STI using satellite data [25]. The estimation accuracy of thermal inertia and SMC can be further increased using the model of Zhang et al. [26] in which surface sensibility and latent fluxes with differential thermal inertia are taken into account. However, the model neglects the impact of vegetation and soil. In the model of Cai et al. [6], the maximum or minimum temperature is used to estimate thermal inertia, which makes it possible that the estimation is not dependent on the time for a satellite overpassing the study area. A relatively new thermal inertia model proposed by Liu et al. [27] introduced surface sensibility, latent fluxes and a parallel two-layer model into a thermal conductivity equation. Minacapilli et al. [28] conducted a laboratory experiment to evaluate the thermal inertia model for estimating surface SMC and found that this model offered great potential for mapping SMC and monitoring its dynamics using visible–near-infrared and TIR data. Although a great deal of research has been conducted and several thermal inertial models have been available for retrieving SMC [15–18], the models are only applicable in bare soil areas and lack the ability of accurately mapping SMC in the vegetated areas due to mixed pixels and vegetation-induced shadows.

The TIR-based methods have advantages of various spatial resolution satellite images available and relate SMC to thermal inertia. However, the weak relationship of TIR images with SMC in the densely vegetated areas impedes the applications of the TIR methods. So far, there have been no reports that deal with how to mitigate the impact of vegetation-induced shadows on estimation of SMC and improving the mapping of SMC in vegetated areas.

This study aimed to develop a novel method to generate the spatial distribution of surface SMC in both bare soil and vegetated areas by modifying the STI model presented by Liu et al. [27] and reducing the effect of vegetation-induced shadows on estimation of SMC. The proposed method was developed by decomposing mixed pixels into three components, bare soil, vegetated soil, and shaded soil, and taking into account the effects of vegetation and corresponding shadows on the estimation of SMC. The improved model was verified by comparing the measured and estimated values of SMC obtained using Advanced Spaceborne Thermal Emission and Reflection Radiometer (ASTER) data in a selected study area.

2. Materials and Methods

In this study, we proposed a modified STI model to improve the estimation of SMC. The flow chart of the method is shown in Figure 1. The methodological framework consists of data collection, data pre-processing, derivation of models, modification of the STI model based on decomposition of mixed pixels and derivation of components at a finer spatial resolution, estimation of soil sensible flux and soil latent flux, and estimation and validation of SMC using the modified model. In addition, a comparison of the modified model with the original model was carried out.

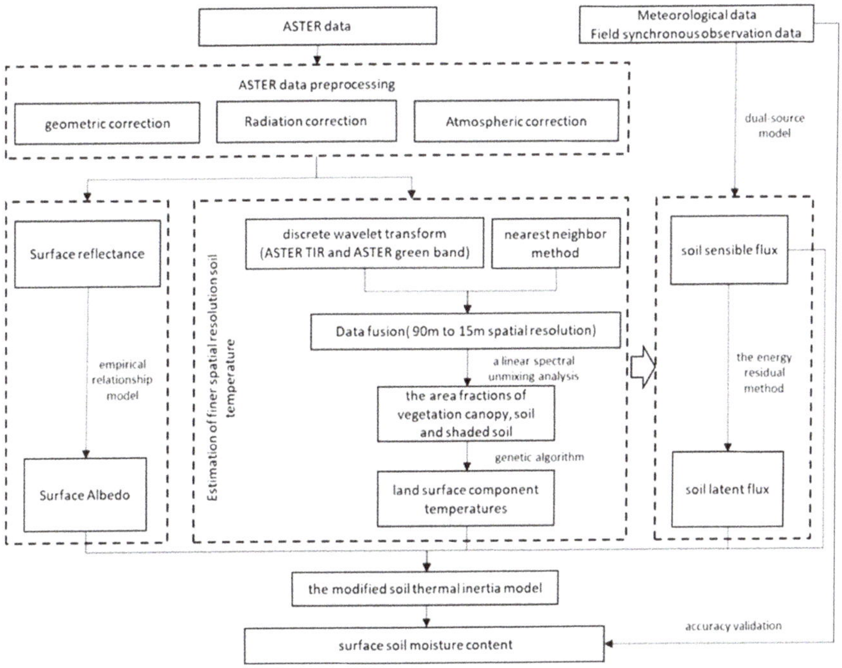

Figure 1. The flow chart of obtaining soil moisture content (SMC) from the modified soil thermal inertia (STI) model.

2.1. Study Site and Data

This study was conducted in an irrigated area located in Yingke of Heihe River Basin, Northwestern China (Figure 2a). This study area falls in the temperate zone characterized by a semi-arid climate, where summers are hot, and winters are cold and dry. The high elevation and aridity lead to great variation of diurnal temperature with the mean annual temperature varying from 7.0 to 10.0 °C and the maximum temperature of 40 °C. The annual precipitation is 140 mm, concentrating in May to October. Due to dry winters, snowing rarely occurs. This area was dominated by wheat, corn, and other crops (Figure 2b).

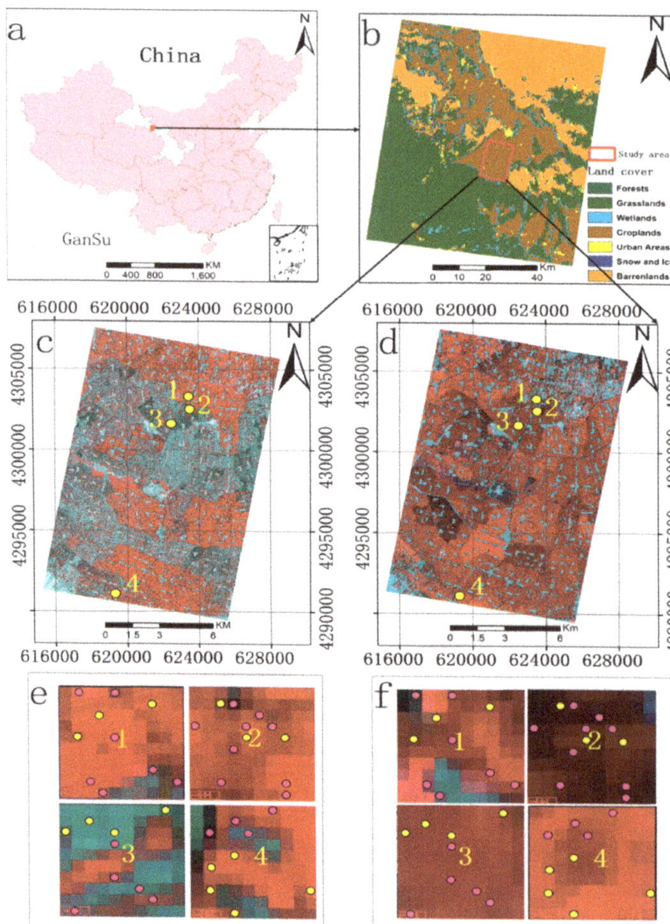

Figure 2. The study area and spatial distribution of sample plots: (**a**) The location of the study area in China; (**b**) a land cover map generated using 500 m spatial resolution MODIS MCD12 product with the study area highlighted. (**c**,**d**) The study area shown using a standard false color composite image from ASTER data dated on June 4th and July 10th of 2008, respectively, and the locations of 4 180 m × 180 m sample blocks. (**e**,**f**) Spatial distributions of the 40 15 m × 15 m sample plots within the sample blocks shown in ASTER images dated on June 4th and July 10th of 2008, respectively (50 purple plots used for developing and selecting the models and 30 yellow plots for validating the accuracy of the models).

In Figure 2c,d, the study area is shown using the ASTER images acquired on June 4th and July 10th of 2008, respectively. In this experiment, a total of 40 sample plots (Figure 2e,f) were selected from 4 sampled blocks of 180 m × 180 m (marked with 1, 2, 3, and 4 in Figure 1c,d). Soil samples were collected on June 4th and July 10th of 2008, respectively, which led to a total of 80 soil samples. The samples fell in in three types of land cover, including wheat, corn, and wheat/corn fields. The sample plots of 15 m × 15 m spatial resolution matched the pixels of the ASTER fusion data mentioned next in size. The locations of the plots were obtained using a global positioning system receiver. Within each of the sample plots, SMC was measured at 5 points in the field with one point at the plot center and other 4 at the middle locations of two diagonal lines from the plot center to four corners and the mean value of the SMC measurements from the 5 points was used as the SMC value of the sample plot. Based on previous experiments [29], a 20 cm depth of soil layer led to the stable measurements of SMC. At each point, SMC at a depth of 20 cm was thus measured in the field with time-domain reflectometry. The time-domain reflectometry sensor for collecting the measurements of SMC was calibrated according to the results of the study by Wu et al. [1]. There were 50 soil samples (purple plots in Figure 2e,f) chosen based on land cover types and used to compare and select the models that were used to account for the relationship between thermal inertia and SMC using a leave-one cross validation. The left 30 soil samples were used for validation.

Moreover, within each of the sample plots, an Analytical Spectral Device with a spectroradiometer of 15° and 25° field of views was used to measure hyperspectral reflectance values of the wavelength range from 300 nm to 2500 nm for different cover types. A thermo-radiometer working in the interval of 800 nm to 1400 nm and with a field of view of larger than 5° was used to measure the temperatures of various cover types. At the same time, the data for atmospheric, meteorological, and micrometeorological variables including wind speed, air temperature, and net radiation were collected in the field.

In this study, ASTER images dated on June 4th of 2008 and July 10th of 2008 were acquired with a total of 14 bands covering visible through TIR wavelength regions. Five 90 m × 90 m TIR bands and three 15 m × 15 m visible bands were used to estimate land surface temperature. The radiance calibration of each ASTER-TIR band q was carried out based on the equation $L_q(i,j) = Gain_q \times (DN_q(i,j) - 1)$, where i and j were the ith row and the jth column of the location in the image, $Gain_{10} = 0.006882$, $Gain_{11} = 0.006780$, $Gain_{12} = 0.006590$, $Gain_{13} = 0.005693$, and $Gain_{14} = 0.0055$, and $DN_q(i,j)$ is the digital number value of the pixel (i,j) [30]. The atmospheric correction of the ASTER images including VNIR and thermal bands was conducted using MODTRAN4.0. The corrected ASTER-TIR bands at the spatial resolution of 90 m × 90 m were re-sampled using a nearest neighbor method (NNM) to create ASTER-TIR images at a spatial resolution of 15 m × 15 m, which kept the original ASTER-TIR values at the 90 m resolution.

2.2. Methods

2.2.1. Multiscale Soil Thermal Inertia Model

As a physical variable, STI characterizes the impedance to the variation of temperature. If a heat transfer is given, usually, the higher the STI values, the smaller the changes of temperature and vice versa. The STI is calculated as follows:

$$P = \sqrt{kc\rho} \quad (1)$$

where P, k, c, and ρ are the STI ($J \cdot m^{-2} \cdot s^{-1/2} \cdot K^{-1}$), the thermal conductivity ($J \cdot m^{-1} \cdot s^{-1} \cdot K^{-1}$), the specific heat of the material ($J \cdot kg^{-1} \cdot K^{-1}$), and the density ($kg \cdot m^{-3}$), respectively. Though STI in Equation (1) can be obtained from in situ measurements, it is difficult to be acquired for a large region. Remote sensing can provide the potential of obtaining large-scale STI values based on a conductive heat transfer equation.

The existing thermal models are based on the assumption of one-dimensional periodic heating of a uniform half-space. A diffusion equation (Equation (2)) with boundary conditions is utilized to estimate the soil temperature [19]:

$$\begin{cases} \frac{\partial T_{sur}(x,t)}{\partial t} = D_H \frac{\partial^2 T_{s1}(x,t)}{\partial x^2} \\ -k\frac{\partial T_{sur}(x,t)}{\partial x}\Big|_{x=0} = (1-A)S_O C_\tau \cos\theta - [A_c + BT_s(0,t)] \end{cases} \tag{2}$$

where $T_{sur}(x,t)$ and $T_{sur}(0,t)$ are the soil temperatures at the depth x and 0 below the surface and time t. The $D_H(=\frac{k}{c\rho})$ represents the thermal diffusivity of the half-space. The A, S_O, C_τ, and θ indicate the surface albedo, the solar constant, the atmospherically optical transmission, and the solar zenith, respectively. The A_c and B are dependent on the conditions of the atmosphere and surface with a linear relationship with the surface soil temperature: $A_c + BT_{sur}(0,t)$. In Equation (2), an assumption of a flat Lambertian surface underneath a horizontally homogeneous atmosphere is made.

Carslow et al. [31] provided a solution of the thermal conduction Equation (2) for the temperature of the ground surface. Based on the ground surface temperature equation, Liu and Zhao [27] proposed a new STI model expressed as

$$P = \left[\sqrt{\left(\frac{A_1(1-A)S_O C_\tau}{T_{sur}(0,t) + (LE+H)/B}\right)^2 - \frac{B^2}{2}} - \frac{B}{\sqrt{2}}\right] / \sqrt{w} \tag{3}$$

where LE, H, and w are, respectively, the bare soil latent, sensible fluxes, and the rate of the earth's rotation. $T_{sur}(0,t)$ is the surface bare soil temperature.

Liu and Zhao [27] considered that each pixel consisted of the bare soil area and vegetated area. A parallel dual-source model was utilized to estimate the soil latent and sensible fluxes in bare soil areas, which neglects the shaded soil flux in the vegetated area. Thus, to improve the STI model so as to apply it to the vegetated areas, in this study, each pixel was decomposed into bare, vegetated, and shaded soil areas. A serial dual-source model was employed to reduce the effects of vegetation and obtain the estimates of the shaded soil latent and sensible fluxes in the vegetated areas. At the same time, to improve the estimation accuracy of the STI model for SMC, a discrete wavelet transform (DWT) was used to obtain a finer spatial resolution thermal inertia image. The STI could be written as

$$P = \left[\sqrt{\left(\frac{A_1(1-A_s)S_O C_\tau}{T_s(0,t) + (LE_s+H_s)/B}\right)^2 - \frac{B^2}{2}} - \frac{B}{\sqrt{2}}\right] / \sqrt{w} \tag{4}$$

where LE_s and H_s are, respectively, the soil latent and sensible fluxes in the soil and vegetated area, A_s is the surface soil albedo in the soil and vegetated area, and $T_s(0,t)$ is the surface soil temperature.

2.2.2. Estimation of Finer Spatial Resolution Soil Temperature

Wavelet Transform of ASTER Thermal Data

The LST is a key physical parameter. To increase the estimation accuracy of the surface component temperatures, the DWT of ASTER TIR images was conducted, which led to finer spatial resolution TIR data. The DWT included the wavelet decomposition of the images and an inverse wavelet transform that resulted in a new image. The wavelet decomposition of an image means that the image is first divided into four components, including the localized high-frequency point features, the localized vertical features, the localized horizontal details, and the lower frequency features. The lower frequency features are then further split at higher levels of the decomposition. For the detail of DWT, readers can be referred to Chang et al. [32].

In this research, as examples three high-frequency components of the ASTER TIR band 14 and green band 1 are shown in Figure 3. Figure 3a,b for green band 1 and Figure 3g,h for TIR band 14 represent the localized high-frequency point features of the June 4th and July 10th images, respectively. Figure 3c,d and Figure 3i,j show the localized vertical features of the June 4th and July 10th images, respectively. Figure 3e,f and Figure 3k,l provide the localized horizontal features of the June 4th and July 10th images, respectively. The ratios of three high-frequency components between the visible green band 1 and TIR band 14 were, respectively, close to the constants of −0.178147, −1.302275, and −0.611796, implying that the high-frequency information of the ASTER TIR data was very similar to that of the ASTER green band data. In fact, the high-frequency components transformed from the 15 m ASTER green band were the same as those from the 90 m ASTER TIR band because the high-frequency components represent the detailed surface features in the whole study area. The 15 m spatial resolution ASTER-TIR products were obtained by fusing the three high-frequency data of the 15 m ASTER green band and the low frequency data of the 90 m ASTER-TIR image (Figure 4). Compared with the original ASTER data, the fusion image showed more details of the surface features.

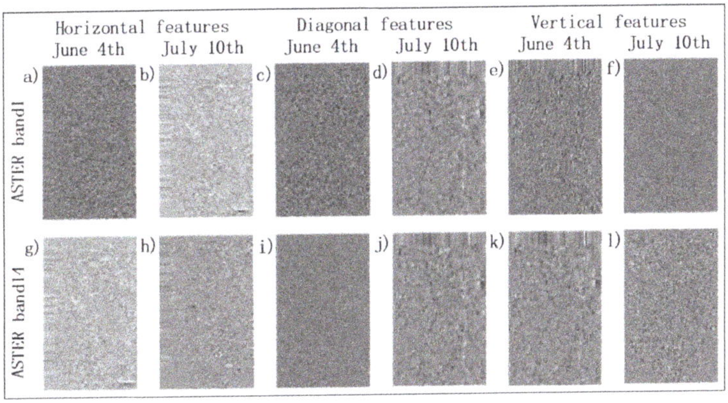

Figure 3. The wavelet decomposition results of ASTER band 1 and band 14 images from 4 June and 10 July 2008 at three directions by the discrete wavelet transform (DWT): (**a**,**b**,**g**,**h**) horizontal features; (**c**,**d**,**i**,**j**) diagonal features; and (**e**,**f**,**k**,**l**) vertical features.

Retrieval of Land Surface Component Temperatures

Mixed pixels often contain multiple land cover types, which impedes the improvement of estimation accuracy of SMC. Although the ASTER-TIR images were scaled down from a 90 m spatial resolution to a 15 m spatial resolution, the impact of mixed pixels in the images on mapping SMC cannot be ignored. The temperatures of land cover types within a mixed pixel *i* would be integrated and determine the average temperature of the mixed pixel. The integration could be linear or non-linear. For simplification, in this study it was assumed that the fluxes of the vegetated area (L_{vi}), bare soil (L_{bi}) and shaded soil (L_{di}) in each mixed pixel had linear contributions to the heat flux (L_i) of the mixed pixel. The area fractions of vegetation, bare soil, and shadow were derived using a linear spectral unmixing analysis. The LST was then obtained from the following equation [33]:

$$L_i = f_{vi}\sigma\varepsilon_{vi}T_{vi}^4 + f_{bi}\sigma\varepsilon_{bi}T_{bi}^4 + (1 - f_{vi} - f_{bi})\sigma\varepsilon_{di}T_{di}^4 \quad (5)$$

where ε_{vi}, ε_{bi}, and ε_{di}, respectively, represent vegetation, bare soil, and shaded soil emissivity at each mixed pixel. The T_{vi}, T_{bi}, and T_{di} are the temperatures of pure vegetation, bare soil, and shaded soil within the mixed pixel, respectively. σ is the Boerziman constant ($\sigma = 5.67 \times 10^{-8} Wm^{-2}K^{-4}$). f_{vi} and f_{bi} are the fractions of the vegetated and bare soil areas within the mixed pixel. In this study, estimating the land surface component values of temperature and emissivity was conducted using a genetic algorithm (GA) as a nonlinear multi-parameter optimization problem. For the detail of GA, readers can be referred to Liu et al. [34].

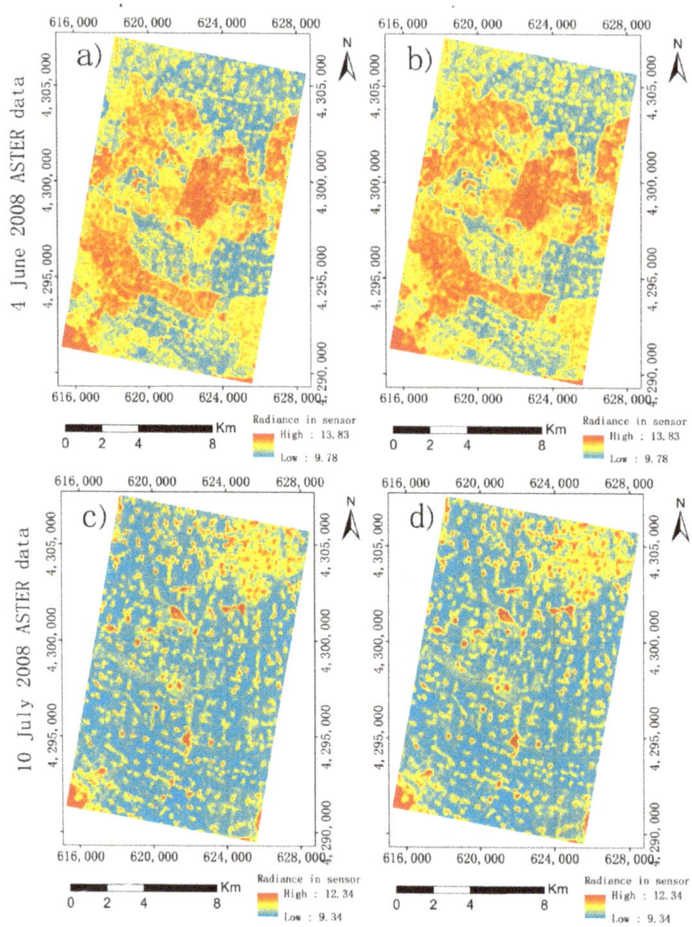

Figure 4. (**a**,**c**) the 15 m spatial resolution ASTER-TIR images from 4 June and 10 July 2008 after the data resampled using the nearest neighbor method (NNM); (**b**,**d**) the 15 m spatial resolution ASTER-TIR images from 4 June and 10 July 2008 after the data fusion using the discrete wavelet transform (DWT).

2.2.3. Estimation of Soil Sensible Flux

In the vegetated area, a series dual-source energy balance model was driven to estimate soil latent and sensible fluxes. In this model, we considered water vapor and heat as two sources and supposed that the bottom water and heat only come through the top layer [35]. The water and heat were then superimposed on each other. Moreover, the surface net radiance flux, the surface sensible heat flux and the surface latent flux in the reference level were respectively obtained by summing the corresponding components at each layer (Figure 5).

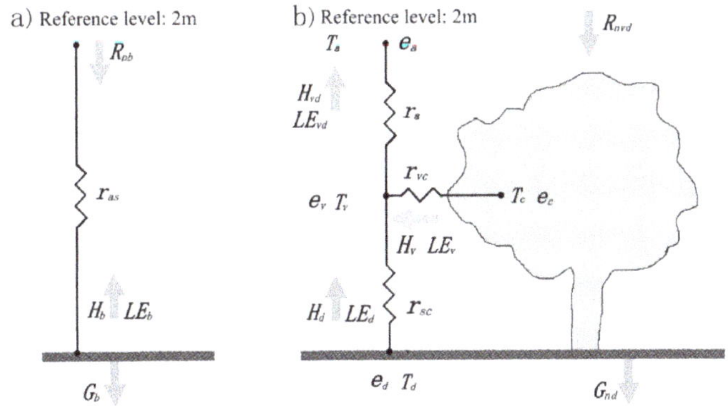

Figure 5. The demonstration of resistance network and important energy balance variables used in (**a**) bare soil and (**b**) the two-source energy balance (TSEB) model (Note: R_{nb} and R_{nvd} (R_{nd} + R_{nv}) represent, respectively, surface net radiance flux under bare soil and vegetation cover; R_{nd} and R_{nv} are the surface net radiance flux of the shaded soil and vegetation. H_b, H_d, and H_v represent, respectively, the surface sensible heat flux of bare soil, shaded soil, and vegetation. H_{vd} is the sum of H_v and H_d. LE_b, LE_d, and LE_v represent, respectively, the surface latent heat flux of bare soil, shaded soil, and vegetation. LE_{vd} is the sum of LE_v and LE_d. G_b and G_d represent, respectively, the surface flux exchange of bare soil and shaded soil. T_a, T_d, and T_c are the atmospheric temperature at the reference level, the surface soil temperature, and the vegetation aerodynamic temperatures, respectively. The r_{as}, r_{vc}, and r_{sc} represent, respectively, the aerodynamic resistance, the resistance of the whole canopy boundary layer, and the aerodynamic resistance between canopy source and soil surface. e_a, e_c, and e_d represent, respectively, the water vapor pressure at the reference level, canopy height, and soil surface).

In the bare soil area (Figure 5a), the soil sensible heat flux (H_b) was written as [36]

$$H_b = \rho_a \cdot C_p \cdot (T_b - T_a)/r_{as} \tag{6}$$

where T_b is the bare soil temperature, T_a is the atmospheric temperature at the reference level, ρ_α and C_p are, respectively, the air density (1.2 kg·m^{-3}) and the air heat capacity (1003.2 J·kg^{-1}), and r_{as} is the aerodynamic resistance obtained using the empirical model of Sauer et al. [37].

In the series dual-source energy balance model, H is obtained by adding the soil sensible flux (H_b) and canopy sensible flux (H_v). Based on the resistance network shown in Figure 5, the soil sensible flux under vegetation cover (H_{vd}) is [38]

$$H_{vd} = H_v + H_d \tag{7}$$

$$H_v = \rho_\alpha \cdot C_p \cdot (T_v - T_c)/r_{vc} \tag{8}$$

$$H_d = \rho_\alpha \cdot C_p \cdot (T_d - T_c)/r_{sc} \tag{9}$$

where ρ_α and C_p are, respectively, the air density (1.2 kg·m^{-3}) and the air heat capacity (1003.2 J·kg^{-1}). H_d and H_v present, respectively, the canopy-covered soil sensible flux and the vegetation sensible flux. T_c, T_d, and T_v are, respectively, the air aerodynamic temperature in the canopy height, the canopy-covered soil, and vegetation aerodynamic temperatures. r_{vc} and r_{sc} represent, respectively, the resistance of the whole canopy boundary-layer and the aerodynamic resistance between canopy source and soil surface. Finally, the soil sensible flux (H_s) was acquired by the following equation: $H_s = H_b + H_d$.

2.2.4. Estimation of Soil Latent Flux

The soil latent heat flux LE_S $(= LE_b + LE_d)$ was derived using the energy residual method, Equation (10) (Figure 5) [35].

$$LE_S = R_{ns} - H_S - G \tag{10}$$

where the soil net radiance flux R_{ns} $(= R_{nb} + R_{nd})$ was estimated by

$$R_{ns} = R_n(1 - f_v) = (1 - f_v)((1 - A)S_O C_\tau \cos\theta + R_{ld} - \varepsilon_s \sigma T_S^4) \tag{11}$$

where R_n and A are the surface net radiance flux and the surface albedo. S_0 and C_τ are, respectively, a solar constant (1367 Wm^{-2}) and the atmospheric optical transmission (0.562). The θ is the solar zenith (20°); R_{ld} is the incident longwave radiation (Wm^2) and is calculated according to the biomass equation ($R_{ld} = aT_a^6$) [39], where $a = 5.3 \times 10^{-13} Wm^{-2}K^{-6}$, and T_a is the atmospheric temperature (= 29°). ε_s $(= \varepsilon_b + \varepsilon_d)$ is the surface emissivity . ε_b and ε_d are, respectively, the emissivity of bare soil and shaded soil. T_s $(= T_b + T_d)$ is the surface soil temperature. The surface flux exchange G $(= G_b + G_{nd})$ can be calculated with the model [40] $G = \beta \times R_{ns}$, and $\beta = 0.2$ [41] in this study.

2.2.5. Estimation and Evaluation of Soil Moisture

In this study, the STI model (Equation (3)) was modified to accurately derive the spatial distribution of SMC in both bare soil and vegetated areas, which led to the modified model (Equation (4)). The modification was conducted by taking into account the effect of vegetation-induced shadows on the estimation of surface SMC. The original model and the modified model were, respectively, used to generate the estimates of STI. On the other hand, the change of STI caused the change of SMC. Thus, the relationship of SMC with STI derived from ASTER images can be developed and used to create the estimates of SMC. In this study, four models, including a linear, logarithm, power, and exponential model [42], were compared and selected using a leave-one cross validation method [43] based on the field measurements of 50 sample plots in the study area. The derivation of STI and its relationships with SMC was conducted using the ASTER TIR images at the spatial resolution of 15 m × 15 m by the NNM re-sample method and the DWT data fusion, respectively.

The accuracies of SMC estimates obtained from the original and modified models with two images at the 15 m spatial resolution obtained using NNM and DWT, respectively, were evaluated. Integrating the original and modified models with two images led to four combinations, including (1) the original-model-based bare STI using the 15 m resolution TIR bands from NNM (OBSTI_NNM); (2) the modified-model-based STI using the 15 m resolution TIR bands from the NNM (MSTI_ NNM); (3) the original-model-based bare STI using the 15 m resolution TIR bands from the DWT (OBSTI_DWT); and (4) the modified-model-based STI using the 15 m resolution TIR bands from the DWT (MSTI_ DWT). The evaluations and comparisons were conducted using the coefficient of determination (R^2), RMSE, relative RMSE (RRMSE), and mean absolute error (MAE) between the estimated and observed values of SMC. In addition, the mean value of SMC predictions was also utilized. The mean absolute error was defined as the mean value of the absolute differences of the estimates from the observations. The RMSE was defined as the square root of the average squared differences between the estimated and observed values. The RRMSE indicated the per-unit RMSE in percentage.

3. Results

3.1. Surface Albedo

The broadband albedo of land surface is an important variable in this study. At present, the conversion models from narrowband to broadband albedos [44] are not available. There are two alternatives to derive the total shortwave albedo, that is, using an empirical relationship of the surface total shortwave albedo with the spectral variables from remotely sensed images and utilizing the radiative transfer simulation with a number of surface reflectance spectra. In this study, the surface albedo was defined as the ratio of irradiance fluxes based on an assumption that narrowband albedos are linearly correlated with broadband ones. The surface narrowband albedos were first derived using the MODTRAN4.0 model and then converted to broadband albedos by taking the spectral distribution solar irradiance at the surface as a weighted function [33]. The obtained broadband albedo, α, was expressed as follows:

$$\alpha = 0.468594\alpha_1 + 0.303217\alpha_2 + 0.228189\alpha_3 \tag{12}$$

where α_1, α_2, and α_3 are, respectively, the narrowband albedos for ASTER band 1, band 2, and band 3. The albedo results are shown in Figure 6.

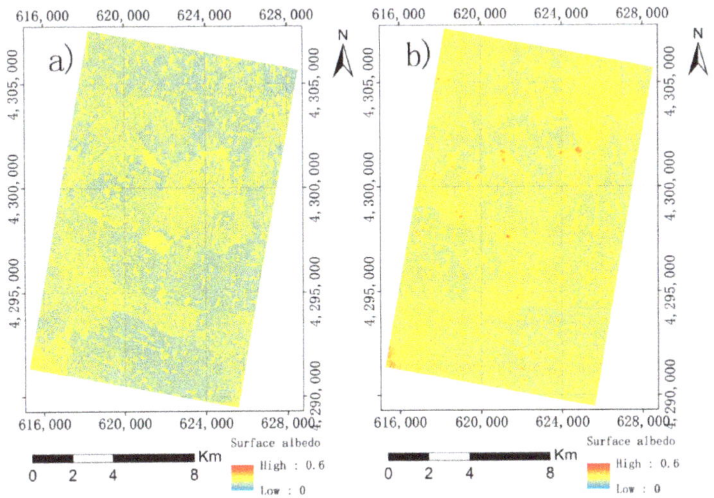

Figure 6. Surface albedo of the study area for (**a**) the June 4th image and (**b**) the July 10th image of 2008.

3.2. Soil Temperature

In Equation (5), the vegetation canopy fraction was estimated using ASTER band 1, band 2, and band 3 as follows [34]:

$$\rho_i = f_v \cdot \rho_{vi} + f_b \cdot \rho_{bi} + (1 - f_v - f_b) \cdot \rho_{di} \tag{13}$$

where ρ_i is the spectral reflectance of a mixed pixel in the ith band, and ρ_{vi}, ρ_{bi}, and ρ_{di} are, respectively, the spectral reflectance values of vegetation, bare soil, and shaded soil under vegetation canopy for the ith band. In order to obtain more accurate spectral reflectance values, atmospheric corrections for the ASTER band 1, band 2, and band 3 were carried out by the Modtran model. Then, a linear spectral unmixing analysis was applied to derive the area fractions (f_v, f_b, and f_d) of vegetation canopy, soil, and shaded soil (Figure 7). Due to the lack of field measurements of three area fractions, a total of 32 sample areas was randomly selected. The spectral reflectance values of the mixed pixels were estimated based on the linear combination of the area fractions with the spectral values of the pure pixels as the training samples for three endmembers. The obtained estimates were compared with the actual values of the mixed pixels, and the determination coefficients were used as an indirect validation of the area fractions. It was found that the mean coefficients of determination varied from 0.78 to 0.85 for the ASTER band 1, band 2, and band 3 dated on June 4th and July 10th of 2008. This indicated that the area fractions obtained using the linear spectral unmixing analysis were potentially reliable.

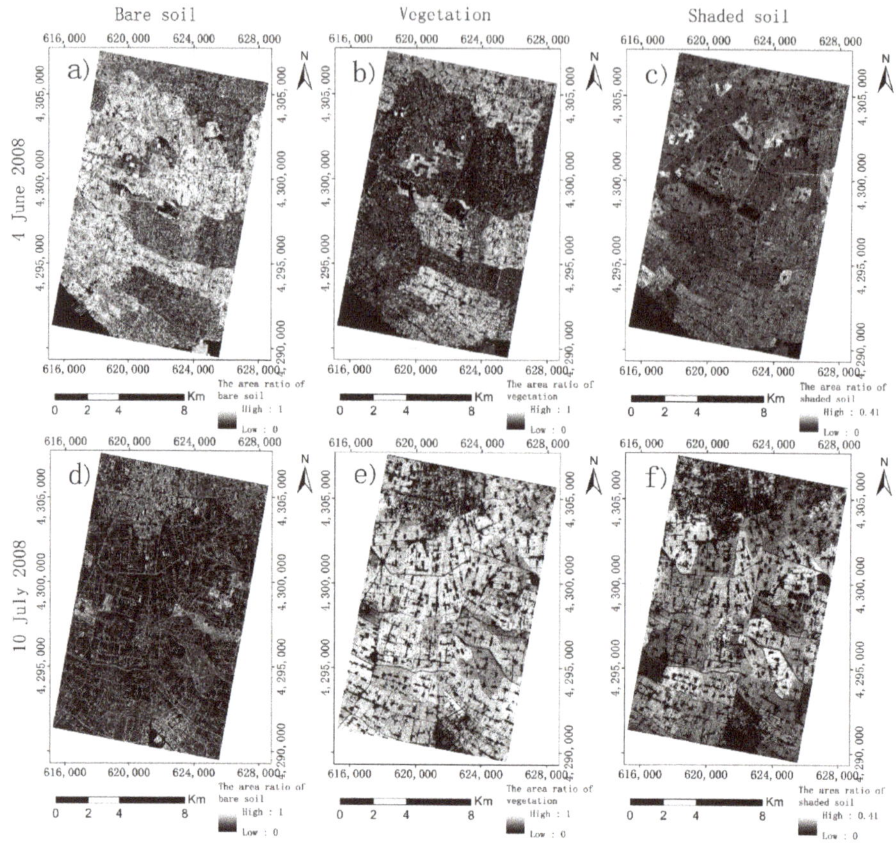

Figure 7. The area fraction maps of three components derived from ASTER data using a spectral unmixing analysis for the study area: (**a**,**d**) bare soil; (**b**,**e**) vegetation; and (**c**,**f**) shaded soil for the June 4th and July 10th images of 2008, respectively.

In order to obtain the unknown parameters (T_{vi}, T_{bi}, T_{di}, ε_{vi}, ε_{bi}, and ε_{di}) in Equation (5), the GA that solves nonlinear optimization problems was driven. The convergence property and speed of GA depended mainly on the given parameters, including the population size, mutation probability, crossover, etc. Because the input parameters of GA are not biophysically and mathematically meaningful, a random trial was utilized to search for the most accurate parameters [34]. In this study, the initial parameters of GA were set as follows: $\varepsilon_{bi} \in [0.85, 0.92]$, $\varepsilon_{di} \in [0.80, 0.95]$,$\varepsilon_{vi} \in [0.95, 1.00]$, $T_{bi} \in [273, 326]$,$T_{di} \in [273, 306]$, and $T_{vi} \in [273, 310]$. The number of the maximum runs and the input parameters for the population size, crossover probability, and mutation probability were 250, 128, 0.9, and 0.02, respectively. The resulting temperatures are demonstrated in Figure 8.

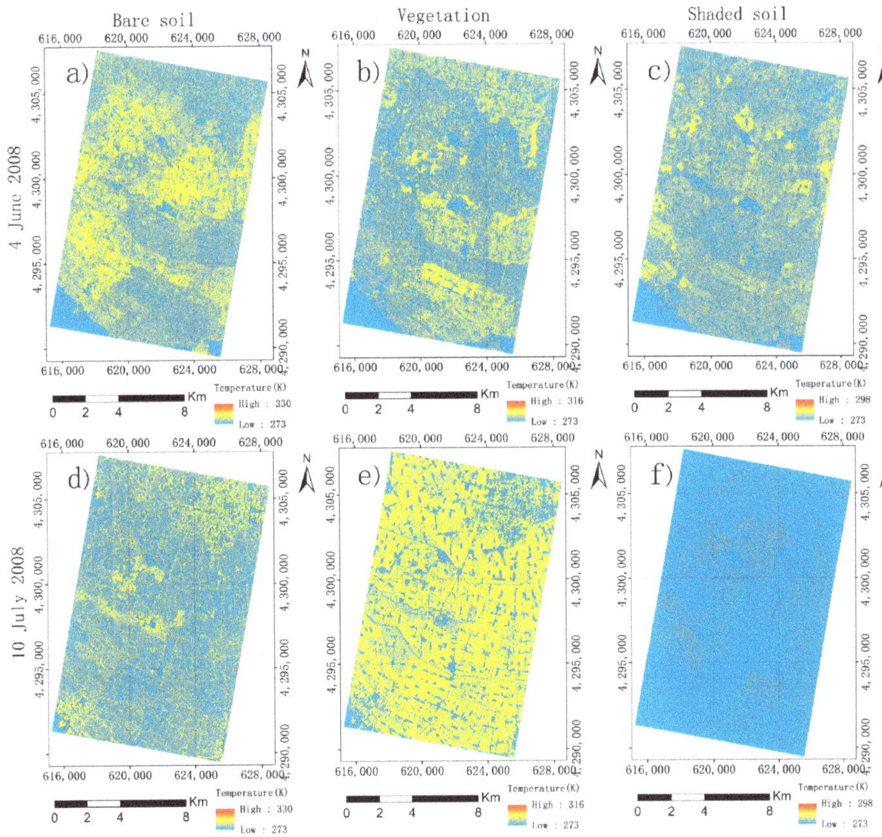

Figure 8. The spatial distributions of temperatures obtained based on three component fraction maps derived from ASTER data using a spectral unmixing analysis for the study area: (**a**,**d**) bare soil; (**b**,**e**) vegetation; and, (**c**,**f**) shaded area for the June 4th and July 10th images of 2008, respectively.

3.3. Soil Latent and Sensible Fluxes

The soil thermal transfer equation was driven to obtain the surface soil inertia, which needed the soil sensible flux, soil latent flux, and net radiant flux. In the bare soil, the soil sensible flux was obtained from Equation (6). Here, r_{as} was estimated from an experimental model [37]:

$$r_{as} = 4.27\frac{\ln^2\left(\frac{h}{z_{0m}}\right)}{(1+0.54u)} \tag{14}$$

where h is the height for measuring air temperature and wind speed u above the canopy height of 2 m. As the roughness length for momentum transfer, z_{0m} equals to 0.01 m. The measured wind speed above the canopy height is 3.92 m/s. The bare soil sensible flux is shown in Figure 9a.

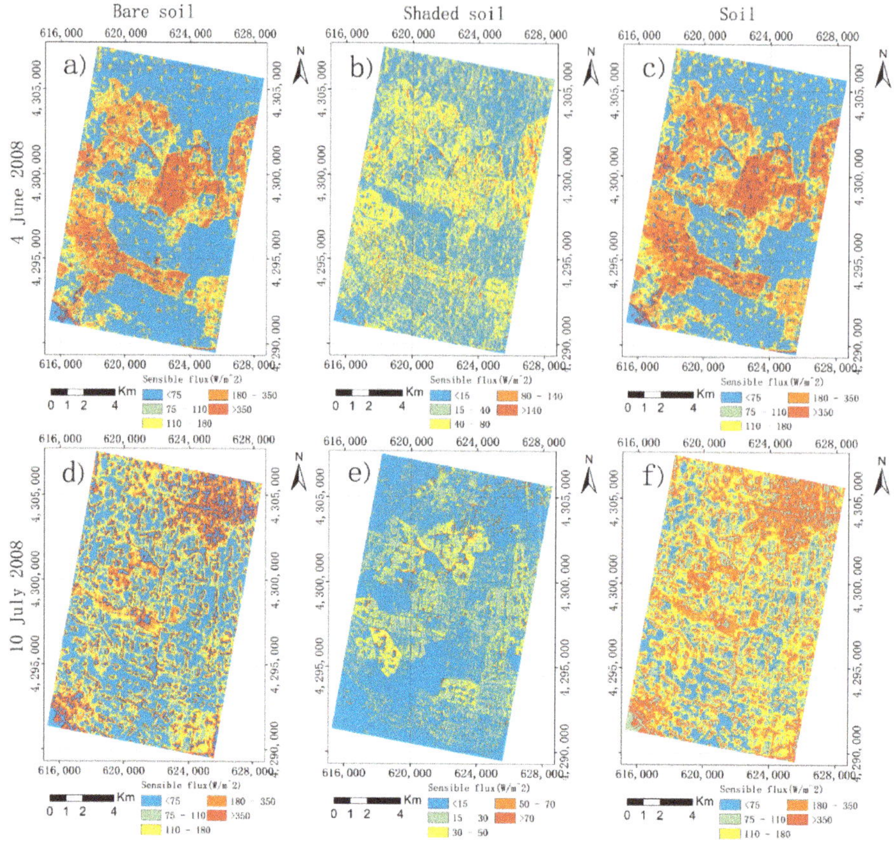

Figure 9. Spatial distributions of the bare soil, shaded soil, and total soil sensible fluxes for the study area: (**a**,**d**) bare soil; (**b**,**e**) shaded soil; and (**c**,**f**) total soil sensible flux using the June 4th image and July 10th image of 2008, respectively.

In the vegetated area, the shaded soil sensible flux was obtained from Equation (9) based on the TSEB described by Cammalleri et al. [35]. The r_{sc} in Equation (9) was estimated using the model in the studies of Kustas and Norman and Kondo and Ishida [40,45]:

$$r_{sc} = \frac{1}{c(T_s - T_c)^{1/3} + b' u_s} \quad (15)$$

where b' was set up to be 0.012 for natural surfaces, c is the coefficient and equal to 0.0025 $m \cdot s^{-1} \cdot K^{-1/3}$ [46,47], and the measured surface wind speed (u_s) is 1.2 m/s in the experiment field. The shaded soil sensible flux is shown in Figure 9b. Thus, the total soil flux of the sensible heat for the whole surface was estimated as the sum of the shaded soil sensible flux and the bare soil sensible flux in Figure 9c. Moreover, Equations (10) and (11) were respectively used to derive the values of soil latent flux and the soil net radiant flux that are shown in Figures 10 and 11.

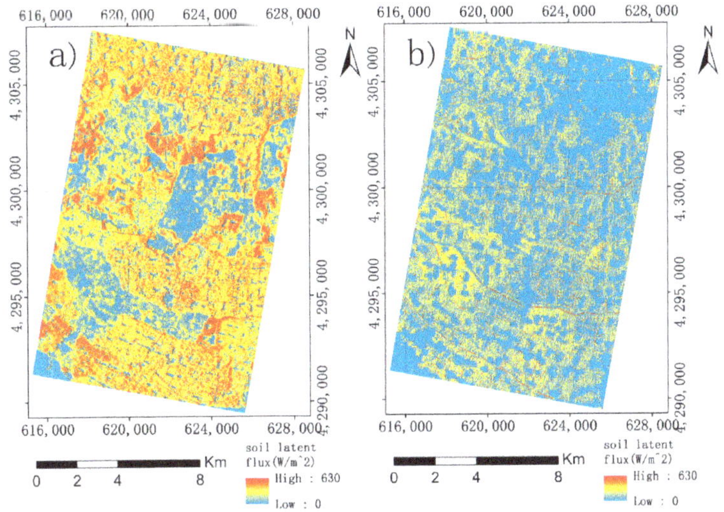

Figure 10. Spatial distribution of the soil latent flux using: (**a**) the June 4th image and (**b**) the July 10th image 2008, respectively.

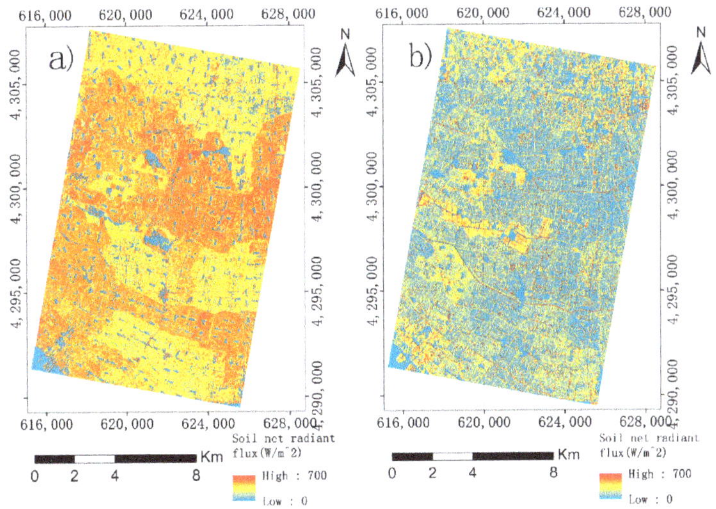

Figure 11. Spatial distribution of the soil net radiant flux using: (**a**) the June 4th image and (**b**) the July 10th image 2008, respectively.

3.4. The Estimation of Surface Soil Moisture Content

Based on the obtained spectral albedo, sensible flux, latent flux, and temperatures, Equation (7) for soil sensible flux under vegetation cover and Equation (8) for canopy sensible flux were used to estimate the STI and the bare STI using two 15 m spatial resolution TIR images by NNM and DWT. In this study, the thermal inertia varied from 0 to 1. The spatial distributions of the normalized STI for four combinations formed by the original and modified STI models and two 15 spatial resolution images are shown in Figure 12.

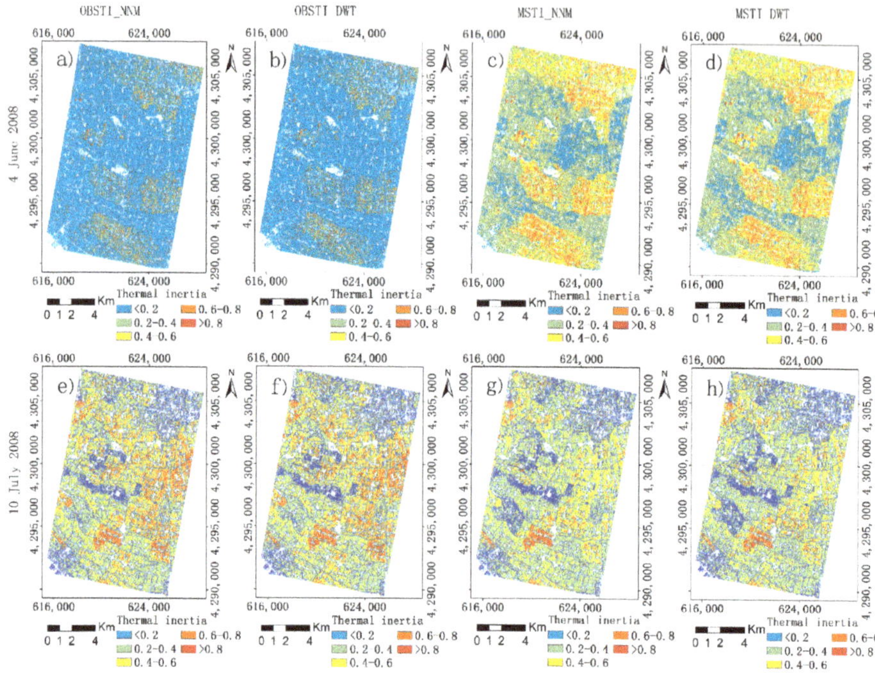

Figure 12. Spatial distributions of four combinations with images dated on June 4th and July 10th 2008: OBSTI_NNM—the original-model-based bare STI using the 15 m resolution thermal infrared bands by re-sampling using the nearest neighbor method; OBSTI_DWT—the original-model-based bare STI using the 15 m resolution fusion thermal infrared bands; MSTI_NNM—the modified-model-based STI using the 15 m resolution thermal infrared bands by re-sampling using the nearest neighbor method; and MSTI_DWT—the modified-model-based STI using the 15 m resolution fusion thermal infrared bands. Spatial distributions of thermal inertia derived using (**a**,**e**) OBSTI_NNM; (**b**,**f**) OBSTI_DWT; (**c**,**g**) MSTI_NNM; and, (**d**,**h**) MSTI_DWT on June 4th and July 10th 2008, respectively.

After STI was obtained using the downscaled images, the regional-scale SMC was generated using the obtained relationship of STI with SMC. In this study, the relationship was modeled and optimized by comparing a linear model, a logarithm model, a power model, and an exponential model in Table 1. A leave-one cross validation method was utilized to select the most accurate model.

Table 1. The comparison of four models to estimate soil moisture content (SMC) using the 50 soil samples and leave-one cross validation for four combinations: OBSTI_NNM—the original-model-based bare STI using the 15 m resolution thermal infrared bands by re-sampling using the nearest neighbor method; OBSTI_DWT—the original-model-based bare STI using the 15 m resolution fusion thermal infrared bands; MSTI_NNM—the modified-model-based STI using the 15 m resolution thermal infrared bands by re-sampling using the nearest neighbor method; and MSTI_DWT—the modified-model-based STI using the 15 m resolution fusion thermal infrared bands. R^2 and RMSE ($J \cdot m^{-2} \cdot s^{-1/2} \cdot K^{-1}$): coefficient of determination and root mean square error between the estimates and observations of SMC.

Model	Linear Model		Exponential Mode		Logarithm Model		Power Model	
Thermal inertia	R^2	RMSE	R^2	RMSE	R^2	RMSE	R^2	RMSE
OBSTI_NNM	0.72	3.46	0.71	3.75	0.59	3.67	0.63	3.49
OBSTI_DWT	0.73	3.41	0.72	3.75	0.58	3.66	0.63	3.47
MSTI_NNM	0.83	2.73	0.81	2.57	0.70	3.76	0.72	3.34
MSTI_DWT	0.85	2.67	0.82	2.47	0.71	3.78	0.72	3.37

For OBSTI_NNM and OBSTI_DWT, there were no statistically significant differences of RMSE among the four models, and the linear model led to slightly smaller RMSE value (Table 1). On the other hand, the exponential model resulted in the most accurate estimates of SMC for both MSTI_NNM and MSTI_DWT, and their RMSE values were statistically significantly smaller than those using the logarithm and power models but not than that using the linear model. Thus, the exponential mode most accurately accounted for the relationship of SMC with the modified-model-based STI, while the linear model most accurately explained the relationship of SMC with the original-model-based bare STI. The results of the models are shown in Figure 13. Given the same model, the image obtained by DWT only slightly reduced the RMSE values compared with that by NNM.

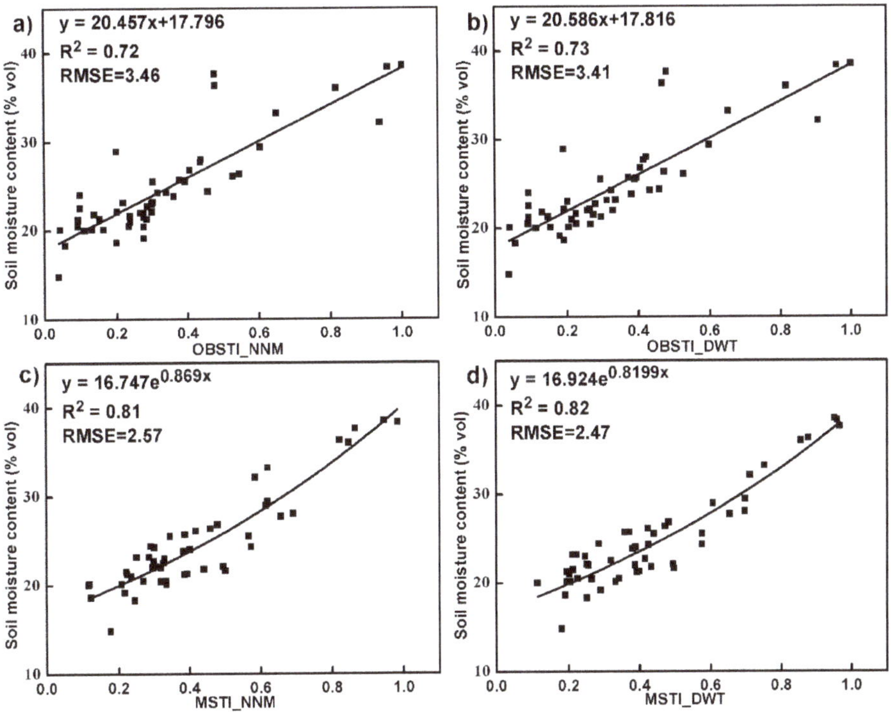

Figure 13. The selected models for accounting for the relationship of soil moisture content (SMC) with soil and bare soil thermal inertia (STI): (**a**) The linear model of SMC with the original-model-derived bare STI at the 15 m resampling spatial resolution (OBSTI_NNM); (**b**) The linear model of SMC with the original-model-derived bare STI at the 15 m fusion spatial resolution (OBSTI_DWT); (**c**) The exponential model of SMC with the modified-model-derived STI at the 15 m resampling spatial resolution (MSTI_NNM); and (**d**) The exponential model of SMC with the modified-model-derived STI at the 15 m fusion spatial resolution (MSTI_DWT). The R^2 and RMSE ($J{\cdot}m^{-2}{\cdot}s^{-1/2}{\cdot}K^{-1}$) are the coefficient of determination and root mean square error between the estimates and observations of SMC.

The spatial distributions of SMC estimates derived using the original and modified models with 15 m resampling and fusion spatial resolution TIR bands were compared in Figure 14. Given the original model, the SMC estimates from the image dated on July 10th (Figure 14e,f) are overall greater than those from the image dated on June 4th (Figure 14a,b), while for the modified model, the SMC estimates from the image dated on July 10th (Figure 14g,h) are slightly smaller than those from the image dated on June 4th (Figure 14c,d). In June and July, the study area was dominated by wheat and corn, respectively. The SMC varied very much depending on the canopy density of the crops.

Generally, the crop-dominated areas had greater SMC values. In addition, precipitation also affected the SMC values. Given the same model (the original or modified model) and the same image date (June 4th or July 10th), the spatial patterns of the SMC predictions looked similar (Figure 14a vs. Figure 14b, Figure 14c vs. Figure 14d, Figure 14e vs. Figure 14f, Figure 14g vs. Figure 14h,). However, the 15 m fusion image by DWT resulted in slightly more detailed information of SMC prediction.

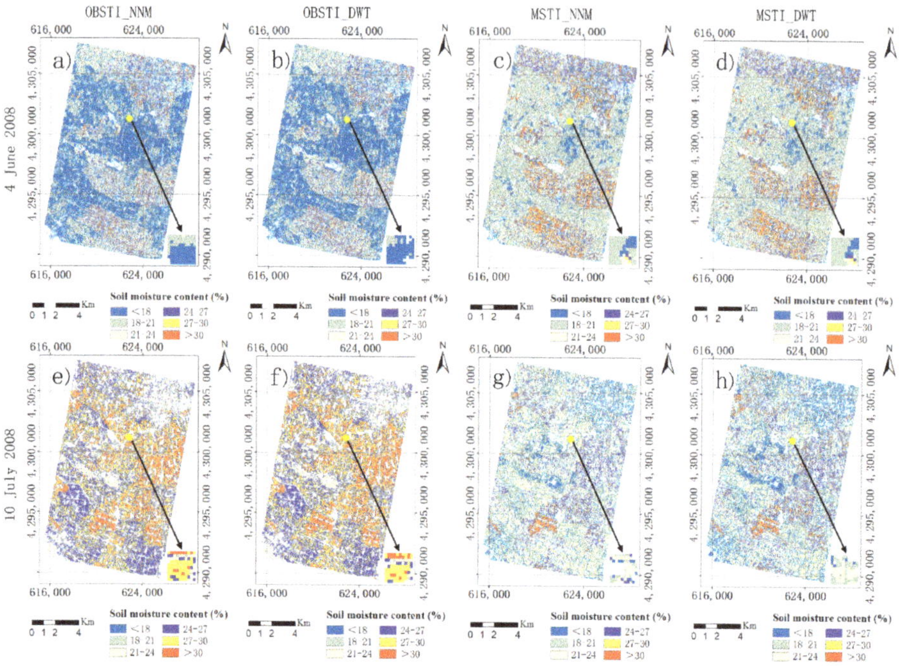

Figure 14. Comparison of spatial distributions of predicted soil moisture contents derived using (**a**,**e**) the original-model-derived bare STI at the 15 m resampling spatial resolution (OBSTI_NNM); (**b**,**f**) the modified-model-derived STI at the 15 m resampling spatial resolution (MSTI_NNM); (**c**,**g**) the original-model-derived bare STI at the 15 m fusion spatial resolution (OBSTI_DWT); and, (**d**,**h**) the modified-model-derived STI at the 15 m fusion spatial resolution (MSTI_DWT) on June 4th and July 10th 2008, respectively.

A total of 30 test soil samples was utilized to evaluate the SMC estimates from the ASTER data by calculating MAE and RRMSE between the observations and estimates in Figure 15. The sample mean of SMC was 23.57% with a standard deviation of 5.80% and a confidence interval of 21.56–25.58% at the significance level of 0.05. All the mean estimates of SMC statistically did not significantly differ from the sample mean, but the mean estimates of 23.85% and 23.58% from the modified model with the 15 m spatial resolution images by NNM and DWT, respectively, were closer to the sample mean than those from the original model. Moreover, the original model without the consideration of the effect of vegetation-induced shadows on estimation of SMC led to an MAE of 3.51% and 3.42% and RRMSE values of 18.95% and 18.60% for the images by NNM and DWT, respectively, while the modified model resulted in a corresponding MAE of 1.30% and 1.26% and RRMSE values of 6.92% and 6.89%. The MAE and RRMSE values from the original model were statistically significantly larger than those from the modified model at a significant level of 0.05. This implied that the modified model significantly increased the estimation accuracy of SMC compared with the original model. Given the same model (the original or the modified model), however, the estimation accuracies of SMC by the images from NNM and DWT were statistically similar to each other.

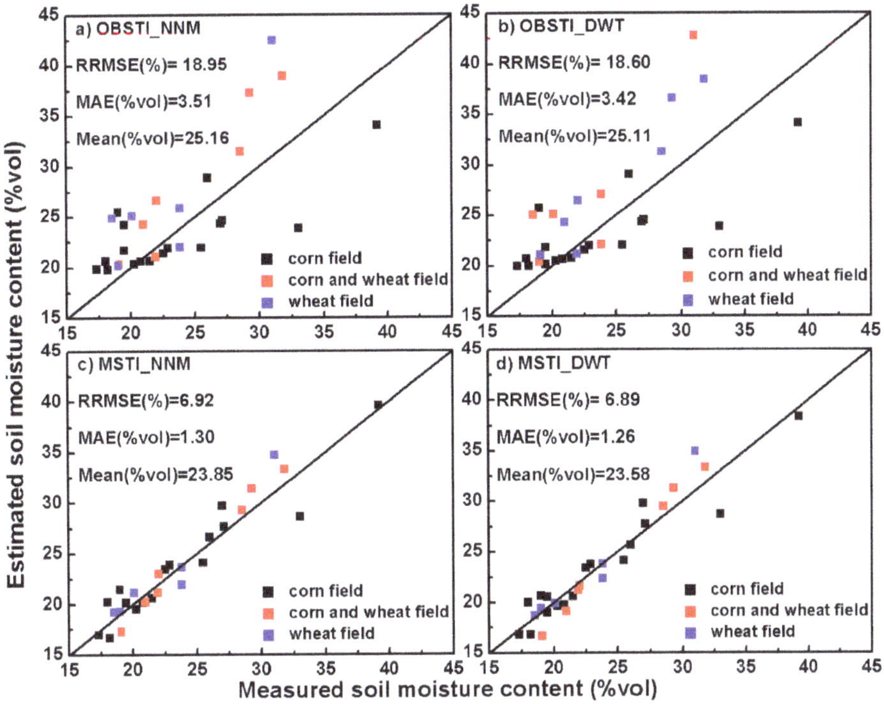

Figure 15. The field measurements of soil moisture content (SMC) versus the estimates: (**a**) OBSTI_NNM: the original-model-based bare STI using 15 m resolution thermal infrared bands by re-sampling using the nearest neighbor method; (**b**) OBSTI_DWT: the original-model-based bare STI using 15 m resolution fusion thermal infrared bands; (**c**) MSTI_NNM: the modified-model-based STI using 15 m resolution thermal infrared bands by re-sampling using the nearest neighbor method; (**d**) MSTI_DWT: the modified model based thermal inertia using 15 m resolution fusion thermal infrared bands.

4. Discussion

The widely used Apparent Thermal Inertia model was developed to estimate SMC using remotely sensed images in bare soil or scarcely vegetated regions [18,23]. Due to its ignorance of the effect from vegetation canopy-induced shadows, it cannot be applied to accurately estimate SMC in the densely vegetated areas. The STI model proposed in this study significantly contributed to the effort being made to improve the applications of the Apparent Thermal Inertia model from bare soil and sparsely vegetated areas to densely vegetated areas by introducing a vegetation-induced shadow component into the modified model using a spectral unmixing analysis. Some other models (e.g., real thermal model) were also devised to acquire the estimates of SMC in vegetated areas [24,26,27]. However, the models did not consider the serial dual-layer effectiveness for vegetated areas, and thus, the estimation accuracy of SMC retrieved by the models for vegetated areas was limited. The new approach provides a promising solution because it does not only take into account the effect of vegetation canopy-induced shadows, but also considers the serial dual-layer effectiveness for vegetated areas and used more detailed soil temperature information derived from the downscaled thermal images.

Moreover, Zhang et al. [4] used the information of red and near-infrared bands to develop a normalized difference vegetation index (NDVI)-albedo model and to retrieve the estimates of SMC, and the results were highly consistent with the field measurements at 5 cm soil depth. However, the limited penetration ability of the spectral reflectance might degrade the quality of SMC estimates. Because the estimates of SMC from the NDVI-albedo model were related to NDVI change, it was difficult to accurately estimate SMC in the irrigated areas using the NDVI-albedo model according to the widely spread triangular scatter distribution of the albedo values against the NDVI values (Figure 16). The approach proposed in this study improved the estimation of SMC in the irrigated soil areas because of using the thermal images and the consideration of the shadows under the vegetation canopies.

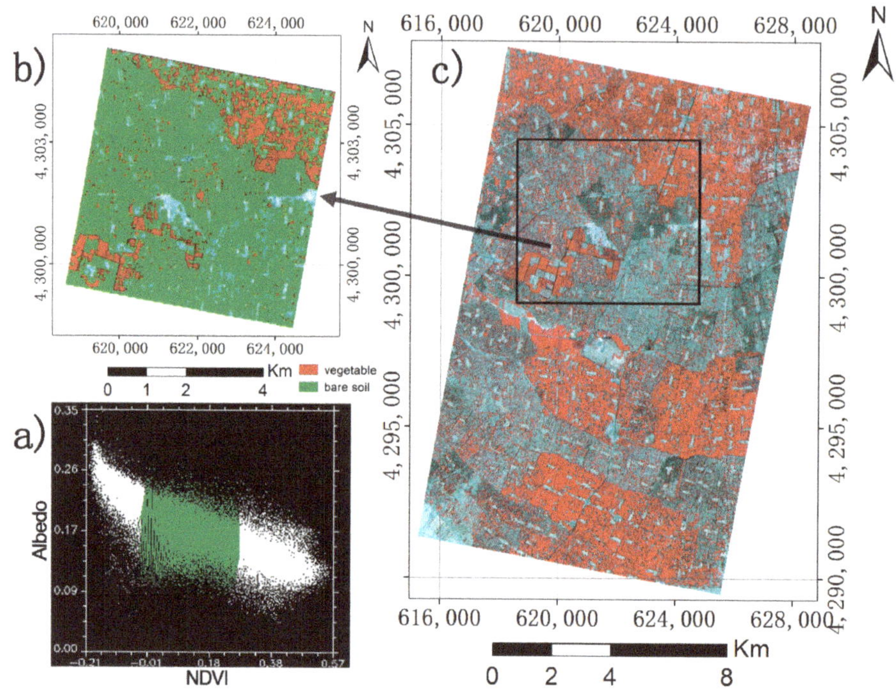

Figure 16. (**a**) The normalized difference vegetation index (NDVI)-albedo scatter plot of a small square extracted from (**b**) an enlarged and bare soil dominated area of (**c**) the study area.

There were also some simple methods for the estimation of SMC using the relationships of thermal image-derived LST with NDVI and other vegetation indices [4,48]. However, the LST-NDVI-based methods require a large number of SMC values that well characterize various land cover types, and the scatter plot of the LST against NDVI often has a wide spread (i.e., Figure 17) that cannot be utilized to accurately obtain the estimates of SMC. The proposed approach in this study modeled the relationship of SMC with STI and the values of STI were derived using the modified model in which the shadows of vegetation canopies were extracted by the shade fraction from the spectral unmixing analysis. That is, the improved inertial model (Equation (4)) theoretically takes into account bare soil, vegetated soil, and shaded soil to obtain the estimates of shaded soil latent and sensible fluxes in the vegetated areas. Thus, it has no limitations of empirical model data and can be applied to other areas with any kind of vegetation types and canopy densities.

In this study, the values of SMC obtained by four inertial methods in bare soil regions varied relatively little. However, in the vegetated area, there was a great variation, and the values of SMC from the modified models MSTI_NNM and MSTI_DWT were closer to the field measurements than those from the original models OBSTI_NNM and OBSTI_DWT, which attributes to the consideration of shaded soil effects. This further indicated that the modified model was more reliable to retrieve SMC. Moreover, although the 15 m spatial resolution thermal bands from DWT showed more detailed information than those from NNM, given the modified or original model, the former only slightly decreased the values of MAE and RRMSE compared with those from the latter (Figure 15). That is, the 15 m spatial resolution thermal bands obtained by DWT did not significantly improve the estimation accuracy of SMC compared to the 15 m spatial resolution thermal bands by NNM. The main reason might be due to the flat study area with relatively simple and uniform topographic and crop canopy features, in which most of the wheat and corn areas had sizes of larger than 90 m × 90 m, and thus, the DWT made the spatial information more available but did not significantly change the values of the thermal band pixels compared with those from NNM.

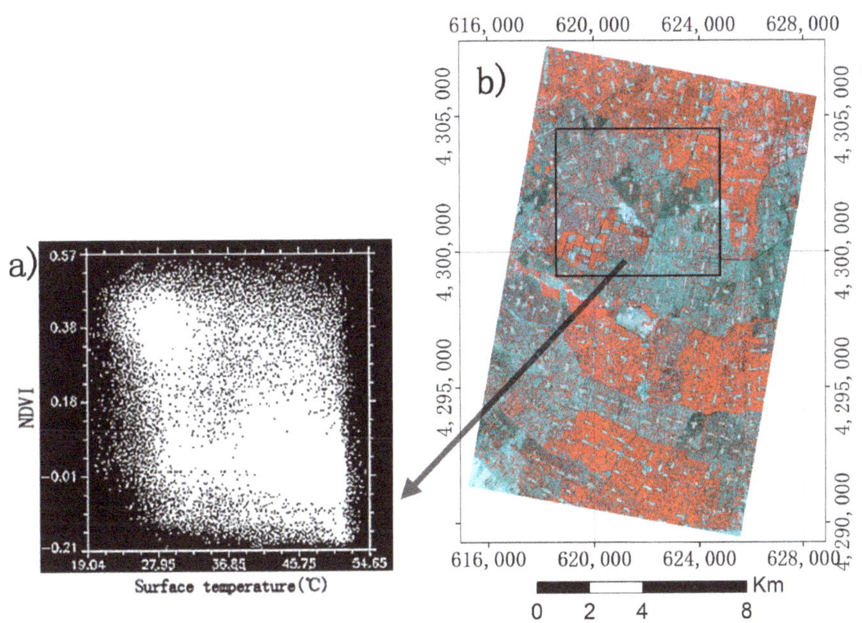

Figure 17. (**a**) The scatter plot of the normalized difference vegetation index (NDVI) values against the land surface temperature (LST) values of a square area extracted from (**b**) the study area shown with an infrared color composite image.

It had to be pointed out that in the modified model of this study, the relationship of soil moisture with the modified model derived STI was most accurately fit by an exponential model. The exponential model showed a statistically significant relationship of using STI to estimate SMC, but lacked a clearly biophysical mechanism, which might have limited the capacity of its predictions in the case of no field measurements available. On the other hand, when the modified model is applied to other regions, its calibration based on some field measurements is needed. Moreover, the application of the proposed method was limited to wheat, corn, and wheat/corn fields, and the application of a future study should be expanded to other land cover types. In addition, to ensure the reliability of the data collected, the collection of data (such as spectroscopy and temperature measurements) was performed within a half hour before and after the satellite's pass. Thus, in the experiment we only measured a total of 30 sample points used for validation within the limited time. In the future, more field measurements should be collected to further validate this method.

5. Conclusions

In this study, the modified STI model was developed to improve the estimation of SMC in the vegetated and irrigated region located in the basin of Heihe River, Northwestern China. The modified model reduced the impact of vegetation on the estimation of SMC by extracting the fraction of shadows due to vegetation canopies using a spectral unmixing analysis, which overcame the gap that currently exists in the original STI model. The results showed that compared to those from the original STI model, the modified STI model significantly decreased the MAE and RRMSE values of SMC by 63.0–63.2% for MAE and 63.0–63.5% for RRMSE (Figure 15). The 15 m spatial resolution thermal bands obtained by DWT data fusion provided more detailed information of SMC but did not significantly improve the estimation of SMC compared to the 15 m spatial resolution thermal bands re-sampled by NNM. This study implied that the new approach can offer the great potential to increase the accuracy of retrieving SMC estimates in the vegetated areas.

Author Contributions: Z.L.: conceptualization, methodology, performance of experiments and writing; G.W.: supervision, review, and revision; L.Z.: performance of experiments and methodology; Y.P.: performance of experiments and validation; Y.H.: project administration. All authors have read and agreed to the published version of the manuscript.

Funding: This research was funded by the National Natural Science Foundation of China (U1901601); Guangdong Province Agricultural Science and Technology Innovation and Promotion Project (No. 2019KJ102); Guangdong Provincial Science and Technology Project of China (2017B030314155); and Qinghai Provincial Science and Technology Project of China (2019-HZ-801).

Acknowledgments: We gratefully acknowledge the experimental assistance of Ting Wang and Ziqing Xia.

Conflicts of Interest: The authors declare no conflict of interest.

References

1. Wu, G.L.; Zhang, Z.N.; Wang, D.; Shi, Z.H.; Zhu, Y.J. Interactions of soil water content heterogeneity and species diversity patterns in semi-arid steppes on the Loess Plateau of China. *J. Hydrol.* **2014**, *519*, 1362–1367. [CrossRef]
2. Zhao, W.; Li, Z.L. Sensitivity study of soil moisture on the temporal evolution of surface temperature over bare surfaces. *Int. J. Remote Sens.* **2013**, *34*, 3314–3331. [CrossRef]
3. Li-Ge, B. Research on the Drought during the Vegetation Growth Period of Inner Mongolia Based on TVDI. *J. Anhui Agric. Sci.* **2011**, *39*, 5945–5948.
4. Zhang, D.; Tang, R.; Tang, B.; Wu, H.; Li, Z. A Simple Method for Soil Moisture Determination From LST–VI Feature Space Using Nonlinear Interpolation Based on Thermal Infrared Remotely Sensed Data. *IEEE J. Sel. Top. Appl. Earth Obs. Remote Sens.* **2015**, *8*. [CrossRef]
5. Escorihuela, M.J.; Chanzy, A.; Wigneron, J.P.; Kerr, Y.H. Effective soil moisture sampling depth of L-band radiometry: A case study. *Remote Sens. Environ.* **2010**, *114*, 995–1001. [CrossRef]
6. Cai, G.; Xue, Y.; Hu, Y.; Wang, Y.; Guo, J.; Luo, Y.; Wu, C.; Zhong, S.; Qi, S. Soil moisture retrieval from MODIS data in Northern China Plain using thermal inertia model. *Int. J. Remote Sens.* **2007**, *28*, 3567–3581. [CrossRef]
7. Peischl, S.; Walker, J.P.; Ye, N.; Ryu, D.; Kerr, Y. Sensitivity of multi-parameter soil moisture retrievals to incidence angle configuration. *Remote Sens. Environ.* **2014**, *143*, 64–72. [CrossRef]
8. Li, B.; Rodell, M. Spatial variability and its scale dependency of observed and modeled soil moisture over different climate regions. *Hydrol. Earth Syst. Sci. Discuss.* **2012**, *9*, 10245–10276. [CrossRef]
9. Lu, Y.; Horton, R.; Zhang, X.; Ren, T. Accounting for soil porosity improves a thermal inertia model for estimating surface soil water content. *Remote Sens. Environ.* **2018**, *212*, 79–89. [CrossRef]
10. Doninck, J.V.; Peters, J.; Baets, B.D.; Clercq, E.M.D.; Ducheyne, E.; Verhoest, N.E.C. The potential of multitemporal Aqua and Terra MODIS apparent thermal inertia as a soil moisture indicator. *Int. J. Appl. Earth Obs. Geoinf.* **2011**, *13*, 941. [CrossRef]
11. Zeng, J.; Chen, K.; Cui, C.; Bai, X. A Physically Based Soil Moisture Index from Passive Microwave Brightness Temperatures for Soil Moisture Variation Monitoring. *IEEE Trans. Geosci. Remote Sens.* **2020**, *58*, 2782–2795. [CrossRef]
12. Kim, S.B.; Moghaddam, M.; Tsang, L.; Burgin, M.; Xu, X.L.; Njoku, E.G. Models of L-Band Radar Backscattering Coefficients Over Global Terrain for Soil Moisture Retrieval. *IEEE Trans. Geosci. Remote Sens.* **2014**, *52*, 1381–1396. [CrossRef]
13. Gao, Z.; Xu, X.; Wang, J.; Yang, H.; Huang, W.; Feng, H. A method of estimating soil moisture based on the linear decomposition of mixture pixels. *Math. Comput. Model.* **2013**, *58*, 606–613. [CrossRef]

14. Nocita, M.; Stevens, A.; Noon, C.; Wesemael, B.V. Prediction of soil organic carbon for different level of soil moisture using Vis-NIR spectroscopy. *Geoderma* **2012**, *199*. [CrossRef]
15. Kang, J.; Jin, R.; Li, X.; Ma, C.; Qin, J.; Zhang, Y. High spatio-temporal resolution mapping of soil moisture by integrating wireless sensor network observations and MODIS apparent thermal inertia in the Babao River Basin, China. *Remote Sens. Environ.* **2017**, *191*, 232–245. [CrossRef]
16. Taktikou, E.; Bourazanis, G.; Papaioannou, G.; Kerkides, P. Prediction of Soil Moisture from Remote Sensing Data. *Procedia Eng.* **2016**, *162*, 309–316. [CrossRef]
17. Verstraeten, W.W.; Veroustraete, F.; van der Sande, C.; Grootaers, I.; Feyen, J. Soil moisture retrieval using thermal inertia, determined with visible and thermal spaceborne data, validated for European forests. *Remote Sens. Environ.* **2006**, *101*, 299–314. [CrossRef]
18. Palombo, A.; Pascucci, S.; Loperte, A.; Lettino, A.; Castaldi, F.; Muolo, M.R.; Santini, F. Soil Moisture Retrieval by Integrating TASI-600 Airborne Thermal Data, WorldView 2 Satellite Data and Field Measurements: Petacciato Case Study. *Sensors* **2019**, *19*, 1515. [CrossRef]
19. Watson, K. Application of thermal modeling in the geologic interpretation of IR images. *Proc. Seventh Int. Symp. Remote Sens. Environ.* **1971**, *3*, 2017–2041.
20. Price, J.C. Thermal inertia mapping: A new view of the Earth. *J. Geophys. Res.* **1977**, *18*, 2582–2590. [CrossRef]
21. Pratt, D.A.; Ellyett, C.D. The thermal inertia approach to mapping of soil moisture and geology. *Remote Sens. Environ.* **1979**, *8*, 151–168. [CrossRef]
22. Price, J.C. On the analysis of thermal infrared imagery: The limited utility of apparent thermal inertia. *Remote Sens. Environ.* **1985**, *18*, 59–73. [CrossRef]
23. Scheidt, S.; Lancaster, N.; Ramsey, M. Eolian dynamics and sediment mixing in the Gran Desierto, Mexico, determined from thermal infrared spectroscopy and remote-sensing data. *Geol. Soc. Am. Bull.* **2011**, *123*, 1628–1644. [CrossRef]
24. Xue, Y.; Cracknell, A.P. Advanced thermal inertia modeling. *Int. J. Remote Sens.* **1995**, *16*, 431–446. [CrossRef]
25. Tao, Y. The Application of Thermal Inertia Method the Monitoring of Soil Moisture of North China Plain Based on NOAA-AVHRR Data. *J. Remote Sens.* **1997**, *1*, 24–31.
26. Zhang, R.; Sun, X.; Zhu, Z.; Su, H.; Tang, X. A remote sensing model for monitoring soil evaporation based on differential thermal inertia and its validation. *Sci. China Ser. D Earth Sci.* **2003**, *46*, 342–355.
27. Liu, Z.-h.; Zhao, Y.-s. Research on the method for retrieving soil moisture using thermal inertia model. *Sci. China Ser. D* **2006**, *49*, 539–545. [CrossRef]
28. Minacapilli, M.; Cammalleri, C.; Ciraolo, G.; D'Asaro, F.; Lovino, M.; Maltese, A. Thermal Inertia Modeling for Soil Surface Water Content Estimation: A Laboratory Experiment. *Soil Sci. Soc. Am. J.* **2012**, *76*, 92. [CrossRef]
29. Chen, H.L.; Feng, D.Y.; Zou, C.H. The study of method on monitoring soil moisture in wheat field by NOAA/AVHRR data. *Remote Sens. Technol. Appl.* **1998**, *13*, 27–35.
30. Liu, J.Y.; Zhang, G.; Li, D.R. Detection of anomaly temperature based on ASTER and ETM+ thermal infrared image. *Infrared Laser Eng.* **2008**, *37*, 677–680.
31. Carslow, H.S.; Jaeger, J.C.; Morral, J.E. Conduction of Heat in Solids, Second Edition. *J. Eng. Mater. Technol.* **1986**, *108*, 378. [CrossRef]
32. Chang, S.G.; Yu, B. Adaptive wavelet thresholding for image denoising and compression. *IEEE Trans. Image Process.* **2000**, *9*, 1532–1546. [CrossRef] [PubMed]
33. Liu, Z.-H.; Zhao, Y.-S.; Song, X.-N. A Simplified Surface Albedo Inverse Model with MODIS Data. *Remote Sens. Technol. Appl.* **2004**, *7*, 508–511. [CrossRef]
34. Liu, Z.-H.; Zhao, Y.-S. Retrieval of plant and soil component temperature under different light conditions based on genetic algorithm. *Trans. Chin. Soc. Agric. Eng.* **2012**, *28*, 161–166.
35. Cammalleri, C.; Anderson, M.; Ciraolo, G.; D'Urso, G.; Kustas, W.P.; La Loggia, G.; Minacapilli, M. The impact of in-canopy wind profile formulations on heat flux estimation in an open orchard using the remote sensing-based two-source model. *Hydrol. Earth Syst. Sci.* **2010**. [CrossRef]
36. Chehbouni, A.; Nouvellon, Y.; Lhomme, J.P.; Watts, C.; Boulet, G.; Kerr, Y.H.; Moran, M.S.; Goodrich, D.C. Estimation of surface sensible heat flux using dual angle observations of radiative surface temperature. *Agric. For. Meteorol.* **2001**, *108*, 65. [CrossRef]
37. Sauer, T.J.; Hatfield, J.L.; Prueger, J.H. Aerodynamic Characteristics of Standing Corn Stubble. *Agron. J.* **1996**, *5*, 733–739. [CrossRef]

38. Shuttleworth, W.J.; Wallace, J.S. Evaporation from sparse crops-an energy combination theory. *Q. J. Royal Meteorol. Soc.* **1985**, *111*, 839–855. [CrossRef]
39. Swinbank, W.C. Long-wave radiation from clear skies. *Q. J. Royal Meteorol. Soc.* **2010**, *89*, 339–348. [CrossRef]
40. Kustas, W.P.; Norman, J.M. Evaluation of soil and vegetation heat flux predictions using a simple two-source model with radiometric temperatures for partial canopy cover. *Agric. For. Meteorol.* **1999**, *94*, 29. [CrossRef]
41. Choudhury, B.J.; Idso, S.B.; Reginato, R.J. Analysis of an empirical model for soil heat flux under a growing wheat crop for estimating evaporation by infrared-temperature based energy balance equation. *Agric. For. Meteorol.* **1987**, *39*, 283–297. [CrossRef]
42. Claps, P.; Laguardia, G. Assessing spatial variability of soil water content through thermal inertia and NDVI. *Int. Soc. Opt. Eng.* **2004**, *5232*, 378–387.
43. Wong, T.-T. Performance evaluation of classification algorithms by k-fold and leave-one-out cross validation. *Pattern Recognit.* **2015**, *48*, 2839–2846. [CrossRef]
44. Riihelä, A.; Manninen, T.; Laine, V.; Andersson, K.; Kaspar, F. CLARA-SAL: A global 28 yr timeseries of Earth's black-sky surface albedo. *Atmos. Chem. Phys.* **2013**, *13*, 3743–3762. [CrossRef]
45. Kondo, J.; Ishida, S. Sensible Heat Flux from the Earth's Surface under Natural Convective Conditions. *J. Atmos. Sci.* **1997**, *54*, 498–509. [CrossRef]
46. Sauer, T.J.; Norman, J.M. Simulated canopy microclimate using estimated below-canopy soil surface transfer coefficients. *Agric. For. Meteorol.* **1995**, *75*, 135–160. [CrossRef]
47. Sauer, T.J. *Sensible and Latent Heat Exchange at the Soil Surface Beneath a Maize Canopy*; ProQuest Dissertations Publishing: Madison, WI, USA, 1993.
48. Wan, Z.; Wang, P.; Li, X. Using MODIS Land Surface Temperature and Normalized Difference Vegetation Index products for monitoring drought in the southern Great Plains, USA. *Int. J. Remote Sens.* **2004**, *25*, 61–72. [CrossRef]

remote sensing

MDPI

Article

High-Precision Soil Moisture Mapping Based on Multi-Model Coupling and Background Knowledge, Over Vegetated Areas Using Chinese GF-3 and GF-1 Satellite Data

Leran Han [1,2], Chunmei Wang [1,*], Tao Yu [1], Xingfa Gu [1] and Qiyue Liu [1,2]

1 Aerospace Information Research Institute, Chinese Academy of Sciences, Beijing 100094, China; hanlr@radi.ac.cn (L.H.); yutao@radi.ac.cn (T.Y.); guxf@radi.ac.cn (X.G.); liuqy@radi.ac.cn (Q.L.)

2 University of Chinese Academy of Sciences, Beijing 100049, China

* Correspondence: wangcm@aircas.ac.cn

Received: 18 June 2020; Accepted: 29 June 2020; Published: 2 July 2020

Abstract: This paper proposes a combined approach comprising a set of methods for the high-precision mapping of soil moisture in a study area located in Jiangsu Province of China, based on the Chinese C-band synthetic aperture radar data of GF-3 and high spatial-resolution optical data of GF-1, in situ experimental datasets and background knowledge. The study was conducted in three stages: First, in the process of eliminating the effect of vegetation canopy, an empirical vegetation water content model and a water cloud model with localized parameters were developed to obtain the bare soil backscattering coefficient. Second, four commonly used models (advanced integral equation model (AIEM), look-up table (LUT) method, Oh model, and the Dubois model) were coupled to acquire nine soil moisture retrieval maps and algorithms. Finally, a simple and effective optimal solution method was proposed to select and combine the nine algorithms based on classification strategies devised using three types of background knowledge. A comprehensive evaluation was carried out on each soil moisture map in terms of the root-mean-square-error (RMSE), Pearson correlation coefficient (PCC), mean absolute error (MAE), and mean bias (bias). The results show that for the nine individual algorithms, the estimated model constructed using the AIEM (m_{v1}) was significantly more accurate than those constructed using the other models (RMSE = 0.0321 cm^3/cm^3, MAE = 0.0260 cm^3/cm^3, and PCC = 0.9115), followed by the Oh model (m_v5) and LUT inversion method under HH polarization (m_{v2}). Compared with the independent algorithms, the optimal solution methods have significant advantages; the soil moisture map obtained using the classification strategy based on the percentage content of clay was the most satisfactory (RMSE = 0.0271 cm^3/cm^3, MAE = 0.0225 cm^3/cm^3, and PCC = 0.9364). This combined method could not only effectively integrate the optical and radar satellite data but also couple a variety of commonly used inversion models, and at the same time, background knowledge was introduced into the optimal solution method. Thus, we provide a new method for the high-precision mapping of soil moisture in areas with a complex underlying surface.

Keywords: soil moisture; multi-model coupling; optimal solution method

---

1. Introduction

The soil moisture is a main component of the earth system, with the first 0–5 cm of the soil layer playing an important role in the exchange of substances and energy between the land and atmosphere; moreover, it is an important parameter in the fields of agriculture, meteorology, and hydrology [1–6]. Although conventional in situ experiments can provide highly accurate data, the regional-scale monitoring via this method has drawbacks because of the poor representativeness of the surface soil moisture in the limited spatial scale of strong heterogeneity areas [7–9]. Compared with the cumbersome and time-consuming in situ experiments, the development of remote sensing technology has led to promising methods for soil moisture monitoring, extending the "point" conventional measurement to objective regional-scale soil moisture information [10–15].

Optical remote sensing has been successfully applied to soil moisture inversion by linking the changes in the spectral characteristics with the soil moisture content. Optical methods are straightforward to implement but are easily affected by weather conditions because of their weak penetration [16]. Most of the thermal infrared methods are applicable only for monitoring the soil moisture in bare soil and sparsely vegetated areas and under cloud-free conditions, based on the estimation of soil thermal characteristics [17]. By contrast, the microwave remote sensing technology is unaffected by cloud or weather conditions owing to the penetration to the vegetation canopy and sensitivity to soil permittivity [18–22]. Therefore, the microwave remote sensing method, owing to the synergy between microwaves and other information acquired from the electromagnetic radiation spectrum, has a good application prospect for estimating the soil moisture of bare soil and vegetation-covered areas [2,14,18].

In recent years, studies have shown that active microwave remote sensing could make up for the low spatial resolution and long revisit period of passive microwave remote sensing and has become a research hotspot in regional-scale soil moisture monitoring [22–24]. The theoretical basis of active microwave remote sensing in extracting soil moisture information is the dielectric nature of the soil. Depending on the soil texture and soil moisture, the dielectric constant of soil may vary significantly even if the soil moisture content changes only slightly, thereby affecting the backscattering coefficient obtained by active microwave remote sensing [25].

In addition to radar system parameters, the backscattering coefficient obtained by active microwave remote sensing is mainly determined by the soil dielectric constant, surface roughness parameters, and vegetation canopy water content [26]. Therefore, under the configuration of a known radar system, the difficulty in estimating the soil moisture at the regional scale lies in effectively removing the influences of the vegetation canopy and surface roughness [27–30]. In the process of vegetation impact correction, the water-cloud model (WCM) and Michigan Microwave Canopy Scattering Model (MIMICS) model are commonly used to separate the contributions of the vegetation backscattering and soil backscattering from the radar data [27,31,32]. The vegetation canopy water content is a vital input parameter in the above models; its value can be obtained by establishing empirical models based on optical remote sensing data. In practical applications, the WCM is more widely used for its simplified physical theory, simple model structure, and fewer input parameters [27,33–35]. Several models have been proposed to develop the relationship between the surface parameters and the backscattering coefficient and then realize the inversion of the soil moisture, including the integral equation model (IEM) [36,37], advanced integral equation model (AIEM) [37,38], Oh model [37,39], Dubois model [37,40,41], and some regression empirical models [42,43]. In recent years, several SAR satellites (Radarsat-2, Sentinel-1, ALOS-2, and TerraSAR-X) have been used for soil moisture monitoring at the L/C/X-bands, and studies based on these remote sensing data have achieved satisfactory results [2,44–52]. The domestic GF-3 satellite, launched on 10 August 2016, has shown good reliability and application prospects in the field of soil moisture retrieval [53].

In conclusion, many efforts have been made for an efficient and accurate soil moisture monitoring, with significant progress in terms of algorithm development and applications. However, there remain several gaps in existing literature: (1) Most of the soil moisture retrieval studies selected only one inversion model, and studies on the coupling of several commonly used models are limited; (2) In the few studies on coupled models, parallel comparisons of these models were emphasized, and only a few have comprehensively utilized the results of each model to obtain an optimal solution soil moisture map with higher accuracy than any of the independent models; (3) In terms of remote sensing data sources, there are few cases of a synergistic application of radar and optical data from high-resolution satellites launched by China; (4) The combination of background knowledge and commonly used inversion models was rarely considered.

The purpose of this study was to integrate multi-source data, realize the coupling of commonly used models, and introduce background knowledge into the optimal solution method to achieve high-precision soil moisture mapping.

2. Study Area and Available Datasets

2.1. Study Area

The study area is located in the southwest of Jiangsu Province, covering an area of approximately 25 km × 25 km in the Yangtze River Delta (see Figure 1). The subtropical monsoon climate here is characterized by mild and humid weather and four distinct seasons. The average temperature and precipitation are approximately 15.5 °C and 1152.1 mm, respectively. The dominant wind direction in this region is easterly. The land cover map of the study area was based on the Finer Resolution Observation and Monitoring of Global Land Cover (FROM-GLC) developed using Thematic Mapper (TM) and Enhanced Thematic Mapper Plus (ETM+) data (see Figure 1).

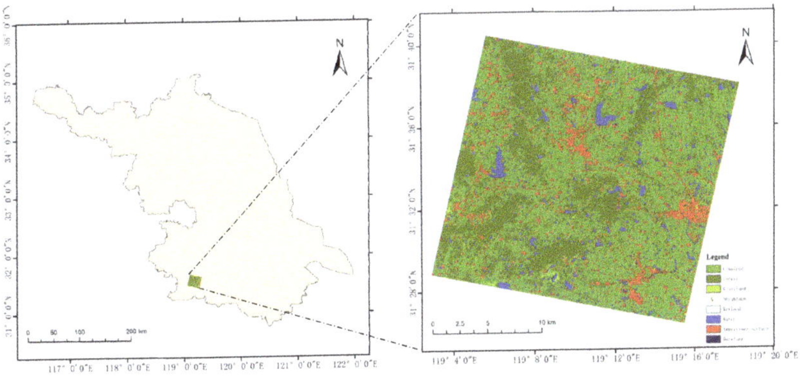

Figure 1. (**a**) Geographical location of the study area and (**b**) major land-cover types.

The study area contains several types of land covers, including cropland, forest, grassland, water, impervious surface, and bare land. Among them, farmland is the main land-cover type, accounting for 59.02% of the study area, and the main crop is rice. During the in-situ experiment, the rice plants were in the mature stage, and there was no water in the fields. There were many fragmented water bodies, accounting for 4.73% of the total area therein. During the field experiments, we found that most of the forest land marked in the land cover map had been planted with shrubs. Therefore, we combined the forest and shrub lands under the same land use type. Although most of the research area is characterized by farmland land cover, the underlying surface of the farmland is relatively complex. From the perspective of planting types, there are rich vegetable varieties, mature rice, and a few rice fields that were still covered with stubble and weeds after harvesting.

2.2. GF-1 Data

To eliminate the effect of vegetation water content (VWC) on the soil moisture estimation, the GF-1 data was used in this study. The GF-1 satellite was launched by China on 26 April 2013. The CCD sensors include four spectral bands (blue: 0.45–0.52 μm, green: 0.52–0.59 μm, red: 0.63–0.69 μm, and near-infrared: 0.77–0.89 μm). The two images used in this study acquired on 17 and 18 of October 2018 were downloaded from the China Center for Resource Satellite Data and Application website (http://218.247.138.119:7777/DSSPlatform/index.html); these have a spatial resolution of 8 m.

The reprocessing of the GF-1 data included radiation correction, atmospheric correction, and geometrical correction. Th radiation correction for the four bands was performed using the ENVI 5.3 software to convert the digital number (DN) of the images to the surface spectral reflectance. The atmospheric correction was conducted using the FLAASH Atmospheric Correction toolbox using the ENVI software [44,47,52–55]. After atmospheric correction, the images were geo-referenced based on 25 ground control points. After the preprocessing procedures, the four processed bands of the GF-1 data were used to calculate the vegetation index during the in-situ experiments.

2.3. GF-3 Data

The C-band GF-3 satellite is a multi-polarization high-resolution commercial SAR satellite, launched by China on 10 August 2016. The GF-3 SAR sensors have 12 imaging modes with a fine spatial resolution of up to 1 m, and the data used in this study were quad-polarization images of the Quad-Polarization StripI (QPSI) mode acquired on 17 and 19 of October. The SAR images have a nominal spatial resolution of 8 m. The incidence angle ranged from 35.38 to 37.21°, at a frequency of 5.4 GHz.

A series of preprocessing procedures were performed on the single-look complex images using SARscape 5.5 module developed by SARmap using the ENVI 5.5 software. First, Multi-Look processing was applied to make the texture of the original images close to the real condition and to reduce the speckle noise. The enhanced Lee filter with a window size of 5 pixels by 5 pixels was applied to reduce the speckle noise as the filtering and denoising process, providing the best results compared to using the other filters tested in this study [56]. The obtained data were geocoded using digital elevation maps for geometric fine correction. Finally, radiometric calibration was conducted to transform the DN of the pixel into the backscattering coefficient in the multi-polarization mode.

2.4. In Situ Measurements

In situ measurements were conducted from 17 to 19 of October 2018, simultaneously with the GF-3 and GF-1 acquisitions. During the in-situ measurements, no evident changes in the temperature or precipitation were found in the study area. A total of 94 soil samples were selected from this area. The position of each sample field was identified and recorded by Global Positioning System (GPS). Figure 2 shows the map of the sampling sites and details regarding the soil moisture samples.

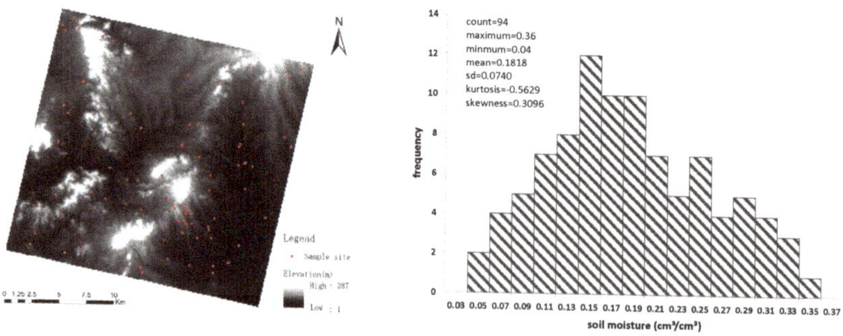

(a) (b)

Figure 2. Map of sampling sites and elevation (**a**) and statistical characteristics of measured soil moisture samples (**b**).

The entire study area was classified on the basis of three strategies, each containing 4–5 classes (listed in Table 1). The number and proportion of sample points distributed in each class were counted to observe the difference between the proportion of sample points contained in a class and the proportion of this class in the total area of the study area [57]. As listed in the table, the proportion of sample points in a certain class is similar to the proportion of the class in the total area of the study area, indicating that the sampling is largely representative.

Table 1. Detailed information of the 94 samples taken from a study area in Jiangsu Province, China.

	Type Class	Number of Samples	Proportion of Total Samples (%)	Proportion of This Class in the Entire Study Area (%)
Land Use Type	Farmland	63	67.02	68.62
	Forest and shrub	25	26.60	30.51
	Grassland	4	4.26	0.84
	Bare land	2	2.13	0.03
Percentage Content of Clay (%)	14	1	1.06	0.67
	26	37	39.36	35.04
	28	26	27.66	35.94
	29	28	29.79	26.12
	37	2	2.13	2.23
Slope (°)	≤1	36	38.30	41.53
	1–2	32	34.04	28.47
	2–3	12	12.77	10.34
	3–4	7	7.45	5.02
	>4	7	7.45	14.64

The soil moisture measurements were mainly acquired using time-domain reflectometry (TDR) probes for the top 5 cm in all the 94 sample fields, on the basis of the hypothesis that the factors affecting radar signal penetration, such as vertical inhomogeneity of soil moisture content at a depth of 0–5cm, could be ignored [42,56,58]. For each sample field, the probes were placed vertically, and three close-distance measurements (few meters apart) of the near-surface volumetric soil moisture were collected and averaged. Furthermore, a total of 29 samples were simultaneously collected, uniformly mixed, and placed in cutting rings and aluminum specimen boxes using the oven-drying method for the top 5 cm. These cutting rings and aluminum specimen boxes were weighed and recorded, and then transported back to the laboratory for drying in an oven at 108 °C for 24 h, until the boxes were completely dehydrated. The obtained gravimetric soil moisture (g/g) and soil bulk density (g/cm^3) were used to calculate the volumetric soil moisture (cm^3/cm^3), which was used to calibrate the TDR measurements. A linear model was used to calibrate the volumetric soil moisture acquired by the TDR probes, and the result is shown in Figure 3. The calibrated values of the volumetric soil moisture ranged from 0.03 to 0.35 cm^3/cm^3, averaging at 0.19 cm^3/cm^3. A total of 60 soil moisture samplings were randomly selected as independent testing data, and the remaining samplings served as training data in different models and algorithms.

In this study, the root-mean-square (RMS) height and correlation length were measured as the roughness parameters using a needle profiler made of a 1 m-long board with 100 pins digitized at an interval of 1 cm, and a digital camera. At the sample fields, the needle profiler was inserted onto the soil surface in the east-west and north-south directions, and then two images were taken to record the roughness characteristics. The roughness values were calculated from the digitized curve based on the roughness images using a developed MATLAB program. The RMS height ranged from 0.3 to 1.7 cm, averaging at 1 cm. The correlation length ranged from 5 to 25 cm, with the average value of 15 cm. The results show that the microtopography of the study area is relatively flat. A total of 57 vegetation samples were collected and sealed in plastic black bags, in order to minimize the effects of photosynthesis and transpiration on the water content of the vegetation samples. The VWC (kg/m^2) was calculated from the fresh and oven dried sampling measurements using the oven-drying method.

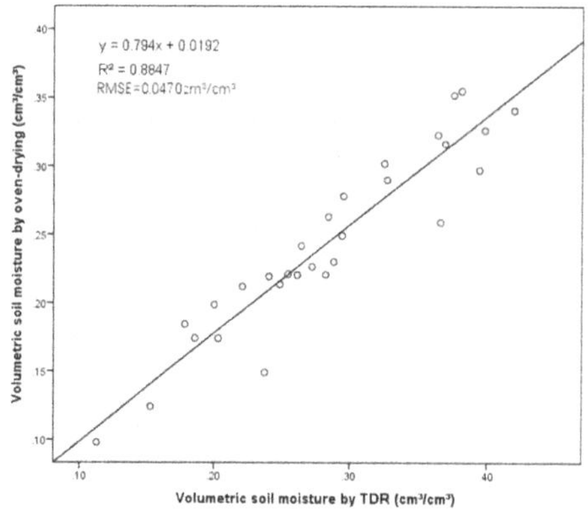

Figure 3. Calibration of volumetric soil moisture acquired by TDR probes.

After fixing the drying process, the vegetation samples were dried in an oven at 65 °C, until the samples were completely dehydrated.

3. Methodology

In this study, a set of combined approach comprising a set of methods to obtain high-precision soil moisture, which integrate multi-source data, realize the coupling of commonly used models, and introduce background knowledge into the optimal solution method. The procedure of the methodology for soil moisture mapping, illustrated in Figure 4, is explained as follows:

In the first step, the influence of the vegetation canopy water content was removed from the total backscattering coefficients, and the backscattering coefficient of the bare soil was obtained. The specific operations were as follows: first, an empirical VWC model was established using four vegetation indices extracted from the GF-1 satellite data and measured vegetation canopy water content data, and the accuracy of this empirical model was then compared with that of the conventional one-variable quadratic model; second, a WCM with localized parameters was established using the HH and VV polarization backscattering coefficients extracted from the GF-3 satellite data and in situ experimental data, and the accuracy of this model was compared with that of a model that uses empirical parameters.

In the second step, four commonly used soil moisture inversion models were coupled, and a total of nine soil moisture maps were obtained. The specific operations were as follows: five maps of the soil moisture and four RMS height maps were obtained using the AIEM model, Oh model, and look-up table (LUT) method; the four RMS height maps were then taken as input data for the Dubois model, and four soil moisture maps were obtained.

In the third step, based on the above nine soil moisture maps, a strategy of pixel classification and algorithm selection was applied to draw the optimal solution soil moisture map based on background knowledge. The specific operations were as follows: first, the study area was divided into several categories based on a certain background knowledge, each of which was assigned one of the nine algorithms with best precision. The results of all the categories were then combined to form a soil moisture map of the entire study area; finally, the three soil moisture maps based on the three sets of background knowledge were compared, and one with the best precision was selected as the optimal solution soil moisture map.

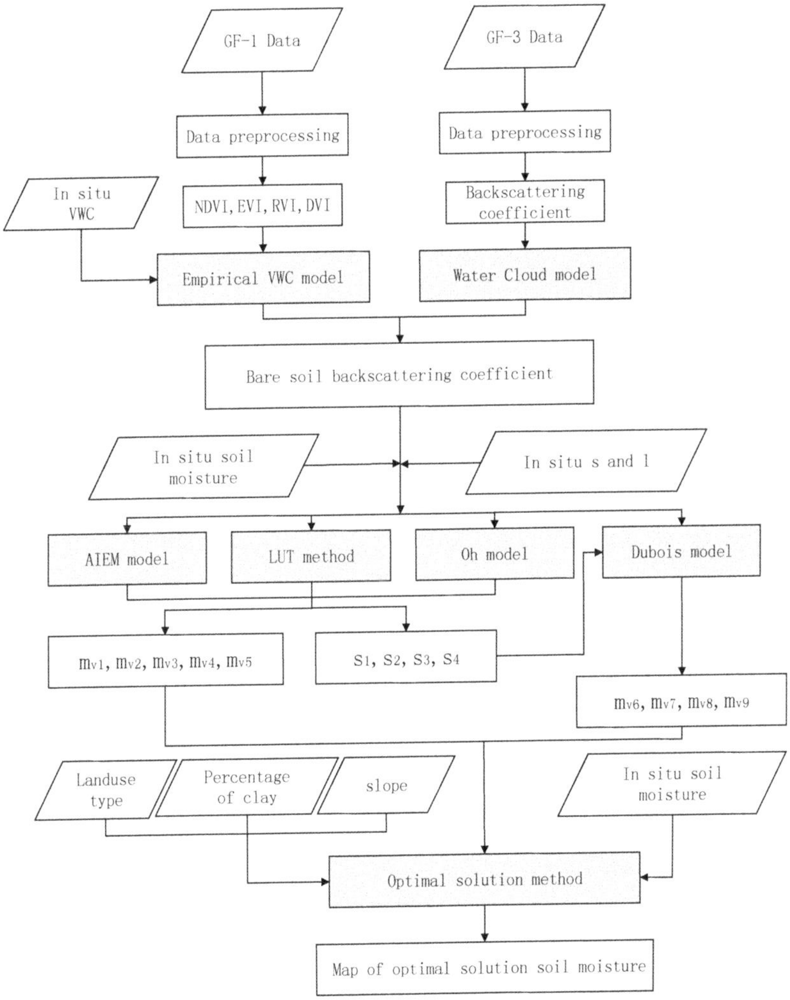

Figure 4. Flowchart of the proposed soil moisture mapping. VWC, vegetation water content; m_{vn}, soil moisture content obtained by four commonly used models; s and l: the root-mean-square (RMS) height and correlation length.

3.1. Vegetation Effect Correction Models

The empirical VWC model and WCM with localized parameters were applied to remove the vegetation effects from the total backscattering coefficient. The reason for constructing an empirical VWC model considering four vegetation indices is as follows: Although most of the study area is farmland, the underlying surface of the farmland is relatively complex, so it is likely that the VWC of the complex vegetation types cannot be accurately described by any single vegetation index; furthermore, a total of 57 VWC samples were collected during the in situ experiment, and the measured data were fully utilized in the process of constructing an empirical model that was more suitable for the study area. Similarly, the WCM with localized parameters makes full use of the in- situ experiment. To make the results more convincing, the accuracies of the empirical VWC model and WCM with localized parameters were compared with those of the conventional one-variable quadratic model and WCM with empirical parameters.

3.1.1. Vegetation Water Content Estimation Model

The VWC is an important parameter in the WCM. To eliminate the influence of vegetation canopy on the total backscattering coefficient, the VWC should be estimated. Because of the time and economic costs, it is difficult to acquire the VWC via ground tests on a large scale. Fortunately, the development of remote sensing technology has led to an effective solution to this problem. Published studies confirmed that the VWC has a significant correlation with the vegetation indices obtained from remote sensing datasets [55–60]. The vegetation indices often used to estimate the VWC include the normalized difference vegetation index (NDVI), normalized difference water index (NDWI), enhanced vegetation index (EVI), difference vegetation index (DVI), and ratio vegetation index (RVI) [61–64].

The GF-1 wide-field-view data synchronized with the GF-3 data passing time were used to calculate the vegetation indices. Considering the configuration of the GF-1 satellite, four vegetation indices were selected for the calculation: the NDVI, EVI, RVI, and DVI. These were calculated as follows:

$$\mathrm{NDVI} = \frac{NIR - RED}{NIR + RED} \tag{1}$$

$$\mathrm{EVI} = 2.5 \times \frac{NIR - RED}{NIR + 6.0RED - 7.5BLUE + 1} \tag{2}$$

$$\mathrm{RVI} = \frac{NIR}{RED} \tag{3}$$

$$\mathrm{DVI} = NIR - RED \tag{4}$$

where NIR is the reflectivity at the near-infrared band; RED is the reflectivity at the red band; $BLUE$ is the reflectivity at the blue band.

Previous works have indicated that the one-variable quadratic models, along with the linear and exponential models, are suitable for describing the correlation between the vegetation indices and VWC [55,61–63]. However, in most studies, only one of the vegetation indices (such as the NDVI) with the strongest correlation with the VWC was used in the calculation [61]. For medium-scale research areas with a complex surface cover, a combination of multiple vegetation indices may be more effective and accurate.

In this study, we developed a new empirical model in which all the four vegetation indices were incorporated in the calculation. Moreover, the accuracy of the new model was compared with those of the four one-variable quadratic models with each vegetation index, so as to verify the efficiency of the new model. The one-variable quadratic models used for accuracy comparison and the new model can be described as follows:

$$VWC_{VI} = \mathrm{a} \times VI^2 + b \times VI + c \tag{5}$$

where VWC_{VI} is the VWC calculated using the one-variable quadratic model (kg/m^2); VI is the vegetation index obtained from the GF-1 satellite data; a, b, and c are the empirical parameters obtained using the least-squares method.

Each vegetation index has a certain scope of application and works well only under a certain growth condition, so a VWC model constructed using only a single vegetation index may not effectively reflect the condition of the entire study area. Considering the diversity of the vegetation types in the study area, a combined model incorporating the four vegetation indexes was developed. The 1stOpt software was used in the process of constructing the combined model, and the fitting accuracy and number of parameters were considered. The combined model can be expressed as follows [59]:

$$\mathrm{VWC} = d + e \times NDVI \times RVI + f \times NDVI \times EVI + g \times DVI \times EVI + h \times RVI \times EVI \tag{6}$$

where *VWC* is the *VWC* calculated using the new empirical model (kg/m^2); *NDVI*, *EVI*, *DVI*, and *RVI* are the vegetation indices obtained from the GF-1 satellite data; d, e, f, g, and h are the empirical parameters obtained using the least-squares method.

3.1.2. Water-Cloud Model

The backscattering coefficient obtained from the study area is affected by the land surface and radar configuration parameters [27,31]. In this study, the radar configurations were obtained from the head file of the GF-3 SAR sensor. Therefore, only the influences of land surface parameters were considered, including the soil moisture, roughness, and vegetation. To extract the backscattering of the bare soil surface, which is directly correlated to the soil moisture, the effect of vegetation canopy should be removed.

To describe the backscattering of the soil and vegetation in the vegetation-covered soil surfaces, a semi-empirical WCM, first developed by Attema and Ulaby [31], has been widely used as the vegetation impact correction model. The WCM is based on the following basic assumptions: (1) The vegetation canopy is represented as a homogeneous horizontal water cloud of identical and uniformly water spheres. (2) The multiple scattering between the canopy and soil is ignored. (3) The only significant variables in the model are the height of the canopy layer and cloud density, and the latter is assumed to be proportional to the volumetric VWC of the canopy [31].

For a given incidence angle, the classic WCM can be expressed as follows:

$$\sigma^0 = \sigma^0_{veg} + \tau^2 \times \sigma^0_{soil} \tag{7}$$

$$\tau^2 = \exp(-2 \times B \times VWC / \cos\theta) \tag{8}$$

$$\sigma^0_{veg} = \mathrm{A} \times VWC \times \cos\theta \times \left(1 - \tau^2\right) \tag{9}$$

where σ^0 is the total backscattering coefficient, considering the incoherent sum of the contributions of the vegetation and underlying soil (dB), σ^0_{veg} is the backscattering coefficient of the vegetation canopy (dB), σ^0_{soil} is the direct backscattering coefficient of the underlying soil surface (dB), τ^2 is the two-way transmissivity of the vegetation, VWC is the vegetation water content (kg/m^2), θ is the incident angle (°), and A and B are parameters depending on the vegetation type. There is no general theoretical basis to obtain the values of A and B, which are typically determined by fitting the WCM against experimental datasets.

The original WCM has been subsequently modified or extended by various researchers [27]. Bindlish and Barros showed that the accuracy of the WCM could be improved by considering the radar-shadow effect. The concept of a dimensionless vegetation correlation length was used to better describe the geometric effect of the vegetation spacing; this parameter is similar to the surface correlation length used in the AIEM [59]. It can be interpreted as the distance at which the plants are considered independent scatterers [60]. Based on the improved WCM, Equations (10) and (11) must be modified to correct for the radar-shadow effect. The modified total backscattering coefficient and vegetation effect can be expressed as follows:

$$\sigma^0 = \sigma^*_{veg} + \tau^2 \times \sigma^0_{soil} \tag{10}$$

$$\sigma^*_{veg} = \sigma^0_{veg} \times [1 - exp(-\alpha)] \tag{11}$$

where σ^*_{veg} is the modified vegetation contribution, and α is the radar-shadow coefficient parameter, which is also considered the dimensionless vegetation correlation length.

In this study, three different WCM experiments were conducted: in the first case, the classic WCM was used, and the empirical parameters A and B were recorded as 0.0012 and 0.091, respectively, based on the values reported by Bindlish [60] (WCM_1). In the second case, the same classic WCM was used; however, the empirical parameters A and B were obtained using the least-squares method with the in situ experimental datasets (WCM_2). In the third case, an improved WCM considering the radar-shadow effects was selected; the empirical parameters A, B, and α were calculated using the least-squares method (WCM_3) with the in situ experimental datasets.

3.2. Soil Moisture Content Retrieval Model

To convert the bare soil backscattering coefficient to the soil moisture content, four commonly used inversion models (AIEM model, LUT method, Oh model, and Dubois model) were coupled. In this section, only the general principles and formulae of these models are introduced. A more detailed discussion of the relevant techniques and theory can be found in the references mentioned in this section.

The process of coupling the four models involved five steps. Only the general principles of these models are introduced in this section, and since steps 2 and 3 involve the analysis and application of the results generated in the intermediate process, the details of these are described in Section 4.2.

Step 1: The AIEM-simulated database was generated using the AIEM model, based on the in situ datasets and GF-3 configuration.

Step 2: Based on the AIEM-simulated database obtained in step 1, the responses of the bare soil backscattering coefficients and input parameters were analyzed, and the relationship structure between the soil moisture and input parameters was then determined. Thereafter, the empirical model was constructed using the measured soil moisture and bare soil backscattering coefficient; this model was used as an algorithm to map the soil moisture (m_{v1}).

Step 3: Similarly, based on the AIEM-simulated database, the LUT method was established using tree cost functions, which generate three soil moisture maps (m_{v2}, m_{v3}, and m_{v4}) and three RMS height results (s1, s2, and s3).

Step 4: In parallel with the above three steps, based on the in-situ database and backscattering coefficient, the Oh model was used to map the soil moisture (m_{v5}), along with the RMS height result (s4).

Step 5: After the above four steps, a total of four RMS height results were obtained, which were used as input data to the Dubois model. Based on the backscattering coefficient and measured soil moisture, four soil moisture maps (m_{v6}, m_{v7}, m_{v8}, and m_{v9}) were obtained corresponding to the four RMS height results.

3.2.1. Estimated Model Constructed by AIEM

The AIEM model was developed by Chen and Wu in 2003 based on the classic IEM [35–37]. The AIEM is a well-established theoretical model in which the discontinuities in the surface roughness and Fresnel reflection coefficient were resolved by replacing the Fresnel reflection coefficients with a transition function [37,38,65]. Extensive experimental datasets were used to analyze the accuracy and validity of this model, the results of which indicated that the AIEM could accurately simulate the backscattering coefficients under a wide range of conditions [38,66].

In this study, the AIEM was adopted to establish a simulated database, and the responses of the bare soil backscattering coefficients, incident angle, roughness parameter, soil dielectric constant, and frequency were simulated under the configuration of the GF-3 sensor. The soil dielectric constant simulated from the AIEM was converted to the soil moisture via the dielectric-mixed Dobson model [67]. The frequency of the GF-3 sensor was fixed at 5.4 GHz, and other input parameters based on in situ data are as listed in Table 2.

Table 2. Values of the input parameters to the AIEM model.

Input Parameters	Minimum	Maximum	Interval
Incidence angle (°)	30	60	1
RMS height (cm)	0.3	1.8	0.1
Correlation length (cm)	5	25	1
Soil moisture (vol. cm^3/cm^3)	0.03	0.36	0.01

Considering the difficulties in simulating the bare soil backscattering coefficients in cross-polarization using the AIEM, only the backscattering coefficients simulated in HH and VV polarizations were used in this study. Based on the AIEM and Dobson model, the equations of the bare soil backscattering coefficients can be expressed as follows:

$$\sigma^0_{PP} = f(\theta, f, s, l, m_v) \tag{12}$$

where θ is the incidence angle, f is the frequency of the GF-3 sensor, s is the RMS height, l is the correlation length, and m_v is the bare soil moisture content.

3.2.2. LUT Inversion Method

The LUT of the HH and VV polarization backscattering coefficients was obtained from previous studies [37,38]. In this study, the LUT method was used to obtain soil moisture and root mean square (RMS) height at the same time. It was based on the AIEM-simulated database, and under the configuration of GF-3 satellite, each record of the database contains five interrelated parameter values, including backscattering coefficient, root mean square (RMS) height, correlation length and soil moisture. For a particular pixel of the bare soil backscattering coefficient map, a specific record in the simulated database could be located by minimizing the cost function, and the soil moisture and root mean square (RMS) height in this record were extracted as the best-fit inversion value corresponding to the backscattering coefficient of the pixel.

Three cost functions were established to retrieve the bare soil moisture in this study:

$$Z_1 = min\left(\sqrt{\left(\sigma^0_{HH} - \sigma^0_{HH\_AIEM}\right)^2}\right) \tag{13}$$

$$Z_2 = min\left(\sqrt{\left(\sigma^0_{VV} - \sigma^0_{VV\_AIEM}\right)^2}\right) \tag{14}$$

$$Z_3 = min\left(\sqrt{\left(\sigma^0_{HH} - \sigma^0_{HH\_AIEM}\right)^2 + \left(\sigma^0_{VV} - \sigma^0_{VV\_AIEM}\right)^2}\right) \tag{15}$$

where σ^0_{HH} and σ^0_{VV} are the HH and VV polarization backscattering coefficients (dB) obtained from the bare soil backscattering images, respectively, and $\sigma^0_{HH\_AIEM}$ and $\sigma^0_{VV\_AIEM}$ are the HH and VV polarization backscattering coefficients (dB) in the AIEM-simulated database, respectively.

3.2.3. Semi-Empirical Oh Model

Oh et al. proposed several versions of the semi-empirical polarimetric models between 1992 and 2004 [37,39]. The Oh model is based on physical theoretical models, and the relevant parameters in this model were obtained as field experimental data using several polarimeter radar scatterometers [39,68]. The latest version of the Oh model modified in 2004 contains a new equation that ignores the correlation length, built on the previous version [37]. This improvement was made on the basis of the results of a published research, which indicated that the ratio q ($\sigma^0_{HV}/\sigma^0_{VV}$) is not sensitive enough to the roughness parameters and that the measurement of the correlation length may vary depending on the scale of the instrument used [39].

In this study, a new expression for the Oh model (2004) was used to estimate the soil moisture of the bare soil. The input parameters of the Oh model (2004) include the incidence angle, wavenumber, RMS height, and soil moisture or dielectric constant. The 2004 version of the Oh model can be expressed as [37,39]:

$$\sigma^0_{VH} = 0.11 \times m_v^{0.7} \times (cos\theta)^{2.2} \times \left[1 - exp\left(-0.32 \times (ks)^{1.8}\right)\right] \tag{16}$$

$$p \equiv \frac{\sigma_{HH}^0}{\sigma_{VV}^0} = 1 - \left(\frac{\theta}{90^\circ}\right)^{0.35 \times m_v^{-0.65}} \times exp\left[-0.4 \times (ks)^{1.4}\right] \tag{17}$$

$$q \equiv \frac{\sigma_{VH}^0}{\sigma_{VV}^0} = 0.095 \times (0.13 + sin1.5\theta)^{1.4} \times \left\{1 - exp\left[-1.3 \times (ks)^{0.9}\right]\right\} \tag{18}$$

where p is the co-polarized ratio, q is the cross-polarized ratio, σ_{VH}^0 represents the backscattering coefficients for VH polarization (dB), and ks is the RMS height normalized to the wavelength (k is the wavenumber). The dataset in this study satisfies the validity range of the Oh model (2004) (0.04–0.29 cm^3/cm^3 for m_v, 0.13–6.98 for ks, and 10–70° for θ) [37].

3.2.4. Semi-Empirical Dubois Model

The Dubois model was developed in 1995 using scatterometer data for retrieving the soil moisture content and RMS height from the remote sensing radar data [37]. The Dubois model has been applied in some studies, with satisfactory accuracy [40,41]. The validity domain of the model is $m_v < 0.35$ cm^3/cm^3 and $ks \leq 2.5$, $\theta \geq 30°$. The equations applicable to HH and VV polarization data are:

$$\sigma_{VV}^0 = 10^{-2.35} \times \frac{\cos^3\theta}{\sin^3\theta} \times 10^{0.046 \times \varepsilon \times \tan\theta} \times (ks \times \sin\theta)^{1.1} \times \lambda^{0.7} \tag{19}$$

$$\sigma_{HH}^0 = 10^{-2.75} \times \frac{\cos^{1.5}\theta}{\sin^5\theta} \times 10^{0.028 \times \varepsilon \times \tan\theta} \times (ks \times \sin\theta)^{1.4} \times \lambda^{0.7} \tag{20}$$

where ε is the dielectric constant, and λ is the radar wavelength.

In this study, the Dubois and Dobson models were combined to complete the following processes: the RMS height map obtained using the Oh model and LUT inversion method was used as the input data, and the soil dielectric constant was generated using the Dubois model based on the HH and VV polarization backscattering coefficients, which were converted to the soil moisture content using the dielectric-mixed Dobson model.

3.3. Optimal Solution Method

This section presents a simple and effective optimal solution method used to select and combine the nine soil moisture mapping algorithms proposed above based on three types of background knowledge. Figure 5 shows the overall process of the optimal solution method.

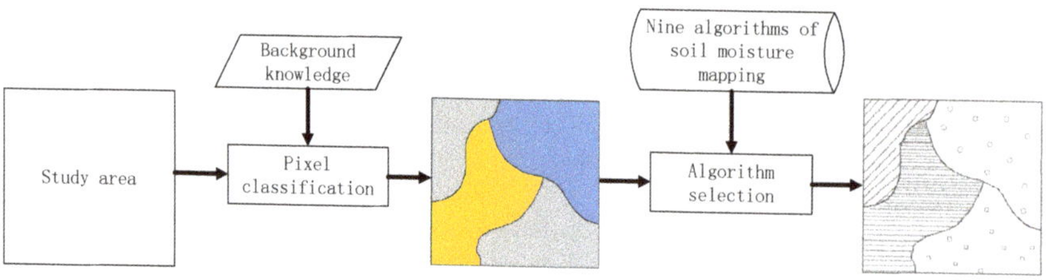

Figure 5. Schematic of the optimal solution method based on a specific background knowledge. Color: pixel classification results; Pattern: soil moisture results after model.

For a specific background knowledge, the pixels in the study area were classified into several categories using the classification strategy constructed based on this knowledge, and the algorithm with the highest accuracy was selected in terms of the evaluation index. The algorithm results corresponding to each category were calculated independently. Thereafter, the calculation results of all the categories were combined to form the soil moisture map of the entire study area.

A total of three spliced maps of the soil moisture were obtained using three classification strategies. Finally, these maps were compared, and the one with the best accuracy was selected as the optimal solution soil moisture map of the study area.

Three optimal solution classification strategies were proposed as follows: (1) Based on the four main land use types (farmland, grassland, shrub, and bare land) in the study area, a statistical analysis was made on the accuracy of the nine soil moisture inversion maps, and the best inversion method for each type was selected; (2) Based on the percentage of clay, nine soil moisture inversion accuracies were calculated corresponding to the five clay percentage values (14, 26, 28, 29, and 37%), and the best inversion method was selected for each type; (3) Similarly, based on the slope levels of the study area, a statistical analysis was made on the accuracies of the nine soil moisture inversion maps under five slope levels (Level 1: ≤1°; Level 2: 1–2°; Level 3: 2–3°; Level 4: 3–4°; Level 5: ≥4°), and the inversion method with the highest accuracy for each slope level was selected as the optimal solution method.

The equation of the optimal solution methods is expressed as follows:

$$m_v = f(m_{v1}, m_{v2}, \ldots\ldots, m_{v9}) \tag{21}$$

where m_{vn} is the retrieved moisture based on the four models.

4. Results and Discussion

4.1. Vegetation Effect Correction

This section presents the results of the calculation parameters of the empirical VWC model and WCM with localized parameters. To prove the superiority of the two models compared with the commonly used empirical models and empirical parameters, we developed conventional one-variable quadratic models and WCM with empirical parameters for comparative experiments. The comparison results are given in this section.

The developed empirical model was used to retrieve the VWC from the GF-1 vegetation indices, compared with the one-variable quadratic model. The agreement between the measured VWC and the modeled VWC was evaluated on the basis of the Pearson correlation coefficient (PCC). Table 3 lists the coefficient of determination values based on 19 validation sample datasets.

Table 3. Parameters of the VWC model and the PCC based on the 19 validation sample datasets.

VWC_{model}	Empirical Parameters					PCC
	a		b	c		
VWC_{NDVI}	26.0820		−1.8558	1.0915		0.9158
VWC_{EVI}	4.6957		1.4673	0.9790		0.9054
VWC_{RVI}	1.3854		−1.7960	1.3724		0.9124
VWC_{DVI}	46.9080		5.8157	0.9459		0.8805
	d	e	f	g	h	
VWC	1.1604	4.3773	18.2684	14.1764	−4.8787	0.9442

From Table 3, we find that by comparing the performances of the four vegetation indices in the one-variable quadratic models, VWC_{NDVI} provides a higher accuracy than the other vegetation index models (PCC = 0.9158). However, the accuracy of the VWC estimated from the combined vegetation model was significantly higher than those estimated from any one of the one-variable quadratic models (PCC = 0.9442).

The VWC was estimated using GF-1 satellite data. As shown in Figure 6, the PCC between the estimated and measured VWC values is 0.9440, the root-mean-square error (RMSE) of the estimated VWC is 0.1651 kg/m^2, and the coefficient of determination (R^2) is 0.8915. The correlation between the measured and estimated VWC values was significant at the 0.01 level.

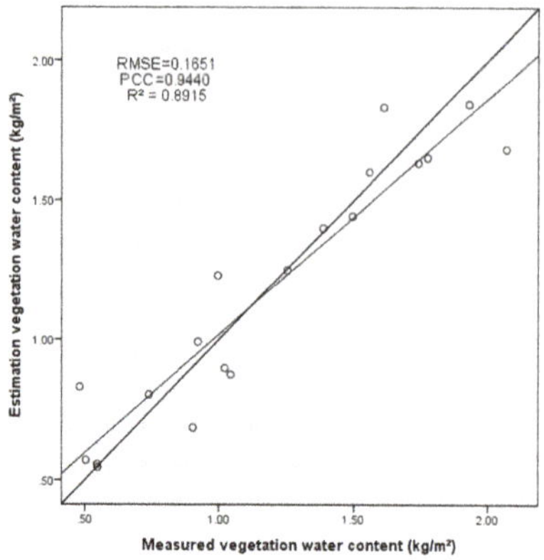

Figure 6. Scatter plot between the measured and estimated vegetation water content values using the new empirical model based on 19 sample datasets. The 1:1 line is included in the plot.

Three different WCM experiments were conducted using the results of the above VWC. As mentioned previously, the classic WCM was used, and the empirical parameters *A* and *B* were recorded as 0.0012 and 0.091, respectively (WCM_1); in the second case, the same classic WCM was used; however, the empirical parameters *A* and *B* were obtained using the least-squares method with the in situ experimental datasets (WCM_2); in the third case, the improved WCM considering the radar-shadow effects was selected, and the empirical parameters *A*, *B*, and α were calculated using the least-squares method (WCM_3) with the in situ experimental datasets. The values of the empirical parameters used in the WCM were recorded by referring to published works or calculated based on the in situ experimental datasets. The AIEM model was used to simulate the bare soil backscattering at each sampling location where in situ experimental data of the soil moisture and auxiliary data were available.

To evaluate the validity of the three WCM experiments conducted on the vegetation effect correction in the GF-1 and GF-3 satellite configurations, the bare soil backscattering simulated using the WCM was compared with that simulated using the AIEM, in Figure 7.

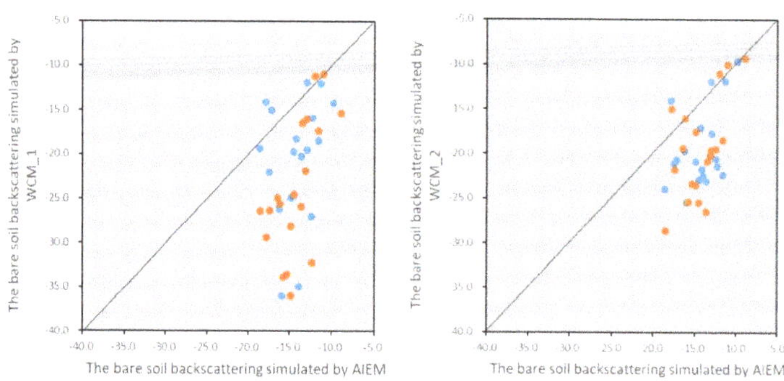

(a) (b)

Figure 7. *Cont.*

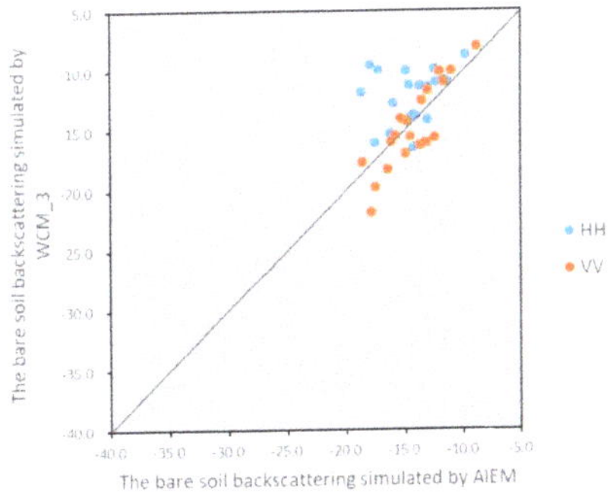

(**c**)

Figure 7. Scatter plots between the bare soil backscattering (dB) simulated by AIEM and three different water-cloud model strategies based on experimental data. The 1:1 line is included in the plot. (**a**) WCM_1; (**b**) WCM_2; (**c**) WCM_3.

Based on the analysis of the comparative experimental results, we find that the empirical VWC model and WCM with localized parameters considering the radar-shadow effects do have evident accuracy advantages, confirming that the accuracy of the calculated bare soil backscattering coefficient could be ensured.

4.2. Bare Soil Moisture Estimation Based on Four Commonly Used Models

The process of coupling the four commonly used models included the five steps described in Section 3.2. In this section, Sections 4.2.1–4.2.4 correspond to steps 2–5, respectively. Because steps 2 and 3 involve the analysis and application of the results generated in the intermediate process, they will be analyzed in more detail.

4.2.1. Bare Soil Moisture Estimation Based on AIEM

To build the bare soil moisture empirical estimation model based on the AIEM and map the soil moisture content, three steps were carried out. First, the responses of the backscattering coefficient and input parameters were analyzed to support the judgment of the structure of the inversion model; second, the parameters of this model were calculated based on the in situ experimental data; third, the soil moisture content in the entire area was mapped based on the model.

The input parameters of the AIEM include the radar system parameters and surface parameters of the study area. In the GF-3 radar sensor configuration, the frequency and incidence angle of the radar system parameters can be obtained from the header file. Therefore, only the responses of the bare soil backscattering coefficients and surface parameters of the study area were mainly considered in this study.

At a frequency of 5.4 GHz and an incident angle of 37°, the AIEM was used to generate a simulated database of the changed soil moisture conditions with a fixed roughness parameter value. Two conditions were required to be analyzed: first, the correlation length was considered a fixed value (15 cm), and the RMS height was considered a fixed-interval parameter (0.3, 0.6, 0.9, 1.2, 1.5, and 1.7 cm); second, the RMS height was considered a fixed value (1.0 cm), and the correlation length was considered a fixed-interval parameter (5, 10, 15, 20, and 25 cm). Figure 8 shows the results. The following rules can be summarized from this figure: First, When the roughness parameter is fixed, there is an evident nonlinear relationship between the soil moisture and the soil backscattering coefficients, the trends of which do not vary with the roughness interval parameter. With the increase in the soil moisture, the trend in the soil backscattering coefficient is upward, and the growth rate gradually decreases with an increase in the soil moisture.

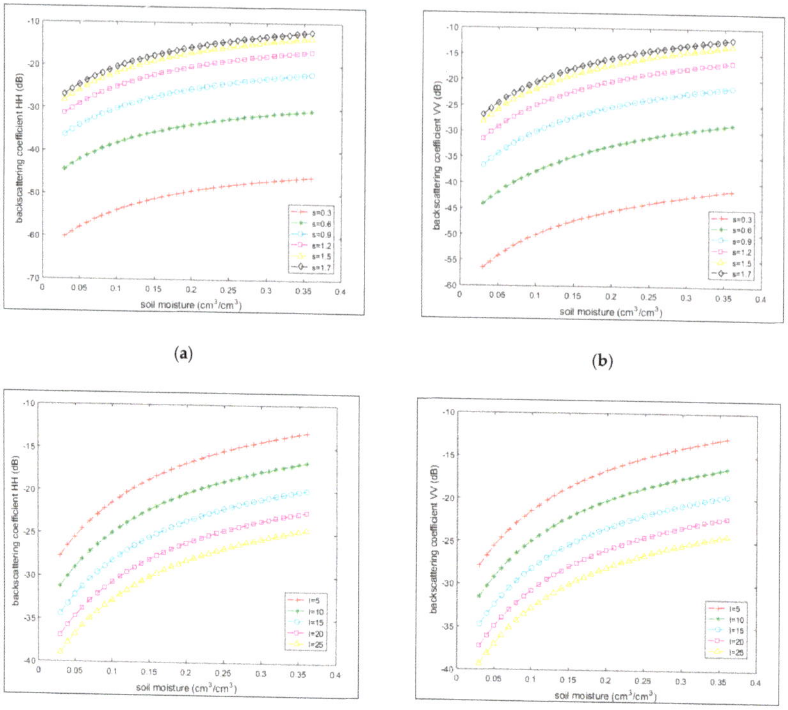

(**a**) (**b**)

(**c**) (**d**)

Figure 8. Responses of the soil moisture and soil backscattering coefficient: (**a**,**c**) HH polarization; (**b**,**d**) VV polarization.

Second, when the correlation length is fixed, the soil backscattering coefficient increases with the increase in the RMS height, and the increase amplitude gradually flattens with the increase in the RMS height. Third, at a fixed RMS height value, the soil backscattering coefficient decreases with the increase in the correlation length; however, the change trend is not as evident as the trends in the soil backscattering coefficients and RMS height. Fourth, under the same parameter conditions, the soil backscattering coefficient value for the HH polarization is slightly lower than that for the VV polarization.

At a frequency of 5.4 GHz and an incident angle of 37°, the AIEM was used to create a database for two conditions: first, the correlation length was considered a fixed value (15 cm), and the soil moisture was considered a fixed-interval parameter (0.03, 0.09, 0.15, 0.21, 0.27, and 0.33 cm^3/cm^3); second, the RMS height was considered a fixed value (1.0 cm), and the soil moisture condition was the same as in the first case. The responses of the roughness parameter and bare soil backscattering coefficient could be analyzed based on the database.

The following rules can be summarized from Figure 9: First, at a fixed soil moisture value, the backscattering coefficient decreases with the increase in the correlation length, whereas it increases with the increase in the RMS height.

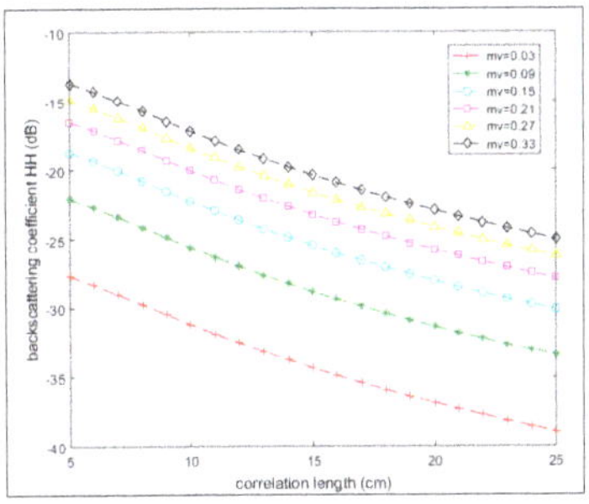

(**a**)

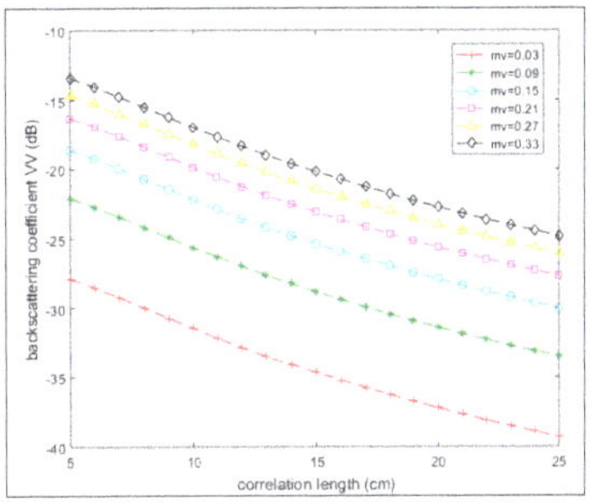

(**b**)

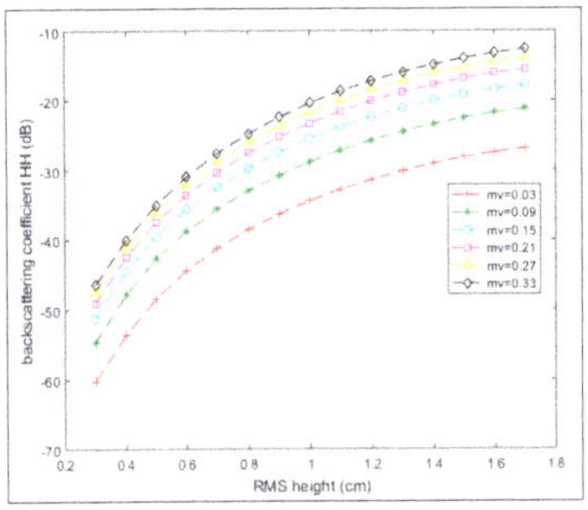

(**c**)

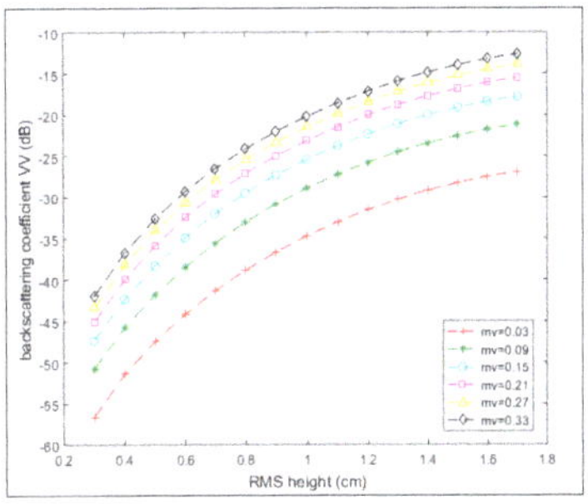

(**d**)

Figure 9. Responses of the roughness parameters and soil backscattering coefficient: (**a**,**c**) HH polarization; (**b**,**d**) VV polarization.

Second, the sensitivity of the backscattering coefficient to the correlation length declines with the increase in the soil moisture.

Third, the soil backscattering coefficients are highly sensitive to the RMS height and tend to decrease with the increase in the soil moisture; however, this trend is not as evident as the change tendency in the correlation length.

Fourth, at a fixed soil moisture value, the soil backscattering coefficients gradually decrease with the increase in the RMS height; the change range gradually decreases, and the curve tends to flatten.

For a fixed sensor configuration, the soil backscattering coefficient is mainly affected by the roughness parameter and soil moisture. As shown in Figure 9, there is a significant nonlinear relationship between the soil backscattering coefficient and surface parameter for both the polarizations. In this section, based on the above qualitative analysis, a quantitative estimation model is presented and verified.

The effects of the roughness parameter and soil moisture on the soil backscattering coefficient are considered independent of each other, based on published studies [9,42]. In this study, the quantitative relationship between the soil backscattering coefficient and surface parameter could be described as the sum of the contributions of the roughness and soil moisture.

The first step is to establish a quantitative relationship between the soil backscattering coefficients and two roughness parameters based on the AIEM-simulated database for an average value (0.19) of the soil moisture from the in-situ datasets. For a quantitative model, the greater the number of roughness parameters, the greater the uncertainty in the model [64]. Therefore, a combined roughness parameter that combines the correlation length and the RMS height into one comprehensive parameter is selected. The combined roughness parameter has been widely used in previous studies for soil moisture estimation [9,37,69]. The results of these studies show that the new roughness parameter can help characterize the natural surface roughness conditions, while reducing the uncertainty of the estimation model. Under the different conditions of study areas and sensor configurations, the form of the combined roughness parameter can be different. In this study, three forms of the combined roughness parameter were tested to select the most suitable one for the surface conditions. Based on the database simulated using the AIEM, a multivariate regression analysis of the three forms of the combined roughness parameter and the soil backscattering coefficient was performed using the 1stOpt software. Table 4 lists the results. From Table 4, we find that there is a significant logarithmic relationship between the three forms of the combined roughness parameter and the soil backscattering coefficients. The PCC of all the constructed equations is greater than 0.8.

Table 4. Quantitative relationship between the three forms of combined roughness parameters and the soil backscattering coefficient.

Combined Roughness Parameter	Formula	PCC	References
$Z_{S1} = S/L$	$\sigma^0_{HH} = 9.0421 + 13.0996 \times \ln(Z_{S1})$ $\sigma^0_{VV} = 6.8906 + 12.0569 \times \ln(Z_{S1})$	0.8070 0.8339	[9]
$Z_{S2} = S^2/L$	$\sigma^0_{HH} = -1.7147 + 8.7825 \times \ln(Z_{S2})$ $\sigma^0_{VV} = -3.2885 + 7.9854 \times \ln(Z_{S2})$	0.9579 0.9659	[70]
$Z_{S3} = S^3/L$	$\sigma^0_{HH} = -8.3608 + 6.1922 \times \ln(Z_{S3})$ $\sigma^0_{VV} = -9.4118 + 5.6029 \times \ln(Z_{S3})$	0.9629 0.9680	[71]

In contrast, the equations based on Z_{S3} give the highest precision for both HH (0.9629) and VV (0.9680) polarizations. Therefore, the form of Z_{S3} was used as Z_S in this study, to characterize the combined effect of the correlation length and RMS height. Figure 10 shows the corresponding relationship between Z_S and the soil backscattering coefficient under different soil moisture conditions.

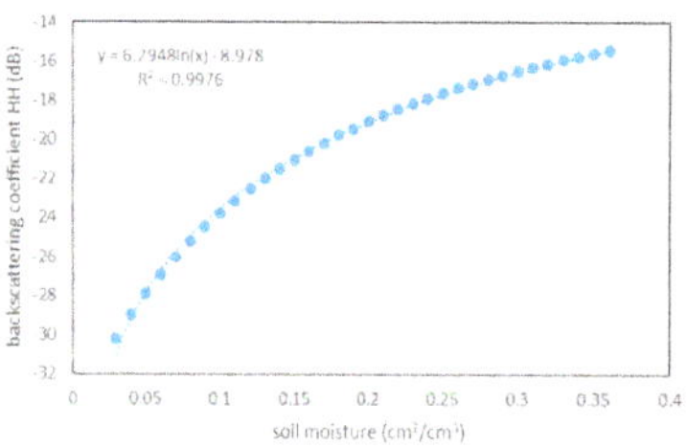

(a)

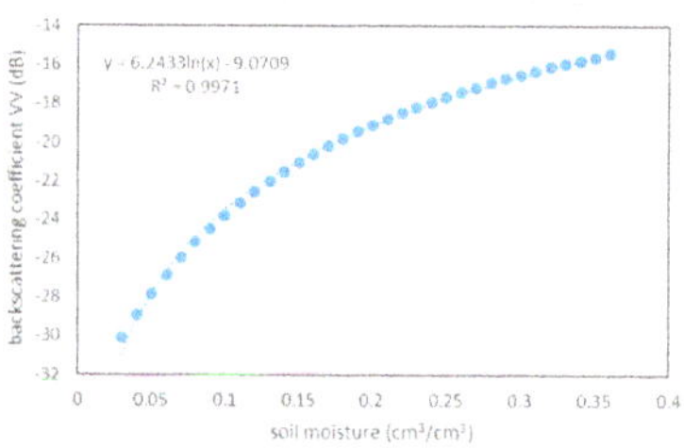

(b)

Figure 10. Responses of soil moisture and soil backscattering coefficients. (**a**) HH polarization; (**b**) VV polarization.

The functional expressions between the soil moisture and the backscattering coefficients for a correlation length of 15 cm and an RMS height of 1 cm based on the simulated database are shown in Figure 10. The R-square of the regression equations for both HH and VV polarizations is greater than 0.99.

According to Figure 10, the effect of soil moisture on soil backscattering coefficients under the condition of fixed roughness parameters can be expressed as follows:

$$\sigma^0_{HH} = -8.9780 + 6.2948 \times \ln(m_v) \quad (22)$$

$$\sigma^0_{VV} = -9.0709 + 6.2433 \times \ln(m_v) \quad (23)$$

Based on the above analysis, for a fixed frequency and incident angle, the expression of the quantitative model can be written as:

$$\sigma^0_{HH} = -17.3388 + 6.2948 \times \ln(m_v) + 6.1922 \times \ln(Z_s) \tag{24}$$

$$\sigma^0_{VV} = -18.4827 + 6.2433 \times \ln(m_v) + 5.6029 \times \ln(Z_s) \tag{25}$$

Combining Equations (24) and (25), the soil moisture can be estimated mathematically, and the general expression is as follows:

$$m_v = \exp\left(i \times \sigma^0_{VV} + j \times \sigma^0_{HH} + k\right) \tag{26}$$

where i, j, and k are the empirical parameters. In this study, the empirical parameters were obtained using the least-squares method based on the in-situ datasets. The bare soil moisture estimation equation based on the in-situ experiments is as follows:

$$m_v = \exp\left(-0.0407 \times \sigma^0_{VV} + 0.0236 \times \sigma^0_{HH} - 2.0599\right) \tag{27}$$

The soil moisture map of the entire study area was obtained using this estimation model based on the backscattering coefficients for HH and VV polarizations.

4.2.2. Bare Soil Moisture Estimation Based on the Oh Model

The backscattering coefficients for the VV, HH, and VH polarizations used in the Oh model were obtained from the results of the vegetation effect correction model described in Section 4.1. The backscattering coefficients of urban land and water bodies are significantly different from those of bare soil [69,72,73]. Therefore, the threshold value summarized based on the datasets can be used to effectively mask the urban land and water bodies. For bare soil, the backscattering coefficient for VV polarization is always greater than that for HH polarization, and the latter is greater than that for VH polarization [37,39]. Based on the statistical analysis of the backscattering coefficients, we found the existence of pixels whose backscattering coefficients for HH polarization are higher than that for VV polarization. The reason for this phenomenon may be that, despite the high accuracy of the vegetation correction model, some pixels are still affected by the influence of sparse vegetation [37].

In this study, the pixels in the Oh model were screened based on the following conditions: First, the backscattering coefficients of the pixel are above the threshold value set for urban land and water bodies, summarized based on the datasets. Second, the backscattering coefficient of the pixel for VV polarization is greater than that for HH polarization. Lastly, the estimated soil moisture and ks should meet the range of soil moisture condition of the Oh model: 0.04 $cm^3/cm^3 < mv < 0.29$ cm^3/cm^3, and $0.13 < ks < 6.98$.

4.2.3. Bare Soil Moisture Estimation Based on LUT

The LUT in this study was based on the AIEM-simulated database, described in Section 3.2.1, with fixed incidence angle (37°) and frequency (5.4 GHz) values under the configuration of GF-3 images. This LUT indicates the one-to-one correspondence relationship between the bare soil backscattering coefficients for HH and VV polarizations, roughness parameters, and soil moisture. The pixels of the bare soil backscattering-coefficient images, obtained in Section 4.1, were calculated using the cost functions one by one. When the cost function value Z_n (n = 1, 2, and 3) reached the minimum, the soil moisture value corresponding to $\sigma^0_{HH\_AIEM}$ and $\sigma^0_{VV\_AIEM}$ in the LUT was used as the best-fit inversion soil moisture value under the HH or VV polarization.

The results of the retrieved soil moisture content based on the cost functions Z_1, Z_2, and Z_3 show that the average soil moisture contents were 0.32, 0.29, and 0.30, respectively, with standard deviations of 0.05, 0.07, and 0.07, respectively.

4.2.4. Soil Moisture Estimation Based on Dubois Model

The four RMS height results retrieved from the LUT and Oh method were used in the Dubois model as input data to indirectly estimate the soil moisture. The inversion results of the soil moisture under the two polarization conditions could be obtained simultaneously in the Dubois model. After comparing their accuracies, we selected the HH polarization mode results of this model. Considering the limitation of the application conditions of the Dubois model, the pixels of the input and output images that do not meet the conditions were set as null values. Consequently, four soil moisture inversion maps were obtained.

4.2.5. Validation and Analysis of the Individual Soil Moisture Inversion Models

To evaluate the performance of the nine inversion soil moisture maps obtained using the most commonly used soil moisture models (AIEM, Oh model, LUT inversion model, and Dubois model), a validation experiment was conducted to confirm the agreement between the modeled soil moisture and the measured soil moisture in terms of the RMSE, PCC, mean absolute error (MAE), and bias. The RMSE and MAE could reflect the differences between the estimated and measured soil moisture values, where a low index value represents a high prediction accuracy. The positive and negative values of the bias could reflect the deviation direction. The PCC could reflect the linear correlation between the estimated and measured soil moisture values, where a value close to 1 represents a strong linear correlation.

Table 5 lists the RMSE, MAE, PCC, and bias between the measured and estimated soil moisture values based on the 60 sample values from the validation dataset acquired during the in situ measurements. It should be noted that m_{v1} represents the result of the estimation model based on the AIEM; m_{v2}, m_{v3}, and m_{v4} represent the results of the LUT inversion model; m_{v5} represents the results of the Oh semi-empirical model; m_{v6}, m_{v7}, m_{v8}, and m_{v9} represent the results of the indirect soil moisture estimation method based on the Dubois model.

Table 5. RMSE, MAE, PCC, and bias between the measured and estimated soil moisture values based on the validation dataset.

Inversion Models	RMSE (cm^3/cm^3)	MAE (cm^3/cm^3)	PCC	Bias (cm^3/cm^3)
m_{v1}	0.0321	0.0260	0.9115	−0.0150
m_{v2}	0.0538	0.0338	0.7441	−0.0117
m_{v3}	0.0769	0.0563	0.7287	−0.0426
m_{v4}	0.1059	0.0811	0.5679	0.0230
m_{v5}	0.0428	0.0306	0.8843	0.0174
m_{v6}	0.0710	0.0567	0.6551	−0.0420
m_{v7}	0.0687	0.0529	0.4114	0.0266
m_{v8}	0.0761	0.0608	0.6278	0.0107
m_{v9}	0.0676	0.0564	0.4599	−0.0141

In terms of the RMSE and MAE, the estimation model constructed using the AIEM (m_{v1}: RMSE = 0.0321 cm^3/cm^3, MAE = 0.0260 cm^3/cm^3) was more accurate than the other models, followed by the Oh model (m_{v5}: RMSE = 0.0428 cm^3/cm^3, MAE = 0.0306 cm^3/cm^3) and the LUT inversion method under HH polarization (m_{v2}: RMSE = 0.0538 cm^3/cm^3, MAE = 0.0338 cm^3/cm^3). The indirect Dubois model combined with the Oh model (m_{v9}), combined with the LUT inversion method under VV polarization (m_{v7}) and the direct LUT inversion method under VV polarization (m_{v3}) gave moderate accuracy results, with an RMSE value lower than 0.077 cm^3/cm^3 and MAE value lower than 0.056 cm^3/cm^3. The accuracy of the LUT model for both HH and VV polarizations (m_{v4}) is relatively lower than those of the other inversion models.

The PCC could directly reflect the linear correlation between the estimated and measured soil moisture values. As listed in Table 5, the soil moisture estimated from the inversion model constructed using the AIEM (m_{v1}) had a strong linear relationship, with the PCC reaching 0.9115. The soil moisture calculated using the Oh model (m_{v5}) and LUT inversion method under HH or VV polarizations (m_{v2} and m_{v3}) also had a good linear relationship with the measured values (PCC higher than 0.70). In contrast, there is no significant linear correlation between the results estimated using the indirect Dubois model combined with the Oh model (m_{v9}) and LUT inversion method under VV polarization (m_{v7}) and the measured soil moisture (PCC lower than 0.50).

The statistical analysis of the bias results shows that, among the nine inversion soil moisture maps, five models (m_{v1}, m_{v2}, m_{v3}, m_{v6}, and m_{v9}) show negative bias, indicating that the inversion results of these five models are generally higher than the measured soil moisture values; at the same time, four models (m_{v4}, m_{v5}, m_{v7}, and m_{v8}) show positive bias, indicating that these models give an overall underestimated result.

In summary, the evaluation and analysis results of the nine inversion maps on the validation dataset are as follows:

(1) The estimation model constructed using the AIEM (m_{v1}) gives the highest accuracy for all the four verification indexes, followed by the Oh model (m_{v5}) and LUT inversion method under HH polarization (m_{v2}), whereas the accuracy of the LUT model for both HH and VV polarizations (m_{v4}) is relatively poor. The good performance of the estimation model constructed using the AIEM is attributed to the well-established theoretical model describing the radar scattering process, and at the same time, 37 sample datasets from the in-situ measurements were used as modeling data in establishing the regression model [9,42]. Previous studies have also shown that the Oh model is a highly accurate semiempirical model and that it is more favorable in the case of cross-polarized data for the inversion of complex surfaces [37,48]. However, because the model cannot express the soil moisture precisely for pixels with soil moisture below 0.04 or greater than 0.29, the precision of the model is compromised [69,70,74].

(2) The RMS height, obtained from the LUT inversion method or Oh model, was used as input data for the Dubois model to estimate the soil moisture indirectly. In general, the indirect Dubois model (m_{v6}, m_{v7}, m_{v8}, and m_{v9}) has a relatively lower inversion accuracy than the direct inversion models (m_{v1}, m_{v2}, m_{v3}, m_{v4}, and m_{v5}). There are two possible reasons for this: first, significant errors exist in the process of RMS height inversion, which will be transmitted to the indirect inversion Dubois model; second, some factors, such as the residual vegetation effects and the row directional effects, may be taken as soil moisture fluctuation in the Dubois model [72,75].

(3) Among the three LUT direct inversion methods (m_{v2}, m_{v3}, and m_{v4}), the differences between the estimated and measured soil moisture values are relatively minimum under HH polarization (m_{v2}: RMSE = 0.0538, MAE = 0.0338), followed by the inversion method under VV polarization (m_{v3}: RMSE = 0.0769, MAE = 0.0563). The linear correlation between the estimated and measured soil moisture values is acceptable under single polarization (m_{v2} and m_{v3}), and the PCC values are greater than 0.7. This shows that the LUT inversion under single polarization is feasible to retrieve the soil moisture, consistent with previous results [37,40]. However, the accuracy of the direct or indirect LUT inversion under dual-polarization (m_{v4} and m_{v8}) is unsatisfactory, inconsistent with a previous research by Zhang et al. [56]. The possible reason for the inconsistency is that the surface coverage of the study area in the previous research was monotonous, whereas in this study, the surface coverage was complex, and the vegetation types were varied, including a certain proportion of nurseries with evident body scattering effect.

4.3. Optimal Solution Method

Based on the nine soil moisture mapping algorithms, a simple and effective optimal solution method of pixel classification and algorithm selection was implemented, in order to obtain the most accurate soil moisture map. The principle and main steps of the optimal solution method have been described in detail in Section 3.3. In this section, the process and results of the algorithm selection, as well as the verification process of the accuracy of the spliced soil moisture maps, are described in detail.

In the last section, the accuracy of all the methods is verified and analyzed for the entire validation dataset. The overall analysis shows that the estimation model constructed using the AIEM ($m_{v1,}$) is the best, followed by the Oh model (m_{v5}) and LUT inversion method under HH polarization (m_{v2}). However, if the study area is divided into different categories based on background knowledge, the accuracy of the nine inversion maps in each category could be estimated differently, probably different from that of the overall analysis. In this study, three types of background knowledge data were used, including land use types, percentage content of clay, and slope levels.

The 30-m resolution land use data were downloaded from the website of Finer Resolution Observation and Monitoring Global Land Cover Products (http://data.ess.tsinghua.edu.cn/). The land use cover product is the first 30-m resolution global land cover map produced using Landsat Thematic Mapper and Enhanced Thematic Mapper Plus data. The soil texture data were downloaded from the Geographic Data Sharing Infrastructure, College of Urban and Environmental Science, Peking University (http://geodata.pku.edu.cn), and divided on the basis of three categories: the percentage of sand, percentage of silt, and percentage of clay. The slope levels were derived from the Shuttle Radar Topography Mission digital elevation model with a resolution of 30 m using the ArcGIS 10.2 software.

4.3.1. Construction of the Optimal Solution Method under Three Strategies

On the premise of introducing background knowledge to classify the study area, the soil moisture inversion method with the highest accuracy under each category was selected as the optimal solution method. In this study, three construction strategies for the optimal solution method were proposed based on the modeling dataset.

In strategy 1, the study area was divided into four categories based on the four main land use types. The soil moisture inversion map with the highest accuracy was selected as the optimal solution model for this certain land use type (Table 6). From Table 6, we find that, although the estimation model constructed using the AIEM (m_{v1}) has the highest accuracy in the overall evaluation process, it does not have the highest accuracy in all the four land use types. The Oh model (m_{v5}) and LUT inversion method under HH polarization (m_{v2}) shows the best precision in the case of grassland and bare land, respectively.

Table 6. Optimal solution strategy based on land use types and the accuracy of each algorithm (strategy 1).

Land Use Types	Optimal Model	RMSE (cm^3/cm^3)	MAE (cm^3/cm^3)	PCC	BIAS (cm^3/cm^3)
farmland	m_{v1}	0.0327	0.0264	0.9436	−0.0187
grassland	m_{v5}	0.0171	0.0128	0.8984	0.0128
shrub	m_{v1}	0.0340	0.0273	0.8763	0.0065
bare land	m_{v2}	0.0231	0.0201	0.9033	0.0134

The study area was divided into five categories based on the value of the percentage content of clay, in strategy 2. The soil moisture inversion map with the highest accuracy was selected as the optimal solution method for this certain category (Table 7). In the category with the highest percentage of clay content, the LUT inversion method under HH polarization (m_{v2}) gave the best precision; in the category with the lowest percentage of clay content, the estimation model constructed using the AIEM (m_{v1}) was selected as the optimal solution method; in the other three categories, respectively, the LUT inversion method under VV polarization (m_{v3}), estimation model constructed using the AIEM (m_{v1}), and Oh model (m_{v5}) gave the highest precision.

Table 7. Optimal solution method strategy based on the percentage content of clay and the accuracy of each algorithm (strategy 2).

Percentage Content of Clay (%)	Optimal Model	RMSE (cm^3/cm^3)	MAE (cm^3/cm^3)	PCC	Bias (cm^3/cm^3)
14	m_{v1}	0.0468	0.0338	0.8409	−0.0296
26	m_{v3}	0.0354	0.0291	0.9145	−0.0119
28	m_{v1}	0.0231	0.0193	0.9673	−0.0141
29	m_{v5}	0.0251	0.0213	0.9555	0.0193
37	m_{v2}	0.0261	0.0269	0.9376	−0.0196

In strategy 3, the study area was divided into five categories based on five slope levels, as listed in Table 8. The LUT inversion method under VV polarization (m_{v3}) was selected as the optimal solution method in Levels 1 and 4. The Oh model (m_{v5}) gave the highest accuracy in Levels 3 and 5.

Table 8. Optimal solution modeling strategy based on the slope levels and the accuracy of each algorithm (strategy 3).

Levels of Slope	Optimal Model	RMSE (cm^3/cm^3)	MAE (cm^3/cm^3)	PCC	BIAS (cm^3/cm^3)
Level 1 (≤1°)	m_{v3}	0.0438	0.0379	0.8419	−0.0156
Level 2 (1–2°)	m_{v1}	0.0216	0.0204	0.9114	−0.0125
Level 3 (2–3°)	m_{v5}	0.0387	0.0291	0.8871	0.0102
Level 4 (3–4°)	m_{v3}	0.0332	0.0254	0.8933	−0.0144
Level 5 (>4°)	m_{v5}	0.0380	0.0288	0.8674	0.0135

4.3.2. Validation and Analysis of the Three Strategies

The scatter plot could directly reflect the correlation between the estimated and measured soil moisture values and the corresponding error between them. In this study, a set of scatter plots (Figure 11) were used to show the results of the optimal solution methods under the three different strategies.

Table 9 lists the validation results under the three strategies. The RMSE, MAE, PCC, and bias between the measured soil moisture and the one estimated using the optimal solution method were calculated based on the 60 sample values from the validation dataset acquired during the in-situ measurements. It should be noted that the modeling strategies represent the three strategies for building the optimal solution method, as described in the previous section, based on land use types, percentage content of clay, and slope levels.

Similarly, the optimal solution method under strategies 2 and 1 exhibited a higher precision than the nine individual inversion soil moisture maps (the highest precision was: PCC = 0.9115 cm^3/cm^3). As listed in Table 9, the soil moisture estimated using the optimal solution method under strategy 2 had a strong linear relationship, with the PCC value reaching 0.9364. The soil moisture calculated using the optimal solution method under strategy 1 also had a good linear relationship with the measured values (PCC value higher than 0.90).

Table 9. RMSE, MAE, PCC, and bias between the measured and estimated soil moisture based on the validation dataset.

Modeling Strategies	RMSE (cm^3/cm^3)	MAE (cm^3/cm^3)	PCC	BIAS (cm^3/cm^3)
Strategy 1	0.0314	0.0249	0.9186	−0.0182
Strategy 2	0.0271	0.0225	0.9364	−0.0133
Strategy 3	0.0502	0.0351	0.7703	−0.0231

In terms of the RMSE and MAE, the soil moisture obtained using the optimal solution method under strategies 2 and 1 was more accurate than that obtained from the nine individual inversion soil moisture maps (the highest precision was: RMSE = 0.0321 cm^3/cm^3, MAE = 0.0260 cm^3/cm^3). The optimal solution method under strategy 2 (RMSE = 0.0271 cm^3/cm^3, MAE = 0.0225 cm^3/cm^3)

exhibited a higher accuracy than the other two models, followed by optimal solution strategy 1 (RMSE = 0.0314 cm^3/cm^3, MAE = 0.0249 cm^3/cm^3).

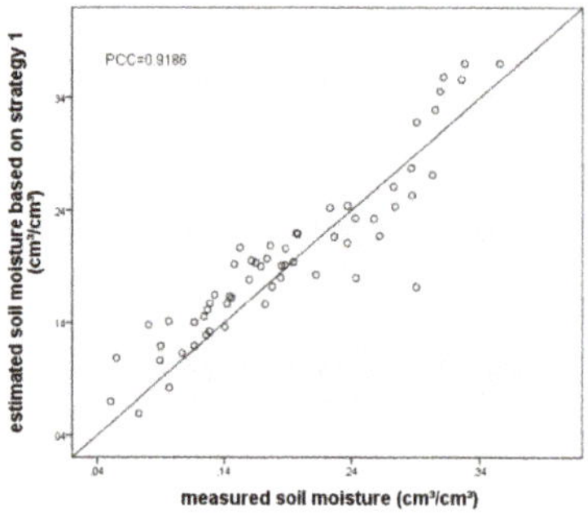

(**a**)

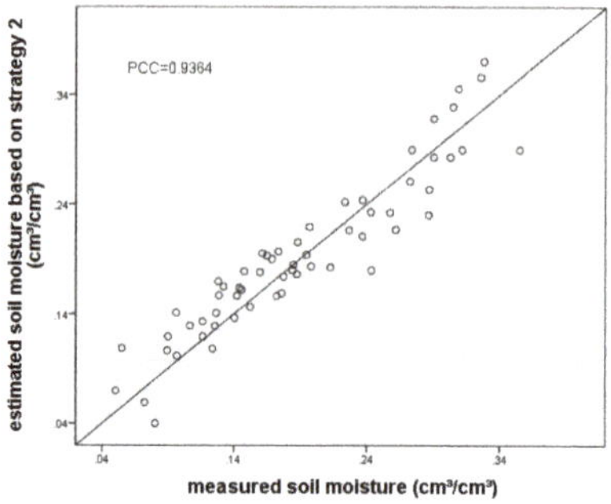

(**b**)

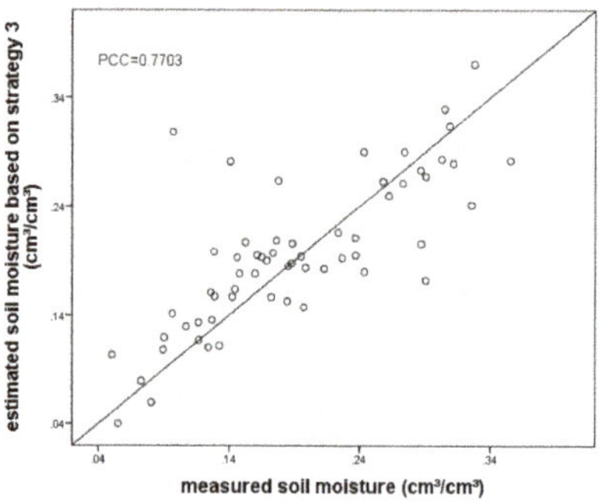

(**c**)

Figure 11. Scatter plots between the measured and estimated soil moisture values obtained using the optimal solution methods based on the 60 in situ experimental datasets. A 1:1 line is included in this plot. (**a**) strategy 1; (**b**) strategy 2; (**c**) strategy 3.

Interestingly, the bias values of the soil moisture estimated using the optimal solution methods under the three strategies were negative, indicating that all the strategies gave an overall overestimated result.

In this study, a mask file built using the threshold value was used to mask the urban land and water bodies to improve the accuracy of the bare soil moisture estimation. Figure 12 shows the soil moisture map retrieved using the optimal solution method under strategy 2.

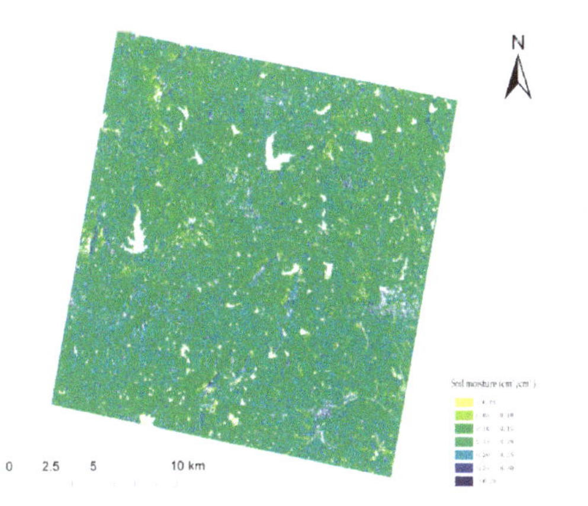

Figure 12. Map of retrieved soil moisture content based on the optimal solution method under strategy 2.

In summary, the evaluation and analysis results on the validation dataset are as follows:

(1) The optimal solution method showed advantages, confirming the assumptions made in this study. Among them, the soil moisture calculated using the optimal solution method under strategies 2 and 1 was more accurate than those obtained from the nine individual inversion soil moisture maps. The RMSE index for the optimal solution method under strategy 2 was 0.005 cm^3/cm^3 lower than that for the individual estimation model constructed using the AIEM (m_{v1}: the only model with the highest accuracy among the nine individual inversion maps), and the RMSE index for the optimal solution method under strategy 1 was 0.0007 cm^3/cm^3 lower than that for the individual estimation model. Similarly, the PCC index for the optimal solution method under strategy 2 was 0.0249 higher than that for the individual estimation model constructed using the AIEM, and the optimal solution method under strategy 1 was 0.0071 higher than that for the individual estimation model. This means that the error between the estimated results of the optimal solution methods under most strategies and the measured soil moisture was lower and that their linear correlation was stronger compared with the individual inversion models.

(2) Among the strategies used for constructing the optimal solution methods, strategy 2 gave the best accuracy, followed by strategy 1. Compared with the former two strategies, the accuracy of the optimal solution method under strategy 3 was relatively low. Nevertheless, it was still higher than the accuracy of the nine individual models (77.78%), thus demonstrating its superiority. To some extent, this inconsistent result reflects the sensitivity of the soil moisture to the different background knowledge. For the study area, the optimal solution construction strategy with the clay percentage content as the background knowledge exhibited the best performance, followed by the optimal solution construction strategy with the land use type as the background knowledge. The optimal solution strategy with the slope level as the background knowledge exhibited the lowest performance.

5. Conclusions

To accurately map the soil moisture of a study area in Jiangsu Province, China, in this study, we used both optical and radar remote sensing data in a combined manner, improved and coupled several commonly used inversion models, fully utilized in situ ground experimental datasets, and introduced background knowledge into the optimal solution method.

For study areas with a complex coverage, it is important to effectively remove the influence of vegetation canopy. In the process of removing the influence of vegetation canopy, we made two improvements based on the conventional WCM. A combined VWC model was proposed, and its accuracy (PCC = 0.9442) was found to be significantly higher than that of the one-variable quadratic models. The empirical parameters A, B, and α calculated based on the in situ experimental datasets (WCM_3) were selected as the WCM parameters to remove the effect of vegetation canopy.

Based on an individual inversion algorithm constructed using four commonly used models, nine soil moisture inversion maps were obtained. The estimated model constructed using the AIEM (m_{v1}) exhibited a higher accuracy than the other models, followed by the Oh model (m_{v5}) and LUT inversion method under HH polarization (m_{v2}). In general, the indirect Dubois model (m_{v6}, m_{v7}, m_{v8}, and m_{v9}) exhibited a relatively lower inversion accuracy than the direct inversion models (m_{v1}, m_{v2}, m_{v3}, m_{v4}, and m_{v5}). The possible reasons for these were analyzed. The influencing factors may have included the inherent characteristics of these models, the sensitivity of the radar polarization mode to the soil moisture, and the error transmission of the intermediate parameters of the indirect models.

Notably, we attempted to retrieve the RMS height based on the LUT method; this is a relatively new approach. The RMS height was retrieved from the LUT method and Oh model as input to estimate the soil moisture using the Dubois model. Although the accuracy of these indirect algorithms was relatively lower than that of the direct inversion algorithm, the attempts made gave feasible and valuable results to some extent.

Optimal solution methods were proposed by analyzing the nine individual inversion algorithms with background knowledge. The superiority of the optimal solution method was significant. The model under strategy 2 (RMSE = 0.0271 cm^3/cm^3, MAE = 0.0225 cm^3/cm^3, PCC = 0.9364) gave the best accuracy, followed by strategy 1. Compared with the former two strategies, the accuracy of the optimal solution method under strategy 3 was relatively low, though it was still higher than the accuracy of the nine individual models (77.78%). The superiority of the optimal solution methods was attributed to the integration of various categories of the nine individual algorithms with the best accuracy based on background knowledge.

In this study, various efforts were made to retrieve the soil moisture with high accuracy. Some of the individual inversion algorithms and most of the optimal solution methods gave satisfactory results. However, as with any study, there remain some shortcomings: (1) The surface coverage of the study area was complex and inconsistent with the basic assumptions of the WCM. (2) Although the influence of vegetation water was reduced as much as possible, the polarization backscattering coefficient is significantly affected by the structure of the vegetation canopy, which would introduce some errors in the calculation process. (3) There were uncertainties in the in situ measurement of the RMS height, which may have caused errors in the inversion processes. (4) Only a few types of background knowledge data were used in the optimal solution methods, and the construction methods were relatively simple. Future research can be carried out from the following aspects: GF-1 could be replaced by GF-6 to obtain more spectral information; this may give a better accuracy in removing the influence of the vegetation canopy. In the process of bare soil moisture inversion, more diversified and comprehensive multi-model coupling methods could be tried to explore the potential of each model and improve the coupling algorithms. More diverse and representative background knowledge could be introduced in the process of constructing the optimal solution method, and the algorithm used for the construction process can be further improved.

Author Contributions: Conceptualization, L.H., C.W.; Methodology, L.H., C.W.; Formal analysis, L.H., C.W.; Data curation, L.H.; Writing—original draft, L.H.; Writing—review & editing, C.W., T.Y., Q.L.; Valuable advice, T.Y., Q.L.; Funding acquisition, T.Y. and X.G.; All authors have read and agreed to the published version of the manuscript.

Funding: This research was funded by the China's 12th Five-Year Plan Civil Space Pre-Research Project (Grant No. Y930280A2F), National Natural Science Foundation of China (Grant No. 41501400), and the Youth fund project (Y5SJ0600CX).

Acknowledgments: The authors sincerely thank the editors and reviewers.

Conflicts of Interest: The authors declare no conflict of interest.

References

1. Anagnostopoulos, V.; Petropoulos, G.P.; Ireland, G.; Carlson, T.N. A modernized version of a 1D soil vegetation atmosphere transfer model for improving its future use in land surface interactions studies. *Environ. Model. Softw.* **2017**, *90*, 147–156. [CrossRef]
2. Gherboudj, I.; Magagi, R.; Berg, A.A.; Toth, B. Soil moisture retrieval over agricultural fields from multi-polarized and multi-angular RADARSAT-2 SAR data. *Remote Sens. Environ.* **2011**, *115*, 33–43. [CrossRef]
3. She, D.; Liu, D.; Liu, Y.; Liu, Y.; Xu, C.; Qu, X.; Chen, F. Profile characteristics of temporal stability of soil water storage in two land uses. *Arab. J. Geosci.* **2014**, *7*, 21–34.
4. Taibi, A.E.O.; Elfeki, A.; Zhang, G. Modeling hydrologic responses of the Zwalm catchment using the REW approach: Propagation of uncertainty in soil properties to model output. *Arab. J. Geosci.* **2011**, *4*, 1005–1018. [CrossRef]
5. Kornelsen, K.C.; Coulibaly, P. Advances in soil moisture retrieval from synthetic aperture radar and hydrological application. *J. Hydrol.* **2013**, *476*, 460–489. [CrossRef]
6. Seneviratne, S.I.; Corti, T.; Davin, E.L.; Hirschi, M.; Jaeger, E.B.; Lehner, I.; Orlowsky, B.; Teuling, A.J. Investigating soil moisture–climate interactions in a changing climate: A review. *Earth Sci. Rev.* **2010**, *99*, 125–161. [CrossRef]
7. Fisher, J.B.; Melton, F.; Middleton, E.; Hain, C.; Anderson, M.; Allen, R.; McCabe, M.F.; Hook, S.; Baldocchi, D.; Townsend, P.A.; et al. The future of evapotranspiration: Global requirements for ecosystem functioning, carbon and climate feedbacks, agricultural management, and water resources. *Water Resour. Res.* **2017**, *53*, 2618–2626. [CrossRef]
8. Qiaozhen, M.U.; Zhao, M.; Steven, W.J. Improvements to a MODIS global terrestrial evapotranspiration algorithm. *Remote Sens. Environ.* **2011**, *115*, 1781–1800.
9. Kong, J.; Yang, J.; Zhen, P.; Li, J.; Yang, L. A Coupling Model for Soil Moisture Retrieval in Sparse Vegetation Covered Areas Based on Microwave and Optical Remote Sensing Data. *IEEE Trans. Geosci. Remote Sens.* **2018**, *56*, 7162–7173. [CrossRef]
10. Amazirh, A.; Merlin, O.; Er-Raki, S.; Gao, Q.; Rivalland, V.; Malbeteau, Y.; Khabba, S.; JoséEscorihuelac, M. Retrieving surface soil moisture at high spatio-temporal resolution from a synergy between Sentinel-1 radar and Landsat thermal data: A study case over bare soil. *Remote Sens. Environ.* **2018**, *211*, 321–337. [CrossRef]
11. Du, Y.; Ulaby, F.T.; Dobson, M.C. Sensitivity to soil moisture by active and passive microwave sensors. *IEEE Trans. Geosci. Remote Sens.* **2000**, *38*, 105–114. [CrossRef]
12. Bruckler, L.; Witono, H.; Stengel, P. Near surface soil moisture estimation from microwave measurements. *Remote Sens. Environ.* **1988**, *26*, 101–121. [CrossRef]
13. Bao, Y.; Zhang, Y.; Wang, J.; Min, J. Surface soil moisture estimation over dense crop using Envisat ASAR and Landsat TM imagery: An approach. *Int. J. Remote Sens.* **2014**, *35*, 6190–6212. [CrossRef]
14. Baghdadi, N.; Hajj, M.E.; Zribi, M. Coupling SAR C-band and optical data for soil moisture and leaf area index retrieval over irrigated grasslands. *IEEE J. Sel. Top. Appl. Earth Obs. Remote Sens.* **2016**, *9*, 3551–3554. [CrossRef]
15. Dianjun, Z.; Guoqing, Z. Estimation of Soil Moisture from Optical and Thermal Remote Sensing: A Review. *Sensors* **2016**, *16*, 1308.
16. Verstraeten, W.W.; Veroustraete, F.; van der Sande, C.J.; Grootaers, I.; Feyen, J. Soil moisture retrieval using thermal inertia, determined with visible and thermal spaceborne data, validated for European forests. *Remote Sens. Environ.* **2006**, *101*, 299–314. [CrossRef]
17. Yansong, B.; Libin, L.; Shanyu, W.; Deng, K.A.K.; Petropoulos, G.P. Surface soil moisture retrievals over partially vegetated areas from the synergy of Sentinel-1 and Landsat 8 data using a modified water-cloud model. *Int. J. Appl. Earth Obs. Geoinf.* **2018**, *72*, 76–85.
18. Zhang, X.; Chen, B.; Fan, H.; Huang, J. The Potential Use of Multi-Band SAR Data for Soil Moisture Retrieval over Bare Agricultural Areas: Hebei, China. *Remote Sens.* **2015**, *8*, 7. [CrossRef]

19. Mattar, C.; Wigneron, J.P.; Sobrino, J.A.; Novello, N.; Calvet, J.C.; Albergel, C.; Richaume, P.; Mialon, A.; Guyon, D.; Jiménez-Muñoz, J.C. A Combined Optical-Microwave Method to Retrieve Soil Moisture Over Vegetated Areas. *IEEE Trans. Geosci. Remote Sens.* **2012**, *50*, 1404–1413. [CrossRef]
20. Santamaria-Artigas, A.; Mattar, C.; Wigneron, J.P. Application of a Combined Optical-Passive Microwave Method to Retrieve Soil Moisture at Regional Scale Over Chile. *IEEE J. Sel. Top. Appl. Earth Obs. Remote Sens.* **2016**, *9*, 1493–1504. [CrossRef]
21. Miernecki, M.; Wigneron, J.P.; Lopez-Baeza, E.; Kerr, Y.; De Jeu, R.; De Lannoy, G.J.M.; Jackson, T.J.; O'Neill, P.E.; Schwankh, M.; Moran, R.F.; et al. Comparison of SMOS and SMAP soil moisture retrieval approaches using tower-based radiometer data over a vineyard field. *Remote Sens. Environ.* **2014**, *154*, 89–101. [CrossRef]
22. Yan, F.; Qin, Z.; Li, M.; Li, W. Progress in soil moisture estimation from remote sensing data for agricultural drought monitoring. *Proc. SPIE Int. Soc. Opt. Eng.* **2006**, *6366*, 636601.
23. Zeng, J.; Li, Z.; Chen, Q.; Bi, H. Method for Soil Moisture and Surface Temperature Estimation in the Tibetan Plateau Using Spaceborne Radiometer Observations. *IEEE Geosci. Remote Sens. Lett.* **2015**, *12*, 97–101. [CrossRef]
24. Chen, Q.; Zeng, J.; Cui, C.; Li, Z.; Chen, K.; Bai, X.; Xu, J. Soil Moisture Retrieval From SMAP: A Validation and Error Analysis Study Using Ground-Based Observations Over the Little Washita Watershed. *IEEE Trans. Geosci. Remote Sens.* **2018**, *56*, 1394–1408. [CrossRef]
25. Ulaby, F.T.; Batlivala, P.P.; Dobson, M.C. Microwave Backscatter Dependence on Surface Roughness, Soil Moisture, and Soil Texture: Part I-Bare Soil. *IEEE Trans. Geosci. Electron.* **1978**, *GE-16*, 286–295. [CrossRef]
26. Zhao, X.; Huang, N.; Song, X.F.; Niu, Z.; Li, Z.Y. A new method for soil moisture inversion in vegetation-covered area based on Radarsat 2 and Landsat 8. *J. Infrared Millim. Waves* **2016**, *35*, 609–616.
27. Lievens, H.; Verhoest, N.E.C. On the Retrieval of Soil Moisture in Wheat Fields from L-Band SAR Based on Water Cloud Modeling, the IEM, and Effective Roughness Parameters. *IEEE Geosci. Remote Sens. Lett.* **2011**, *8*, 740–744. [CrossRef]
28. Wang, C.; Qi, J.; Moran, S.; Marsett, R. Soil moisture estimation in a semiarid rangeland using ERS-2 and TM imagery. *Remote Sens. Environ.* **2004**, *90*, 178–189. [CrossRef]
29. Holah, N.; Baghdadi, N.; Zribi, M.; Bruand, A.; King, C. Potential of ASAR/ENVISAT for the characterization of soil surface parameters over bare agricultural fields. *Remote Sens. Environ.* **2005**, *96*, 78–86. [CrossRef]
30. Joseph, A.T.; Velde, R.V.D.; O'Neill, P.E.; Lang, R.; Gish, T. Effects of corn on C- and L-band radar backscatter: A correction method for soil moisture retrieval. *Remote Sens. Environ.* **2010**, *114*, 2417–2430. [CrossRef]
31. Attema, E.P.W.; Ulaby, F.T. Vegetation modeled as a water cloud. *Radio Sci.* **1978**, *13*, 357–364. [CrossRef]
32. Ulaby, F.T.; Mcdonald, K.; Sarabandi, K.; Dobson, M.C. Michigan microwave canopy scattering model. *Int. J. Remote Sens.* **1990**, *11*, 1223–1253. [CrossRef]
33. Hu, D.; Guo, N.; Sha, S.; Wang, L. Soil Moisture Retrieved Using Radarsat-2/SAR and MODIS Remote Sensing Data in Vegetated Areas of Loess Plateau Soil Moisture Retrieved Using Radarsat-2/SAR and MODIS Remote Sensing Data in Vegetated Areas of Loess Plateau. *Remote Sens. Technol. Appl.* **2015**, *30*.
34. Baghdadi, N.; King, C.; Chanzy, A.; Wigneron, J.P. An empirical calibration of the integral equation model based on SAR data, soil moisture and surface roughness measurement over bare soils. *Int. J. Remote Sens.* **2002**, *23*, 4325–4340. [CrossRef]
35. He, B.; Xing, M.; Bai, X. A Synergistic Methodology for Soil Moisture Estimation in an Alpine Prairie Using Radar and Optical Satellite Data. *Remote Sens.* **2014**, *6*, 10966–10985. [CrossRef]
36. Fung, A.K.; Li, Z.; Chen, K.S. Backscattering from a randomly rough dielectric surface. *IEEE Trans. Geosci. Remote Sens.* **1992**, *30*, 369. [CrossRef]
37. Baghdadi, N.; Zribi, M. Evaluation of radar backscatter models IEM, OH and Dubois using experimental observations. *Int. J. Remote Sens.* **2006**, *27*, 3831–3852. [CrossRef]
38. Wu, T.D.; Chen, K.S.; Fung, A.K.; Tsay, M.K. A transition model for the reflection coefficient in surface scattering. *IEEE Trans. Geosci. Remote Sens.* **2001**, *39*, 2040–2050.
39. Oh, Y.; Sarabandi, K.; Ulaby, F.T. Semi-empirical model of the ensemble-averaged differential Mueller matrix for microwave backscattering from bare soil surfaces. *IEEE Trans. Geosci. Remote Sens.* **2002**, *40*, 1348–1355. [CrossRef]
40. Dubois, P.C.; Van Zyl, J.; Engman, T. Measuring soil moisture with imaging radars. *IEEE Trans. Geosci. Remote Sens.* **1995**, *33*, 926. [CrossRef]

41. Khabazan, S.; Motagh, M.; Hosseini, M. Evaluation of Radar Backscattering Models IEM, OH, and Dubois using L and C-Bands SAR Data over different vegetation canopy covers and soil depths. *ISPRS Int. Arch. Photogramm. Remote Sens. Spat. Inf. Sci.* **2013**, *XL-1/W3*, 225–230. [CrossRef]
42. Huang, S.; Ding, J.; Zou, J.; Liu, B.; Zhang, J.; Chen, W. Soil Moisture Retrival Based on Sentinel-1 Imagery under Sparse Vegetation Coverage. *Sensors* **2019**, *19*, 589. [CrossRef]
43. Biftu, G.F.; Gan, T.Y. Retrieving near-surface soil moisture from Radarsat SAR data. *Water Resour. Res.* **1999**, *35*, 1569–1579. [CrossRef]
44. Chai, X.; Zhang, T.; Shao, Y.; Gong, H.; Liu, L.; Xie, K. Modeling and Mapping Soil Moisture of Plateau Pasture Using RADARSAT-2 Imagery. *Remote Sens.* **2015**, *7*, 1279–1299. [CrossRef]
45. Chiara, P.; Brian, B.; Alexander, G.; Dwyer, E. Quality Assessment of the CCI ECV Soil Moisture Product Using ENVISAT ASAR Wide Swath Data over Spain, Ireland and Finland. *Remote Sens.* **2015**, *7*, 15388–15423.
46. Dabrowska, K.; Budzynska, M.; Tomaszewska, M.; Malińska, A.; Gatkowska, M.; Bartold, M.; Malek, I. Assessment of Carbon Flux and Soil Moisture in Wetlands Applying Sentinel-1 Data. *Remote Sens.* **2016**, *8*, 756. [CrossRef]
47. Alexakis, D.D.; Mexis, F.K.; Vozinaki, A.K.; Daliakopoulos, I.N.; Tsanis, I.K. Soil Moisture Content Estimation Based on Sentinel-1 and Auxiliary Earth Observation Products. A Hydrological Approach. *Sensors* **2017**, *17*, 1455. [CrossRef] [PubMed]
48. Gao, Q.; Zribi, M.; Escorihuela, M.J.; Baghdadi, N. Synergetic Use of Sentinel-1 and Sentinel-2 Data for Soil Moisture Mapping at 100 m Resolution. *Sensors* **2017**, *17*, 1966. [CrossRef]
49. Rahman, M.S.; Di, L.; Yu, E.; Lin, L.; Zhang, C.; Tang, J. Rapid Flood Progress Monitoring in Cropland with NASA SMAP. *Remote Sens.* **2019**, *11*, 191. [CrossRef]
50. Meyer, T.; Weihermüller, L.; Vereecken, H.; Jonard, F. Vegetation Optical Depth and Soil Moisture Retrieved from L-Band Radiometry over the Growth Cycle of a Winter Wheat. *Remote Sens.* **2018**, *10*, 1637. [CrossRef]
51. Wang, C.; Zhang, Z.; Paloscia, S.; Zhang, H.; Wu, F.; Wu, Q. Permafrost Soil Moisture Monitoring Using Multi-Temporal TerraSAR-X Data in Beiluhe of Northern Tibet, China. *Remote Sens.* **2018**, *10*, 1577. [CrossRef]
52. Leconte, R.; Brissette, F.; Galarneau, M.; Rousselle, J. Mapping near-surface soil moisture with RADARSAT-1 synthetic aperture radar data. *Water Resour. Res.* **2004**, *40*. [CrossRef]
53. Zhang, L.; Lv, X.; Chen, Q.; Sun, G.; Yao, J. Estimation of Surface Soil Moisture during Corn Growth Stage from SAR and Optical Data Using a Combined Scattering Model. *Remote Sens.* **2020**, *12*, 1844. [CrossRef]
54. Hoskera, A.K.; Nico, G.; Ahmed, M.I.; Whitbread, A. Accuracies of Soil Moisture Estimations Using a Semi-Empirical Model Over Bare Soil Agricultural Croplands from Sentinel-1 SAR Data. *Remote Sens.* **2020**, *12*, 1664. [CrossRef]
55. Page, M.L.; Jarlan, L.; El Hajj, M.M.; Zribi, M.; Baghdadi, N.; Boone, A. Potential for the Detection of Irrigation Events on Maize Plots Using Sentinel-1 Soil Moisture Products. *Remote Sens.* **2020**, *12*, 1621. [CrossRef]
56. Zhang, L.; Meng, Q.Y.; Yao, S.; Wang, Q.; Zeng, J.; Zhao, S.; Ma, J. Soil Moisture Retrieval from the Chinese GF-3 Satellite and Optical Data over Agricultural Fields. *Sensors* **2018**, *18*, 2675. [CrossRef]
57. He, L.; Chen, J.M.; Chen, K.S. Simulation and SMAP Observation of Sun-Glint Over the Land Surface at the L-Band. *IEEE Trans. Geosci. Remote Sens.* **2017**, *55*, 2589–2604. [CrossRef]
58. Zribi, M.; Gorrab, A.; Baghdadi, N.; Lili-Chabaane, Z. Influence of Radar Frequency on the Relationship between Bare Surface Soil Moisture Vertical Profile and Radar Backscatter. *IEEE Geosci. Remote Sens. Lett.* **2014**, 848–852. [CrossRef]
59. Gong, P.; Wang, J.; Yu, L.; Zhao, Y.; Zhao, Y.; Liang, L. Finer resolution observation and monitoring of global land cover: First mapping results with Landsat TM and ETM+ data. *Int. J. Remote Sens.* **2013**, *34*, 2607–2654. [CrossRef]
60. Bindlish, R.; Barros, A.P. Parameterization of vegetation backscatter in radar-based, soil moisture estimation. *Remote Sens. Environ.* **2001**, *76*, 130–137. [CrossRef]
61. Hansan, Z.; Quan, J. Accurate Measurement of Key Parameters of Film Capacitors for EV Power Control Unit. In Proceedings of the 2019 IEEE 4th Int. Future Energy Electronics Conference (IFEEC), Singapore, 25–28 November 2019; Volume 10, p. 1109.
62. Eom, H.J.; Fung, A.K. A scatter model for vegetation up to Ku-band. *Remote Sens. Environ.* **1984**, *15*, 185–200. [CrossRef]
63. Zhang, J.H.; Xu, Y.; Yao, F.M.; Wng, P.J.; Gou, W.J.; Yang, L.M. Advances in estimation methods of vegetation water content based on optical remote sensing techniques. *Sci. China Technol. Sci.* **2010**, *5*, 5–13. [CrossRef]

64. Prakash, R.; Singh, D.; Pathak, N.P.A. Fusion approach to retrieve soil moisture with SAR and optical data. *Remote Sens.* **2012**, *5*, 196–206. [CrossRef]
65. Bai, X.; He, B.; Li, X. Optimum Surface Roughness to Parameterize Advanced Integral Equation Model for Soil Moisture Retrieval in Prairie Area Using Radarsat-2 Data. *IEEE Trans. Geosci. Remote Sens.* **2016**, *54*, 2437–2449. [CrossRef]
66. Zheng, X.M.; Ding, Y.L.; Zhao, K.; Jiang, T.; Li, X.F.; Zhang, S.Y.; Li, Y.Y.; Wu, L.L.; Sun, J.; Ren, J.H.; et al. Estimation of Vegetation Water Content from Landsat 8 OLI Data. *Spectrosc. Spectr. Anal.* **2014**, *34*, 3385–3390.
67. Zeng, J.; Chen, K.S.; Liu, Y.; Bi, H.; Chen, Q. Radar Response of Off-Specular Bistatic Scattering to Soil Moisture and Surface Roughness at L-Band. *IEEE Geosci. Remote Sens. Lett.* **2016**, *13*, 1–5. [CrossRef]
68. Dobson, M.C. Microwave dielectric behavior of wet soil-Part II: Dielectric-mixing models. *IEEE Trans. Geosci. Remote Sens* **1985**, *23*, 35–46. [CrossRef]
69. Barber, M.; Grings, F.; Perna, P.; Piscitelli, M.; Maas, M.; Bruscantini, C.A.; Jacobo-berlles, J.; Karszenbaum, H. Speckle Noise and Soil Heterogeneities as Error Sources in a Bayesian Soil Moisture Retrieval Scheme for SAR Data. *IEEE J. Sel. Top. Appl. Earth Obs. Remote Sens.* **2012**, *5*, 942–951. [CrossRef]
70. Zribi, M.; Dechambre, M. A new empirical model to retrieve soil moisture and roughness from C-band radar data. *Remote Sens. Environ.* **2003**, *84*, 42–52. [CrossRef]
71. Shi, J.; Wang, J.; Hsu, A.Y.; O'Neill, P.E.; Engman, E.T. Estimation of bare surface soil moisture and surface roughness parameter using L-band SAR image data. *IEEE Trans. Geosci. Remote Sens.* **2002**, *35*, 1254–1266.
72. Chen, J.; Jia, Y.; Yu, F. Soil moisture inversion by radar with dual-polarization. *Trans. Chin. Soc. Agric. Eng.* **2013**, *5*, 109–115.
73. Gou, S.; Bai, X.; Chen, Y.; Zhang, S.; Zhu, Q.; Du, P. An Improved Approach for Soil Moisture Estimation in Gully Fields of the Loess Plateau Using Sentinel-1A Radar Images. *Remote Sens.* **2019**, *11*, 349.
74. Brunner, D.; Bruzzone, L.; Ferro, A.; Fortuny, J.; Lemoine, G. Analysis of the double bounce scattering mechanism of buildings in VHR SAR data. *Proc. SPIE Int. Soc. Opt. Eng.* **2008**, *7109*. [CrossRef]
75. Al-Bakri, J.; Suleiman, A.; Berg, A. A comparison of two models to predict soil moisture from remote sensing data of RADARSAT II. *Arab. J. Geosci.* **2014**, *7*, 4851–4860. [CrossRef]

remote sensing

MDPI

Article

Deriving Field Scale Soil Moisture from Satellite Observations and Ground Measurements in a Hilly Agricultural Region

Luca Zappa [1,*], **Matthias Forkel** [1,2], **Angelika Xaver** [1] and **Wouter Dorigo** [1]

[1] Department of Geodesy and Geoinformation, TU Wien, 1040 Vienna, Austria; matthias.forkel@tu-dresden.de (M.F.); angelika.xaver@geo.tuwien.ac.at (A.X.); wouter.dorigo@geo.tuwien.ac.at (W.D.)

[2] Institute of Photogrammetry and Remote Sensing, Technische Universität Dresden, 01069 Dresden, Germany

* Correspondence: luca.zappa@geo.tuwien.ac.at

Received: 3 September 2019; Accepted: 4 November 2019; Published: 6 November 2019

Abstract: Agricultural and hydrological applications could greatly benefit from soil moisture (SM) information at sub-field resolution and (sub-) daily revisit time. However, current operational satellite missions provide soil moisture information at either lower spatial or temporal resolution. Here, we downscale coarse resolution (25–36 km) satellite SM products with quasi-daily resolution to the field scale (30 m) using the random forest (RF) machine learning algorithm. RF models are trained with remotely sensed SM and ancillary variables on soil texture, topography, and vegetation cover against SM measured in the field. The approach is developed and tested in an agricultural catchment equipped with a high-density network of low-cost SM sensors. Our results show a strong consistency between the downscaled and observed SM spatio-temporal patterns. We found that topography has higher predictive power for downscaling than soil texture, due to the hilly landscape of the study area. Furthermore, including a proxy of vegetation cover results in considerable improvements of the performance. Increasing the training set size leads to significant gain in the model skill and expanding the training set is likely to further enhance the accuracy. When only limited in-situ measurements are available as training data, increasing the number of sensor locations should be favored over expanding the duration of the measurements for improved downscaling performance. In this regard, we show the potential of low-cost sensors as a practical and cost-effective solution for gathering the necessary observations. Overall, our findings highlight the suitability of using ground measurements in conjunction with machine learning to derive high spatially resolved SM maps from coarse-scale satellite products.

Keywords: soil moisture; downscaling; advanced scatterometer (ASCAT); soil moisture active passive (SMAP); random forest; low-cost sensor

1. Introduction

By modulating the water, carbon, and energy fluxes between the soil, the vegetation, and the atmosphere, soil moisture is a key variable in climatological and hydrological processes and a regulator of productivity in agricultural systems. It affects the partitioning of water between infiltration, runoff, and evapotranspiration, thus governing the water available for photosynthesis. Accurate knowledge of the spatial and temporal patterns of moisture conditions is required for crop yield estimation [1,2], drought monitoring [3–6], weather and climatic prediction [7–9], rainfall-runoff response [10–12], and landslide forecasting [13,14].

Different factors govern and affect soil moisture patterns across scales: Point- and field-scale variations in space are mainly caused by topography, soil texture, and vegetation, while regional- and continental-scale patterns are controlled by meteorological forcing [15,16]. Hence, monitoring soil moisture is challenging because of its high variability both in space and in time. In-situ sensors provide accurate and reliable measurements at the point scale, however they are costly, and their installation and maintenance are time consuming. Therefore, in-situ sensors are limited in number on a global basis [17–19]. Due to their sparsity, available ground stations can only provide an incomplete picture of the moisture conditions over large areas, making their use impractical for large-scale monitoring [20]. On the other hand, remotely sensed data from microwave sensors have been successfully used to retrieve soil moisture globally [21,22]. This is possible because of the high contrast between dielectric properties of liquid water and dry soil, which can be related either to the microwave emission or to the backscatter detected by passive and active microwave sensors, respectively. However, a trade-off between the spatial and temporal resolution of remotely sensed soil moisture products exists. Synthetic aperture radar (SAR) systems can retrieve data at (sub-) field scale but are characterized by a long revisit time. For example, the recently launched Sentinel-1 mission is used to retrieve surface soil moisture at 1 km with a revisit time of up to four days over Europe [23]. Consequently, short-term events such as rainfall often cannot be captured. Passive (radiometers) and active (radars) microwave sensors with large footprints can observe the same location daily or sub-daily, but typically have spatial resolutions in the order of tens of km. Therefore, such coarse-scale products can capture the temporal dynamics of soil moisture but are inadequate in providing spatial details.

Several agricultural and hydrological applications would benefit from soil moisture observations with a sub-kilometer spatial resolution while preserving a daily revisit time [24,25]. To meet these requirements, several downscaling approaches have been developed, differing in terms of input data and scaling models. Generally, the attainable resolution of the soil moisture downscaled using auxiliary data (e.g., land surface attributes, optical and/or radar remote sensing) depends on the resolution of these auxiliary data. The downscaling method used to describe the relationship between the coarse-scale soil moisture and the local variability can be either statistical or physically based. A comprehensive review of downscaling methods for remotely sensed soil moisture products, discussing assumptions, advantages, and disadvantages associated with each method, is given in Peng et al. [26] and Sabaghy et al. [27].

Over the last decade, machine learning methods have been introduced to downscale soil moisture products and were found to be superior to other techniques [27]. A common way to use machine learning algorithms is to build a model between the coarse-scale soil moisture as the response variable and ancillary surface parameters aggregated to the same coarse resolution as explanatory variables. The model is then applied to the native resolution of those explanatory variables [28], thus assuming that the relationship between soil moisture and ancillary variables remains constant among scales. However, uncertainties and errors arise when aggregating high-resolution surface variables to coarse scale, because of smoothing effects [29]. Zhao et al. [30] clearly presented this issue by taking the elevation as an example: At 1 km resolution, it varied between 4 m and 2418 m, while after the spatial aggregation to 36 km, elevation ranged between 152 m and 1507 m. Extreme values are smoothed after spatial averaging, thus are not included in the training data. To avoid such shortcoming, one could derive robust relationships between coarse-scale soil moisture products and high-resolution ancillary variables using in-situ measurements. However, extensive in-situ observations would be a prerequisite for training the model. Moreover, the accuracy of model predictions increases with increasing the number of samples used as the training set [31]. The limited number of in-situ observations available has been a limiting factor for the application of such a downscaling approach [26].

An appealing solution is offered by low-cost soil moisture sensors, which have gained great attention in the last decade. A large variety of sensors with different measuring techniques and objectives has been developed recently [32–34]. In particular, capacitance sensors found widespread use because they are relatively inexpensive and easy to operate. The significantly lower cost of these sensors compared to traditional probes makes them suitable for high-density and/or large-scale monitoring of soil moisture [35,36]. Increasing the number of sensors in a network further reduces the sampling error due to the high spatial variability of soil moisture [37].

Here, we aim to downscale coarse resolution (25–236 km) satellite soil moisture products to high spatial (30 m) resolution, in order to meet the requirements of agricultural and hydrological applications. The downscaling framework is based on the random forest machine learning algorithm trained against in-situ soil moisture measurements. Coarse-scale soil moisture products and ancillary information on soil texture, topography, and temporal dynamics of vegetation are used as model predictors. A high-density network of low-cost sensors was installed in an agriculturally dominated catchment in Austria to develop and test the proposed downscaling approach. The main objectives of this study were to (i) assess the robustness of the downscaling model, in relation to environmental conditions and physical properties of the study area, (ii) evaluate the impact of different input variables on the model accuracy, and finally, (iii) estimate the effect of various training sets, differing in terms of size and sampling schemes (spatially limited and temporally limited scenarios), on the model skill.

2. Materials and Methods

2.1. Study Area

The hydrological open air laboratory (HOAL, [38]) is an agricultural catchment (66 ha) located in Petzenkirchen, lower Austria (48°9′N, 15°9′E) (Figure 1a). The area is characterized by a humid climate with higher precipitation in summer than in winter. The mean annual temperature and precipitation measured during the period 1990–2014 are 9.5 °C and 823 mm year^{-1}, respectively. The mean annual evapotranspiration estimated for the same period is 628 mm year^{-1} [38]. The elevation ranges from 268 to 323 m a.s.l. with a mean slope of 8%. The dominant soil types in the catchment are Cambrisols and Planosols, characterized by medium to poor infiltration capacities [38].

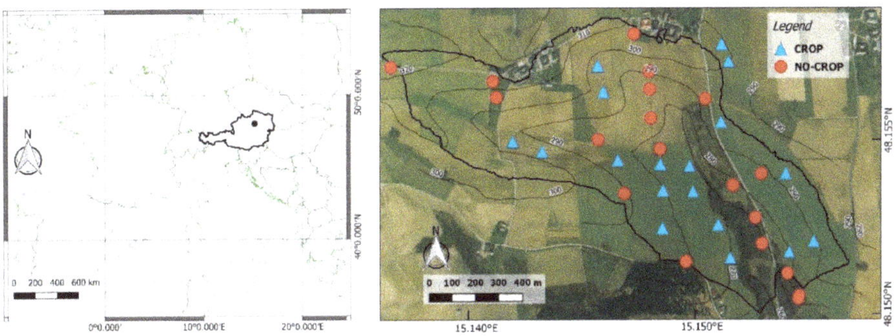

(a) (b)

Figure 1. Location of the study area in Petzenkirchen, Austria (**a**) and distribution of the low-cost sensors within the study area (**b**). Map data ©2019 Bing.

Most of the catchment area is arable land (87%), while the rest is forested (6%), used as pasture (5%), or paved (2%). Usually, winter crops (mainly wheat) are planted in November and harvested in June, while summer crops (mainly maize) are planted in April and harvested in October. Depending on the crop and the weather conditions, these dates might vary by a few weeks.

2.2. In-Situ Measurements

Since April 2017, an in-situ network of low-cost sensors (Flower Power, Parrot [39]) measuring both soil moisture and incoming solar radiation has been installed in the catchment [36]. The selection of the sensor locations follows the design employed by Vreugdenhil et al. [40] for the validation of coarse-scale soil moisture products over the same area [41]. In this study, we used 38 low-cost sensors covering a wide range of soil texture and topographic conditions (Supplementary Figures S1 and S2). Nineteen sensors were installed within agricultural fields and temporarily removed during field management practices (planting, harvesting, ploughing, etc.) and are referred to as CROP (Supplementary Figure S3). The remaining 19 sensors were permanently positioned in forest and grassland sites, as well as at the edges of agricultural fields (NO-CROP, Figure 1b). The original readings of the low-cost sensors, taken every 15 min, were resampled to daily averages.

2.2.1. Soil Moisture

A capacitance sensor inserted vertically in the top layer of the soil (approximately 7 cm) measures the soil capacitance, which is related to the dielectric permittivity. The latter is then converted to volumetric soil moisture using a predefined calibration equation [36]. An extensive evaluation of the low-cost sensor performances was carried out both in the laboratory and in the field, proving the capability of this sensor to reliably observe soil moisture [36]. Measurements from the low-cost sensors have been evaluated against gravimetric observations in the lab for the dominant soil type present in the catchment [36]. Several water content levels have been investigated, ranging from air dry to full saturation. For each soil moisture level considered, five low-cost sensors have been used to record data. Overall, this comparison showed a slight overestimation (up to 0.08 $m^3\ m^{-3}$) of the low-cost sensors for dry conditions (soil moisture < 0.20 $m^3\ m^{-3}$), but a high correlation was found ($R > 0.90$, unbiased root mean square deviation (uRMSD) < 0.035 $m^3\ m^{-3}$). The inter-sensor variability was also negligible. Furthermore, Xaver et al. [36] compared the measurements from the low-cost sensors against those of professional probes already present in the study site [38,40]. The analysis was performed using the top layer sensor of the professional stations, installed horizontally at 5 cm depth. Results of a 10-month field assessment involving 33 pairs of low-cost and professional sensors showed a good agreement ($R = 0.80$, uRMSD = 0.05 $m^3\ m^{-3}$). Some differences in the readings could be attributed to the different volumes of soil investigated by the two sensor types because of the different positioning of low-cost and professional probes.

2.2.2. Incoming Solar Radiation and Fraction of Absorbed Radiation

The low-cost devices are equipped with a light sensor measuring the incoming solar radiation centered at 550 nm. The accuracy of such observations was tested by comparing measurements from a low-cost sensor placed on top of a 2 m pole, i.e., undisturbed from shadowing sources, with the incoming radiation integrated between 300 and 2800 nm recorded from a weather station in close proximity (<5 m). Also, for the solar radiation, we found a high correlation ($R > 0.90$) between observations from low-cost and professional sensors.

The sensor on top of the 2 m pole is used as reference of the top of canopy radiation (ToC), which we assumed homogeneous over the study area. The other 38 low-cost sensors installed in the ground, in addition to soil moisture, monitor the solar radiation at the bottom of the vegetation canopy (BoC). Similar as for the fraction of absorbed photosynthetically active radiation (fAPAR), we calculated the fraction of absorbed green radiation (fAGR) as:

$$fAGR_{i,t} = \frac{ToC_t - BoC_{i,t}}{ToC_t} \tag{1}$$

where fAGR is a proxy of vegetation cover. The subscripts i and t denote the radiation measured at a specific site and on a certain day, respectively. fAGR ranges between 0 and 1, indicating the entire radiation (i.e., no vegetation) and no radiation (i.e., dense vegetation) reaching the ground, respectively. After visual inspection of the fAGR timeseries, we applied a moving average of 20 days to smooth the data and avoid artifacts such as spikes and drops.

2.3. Remotely Sensed Soil Moisture

Here, we employed two state-of-the-art remotely sensed soil moisture products that guarantee adequate temporal frequency for agricultural and hydrological applications. We selected active C-band (advanced scatterometer (ASCAT)) and passive L-band (soil moisture active passive (SMAP)) soil moisture products, to assess the robustness of the downscaling approach with respect to different observation frequencies and sensing techniques. An assessment of the ASCAT and SMAP products using ground-based measurements is given, e.g., in [42,43], respectively. Figure 2 shows the area (approximately) covered by the ASCAT and SMAP footprints compared to the study catchment. Notwithstanding the considerably larger area observed by the satellite products, the HOAL is representative of the topographic conditions and land cover classes captured by ASCAT and SMAP [44].

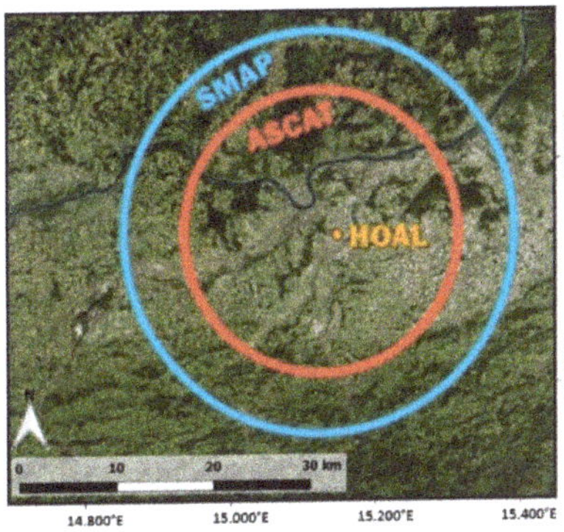

Figure 2. Schematic overview of the sensor footprints of advanced scatterometer (ASCAT) and soil moisture active passive (SMAP) over the study area (hydrological open air laboratory (HOAL)). Map data ©2019 Bing.

2.3.1. ASCAT

Backscatter observations from the advanced scatterometer (ASCAT) sensor onboard Metop-A are used to retrieve soil moisture via the TU Wien change detection method [45,46]. In short, the observed backscatter is scaled between the historically lowest and highest backscatter for each pixel, corresponding to the driest and the wettest observations, respectively [46]. Thus, the ASCAT soil moisture is expressed as percentage of saturation. The ASCAT product accounts for the top 2 cm soil layer and is provided at 25 km resolution [42]. Recently, Pfeil et al. [44] improved the soil moisture retrieval algorithm by changing the parameterization of the cross-over angles (10°/30° instead of 25°/40°). We selected this product for the analysis and calculated the daily average between morning and evening overpasses.

2.3.2. SMAP

The soil moisture active passive (SMAP) L-band radiometer measures brightness temperature [47], from which soil moisture is derived by inversion of a tau-omega model [22]. Surface soil moisture (top 5 cm) is expressed in $m^3\ m^{-3}$ and is provided at 36 km resolution [41]. The SMAP Level-3 version V005 was chosen for this study. Only the morning overpasses (06:00) were included in the analysis as in Alemohammad et al. [28], because the 18:00 ascending passes show degradation in quality [48].

2.4. Topography and Soil Texture

A digital elevation model (DEM) of the study area is available at 0.5 m resolution [38]. We down-sampled the DEM to 30 m, to simulate the resolution of freely available data, e.g., EU-DEM [49]. Then, we extracted five topographic indices ensuring that different hydrological processes affecting the spatial distribution of soil moisture were considered [50]. In particular, the DEM was used to compute slope, upslope area, topographic wetness index (TWI) [51], total insolation [52], and general curvature [53].

Percentages of different soil particle sizes (clay, silt, and sand) in the top 5 cm are available from a soil survey campaign conducted over the study area on a regular 50 × 50 m grid [38]. Data gathered at these point locations were interpolated to 30 m resolution using the inverse distance weighting method, to spatially match the DEM map.

3. Methods

3.1. Downscaling Framework

In order to estimate soil moisture at fine spatial resolution (30 m), we trained random forest (RF) regression models [54] against in-situ soil moisture measurements from low-cost sensors (SSM_{HR}), by using coarse-resolution soil moisture products (SSM) and ancillary data related to soil texture, topography, and fAGR as predictor variables:

$$SSM_{HR} = RF(SSM,\ Soil\ texture,\ Topography,\ fAGR). \quad (2)$$

SSM is assumed to represent the large (i.e., catchment) scale wetness conditions, mainly regulated by meteorological factors, while the other parameters are drivers of small-scale spatial patterns of surface water content [16,50]. Note that the spatial resolution of the downscaled soil moisture is equal to the resolution of the soil texture and topography data employed, i.e., 30 m. The temporal resolution corresponds to that of satellite observations, i.e., quasi-daily. RF was selected as downscaling algorithm because of its ability to capture and model complex non-linear interactions. We employed the *RandomForestRegressor* implementation in Python [55] using 1000 trees for each RF model, as a large number of trees tends to improve the estimation accuracy [54,56].

In order to assess the role of different input variables on the downscaling accuracy, we trained seven RF models with various combinations of predictors (Table 1). A coarse-scale surface soil moisture (SSM) dataset, representing the catchment scale wetness condition, was always included as model input. The SSM source was either a satellite-derived product (ASCAT or SMAP) or the spatially averaged soil moisture calculated from the low-cost sensors (AVG_insitu). The latter was calculated as average from all the in-situ measurements available at day *t* and exemplifies the scenario where a remotely sensed product optimally represents the conditions of the study area. Therefore, AVG_insitu is used as reference to assess the impact of the SSM source used in the downscaling framework.

Table 1. Summary of model combinations tested in this study.

		Predictors			
		Coarse Scale Surface Soil Moisture (SSM)	**Soil Texture (S)**	**Topography (T)**	**Vegetation (V)**
	Variables	AVG_insitu or ASCAT or SMAP	Clay, silt, sand	Slope, TWI, Upslope_area, Total_insolation, General_curvature	fAGR
Model combinations	**SSM+V**	✔			✔
	SSM+S	✔	✔		
	SSM+T	✔		✔	
	SSM+S+T	✔	✔	✔	
	SSM+S+V	✔	✔		✔
	SSM+T+V	✔		✔	✔
	SSM+S+T+V	✔	✔	✔	✔

For this analysis, we considered days when soil moisture observations were available for the low-cost sensors and both the satellite-derived products. Furthermore, to exclude unreliable measurements due to frozen soil, only dates with daily air temperature (measured by a weather station in the catchment) higher than 3 °C were used, resulting in 237 days between May 2017 and June 2018. Overall, 3494 observations were available for the analysis.

3.2. Evaluation Strategy

3.2.1. Model Comparison and Evaluation

To evaluate the predictive performance of each model combination (Table 1), a K-fold cross validation was performed because of the relatively limited data available, i.e., 3494 observations. The original data are divided into K subsets, i.e., the folds, consisting of predictors and response variable. We defined K = 10 subsets, i.e., each training set is made up of 90% of all available observations and the remaining 10% are used for validation. Therefore, the training sets comprised a vast range of moisture values, thus reducing errors and uncertainties arising from extrapolation (i.e., using values beyond the soil moisture range within the training data). In order to maximize the temporal independence between training and testing sets, each fold contained observations sampled from contiguous dates. For each of the 10 iterations of the cross validation, the Pearson correlation (R) and the unbiased root mean square deviation (uRMSD) were calculated [57]. The uRMSD accounts for the bias as the difference of the long-term mean between the measured and predicted soil moisture [57]. We selected the uRMSD instead of the RMSD, because the latter might be severely compromised in the presence of biases [58]. Moreover, the uRMSD is the target metric for evaluating the soil moisture products of various satellite missions, e.g., [41].

We further investigated the robustness of the downscaled soil moisture by assessing if differences in the skill could be ascribed either to static physical characteristics of the landscape or to dynamic environmental conditions. Therefore, we related the statistical metrics obtained for each sensor location to soil texture and topography, and for each time step to the overall (i.e., catchment scale) soil moisture and vegetation cover. Additionally, we explored the models' performances with respect to the vegetation type (i.e., CROP and NO-CROP).

3.2.2. Testing the Effect of the Training Set Size

Considering that the number of in-situ observations used for training machine learning models strongly affects their accuracy, another objective of this study was to assess the model performance in response to various data-limited scenarios. In particular, we investigated how many sensors are needed, and for how long sensors should be installed, to develop robust RF models and produce accurate estimates of soil moisture.

Based on the 10-fold cross validation, the "original" training sets consisted of observations from all available sensor locations and a time span covering 90% of the observation period. We set up a spatially limited scenario and a temporally limited scenario. In the spatially limited case, training sets were created by randomly selecting 25%, 50%, and 75% of sensors (i.e., locations) from the original training sets. Therefore, the spatially limited training sets consisted of observations from 10, 20, and 30 sensors, covering the same time interval of the original training sets. In the temporally limited scenario, new training sets were obtained by removing m observations from adjacent dates, with m corresponding to 25%, 50%, and 75% of the original number of observations (i.e., days). Thus, the temporally limited training sets consisted of data from all the available sensors covering a shorter period than the original training sets. Note that for each scenario (2) and training set size (4) considered, the evaluation was repeated for 10 random combinations.

4. Results

4.1. Impact of Predictors on Model Performance

Results of the 10-fold cross validation show that the choice of the coarse-scale soil moisture source (SSM) plays a major role in the model accuracy (Figure 3). Statistical metrics obtained using the average soil moisture from the low-cost sensors (AVG_insitu) as SSM predictor are significantly better than those obtained employing satellite-derived soil moisture products. This finding is consistent among all the model combinations investigated. As an example, considerable differences in the accuracy are found among models trained using soil texture information (SSM+S): If AVG_insitu is the SSM source, Pearson R is 0.18 (0.14) higher and uRMSD is 0.019 (0.017) $m^3\ m^{-3}$ lower as compared to providing ASCAT (SMAP) as the predictor. Similar results are found for the other model combinations.

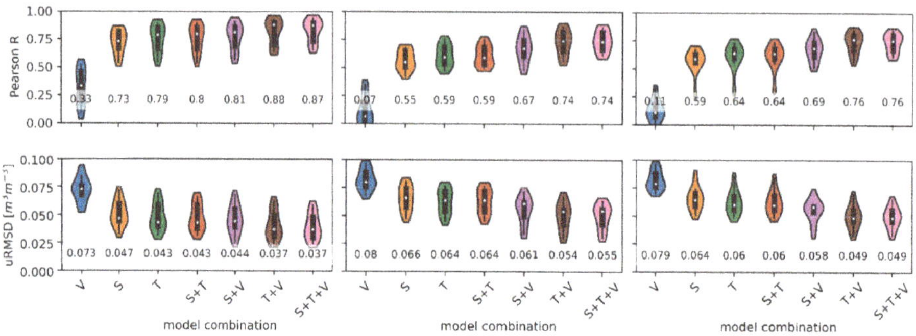

(a) AVG_insitu **(b)** ASCAT **(c)** SMAP

Figure 3. Violin plots of Pearson R (top) and unbiased root mean square deviation (uRMSD) (bottom) between measured and predicted soil moisture (see Table 1). The soil moisture source used as predictor (surface soil moisture (SSM)) is displayed above each graph: AVG_insitu (**a**), ASCAT (**b**), and SMAP (**c**). The different model combinations are reported on the X-axis: S indicates "soil texture", T indicates "topography", and V indicates "vegetation". The boxplots within the violins depict quartiles, and the white dots indicate the median values (also reported below the violins).

Furthermore, Figure 3 indicates that models employing only topographic indices (SSM+T) have better skill than models trained using only soil texture information (SSM+S). If a proxy of vegetation cover (i.e., fAGR) is the only model predictor (SSM+V), the model accuracy is very poor. For example, when AVG_insitu is the model SSM source, the median correlation is equal to 0.79, 0.73, and 0.33 while the uRMSD is 0.043, 0.047, and 0.073 $m^3\ m^{-3}$ for the SSM+T, SSM+S, and SSM+V cases, respectively. An improvement of model performances is obtained if topography-derived indices are added to soil texture data (SSM+S+T). Interestingly, the synergetic use of both soil texture and topographic information does not improve the accuracy of the model as compared to using topography alone (SSM+T).

Models including fAGR as predictor always outperform the counterparts based on static data alone (SSM+S, SSM+T, and SSM+S+T). For instance, the combination SSM+S is subject to a substantial improvement if a vegetation proxy is added to the model (SSM+S+V): When AVG_insitu is the SSM source, the correlation increases from 0.73 to 0.81, while the uRMSD decreases from 0.047 to 0.044 $m^3\ m^{-3}$. For the same model combinations, if ASCAT (SMAP) is the predictor, Pearson R increases from 0.55 to 0.67 (from 0.59 to 0.69) and the uRMSD drops from 0.066 to 0.061 (from 0.064 to 0.058) $m^3\ m^{-3}$, respectively. Overall, the highest accuracy is achieved by the SSM+S+T+V combination, which includes both static (soil texture and topography) and dynamic (vegetation) information. Therefore, further analyses are conducted only for this model combination.

4.2. Spatial and Temporal Evaluation

Figure 4 shows the cumulative distribution of Pearson R and uRMSD, allowing for quick identification of the recurrence of poor agreement between the measured and predicted soil moisture. In order to quantify the model skill in estimating the temporal dynamics, we calculated the statistical metrics between the downscaled and measured timeseries at each sensor location (Figure 4a). As expected, using AVG_insitu as SSM source generates better results than providing satellite-derived products. When using ASCAT or SMAP as predictors, the accuracy is poor (R < 0.5, uRMSD > 0.04 $m^3 m^{-3}$) at few locations. Further investigations with the aim to identify a relationship between the model accuracy and static properties of the study area, i.e., soil texture and topography, showed the ability of the proposed method to estimate soil moisture independently from these factors. Similarly, we calculated statistical metrics between the downscaled soil moisture and in-situ measurements at each time step, thus investigating the skill of the model to reproduce the observed spatial patterns (Figure 4b). In this case, the models trained with coarse-scale satellite products perform comparably as employing AVG_insitu as SSM source. Additional analysis revealed that no relation exists between the model accuracy and catchment scale wetness or vegetation cover.

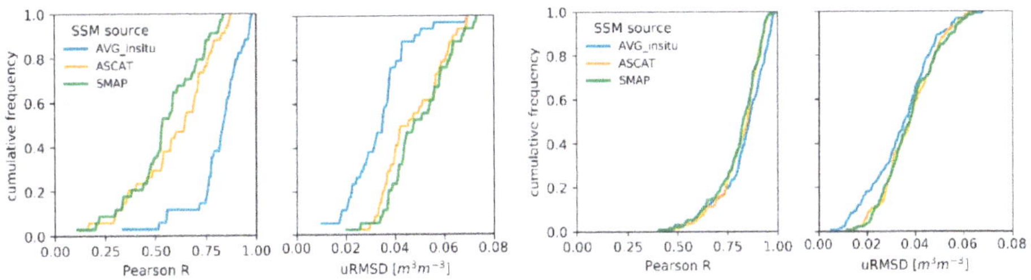

(a) Temporal accuracy **(b)** Spatial accuracy

Figure 4. Cumulative frequency of Pearson R and uRMSD between measured and predicted soil moisture (model combination SSM+S+T+V). The statistical metrics were calculated for each sensor location, thus representing the ability of the model to capture temporal dynamics (**a**), and for each time-step, accounting for the model skill to reproduce spatial patterns (**b**).

We further analyzed the accuracy of the downscaled soil moisture depending on the two main land cover types within the study site, i.e., CROP and NO-CROP. Independently from the SSM source used, a higher correlation is achieved for locations in grassland and forest (i.e., NO-CROP) (Figure 5). We found a substantial difference between NO-CROP (R = 0.87) and CROP (R = 0.70) locations if AVG_insitu is the model predictor. Likewise, when satellite-derived products are used as SSM source, NO-CROP locations outperform CROP sites (e.g., correlation equal to 0.74 and 0.64, respectively, when ASCAT is the model predictor). Interestingly, when considering the uRMSD, we found a better skill for locations in agricultural fields: If AVG_insitu is provided as model input, the uRMSD is 0.037 for NO-CROP and 0.036 $m^3 m^{-3}$ for CROP locations. Given ASCAT as SSM source, results in uRMSD equal to 0.055 and 0.040 $m^3 m^{-3}$ for NO-CROP and CROP, similar to SMAP (0.049 and 0.036 $m^3 m^{-3}$, respectively). A reason for this finding can be ascribed to the strong impact of outliers on the calculation of uRMSD, as the difference between predicted and measured values is squared. Indeed, soil moisture differences larger than 0.10 $m^3 m^{-3}$ were found for few pairs (predicted/measured) belonging to NO-CROP locations.

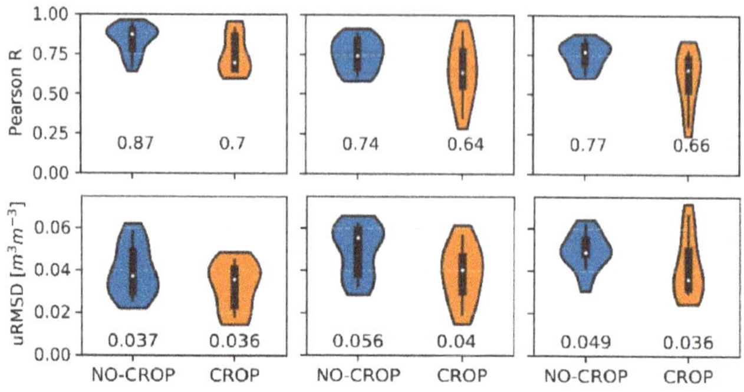

(a) AVG_insitu **(b)** ASCAT **(c)** SMAP

Figure 5. Violin plots of Pearson R (top) and uRMSD (bottom) between measured and predicted soil moisture (model combination SSM+S+T+V) depending on the vegetation type. CROP indicates agricultural fields, while NO-CROP includes grassland, forest, and field edges. The SSM source used as model predictor is displayed above each graph. The boxplots within the violins indicate quartiles and the white dots depict the median values (also reported below the violins). (**a**) AVG_insitu, (**b**) ASCAT, (**c**) SMAP.

4.3. Comparison of Coarse-Scale and Downscaled Products

Here, we evaluate the improvement of statistical metrics achieved by applying the proposed downscaling framework in comparison to using the original coarse-scale products (SSM). The latter, especially ASCAT and SMAP, show low correlation and high uRMSD (Figure 6, top). On the other hand, the downscaled soil moisture is closely distributed around the 1:1 line, resulting in a considerable improvement of the statistical metrics (Figure 6, bottom). Pearson R and uRMSD between individual in-situ measurements and the catchment average (AVG_insitu) are equal to 0.68 and 0.061 m^3 m^{-3}, while for the downscaled soil moisture using AVG_insitu as SSM source, these metrics are 0.86 and 0.042 m^3 m^{-3}, respectively. Even more remarkable is the improvement observed for satellite-derived products: Pearson R increases from 0.50 to 0.76 (from 0.38 to 0.74) and uRMSD decreases from 0.0872 to 0.054 (from 0.080 to 0.056) m^3 m^{-3} when considering the original ASCAT (SMAP) product and the downscaled soil moisture, respectively. As expected, soil moisture downscaled with remotely sensed products as SSM source shows a more dispersed cloud of points compared to using AVG_insitu (Figure 6, bottom). When employing the latter as model predictor, 74% of the downscaled soil moisture fell within an absolute accuracy of 0.04 m^3 m^{-3}, while this percentage is 59% when either ASCAT or SMAP is the SSM source.

The density distribution of the original coarse-scale predictors (SSM) and of the downscaled soil moisture is further compared in Figure 7. We found that the coarser the spatial resolution (i.e., support) of the data, the higher the occurrence of moderate (0.22 m^3 m^{-3}) soil moisture values (green lines). The original SMAP soil moisture product (36 km) exhibits the highest density peak, followed by the ASCAT product (25 km) and the average soil moisture from ground measurements, i.e., AVG_insitu (catchment scale, approximately 2 km). In addition, the coarse-scale products underestimate the occurrence of extreme high and low soil moisture conditions. These results agree with the scaling theory proposed by Western et al. [59], which states that more and more small-scale features are averaged and thus disappear at coarser resolutions because moisture conditions are assumed homogeneous at the support scale. The density distributions of the downscaled soil moisture (orange lines) are very close to that of in-situ measurements (blue line), regardless of the SSM predictor. For instance, the first quartiles of the original ASCAT product, the downscaled soil moisture from ASCAT+S+T+V, and in-situ measurements are equal to 0.23, 0.20, and 0.19 m^3 m^{-3}, respectively. The third quartiles for the same data are 0.27, 0.30, and 0.32 m^3 m^{-3}. However, the larger spread visible at very dry/wet states

(Figure 7) suggests that estimating extreme moisture conditions is more challenging if the SSM source is a coarse-scale satellite-derived product. This finding is likely caused by the insufficient number of training data covering such conditions, as observed by Hutengs and Vohland [56] for temperature.

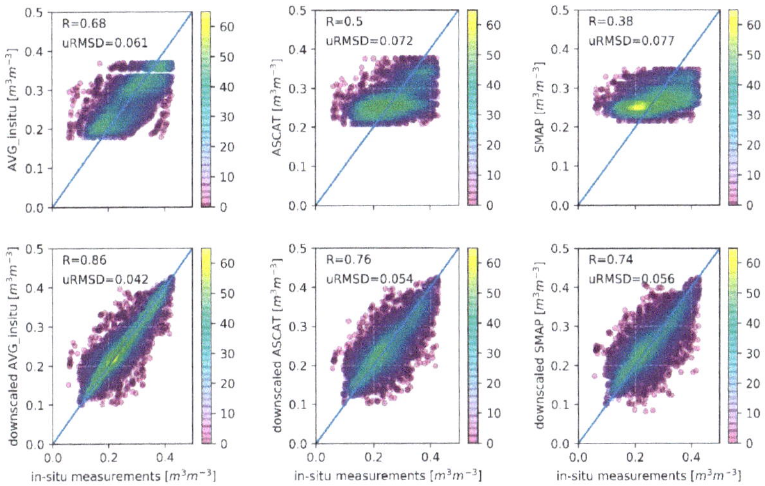

(a) AVG_insitu **(b)** ASCAT **(c)** SMAP

Figure 6. Scatterplots between measured soil moisture and original coarse-scale SSM products (top) and between measured and predicted soil moisture (model combination SSM+S+T+V) (bottom). The color indicates the number of observations. In each graph, the Pearson R and uRMSD are given. (**a**) AVG_insitu, (**b**) ASCAT, (**c**) SMAP.

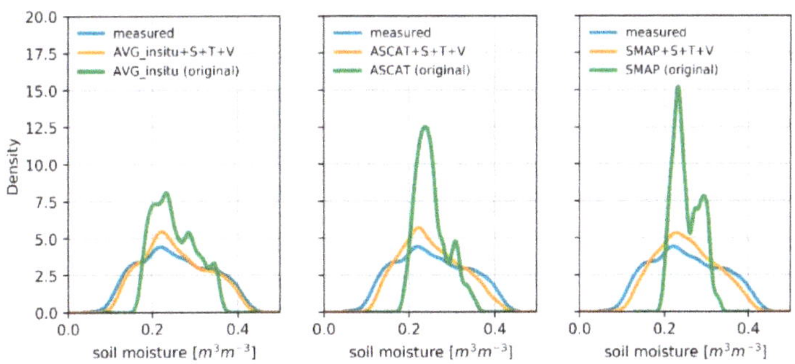

(a) AVG_insitu **(b)** ASCAT **(c)** SMAP

Figure 7. Density distributions of soil moisture obtained for in-situ measurements (blue lines), downscaled soil moisture from the model combination SSM+S+T+V (orange lines), and the original coarse-scale SSM products (green lines). (**a**) AVG_insitu, (**b**) ASCAT, (**c**) SMAP.

4.4. Spatio-Temporal Patterns at the Catchment Scale

To represent the catchment scale temporal pattern, the downscaled soil moisture was averaged over the entire study area and compared to the spatially averaged in-situ measurements (Figure 8). We focus our analysis on the models using remotely sensed products as SSM source. The predicted soil moisture can capture drying and wetting events, regardless of the satellite dataset used (ASCAT or

SMAP). However, the range of soil water content from the predictions is smaller than the observed one. The catchment scale soil moisture from in-situ measurements varies between 0.17 and 0.36 m^3 m^{-3}, while the spatially aggregated estimates using ASCAT (SMAP) as SSM source ranges between 0.19 (0.20) and 0.33 (0.33) m^3 m^{-3}. Furthermore, during summer, the predicted catchment scale soil moisture exhibits a persistent positive bias, while a negative bias is found in the winter period, as reported, e.g., in [44,60].

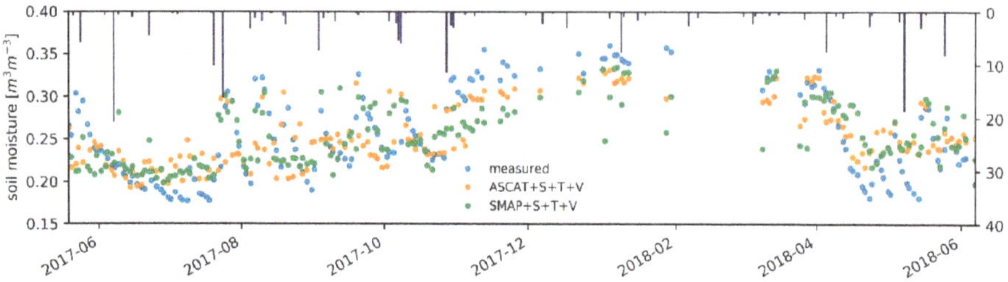

Figure 8. Temporal patterns of catchment scale soil moisture. The predicted soil moisture (model combination SSM+S+T+V) was spatially averaged and compared to the average of in-situ measurements over the catchment (blue dots).

Nevertheless, we found a substantial improvement in the spatially aggregated soil moisture predictions compared to the original satellite products. Pearson R increases from 0.64 to 0.74 (16% increase) when considering the original ASCAT product and the upscaled soil moisture from the ASCAT+S+T+V combination. Similarly, the correlation increases from 0.53 to 0.70 (32% increase) from the original SMAP soil moisture product to the spatially aggregated model predictions. Also, the uRMSD drops from 0.046 to 0.037 m^3 m^{-3} (20% decrease) and from 0.056 to 0.038 m^3 m^{-3} (32% decrease) between the original satellite datasets (ASCAT and SMAP, respectively) and the upscaled soil moisture estimates.

Soil moisture spatial patterns obtained from the model combination ASCAT+S+T for three days with dry, medium, and wet conditions, are shown in Figure 9 (note that these days were not part of the training data). We did not use the best-performing model (i.e., SSM+S+T+V), which also includes information about vegetation cover, because fAGR was available exclusively for the sensor locations and not for the entire catchment. The predicted patterns derived from the SMAP+S+T combination are not shown because they are very similar to those obtained for ASCAT+S+T.

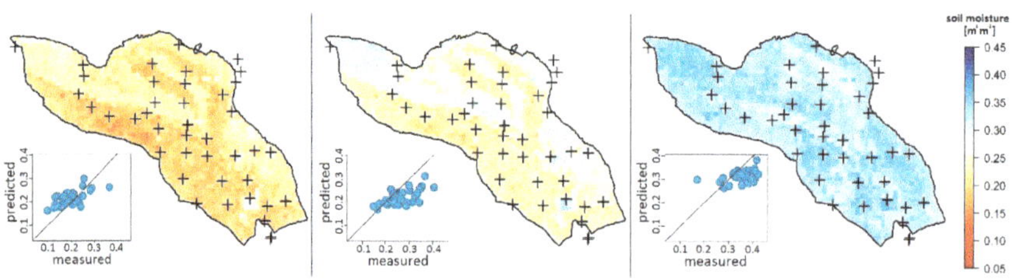

(**a**) Dry (**b**) Medium (**c**) Wet

Figure 9. Spatial patterns of soil moisture over the study site for three days with varying moisture conditions. Each graph shows also the scatterplot between measured and predicted soil moisture for the same day. Soil moisture was obtained from the sub-optimal model combination ASCAT+S+T (similar patterns were found for the SMAP+S+T combination, not shown). Note that a proxy of vegetation cover "V", i.e., fraction of absorbed green radiation (fAGR), was not included because it was available only for the sensor locations (depicted with the cross) but not for the entire study area. (**a**) Dry, (**b**) Medium, (**c**) Wet.

Some clear patterns are visible, e.g., the north-western part of the catchment is generally wetter than the rest. This result can be explained by the higher clay content and the relatively flat topography of this portion of the study area. However, it is important to note that such patterns are only related to static properties, i.e., soil texture and topography. A solution for this constraint is offered by vegetation indices retrieved from optical [61] or microwave [62] satellite sensors, which provide spatially continuous coverage of large areas. Therefore, it is crucial to highlight that great improvements and more representative maps are expected by including information about the spatio-temporal patterns of vegetation (as shown in Figure 3).

4.5. *Effect of the Training Set Size on Model Performance*

In order to quantify how many sensors are needed and for how long they should be installed, we investigated the impact of different training set sizes with varying spatial and temporal information content on the model accuracy (Figures 10 and 11). Note that when the original training sets (ALL) are used, results are equivalent to those shown in Figure 3 for the SSM+S+T+V combination.

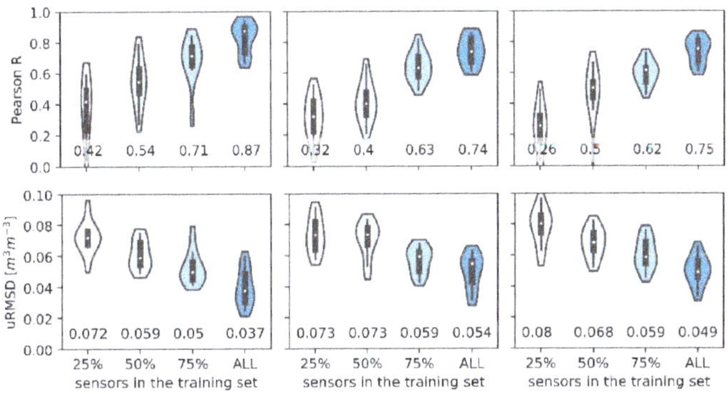

(a) AVG_insitu **(b)** ASCAT **(c)** SMAP

Figure 10. Violin plots of Pearson R (top) and uRMSD (bottom) between measured and predicted soil moisture (SSM+S+T+V) against the number of sensors used to train the models. Training sets consisted of 25%, 50%, and 75% of all available sensors (38), each sensor covering the same time interval of the original training set. For each training set size, we repeated the evaluation for 10 random permutations. The SSM source used is displayed above each graph. The boxplots within the violins indicate quartiles and the white dots depict the median values (also reported below the violins). (**a**) AVG_insitu, (**b**) ASCAT, (**c**) SMAP.

The model performance generally improves with an increasing number of sensors. If AVG_insitu is used as model predictor, the median correlation increases from 0.42 to 0.87 (107% increase) from using data from only 25% of the sensors to using all the available ones. Even more outstanding is the improvement achieved when employing satellite-derived soil moisture products as SSM source. We found Pearson R to increase from 0.32 to 0.74 (131% increase) and from 0.26 to 0.72 (177% increase), if ASCAT and SMAP are the model predictors, respectively. Like the improvement in Pearson correlation, the uRMSD decreases with an increase in the number of sensors used in the training set. If the SSM source is AVG_insitu, the uRMSD drops from 0.072 to 0.037 $m^3\ m^{-3}$, by using observations from 25% of the sensors and all the available ones, respectively. Similarly, when using ASCAT (SMAP) as the model predictor, the uRMSD decreases from 0.073 to 0.055 (from 0.080 to 0.049) $m^3\ m^{-3}$.

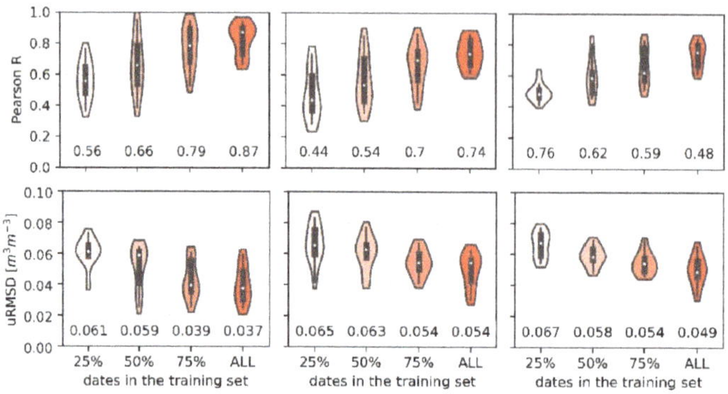

(a) AVG_insitu **(b)** ASCAT **(c)** SMAP

Figure 11. Violin plots of Pearson R (top) and uRMSD (bottom) between measured and predicted soil moisture (SSM+S+T+V) against the number of dates used to train the models. Training sets consisted of 25%, 50%, and 75% of contiguous observations sampled from the original training sets (ALL). Note that all the available sensor locations are included. For each training set size, we repeated the evaluation for 10 random permutations. The SSM source used as model predictor is displayed above each graph. The boxplots within the violins indicate quartiles and the white dots depict the median values (also reported below the violins). (**a**) AVG_insitu, (**b**) ASCAT, (**c**) SMAP.

The model accuracy also improves when increasing the temporal information, i.e., longer duration of the in-situ measurements, in the training set. When AVG_insitu is provided as SSM predictor and only 25% of the observations are used, the correlation is equal to 0.56, while using the original time interval results in a Pearson R of 0.87 (55% increase). If ASCAT is the model predictor, the correlation increases by 68% (from 0.44 to 0.74), and similar results are found when SMAP is used as SSM source (58% increase). Analogously, the uRMSD reduces from 0.061 to 0.037 m^3 m^{-3} for the 25% and all cases, respectively, when AVG_insitu is the SSM source. If ASCAT (SMAP) is the model predictor, we found the uRMSD to decrease from 0.065 to 0.054 m^3 m^{-3} (from 0.067 to 0.049 m^3 m^{-3}) by using only 25% of the observations and the original training set, respectively.

The model skill improves with both an increasing number of sensors and a longer duration of the measurements. Furthermore, Figures 10 and 11 suggest that more accurate estimates could be expected with an even higher number of sensors. However, our results indicate that for a given training set size (within the sizes tested here), the model performance is better when the training data includes a high number of sensors collecting measurements for a short period, rather than consisting of few sensors measuring for a longer period.

5. Discussion

5.1. Role of Model Predictors on Downscaling Performance

The model accuracy obtained in this study is similar to that reported in the review paper of Sabaghy et al. [27] for various downscaling methods based on machine learning (R ≈ 0.80, RMSD = 0.056 m^3 m^{-3} on average). However, we found that the skill greatly improved (R = 0.87, uRMSD = 0.037 m^3 m^{-3}) if more representative SSM sources (i.e., AVG_insitu) are provided as input to the model instead of using coarse-scale remotely sensed products (ASCAT or SMAP). Therefore, our results highlight the importance of the SSM source provided as input to the model to produce satisfactory estimates. Indeed, satellite-derived products are subject to considerable challenges and difficulties, such as the extraction of soil moisture from the retrieved signal (either backscatter or brightness temperature), leading to random and/or systematic errors [63]. In our analysis, an additional source of error is the spatial mismatch between the satellite footprint and the size of the study site. To reduce uncertainties arising from the spatial representativeness, newly developed products could be used as SSM predictor. For example, Sentinel-1 observations allow estimating surface soil moisture at a spatial resolution of 1 km and a revisit time up to four days over Europe [23]. However, the temporal frequency of Sentinel-1 might still be insufficient for some applications. To overcome this issue, novel approaches fusing ASCAT and Sentinel-1 soil moisture products have been developed, preserving the temporal resolution of ASCAT while improving the spatial details [64]. Both products could reduce the spatial mismatch between the study site and the satellite pixel size leading to more accurate soil moisture estimates.

When examining results from different model combinations, it appears that topography (SSM+T) has higher predictive power than soil texture (SSM+S). Furthermore, the addition of soil texture information to topography did not generate any substantial improvement to the model accuracy. This finding can be partially explained by the uncertainties due to the interpolation process of the soil texture data employed in this work. Soil samples were collected on a regular grid, thus locations at the field edges (i.e., NO-CROP) are likely to have different soil properties than inferred by the interpolated map. Furthermore, Picciafuoco et al. [65] suggested that variations in soil properties, such as soil packing and the presence of macropores, might not be captured by soil texture alone. However, the higher predictive power of topography over soil texture is site-specific and different outcomes can be expected for other locations. For instance, Teuling and Troch [66] found that the importance of topography increased with increasing relief and complexity of the study area. Similarly, Brocca et al. [67] concluded that flat areas showed random soil moisture patterns, while over-undulating areas soil moisture patterns were strongly related to topography. Our results confirm the dominant role of topography, compared to soil texture, in controlling the spatial distribution of soil moisture in hilly landscapes.

While the inclusion of soil texture information produced a negligible increase in the model accuracy, adding a proxy of vegetation dynamics yielded significant improvements. We also observed a strong influence due to different vegetation types on the predicted soil moisture, as reported in Picciafuoco et al. [65] for saturated hydraulic conductivity. This can be expected because different vegetation types, as well as phenological stages, strongly influence the local environmental conditions. Various rooting structures and canopy properties (e.g., fractional vegetation cover and vertical structure) affect interception, evapotranspiration, percolation, and runoff. Gomez-Plaza et al. [68] determined that the spatial variability of surface soil moisture content is affected by vegetation, and the controls regulating soil moisture spatial patterns are different in vegetated and non-vegetated zones. Also, Hupet and Vanclooster [69] concluded that root water uptake and evapotranspiration play a considerable role in the spatial organization of soil moisture. Similarly, Baroni et al. [70] highlighted the significant contribution of vegetation both directly and through the interaction with other factors (e.g., soil texture and topography). However, we clearly showed that solely using a proxy of vegetation as the model predictor resulted in poor downscaling accuracy, regardless of the SSM source. This finding supports the conclusions of previous studies, which attempted to identify and quantify the contribution of various controls on the spatial distribution of soil moisture [50,71]. Humid catchments (as our study site) showed strong similarities between soil moisture patterns and topography, however a significant contribution to the total variability was related to other factors, i.e., either soil or vegetation (or both) [50]. For instance, vegetation might reduce the lateral flow driven by topography, thus minimizing the variability due to the landscape. On the other hand, different vegetation types might also lead to varying evapotranspiration fluxes, otherwise mainly controlled by soil properties, hence increasing the spatial variability. Furthermore, Western et al. [71] observed that the processes regulating the spatial patterns of soil moisture are not only site-specific, but also vary depending on the seasonal wetness conditions. In this respect, our findings corroborate the relevant role of vegetation in the spatial distribution of soil moisture, and emphasize the potential of using machine learning methods to infer soil moisture based on its complex relationships with other surface parameters [72].

5.2. Random Forest and Training Data

A wide range of machine learning algorithms exists and has been applied for downscaling purposes. In this work, we selected RF because of its ability to model complex and non-linear relationships while minimizing the risk of overfitting. Furthermore, previous comparative studies identified RF as the most suitable method for downscaling remotely sensed products, ranging from soil moisture [73] to evapotranspiration [74] to temperature [75]. As discussed above, RF is sensitive to the quality of the input data used, like any other machine learning method. While systematic differences between satellite-derived and local moisture conditions can be incorporated and accounted for in the RF model, random errors and sporadic discrepancies that propagate through the model are unavoidable. The same applies to the surface parameters (i.e., soil texture, topography, and vegetation) used as model inputs. Thus, the quality of the predictors is essential for obtaining reliable results. Similarly, a thorough evaluation of the ground measurements used as training data must be performed. We observed that increasing the training set size produced significant improvements in the model accuracy, and larger training sets are likely to further enhance the developed models. However, our results also suggest that if limited resources are available, e.g., computational power or monitoring resources, a preference should be given to use as much spatial information as possible at the expense of the temporal component. Concretely, this finding would translate in collecting and employing observations from many sensors—even if for a shorter period—rather than using the same amount of data from fewer sensors covering a longer time interval. This remark might open new frontiers for a wide adoption of low-cost sensors for monitoring purposes and scientific research. Low-cost sensors could become an established element of environmental networks for monitoring soil moisture at different spatial extents (from field to regional scale). In an envisaged scenario, many (around 30 to 50) low-cost sensors would be distributed over an area of interest, in conjunction with one or few professional sensors. The first would provide detailed information about the spatial patterns, while the latter would allow monitoring the long-term temporal dynamics of soil moisture.

5.3. Limitations, Opportunities, and Transferability

The current analysis was carried out using an in-situ derived vegetation index, i.e., fAGR Equation (1), available at few points in space. Therefore, the application of the proposed downscaling method including a proxy of vegetation density was not possible for the entire study site but was limited to the locations of the low-cost sensors. fAGR is also subject to saturation compared to other indices (e.g., leaf area index) and might not capture important differences in the canopy structure of different vegetation types. Given the essential role played by vegetation in regulating the spatial patterns of soil moisture, we expect more accurate results when employing vegetation indicators that better discriminate between the various vegetation types present in the study site (winter cereals, maize, rapeseed, grassland, and forest). Therefore, we highlight the potential to replace fAGR with canopy observations from remotely sensed observations. Particularly suited for this task would be the Sentinel-2 mission, which can provide several vegetation indices at high spatial resolution (20 m) and high revisit time (five days) [61]. Alternatively, one could use indices derived from Sentinel-1 backscatter, with the advantage that observations can be made under almost all weather conditions [62]. Additional ancillary attributes can be added to the downscaling model, in order to increase the knowledge about the surface conditions. A widely used variable for downscaling purposes is the land surface temperature (LST) [26]. Here, LST was not included because freely available products are characterized either by a (too) coarse spatial resolution (e.g., MODIS MOD21, [76]) or an insufficient revisit frequency (e.g., Landsat, [77]). However, if the target resolution is in the order of, e.g., 1 km, including LST as a model predictor might further increase the model accuracy.

The proposed downscaling framework aims to build a robust model that can be applied to satellite data in the absence of concomitant in-situ observations. The latter are only required for training purposes. Therefore, the resulting downscaling model can be applied to generate high-resolution soil moisture back in the past. The only constraint to the applicability of the method is represented by the duration of the coarse-scale soil moisture (SSM) predictor and, eventually, the availability of the vegetation proxy used. The generation of long-term soil moisture records at high resolution would be possible by employing, e.g., the ESA CCI (climate change initiative) SM [78], and a "static" implementation of the RF model, i.e., SSM+S+T. Similarly, the framework presented in this study can be employed for near real-time monitoring at high resolution, even after the in-situ sensors have been moved or stopped working.

Furthermore, the downscaling approach developed here can be transferred to other sites, ranging from catchment to regional scales, given the presence of in-situ observations for training robust models. Clearly, the main constraint related to such an approach is the need for extensive ground measurements [26]. A possible source of information is offered by the International Soil Moisture Network (https://ismn.geo.tuwien.ac.at/; [17]), a data hosting platform where soil moisture in-situ measurements are publicly available from dozens of networks worldwide, with higher density and coverage in North America and Europe. Alternatively, the recent expansion of low-cost sensors might help to broaden soil moisture monitoring, as in-situ sensors are now available for a fraction of the price of professional probes. Hence, environmental monitoring networks of unprecedented spatial density and coverage might be established [36,79], providing the required source of information for machine learning techniques. Another novel opportunity is offered by crowdsourced observations, i.e., measurements collected by citizens. On the one hand, such observations might be subject to additional sources of errors and uncertainties, but on the other hand they can provide amounts of data not economically sustainable by traditional networks. For instance, the GROW Observatory (https://growobservatory.org/; [80]) is a citizen observatory unique in terms of number and variety of participants involved. Thousands of citizens are recording soil moisture measurements with low-cost sensors, thus monitoring environmental conditions at the European scale. These various sources of information can be either integrated or used independently as training data for downscaling satellite products. Furthermore, downscaling models developed for specific areas can be transferred and tested for other sites with similar environmental and climatic conditions.

6. Conclusions

A downscaling framework based on random forest models has been developed in order to estimate soil moisture at sub-field resolution (30 m), thus meeting the requirements of a growing number of applications. Input variables of the RF models were coarse-scale soil moisture products (ASCAT or SMAP or the spatial average from the in-situ sensors), soil texture, topographic indices, and a proxy of vegetation cover. The models have been trained against in-situ measurements collected by low-cost sensors installed in an agricultural catchment, covering a wide range of edaphic and topographic conditions, as well as vegetation types. Based on our results, the following conclusions can be drawn:

- The accuracy of the downscaling result is strongly related to the quality of the model predictors;
- Topography has higher predictive power than soil texture, which can be explained by the hilly landscape of the study site;
- Vegetation plays a key role in regulating the spatial distribution of soil moisture, and a proxy of vegetation should ideally be added as model predictor;
- The observed spatial patterns are consistently captured by the downscaled soil moisture;
- The training set size strongly affects the model accuracy, and larger training sets are likely to further improve the results;
- When limited training data can be used, priority should be given to increase the number of sensor locations to adequately cover the spatial heterogeneity, rather than expanding the duration of the measurements.

Furthermore, the proposed method can be improved by including a spatially continuous representation of vegetation, i.e., from remotely sensed vegetation indices. Future studies should focus on applying the downscaling framework developed here to other regions characterized by similar environmental and climatic conditions.

Overall, our results demonstrate the potential for employing data gathered from low-cost sensors for scientific applications. Low-cost devices could become an established component of environmental networks, allowing one to monitor soil moisture at an unprecedented density from the field to the continental scale. The extensive amount of data generated through low-cost sensors could provide, among others, the necessary basis for developing robust machine learning models for downscaling purposes.

Supplementary Materials: The following are available online at http://www.mdpi.com/2072-4292/11/22/2596/s1, Figure S1: Digital elevation model at 0.5 m resolution (a), clay and sand percentages at 30 m resolution (b and c, respectively) of the study area. Sensor locations are also shown, Figure S2: Violin plots showing the distribution of the static variables considered over the study area. Blue dots represent the values of such variables at the sensor locations, Figure S3: Maps showing crops grown within individual fields over the study area for 2018 (a) and 2019 (b).

Author Contributions: L.Z. conceived and designed the experiments together with W.D.; L.Z. analyzed the data and wrote the paper; L.Z. and A.X. collected in-situ measurements; M.F. and W.D. contributed with their expertise and provided input on analysis of results. All authors participated in the revision of the manuscript.

Funding: This project has received funding from the European Union's Horizon 2020 research and innovation program under grant agreement No. 690199 and the TU Wien Wissenschaftspreis 2015 awarded to W.D.

Acknowledgments: The authors would like to thank Gerhard Rab for the support with the in-situ data collection, and Isabella Pfeil for kindly providing Metop ASCAT data. Furthermore, the authors are grateful for the valuable feedback from the four anonymous reviewers.

Conflicts of Interest: The authors declare no conflict of interest.

References

1. Gibon, F.; Pellarin, T.; Román-Cascón, C.; Alhassane, A.; Traoré, S.; Kerr, Y.; Lo Seen, D.; Baron, C. Millet yield estimates in the Sahel using satellite derived soil moisture time series. *Agric. For. Meteorol.* **2018**, *262*, 100–109. [CrossRef]
2. Ines, A.V.M.; Das, N.N.; Hansen, J.W.; Njoku, E.G. Assimilation of remotely sensed soil moisture and vegetation with a crop simulation model for maize yield prediction. *Remote Sens. Environ.* **2013**, *138*, 149–164. [CrossRef]
3. AghaKouchak, A.; Farahmand, A.; Melton, F.S.; Teixeira, J.; Anderson, M.C.; Wardlow, B.D.; Hain, C.R. Remote sensing of drought: Progress, challenges and opportunities: Remote sensing of drought. *Rev. Geophys.* **2015**, *53*, 452–480. [CrossRef]
4. Bolten, J.D.; Crow, W.T.; Zhan, X.; Jackson, T.J.; Reynolds, C.A. Evaluating the utility of remotely sensed soil moisture retrievals for operational agricultural drought monitoring. *IEEE J. Sel. Top. Appl. Earth Obs. Remote Sens.* **2010**, *3*, 57–66. [CrossRef]
5. Lorenz, D.J.; Otkin, J.A.; Svoboda, M.; Hain, C.R.; Anderson, M.C.; Zhong, Y. Predicting U.S. drought monitor states using precipitation, soil moisture, and evapotranspiration anomalies. Part I: Development of a Nondiscrete USDM Index. *J. Hydrometeorol.* **2017**, *18*, 1943–1962. [CrossRef]
6. Wang, H.; Vicente-serrano, S.M.; Tao, F.; Zhang, X.; Wang, P.; Zhang, C.; Chen, Y.; Zhu, D.; Kenawy, A.E. Monitoring winter wheat drought threat in Northern China using multiple climate-based drought indices and soil moisture during 2000–2013. *Agric. For. Meteorol.* **2016**, *228–229*, 1–12. [CrossRef]
7. Dirmeyer, P.A.; Wu, J.; Norton, H.E.; Dorigo, W.A.; Quiring, S.M.; Ford, T.W.; Santanello, J.A.; Bosilovich, M.G.; Ek, M.B.; Koster, R.D.; et al. Confronting weather and climate models with observational data from soil moisture networks over the United States. *J. Hydrometeorol.* **2016**, *17*, 1049–1067. [CrossRef] [PubMed]
8. Orth, R.; Seneviratne, S.I. Using soil moisture forecasts for sub-seasonal summer temperature predictions in Europe. *Clim. Dyn* **2014**, *43*, 3403–3418. [CrossRef]
9. Tuttle, S.; Salvucci, G. Empirical evidence of contrasting soil moisture-precipitation feedbacks across the United States. *Science* **2016**, *352*, 825–828. [CrossRef]

10. Brocca, L.; Melone, F.; Moramarco, T.; Wagner, W.; Naeimi, V.; Bartalis, Z.; Hasenauer, S. Improving runoff prediction through the assimilation of the ASCAT soil moisture product. *Hydrol. Earth Syst. Sci.* **2010**, *14*, 1881–1893. [CrossRef]
11. Koster, R.D.; Mahanama, S.P.P.; Livneh, B.; Lettenmaier, D.P.; Reichle, R.H. Skill in streamflow forecasts derived from large-scale estimates of soil moisture and snow. *Nat. Geosci.* **2010**, *3*, 613–616. [CrossRef]
12. Tramblay, Y.; Bouvier, C.; Martin, C.; Didon-Lescot, J.-F.; Todorovik, D.; Domergue, J.-M. Assessment of initial soil moisture conditions for event-based rainfall–runoff modelling. *J. Hydrol.* **2010**, *387*, 176–187. [CrossRef]
13. Brocca, L.; Ponziani, F.; Moramarco, T.; Melone, F.; Berni, N.; Wagner, W. Improving landslide forecasting using ASCAT-derived soil moisture data: a case study of the torgiovannetto landslide in central Italy. *Remote Sens.* **2012**, *4*, 1232–1244. [CrossRef]
14. Ray, R.L.; Jacobs, J.M.; Cosh, M.H. Landslide susceptibility mapping using downscaled AMSR-E soil moisture: A case study from Cleveland Corral, California, US. *Remote Sens. Environ.* **2010**, *114*, 2624–2636. [CrossRef]
15. Famiglietti, J.S.; Ryu, D.; Berg, A.A.; Rodell, M.; Jackson, T.J. Field observations of soil moisture variability across scales: Soil moisture variability across scales. *Water Resour. Res.* **2008**, *44*. [CrossRef]
16. Blöschl, G.; Sivapalan, M. Scale issues in hydrological modelling: A review. *Hydrol. Process.* **1995**, *9*, 251–290. [CrossRef]
17. Dorigo, W.A.; Wagner, W.; Hohensinn, R.; Hahn, S.; Paulik, C.; Xaver, A.; Gruber, A.; Drusch, M.; Mecklenburg, S.; van Oevelen, P.; et al. The International Soil Moisture Network: A data hosting facility for global in situ soil moisture measurements. *Hydrol. Earth Syst. Sci.* **2011**, *15*, 1675–1698. [CrossRef]
18. Ochsner, T.E.; Cosh, M.H.; Cuenca, R.H.; Dorigo, W.A.; Draper, C.S.; Hagimoto, Y.; Kerr, Y.H.; Njoku, E.G.; Small, E.E.; Zreda, M. State of the Art in Large-Scale Soil Moisture Monitoring. *Soil Sci. Soc. Am. J.* **2013**, *77*, 1888. [CrossRef]
19. Gruber, A.; Dorigo, W.A.; Zwieback, S.; Xaver, A.; Wagner, W. Characterizing coarse-scale representativeness of in situ soil moisture measurements from the International Soil Moisture Network. *Vadose Zone J.* **2013**, *12*. [CrossRef]
20. Fontanet, M.; Fernàndez-Garcia, D.; Ferrer, F. The value of satellite remote sensing soil moisture data and the DISPATCH algorithm in irrigation fields. *Hydrol. Earth Syst. Sci.* **2018**, *22*, 5889–5900. [CrossRef]
21. Ulaby, F.T.; Moore, R.K.; Fung, A.K. Radar remote sensing and surface scattering and emission theory. In *Microwave Remote Sensing: Active and Passive*; Artech House: Norwood, MA, USA, 1982; Volume 2, p. 624.
22. Njoku, E.G.; Entekhabi, D. Passive microwave remote sensing of soil moisture. *J. Hydrol.* **1996**, *184*, 101–129. [CrossRef]
23. Bauer-Marschallinger, B.; Freeman, V.; Cao, S.; Paulik, C.; Schaufler, S.; Stachl, T.; Modanesi, S.; Massari, C.; Ciabatta, L.; Brocca, L.; et al. Toward global soil moisture monitoring with Sentinel-1: Harnessing assets and overcoming obstacles. *IEEE Trans. Geosci. Remote Sens.* **2019**, *57*, 520–539. [CrossRef]
24. Molero, B.; Merlin, O.; Malbéteau, Y.; Al Bitar, A.; Cabot, F.; Stefan, V.; Kerr, Y.; Bacon, S.; Cosh, M.H.; Bindlish, R.; et al. SMOS disaggregated soil moisture product at 1 km resolution: Processor overview and first validation results. *Remote Sens. Environ.* **2016**, *180*, 361–376. [CrossRef]
25. Walker, J.P.; Houser, P.R. Requirements of a global near-surface soil moisture satellite mission: Accuracy, repeat time, and spatial resolution. *Adv. Water Resour.* **2004**, *27*, 785–801. [CrossRef]
26. Peng, J.; Loew, A.; Merlin, O.; Verhoest, N.E.C. A review of spatial downscaling of satellite remotely sensed soil moisture: Downscale Satellite-Based Soil Moisture. *Rev. Geophys.* **2017**, *55*, 341–366. [CrossRef]
27. Sabaghy, S.; Walker, J.P.; Renzullo, L.J.; Jackson, T.J. Spatially enhanced passive microwave derived soil moisture: Capabilities and opportunities. *Remote Sens. Environ.* **2018**, *209*, 551–580. [CrossRef]
28. Alemohammad, S.H.; Kolassa, J.; Prigent, C.; Aires, F.; Gentine, P. Global downscaling of remotely sensed soil moisture using neural networks. *Hydrol. Earth Syst. Sci.* **2018**, *22*, 5341–5356. [CrossRef]
29. He, L.; Chen, J.M.; Chen, K.-S. Simulation and SMAP Observation of Sun-Glint Over the Land Surface at the L-Band. *IEEE Trans. Geosci. Remote Sens.* **2017**, *55*, 2589–2604. [CrossRef]
30. Zhao, W.; Sánchez, N.; Lu, H.; Li, A. A spatial downscaling approach for the SMAP passive surface soil moisture product using random forest regression. *J. Hydrol.* **2018**, *563*, 1009–1024. [CrossRef]
31. Pelletier, C.; Valero, S.; Inglada, J.; Champion, N.; Dedieu, G. Assessing the robustness of Random Forests to map land cover with high resolution satellite image time series over large areas. *Remote Sens. Environ.* **2016**, *187*, 156–168. [CrossRef]

32. González-Teruel, J.; Torres-Sánchez, R.; Blaya-Ros, P.; Toledo-Moreo, A.; Jiménez-Buendía, M.; Soto-Valles, F. Design and Calibration of a Low-Cost SDI-12 Soil Moisture Sensor. *Sensors* **2019**, *19*, 491. [CrossRef] [PubMed]
33. Kumar, M.S.; Chandra, T.R.; Kumar, D.P.; Manikandan, M.S. Monitoring moisture of soil using low cost homemade Soil moisture sensor and Arduino UNO. In Proceedings of the 2016 3rd International Conference on Advanced Computing and Communication Systems (ICACCS), Coimbatore, India, 22–23 January 2016; pp. 1–4.
34. Chawla, S.; Bachhtey, S.; Gupta, V.; Sharma, S.; Seth, S.; Gandhi, T.; Varshney, S.; Mehta, S.; Jha, R. Low cost soil moisture sensors and their application in automatic irrigation system. In Proceedings of the IEEE International Conference on Advances in Computing, Communications, and Informatics, Jaipur, India, 21–24 September 2016.
35. Bogena, H.R.; Huisman, J.A.; Oberdörster, C.; Vereecken, H. Evaluation of a low-cost soil water content sensor for wireless network applications. *J. Hydrol.* **2007**, *344*, 32–42. [CrossRef]
36. Xaver, A.; Zappa, L.; Rab, G.; Pfeil, I.; Vreugdenhil, M.; Hemment, D.; Dorigo, W. Evaluating the suitability of the consumer low-cost Parrot Flower Power soil moisture sensor for scientific environmental applications. *Geosci. Instrum. Methods Data Syst.* In review.
37. Teuling, A.J.; Uijlenhoet, R.; Hupet, F.; van Loon, E.E.; Troch, P.A. Estimating spatial mean root-zone soil moisture from point-scale observations. *Hydrol. Earth Syst. Sci.* **2006**, *13*, 1447–1485. [CrossRef]
38. Blöschl, G.; Blaschke, A.P.; Broer, M.; Bucher, C.; Carr, G.; Chen, X.; Eder, A.; Exner-Kittridge, M.; Farnleitner, A.; Flores-Orozco, A.; et al. The Hydrological Open Air Laboratory (HOAL) in Petzenkirchen: A hypothesis-driven observatory. *Hydrol. Earth Syst. Sci.* **2016**, *20*, 227–255. [CrossRef]
39. Parrot Flower Power. Available online: https://www.parrot.com/global/support/products/parrot-flower-power (accessed on 5 November 2019).
40. Vreugdenhil, M.; Dorigo, W.; Broer, M.; Haas, P.; Eder, A.; Hogan, P.; Bloeschl, G.; Wagner, W. Towards a high-density soil moisture network for the validation of SMAP in Petzenkirchen, Austria. In Proceedings of the 2013 IEEE International Geoscience and Remote Sensing Symposium—IGARSS, Melbourne, Australia, 21–26 July 2013; pp. 1865–1868.
41. Colliander, A.; Jackson, T.J.; Bindlish, R.; Chan, S.; Das, N.; Kim, S.B.; Cosh, M.H.; Dunbar, R.S.; Dang, L.; Pashaian, L.; et al. Validation of SMAP surface soil moisture products with core validation sites. *Remote Sens. Environ.* **2017**, *191*, 215–231. [CrossRef]
42. Wagner, W.; Hahn, S.; Kidd, R.; Melzer, T.; Bartalis, Z.; Hasenauer, S.; Figa-Saldaña, J.; de Rosnay, P.; Jann, A.; Schneider, S.; et al. The ASCAT soil moisture product: A review of its specifications, validation results, and emerging applications. *Meteorol. Z.* **2013**, *22*, 5–33. [CrossRef]
43. Chan, S.K.; Bindlish, R.; O'Neill, P.E.; Njoku, E.; Jackson, T.; Colliander, A.; Chen, F.; Burgin, M.; Dunbar, S.; Piepmeier, J.; et al. Assessment of the SMAP Passive Soil Moisture Product. *IEEE Trans. Geosci. Remote Sens.* **2016**, *54*, 4994–5007. [CrossRef]
44. Pfeil, I.; Vreugdenhil, M.; Hahn, S.; Wagner, W.; Strauss, P.; Blöschl, G. Improving the Seasonal Representation of ASCAT Soil Moisture and Vegetation Dynamics in a Temperate Climate. *Remote Sens.* **2018**, *10*, 1788. [CrossRef]
45. Wagner, W.; Lemoine, G.; Rott, H. A Method for Estimating Soil Moisture from ERS Scatterometer and Soil Data. *Remote Sens. Environ.* **1999**, *70*, 191–207. [CrossRef]
46. Naeimi, V.; Scipal, K.; Bartalis, Z.; Hasenauer, S.; Wagner, W. An Improved Soil Moisture Retrieval Algorithm for ERS and METOP Scatterometer Observations. *IEEE Trans. Geosci. Remote Sens.* **2009**, *47*, 1999–2013. [CrossRef]
47. Entekhabi, D.; Njoku, E.G.; O'Neill, P.E.; Kellogg, K.H.; Crow, W.T.; Edelstein, W.N.; Entin, J.K.; Goodman, S.D.; Jackson, T.J.; Johnson, J.; et al. The Soil Moisture Active Passive (SMAP) mission. *Proc. IEEE* **2010**, *98*, 704–716. [CrossRef]
48. O'Neill, P.E.; Chan, S.; Njoku, E.G.; Jackson, T.; Bindlish, R. *SMAP L3 Radiometer Global Daily 36 km EASE-Grid Soil Moisture, Version 5*; NASA National Snow and Ice Data Center Distributed Active Archive Center: Boulder, CO, USA.
49. Garcia Gonzalez, J.C.; Redondo, J.A.; Garzon, A. *EU-Hydro/EU-DEM Upgrade*; Indra Sistemas S.A.: Madrid, Spain, 2015.

50. Wilson, D.J.; Western, A.W.; Grayson, R.B. Identifying and quantifying sources of variability in temporal and spatial soil moisture observations: Sources of soil moisture variability. *Water Resour. Res.* **2004**, *40*. [CrossRef]
51. Beven, K.J.; Kirkby, M.J. A physically based, variable contributing area model of basin hydrology/Un modèle à base physique de zone d'appel variable de l'hydrologie du bassin versant. *Hydrol. Sci. Bull.* **1979**, *24*, 43–69. [CrossRef]
52. Hofierka, J.; Šúri, M. The solar radiation model for Open source GIS: Implementation and applications. In Proceedings of the Open Source GIS-GRASS Users Conference, Trento, Italy, 11–13 September 2002.
53. Mitášová, H.; Hofierka, J. Interpolation by regularized spline with tension: II. Application to terrain modeling and surface geometry analysis. *Math. Geol.* **1993**, *25*, 657–669. [CrossRef]
54. Breiman, L. Random forests. *Mach. Learn.* **2001**, *45*, 5–32. [CrossRef]
55. Pedregosa, F.; Varoquaux, G.; Gramfort, A.; Michel, V.; Thirion, B.; Grisel, O.; Blondel, M.; Prettenhofer, P.; Weiss, R.; Dubourg, V.; et al. Scikit-learn: Machine learning in Python. *J. Mach. Learn. Res.* **2011**, *12*, 2825–2830.
56. Hutengs, C.; Vohland, M. Downscaling land surface temperatures at regional scales with random forest regression. *Remote Sens. Environ.* **2016**, *178*, 127–141. [CrossRef]
57. Paulik, C.; Plocon, A.; Baum, D.; Hahn, S.; Mistelbauer, T.; Preimesberger, W.; Schmitzer, M.; Gruber, A.; Teubner, I.; Reimer, C. *Pytesmo: Python Toolbox for the Evaluation of Soil Moisture Observations*; Python Software Foundation: Wilmington, DE, USA, 2019.
58. Albergel, C.; Brocca, L.; Wagner, W.; de Rosnay, P.; Calvet, J.P. Selection of performance metrics for global soil moisture products: The case of ASCAT product. In *Remote Sensing of Energy Fluxes and Soil Moisture Content*; CRC Press: Boca Raton, FL, USA, 2013; pp. 427–444.
59. Western, A.W.; Grayson, R.B.; Blöschl, G. Scaling of soil moisture: A hydrologic perspective. *Annu. Rev. Earth Planet. Sci.* **2002**, *30*, 149–180. [CrossRef]
60. Brocca, L.; Hasenauer, S.; Lacava, T.; Melone, F.; Moramarco, T.; Wagner, W.; Dorigo, W.; Matgen, P.; Martínez-Fernández, J.; Llorens, P.; et al. Soil moisture estimation through ASCAT and AMSR-E sensors: An intercomparison and validation study across Europe. *Remote Sens. Environ.* **2011**, *115*, 3390–3408. [CrossRef]
61. Vuolo, F.; Żółtak, M.; Pipitone, C.; Zappa, L.; Wenng, H.; Immitzer, M.; Weiss, M.; Baret, F.; Atzberger, C. Data service platform for Sentinel-2 surface reflectance and value-added products: System use and examples. *Remote Sens.* **2016**, *8*, 938. [CrossRef]
62. Vreugdenhil, M.; Wagner, W.; Bauer-Marschallinger, B.; Pfeil, I.; Teubner, I.; Rüdiger, C.; Strauss, P. Sensitivity of Sentinel-1 Backscatter to Vegetation Dynamics: An Austrian Case Study. *Remote Sens.* **2018**, *10*, 1396. [CrossRef]
63. Dorigo, W.A.; Scipal, K.; Parinussa, R.M.; Liu, Y.Y.; Wagner, W.; de Jeu, R.A.M.; Naeimi, V. Error characterisation of global active and passive microwave soil moisture datasets. *Hydrol. Earth Syst. Sci.* **2010**, *14*, 2605–2616. [CrossRef]
64. Bauer-Marschallinger, B.; Paulik, C.; Hochstöger, S.; Mistelbauer, T.; Modanesi, S.; Ciabatta, L.; Massari, C.; Brocca, L.; Wagner, W. Soil moisture from fusion of scatterometer and SAR: Closing the scale gap with temporal filtering. *Remote Sens.* **2018**, *10*, 1030. [CrossRef]
65. Picciafuoco, T.; Morbidelli, R.; Flammini, A.; Saltalippi, C.; Corradini, C.; Strauss, P.; Blöschl, G. On the estimation of spatially representative plot scale saturated hydraulic conductivity in an agricultural setting. *J. Hydrol.* **2019**, *570*, 106–117. [CrossRef]
66. Teuling, A.J. Improved understanding of soil moisture variability dynamics. *Geophys. Res. Lett.* **2005**, *32*, L05404. [CrossRef]
67. Brocca, L.; Morbidelli, R.; Melone, F.; Moramarco, T. Soil moisture spatial variability in experimental areas of central Italy. *J. Hydrol.* **2007**, *333*, 356–373. [CrossRef]
68. Gómez-Plaza, A.; Martínez-Mena, M.; Albaladejo, J.; Castillo, V.M. Factors regulating spatial distribution of soil water content in small semiarid catchments. *J. Hydrol.* **2001**, *253*, 211–226. [CrossRef]
69. Hupet, F.; Vanclooster, M. Intraseasonal dynamics of soil moisture variability within a small agricultural maize cropped field. *J. Hydrol.* **2002**, *261*, 86–101. [CrossRef]
70. Baroni, G.; Ortuani, B.; Facchi, A.; Gandolfi, C. The role of vegetation and soil properties on the spatio-temporal variability of the surface soil moisture in a maize-cropped field. *J. Hydrol.* **2013**, *489*, 148–159. [CrossRef]

71. Western, A.W.; Zhou, S.-L.; Grayson, R.B.; McMahon, T.A.; Blöschl, G.; Wilson, D.J. Spatial correlation of soil moisture in small catchments and its relationship to dominant spatial hydrological processes. *J. Hydrol.* **2004**, *286*, 113–134. [CrossRef]
72. Notarnicola, C.; Angiulli, M.; Posa, F. Soil moisture retrieval from remotely sensed data: Neural network approach versus Bayesian method. *IEEE Trans. Geosci. Remote Sens.* **2008**, *46*, 547–557. [CrossRef]
73. Im, J.; Park, S.; Rhee, J.; Baik, J.; Choi, M. Downscaling of AMSR-E soil moisture with MODIS products using machine learning approaches. *Environ. Earth Sci.* **2016**, *75*, 1120. [CrossRef]
74. Ke, Y.; Im, J.; Park, S.; Gong, H. Downscaling of MODIS one kilometer evapotranspiration using landsat-8 data and machine learning approaches. *Remote Sens.* **2016**, *8*, 215. [CrossRef]
75. Eccel, E.; Ghielmi, L.; Granitto, P.; Barbiero, R.; Grazzini, F.; Cesari, D. Prediction of minimum temperatures in an alpine region by linear and non-linear post-processing of meteorological models. *Nonlinear Process.* **2007**, *14*, 211–222. [CrossRef]
76. Hulley, G.; Freepartner, R.; Malakar, N.; Sarkar, S. *Moderate Resolution Imaging Spectroradiometer (MODIS) Land Surface Temperature and Emissivity Product (MxD21) User Guide*; NASA: Washington, DC, USA, 2016.
77. Jimenez-Munoz, J.C.; Sobrino, J.A.; Skokovic, D.; Mattar, C.; Cristobal, J. Land surface temperature retrieval methods from Landsat-8 thermal infrared sensor data. *IEEE Geosci. Remote Sens. Lett.* **2014**, *11*, 1840–1843. [CrossRef]
78. Dorigo, W.; Wagner, W.; Albergel, C.; Albrecht, F.; Balsamo, G.; Brocca, L.; Chung, D.; Ertl, M.; Forkel, M.; Gruber, A.; et al. ESA CCI Soil Moisture for improved Earth system understanding: State-of-the art and future directions. *Remote Sens. Environ.* **2017**, *203*, 185–215. [CrossRef]
79. Bogena, H.; Schulz, K.; Vereecken, H. Towards a network of observatories in terrestrial environmental research. *Adv. Geosci.* **2006**, *9*, 109–114. [CrossRef]
80. Kovács, K.Z.; Hemment, D.; Woods, M.; van der Velden, N.K.; Xaver, A.; Giesen, R.H.; Burton, V.J.; Garrett, N.L.; Zappa, L.; Long, D.; et al. Citizen observatory based soil moisture monitoring—The GROW example. *Hung. Geo. Bull.* **2019**, *68*, 119–139. [CrossRef]

remote sensing

MDPI

Article

A New Retrieval Algorithm for Soil Moisture Index from Thermal Infrared Sensor On-Board Geostationary Satellites over Europe and Africa and Its Validation

Nicolas Ghilain [1,*], Alirio Arboleda [1], Okke Batelaan [2], Jonas Ardö [3], Isabel Trigo [4], Jose-Miguel Barrios [1] and Francoise Gellens-Meulenberghs [1]

1 Royal Meteorological Institute of Belgium, B-1180 Brussels, Belgium
2 College of Science and Engineering, Flinders University, Adelaide SA 5001, Australia
3 Department of Physical Geography and Ecosystem Science, Lund University, SE-221 00 Lund, Sweden
4 Instituto do Mar e Atmosfera, 1749-077 Lisbon, Portugal
* Correspondence: nicolas.ghilain@meteo.be

Received: 23 July 2019; Accepted: 16 August 2019; Published: 21 August 2019

Abstract: Monitoring soil moisture at the Earth'surface is of great importance for drought early warnings. Spaceborne remote sensing is a keystone in monitoring at continental scale, as satellites can make observations of locations which are scarcely monitored by ground-based techniques. In recent years, several soil moisture products for continental scale monitoring became available from the main space agencies around the world. Making use of sensors aboard polar satellites sampling in the microwave spectrum, soil moisture can be measured and mapped globally every few days at a spatial resolution as fine as 25 km. However, complementarity of satellite observations is a crucial issue to improve the quality of the estimations provided. In this context, measurements within the visible and infrared from geostationary satellites provide information on the surface from a totally different perspective. In this study, we design a new retrieval algorithm for daily soil moisture monitoring based only on the land surface temperature observations derived from the METEOSAT second generation geostationary satellites. Soil moisture has been retrieved from the retrieval algorithm for an eight years period over Europe and Africa at the SEVIRI sensor spatial resolution (3 km at the sub-satellite point). The results, only available for clear sky and partly cloudy conditions, are for the first time extensively evaluated against in-situ observations provided by the International Soil Moisture Network and FLUXNET at sites across Europe and Africa. The soil moisture retrievals have approximately the same accuracy as the soil moisture products derived from microwave sensors, with the most accurate estimations for semi-arid regions of Europe and Africa, and a progressive degradation of the accuracy towards northern latitudes of Europe. Although some possible improvements can be expected by a better use of other products derived from SEVIRI, the new approach developped and assessed here is a valuable alternative to microwave sensors to monitor daily soil moisture at the resolution of few kilometers over entire continents and could reveal a good complementarity to an improved monitoring system, as the algorithm can produce surface soil moisture with less than 1 day delay over clear sky and non-steady cloudy conditions (over 10% of the time).

Keywords: soil moisture; geostationary; validation; SEVIRI; thermal infrared; land surface temperature

1. Introduction

Soil moisture at the surface of the Earth affects everyday life of people across the globe. The infiltration and hence soil moisture influence various meteorological processes [1], natural hazards as floodings [2], land slides and droughts, depriving populations of agricultural yield due to lack of soil moisture. Hence, water required to irrigate fields in semi-arid areas determines the long-term natural resources availability at regional level and may impact local [3] to international economics and policy [4,5].

Since late 1980's, efforts, based on numerical meteorological modelling, have resulted in estimation of soil moisture at continental scale [6]. Yet, observation-based systems are required for realistic data without *a priori* knowledge. This aspect is especially important in regions where few ground-based observations are available to confront model results with reality. The launch of the first Earth-observing satellites into space brought hopes and promises in characterizing and monitoring the Earth natural resources [7], especially water [8]. Njoku and Entekhabi [9] were among the first to use sensors aboard polar orbiting satellites for retrieving soil moisture. Recent success stories in near-real time soil moisture monitoring at global scale at 25 km resolution independently from the weather type are with micro-wave sensors [10–12], with a revisiting time of one or two passages per day. Geostationary satellites, initially designed for weather monitoring and forecasting, have long been discarded from this research effort. However, the imagery in the visible and infra-red spectra covers whole continents at a few kilometers resolutions in less than an hour and is potentially powerful in detecting rapid changes of the land surface [13]. They can potentially reduce the noise in data of other satellite sensors by bringing a wealth of observations every day. The Spinning Enhanced Visible and Infrared Imager (SEVIRI) instruments onboard the european Meteosat Second Generation (MSG) satellites provide 96 images per day in the visible and infra-red spectral bands, providing a detailed information on the sub-daily variations of the atmosphere and the surface.

Information from infrared spectral channels of sensors can be converted to a physical land surface temperature [14]. From the early 1980s, the thermal inertia concept, consisting of linking the amplitude of the surface heating over a day with the quantity of water present in the soil [15–21], has been applied to space-based observations [20,22–24]. This concept has been applied with various degrees of success to local and regional studies [25–29], and though most studies based on thermal inertia show potential for soil moisture monitoring, the success of this measurement methodology has not be quantified for large datasets so far. With the increasing interest on dynamical water resources assessment, new international initiatives [30] have brought a wealth of soil moisture ground-based observations to the public through a unique internet-based platform, otherwise locally managed, pushing forward model developments and remote sensing validation exercises. With these new opportunities, we applied a new retrieval algorithm based on the thermal inertia concept to SEVIRI observations over European and African continents, spanning an eight years period (2007–2014). Its success could be quantified in an extensive sample of climate zones, therefore filling some of the gaps left by other studies on thermal inertia by providing quantitative information regarding its accuracy and performance against in-situ ground measurements in a large panel of conditions.

Thanks to its unprecedented temporal and spatial observation resolution, the SEVIRI instument allows an accurate estimation of the daily surface heating rates, allowing a well-sampled tracking of soil moisture, for clear sky and partly cloudy weather over semi-arid areas. In this paper, we present first the new approach of the thermal inertia model. Then, to characterize the accuracy of the resulting soil moisture time series, we present the comparison with in situ measurements and discuss the limitations of both the method and the data.

2. Material and Methods

Thermal infra-red sensitive channels aboard satellites are particularly suited to derive land surface temperature (LST) by clear sky conditions. LST can be derived from two spectral channels [14], or even from a single thermal channel [31]. The impact of soil moisture on the land surface temperature is evident [32] and several attempts have been carried out to derive soil moisture from thermal imagery for various landscapes, at different scales and spatial resolutions, using the thermal inertia property of surface elements. The concept of thermal inertia is the following: a surface element with high thermal inertia will heat and cool more slowly than a surface element with a low thermal inertia, subjected to the same radiation. This concept does apply to compare different surface materials: sandy soil has a low thermal inertia compared to most rocks, and so heat and cool more rapidly. This technique is very popular in geological and planetary sciences (e.g., [33]). Thermal inertia property also applies to changes in the same soil over time, as it depends on soil moisture [32,34]. A soil with high soil moisture has a higher thermal inertia than the same soil when it is dry. Over time, detection of changes of thermal inertia at the same location can help to detect variations of soil moisture [35].

Various approaches have been suggested to derive soil moisture from thermal inertia variations observed from space. Two popular approaches are: the establishment of direct empirical relations between soil moisture and the changes in land surface temperature [26,28,35–37] and physical approaches, based on the estimation of evapotranspiration from land surface temperature changes and the deduction of soil moisture (for example Merlin et al. [38] using polar satellites, and Wetzel et al. [39], Anderson et al. [40], Hain et al. [41], Parinussa et al. [42] using geostationary satellites).

In several studies, at least two polar satellites passes are used to derive a soil moisture index based on land surface temperature, exploiting the diurnal heating or the heating over several days. While polar orbiters offer the possibility of global data coverage, these data are generally exploited for local applications. They are often limited by the relatively low temporal sampling and further hampered by cloud coverage and deficient cloud screening or atmospheric correction. Geostationary satellites offer an interesting alternative to polar satellite because of their high temporal sampling rate at the expense of coarser spatial resolutions. Verstraeten et al. [35] (referred as V2006) made a first attempt to derive a soil moisture index from thermal inertia concept and MSG/SEVIRI data for various forested sites across Europe during a vegetation period spanning April to October. The methodology proposed in V2006 is fairly simple and easily applicable, but does not exploit the full capabilities of the satellite, as it only uses two observations per day, as would be the case for a system of one sun-synchronous polar orbiter like MODIS/Terra or Aqua. The methodology has been further examined in Van Doninck et al. [37], where reconstructed LST signal from MODIS suite of satellites (4 overpasses per day) was used over Southern Africa and also more recently by Garcia et al. [26] with MSG/SEVIRI data over two sites in Mali and Southern Spain. It has been shown that a more robust estimate could be obtained by using a full morning dataset of LST, as Stisen et al. [43] showed for evaporative fraction from the same satellite. The present derivation has been built upon those previous studies to set up a new simple model to retrieve a soil moisture estimate from a single input variable, the MSG/SEVIRI LST produced by the Land Surface Analysis Satellite Application Facility (LSA-SAF) [44].

2.1. A New Thermal Infrared-Based Soil Moisture Retrieval

The new method for soil moisture retrievals presented here is based on the assumption that the algorithm can rely on only thermal infrared (TIR) observations from geostationnary satellites over entire continents. More specifically, we focus on the MSG/SEVIRI land surface temperatures (LST), and the exploitation of its exceptionnal temporal sampling, to derive surface soil moisture, not only for fully clear sky days, but also for days with intermittent cloud cover. In the following section, we describe the satellite data, the advantages and disadvantages of the data, the calibration of the empirical retrieval algorithm, and the processing methodology used to generate daily averaged surface soil moisture maps.

2.1.1. MSG Satellite and Its SEVIRI Sensors

The necessary input variables are derived from thermal infrared data acquired by the SEVIRI sensor onboard the series of MSG satellites of the European Organization for the Exploitation of Meteorological Satellites (EUMETSAT).

SEVIRI is one of the sensors aboard MSG suite of satellites. SEVIRI has 12 spectral channels recording the radiance from the Earth in the visible, near infrared and thermal infrared wavelengths. Every 15 min the whole field of view of MSG, covering Europe, Africa, South America and Middle East, is scanned in all the considered wavelengths. SEVIRI has a spatial footprint of 3.1 km at the sub-satellite location. Because of its position at the equator and of the associated angle of view (Figure 1), the spatial resolution is variable over the field of view and the surface elements are not seen from the same perspective.

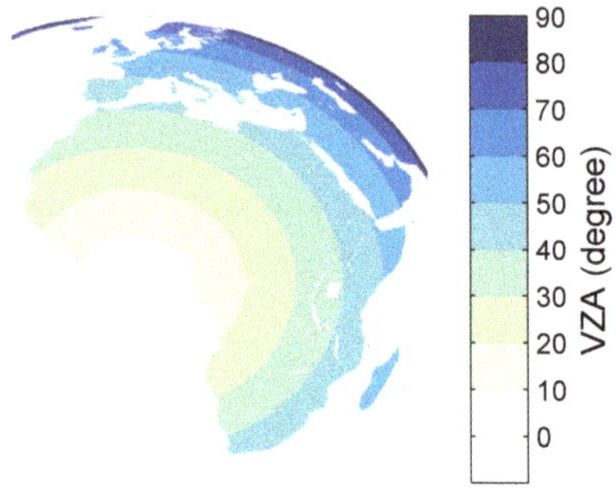

Figure 1. The viewing zenithal angle (VZA) from MSG/SEVIRI, expressed in degrees, is varying over the study area.

Several EUMETSAT decentralized thematic centers, called Satellite Application Facilities (SAF), have been created to generate further exploitations of its satellites, including the MSG suite. One particular component focus on land surface: the Satellite Application Facility on Land Surface Analysis, LSA-SAF [44]. Several variables of interest for the land surface and land-atmosphere transfers are produced in near-real time over the entire SEVIRI field of view using MSG/SEVIRI acquired data. The thermal infra-red channels are particularly exploited for the generation of the land surface temperature (LST). The LST product is estimated every 15 min for clear-sky areas using a split-window algorithm applied to brightness temperatures measured in the 10.8 and 12.0 μm [45] of the SEVIRI instrument. The reported accuracy from comparison over selected locations in Portugal, Senegal, and Namibia, is of the order of 1.7 to 2.7 K, with a bias generally less than 1 K [46–49]. LST is provided with uncertainty bars and a quality indicator reflecting the state of the input data or the nature of the surface. Over the period covered by this study, from 2007 to 2014, consecutive versions of retrieval algorithm have been used to generate LST maps, versions 5.0 to 7.7, mostly including minor changes to files technical specifications (attributes) (http://landsaf.ipma.pt), and 2 changes of satellites, although the longest period was covered by MSG-2: MSG-1 (1 March 2007–10 April 2007, operating from 3.4° longitude west), MSG-2 (11 April 2007–20 January 2013, operating from 0° longitude), MSG-3 (21 January 2013–31 December 2014, operating from 0° longitude).

Because of the specific view of the MSG satellites, LST derived from MSG can be affected by anisotropy effects and be not fully compatible in all areas with devices operating with other viewing geometries. Vegetation density and structure play a role, affecting for example the observation of shaded and sunlit areas [50]. In a *hotspot* configuration, the surface is seen by the sensor within a few degrees of the sunrays direction, and an enhancement in the land surface temperature retrieved is observed [51]. Another cause that can affect the retrieval is the estimation of the humidity in the total thickness of the atmosphere [46].

2.1.2. The Soil Moisture Retrieval Algorithm

Initial Signal from the Land Surface Temperature

In the present study, we compute morning surface heating rates per pixel through linear regression on LST data between local dawn and noon, excluding one hour on each side of the time series [43], therefore aiming at exploiting the daily changes of thermal inertia. A surface submitted to the same water stress has been evidenced to exhibit higher amplitude of LST diurnal cycle under high than under low insulation Garcia et al. [26], Verstraeten et al. [35], Van Doninck et al. [37]. Therefore, in those studies, the daily amplitude of the LST signal was normalized by an astronomical factor accounting for seasonality in sun zenithal angle. This was also supported by Prigent et al. [52], where LST amplitudes were normalized with incoming solar radiation. Here, because we derive heating rates and not amplitude, we already take into account the length of the morning, and no normalization is needed. An error range is calculated indicating the goodness of the linear fit: a poor linear fit can find its roots in a change of atmospheric conditions within the sampling period, a systematically poor cloud masking, or a rain event occurring during the estimation period.

Geometric Correction

Because of the specific viewing geometry of MSG/SEVIRI the heating rates should be corrected from viewing anisotropy effects. The correction we apply follows other studies for correction of LST observations from geostationary satellites (Ermida et al. [53] for MSG/SEVIRI, Vinnikov et al. [54] for GOES). It consists of a multiplicative factor accounting for the effects of surface heterogeneity and land use, see Equation (1), and relies on viewing zenith angle (VZA) from the satellite, θ_{sat} and the solar zenith angle, θ_{sun}. It is composed of two additive kernels, the emissivity and solar kernels, incorporating several effects related to shading and hotspot.

$$\frac{HR_{MSG}}{HR_{VZA=0}} = 1 + A\Phi(\theta_{sat}) + B\Psi(\theta_{sun}, \theta_{sat}) \tag{1}$$

$$\Phi = 1 - cos(\theta_{sat}) \tag{2}$$

$$\Psi = sin(\theta_{sat})cos(\theta_{sun})sin(\theta_{sun})cos(\theta_{sun} - \theta_{sat}) \tag{3}$$

Because the major effect on heating rates is expected to be from surface heterogeneity, we use the set of Equations (1)–(3) to correct the heating rate calculated in the view of the satellite HR_{MSG} to nadir view. The emissivity kernel parameter (Equation (2)) is set constant, $A = -0.2$. A map of the constant of the solar kernel (Equation (3)), B, has been obtained by calculating the ratio of the maximum heating rate obtained per pixel (percentile 97 of the distribution) with the regional average of maxima (60 × 60 pixels), such as effects from terrain complexity are clearly visible (not shown).

Calibration

Because higher heating rates appear in water stress affected regions, the signal is generally inversely correlated to the soil moisture. A linear relation between soil moisture and heating rate or normalized difference of temperatures is usually assumed. However, no evidence of that linearity has been provided over a large set of conditions. A series of adaptations have therefore been brought to the original algorithm to account for the behaviour of the signal locally observed. The selection of the model and its calibration was based on pairing daily soil moisture in-situ measurements with daily heating rates, estimated from at least 40% of daily LST measurements and smoothened by a low-pass exponential filter with a decay time of 3 days, at a single calibration site (Tojal, Portugal) over one year (2007), which present the characteristic to have clear moist and dry seasons, and a low vegetation cover. Soil moisture and heating rates have been both normalized between 0 and 1. Three functions have been tested (linear, exponential, double exponential) and the coefficients optimalized using the bayesian inversion algorithm *Shuffled Complex Evolution Metropolis algorithm*, SCEM-UA [55].

As a result, the relation with the smallest residuals was selected: it is exponential (Equation (4)), increasingly sensitive to variations towards high soil moisture contents (Figure 2), with $k_1 = 1.6$, $k_2 = -1.05$ and $k_3 = -0.6$.

$$SSM_{0-1} = k_1.exp\left(k_2.\frac{HR - HR_{min}}{HR_{max} - HR_{min}}\right) + k_3 \quad (4)$$

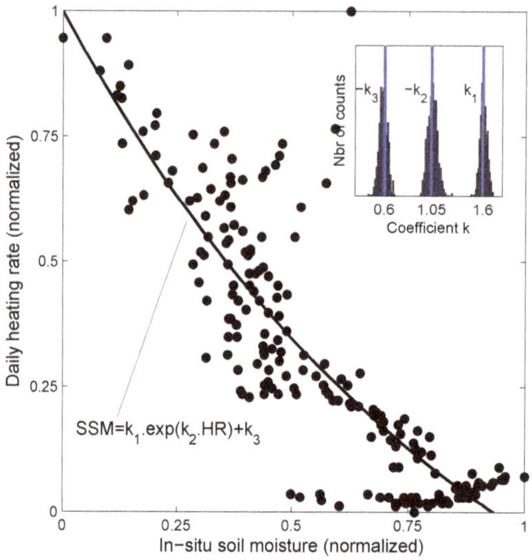

Figure 2. The exponential relation between daily heating rates and ground measurements of surface soil moisture at Tojal, Portugal, has been calibrated with the SCEM-UA algorithm. Each parameters is determined as the median of the probability density function.

Processing Steps

The production of daily averaged surface soil moisture (SSM) from MSG/SEVIRI LST follows the following sequence:

- Daily surface heating rates (*HR*) are computed for each pixel on land, by a linear regression through the LST observations from 1 h after dawn to 1 h before noon. Only the *HR* computed with at least 10% of LST over the morning are retained.
- The heating rates are corrected for the viewing anisotropy effects (Equation (1)). The inclusion of this viewing anisotropy correction has effects, for example, at latitudes above 40°N and in hilly areas, where it forces the signal to exhibit a larger soil moisture in winter, which is not seen without the correction.
- The heating rates are normalized between the minimum *Min* and maximum *Max*, determined pixelwise as the value corresponding to respectively 97% and 3% of the cumulative distribution function of the pixel heating rates over one year. The thresholds have been chosen as to avoid false extremes caused by wrong detection of clouds, dust or any failure in the retrieval of the LST. Because full variability of soil moisture between the wilting point (*pwp*) and the field capacity (*fc*) can be assumed if the series is sufficiently long, the extrema could be set to *fc* and *pwp*, as proposed by V2006.

- The exponential function is applied on the normalized corrected heating rates, resulting in daily surface soil moisture estimates (Equation (4)).
- A low-pass filter is applied, because of day-to-day apparent variability of the thermal inertia making the time series instable for some periods [37], an effect that maybe due to aerosol load variations, wind speed at the surface or fuzzy geolocation of the MSG/SEVIRI signal. It consists of a running exponential filter over the previous 30 days, with a characteristic time $T = 3$ days (Equation (5)), which allows to smoothen the signal with moderate loss of day-to-day information. This has been shown to be also beneficial for ASCAT SM products [56], while low-pass filters have been shown to reduce noise on SMOS soil moisture product [57].

$$SM_{Final}(t_n) = \frac{\sum_{i=n-30}^{n} SM(t_i) \cdot e^{-(\frac{t_n - t_i}{T})}}{\sum_{i=n-30}^{n} e^{-(\frac{t_n - t_i}{T})}} \quad (5)$$

2.2. Validation Material

2.2.1. In-Situ Soil Moisture Measurements

Datasets have been acquired for 8 yearly periods from March 2007 to December 2014 across Europe and Africa. Half-hourly measurements, if available, or hourly measurements otherwise, of soil moisture at depths of 2, 5, 10 or 30 cm have been made available from CarboEurope, CarboAfrica [58], AMMA [59–62], SMOSMANIA [63,64], REMEDHUS [65,66], HOBE [67], UMSUOL, CALABRIA, PERUGIA, VAS, UMBRIA [68–70], COSMOS [71] and UDC-SMOS [72] networks, with 161 station-years in total, sampling Spain, Portugal, Italy, France, Denmark, Mali, Niger, Sudan, Germany, Senegal, South Africa and Kenya (Figure 3).

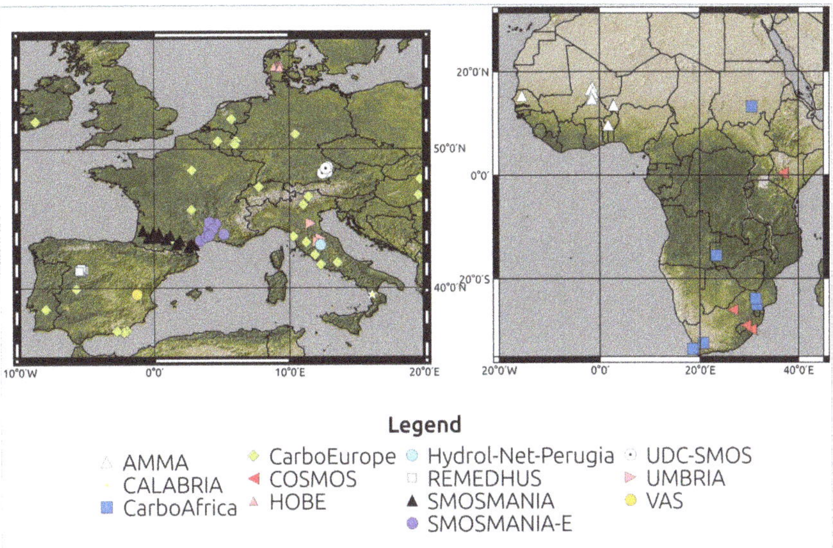

Figure 3. The validation sites of the soil moisture retrievals are spread over Europe and Africa, some are situated in challenging environments or settings for satellite remote sensing in terms of complexity of landscapes or topography. Most of the validation sites are grouped in networks, sharing same measurement methodologies, but not necessarily grouped geographically.

The in-situ set-up of the networks intent to provide local observations of soil moisture under various land cover types (coniferous, broadleaf or mixed forests, grasslands, croplands, wetlands, savannahs) submitted to a large range of climatic regimes. Sites considered for the validation are listed in Table 1, together with information on geographical location, climate regime, ground measurement depth and measurement technique (see Albergel et al. [73] for an overview of the various experimental set-ups). When possible, at least two yearly periods have been selected per site, with a sampling of the first years of operations of the LST product and of more recent years, where overlap is possible with other missions, for instance SMOS. Each time series has been rescaled between 0 and 1, 0 being considered as the permanent wilting point and 1, the field capacity.

2.2.2. Soil Moisture Products from Microwave Sensors

In addition, available products from current soil moisture missions have been used to compare their relative fit to the proposed retrieval during clear sky conditions. Products derived from the operational EUMETSAT MetOp/Advanced Scatterometer (ASCAT), an active sensor, and the research European Satellite Agency Soil Moisture Ocean Salinity Earth explorer mission (SMOS), a passive sensor. Both operate in the microwave domain, C-Band (5.255 GHz) and L-Band (1.4 GHz) respectively, and provide global imagery at 25 to 50 km and 35 km resolution every 1 to 3 days, respectively. The retrieval principles for both ASCAT and SMOS products can be found in Srivastava et al. [74].

Table 1. In-situ soil moisture datasets used in this study.

Soil Moisture Dataset	Soil Layer Depth (cm)	Number of Stations	Years	Measurement Technique/Device
AMMA (West Africa)	5	13	2007–2012	TDR CSC616
REMEDHUS (Spain)	5	21	2007, 2010	TDR CSC616
UMSUOL (Italy)	10	1	2010	TDR 100
UDC-SMOS (Germany)	5	4	2010	IMKO TDR
CALABRIA (Italy)	30	1	2010	ThetaProbe ML2X
Hydrol-Net-Perugia (Italy)	5	1	2010	TDR-SM Eq Corp TRASE-BE
UMBRIA (Italy)	5–15	3	2010	ThetaProbe ML2X
VAS (Spain)	0–5	1	2010	Stevens Hydraprobe
HOBE (Denmark)	0–5	18	2010	Decagon 5TE
Carboltaly (Italy)	2	5	2007, 2008, 2010, 2011	TDR
CarboEurope (Spain & Portugal)	0–5	5	2007, 2010, 2011	ThetaProbe ML2X
CarboEurope (France)	2	2	2007, 2010, 2011	TDR
CarboEurope (Belgium, Netherlands)	0–5	4	2007, 2010, 2011	TDR
CarboEurope (Swiss, Germany, Hungary)	0–5	4	2007, 2008, 2010	TDR
SMOSMANIA (France)	5	12	2007, 2010, 2011	ThetaProbe ML2X
SMOSMANIA-E (France)	5	9	2010, 2011	ThetaProbe ML2X
CarboAfrica	6	1	2007–2011, 2013	TDR
COSMOS	-	5	2007–2009, 2014	CosmicRay Probes

In this study, we have used the ASCAT SSM L3 product delivered as time series granules by the EUMETSAT Hydrology SAF (H25 product, http://hsaf.meteoam.it) for its easy handling of time series. The SMOS SSM L3 product has been obtained for the period starting at 1 January 2010 from the Centre Aval de Traitement des Données SMOS (CATDS) (http://www.catds.fr/). Both products have global coverage. Time series files for ASCAT H25 and daily files for SMOS SSM L3 are provided with the instantaneous retrievals, the time of acquisition and the acquisition mode (descending or ascending). In most files, there are one to two observations per day. The spatial resolution of the products is 25 km, with sampling at 12.5 km, for ASCAT H25 [75] and 25 km for SMOS L3. A buffer zone on land surface at the coast is foreseen by the products. Because SMOS and ASCAT L3 products are instantaneous, the direct comparison with MSG/SEVIRI derived daily soil moisture was not possible. Therefore, we applied a running exponential filter with a decay time of 3 days to smoothen the signal from instataneous values to daily values. It follows a similar strategy as used to produce the soil water index (SWI) product from MetOp/ASCAT SSM in the frame of Copernicus Global Land Services (http://land.copernicus.eu/global/ [56]).

2.3. Benchmarking Protocol

The time series of daily soil moisture derived from TIR remote sensing have been compared to daily averaged local surface soil moisture measurements. Time series are compared to time collocated daily averaged surface soil moisture data. The statistical scores, the standard deviation of both observations and remote sensing retrieval, the root mean square difference between both datasets (RMSD), the bias (in-situ minus satellite) and the correlation coefficient (R), have been calculated for each time series [73].

A synthesis of the statistical scores is presented in a Taylor diagram [76], which allows both a quantitative evaluation of per-time series comparison and a visual interpretation of how the site specific scores compare to each other. It requires the standard deviations from all retrievals to be normalized by the standard deviation of each individual in-situ dataset. Visually, the distance between the observation and the retrieval point gives the relative performance compared to other points. The normalized standard deviation (SDV) indicates if the remote sensing based retrieval can reproduce the locally observed variability: underestimated if less than 1, overestimated otherwise. The correlation coefficient indicates the match in the temporal evolution of the soil moisture, at both diurnal and seasonal scales: 1 is perfect match. In-situ data are represented by a point on the x axis at $R = 1$ and $SDV = 1$, every single time series comparison is displayed in polar coordinates. For the ease of interpretation, we group the results by network, and by climate classes, according to Köppen-Geiger classification [77], reflecting a variety of precipitation and air temperature regimes.

The validation is carried out on the total set of sites, where environmental conditions seem to be the best suitable for MSG/SEVIRI (marked in bold in Table 2), and also encompassing conditions that can be not optimal, but can nevertheless showcase the performance of the new method.

It is clear that a mismatch can be expected between local observations and retrieval from satellite data because of the difference in spatial scale between the observations and remote sensing data [78]. Though this issue has not been considered in Albergel et al. [73], higher correlation between coarser scale soil moisture retrieval and local observations may be found if the local nature of the surface corresponds to the coarse pixel. Several sites are concerned with this representativity issue at MSG/SEVIRI scale. For example, both SMOSMANIA transects in Southern France are concerned in our study: the South-Western transect (SMOSMANIA) consists of sites situated in heavily irrigated areas during summer, and both transects, but mostly the South-Eastern (SMOSMANIA-E), consists of sites nested in hilly to montaineous landscapes. The mapping tool available from french government (http://www.geoportail.gouv.fr/accueil) suggests irrigation fractions within a MSG SEVIRI pixel being around 80% for Montaut and Urgons, between 20% and 80% for Sabrès, 60% around Créon d'Armagnac, 30% around Savenes, between 10% and 20% around Peyrusse-Grande, and less than 5% around Lézignan, Saint-Félix, Lahas and Narbonne. As MSG/SEVIRI is expected to be sensitive to irrigation effects, this will obviously lead to discrepancies. In order to get the most benefits from the SMOSMANIA observations, the statistical scores have been calculated for both the annual time series and on a reduced time series, excluding only the Summer irrigation period, which extends from 20 June to 20 August for maize, the dominant crop type in the area [79], and the results reflected in the comparison with SMOS and ASCAT derived products. For SMOSMANIA-E, some sites are situated at the bottom of steep slopes, leading also potential discrepancies with the satellite signal which encompass terrains at higher altitudes, and slopes (Figure 4).

Table 2. Climate classification of sites and grouping in this study. The number of sites that seem optimal are marked in bold. The reasons for non-optimality are classified in 3 classes: [1] proximity of large water body, [2] proximity to large irrigated areas, [3] set in a location which topography does not reflect the neighbourhood within the 5 km.

Climate	Stations
Tropical Savannah(Aw)	AMMA (**2**), CarboAfrica(**1**), COSMOS (**2**)
Arid Steppe hot (Bsh)	AMMA (**4**), CarboAfrica (**1**)
Arid Steppe Cold (Bsk)	VAS (1), REMEDHUS (**20**), CarboEurope (1 [3])
Arid Desert Hot (Bwh)	AMMA (**7**), CarboAfrica(**1**)
Arid Desert Cold (Bwk)	CarboEurope (**2**)
Temperate Dry Hot Summer (Csa)	SMOSMANIA(-E) (**1**, 1 [1]), Calabria (1 [3]), Hydrol-Net-Perugia (1 [3]), CarboEurope (**2**, 4 [1,2,3])
Temperate Dry warm Summer (Csb)	SMOSMANIA(-E) (**2**,2 [3]), CarboAfrica (**1**)
Temperate without dry season, warm summer (Cfb)	Umbria (3 [3]), SMOSMANIA (**1**, 1), COSMOS (**1**)
Temperate without dry season, cold summer (Cfc)	CarboEurope (**6**), SMOSMANIA(-E) (**1**, 6 [2,3])
Temperate dry winter hot summer (Cwa)	CarboAfrica (**1**), COSMOS (**1**)
Temperate dry winter warm summer (Cwb)	COSMOS (**1**)
Cold without dry season, cold summer (Dfc)	UDC-SMOS (**4**), HOBE (**18**), SMOSMANIA(-E) (5 [2,3])

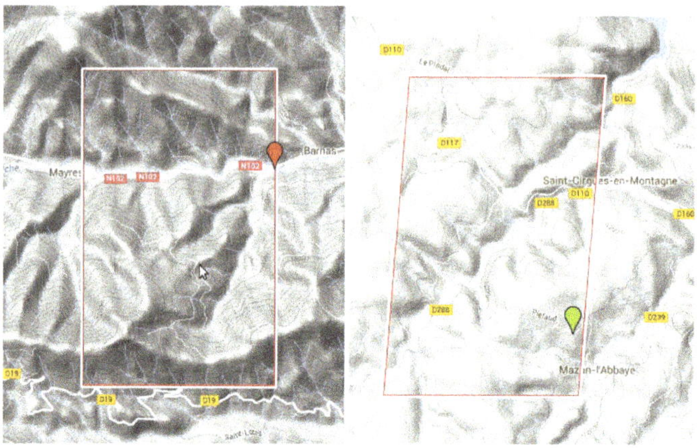

Figure 4. The topography around some SMOSMANIA-E sites is not favourable for MSG/SEVIRI soil moisture validation (two examples shown): the contrasts in topography within one pixel do not allow a fair comparison, as noted in the correlation results obtained in this study. Red and green markers indicate the location of the site within its overlapping MSG/SEVIRI pixel.

3. Results

The comparison of surface soil moisture time series shows that the remote sensing data and model seem well adapted to reproduce the observed signal in regions where vegetation is experiencing water stress: seasonality and shorter fluctuations are very well reproduced over arid to semi-arid areas of Spain and Africa, while the signal tends to be noisier and less correlated for temperate sites of Southern France, with scattering increasing for higher latitudes (Figure 5). The retrieved signal is discontinuous at African sites, mostly during the wet season, and at temperate sites throughout the year because of persitent clouds during entire days. For the temperate sites, the retrieved soil moisture captures the dry spells and wet events, however the signal does not match very well the observed amplitude over most SMOSMANIA sites during summer, e.g., St Felix, possibly due to intensive irrigation in the region, as expected (see another example in Figure 6). For the site in Germany, the estimations are less accurate and sparser, a summertime dry spell is correctly retrieved from the satellite information, but another one in Spring is not detected in the local measured soil moisture. For Spain, soil moisture is globally well retrieved compared to in-situ observations. For sites in Africa, the seasonal course is well reproduced, apart from isolated apparent wet events, e.g., in Agoufou, Mali, which coincide with strong dust storm. In this case, the amplitude of top-of-atmosphere brightness temperatures in the infrared spectral window tends to be attenuated over high aerosol loads, if not identified, these may be wrongly interpreted as a wet event. This will be further discussed in the next section.

The statistical scores for each yearly dataset grouped in averages by network and climate zones are provided in Table 3 and Figure 7, along with additional information about the number of daily observations used. This latter information reflects both the availability of the soil moisture retrieved from the satellite and availability of local observations. Averaged correlations for Sahelian (AMMA network) and Iberian sites (REMEDHUS, CarboEurope ES & PT) are around 0.7 (0.69, 0.75 and 0.68, respectively), and with values as high as 0.87 (Figure 7). Retrievals over Southern France and Italy are less correlated to in-situ data, with averages between 0.45 and 0.55, except for one site in CALABRIA which exhibits even lower correlation. Correlations tend to decrease with colder climates. While high correlations are found for the group of sites with *Aw* climate with an average correlation of 0.81, average correlations between 0.61 and 0.69 are found for the Arid climates (*BSk, BSh, Bwk, Bwh*). Correlations for the temperate climates *C* range between 0.44 and 0.66, except for one site (*Cwb*, 0.88). The cold climate *Dfc* present a low average correlation of 0.36. Retrievals have a bias around 10%, but a variability comparable to observations. In Figure 7, it is clear that the retrievals have a variation fairly comparable to the observations, with a normalized standard deviation between 0.5 and 1.5 and centered around 1. For most climate groups, the variability of correlations is contained in intervals of approximately 0.2, such that the average correlation gives a good idea of the performance in each class.

Highest scores are found in semi-arid and arid areas with a strong annual cycle in the precipitation regime. Less high correlations are found in Southern France and in Italy, where results are more scattered, and low correlations are found in temperate cold Europe, where variability of soil moisture is not dominated by an annual cycle. This result is not in agreement with the findings of Verstraeten et al. [35], who reported high correlations for european forested sites, with an average correlation of 0.556, during the growing season 1997. Possible explanations for this lower correlation for temperate Europe can be a less accurate retrieval of land surface temperature due to a large viewing angle of the surface and associated problems (a long optical path through the atmosphere leading to a weakened signal from the surface compared to noise generated by atmospheric motion, azimuthal anisotropy effects) and to a decreased contrast in the thermal infrared signal in the presence of denser vegetation covers, that, for Europe, increases with latitude.

Compared to the results of Prigent et al. [52], higher correlations have been found in the present study, which can be attributed either to the modifications of the model or to the selected regions. The application of V2006 model with MODIS data by Veroustraete et al. [80] revealed correlations varying between 0.17 and 0.85 with an average of 0.56 against TDR observations at six sites in China at 1 m depth (deeper than in this study). In a recent study, Zhao and Li [27] reported low correlation between soil moisture from SEVIRI and local observations at the REMEDHUS network, presenting poorer validation results than those shown here. Compared to studies combining two sources of satellite data, for example SMOS and MSG/SEVIRI [81] or ESA CCI and MODIS [82], the results from MSG/SEVIRI alone presented here show a better correlation for daily values for both SMOSMANIA and REMEDHUS sites, which shows the potential of such combinations of products.

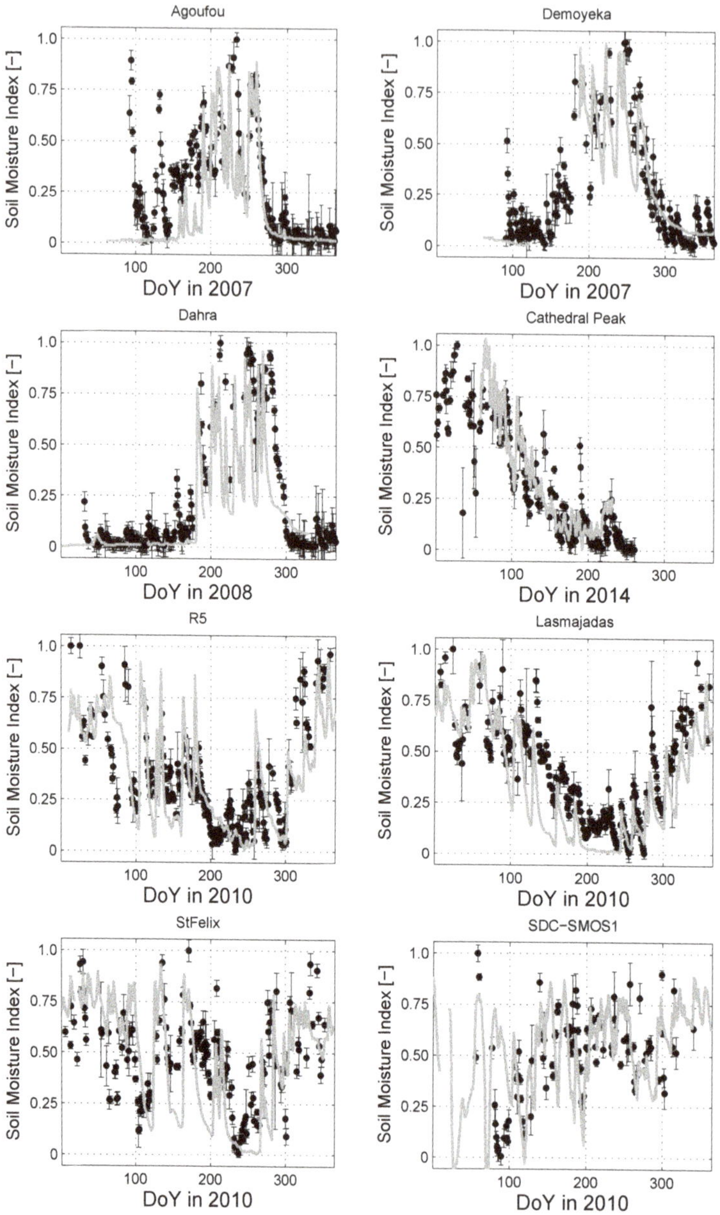

Figure 5. Comparison of daily surface soil moisture retrieved from MSG/SEVIRI LST (black) and measured locally (grey), for a set of 8 sites from different networks and climates: Agoufou (Mali, AMMA, BWh), Demokeya (Sudan, CarboAfrica, BWh), Dahra (Senegal, AMMA, BSh), Cathedral Peak (South Africa, COSMOS, Cwb), Guarena (Spain, REMEDHUS, BSk), Las Majadas del Tietar (Spain, CarboEurope, Csa), St Felix (France, SMOSMANIA, Cfc), Friedling (Germany, UDC-SMOS, Dfc) (from (**top left**) to (**bottom right**)).

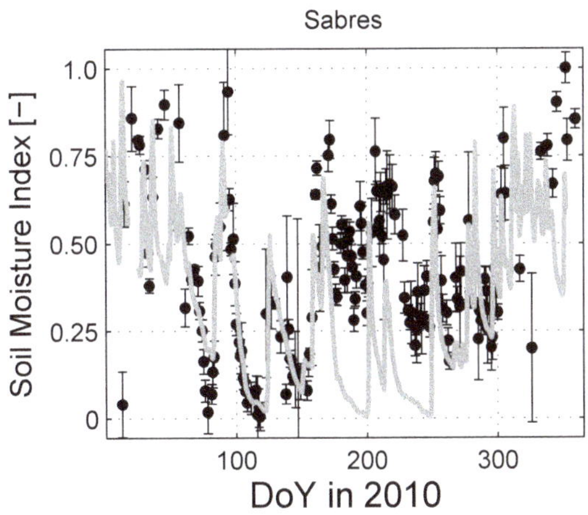

Figure 6. The quality of sites representativity within the SEVIRI pixel affects the statistical results obtained. In Sabrès (France), soil moisture retrievals from SEVIRI (black) compares well with ground observations (grey) except in Summer: intensive irrigation is detected, while the ground observation are taken at a non-irrigated site. Removing the summer improves the statistics.

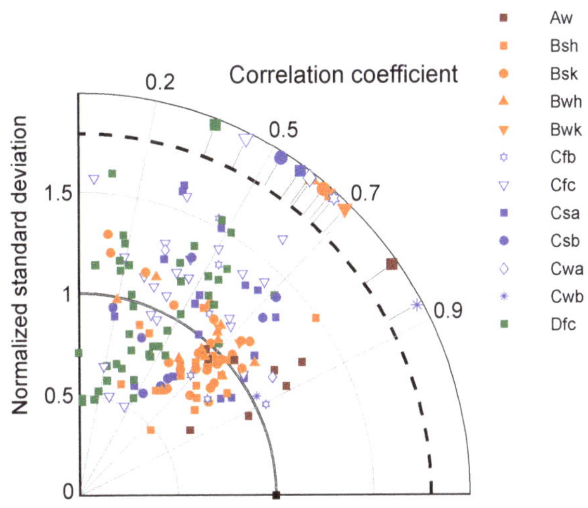

Figure 7. Statistical scores of the comparison between local observations and soil moisture derived from LSASAF LST, presented in a Taylor diagram. Each colored point represents the comparison for a set of data at a in-situ validation site. Codes of colour indicate the climate type region associated to the data. Average of correlations par climate type is represented with bigger symbols next to the correlation curved axis.

The comparison with SMOS and ASCAT products as described in the methodology section allows the observation of similarities and disagreement between the observation techniques from space. In Figure 8, three sites have been chosen to showcase some of those comparisons, which have been extended to all the datasets from 2010 to 2014 and synthetized in Figure 9. The visual comparison of the time series show very good consistency of the satellite retrievals for the site in REMEDHUS network in Spain: a clear marked seasonal pattern and weekly fluctuations are in agreement with the satellite retrievals, although not always exactly in line with the ground observations, especially during summer precipitation events. The comparison over a site in the Southern Spain in 2011 shows a high consistency between surface soil moisture from SMOS, MSG and the ground observation, but a different seasonal pattern from ASCAT SSM, which is quite rare within the datasets examined. The comparison over Agoufou in Mali in 2007 reveals a better consistency of MSG surface soil moisture with observation during the onset of rainy season, while the end of wet season and the beginning of the dry season is equally well reproduced by ASCAT and MSG retrievals. At the end of the dry season (day 100 to 180), MSG retrievals are affected by the attenuation of the signal by aerosols, while ASCAT surface soil moisture shows a consistent dry condition of the soil. These are only visual illustrations of agreements and disagreements between space-based retrieval of surface soil moisture. A more extensive analysis has been conducted to better classify the performance of the three retrieval techniques, and the result presented here.

Table 3. For each network, the average statistical scores of the comparison between the satellite retrievals and the in-situ observations shows the geographical variability of the performance of the developed retrieval strategy. The retrievals are the most accurate over the Mediterranean region and Sahel, and less correlated to soil moisture observations in temperate Europe.

Datasets	R	Bias	RMS	obs/yr (Mean)	st-yr
AMMA	0.69	0.17	0.26	202	29
CALABRIA	0.19	0.02	0.21	48	1
COSMOS	0.88	0.09	0.16	154	5
CarboAfrica	0.54	0.10	0.25	170	7
CarboEurope					
BE&DE&NL	0.39	−0.09	0.30	71	6
FR&IT&CH	0.45	−0.01	0.29	150	9
ES&PT	0.75	0.10	0.21	221	10
HU	0.33	−0.07	0.27	112	2
HOBE	0.24	0.06	0.29	77	18
Hydrol-Perugia	0.56	−0.25	0.35	183	1
REMEDHUS	0.68	0.04	0.21	208	24
SMOSMANIA	0.39	0.08	0.29	139	26
SMOSMANIA-E	0.54	0.04	0.25	171	14
UDC-SMOS	0.48	−0.16	0.21	103	4
UMBRIA	0.50	−0.11	0.25	163	2
Climate					
Aw	0.81	0.12	0.21	171	8
Bsh	0.65	0.17	0.27	192	16
Bsk	0.64	0.03	0.22	202	26
Bwh	0.61	0.18	0.27	223	11
Bwk	0.69	0.12	0.23	244	2
Csa	0.58	0.07	0.25	182	17
Csb	0.53	0.12	0.26	159	8
Cwa	0.60	0.07	0.18	122	2
Cwb	0.88	0.14	0.17	126	1
Cfb	0.66	0.06	0.22	156	6
Cfc	0.44	0.01	0.27	139	22
Dfc	0.36	0.00	0.29	107	41

Scores for various networks are similar to results obtained with ASCAT and SMOS products, which are available at coarser resolution than SEVIRI [73]. The results presented here reveal higher correlations for REMEDHUS, HOBE, UDC-SMOS, but lower correlations for the SMOSMANIA sites. The comparison of the in-situ observations with the datasets from the three different families of sensors aboard satellites, i.e., MSG/SEVIRI, MetOp/ASCAT and SMOS, reveals the high complementarity of the systems for clear to partial cloud cover days over the different climate zones of Europe and Africa (Figure 9). While ASCAT derived daily soil moisture outperforms SMOS for 8 groups over 9 in terms of averaged correlation, SMOS derived soil moisture displays best scores for CarboEurope sites in Italy and for Southern African sites. For the AMMA sites, there was no common timeframe for SMOS and in-situ data, except for Dahra, Senegal, where corrrelation of 0.75 was found.

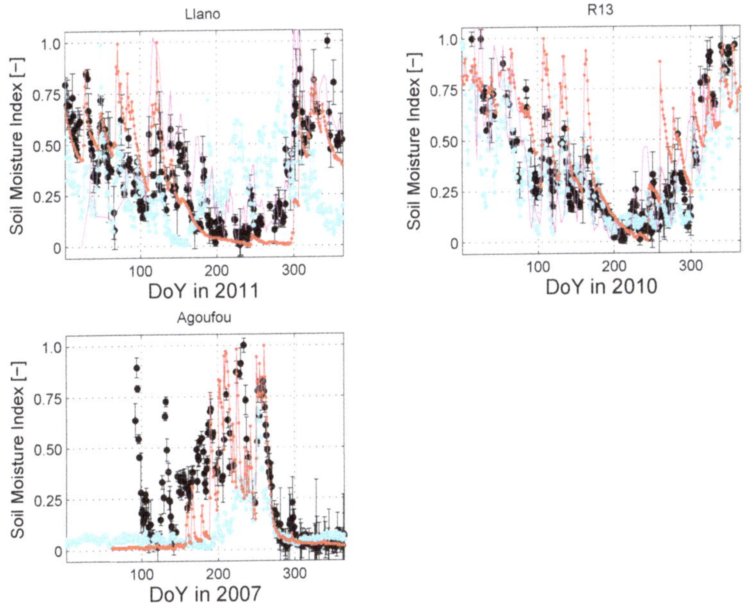

Figure 8. The comparison of MSG/SEVIRI (black), SMOS (magenta) and ASCAT (cyan) products (low-pass filter applied) with observations (red) at sites show sometimes strong consistency (R13, REMEDHUS Spain), or disagreement (Llano de los Juanes, Spain and Agoufou, Mali).

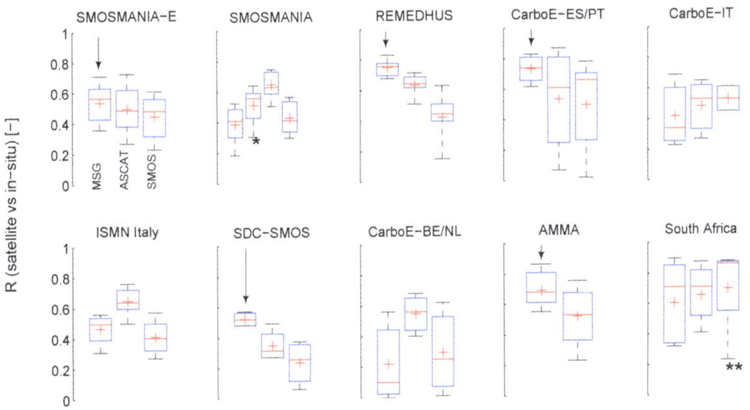

Figure 9. The correlation (average (crosses), distribution (boxplot with median, inter-quartile and extrema)) of the SEVIRI retrieval with ground observations is compared to scores obtained with SMOS L3 and ASCAT H25 products. Highest correlations for SEVIRI are obtained in five groups. * Correlation is improved with SEVIRI when removing summer season due to irrigation in the SMOSMANIA area (second box). ** SMOS only available for a selection of site-years.

The soil moisture derived from MSG/SEVIRI in this study performs best over the two other systems in the REMEDHUS, SMOSMANIA-E, CarboEurope-Spain/Portugal and AMMA sites, suggesting a good capability of the new system to monitor surface soil moisture for semi-arid and arid climate zones with a marked dry season (Aw, BSk, Csb). The exclusion of the irrigation period in the evaluation of the TIR based soil moisture at the SMOSMANIA sites shows an improved correlation, from 0.40 to 0.50 in average, in agreement with the scores obtained at SMOSMANIA-E sites.

4. Discussion

The method used to retrieve daily soil moisture has several limitations, related to the nature of the surface, the composition of the local atmosphere and the geometry of viewing, as most remote sensing data for surface analysis from space do. A dense vegetation cover will probably screen or alter the signal from the soil, daily retrieval is not possible for persistent cloudy situations, while transport of dust by wind induces wrong estimations if not properly accounted for. Similar limitations have been documented for the triangle method in [83]. The two last factors are discussed here to give an insight on limitations.

4.1. Availability of Retrievals

Compared to retrievals from microwave sensors aboard polar satellites, the proposed retrieval algorithm has the advantage of being based on high repetition rate data at spatial resolution of 3 to 5 km. However, no information on the surface can be retrieved under persistent cloudy sky situations, contrarily to microwave sensors such as ASCAT onboard MetOp satellites. This inherent limitation due to the observing system affects the use of the retrieval for continuous monitoring of soil moisture. We have analyzed for each year between 2007 and 2011 the proportion of days that were sampled with a sufficient accuracy by the retrieval algorithm over the five years. The respective seasonal averages (December to February, March to May, June to August, September to November) are shown in Figure 10. This analysis reveals that most temperate Europe is sampled less than 10% during winter. Over most of the African continent, the availability varies between 90% during the dry season and down to 40% during the rain season. The Western coast of central Africa could be only sampled up to 30% due to almost constant cloud coverage even during the dry season [84].

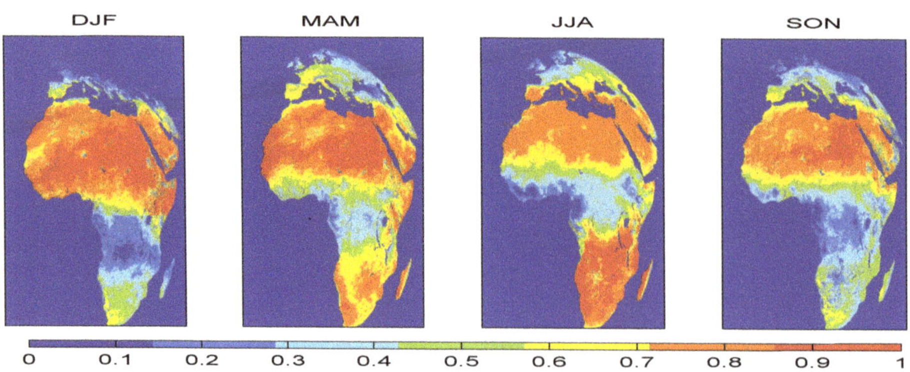

Figure 10. Tri-monthly availability of daily soil moisture retrieval averaged over 5 years. Cloud cover affects the number of available retrieval, especially over the equator and during winter over Europe. Sahelian regions are well sampled.

4.2. Aerosol Loads and Wind Speed

During the annual dry period over Sahel and Sahara desert, apparent wet events are seen in the soil moisture retrieved from remote sensing. However, such events are not supported by local observations of soil moisture. Those apparent wet peaks may be due to either sand or dust storms, leading to a significant increase in aerosol loads, which in turn affect the radiation budget at the surface and may contribute to smaller daily LST amplitudes. Under very high aerosol optical depth, the LST retrievals may also be subject to higher errors: directly associated to the atmospheric correction or indirectly caused by poorer cloud screening [85–88]. A physical effect of cooling of the surface by the wind cannot be rulled out either. Surface heating rates seem sensitive to the wind, but most of its influence is limited to almost bare surfaces as shown on Figure 11, where the relative difference of HR between two distinct wind speed regimes (0 to 3 m·s^{-1} and 6 to 9 m·s^{-1}, obtained from the European Center for Medium-range Weather Forecasts deterministic short-term forecasts of wind speed at 10 m from the surface) is displayed in the two dimensional space delimitated by the possible heating rates obtained by vegetation density, leaf area index obtained from LSA-SAF, for latitudes under 45°N (Figure 11). However, it is still difficult to disentangle the effect of wind speed and aerosol, as aerosol load may affect the retrievals, especially over the bare soil areas.

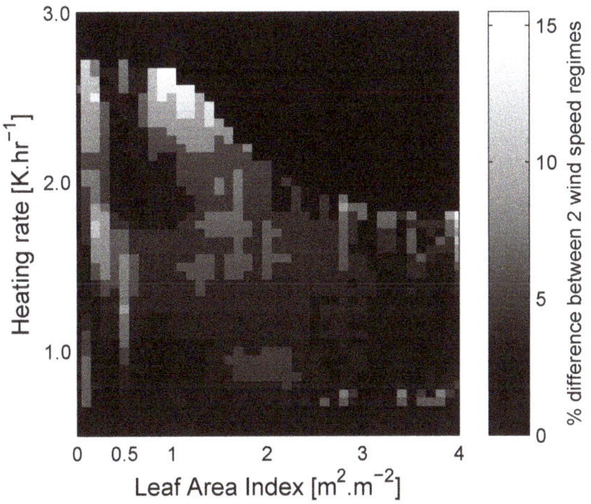

Figure 11. Daily heating rates are increasingly sensitive to wind speed over non vegetated areas with up to 15% of change under another wind speed regime.

In Figure 12, an example of an apparent wet event associated to dust presence in the atmosphere is shown for the 19 February 2008. According to Ben-Ami et al. [89], this released dust was a result of two consecutive days of active emission from the Bodélé depression in Saharian desert, with possible mixing with wild fires. Another example in Bamba, Mali, is shown in Figure 12. The evolution of the retrieved soil moisture seems directly correlated to the evolution of high concentrations of aerosols, as monitored by the MACCII aerosol optical depth product at 1.125° resolution (Addition of 5 types of aerosols: sea salt, dust, organic matter, black carbon and sulphates). Therefore, problems can occur when high aerosol optical depths are not properly taken into account. This is particularly relevant during the agricultural and wild fire season. Whether the poorer performance of soil moisture estimates is associated to enhanced LST errors, associated to deficiencies in the atmospheric correction, poorer cloud screening, or to the impact of those high aerosol loads toghether with near surface wind on the LST dynamics remains to be investigated. It is likeky that all these factors play a role and therefore a proper screening of those retrievals should be conducted with the help of a dedicated aerosol product derived from satellites or assimilation systems. MACC service, now evolved to Copernicus Atmospheric Monitoring Service (CAMS), is an available candidate, however, as its spatial resolution is coarse compared to MSG/SEVIRI, other products could be considered, as AOD products derived from MSG/SEVIRI and produced by the ICARE Data and Service center, AERUS-GEO [90,91].

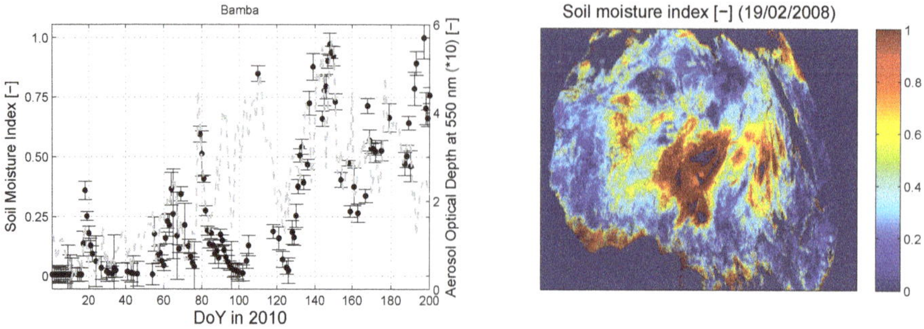

Figure 12. (**Left**) Apparent wet peaks from SSM retrieval in Bamba (Mali) are correlated with AOD forecast from MACCII. (**Right**) Soil moisture retrieved over Africa for 19 February 2008 displays an anomously large wet area in Sahara, due to a large dust storm and local fire emissions.

4.3. Temperate and Cold Climates versus Viewing Angles

In this study, the coldest climate zones sampled are all located in areas where viewing angle is already large. In that case, it is difficult to separate the effects originating from the specific nature of the surface and its response to solar illumination and from the consequences of large viewing angles and long optical paths. Scores presented in the Results section show a poorer performance for the temperate and cold than for warmer climates. It is possible to mitigate the global effect by spatial and temporal averaging. For two sites (BE-Lonzée, BE-Vielsalm, CarboEurope network), we have obtained a time series of heating rates from the average of nine up to twenty five adjacent pixels. The results show a better performance compared to the retrieval based on only one pixel, especially in terms of availability of data: 66 to 70 estimations over 9 months for one pixel and 114 to 120 over the same period for 25 pixels averaged. The correlation is also slightly improved by 10%. Over Belgium in Europe (51°N, 1°E), the new applied resolution is therefore approximately 15 km for nine pixels and 25 km for twenty five pixels, which remains below or equal to present time SSM products. Nevertheless, future studies should concentrate on the disentanglement of the effects to better understand the mechanisms.

5. Conclusions

This study aimed at deriving daily surface soil moisture from land surface temperature data retrieved from geostationary satellites under the assumption that only the observation of variations of land surface temperature were necessary. The new proposed retrieval algorithm relies on the local morning heating rates of the land surface based on the 15 min data provided by the Satellite Application Facility on Land Surface Analysis from EUMETSAT MSG/SEVIRI sensor, and is designed to produce surface soil moisture with less than 1 day delay over clear sky and non-steady cloudy (over 10% of the time) conditions. A dataset of eight years (from 2007 to 2014) has been generated over Europe and Africa at the MSG/SEVIRI variable resolution, being 3 km over equatorial Africa, and around 5 km over northern Europe. An extensive quantitative validation of the retrieved soil moisture against in-situ daily averaged ground measurements over 161 site-years across Europe and Africa reveals a good capability of this monitoring system in semi-arid to arid areas, with correlation as high as 0.85, and around 0.7 in average in Sahel, Spain, and Portugal. In those regions, the seasonal behaviour, as well as the day-to-day variations, are well reproduced. Reduced capabilities have been found over temperate areas of Europe. Although warm temperate climates show relatively fair correlations of 0.5 in average, scores in Belgium, the Netherlands and Denmark show poorer results (between 0.25 and 0.40), but the deterioration of scores could potentially be partly mitigated by some minor adaptations, including averaging the data spatially over mutiple pixels (correlation improved of 10% in Belgium). Effects of irrigation on the retrieved surface soil moisture is observed at that resolution, and as been clearly evidenced in South-West of France.

The limitations of the new method are (1) a viewing geometry of the satellite that is optimal neither for northern regions of Europe nor for areas with high topography complexity, (2) a reduced capability over dense vegetated areas, where the ratio noise over signal is significant, (3) a discontinuous monitoring when the sky is overcast during a whole morning, and (4) a high sensitivity to atmospheric aerosol load, giving spurious moist detection in case of a deficient land surface data screening over semi-desert overpassed by dust storms. While the limitations can be mitigated by solutions proposed in this paper, they should be further investigated.

The comparison of scores with available global products for surface soil moisture derived from passive and active microwave satellite data (SMOS L3 SSM and MetOp/ASCAT H25 SSM) shows the competitiveness of the new TIR method in sampling wide areas over clear and non-steady cloudy sky conditions. Especially, the TIR method allowed better surface soil moisture sampling in the Sahel and in Spain, and similar performances over Southern France at the daily time scale. The comparison revealed that none of the three observational techniques works best everywhere, leaving some more research avenues in studying their complementarity.

The study has pointed out several shortcomings in validating satellite retrievals by direct comparison with in-situ observations. Especially, this is the case for partly irrigated districts and some hilly areas, where further instrumentation equipment could be installed in places with spatially scalable observations.

The method proposed in this study relies on an empirical relation between heating rates and surface soil moisture. Given the relatively high degree of success of this model to sample surface soil moisture in large areas of Africa and Europe, future work should focus on the physical interpretation and modeling to better understand the mechanisms.

The methodology is already recommended to derive daily surface soil moisture at the MSG SEVIRI spatial resolution of 3.1 km over Africa and 4 km over Southern Europe, and can potentially be applied to other geostationary satellites to cover other continents. It is expected to be useful in applications needing continental to global soil moisture states to be assimilated by models, especially in regions where in-situ data are scarse and if details about the surface is needed, as for example irrigation. The results could be useful to drought early warning or evaluation of susceptibility to flooding in regions and periods with not completely overcasts days. Since this observation technique is not valid for totally cloud covered sky conditions, other sources of soil moisture data can potentially complement the information provided by thermal inertia to provide an all-weather daily monitoring of soil moisture.

Further research will concentrate on the complementarity between multiple remote sensing techniques, such as SAR, and sensors to achieve a better accuracy and spatial sampling. In this perspective, the disaggregation of the surface soil moisture from SEVIRI, or from microwave sensors, thanks to their combined use with the fine resolution soil moisture estimations from the Sentinel satellites, for example, could be a leap forward in achieving merged systems of more continuous, accurate and detailed surface soil moisture observations.

Author Contributions: Conceptualization, methodology, software, validation, formal analysis, investigation, N.G.; resources, J.-M.B., A.A., J.A., I.T.; data curation, N.G., I.T., A.A.; writing–original draft preparation, N.G.; writing—review and editing, N.G., A.A., J.-M.B., F.G.-M., J.A., O.B., I.T.; supervision, O.B., I.T., vizualization, N.G., J.-M.B.; project administration, F.G.-M.; funding acquisition, F.G.-M., I.T.

Funding: This research was funded by EUMETSAT and the European Space Agency through the PRODEX programme of the Belgian Science Policy.

Acknowledgments: The authors thank the scientists who have contributed to build the soil moisture databases either in the context of FLUXNET or in ISMN and to share this extremely valuable information freely through accessible platforms.

Conflicts of Interest: The authors declare no conflict of interest.

References

1. Taylor, C.; de Jeu, R.A.M.; Guichard, F.; Harris, P.P.; Dorigo, W. Afternoon rain more likely over drier soils. *Nature* **2012**, *489*, 423–426. [CrossRef] [PubMed]
2. Massari, C.; Brocca, L.; Barbetta, S.; Papathanasiou, C.; Mimikou, M.; Moramarco, T. Using globally available soil moisture indicators for flood modelling in Mediterranean catchments. *Hydrol. Earth Syst. Sci. Discuss.* **2013**, *10*, 10997–11033. [CrossRef]
3. Gil, M.; Garrido, A.; Hernandez-Mora, N. Direct and indirect economic impacts of drought in the agri-food sector in the Ebro River basin (Spain). *Nat. Hazards Earth Syst. Sci.* **2013**, *13*, 2679–2694. [CrossRef]
4. Turner, K.; Georgiou, S.; Clarck, R.; Brouwer, R.; Burke, J. *FAO Water Report 27: Economic Valuation of Water Resources in Agriculture: From the Sectoral to a functional Perspective of Natural Resources Management*; Technical Report; FAO: Rome, Italy, 2004.
5. Hunsberger, C.; Evans, T.P.; Aide, T.M.; Montoro, J.A.; Borras, S.M.; del Valle, H.F.; Devisscher, T.; Jabbour, J.; Kant, S.; Lopez-Carr, D. et al. *The Fifth Global Environmental Outlook Report, Chapter 3: Land*; UNEP: Nairobi, Kenya, 2012; 32p. Available online: http://people.uncw.edu/pricopen/documents/GEO5_report_C3_Land.pdf (accessed on 20 August 2019).
6. Delworth, T.L.; Manabe, S. The influence of potential evaporation on the variabilities of simulated soil wetness and climate. *J. Clim.* **1988**, *1*, 523–547. [CrossRef]
7. Macauley, M.K. Earth Observations in Social Science Research for Management of Natural Resources and the Environment: Identifying the Landsat Contribution. *J. Terr. Obs.* **2009**, *1*, 31–51.
8. Koster, R.D.; Houser, P.R.; Engman, E.T.; Kustas, W.P. Remote sensing may provide unprecedented hydrological data. *Eos Trans. AGU* **1999**, *80*, 156. [CrossRef]
9. Njoku, E.G.; Entekhabi, D. Passive microwave remote sensing of soil moisture. *J. Hydrol.* **1996**, *184*, 101–129. [CrossRef]
10. Njoku, E.G.; Jackson, V.; Lakshmi, V.; Chan, T.; Nghiem, S.V. Soil moisture retrieval from AMSR-E. *IEEE Trans. Geosci. Remote* **2003**, *41*, 215–229. [CrossRef]
11. Bartalis, Z.; Wagner, W.; Naeimi, V.; Hasenauer, S.; Scipal, K.; Bonekamp, H.; Figa, J.; Anderson, C. Initial soil moisture retrieval from the METOP-A Advanced Scatterometer. *Geophys. Res. Lett.* **2007**, *34*, doi:10.1029/2007GL031088. [CrossRef]
12. Owe, M.; de Jeu, R.; Holmes, T. Multisensor historical climatology of satellite-derived global land surface moisture. *J. Geophys. Res.* **2008**, *113*, doi:10.1029/2007JF000769. [CrossRef]
13. Chesters, D. *The Scientific Basis for the Advanced Geosynchronous Studies Program*; Technical Report; NASA: Washington, DC, USA, 1998. Available online: http://goes.gsfc.nasa.gov/text/ags_science.html (accessed on 20 August 2019).
14. Wan, Z.M.; Dozier, J. A generalized split-window algorithm for retrieving land-surface temperature from space. *IEEE Trans. Geosci. Remote* **1996**, *34*, 892–905.
15. Sellers, W.D. *Physical Climatology*; The University of Chicago Press: Chicago, IL, USA, 1965.
16. Blanchard, M.B.; Greeley, R.; Goettelman, R. *Use of Visible, near Infrared and Thermal Infrared Remote Sensing*; NASA Technical Report TM X-62343; NASA: Washington, DC, USA, 1974.
17. Carlson, T.N.; Boland, F.E. Analysis of urban-rural canopy using a surface heat flux/temperature model. *J. Appl. Meteorol.* **1978**, *17*, 998–1013. [CrossRef]
18. Watson, K. Geologic applications of thermal infrared images. *Proc. IEEE* **1975**, *63*, 128–137. [CrossRef]
19. Becker, F. *Thermal Infrared Remote Sensing Principles and Applications*; Ispre Courses; A. A. Balkema: Varese, Italy, 1980.
20. Price, J.C. On the analysis of thermal infrared imagery: The limited utility of apparent thermal inertia. *Remote Sens. Environ.* **1985**, *18*, 59–73. [CrossRef]
21. Abdellaoui, A.; Becker, F.; Olory-Hechinger, E. Use of Meteosat for mapping thermal inertia and evapotranspiration over a limited region of Mali. *J. Clim. Appl. Meteorol.* **1986**, *25*, 1489–1506. [CrossRef]
22. Price, J.C. The Potential of Remotely Sensed Thermal Infrared Data to Infer Surface Soil Moisture and Evaporation. *Water Resour. Res.* **1980**, *16*, 787–795. [CrossRef]
23. Carlson, T.N.; Dodd, J.K.; Benjamin, S.G.; Cooper, J.N. Satellite estimation of the surface energy balance, moisture availability and thermal inertia. *J. Appl. Meteorol.* **1981**, *20*, 67–87. [CrossRef]

24. Gillies, K.K.; Carlson, T.N. Thermal remote sensing of surface soil water content with partial vegetation cover for incorporation into climate models. *J. Appl. Meteorol.* **1995**, *34*, 745–756. [CrossRef]
25. Wagner, W.; Naeimi, V.; Scipal, K.; de Jeu, R.; Martinez-Fernandez, J. Soil moisture from operational meteorological satellites. *Hydrolgeol. J.* **2007**, *15*, 121–131. [CrossRef]
26. Garcia, M.; I. Sandholt, P.C.; Ridler, M.; Mougin, E.; Kergoat, L.; Timouk, F.; Fensholt, R.; Domingo, F. Actual Evapotranspiration in Drylands derived from In-Situ and Satellite Data: Assessing Biophysical Constraints. *Remote Sens. Environ.* **2014**, *131*, 103–118. [CrossRef]
27. Zhao, W.; Li, A. A Downscaling Method for Improving the Spatial Resolution of AMSR-E Derived Soil Moisture Product Based on MSG-SEVIRI Data. *Remote Sens.* **2013**, *5*, 6790–6811. [CrossRef]
28. Song, X.; Leng, P.; Li, X.; Li, X.; Ma, J. Retrieval of daily evolution of soil moisture from satellite-derived land surface temperature and net surface shortwave radiation. *Int. J. Remote Sens.* **2013**, *34*, 3289–3298. [CrossRef]
29. Leng, P.; Song, X.; Li, Z.L.; Ma, J.; Zhou, F.; Li, S. Bare surface soil moisture retrieval from the synergistic use of optical and thermal infrared data. *Int. J. Remote Sens.* **2014**, *35*, 988–1003. [CrossRef]
30. Dorigo, W.A.; Wagner, W.; Hohensinn, R.; Hahn, S.; Paulik, C.; Xaver, A.; Gruber, A.; Drusch, M.; Mecklenburg, S.; van Oevelen, P.; et al. The International Soil Moisture Network: A data hosting facility for global in situ soil moisture measurements. *Hydrol. Earth Syst. Sci.* **2011**, *15*, 1675–1698. [CrossRef]
31. Duguay-Tetzlaff, A.; Bento, V.A.; Göttsche, F.; Stöckli, R.; Martins, J.P.A.; Trigo, I.F.; Olesen, F.; Bojanowski, J.S.; da Camara, C.; Kunz, H. Meteosat Land Surface Temperature Climate Data Record: Achievable Accuracy and Potential Uncertainties. *Remote Sens.* **2015**, *7*, 13139–13156. [CrossRef]
32. Murray, T.; Verhoef, A. Moving towards a more mechanistic approach in the determination of soil heat flux from remote measurements—I. A universal approach to calculate thermal inertia. *Agric. For. Meteorol.* **2007**, *147*, 80–87. [CrossRef]
33. Mitra, D.S.; Majumdar, T.J. Thermal inertia mapping over the Brahmaputra basin, India using NOAA-AVHRR data and its possible geological applications. *Int. J. Remote Sens.* **2004**, *25*, 3245–3260. [CrossRef]
34. Coppola, A.; Basile, A.; Menenti, M.; Buonannno, M.; Colin, J.; Mascellis, R.D.; Esposito, M.; Lazzaro, U.; Magliulo, V.; Manna, P. Spatial distribution and structure of remotely sensed surface water content estimated by a thermal inertia approach. In *Remote Sensing for Environmental Monitoring and Change Detection*; Owe, M., Neale, C., Eds.; IAHS Publisher: Wallingford, UK, 2007; Volume 316; pp. 1–12.
35. Verstraeten, W.W.; Veroustraete, F.; van der Sande, C.J.; Grootaers, I.; Feyen, J. Soil moisture retrieval using thermal inertia, determined with visible and thermal spaceborne data, validated for European forests. *Remote Sens. Environ.* **2006**, *101*, 299–314. [CrossRef]
36. Petropoulos, G.; Carlson, T.N.; Wooster, M.J.; Islam, S. A review of Ts/VI remote sensing based methods for the retrieval of land surface energy fluxes and soil surface moisture. *Prog. Phys. Geogr.* **2009**, *33*, 224–250. [CrossRef]
37. Van Doninck, J.; Peters, J.; De Baets, B.; De Clercq, E.; Ducheyne, E.; Verhoest, N.E.C. The potential of multitemporal Aqua and Terra MODIS apparent thermal inertia as a soil moisture indicator. *Int. J. Appl. Earth Obs.* **2011**, *13*, 934–941. [CrossRef]
38. Merlin, O.; Malbéteau, Y.; Notfi, Y.; Bacon, S.; Khabba, S.R.; Jarlan, L. Performance Metrics for Soil Moisture Downscaling Methods: Application to DISPATCH Data in Central Morocco. *Remote Sens.* **2015**, *7*, 3783–3807. [CrossRef]
39. Wetzel, P.J.; Atlas, D.; Woodward, R.H. Determining soil moisture from geosynchronous satellite infrared data: A feasibility study. *J. Clim. Appl. Meteorol.* **1984**, *23*, 375–391. [CrossRef]
40. Anderson, M.C.; Mecikalski, J.M.N.J.R.; Otkin, J.P.; Kustas, W.P. A climatological study of evapotranspiration and moisture stress across the continental U.S. based on thermal remote sensing: II. Surface moisture climatology. *J. Geophys. Res.* **2007**, *112*. [CrossRef]
41. Hain, C.R.; Mecikalski, J.R.; Anderson, M.C. Retrieval of an available water-based soil moisture proxy from thermal infrared remote sensing. Part I: Methodology and validation. *J. Hydrometeorol.* **2009**, *10*, 663–683. [CrossRef]
42. Parinussa, R.M.; Yilmaz, M.T.; Anderson, M.C.; Hain, C.R.; de Jeu, R.A.M. An intercomparison of remotely sensed soil oisture products at various spatial scales over the Iberian Peninsula. *Hydrol. Process.* **2014**, *28*, 4865–4876. [CrossRef]

43. Stisen, S.; Sandholt, I.; Noergaard, A.; Fensholt, R.; Jensen, K.H. Combining the triangle method with thermal inertia to estimate regional evapotranspiration—Applied to MSG-SEVIRI data in the Senegal River basin. *Rem. Sens. Environ.* **2008**, *112*, 1242–1255. doi:10.1016/j.rse.2007.08.013. [CrossRef]
44. Trigo, I.F.; DaCamara, C.C.; Viterbo, P.; Roujean, J.L.; Olesen, F.; Barroso, C.; de Coca, F.C.; Carrer, D.; Freitas, S.C.; Garcia-Haro, J.; et al. The Satellite Application Facility on Land Surface Analysis. *Int. J. Remote Sens.* **2011**, *32*, 2725–2744. [CrossRef]
45. Trigo, I.F.; Monteiro, I.T.; Olesen, F.; Kabsch, E. An assessment of remotely sensed land surface temperature. *J. Geophys. Res.* **2008**, *113*, doi:10.1029/2008JD010035. [CrossRef]
46. Freitas, S.C.; Trigo, I.F.; Bioucas-Dias, J.M.; Göttsche, F. Quantifying the Uncertainty of Land Surface Temperature Retrievals from SEVIRI/Meteosat. *IEEE Trans. Geosci. Remote* **2010**, *48*, 523–534. [CrossRef]
47. Göttsche, F. Validation of land surface temperature products with 5 years of permanent in-situ measurements in 4 different climate regions. In Proceedings of the EUMETSAT Meteorological Satellite Conference, Vienna, Austria, 16–20 September 2013.
48. Ermida, S.L.; Trigo, I.F.; DaCamara, C.C.; Göttsche, F.M.; Olesen, F.S.; Hulley, G. Validation of remotely sensed surface temperature over an oak woodland landscape—The probel of viewing and illumination geometries. *Remote Sens. Environ.* **2014**, *148*, 16–27. [CrossRef]
49. Göttsche, F.; Olesen, F.; Trigo, I.F.; Bork-Unkelbach, A.; Martin, M.A. Long term validation of land surface temperature retrieved from MSG/SEVIRI with continuous in-situ measurements in Africa. *Remote Sens.* **2016**, *8*, 410. [CrossRef]
50. Rasmussen, M.O.; Pinheiro, A.C.; Proud, S.R.; Sandholt, I. Modeling Angular Dependences in Land Surface Temperatures From the SEVIRI Instrument Onboard the Geostationary Meteosat Second Generation Satellites. *IEEE Trans. Geosci. Remote Sens.* **2010**, *48*, 3123–3133. [CrossRef]
51. Grant, I.; Heyraud, C.; Breon, F.M. Continental scale hotspot observations of Australia at sub-degree angular resolution from POLDER. *Int. J. Remote Sens.* **2004**, *25*, 3625–3636. [CrossRef]
52. Prigent, C.; Aires, F.; Rossow, W.B.; Robock, A. Sensitivity of satellite microwave and infrared observations to soil moisture at a global scale: Relationship of satellite observations to in situ soil moisture measurements. *J. Geophys. Res.* **2005**, *110*, doi:10.1029/2004JD005094. [CrossRef]
53. Ermida, S.; Pires, A.; Trigo, I.F.; da Camara, C. Towards a Harmonized LST Product—The problem of angular anisotropy of LST. In Proceedings of the 6th LSA-SAF Workshop, Reading, UK, 8–10 June 2015.
54. Vinnikov, K.Y.; Yu, Y.; Goldberg, M.D.; Tarpley, D.; Romanov, P.; Laszlo, I.; Chen, M. Angular anisotropy of satellite observations of land surface temperature. *Geophys. Res. Lett.* **2012**, *39*. [CrossRef]
55. Vrugt, J.A.; Gupta, H.V.; Bouten, W.; Sorooshian, S. A Shuffled Complex Evolution Metropolis algorithm for optimization and uncertainty assessment of hydrologic model parameters. *Water Resour. Res.* **2003**, *39*. [CrossRef]
56. Paulik, C.; Dorigo, W.; Wagner, W.; Kidd, R. Validation of the ASCAT Soil Water Index using in situ datafrom the International Soil Moisture Network. *Int. J. Appl. Earth Obs. Geoinf.* **2014**, *30*, 1–8. [CrossRef]
57. Su, C.H.; Ryu, D.; Western, A.W.; Wagner, W. De-noising of passive and active microwave satellite soil moisture time series. *Geophys. Res. Lett.* **2013**, *40*, 3624–3630. [CrossRef]
58. Merbold, L.; Ardö, J.; Arneth, A.; Scholes, R.J.; Nouvellon, Y.; de Grandcourt, A.; Archibald, S.; Bonnefond, J.M.; Boulain, N.; Brueggemann, N.; et al. Precipitation as driver of 20 carbon fluxes in 11 African ecosystems. *Biogeosciences* **2009**, *6*, 1027–1041. [CrossRef]
59. Pellarin, T.; Laurent, J.; Cappelaere, B.; Decharme, B.; Descroix, L.; Ramier, D. Hydrological modelling and associated microwave emission of a semi-arid region in South-western Niger. *J. Hydrol.* **2009**, *375*, 262–272. [CrossRef]
60. Mougin, E.; Hiernaux, P.; Kergoat, L.; Grippa, M.; de Rosnay, P.; Timouk, F.; Dantec, V.L.; Demarez, V.; Lavenu, F.; Arjounin, M.; et al. The AMMA-CATCH Gourma observatory site in Mali: Relating climatic variations to changes in vegetation, surface hydrology, fluxes and natural resources. *J. Hydrol.* **2009**, *375*, 14–33. doi:10.1016/j.jhydrol.2009.06.045. [CrossRef]
61. Cappelaere, B.; Descroix, L.; Lebel, T.; Boulain, N.; Ramier, D.; Laurent, J.P.; Favreau, G.; Boubkraoui, S.; Boucher, M.; Moussa, I.B.; et al. The AMMA Catch observing system in the cultivated Sahel of South West Niger- Strategy, Implementation and Site conditions. *J. Hydrol.* **2009**, *375*, 34–51. [CrossRef]
62. de Rosnay, P.; Gruhier, C.; Timouk, F.; Baup, F.; Mougin, E.; Hiernaux, P.; Kergoat, L.; LeDantec, V. Multi-scale soil moisture measurements at the Gourma meso-scale site in Mali. *J. Hydrol.* **2009**, *375*, 241–252. [CrossRef]

63. Albergel, C.; Rüdiger, C.; Pellarin, T.; Calvet, J.C.; Fritz, N.; Froissard, F.; Suquia, D.; Petitpa, A.; Piguet, B.; Martin, E. From near-surface to root-zone soilmoisture using an exponential filter: An assessment of the method based on insituobservations and model simulations. *Hydrol. Earth Syst. Sci.* **2008**, *12*, 1323–1337. [CrossRef]
64. Calvet, J.C.; Fritz, N.; Froissard, F.; Suquia, D.; Petitpa, A.; Piguet, B. In situ soil moisture observations for the CAL/VAL of SMOS: The SMOSMANIA network. In Proceedings of the International Geoscience and Remote Sensing Symposium, Barcelona, Spain, 23–28 July 2007; pp. 1196–1199.
65. Sánchez, N.; Martinez-Fernandez, J.; Scaini, A.; Perez-Gutierrez, C. Validation of the SMOS L2 Soil Moisture Data in the REMEDHUS Network (Spain). *IEEE Trans. Geosci. Remote Sens.* **2012**, *50*, 1602–1611. [CrossRef]
66. Sánchez-Ruiz, S.; Piles, M.; Sánchez, N.; Martinez-Fernandez, J.; Vall-llossera, M.; Camps, A. Combining SMOS with visible and near/shortwave/thermal infrared satellite data for high resolution soil moisture estimates. *J. Hydrol.* **2014**, *516*, 273–283. [CrossRef]
67. Bircher, S.; Skou, N.; Jensen, K.; Walker, J.; Rasmussen, L. A soil moisture and temperature network for SMOS validation in Western Denmark. *Hydrol. Earth Syst. Sci. Discuss.* **2011**, *8*, 991–10006. [CrossRef]
68. Brocca, L.; Hasenauer, S.; Lacava, T.; Melone, F.; Moramarco, T.; Wagner, W.; Dorigo, W.; Matgen, P.; Martinez-Fernandez, J.; Llorens, P.; et al. Soil moisture estimation through ASCAT and AMSR-E sensors: An intercomparison and validation study across Europe. *Remote Sens. Environ.* **2011**, *115*, 3390 – 3408. [CrossRef]
69. Brocca, L.; Melone, F.; Moramarco, T. On the estimation of antecedent wetness condition in rainfall-runoff modelling. *Hydrol. Process.* **2008**, *22*, 629–642. [CrossRef]
70. Brocca, L.; Melone, F.; Moramarco, T.; Morbidelli, R. Antecedent wetness conditions based on ERS scatterometer data. *J. Hydrol.* **2009**, *364*, 73–87. [CrossRef]
71. Zreda, M.; Shuttleworth, W.J.; Zeng, X.; Zweck, C.; Desilets, D.; Franz, T.; Rosolem, R.; Ferre, T.P. COSMOS: The cosmic-ray soil moisture observing system. *Hydrol. Earth Syst. Sci.* **2012**, *9*, 4505–4551. [CrossRef]
72. Loew, A.; Dall'Amico, J.T.; Schlenz, F.; Mauser, W. The Upper Danube soil moisture validation site: Measurements and activities. In Proceedings of the Earth Observation and Water Cycle Conference, Frascati, Italy, 18 November 2009.
73. Albergel, C.; de Rosnay, P.; Gruhier, C.; Munoz-Sabater, J.; Hasenauer, S.; Isaksen, L.; Kerr, Y.; Wagner, W. Evaluation of remotely sensed and modelled soil moisture products using global ground-based in situ observations. *Remote Sens. Environ.* **2012**, *118*, 215–226. [CrossRef]
74. Srivastava, P.; Petropoulos, G.P.; Kerr, Y. *Satellite Soil Moisture Retrieval*, 1st ed.; Techniques and Applications; Elsevier: Amsterdam, The Netherlands, 2016; pp. 1–448.
75. Hahn, S. Product Validation Report: H-SAF H25 Metop ASCAT DR2015 SSM Time Series 12.5 km Sampling. 2016. Available online: http://hsaf.meteoam.it/documents/PVR/H25_ASCAT_SSM_CDR_PVR_v0.1.pdf (accessed on 20 August 2019).
76. Taylor, K.E. Summarizing multiple aspects of model performance in a single diagram. *J. Geophys. Res.* **2001**, *106*, doi:10.1029/2000JD900719. [CrossRef]
77. Peel, M.C.; Finlayson, B.L.; McMahon, T.A. Updated world map of the Köppen-Geiger climate classification. *Hydrol. Earth Syst. Sci.* **2007**, *11*, 1633–1644. [CrossRef]
78. Crow, W.; Berg, A.A.; Cosh, M.H.; Loew, A.; Mohanty, B.P.; Panciera, R.; De Rosnay, P.; Ryu, D.; Walker, J. Upscaling sparse ground-based soil moisture observations for the validation of coarse-resolution satellite soil moisture products. *Rev. Geophys.* **2012**, *50*, doi:10.1029/2011RG000372. [CrossRef]
79. Maton, L. Représentation et Simulation des Pratiques Culturales des Agriculteurs à L'échelle Régionale pour Estimer la Demande en eau D'irrigation. Ph.D. Thesis, Institut National Polytechnique de Toulouse, Toulouse, France, 2006.
80. Veroustraete, F.; Li, Q.; Verstraeten, W.W.; Chen, X.; Bao, A.; Dong, Q.; Liu, T.; Willems, P. Soil moisture content retrieval based on apparent thermal inertia for Xinjiang province in China. *Int. J. Remote Sens.* **2012**, *33*, 3870–3885. [CrossRef]
81. Piles, M.; Petropoulos, G.P.; Sanchez, N.; Gonzalez-Zamora, A.; Ireland, G. Towards improved spatio-temporal resolution soil moisture retrievals from the synergy of SMOS and MSG SEVIRI spaceborne observations. *Remote Sens. Environ.* **2016**, *180*, 403–417. [CrossRef]
82. Peng, J.; Niesel, J.; Loew, A. Evaluation of soil moisture downscaling using a simple thermal-based proxy—The REMEDHUS network (Spain) example. *Hydrol. Earth Syst. Sci.* **2015**, *19*, 4765–4782. [CrossRef]

83. Maltese, A.; Capodici, F.; Ciraolo, G.; Loggia, G.L. Soil Water Content Assessment: Critical Issues Concerning the Operational Application of the Triangle Method. *Remote Sens.* **2015**, *15*, 6699–6718. [CrossRef]
84. Castaldi, S.; de Grandcourt, A.; Rasile, A.; Skiba, U.; Valentini, R. CO_2, CH_4 and N_2O fluxes from soil of a burned grassland in Central Africa. *Biogeosciences* **2010**, *7*, 3459–3471. [CrossRef]
85. Legrand, M.; Plana-Fattori, A.; N'doumé, C. Satellite detection of dust using the IR imagery of Meteosat 1. Infrared difference dust index. *J. Geophys. Res.* **2001**, *106*, 18251–18274. [CrossRef]
86. Kaurila, T.; Hagard, A.; Persson, R. Aerosol extinction models based on measurements at two sites in Sweden. *Appl. Opt.* **2006**, *45*, 6750–6761. [CrossRef] [PubMed]
87. Mogili, P.K.; Yang, K.H.; Young, M.A.; Kleiber, P.D.; Grassian, V.H. Extinction spectra of mineral dust aerosol components in an environmental aerosol chamber: IR resonance studies. *Atmos. Environ.* **2008**, *42*, 1752–1761. [CrossRef]
88. Laskina, O.; Young, M.A.; Kleiber, P.D.; Grassian, V.H. Infrared extinction spectra of mineral dust aerosol: Single components and complex mixtures. *J. Geophys. Res.* **2012**, *117*. [CrossRef]
89. Ben-Ami, Y.; Koren, I.; Rudich, Y.; Artaxo, P.; Martin, S.T.; Andreae, M.O. Transport of North African dust from the Bodélé depression to the Amazon Basin: A case study. *Atmos. Chem. Phys.* **2010**, *10*, 7533–7544. [CrossRef]
90. Carrer, D.; Roujean, J.L.; Hautecoeur, O.; Elias, T. Daily estimates of aerosol optical thickness over land surface based on a directional and temporal analysis of SEVIRI MSG visible observations. *J. Geophys. Res.* **2010**, *115*. [CrossRef]
91. Carrer, D.; Ceamenos, X.; Six, B.; Roujean, J.L. AERUS-GEO: A newly available satellite-derived aerosol optical depth product over Europe and Africa. *Geophys. Res. Lett.* **2014**, *41*, 7731–7738. [CrossRef]

remote sensing

MDPI

Article

Stepwise Disaggregation of SMAP Soil Moisture at 100 m Resolution Using Landsat-7/8 Data and a Varying Intermediate Resolution

Nitu Ojha [1,*], Olivier Merlin [1], Beatriz Molero [1], Christophe Suere [1], Luis Olivera-Guerra [1], Bouchra Ait Hssaine [2,3], Abdelhakim Amazirh [2,3], Ahmad Al Bitar [1], Maria Jose Escorihuela [4] and Salah Er-Raki [2,3]

1 Centre d'Etudes Spatiales de la BIOsphère (CESBIO, Université de Toulouse (CNES/CNRS/INRA,IRD/UPS), 31013 Toulouse, France
2 Faculty of Sciences and Techniques, Cadi Ayyad University, Marrakech 40000, Morocco
3 Center for Remote Sensing Application (CRSA), University Mohammed VI Polytechnic (UM6P), Benguerir 43150, Morocco
4 isardSAT S.L, Parc Tecnologic Barcelona Activa, 08042 Barcelona, Spain
* Correspondence: ojhan@cesbio.cnes.fr or nituojha11@gmail.com

Received: 25 June 2019; Accepted: 4 August 2019; Published: 9 August 2019

Abstract: Global soil moisture (SM) products are currently available from passive microwave sensors at typically 40 km spatial resolution. Although recent efforts have been made to produce 1 km resolution data from the disaggregation of coarse scale observations, the targeted resolution of available SM data is still far from the requirements of fine-scale hydrological and agricultural studies. To fill the gap, a new disaggregation scheme of Soil Moisture Active and Passive (SMAP) data is proposed at 100 m resolution by using the disaggregation based on physical and theoretical scale change (DISPATCH) algorithm. The main objectives of this paper is (i) to implement DISPATCH algorithm at 100 m resolution using SMAP SM and Landsat land surface temperature and vegetation index data and (ii) to investigate the usefulness of an intermediate spatial resolution (ISR) between the SMAP 36 km resolution and the targeted 100 m resolution. The sequential disaggregation approach from 36 km to ISR (ranging from 1 km to 30 km) and from ISR to 100 m resolution is evaluated over 22 irrigated field crops in central Morocco using in-situ SM measurements collected from January to May 2016. The lowest root mean square difference (RMSD) between the 100 m resolution disaggregated and in-situ SM is obtained when the ISR is around 10 km. Therefore, the two-step disaggregation is more efficient than the direct disaggregation from SMAP to 100 m resolution. Moreover, we propose a moving average window algorithm to increase the accuracy in the 100 m resolution SM as well as to reduce the low-resolution boxy artifacts on disaggregated images. The correlation coefficient between 100 m resolution disaggregated and in situ SM ranges between 0.5–0.9 for four out of the six extensive sampling dates. This methodology relies solely on remote sensing data and can be easily implemented to monitor SM at a high spatial resolution over irrigated regions.

Keywords: disaggregation; soil moisture; DISPATCH; Intermediate spatial resolution; SMAP

---

1. Introduction

Knowledge of soil moisture provides key information about the coupling between the land surface and atmosphere. By controlling the partitioning of water inputs (precipitation, irrigation) into evaporation, infiltration, and runoff, the soil water content is related to the crop water consumption [1], hydrological fluxes [2], weather predictions [3] and climate projections [4].

L-band (1.4 GHz) microwave radiometry is currently the most adapted remote sensing technique for the estimation of near-surface soil moisture (SM) from space [5–10]. Microwave observations at L-band, as compared to higher microwave frequencies, are more sensitive to SM and less sensitive to the soil surface roughness and vegetation optical depth [11]. In this context, European Space Agency (ESA) and National Aeronautics and Space Administration (NASA) have launched the SMOS [6,12] and SMAP [5] satellites in 2009 and 2014, respectively. Both satellites embark an L-band radiometer to retrieve the 3–5 cm SM with a repeat cycle of less than 3 days globally. The spatial resolution of both radiometers is approximately 40 km [5,6].

Despite the high radiometric accuracy achieved by L-band radiometers, the data provided globally have a low spatial resolution, which makes the validation of remote sensing products difficult and limits their application to large scale studies only [13]. For hydro-agriculture purposes, there is a crucial need for SM data at a higher spatial resolution [14,15]. Consequently, disaggregation techniques have been proposed to improve the spatial resolution of the SM data available at a high temporal frequency [16–22]. Existing downscaling methods for SM can be classified into three major groups (1) satellite-based methods; (2) methods using Geo-information data and (3) model-based methods [23,24]. The satellite methods combine the use of radar and optical data to coarse scale microwave radiometry [24]. Among optical-based methods, many studies have used as fine-scale information the fractional vegetation cover and land surface temperature (LST) derived from high-resolution (1 km to 100 m) optical/thermal sensors. The general idea of these methods is to relate LST to SM via the evapotranspiration process [20,25].

Relying on this principle, the disaggregation based on physical and theoretical scale change (DISPATCH) method [26,27] estimates the soil evaporative efficiency (SEE defined as the ratio of actual to potential soil evaporation), and implements a downscaling relationship that links the disaggregated SM to the low-resolution (LR) observation and the high-resolution (HR) SEE. The optical-derived SEE is expressed as a linear function of the retrieved soil temperature [28] and the minimum and maximum soil temperatures observed at HR within the LR pixel [26], according to the so-called contextual approach [29]. Based on the DISPATCH algorithm, the CATDS Level-4 Disaggregation (C4DIS) processor [30] was implemented at the Centre Aval de Traitement des Donnees SMOS (CATDS) as a level 4 product. C4DIS produces 1 km resolution SM data at the quasi-global scale from SMOS level 3 and Moderate resolution Imaging Spectroradiometer (MODIS) data.

The 1 km resolution SM disaggregated from SMOS produts are currently used in a range of disciplines including root-zone soil moisture monitoring [31], detecting irrigated areas at the perimeter scale [32,33], retrieving soil properties from space [34], preventing the spread of desert locust swarms [35], evapotranspiration monitoring over rainfed areas [36], flood forecasting over large basins [37], estimating crop yield [38], and the methods to produce them are continuously evolving and maturing (Merlin et al. 2017). Note that few studies have applied the DISPATCH method to SMAP SM using MODIS data [39]. However, the 1 km resolution is often insufficient for many other fine-scale applications and areas where the surface is highly heterogeneous (e.g., [1]). SM data at the sub-kilometric (typically hectometric) resolution are especially required in agriculture for early crop detection, irrigation scheduling, water stress and yield monitoring [40] and in fine-scale hydrological studies for flood risk prevention, drought monitoring and groundwater level assessment [41], among other potential applications.

In fact, there is still no routine application of DISPATCH to Landsat data, which yet would be useful to increase the spatial resolution of available SM products up to 100 m [27,42]. One difficulty is that Landsat data do not provide global coverage at the daily scale (like MODIS data) so that a sequential approach is needed to "delineate" the 1 km resolution disaggregated SMAP data over each Landsat scene separately before DISPATCH can be implemented at 100 m resolution. Another difficulty is the contextual nature of DISPATCH, which relies on the extreme wet and dry conditions present within the LR pixel to calibrate the SEE model. Especially, the accuracy in temperature endmembers is expected to vary with the spatial extent over which DISPATCH is applied. A third difficulty is the presence of boxy artifacts visible at LR when combining multi-source/multi-resolution remote sensing data within a disaggregation methodology. Boxy artifacts are common problems with downscaling methods [26,27,43].

In this context, this paper presents a new methodology to disaggregate optimally the 36 km resolution SMAP SM to 100 m resolution using Landsat data. The main objective is to assess the usefulness of an optimal intermediate spatial resolution (ISR) between the SMAP and Landsat resolutions. In practice, the 1 km resolution SM disaggregated from SMAP data using MODIS data (similar to C4DIS product for SMAP) is aggregated at ISR ranging from 1 km to 30 km, and DISPATCH is applied to ISR SM. The novelty of this paper thus lies in: (1) the application of DISPATCH to SMAP data at 100 m resolution, (2) the use of an ISR between SMAP and Landsat resolutions and (3) the removal of boxy artifacts on the 100 m resolution disaggregated images using a new technique.

Herein, the stepwise disaggregation approach is tested over an experimental area in central Morocco, comprised of 22 irrigated field crops over which the 0–5 cm SM has been monitored during the 2015–2016 season.

2. Materials and Methods

2.1. Study Area

The study focuses on a 30 km by 30 km area of the R3 irrigated zone (31.70N, 7.35W) located 40 km east of Marrakesh city in the Haouz plain, central Morocco (see Figure 1). To assess the performance of DISPATCH at the 100 m resolution, a set of 22 irrigated wheat fields, covering 3–4 ha each (Figure 1), were selected within a 1 km resolution MODIS pixel. SM sampling was undertaken on clear sky dates with almost simultaneous SMAP/MODIS/Landsat data over the 22 crop fields and repeated measurements were made along the agricultural season to cover all the phenological stages of wheat. Climate of the study area is mainly semi-arid with an annual average precipitation of 250 mm [44,45]. The soil texture in the R3 perimeter is mainly clayey. Flood irrigation is the most widely used method in this district. Wheat is generally sown in November-December and a mean total of 6 irrigations is applied to wheat crops from February till April, typically every 3 weeks. Harvesting is done in late May or early June [46,47].

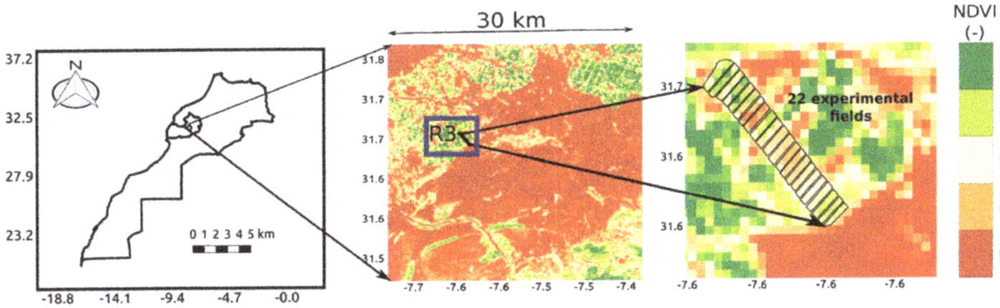

Figure 1. Location of Marrakesh, the Haouz plain in central Morocco and the 22 experimental crop fields in the R3 perimeter.

2.2. In-Situ Data

The 0–5 cm SM was measured manually from January to May months during the 2015–2016 agricultural season. The sampling strategy was to use theta probes to collect 10 distinct measurements for each of the 22 crop fields [48]. In practice, as the 22 parcels were chosen to be aligned and contiguous, two transects along both sides of the crop fields were walked and 5 theta probes measurements were taken on each crop side, at least 5 m from the field border. Theta probe measurements were calibrated using the gravimetric method, based on soil samples collected on each sampling date. In this study, the field-scale SM corresponds to the mean 0–5 cm calibrated theta probe measurements as in Amazirh et al. [49]. Among the 7 available sampling dates, only 6 are used herein in order to satisfy the following criteria: SM measurements are available on a clear-sky Landsat overpass date, and the time difference between SMAP and Landsat overpasses is one day at maximum. The clear-sky dates with quasi-concurrent in-situ sampling and the satellite overpasses of SMAP, SMOS, MODIS and Landsat were on DOY 6, 14, 30, 38, 62 and 78. The remaining sampling dates are suitable for our study and help us identify the variability in SM along the growth of crops.

2.3. Remote Sensing Data

2.3.1. SMAP

SMAP was launched by NASA on 31 January 2015. SMAP is the first L-band mission combining both radar (active) and radiometer (passive) data to provide SM at a range of resolutions from 3 km (active) to 36 km (passive) with a revisit cycle of 2–3 days. But due to the failure of the SMAP radar, the SM produced from SMAP is currently provided on a ∼36 km and ∼9 km (by using a re-sampling technique) resolution grid. Note that a product combining SMAP and C-Band Sentinel-1 data has been recently provided by the mission [50]. SMAP has a near-polar sun-synchronous orbit at an altitude of 658 km with 6:00 a.m./p.m. local time descending/ascending overpass. SMAP works on multi-polarization with a fixed incidence angle at 40 degrees and a swath of ∼1000 km [5]. In this paper, the SMAP level-3 (product name SPL3SMP A/D, version 005, Colliander et al. [51]) is used. The product is provided in HDF format on the version 2 cylindrical EASE grid at 36 km resolution. Data can be downloaded from https://nsidc.org/data/SPL3SMP/versions/5.

2.3.2. MODIS

MODIS LST and NDVI data are used by the C4DIS processor [30] to provide the 1 km resolution SM from the disaggregation of SMAP level-3 SM data. LST is extracted from version 5 MOD11A1, Terra overpass (10:30 a.m.) on ascending node and MYD11A, Aqua overpass (1 p.m.) on descending node. For each (ascending or descending) SMAP overpass, there are 6 MODIS LST products taken as input to C4DIS (one day before, same day and one day after SMAP overpass for both Aqua and Terra platforms). NDVI is extracted from version 5 MOD13, only for Terra overpass with an interval of 16 days [19,52].

2.3.3. Landsat

Landsat-7 and Landsat-8 were launched by NASA in April 1999 and February 2013, respectively. The images were downloaded from the USGS website, which provides surface reflectance and thermal radiances data in different spectral bands. The revisit time of each sensor is 16 days and there is an 8-day lag between Landsat-7 and Landsat-8 so that the Landsat constellation potentially provides (in cloud-free conditions) optical/thermal data every 8 days globally. The Landsat-7/8 30 m resolution reflectance data are aggregated at 100 m resolution and used to derive the fractional vegetation cover. The Landsat NDVI is calculated as the ratio of the re-sampled near-infrared reflectance to re-sampled red reflectance difference divided by their sum, and the fractional vegetation (f_v) is estimated as:

$$f_v = \frac{NDVI_{HR} - NDVI_s}{NDVI_v - NDVI_s} \tag{1}$$

where, $NDVI_{HR}$ represents the NDVI at high (100 m) resolution, $NDVI_s$ the NDVI at bare soil and $NDVI_v$ the NDVI at full cover vegetation. For this study, $NDVI_s$ and $NDVI_v$ are set to 0.1 and 0.9, respectively. Landsat-7 and Landsat-8 provide thermal infrared (TIR) data with a spatial resolution of 60 m and 100 m, respectively. LST is derived by using the single channel (SC) algorithm [53] from Landsat-7 band-6 and Landsat-8 band-7 as:

$$LST = \gamma \left[\frac{1}{\varepsilon} \left(\varphi_1 . L_{sen} + \varphi_2 \right) + \varphi_3 \right] + \delta \tag{2}$$

where, ε is the surface emissivity, (γ, δ) are parameters depending on the radiance and brightness temperature of the Landsat thermal band and $\varphi_1, \varphi_2, \varphi_3$ are atmospheric variables function of the atmospheric water vapor content (ω) and derived from radiative transfer simulations using the GAPRI database [54]. The ω variable is obtained from the MODIS product MOD05.

2.3.4. SRTM

SRTM (Shuttle Radar Topography Mission) 1 arc second global data are used to correct Landsat LST for topographic effects [27]. Although the study area is rather flat, the topographic correction is applied by default in DISPATCH. The 30 m resolution SRTM data are aggregated to 100 m resolution, consistent with the Landsat LST resolution.

2.4. DISPATCH

2.4.1. General Equations

The main equations of the DISPATCH method implemented at both 1 km and 100 m resolutions are reminded in this subsection. The SM downscaled at HR (refers to either 1 km or 100 m resolution) is written as:

$$SM_{HR} = SM_{LR} + \left(\frac{\delta SEE}{\delta SM}\right)^{-1}_{LR} * \left(SEE_{HR} - SEE_{LR}\right) \quad (3)$$

where, SM_{HR} represents the disaggregated SM at HR, SM_{LR} the SM at LR (refers to either SMAP or ISR resolution) derived from SMAP data or from their disaggregation to 1 km resolution, SEE_{HR} the SEE at HR derived from MODIS or Landsat, SEE_{LR} the HR SEE aggregated at LR, and $\left(\frac{\delta SEE}{\delta SM}\right)^{-1}_{LR}$ the inverse of the partial derivative of the SEE(SM) model evaluated at LR. SEE is assumed to follow a linear relationship with the soil temperature [28] and is thus expressed as:

$$SEE_{HR} = \frac{T_{s,dry} - T_{s,HR}}{T_{s,dry} - T_{s,wet}} \quad (4)$$

where T_s is the soil surface temperature, $T_{s,dry}$ and $T_{s,wet}$ the soil temperature in fully dry (SEE = 0) and water-saturated (SEE = 1) conditions, respectively. Temperature endmembers $T_{s,dry}$ and $T_{s,wet}$ are calculated from a graph between LST and f_v derived at HR from MODIS or Landsat data. The soil temperature is derived from a linear decomposition of LST into soil and vegetation temperature. The trapezoidal method [26,55] is used to estimate the vegetation temperature, and the soil temperature is expressed as the residual term:

$$T_{s,HR} = \frac{T_{HR} - f_{v,HR} * T_{v,HR}}{1 - f_{v,HR}} \quad (5)$$

where T_{HR} represents the LST at HR, $T_{v,HR}$ the vegetation temperature at HR and $f_{v,HR}$ the fractional vegetation cover at HR.

The downscaling relationship of Equation (3) is hence based on two SEE models: SEE as a function of SM to estimate the first derivative at LR, and SEE as a function of LST (expressed in Equations (4) and (5)) to estimate the spatial variability of SM at HR.

2.4.2. DISPATCH at 1 km Resolution

Note that C4DIS is labeled as DISPATCH$_{Lin}$ (for linear SEE model) in this study to distinguish the methodologies applied at 1 km and 100 m resolution. The current version of C4DIS/DISPATCH$_{Lin}$

is based on the various studies that have been done in the past using 1 km resolution MODIS data (Merlin et al. 2013, 2012b, 2010, 2009, 2008). In DISPATCH$_{Lin}$ algorithm, the SEE(SM) model is linear:

$$SEE = \frac{SM}{SM_p} \tag{6}$$

where, SM_p is a soil moisture parameter (in soil moisture unit), which depends on soil properties and atmospheric conditions. It is calibrated at the SMAP pixel scale at the satellite overpass time using LR SEE and SM estimates ($SM_p = \frac{SM_{LR}}{SEE_{LR}}$). In this case, the derivative in Equation (3) is simply SM_p. The SEE(LST) model implemented in DISPATCH$_{Lin}$ is based on Equation (4) using the temperature endmembers calculated by the simplest extrapolation method within the LST-f_v feature space: $T_{s,dry}$ and $T_{s,wet}$ are set to the maximum and minimum soil temperature within a given LR pixel.

This approach is implemented for C4DIS/DISPATCH$_{Lin}$ and has provided favorable results at 1 km resolution for arid and semi-arid areas [26,30].

2.4.3. DISPATCH at 100 m Resolution

When applying the DISPATCH methodology at 100 m resolution over extremely heterogeneous areas like irrigated perimeters, one expects two main differences with the 1 km case. First, the range of LST values should increase at 100 m resolution, thereby enabling a more accurate definition of temperature endmembers, if the effect of outliers can be removed [56]. The second difference is that the full SM range (from the residual SM to the SM at saturation) is likely to be present within each LR pixel. Such extreme heterogeneity requires a robust representation of the SEE(SM) relationship over the full SM range. Especially, given that the SEE(SM) is known to be nonlinear [28,57,58], the linear approximation made in the 1 km case (Equation (6) is no more valid. Both differences between the 100 m and 1 km case involve two changes in the disaggregation algorithm: (i) the SEE model in Komatsu [57] replaces the linear SEE model, and (ii) the method in Tang et al. [56] is used to robustly determine the wet and dry edges. For clarity, the implementation of DISPATCH at 100 m resolution is labeled DISPATCH$_{Exp}$ (for nonlinear SEE model).

In DISPATCH$_{Exp}$, the SEE(SM) model is expressed as [57]:

$$SEE = 1 - exp\left(\frac{SM}{SM_p}\right) \tag{7}$$

where SM_p is calculated from LR SM (the 1 km resolution disaggregated SM aggregated at ISR) and HR (Landsat-derived) SEE aggregated at ISR:

$$SM_p = \frac{SM_{LR}}{-ln\left(1 - SEE_{LR}\right)} \tag{8}$$

Note that the derivative in Equation (3) of the SEE(SM) model of Equation (7) can be computed in two different ways, as a function of LR SM:

$$\left(\frac{\delta SEE}{\partial SM}\right)_{LR}^{-1} = SM_p * exp\left(\frac{SM_{LR}}{SM_p}\right) \tag{9}$$

or as a function of LR SEE:

$$\left(\frac{\delta SEE}{\delta SM}\right)_{LR}^{-1} = \frac{SM_p}{1 - SEE_{LR}} \tag{10}$$

As both expressions of the derivative are valid, the average of both estimates is implemented in DISPATCH$_{Exp}$ in order to stabilize the slope estimation with respect to uncertainties in both LR SM and SEE.

Regarding the SEE(LST) model, the algorithm in Tang et al. [56] automatically calculates the temperature endmembers by removing outliers. It processes pixels in an iterative manner to calculate the highest temperature for each f_v interval. A linear approximation of highest temperatures is used to estimate the dry edge. In Tang et al. [56], the wet edge is assumed to be parallel to x-axis with constant surface temperature. Herein, a slight modification is done to estimate the wet edge similar to the dry edge (in that case, the wet edge temperature is not kept as a constant) by removing outliers. This process thus removes specious dry and wet points before determining the dry and wet edges and their corresponding temperature endmembers.

Figure 2 gives an illustration of the calculation of wet and dry edges from LST-f_v graph by using two different algorithms. For DISPATCH$_{Lin}$, the temperature endmembers ($T_{s,dry}$ and $T_{s,wet}$) are the minimum and maximum temperature within a given SMAP pixel. Dry and wet edges are calculated independently for every SMAP pixel in the image. It can be seen that if we apply the same algorithm for the calculation of temperature endmembers at 100 m resolution, the temperature endmember calculation may overestimate the dry edge and underestimate the wet edge. When removing the outliers from the temperature endmember calculation [56], the dry and wet edges follow more closely the contour of data points, and the estimated temperature endmembers are supposedly more accurate.

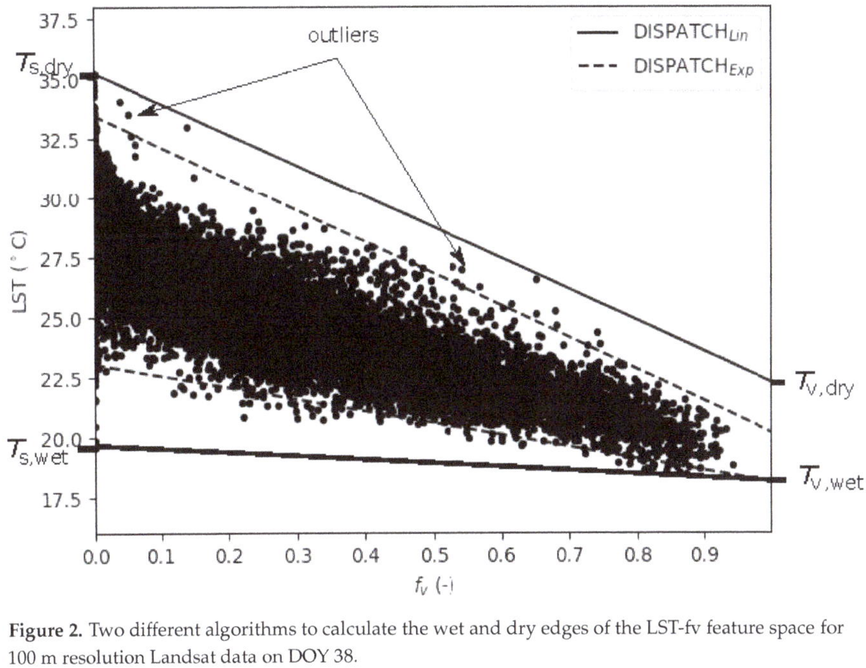

Figure 2. Two different algorithms to calculate the wet and dry edges of the LST-fv feature space for 100 m resolution Landsat data on DOY 38.

2.5. Sequential Downscaling

Figure 3 presents a flow chart of the sequential disaggregation in 3 successive steps. SMAP SM is first disaggregated from 36 km to 1 km resolution using MODIS data and DISPATCH$_{Lin}$ algorithm. Then the 1 km resolution SM is aggregated at ISR. Next, the ISR SM is further disaggregated at 100 m resolution using Landsat data and DISPATCH$_{Exp}$ algorithm.

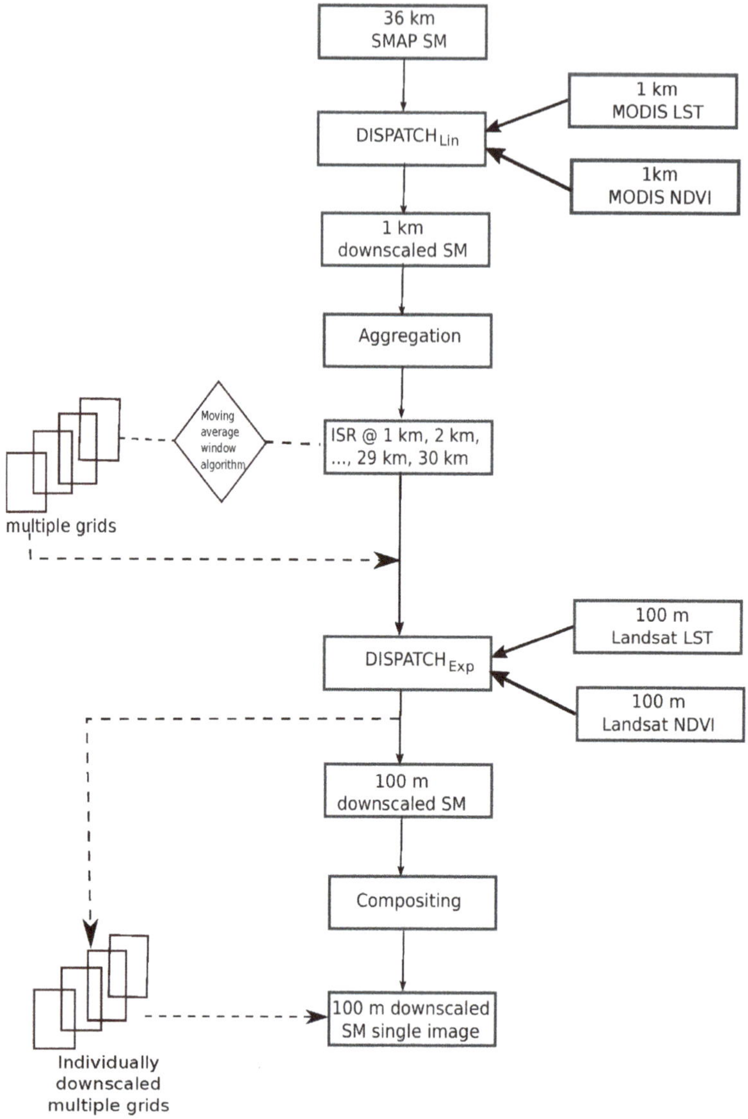

Figure 3. Flowchart of the stepwise sequential downscaling approach from 36 km resolution to 1 km (DISPATCH$_{Lin}$), from 1 km to ISR (aggregation to variable ISR), and from ISR to 100 m (DISPATCH$_{Exp}$).

The reason for the selection of a range of ISRs is associated with the contextual nature of DISPATCH, which makes the determination of temperature endmembers i.e., $T_{s,dry}$ and $T_{s,wet}$, and hence the 100 m resolution Landsat-derived SEE, dependent on the spatial extent [59]. In particular, the larger the spatial extent, the more heterogeneous the surface becomes. Therefore, the accuracy in temperature endmembers should increase with the spatial extent, as long as the meteorological forcing data remain relatively uniform (underlying assumption of the contextual analysis). Therefore, an optimal ISR in terms of SM accuracy at the 100 m resolution appears to be a compromise between (i) the accuracy in temperature endmembers and (ii) the gap between LR and HR. One major objective of this paper is to check the sensitivity of the sequential approach to ISR and to propose an optimal ISR for routine application.

Figure 4 plots the LST-f_v feature space obtained for 100 m resolution Landsat data collected on DOY 38 and for 4 distinct spatial extents within the study area: 1 km, 3 km, 10 km, and 30 km. It can be seen that the range of LST and f_v values increases with the spatial extent over which the temperature endmembers are estimated. The mean LST is also larger for the smaller ISR values; which may induce bias in the disaggregation. If we consider that the temperature endmembers retrieved from the 10 km pixel resemble the real dry and wet soil temperatures, then more uncertainty can be seen while decreasing the spatial extent. An inaccurate representation of the wet/dry and bare/vegetated soil conditions within the spatial extent (ISR pixel) will directly affect the calculation of temperature endmembers and hence the thermal-derived SEE, and finally downscaling. Conversely, when extending (too much) the spatial extent, the spatial variability of air temperature (and wind speed notably) may reach a critical level that invalidates the contextual approach's assumption regarding the uniformity of meteorological forcing. In this respect, the LST-f_v feature space plotted in Figure 4 for a spatial extent of 30 km indicates that two trapezoidal shapes appear separately, which may be a signature of sub-areas having different meteorological forcing. Both constraints (heterogeneity of surface conditions and homogeneity of meteorological conditions) actually represent an important rationale for implementing a sequential method with an ISR between the 1 km and SMAP resolution.

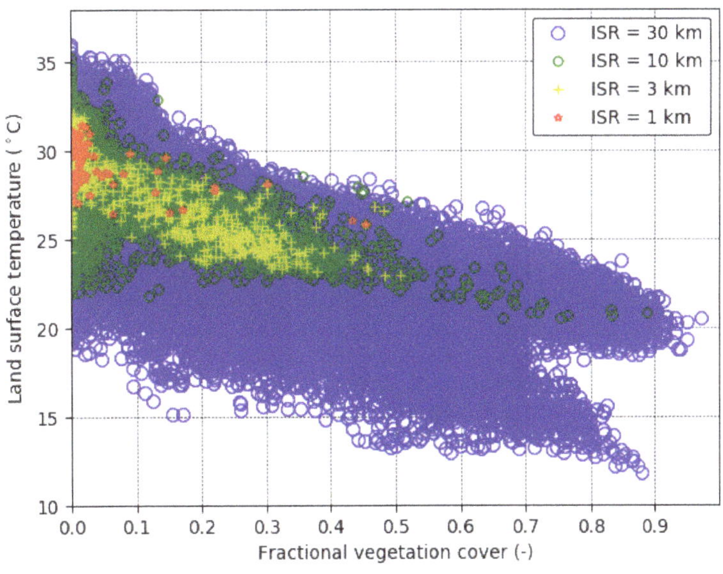

Figure 4. The LST-f_v feature space is plotted for 100 m resolution pixels within a 30 km, 10 km, 3 km and 1 km ISR pixel.

Once the spatial extent of the LST-f_v feature space and the algorithm for estimating temperature endmembers have been defined, one can check the linearity or non-linearity of the SEE(SM) model and its consistency with the SEE(LST) model. Figure 5 plots Landsat-derived SEE as a function of in-situ SM for DOY 14. The predictions of the SEE(SM) model of Equation (7) is also plotted, with SM_p estimated by setting in Equation (8) LR SM and LR SEE to the mean in situ SM and the mean Landsat-derived SEE, respectively. The correlation coefficient and slope between SEE(SM) and SEE(LST) estimates are 0.87 and 0.41 for DISPATCH$_{Exp}$ algorithm. For comparison purposes, Figure 5 also plots SEE(SM) and SEE(LST) predictions for the DISPATCH$_{Lin}$ algorithm. The correlation coefficient and slope between both models are 0.87 and 0.17 respectively. Even though the correlation coefficient shows similar values for both DISPATCH$_{Lin}$ and DISPATCH$_{Exp}$, the slope is significantly lower than that for DISPATCH$_{Exp}$. Those results are fully consistent with our approach to make the SEE(SM) non-linear (using the exponential form of [57]) and to improve the temperature endmembers algorithm [56] of the SEE(LST) model within the new DISPATCH$_{Exp}$ downscaling algorithm.

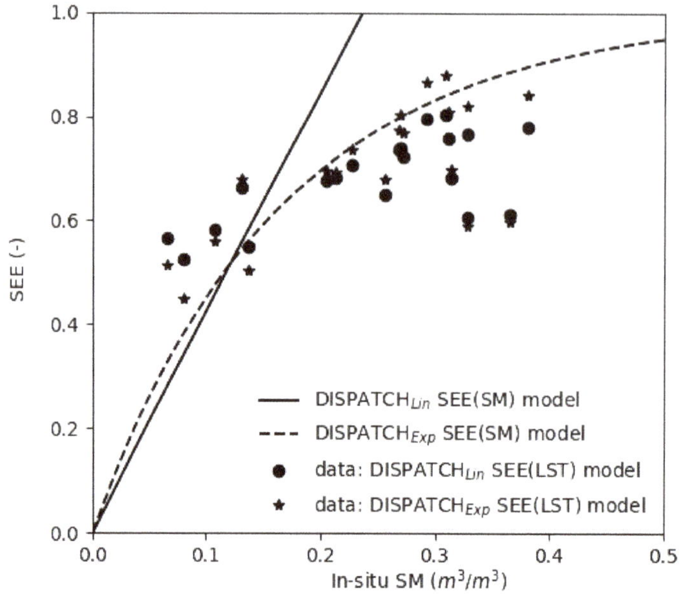

Figure 5. Landsat-derived SEE as a function of in situ SM superimposed with the SEE(SM) model for both DISPATCH$_{Lin}$ and DISPATCH$_{Exp}$ algorithms, for data on DOY 14.

2.6. Inclusion of Multiple ISR Grids

The use of multiple LR grids as input to disaggregation approaches has been proposed in Hoehn et al. [60] and Merlin et al. [26]. Hoehn et al. [60] compared downscaling results obtained from single coarse resolution grid (using fixed window) and using multiple overlapping coarse resolution grids (by shifting windows). Shifting windows using multiple grids showed better performance as compared to the fixed window case with respect to error and smoothness.

In this paper, we propose to define multiple ISR grids as an input to DISPATCH$_{Exp}$ as in Merlin et al. [26] and Hoehn et al. [60]. However, one difference herein is that the multiple ISR grids are built from actual observations at the (higher) 1 km resolution and consequently, they are derived neither from the LR overlapped observations [26], nor from the oversampling of LR (SMAP) observations [60].

Figure 6 gives an illustration of the moving average window algorithm, over the 1 km resolution grid of the DISPATCH$_{Lin}$ output data. The algorithm successively shifts an ISR grid in both directions (east-west and north-south) of a predefined constant distance. In the diagram of Figure 6, ISR is set to 10 km and the distance separating the so generated ISR grids is set to 2 km. Once the multiple grids have been generated, they are independently used as input to DISPATCH$_{Exp}$. As a first step, multiple ISR SM grids are thus overlapped with 100 m resolution Landsat data and disaggregated separately to get multiple 100 m resolution disaggregated SM images. As a second step, the separate

100 m resolution downscaled SM images are composited (simple average) to produce a single 100 m resolution SM disaggregated image.

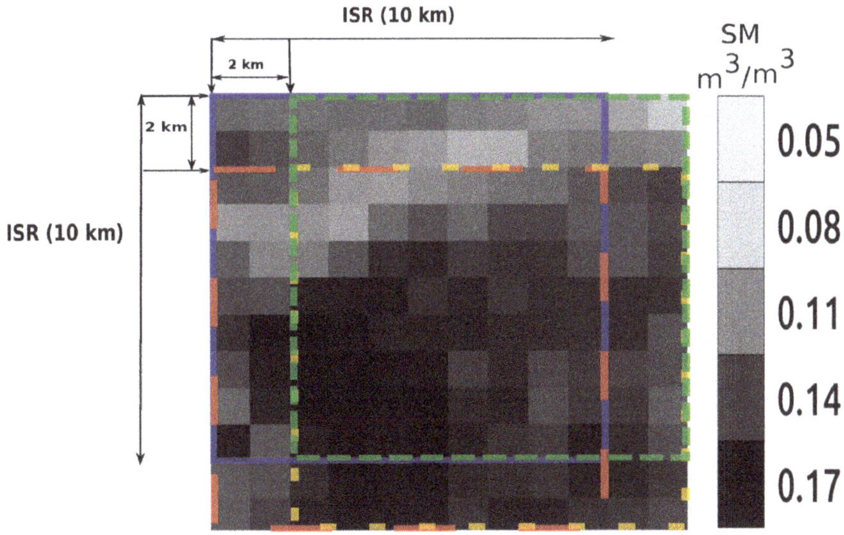

Figure 6. Illustration of the moving average window algorithm applied to the 1 km resolution $DISPATCH_{Lin}$ SM with a shift of ISR (set to 10 km in this case) grids in both directions (east-west and north-south) with a constant spacing of 2 km between ISR grids.

3. Results

This section analyzes the potential of the sequential downscaling approach by investigating (1) the method calibration (2) the method accuracy for a range of ISR values using the single grid algorithm, and (3) the usefulness of the multiple grid (compared to the single grid) algorithm.

3.1. Calibration

The calibration of SM_p parameter in Equation (8) is undertaken using the $DISPATCH_{Lin}$ data sets derived from SMAP, on each date when Landsat data are available. Figure 7 plots SM_p as a function of ISR for each Landsat overpass date. The mean and standard deviation of retrieved SM_p are computed within the 30 km by 30 km study domain. It can be seen that for all dates, the retrieved SM_p behaves quite similarly with respect to ISR. It sharply increases for an ISR increasing from 1 km to 3–4 km and then keeps a relatively stable value for ISR values ranging between 3–4 km and 30 km. Note that significant fluctuations of SM_p are observed for ISR values larger than 15 km, due to the bounded extent of the study area i.e., the mean SM_p is computed using a single retrieved value. However, the SM_p value after convergence is not fully consistent for different dates. In fact, the estimation of SM_p in Equation (8) mainly depends on LR SM and SEE data, so that any error in SMAP SM and Landsat-derived SEE estimates leads to temporal variabilities in retrieved SM_p.

The standard deviation of the retrieved SM_p values within the 30 km by 30 km area is also plotted as a function of ISR ranging from 1 km to 30 km. It can be seen that the spatial variability in retrieved SM_p significantly decreases in the higher ISR range to reach a minimum for ISR values larger than 10 km. Note that the standard deviation becomes zero for ISR equal or larger than 15 km (not shown in the graph) because in such cases, a single ISR pixel is obtained within the whole extent of the (30 km wide) study area. Figure 8 presents the images of SM_p retrieved for ISR equal to 1 km, 3 km, 10 km, and 30 km. The spatial variability of SM_p strongly increases when ISR decreases and tends to 1 km and the average of 1 km resolution SM_p is significantly different from the SM_p

retrieved over 10 km ISR pixels. Such behavior is explained by the non-linear impact of LR SEE on SM_p (see Equation (8), and by the non-representativeness of temperature endmembers for ISR lower than 5 km.

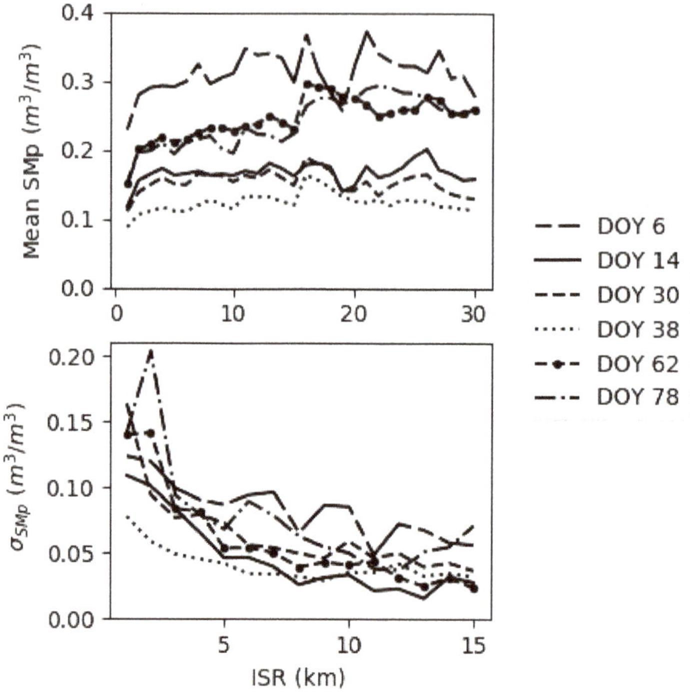

Figure 7. Mean (**top**) and standard deviation (**bottom**) of the parameter SM_p (Equation (8) plotted as function of ISR ranging from 1 to 30 km for each Landsat overpass date separately.

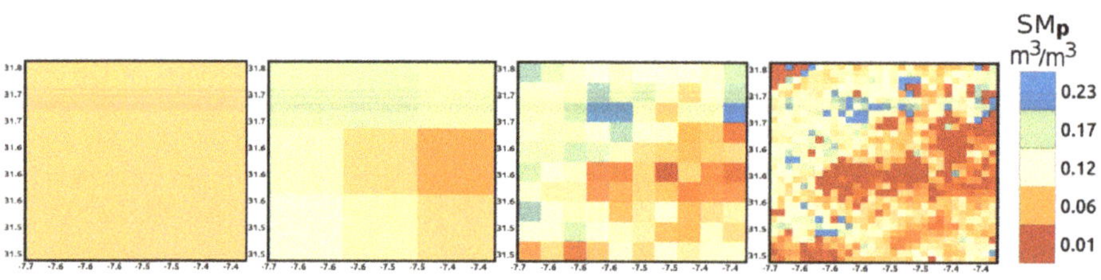

Figure 8. SM_p parameter images derived from SMAP data on DOY 38 for ISR equal to 30 km, 10 km, 3 km and 1 km from left to right, respectively

From the results presented in Figures 7 and 8, it can be concluded that (1) the retrieved SM_p is spatially and temporally representative for ISR equal to or larger than 10 km, and (2) significant spatial/temporal variabilities of SM_p (associated with uncertainties in temperature endmembers) and non linear effects (associated with the non linear SEE(SM) relationship) are obtained for ISR lower than 5 km.

3.2. Evaluation of 100 m Disaggregated SM

The SM_p retrieved from Equation (8) is first used to calculate the derivative from the average of Equations (9) and (10). A range of different ISR values is then chosen to evaluate the sensitivity of

100 m resolution disaggregated SM to ISR. To do so, the 1 km resolution disaggregated SM (the output of DISPATCH$_{Lin}$) is aggregated to 1, 2, 3, ..., and 30 km and in each case, the aggregated ISR SM together with its associated spatial extent is used as an input to DISPATCH$_{Exp}$. Such a sensitivity analysis is undertaken for each SMAP overpass date, separately.

The statistical comparison in terms of correlation coefficient (R), slope of the linear regression (slope), root mean square deviation (RMSD) and absolute mean bias (MB) between DISPATCH$_{Exp}$ disaggregated SM and in-situ SM is illustrated in Figure 9 for ISR ranging from 1 to 30 km for each sampling date. The temporal variability (standard deviation) of R and of the slope of the linear regression is relatively large in the lower range of ISR values. The slope gets even negative values for ISR lower than 5 km on several dates, while the slope is always positive for ISR larger than 10 km. This result is consistent with the stability of SM_p retrievals observed previously for ISR larger than 5 km.

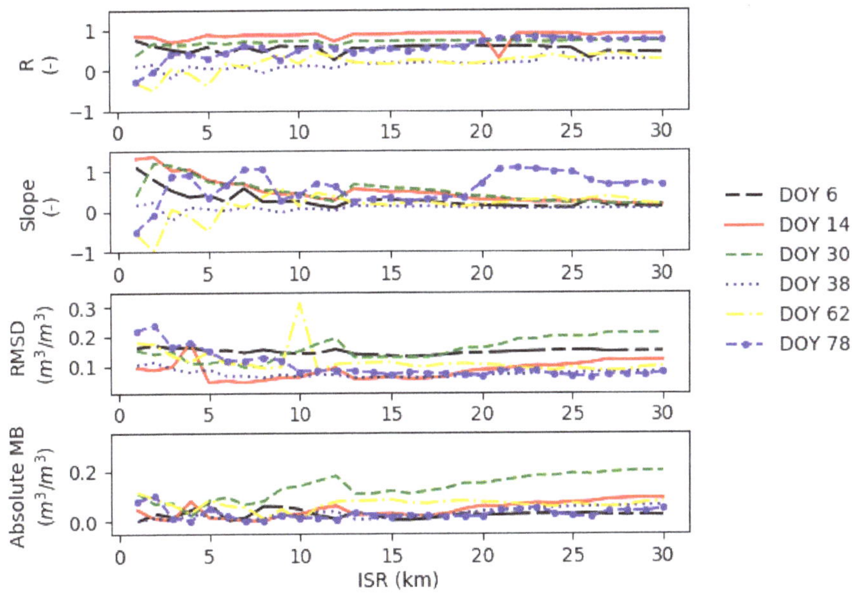

Figure 9. Correlation coefficient (R), slope of the linear regression, absolute mean bias (MB) and root mean square difference (RMSD) between 100 m resolution disaggregated and in situ SM for a range of ISR values (from 1 km to 30 km) for each sampling date separately.

When considering the full ISR range (1–30 km), and despite the date-to-date variability, a slight general increase of R is obtained in the 1–10 km range, whereas it keeps an approximately constant value for larger ISRs. Regarding the slope of the linear regression, an opposite finding is obtained. For ISR values larger than 5 km, the slope keeps decreasing with a value at ISR = 30 km mostly very close to zero. Note that the sudden increase of the slope for ISR = 20 km on DOY 78 is due to the fact that statistical results are obtained from a single (unrepresentative) ISR pixel that fits into the 30 km by 30 km study area. The decrease of the slope is attributed to the gap between the LR and the HR, which increases with ISR. In fact, the disaggregation efficiency (as defined in [61]) is expected to decrease with the LR to HR ratio, due to the decrease of the spatial variability represented at HR by the LR observation. The slope of the linear regression was actually found to be a good indicator of the disaggregation efficiency [61], consistent with the results presented in Figure 9. A second important impact of ISR is the increase of the absolute MB between the 100 m resolution disaggregated and in situ SM, especially in the 10–30 km range. The worsening of downscaling performances (in terms of the slope of the linear regression and MB) in the 10–30 km range is due to the linear approximation of the downscaling relationship (Equation (3). An optimal ISR is thus found at around 10 km. Optimal results in terms of RMSD between 100 m resolution and in-situ SM is actually obtained for ISR close to 10 km. Therefore, ISR set to 10 km throughout the rest of the paper.

For illustration purposes, Figure 10 represents the sequential downscaling of SM from SMAP data collected on DOY 38: the disaggregation of SMAP SM to 1 km resolution, the aggregation of the 1 km resolution disaggregated SM to ISR (10 km), and the disaggregation of ISR SM to 100 m resolution. It is reminded that the extra aggregation step is undertaken (i) to increase the representativeness/accuracy of the temperature endmembers extrapolated from the LST-f_v feature space, (ii) to increase the stability of the disaggregation calibration (via the SM_p retrieval) and (iii) to reduce random uncertainties in the ISR SM used as input to DISPATCH$_{Exp}$.

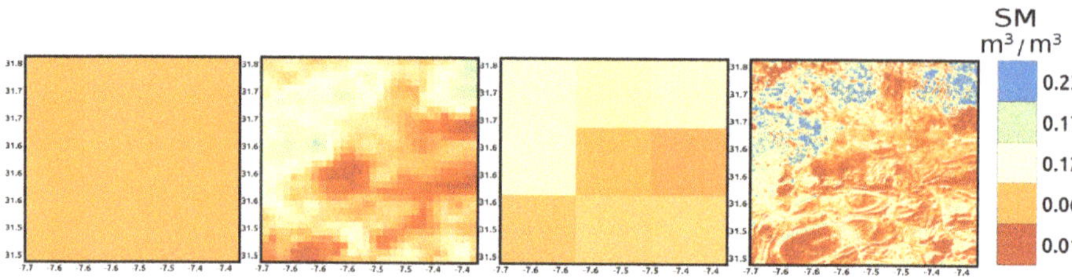

Figure 10. From left to right: images of 36 km resolution SMAP SM, 1 km resolution DISPATCH$_{Exp}$ SM, 10 km resolution aggregated DISPATCH$_{Exp}$ SM and 100 m resolution disaggregated SM on DOY 38.

As a first evaluation of the disaggregation at 100 m resolution independently from the uncertainty in SMAP data, DISPATCH$_{Exp}$ is run for all sampling dates (DOY 6, 14, 30, 38, 62 and 78) by setting the ISR observation to a fraction of the mean (daily areal average of) in-situ SM. By considering that the average of all in-situ SM measurements is representative of the SM over the irrigated area and that the SM over dry land is about 0, a rough estimate of the LR SM is derived as half the mean in situ measurements (the fraction of dry land in the 10 km ISR pixel covering the experimental fields is about 50%). Figure 11 plots the 100 m resolution disaggregated SM versus in situ measurements. Statistical results in terms of R, slope of the linear regression, absolute MB and RMSD are reported in Table 1 for synthetic LR. R is in the range 0.6–0.9 for four dates (DOY 6, 14, 30 and 78), while it is in the range 0.1–0.2 on two dates (DOY 38 and 62). In terms of correlation, better results are obtained on the sampling dates with a larger spatial variability in SM measurements, and reciprocally, poorer results are obtained when SM is relatively uniform at the sub-pixel scale. In terms of bias, however, relatively low absolute MB (lower than 0.03 m^3/m^3) is obtained except for DOY 6, 62 and 78 with an absolute MB of 0.07, 0.08 and 0.11 m^3/m^3, respectively. The reason is that the mean in situ measurements (weighted by the fraction of irrigated land) may not be fully representative of the real SM at the ISR (10 km) scale, as irrigation is not applied uniformly within the irrigated perimeter. Nevertheless, the application of DISPATCH$_{Exp}$ to synthetic LR (ISR) SM data allows for assessing the performance of the downscaling methodology independently of SMAP data and DISPATCH$_{Lin}$ algorithm. We conclude that DISPATCH$_{Exp}$ is relatively efficient when the sub-pixel variability is larger than 0.06 m^3/m^3.

Next, DISPATCH$_{Exp}$ is tested using SMAP data (ISR is still set to 10 km in the sequential downscaling). Figure 12 represents the comparison between DISPATCH$_{Exp}$ and in situ SM and Table 1 for SMAP single grid reports the associated statistical results. It can be seen that results are not significantly degraded in terms of R compared to the case when using synthetic LR observation as input to DISPATCH$_{Exp}$ (see Table 1). In fact, the sub-pixel variability of SM is represented by the Landsat-derived SEE in both real and synthetic cases, which explains similar R results. However, the DISPATCH$_{Lin}$ data derived from SMAP data may involve LR differences in terms of MB and RSMD at 100 m resolution.

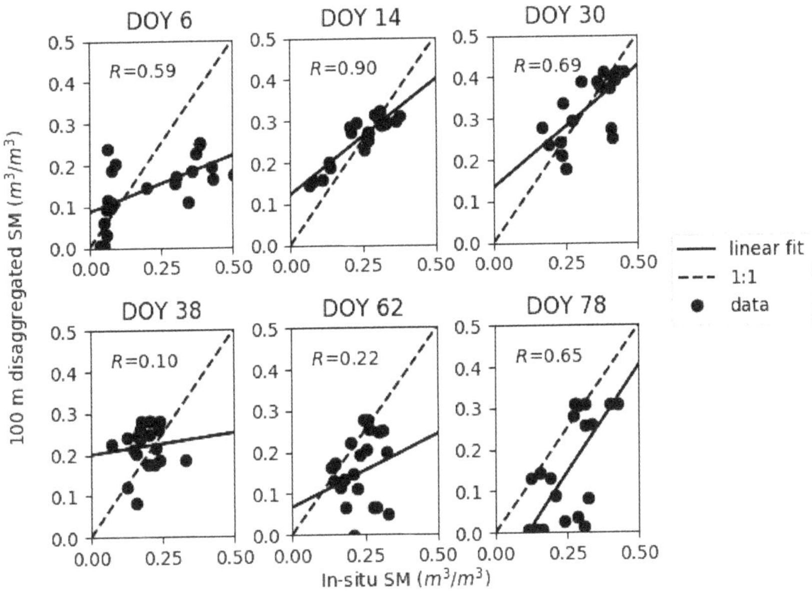

Figure 11. Graph plotting 100 m resolution disaggregated versus in-situ SM with the LR SM set to a fraction of the mean in-situ SM.

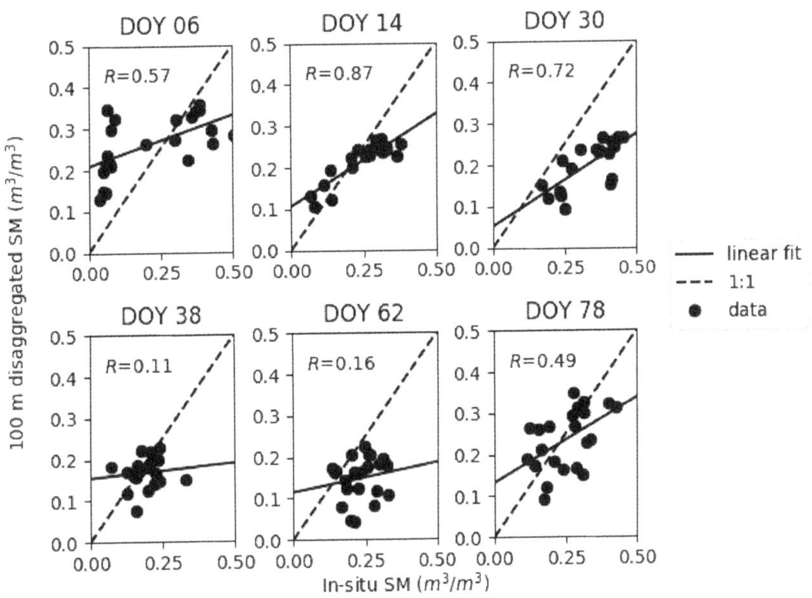

Figure 12. Same as Figure 11 but with LR SM set to the DISPATCH$_{Lin}$ SM obtained from SMAP and aggregated at 10 km resolution (single grid).

Table 1. Statistical results in terms of correlation coefficient (R), slope of the linear regression, absolute mean bias (MB) and root mean square difference (RMSD) between 100 m resolution disaggregated and in-situ SM for Synthetic, SMAP single grid and SMAP multiple grid LR SM cases separately (ISR is set to 10 km).

Day of Year (DOY)	Synthetic				SMAP Single Grid				SMAP Multiple Grid			
	R (-)	Slope (-)	Absolute MB (m^3/m^3)	RMSD (m^3/m^3)	R (-)	Slope (-)	Absolute MB (m^3/m^3)	RMSD (m^3/m^3)	R (-)	Slope (-)	Absolute MB (m^3/m^3)	RMSD (m^3/m^3)
6	0.59	0.27	0.069	0.15	0.57	0.24	0.05	0.14	0.54	0.23	0.01	0.14
14	0.90	0.55	0.014	0.049	0.87	0.44	0.03	0.06	0.90	0.47	0.03	0.06
30	0.69	0.59	0.006	0.066	0.72	0.44	0.14	0.15	0.70	0.52	0.12	0.14
38	0.10	0.10	0.03	0.08	0.11	0.07	0.02	0.07	0.12	0.08	0.03	0.07
62	0.22	0.35	0.08	0.13	0.16	0.14	0.002	0.31	0.20	0.21	0.07	0.10
78	0.65	1.04	0.11	0.15	0.49	0.40	0.02	0.08	0.54	0.31	0.12	0.14
All	0.53	0.48	0.052	0.104	0.55	0.34	0.05	0.09	0.57	0.35	0.08	0.10

3.3. Reducing Boxy Artifact

As demonstrated and discussed above, setting an ISR between the SMAP and Landsat resolutions has many advantages in terms of accuracy and robustness of $DISPATCH_{Exp}$. However, one drawback with an ISR equal to 10 km, is that the ISR grid may be still apparent in the 100 m resolution disaggregated image. Such effects are called boxy artifacts [62]. To reduce these boxy artifacts and to potentially increase the accuracy in 100 m disaggregated SM, a Monte-Carlo sampling method is proposed as an extra step in the pre and post-processing of input/output data of $DISPATCH_{Exp}$.

The preprocessing steps include: (i) selecting 10 km resolution SM pixels such that an equal number of HR (Landsat) pixels falls within each ISR pixel (ii) shifting the 10 km ISR pixels with a distance of 2 km in east-west and north-south directions, so as to generate a set of 25 ISR SM images, (iii) overlapping each image with HR Landsat optical/thermal data, and (iv) disaggregating individually each ISR image to 100 m resolution. Therefore, a set of 25 possible disaggregated 100 m SM images is obtained. The post-processing step consists in combining the 25 disaggregated SM images. The simple averaging is used to produce a single 100 m disaggregated image.

The multiple-grid procedure illustrated in Figure 6 is applied over our 30 km by 30 km study area for SMAP data and for each date separately. Figure 13 presents the 100 m resolution SM disaggregated images by applying the single grid and multiple grid algorithms for each date separately. It can be seen on DOY 6, 38 and 78 that the boxy artifacts at 10 km resolution present on the image obtained using the single grid algorithm have completely disappeared in the multiple grids application. Note that the boxy artifacts are not visible for the other dates due to strips (data gaps) present in the Landsat 7 images. The moving window algorithm also smoothens the disaggregated image at the image borders. Especially, the errors that generally occur at the corners of the image due to sudden changes in temperature endmembers and coarse scale SM are reduced. The composited image is of better quality by reducing the random errors associated with the uncertainty in LR observations and the disaggregation methodology (involving non-linear relationships between SEE and SM), which make the disaggregated image more realistic than using the single grid algorithm.

Table 2 reports the standard deviation of 100 m resolution disaggregated SM within each image for the single and multiple grid algorithms, separately. The standard deviation is systematically lower when applying multiple grids for all the dates. It means that the multiple grid application significantly reduces the variabilities attributed to random uncertainties in $DISPATCH_{Exp}$ input data. A quantitative comparison between 100 m disaggregated SM and in situ measurements is also proposed in Table 1 for SMAP multiple grid. By comparing the statistical results from the single grid and multiple grid (Table 1) algorithm, it can be seen that both the R and slope of the linear regression between 100 m resolution disaggregated and in situ SM are generally increased by applying multiple grids. Therefore, the proposed moving window method not only provides continuous SM images but also increases the efficiency of the disaggregation approach at 100 m resolution.

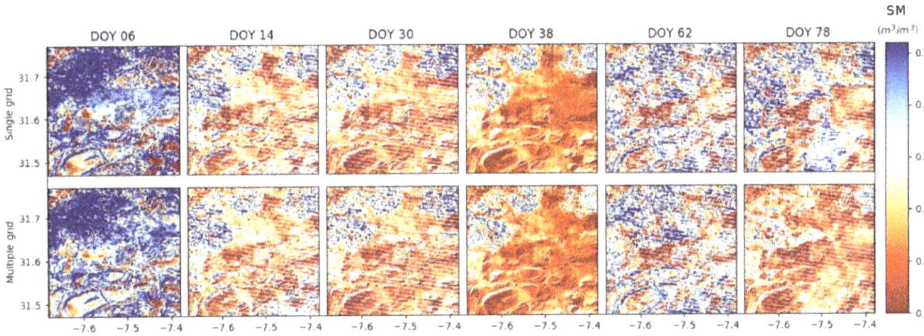

Figure 13. 100 m resolution disaggregated SMAP SM images when using an ISR set to 10 km for single grid (**top**) and moving average window (**bottom**) algorithm for each Landsat overpass date separately.

Table 2. Standard deviation of 100 m resolution disaggregated SM within each $DISPATCH_{Exp}$ image for the single grid and multiple grid algorithm.

Day of Year (DOY)	Single Grid (m^3/m^3)	Multiple Grid (m^3/m^3)
6	0.135	0.115
14	0.075	0.069
30	0.075	0.068
38	0.058	0.055
62	0.096	0.092
78	0.094	0.052
All	0.089	0.075

4. Discussion

High spatial resolution soil moisture data are fundamental for hydro-agricultural purposes as well as for other kind of applications. The DISPATCH method sequentially applied to SMAP data at 1 km resolution (using MODIS) and at 100 m resolution (using Landsat) has potential for providing such data. However, the performance of the approach may depend on the surface and atmospheric conditions. In addition, the temporal resolution of 100 m resolution DISPATCH data is currently limited by (i) the repeat cycle (16 days) of Landsat and (ii) the cloud free conditions required to use optical/thermal data. This section thus discusses the applicability and expected performance of DISPATCH in a context wider than our semi-arid irrigated study area.

The disaggregation of coarse scale soil moisture data is still a relatively recent research avenue [24], and consequently, few studies have compared the performance of existing methods. Sabaghy et al. [23] undertook the first comprehensive and systematic comparison study of several radar-based and optical/thermal-based SM downscaling methods. The SM downscaled from SMOS and SMAP data were evaluated against in situ as well as airborne SM estimates using the AACES data set in Southeastern Australia. DISPATCH was among the most efficient downscaling methods, especially when evaluating the spatial representation at 1 km resolution. The results presented in this paper are consistent with Sabaghy et al. [23] and previous validation exercises of DISPATCH. However, several intrinsic limitations common to optical/thermal-based downscaling approaches needs to be acknowledged, while several weaknesses specific to DISPATCH could be addressed in the future.

In this study, the mean RMSD (about 0.10 m^3/m^3) between disaggregated SMAP and in situ SM is relatively large and need to be interpreted in terms of bias and precision and to be compared with the spatio-temporal variability of SM existing within the study area. First, the mean RMSD is mostly explained by daily biases (mean bias of about 0.08 m^3/m^3) while the slope of the linear regression between disaggregated and in-situ SM is systematically and significantly positive. Second, the spatial variability of SM at 100 m resolution is extreme over the irrigated area with surface conditions ranging from bone dry to soil fully saturated. Over the sampled area, the minimum and maximum measured SM value was 0.03 to 0.45 m^3/m^3, respectively. In such highly heterogeneous areas, relatively large errors in SM estimates are thus expected. It is however reminded that the disaggregation error should be smaller than the actual SM variability, as an indicator of the relevance of such disaggregated SM data [63].

It is reminded that the DISPATCH methodology relies on the relationship between moisture and the LST. LST is in fact a signature of the surface energy balance, which is highly linked to the evaporation flux and the associated soil water availability. NDVI data are used in DISPATCH to partition the LST into its soil and vegetation components, given that the soil temperature is more directly linked to the top SM while the vegetation temperature is related to the deeper root zone SM [26]. Therefore, the application of the DISPATCH method over irrigated regions is fully relevant. In our study area, flood irrigation consists in applying about 60 mm in several (typically 4) hours so as to flood the entire field. However, the irrigation water rapidly infiltrates into the soil so that there is little chance that the Landsat satellite actually "sees" any inundated field (although it may potentially happen), all the more as the typical frequency of flood irrigations is one every 3 weeks. The flood irrigation technique is still widely applied in Morocco and in many developing countries where traditional practices persist. Note that the application of DISPATCH over drip irrigated crops would require thermal data with a high repeat cycle, consistent with the irrigation frequency for the drop-by-drop technique.

One major limitation of DISPATCH is the availability of optical/thermal data. DISPATCH has been tested mostly under arid or semi-arid regions where the cloud cover is rather small. In particular, the cloud cover of the Haouz plain is about 40% from January to May while its overall yearly percentage is 30% [64]. The 100 m resolution downscaled SM time series is thus expected to be much less dense over other regions having a larger cloud cover. In addition, as DISPATCH relies on the LST-moisture relationship, the performance of downscaling depends on the atmospheric evaporative demand. Therefore, testing its applicability to different climatic conditions will be needed in the future, notably by identifying moisture-limited and energy-limited regions.

Other limitations specific to DISPATCH includes the non-linear behavior of the SEE(SM) relationship and the so-called "boxiness" within the downscaled image. From Figures 11 and 12, relationships between disaggregated and in-situ SM appear to be non-linear on several dates (DOY 6 notably). This is explained by (i) the non-linear behavior of SEE for a range of SM values and (ii) the linear approximation of DISPATCH around the LR SM (Equation (3). It seems that using a non-linear SEE(SM) model in Equation (3) is not sufficient to represent the nonlinear SEE(SM) relationship when the SM variability is extreme. Future studies will address this issue by, for instance, correcting the SEE(SM) relationship when the sub-pixel SM variability is larger than a given threshold. Regarding the "boxiness" within the image, it is a relatively small effect compared to the spatial variability of SM represented by the DISPATCH method. Quantitatively, the RMSD between the SM produced from single grid and multiple grid applications is 0.019 m^3/m^3, while the standard deviation of disaggregated SM within the study area is 0.089 and 0.075 m^3/m^3 for the single and multiple grid case, respectively. In fact, as shown in Figure 13 the single grid application (without removing the boxiness within the image) already provides 100 m resolution SM images with borders between ISR pixels very consistent from one ISR pixel to another adjacent ISR pixel. Therefore, the stepwise method proposed in this paper does transfer in a satisfying manner the SM information from the 36 km SMAP resolution to the targeted 100 m resolution.

5. Conclusions

A stepwise disaggregation approach of SMAP SM is developed at 100 m resolution using the DISPATCH methodology and Landsat data. SMAP SM is first disaggregated from 36 km to 1 km resolution using MODIS data and $DISPATCH_{Lin}$ algorithm. Then the 1 km resolution SM is aggregated as ISR. Next, the ISR SM is further disaggregated at 100 m resolution using Landsat data and $DISPATCH_{Exp}$ algorithm. In order to take into account the increase of resolution, the new

$DISPATCH_{Exp}$ algorithm brings several innovations compared to the DISPATCH version currently implemented at CATDS ($DISPATCH_{Lin}$) in two aspects: (i) the SEE is a non-linear function of SM, and (ii) the SEE(LST) model is improved by better constraining the determination of temperature endmembers. The approach is evaluated using in situ measurements collected on the dates with concurrent SMAP, MODIS, and Landsat overpasses.

ISR is varied between 1 km and 30 km with a 1 km step, and sensitivity of the calibration parameter (SMp) of DISPATCHExp to ISR is analyzed. The retrieved SM_p is spatially and temporally representative for ISR equal or larger than 10 km, while significant spatial variabilities of SM_p (associated with uncertainties in temperature endmembers) and non-linear effects (associated with the non-linear SEE(SM) relationship) are obtained for ISR lower than 5 km. Optimal results in terms of RMSD between 100 m resolution and in situ SM are obtained for ISR close to 10 km. Therefore, the two-step disaggregation is more efficient than the direct disaggregation from SMAP to 100 m resolution. This is due to the trade-off existing between the performance (increasing with the ISR and its sub-pixel variability) of the contextual-based DISPATCH method and the statistical match (decreasing with ISR) between ISR remotely sensed and field-scale SM estimates. The correlation coefficient between 100 m resolution disaggregated and in situ SM ranges between 0.5–0.9 for four out of the six sampling dates. Better results are obtained on the sampling dates with a larger spatial variability in SM measurements, and reciprocally, poorer results are obtained when SM is relatively uniform at the sub-pixel scale.

Finally, a new method is proposed to reduce boxy artifacts at 10 km resolution in 100 m resolution disaggregated SM images. The multiple grid application perfectly smoothens the composited 100 m resolution disaggregated SM image and, in addition, quantitatively improves the efficiency of the downscaling approach by increasing the correlation coefficient and slope of the linear regression between 100 m resolution disaggregated and in situ SM.

The DISPATCH-based sequential disaggregation scheme has the advantage of being independent on ground-based measurements, as all input parameters (i.e., temperature endmembers and SM_p) are calibrated using remote sensing data. However, the unavailability of optical/thermal (MODIS/Landsat) data in cloudy conditions is still a severe limitation for operational applications. One key avenue for producing SM data sets at high spatial-temporal resolution could be the synergy with radar-based approaches [18,24]. Recently, Amazirh et al. [49] calibrated the main parameters of a radar-based SM retrieval method using a thermal-derived SM proxy. In the same vein, the 100 m resolution $DISPATCH_{Exp}$ SM data sets obtained from SMAP data on MODIS/Landsat clear sky days could represent a cornerstone in the construction of synergies between passive/active microwave and optical/thermal data.

Author Contributions: Conceptualization, N.O. and O.M.; methodology, N.O. and O.M.; validation N.O. and O.M.; supervision O.M.; investigation, N.O.; data curation, N.O., L.O.-G., B.A.H. and A.A.; writing—original draft preperation, N.O., O.M.; writing—review and editing, N.O., O.M., B.M., C.S., A.A.B., M.J.E. and S.E.-R.

Funding: This study was supported by the European Commission Horizon 2020 Programme for Research and Innovation (H2020) in the context of the Marie Sklodowska-Curie Research and Innovation Staff Exchange (RISE) action (REC project, grant agreement no: 645642 followed by ACCWA project, grant agreement no.: 823965).

Acknowledgments: We would like to thank the National Aeronautics and Space Administration (NASA) for freely providing the data. We would also like to thank the reviewers for their valuable comments.

Conflicts of Interest: The authors declare no conflict of interest.

References

1. Bisquert, M.; Sánchez, J.; López-Urrea, R.; Caselles, V. Estimating high resolution evapotranspiration from disaggregated thermal images. *Remote Sens. Environ.* **2016**, *187*, 423–433. [CrossRef]
2. Maurer, E.P.; Wood, A.; Adam, J.; Lettenmaier, D.P.; Nijssen, B. A long-term hydrologically based dataset of land surface fluxes and states for the conterminous United States. *J. Clim.* **2002**, *15*, 3237–3251. [CrossRef]

3. Koster, R.D.; Dirmeyer, P.A.; Guo, Z.; Bonan, G.; Chan, E.; Cox, P.; Gordon, C.; Kanae, S.; Kowalczyk, E.; Lawrence, D.; et al. Regions of strong coupling between soil moisture and precipitation. *Science* **2004**, *305*, 1138–1140. [CrossRef] [PubMed]
4. Wang, K.; Dickinson, R.E. A review of global terrestrial evapotranspiration: Observation, modeling, climatology, and climatic variability. *Rev. Geophys.* **2012**, *50*. [CrossRef]
5. Entekhabi, D.; Njoku, E.G.; O'Neill, P.E.; Kellogg, K.H.; Crow, W.T.; Edelstein, W.N.; Entin, J.K.; Goodman, S.D.; Jackson, T.J.; Johnson, J.; et al. The soil moisture active passive (SMAP) mission. *Proc. IEEE* **2010**, *98*, 704–716. [CrossRef]
6. Kerr, Y.H.; Waldteufel, P.; Wigneron, J.P.; Delwart, S.; Cabot, F.; Boutin, J.; Escorihuela, M.J.; Font, J.; Reul, N.; Gruhier, C.; et al. The SMOS mission: New tool for monitoring key elements ofthe global water cycle. *Proc. IEEE* **2010**, *98*, 666–687. [CrossRef]
7. Wagner, W. Operational Readiness of Microwave Remote Sensing of Soil Moisture: An Update. In Proceedings of the EGU General Assembly 2010, Vienna, Austria, 2–7 May 2010; Volume 12, p. 2636.
8. Kerr, Y.H.; Waldteufel, P.; Wigneron, J.P.; Martinuzzi, J.; Font, J.; Berger, M. Soil moisture retrieval from space: The Soil Moisture and Ocean Salinity (SMOS) mission. *IEEE Trans. Geosci. Remote Sens.* **2001**, *39*, 1729–1735. [CrossRef]
9. Schmugge, T. Applications of passive microwave observations of surface soil moisture. *J. Hydrol.* **1998**, *212*, 188–197. [CrossRef]
10. Schmugge, T.; Jackson, T. Mapping surface soil moisture with microwave radiometers. *Meteorol. Atmos. Phys.* **1994**, *54*, 213–223. [CrossRef]
11. Bindlish, R.; Crow, W.T.; Jackson, T.J. Role of passive microwave remote sensing in improving flood forecasts. *IEEE Geosci. Remote Sens. Lett.* **2009**, *6*, 112–116. [CrossRef]
12. Kerr, Y.H.; Waldteufel, P.; Richaume, P.; Wigneron, J.P.; Ferrazzoli, P.; Mahmoodi, A.; Al Bitar, A.; Cabot, F.; Gruhier, C.; Juglea, S.E.; et al. The SMOS soil moisture retrieval algorithm. *IEEE Trans. Geosci. Remote Sens.* **2012**, *50*, 1384–1403. [CrossRef]
13. Malbéteau, Y.; Merlin, O.; Molero, B.; Rüdiger, C.; Bacon, S. DisPATCh as a tool to evaluate coarse-scale remotely sensed soil moisture using localized in situ measurements: Application to SMOS and AMSR-E data in Southeastern Australia. *Int. J. Appl. Earth Obs. Geoinf.* **2016**, *45*, 221–234. [CrossRef]
14. Walker, J.P.; Houser, P.R. Requirements of a global near-surface soil moisture satellite mission: Accuracy, repeat time, and spatial resolution. *Adv. Water Resour.* **2004**, *27*, 785–801. [CrossRef]
15. Hawley, M.E.; Jackson, T.J.; McCuen, R.H. Surface soil moisture variation on small agricultural watersheds. *J. Hydrol.* **1983**, *62*, 179–200. [CrossRef]
16. Fang, B.; Lakshmi, V.; Bindlish, R.; Jackson, T.J. Downscaling of SMAP soil moisture using land surface temperature and vegetation data. *Vadose Zone J.* **2018**, *17*, 1. [CrossRef]
17. Chen, N.; He, Y.; Zhang, X. NIR-Red Spectra-Based Disaggregation of SMAP Soil Moisture to 250 m Resolution Based on OzNet in Southeastern Australia. *Remote Sens.* **2017**, *9*, 51. [CrossRef]
18. Tomer, S.K.; Al Bitar, A.; Sekhar, M.; Zribi, M.; Bandyopadhyay, S.; Kerr, Y. MAPSM: A spatio-temporal algorithm for merging soil moisture from active and passive microwave remote sensing. *Remote Sens.* **2016**, *8*, 990. [CrossRef]
19. Piles, M.; Camps, A.; Vall-Llossera, M.; Corbella, I.; Panciera, R.; Rudiger, C.; Kerr, Y.H.; Walker, J. Downscaling SMOS-derived soil moisture using MODIS visible/infrared data. *IEEE Trans. Geosci. Remote Sens.* **2011**, *49*, 3156–3166. [CrossRef]
20. Merlin, O.; Chehbouni, A.; Walker, J.P.; Panciera, R.; Kerr, Y.H. A simple method to disaggregate passive microwave-based soil moisture. *IEEE Trans. Geosci. Remote Sens.* **2008**, *46*, 786–796. [CrossRef]
21. Panciera, R.; Walker, J.P.; Kalma, J.D.; Kim, E.J.; Hacker, J.M.; Merlin, O.; Berger, M.; Skou, N. The NAFE'05/CoSMOS data set: Toward SMOS soil moisture retrieval, downscaling, and assimilation. *IEEE Trans. Geosci. Remote Sens.* **2008**, *46*, 736–745. [CrossRef]
22. Kim, G.; Barros, A.P. Downscaling of remotely sensed soil moisture with a modified fractal interpolation method using contraction mapping and ancillary data. *Remote Sens. Environ.* **2002**, *83*, 400–413. [CrossRef]
23. Sabaghy, S.; Walker, J.P.; Renzullo, L.J.; Jackson, T.J. Spatially enhanced passive microwave derived soil moisture: Capabilities and opportunities. *Remote Sens. Environ.* **2018**, *209*, 551–580. [CrossRef]
24. Peng, J.; Loew, A.; Merlin, O.; Verhoest, N.E. A review of spatial downscaling of satellite remotely sensed soil moisture. *Rev. Geophys.* **2017**, *55*, 341–366. [CrossRef]

25. Taconet, O.; Bernard, R.; Vidal-Madjar, D. Evapotranspiration over an agricultural region using a surface flux/temperature model based on NOAA-AVHRR data. *J. Clim. Appl. Meteorol.* **1986**, *25*, 284–307. [CrossRef]
26. Merlin, O.; Rudiger, C.; Al Bitar, A.; Richaume, P.; Walker, J.P.; Kerr, Y.H. Disaggregation of SMOS soil moisture in Southeastern Australia. *IEEE Trans. Geosci. Remote Sens.* **2012**, *50*, 1556–1571. [CrossRef]
27. Merlin, O.; Escorihuela, M.J.; Mayoral, M.A.; Hagolle, O.; Al Bitar, A.; Kerr, Y. Self-calibrated evaporation-based disaggregation of SMOS soil moisture: An evaluation study at 3 km and 100 m resolution in Catalunya, Spain. *Remote Sens. Environ.* **2013**, *130*, 25–38. [CrossRef]
28. Merlin, O.; Stefan, V.G.; Amazirh, A.; Chanzy, A.; Ceschia, E.; Er-Raki, S.; Gentine, P.; Tallec, T.; Ezzahar, J.; Bircher, S.; et al. Modeling soil evaporation efficiency in a range of soil and atmospheric conditions using a meta-analysis approach. *Water Resour. Res.* **2016**, *52*, 3663–3684. [CrossRef]
29. Sandholt, I.; Rasmussen, K.; Andersen, J. A simple interpretation of the surface temperature/vegetation index space for assessment of surface moisture status. *Remote Sens. Environ.* **2002**, *79*, 213–224. [CrossRef]
30. Molero, B.; Merlin, O.; Malbéteau, Y.; Al Bitar, A.; Cabot, F.; Stefan, V.; Kerr, Y.; Bacon, S.; Cosh, M.; Bindlish, R.; et al. SMOS disaggregated soil moisture product at 1km resolution: Processor overview and first validation results. *Remote Sens. Environ.* **2016**, *180*, 361–376. [CrossRef]
31. Dumedah, G.; Walker, J.P.; Merlin, O. Root-zone soil moisture estimation from assimilation of downscaled Soil Moisture and Ocean Salinity data. *Adv. Water Resour.* **2015**, *84*, 14–22. [CrossRef]
32. Escorihuela, M.J.; Quintana-Seguí, P. Comparison of remote sensing and simulated soil moisture datasets in Mediterranean landscapes. *Remote Sens. Environ.* **2016**, *180*, 99–114. [CrossRef]
33. Malbeteau, Y.; Merlin, O.; Balsamo, G.; Er-Raki, S.; Khabba, S.; Walker, J.; Jarlan, L. Toward a Surface Soil Moisture Product at High Spatiotemporal Resolution: Temporally Interpolated, Spatially Disaggregated SMOS Data. *J. Hydrometeorol.* **2018**, *19*, 183–200. [CrossRef]
34. Bandara, R.; Walker, J.P.; Rüdiger, C.; Merlin, O. Towards soil property retrieval from space: An application with disaggregated satellite observations. *J. Hydrol.* **2015**, *522*, 582–593. [CrossRef]
35. Escorihuela, M.J.; Merlin, O.; Stefan, V.; Moyano, G.; Eweys, O.A.; Zribi, M.; Kamara, S.; Benahi, A.S.; Ebbe, M.A.B.; Chihrane, J.; et al. SMOS based high resolution soil moisture estimates for Desert locust preventive management. *Remote Sens. Appl. Soc. Environ.* **2018**, *11*, 140–150.
36. Hssaine, B.A.; Merlin, O.; Ezzahar, J.; Ojha, N.; Er-raki, S.; Khabba, S. An evapotranspiration model self-calibrated from remotely sensed surface soil moisture, land surface temperature and vegetation cover fraction: Application to disaggregated SMOS and MODIS data. *Hydrol. Earth Syst. Sci.* **2019**, under review. [CrossRef]
37. Lievens, H.; Tomer, S.K.; Al Bitar, A.; De Lannoy, G.J.; Drusch, M.; Dumedah, G.; Franssen, H.J.H.; Kerr, Y.; Martens, B.; Pan, M.; et al. SMOS soil moisture assimilation for improved hydrologic simulation in the Murray Darling Basin, Australia. *Remote Sens. Environ.* **2015**, *168*, 146–162. [CrossRef]
38. Leroux, L.; Baron, C.; Castets, M.; Escorihuela, M.J.; Diouf, A.; Bégué, A.; Lo Seen, D. Estimating maize grain yield in scarce field-data environment: An approach combining remote sensing and crop modeling in Burkina Faso. In Proceedings of the ABSTRACT AFRICAGIS 2017, Addis Ababa, Ethiopia, 20–24 November 2017.
39. Colliander, A.; Fisher, J.B.; Halverson, G.; Merlin, O.; Misra, S.; Bindlish, R.; Jackson, T.J.; Yueh, S. Spatial downscaling of SMAP soil moisture using MODIS land surface temperature and NDVI during SMAPVEX15. *IEEE Geosci. Remote Sens. Lett.* **2017**, *14*, 2107–2111. [CrossRef]
40. French, A.N.; Hunsaker, D.J.; Thorp, K.R. Remote sensing of evapotranspiration over cotton using the TSEB and METRIC energy balance models. *Remote Sens. Environ.* **2015**, *158*, 281–294. [CrossRef]
41. Loheide , S.P., II; Gorelick, S.M. A local-scale, high-resolution evapotranspiration mapping algorithm (ETMA) with hydroecological applications at riparian meadow restoration sites. *Remote Sens. Environ.* **2005**, *98*, 182–200. [CrossRef]
42. Merlin, O.; Al Bitar, A.; Walker, J.P.; Kerr, Y. A sequential model for disaggregating near-surface soil moisture observations using multi-resolution thermal sensors. *Remote Sens. Environ.* **2009**, *113*, 2275–2284. [CrossRef]
43. Fang, B.; Lakshmi, V. Soil moisture at watershed scale: Remote sensing techniques. *J. Hydrol.* **2014**, *516*, 258–272. [CrossRef]
44. Jarlan, L.; Khabba, S.; Er-Raki, S.; Le Page, M.; Hanich, L.; Fakir, Y.; Merlin, O.; Mangiarotti, S.; Gascoin, S.; Ezzahar, J.; et al. Remote sensing of water resources in semi-arid mediterranean areas: The joint international laboratory TREMA. *Int. J. Remote Sens.* **2015**, *36*, 4879–4917. [CrossRef]

45. Er-Raki, S.; Chehbouni, A.; Guemouria, N.; Duchemin, B.; Ezzahar, J.; Hadria, R. Combining FAO-56 model and ground-based remote sensing to estimate water consumptions of wheat crops in a semi-arid region. *Agric. Water Manag.* **2007**, *87*, 41–54. [CrossRef]
46. Duchemin, B.; Hagolle, O.; Mougenot, B.; Benhadj, I.; Hadria, R.; Simonneaux, V.; Ezzahar, J.; Hoedjes, J.; Khabba, S.; Kharrou, M.; et al. Agrometerological study of semi-arid areas: An experiment for analysing the potential of time series of FORMOSAT-2 images (Tensift-Marrakech plain). *Int. J. Remote Sens.* **2008**, *29*, 5291–5299. [CrossRef]
47. Duchemin, B.; Hadria, R.; Erraki, S.; Boulet, G.; Maisongrande, P.; Chehbouni, A.; Escadafal, R.; Ezzahar, J.; Hoedjes, J.; Kharrou, M.; et al. Monitoring wheat phenology and irrigation in Central Morocco: On the use of relationships between evapotranspiration, crops coefficients, leaf area index and remotely-sensed vegetation indices. *Agric. Water Manag.* **2006**, *79*, 1–27. [CrossRef]
48. Miller, J.; Gaskin, G. ThetaProbe ML2x. In *Principles of Operation and Applications*; MLURI Technical Note; The Macaulay Land Use Research Institute: Aberdeen, UK, 1999.
49. Amazirh, A.; Merlin, O.; Er-Raki, S.; Gao, Q.; Rivalland, V.; Malbeteau, Y.; Khabba, S.; Escorihuela, M.J. Retrieving surface soil moisture at high spatio-temporal resolution from a synergy between Sentinel-1 radar and Landsat thermal data: A study case over bare soil. *Remote Sens. Environ.* **2018**, *211*, 321–337. [CrossRef]
50. Das, N.N.; Entekhabi, D.; Kim, S.; Jagdhuber, T.; Dunbar, S.; Yueh, S.; Colliander, A. High-resolution enhanced product based on SMAP active-passive approach using Sentinel 1 data and its applications. In Proceedings of the 2017 IEEE International Geoscience and Remote Sensing Symposium (IGARSS) 2017, Fort Worth, TX, USA, 23–28 July 2017; pp. 2493–2494.
51. Colliander, A.; Jackson, T.J.; Bindlish, R.; Chan, S.; Das, N.; Kim, S.; Cosh, M.; Dunbar, R.; Dang, L.; Pashaian, L.; et al. Validation of SMAP surface soil moisture products with core validation sites. *Remote Sens. Environ.* **2017**, *191*, 215–231. [CrossRef]
52. Chen, D.; Huang, J.; Jackson, T.J. Vegetation water content estimation for corn and soybeans using spectral indices derived from MODIS near-and short-wave infrared bands. *Remote Sens. Environ.* **2005**, *98*, 225–236. [CrossRef]
53. Jiménez-Muñoz, J.C.; Sobrino, J.A.; Skoković, D.; Mattar, C.; Cristóbal, J. Land surface temperature retrieval methods from Landsat-8 thermal infrared sensor data. *IEEE Geosci. Remote Sens. Lett.* **2014**, *11*, 1840–1843. [CrossRef]
54. Mattar, C.; Durán-Alarcón, C.; Jiménez-Muñoz, J.C.; Santamaría-Artigas, A.; Olivera-Guerra, L.; Sobrino, J.A. Global atmospheric profiles from reanalysis information (GAPRI): A new database for earth surface temperature retrieval. *Int. J. Remote Sens.* **2015**, *36*, 5045–5060. [CrossRef]
55. Moran, M.; Clarke, T.; Inoue, Y.; Vidal, A. Estimating crop water deficit using the relation between surface-air temperature and spectral vegetation index. *Remote Sens. Environ.* **1994**, *49*, 246–263. [CrossRef]
56. Tang, R.; Li, Z.L.; Tang, B. An application of the Ts–VI triangle method with enhanced edges determination for evapotranspiration estimation from MODIS data in arid and semi-arid regions: Implementation and validation. *Remote Sens. Environ.* **2010**, *114*, 540–551. [CrossRef]
57. Komatsu, T.S. Toward a robust phenomenological expression of evaporation efficiency for unsaturated soil surfaces. *J. Appl. Meteorol.* **2003**, *42*, 1330–1334. [CrossRef]
58. Chanzy, A.; Bruckler, L. Significance of soil surface moisture with respect to daily bare soil evaporation. *Water Resour. Res.* **1993**, *29*, 1113–1125. [CrossRef]
59. Wang, J. Scale Effects on the Remote Estimation of Evapotransportation. Ph.D. Thesis, Texas State University, San Marcos, TX, USA, 2012.
60. Hoehn, D.C.; Niemann, J.D.; Green, T.R.; Jones, A.S.; Grazaitis, P.J. Downscaling soil moisture over regions that include multiple coarse-resolution grid cells. *Remote Sens. Environ.* **2017**, *199*, 187–200. [CrossRef]
61. Merlin, O.; Chehbouni, A.G.; Kerr, Y.H.; Njoku, E.G.; Entekhabi, D. A combined modeling and multispectral/multiresolution remote sensing approach for disaggregation of surface soil moisture: Application to SMOS configuration. *IEEE Trans. Geosci. Remote Sens.* **2005**, *43*, 2036–2050. [CrossRef]
62. Agam, N.; Kustas, W.P.; Anderson, M.C.; Li, F.; Colaizzi, P.D. Utility of thermal sharpening over Texas high plains irrigated agricultural fields. *J. Geophys. Res. Atmos.* **2007**, *112*, D19. [CrossRef]

63. Merlin, O.; Malbéteau, Y.; Notfi, Y.; Bacon, S.; Khabba, S.; Jarlan, L. Performance metrics for soil moisture downscaling methods: Application to DISPATCH data in central Morocco. *Remote Sens.* **2015**, *7*, 3783–3807. [CrossRef]
64. Wilson, A.M.; Jetz, W. Remotely sensed high-resolution global cloud dynamics for predicting ecosystem and biodiversity distributions. *PLoS Biol.* **2016**, *14*, e1002415. [CrossRef]

remote sensing

MDPI

Article

Evaluation and Analysis of AMSR2 and FY3B Soil Moisture Products by an In Situ Network in Cropland on Pixel Scale in the Northeast of China

Haoyang Fu [1,2], Tingting Zhou [1] and Chenglin Sun [1,*]

1 Coherent Light and Atomic and Molecular Spectroscopy Laboratory, College of Physics, Jilin University, Changchun 130012, China; fuhy16@mails.jlu.edu.cn (H.F.); ttzhou18@mails.jlu.edu.cn (T.Z.)
2 College of Electronic Science and Engineering, Jilin University, Changchun 130012, China
* Correspondence: chenglin@jlu.edu.cn; Tel.: +86-150-4307-6988

Received: 22 February 2019; Accepted: 5 April 2019; Published: 10 April 2019

Abstract: An in situ soil moisture observation network at pixel scale is constructed in cropland in the northeast of China for accurate regional soil moisture evaluations of satellite products. The soil moisture products are based on the Japan Aerospace Exploration Agency (JAXA) algorithm and the Land Parameter Retrieval Model (LPRM) from the Advanced Microwave Scanning Radiometer 2 (AMSR2), and the products from the FengYun-3B (FY3B) satellite are evaluated using synchronous in situ data collected by the EC-5 sensors at the surface in a typical cropland in the northeast of China during the crop-growing season from May to September 2017. The results show that the JAXA product provides an underestimation with a bias (b) of -0.094 cm^3/cm^3, and the LPRM soil moisture product generates an overestimation with a b of 0.156 cm^3/cm^3. However the LPRM product shows a better correlation with the in situ data, especially in the early experimental period when the correlation coefficient is 0.654, which means only the JAXA product in the early stage, with an unbiased root mean square error (ubRMSE) of 0.049 cm^3/cm^3 and a b of -0.043 cm^3/cm^3, reaches the goal accuracy (± 0.05 cm^3/cm^3). The FY3B has consistently obtained microwave brightness temperature data, but its soil moisture product data in the study area is seriously missing during most of the experimental period. However, it recovers in the later period and is closer to the in situ data than the JAXA and LPRM products. The three products show totally different trends with vegetation cover, soil temperature, and actual soil moisture itself in different time periods. The LPRM product is more sensitive and correlated with the in situ data, and is less susceptible to interferences. The JAXA is numerically closer to the in situ data, but the results are still affected by temperature. Both will decrease in accuracy as the actual soil moisture increases. The FY3B seems to perform better at the end of the whole period after data recovery.

Keywords: regional soil moisture; in situ network; AMSR2; FY3B; evaluation; EVI; SST

---

1. Introduction

Soil moisture is vital to the earth's water cycle, energy cycle, ecological environment, and agriculture. It is a critical boundary between the land surface and the atmosphere and a key medium surface evapotranspiration [1–7]. Satellite microwave remote sensing technology can be used to monitor surface soil moisture changes in near real time at regional and global scales. Therefore, the evaluation of the accuracy of onboard soil moisture products is of great significance for the calibration of products and the future scientific research on the global water cycle.

In recent decades, satellite remote sensing technology has been continuously developed, and many satellites have been used to monitor various parameters of surface soil. Compared with the visible light band, microwave remote sensing has long wavelength, strong penetrability, and is not affected by cloud layer and weather conditions. It can realize global all-weather monitoring and ground observation, and is widely used for retrieval of surface soil moisture [8–11] and temperature [12].

In recent years, the L-band is considered to be the most suitable band for soil moisture observation because of its longer wavelength and deeper penetration depth. The Soil Moisture and Sea Salinity (SMOS) mission of the European Space Agency (ESA) can achieve the soil moisture observation at multi-angles [13]. The Soil Moisture Active and Passive (SMAP) mission of the National Aeronautics and Space Administration (NASA) is equipped with a RADAR (stopped transmitting on 7 July 2015) and a radiometer, and it could improve the retrieval accuracy and spatial resolution [14]. Compared to the L-band, the X-band has a much longer temporal sequence of soil moisture observations. The AMSR2 was mounted on the Global Change Observation Mission 1-Water (GCOM-W1) satellite launched on 18 May 2012 and started to acquire observed data on 3 July 2012 [15]. It is the successor to the AMSR-E, which successfully operated for almost ten years from June 2002 to October 2011 [16]. The FY-3B satellite, launched on 5 November 2010, is the second satellite of FY3 (Feng Yun 3) series, a member of China's second generation of polar-orbiting meteorological satellites. It provides measurements of terrestrial, oceanic, and atmospheric parameters, including precipitation rate, sea ice concentration, snow water equivalent, soil moisture, atmospheric cloud water, and water vapor [17]. There was a gap of about ten months between the AMSR-E ceased and the AMSR2 operated, and the Microwave Radiation Imager (MWRI) onboard the FY3B had been running successfully during this period. The AMSR-E, the AMSR2 and the FY3B/MWRI all provide soil moisture products based on X-band. The difference between equatorial local crossing times (the GCOM-W1 at 1:30 a.m./p.m. and the FY3B at 1:38 a.m./p.m.) is within 10 minutes. FY3B/MWRI can fill up the window period if the consistency can be confirmed. Therefore, the evaluation of AMSR2 and FY3B soil moisture products to obtain continuous data on global soil moisture monitoring by the same type of microwave radiometer is of great significance for global water cycle monitoring and long-term continuous monitoring of climate change [18].

In recent years, there have been many studies on the evaluation of soil moisture products. They use a variety of error analysis methods to compare the performance of various products and algorithms based on single site or multiple site data at local or global scales [19–24]. However, due to the complexity of surface soil moisture in the temporal and spatial changes of the surface, experiments at this stage are still insufficient to confirm the superiority and inferiority of various soil moisture products. As the environmental factors change, the performance of soil moisture products may reverse the change results [25].

An in situ soil moisture observation network at pixel scale was constructed in the corn cropland located in the northeast of China, and the experimental period was from early May to late September 2017 which was the only frost-free period in this typical area. In addition, the surface soil structure remained naturally stable without artificial damage during this period. The JAXA and the LPRM soil moisture products from AMSR2 and FY3B/MWRI soil moisture product were evaluated by the performance metrics [26] using the up-scaled in situ soil moisture collected synchronously by the EC-5 probes at 2.5 cm depth. All the three products are all based on X-band where Radio Frequency Interference (RFI) issues are less severe for the X-band soil moisture retrievals than lower frequencies [27,28]. Vegetation has always been one of the main attenuation factors in microwave transfer. Compared to the L-band, the X-band is more susceptible to the surface vegetation cover due to its shorter wavelength [29,30]. The surface temperature also affects the calculation of the vegetation optical thickness and the soil surface emissivity. In the LPRM algorithm, the temperatures of vegetation and soil are approximately equal. In the JAXA algorithm, although the multi-frequency is used to overcome the influence of the surface temperature, the previous research has shown that soil moisture retrieval results were still affected by temperature [25]. Vegetation has always been one of the main attenuation factors in microwave transfer. Compared to the L-band, the X-band is more susceptible to the surface vegetation cover due to its shorter wavelength [29,30]. The microwave radiation is affected by the temperature. Although the JAXA algorithm hoped to eliminate the effects by frequency difference, the brightness temperature value itself is directly affected by the surface soil temperature (SST). The ranges of products are different, and the field capacity will also vary depending on the climate, environment, and time change. Therefore, the accuracy of soil moisture products under different actual soil moisture conditions may change. The performance of all the products was discussed according to the effect factors including vegetation cover, SST, and actual soil moisture itself in different time periods.

2. Materials and Methods

2.1. Study Area

As shown Figure 1, the study area is in the north of Changchun City in the northeast of China where the climate is temperate monsoon climate with four distinct seasons. It is a semi-humid area with a flat terrain. The land features are simple, and the main type is cropland with scarce water bodies. The annual sunshine hours are about 2695.2 hours. The average annual precipitation is 520 mm which is mainly concentrated in July and August in summer. The annual average temperature and annual accumulated temperature are 4.4 °C and 2851 °C, respectively. The average daily temperature is below 0 °C from November to March. The temperature difference is large between winter and summer over 50 to 60 °C, and this region is significantly colder than other regions in the same latitude in winter. The frost-free period is about 140 to 150 days from May to September which is also the crop-growing season. The most suitable crop for growing in the region is corn. The study area is a typical and representative cropland in the northeast of China. The research in this paper is of great significance due to the distinctive and specific climate characteristics in the northeast of China unlike the other parts of the world. In this paper, the observation network was established with the SMAP pixel as spatial reference, so that it could include the coverage of pixels of other microwave products. Then the scale of the in situ soil moisture was converted using Thiessen Polygons method to match the pixel size of the target product.

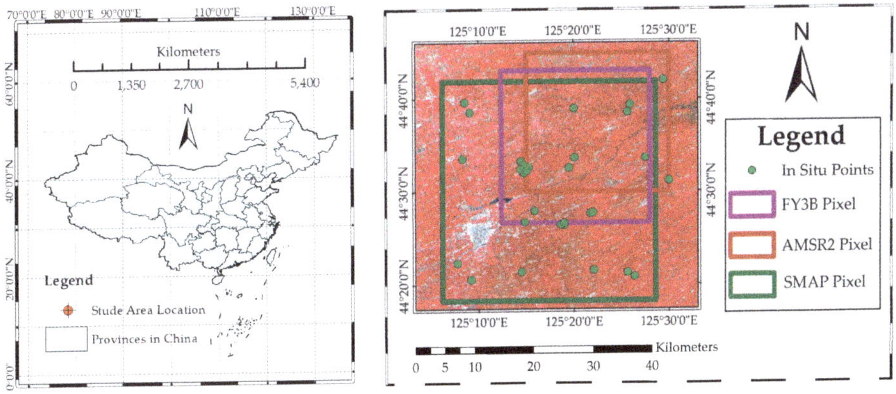

(a) (b)

Figure 1. Study Area. (**a**) shows the geographical location of the study area in the northeast of China; (**b**) shows the distribution of the points from the in situ observation network, the satellite pixels in the study area. The background is a false-color Landsat 8 image at 25 September 2017 with band 5, 4, 3 as the RGB.

2.2. Satellite Soil Moisture Products Based on X-Band

Three X-band soil moisture products, the AMSR2/JAXA product, the AMSR2/LPRM product and the FY3B/MWRI product, were selected in this paper. All the three algorithms use a simple radiative transfer model, the *tau-omega* model [10]. To minimize environmental interference, only the descending products were used while the geophysical conditions were complicated at day times but simple at night times [8,9,31,32].

1. The AMSR2/JAXA L3 Soil Moisture Product

The AMSR2/JAXA Level 3 0.25° global grid soil moisture product used is acquired at the GCOM-W1 Data Providing Service (https://gcom-w1.jaxa.jp/auth.html). The JAXA algorithm uses a forward radiative transfer scheme to calculate brightness temperatures in multiple frequencies and polarizations according to different vegetation and soil conditions and the surface temperature is assumed constant at 293 K. The soil moisture is estimated by a lookup table built up based on the results and the polarization ratio (PI) at 10.65 GHz and index of soil wetness (ISW) at 36.5 and 10.65 GHz horizontal channels [8,33,34].

2. The AMSR2/LPRM L3 Soil Moisture Product

The AMSR2/LPRM Level 3 0.25° global grid soil moisture product used is acquired at the Goddard Earth Sciences Data and Information Services Center (GES DISC) (https://gcmd.gsfc.nasa.gov/). The LPRM algorithm is developed by the Vrije Universiteit (VU) University Amsterdam and NASA for multiple frequencies. It uses brightness temperature at 36.5 GHz V channel to estimate land surface temperature and retrieve the soil moisture vegetation optical depth at the same time by an iteration using PI [9,35,36]. For consistency with the other two products, only the X-band LPRM product was used in this paper.

3. The FY3B/MWRI L2 Soil Moisture Product

The FY3B/MWRI L2 EASE-Grid Soil Moisture Product is acquired at the FENGYUN Satellite Data Center (http://satellite.nsmc.org.cn/). The FY-3B soil moisture retrieval algorithm uses the brightness temperature at 10.65GHz H/V channels based on a parameterized surface emission model (the Qp model) [37] for the bare surface and the empirical relationship between the Normalized Difference Vegetation Index (NDVI) and the vegetation water content to estimate the vegetation optical depth [38] for the vegetation correction.

2.3. The In Situ Observation Network on Pixel Scale

2.3.1. Selection of Each Point Location in the In Situ Observation Network

The in situ soil moisture observation network using Decagon EC-5 sensors at pixel scale was constructed to better represent the real surface soil moisture corresponding to the depth of satellite products. At first, the location of each observation site is a key factor in determining whether the observation network is a good representation of the whole experimental area. As shown in Figure 2, a 36 km × 36 km SMAP pixel grid was used as the spatial coverage reference, and the pixel was then subdivided evenly. According to the spatial heterogeneity factors including soil types and bare soil thermal inertia related to bare soil moisture, representative sub-pixels were selected to represent the overall distribution of soil moisture in the whole experimental area, and to minimize the impact of spatial heterogeneity.

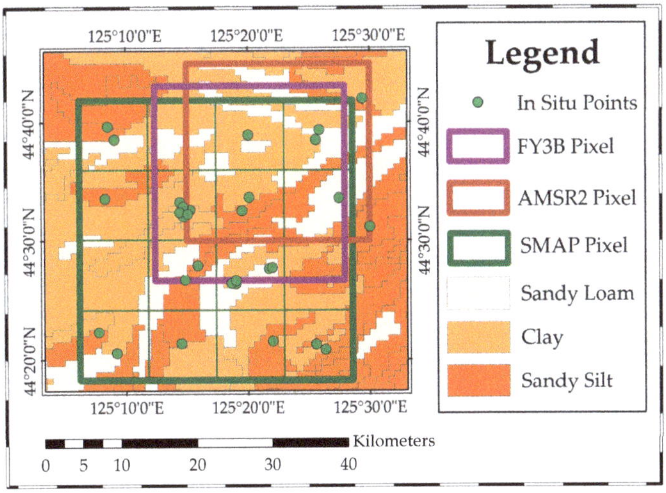

Figure 2. Location of each point in the in situ observation network and the distribution of soil types in the experimental area.

2.3.2. The Sensor Tests of the In Situ Soil Moisture

The Decagon EC-5 probe was selected due to its effective soil moisture measurement at shallow depths as close as possible to satellite data. The measured soil moisture data were hourly collected from in situ points distributed in the study area. The parameters of all the EC-5 probes had been previously tested and confirmed through:

- The sensing boundary test
- The consistency test
- The calibration according to actual soil from study area

The sensing boundary of the EC-5 probe was confirmed in the laboratory with dry sand and water. The experimental apparatus is shown in Figure 3. A plastic cylindrical container with a height of 30 cm and a diameter of 15 cm is filled with dry sand and placed in a big container filled with water surrounding the plastic container. The EC-5 probe was inserted into the dry sand completely, and gradually collected data from the edge to the center of the container. With a movement interval of 0.5 cm, the data were collected for five times at each position, and the time interval between each data collection was one minute.

The experimental results are shown in Figure 4. During the process of the probe moving from the distance of 0.5 cm to 2.5 cm, the voltage measurement value becomes significantly smaller as the distance becomes larger. Between 2.5 cm and 3 cm, the voltage decreases significantly; after 3 cm, the voltage measurement remains stable. This explains that the probe's boundary measurement range is 2.5–3 cm around.

Due to the precisions of the EC-5 sensors, there existed subtle differences in measurement precision and range between the untested probes. To minimize these differences, we performed a consistency test on all EC-5 sensors with ethanol and dry sand. All the EC-5 soil moisture sensor probes synchronously collected data in ethanol then dry sand at the same temperature. The data sampling rate was once every minute, and the total sampling time was 10 to 15 minutes. The measurement results in ethanol and dry sand are shown in Figure 5a,b. Then the data average of each probe in ethanol and dry sand were calculated. Based on the theoretical values of ethanol and dry sand, the probe with the average value closest to the standard value was selected as the standard probe to correct all the other probes.

The results after correction are shown in Figure 5c,d. The error of the measurement baseline of each probe was significantly reduced after the consistency test.

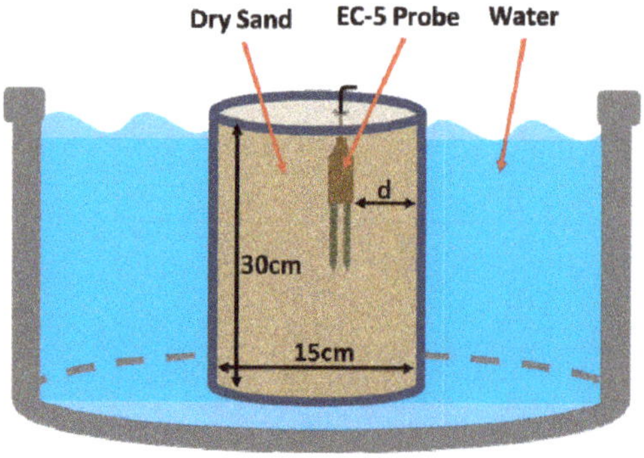

Figure 3. Inductive boundary test of EC-5 sensor. The plastic cylindrical container with a height of 30 cm and a diameter of 15 cm was filled with dry sand and placed in a bigger container filled with water to surround the plastic container. Insert the EC-5 probe into the dry sand completely, and gradually collect data from the edge to the center of the container. d was the distance from the probe to the edge of the small container.

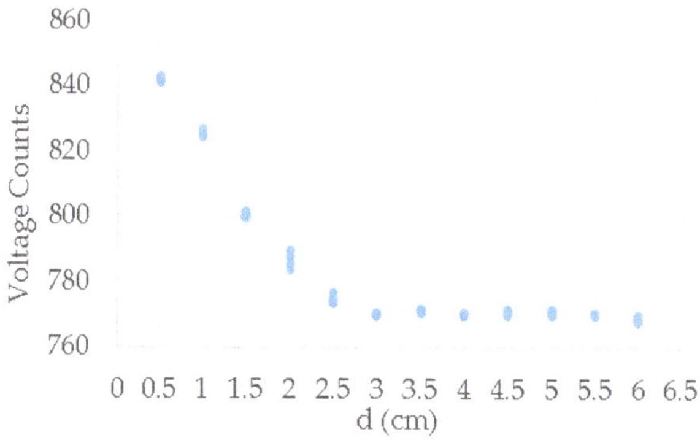

Figure 4. EC-5 sensor probe boundary measurement range experiment, the probe measurement voltage value changes with the probe distance d from the dry sand container boundary.

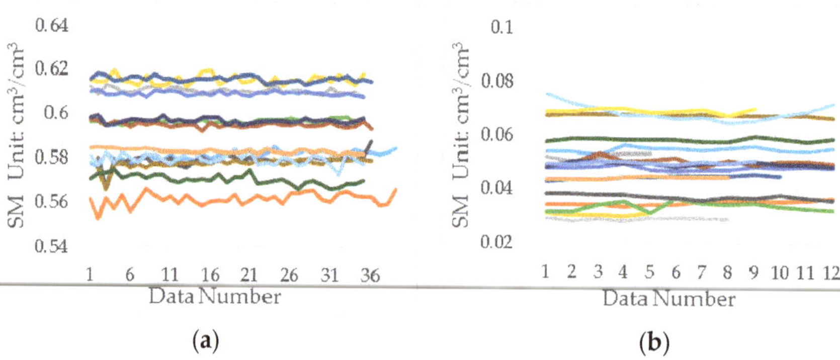

Figure 5. *Cont.*

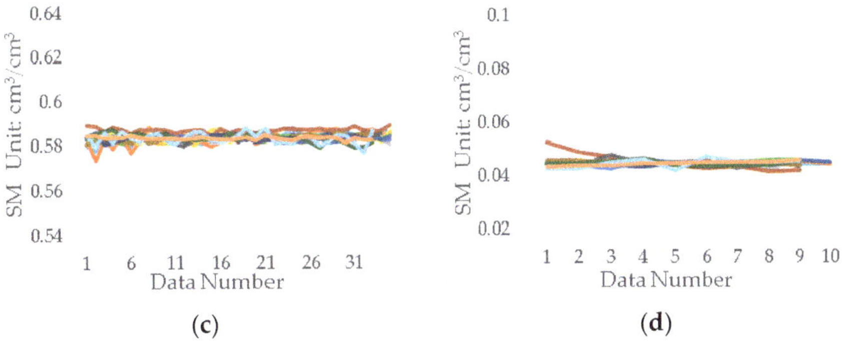

Figure 5. Consistency comparison and correction results of EC-5 sensors in ethyl alcohol and dry sand measurements, and different colored lines represent different sensors. (**a**) shows the data collected in ethanol, (**b**) shows the data collected in dry sand. (**c**) shows the ethanol data after correction, and (**d**) shows the dry sand data after correction.

The consistency-tested EC-5 sensors met the uniform accuracy requirements of measurement, but the same sensor probe's response to soil moisture would still vary attributed to different soil types. To acquire accurate in situ data of soil moisture, we calibrated the sensors according to the actual soil samples collected in the study area. Subject to the soil components based on the Harmonized World Soil Database (described in Section 2.4.3), the soil in the study area was divided into sandy loam soil, clay soil, and sandy silt soil, of which the specific contents are shown in Table 1.

Table 1. Content of each component in different classification of soil samples.

Soil Texture	Clay (%)	Silt (%)	Sand (%)
Sandy Loam Soil	12.41	64.28	23.31
Clay Soil	11.80	57.71	30.48
Sandy Silt Soil	11.81	55.87	32.32

Firstly, various types of soil samples collected in the field need pretreatment, and the soil samples were dried at 105 °C for 48 hours, and the dried soil samples were ground and sieved to remove debris such as stones. The sieve pore size is not less than the specific soil type particle size. Then, the pretreated soil was filled into a container (50 cm × 50 cm × 40 cm in volume) and was slowly sprayed with fresh water about 10% of the weight of the soil. The watered soil was kept being stirred to mix evenly. After that, the uniformly mixed wet soil was filled into a small container (13 cm × 13 cm × 15 cm in volume) in a natural state without pressing. The standard EC-5 probe selected in the consistency test was used for calibration. The probe was vertically inserted into the soil at a position greater than 3 cm from the container wall and totally collected 5 data with a sampling interval of 60 seconds. Then a soil sample was taken using a cutting ring (100 cm^3 in volume) at its adjacent position and was weighed as its fresh weight. After that, the remaining soil was put back into the container used for mixing. Then all the procedures above were repeated until the soil moisture content was saturated. Finally, all the soil samples collected with cutting ring were dried (105 °C, 48 hours) and weighed. We calculated the volumetric water content of the soil samples, then linearly fit the probe readings corresponding to soil moisture to obtain the calibration parameters and equations of the EC-5 sensor to the three soil types in the study area as shown in Figure 6.

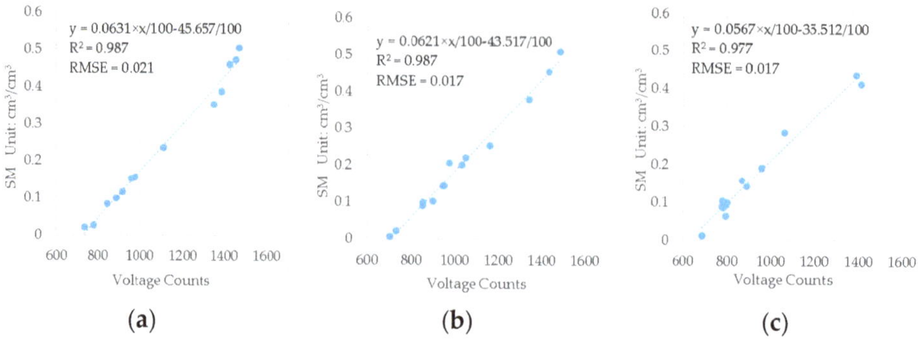

Figure 6. The calibration parameters and equations of the EC-5 sensor to the three soil types in the study area, wherein (**a**) is clay soil, (**b**) is sand silt soil, and (**c**) is sandy loam soil.

The accuracies after calibration for different soil types are 0.021 cm^3/cm^3, 0.017 cm^3/cm^3 and 0.017 cm^3/cm^3, which are better than the standard accuracy (0.02 cm^3/cm^3) of the EC-5 sensor. The specific parameters of the EC-5 sensors after testing and calibration are listed in Table 2.

Table 2. Parameters of the EC-5 sensor after testing and calibration.

Sensing Range (cm)	2.5~3
Operating Temperature (°C)	−40~+60
Measurement Range of SM (%)	0~100
Accuracy (cm^3/cm^3)	0.02

2.3.3. The In Situ SST

The temperature sensors used to measure SST were the DS18B20 soil temperature sensor with a range of −55 °C to +125 °C and an accuracy of ± 0.5 °C. Similarly, the DS18B20 temperature sensors were also installed 2.5 cm below soil surface like EC-5 probes. To avoid measurement interference between them and to ensure that the two measurements represent the same position, the DS18B20 temperature sensor should be installed within 5 to 15 cm from the EC-5 probe.

2.3.4. Placement of Sensors at In Situ Points

All the sensors were laid in the field at the middle of May after all the land in the experimental area was fully cultivated and were retrieved at the end of September before harvesting. It was also in the frost-free period at this time, which ensured the measured data valid. The surface soil structure remained naturally stable without artificial damage. To exclude the influence of other factors, the sensors were placed under the plain surface of pure soil at 2.5 cm depth that was more than 40 cm away from the plant seed position. As shown in Figure 7a,b, a section was dug next to the selected position and measured with a ruler. The sensor probe was horizontally inserted into the soil at 2.5 cm from the upper soil surface so as not to damage the natural structure of the soil in the vertical direction. The host was buried aside as shown in Figure 7c. The specific terrain of the agricultural land in the study area was the alignment arrangement of ditches and ridges. As shown in Figure 7d, two EC-5 sensors were separately placed at the ditch and the ridge of each in situ point to enable the collected soil moisture data to be more representative. The data were recorded once every hour.

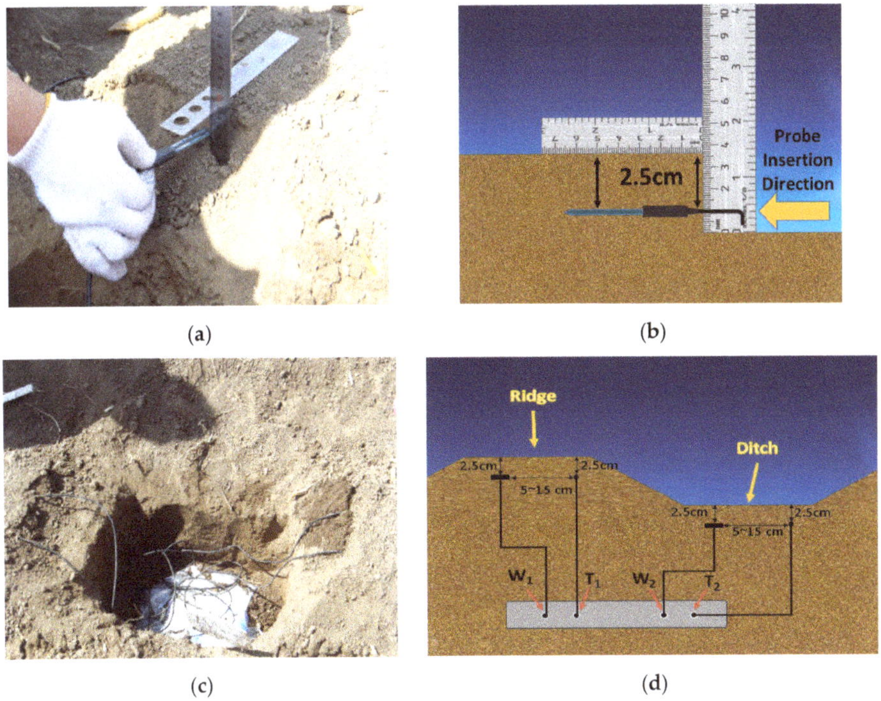

Figure 7. Installation and arrangement of the soil moisture and temperature sensors at in situ points. (**a**) shows the actual installation of the EC-5 probe, (**b**) describes the specific installation details of the EC-5 probe, (**c**) shows the situation of host and probes after installation, and (**d**) shows the position details of the probes and host. W_1 and W_2 are the EC-5 probes, and T_1 and T_2 are the temperature sensors.

2.4. Ancillary Data

2.4.1. Meteorological Data

The global meteorological station's timed observation data were downloaded at the National Meteorological Information Center (http://data.cma.cn/data/cdcdetail/dataCode/A.0013.0001.html), the site data is updated every three hours. With the satellite equatorial crossing time (1:30 A.M.) as the node, all precipitation in the past 24 hours was accumulated as the daily precipitation data.

2.4.2. The Moderate Resolution Imaging Spectroradiometer (MODIS) Vegetation Index

The NASA vegetation index product, MOD13C1 VIs 16-day 0.05deg data, was used to represent the surface vegetation cover situation in the study area [39]. According to different calculation formulas, there are two vegetation indices in this product, the NDVI and the Enhanced Vegetation Index (EVI). The original 0.05° resolution product was resampled to 0.25° by taking the average value. As shown in Figure 8, the trend and amplitude of NDVI and EVI were basically the same, but their change intervals were different, and the correlation between them was significant in the study area.

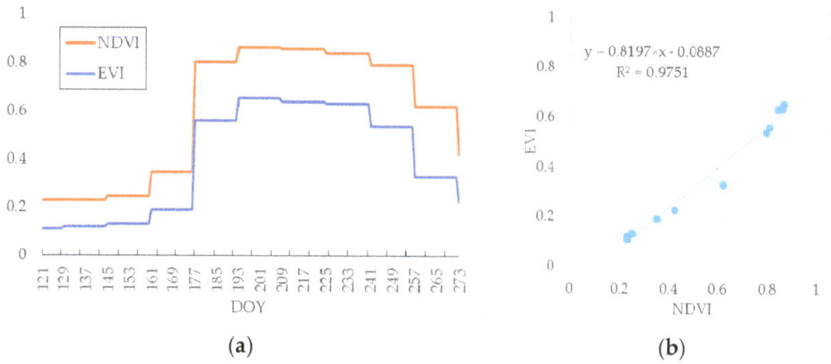

Figure 8. The comparison of NDVI and EVI in the study area during the study period. (**a**) shows the changes of NDVI and EVI during the study period, and (**b**) shows a linear relationship between NDVI and EVI.

The EVI was mainly used for analysis in this paper which exhibited the EVI changing obviously. The EVI was at a low level at first, then sharply increased, and maintained at a high level. At the end of the study period, the EVI showed a gradual decline.

2.4.3. The Harmonized World Soil Database

The Harmonized World Soil Database is the result of a collaboration between the FAO with IIASA, ISRIC-World Soil Information, Institute of Soil Science, Chinese Academy of Sciences (ISSCAS), and the Joint Research Centre of the European Commission (JRC). It is a 30 arc-second raster database with over 15,000 different soil mapping units [8]. In this paper, we used it to identify the soil types in the study area according to its texture and components.

2.5. Methodology

2.5.1. Thiessen Polygons Method for Pixel Scale Matching of the In Situ Data

The pixel size of the passive microwave soil moisture product used in this paper was 25 km × 25 km. To make the point data well represent the actual soil moisture of the whole passive microwave data pixel, the point data, including surface soil moisture and SST, were all up-scaled using Thiessen Polygons (TP) method [40,41] and compared with directly averaged value. As shown in Figure 9, the Thiessen Polygons method divided the entire target pixel area into several polygons according to the position of each point. Every edge of the polygon is in the middle of two points and is perpendicular to the connecting line between the two points. The ratio of the polygon to the total area is the weight of the center point. The sum of all weighted point data is the value of the entire target pixel. The advantages of the Thiessen Polygons method are the simple operation, smooth interpolation results, and basically closed contours generated. Its disadvantage is that it is greatly affected by the known points and only considers the factor of distance. The Thiessen Polygons (TP) approach as an up-scaled method takes the spatial distribution of in situ points into account, while direct average treats the proportion of each point equally. The difference between them is small when the distribution of the points is uniform. However, when some points are gathered together and they have a significant spatial difference from other locations in the pixel, the direct average will be overestimated or underestimated because of the high or low data of the gathered points. However, it could be avoided by the TP approach.

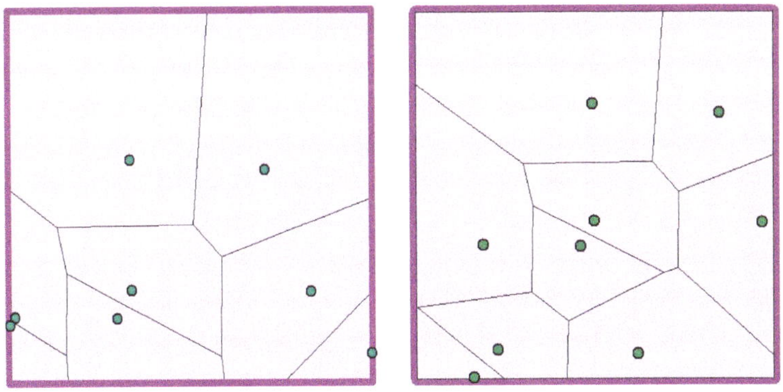

Figure 9. Thiessen Polygons calculated from the spatial distribution of the in situ points of soil moisture in microwave pixels, where (a) is the AMSR2 pixel with a spatial resolution of 0.25 degrees and a total of 8 in situ points, and (b) is the FY3B pixel with a spatial resolution of 0.25 degrees and a total of 9 in situ points.

2.5.2. The Performance Metrics for the Evaluation with In Situ Data

The performance metrics including the root mean square error (RMSE), the unbiased root mean square error (ubRMSE), the bias (b) and the correlation coefficient (R) were used to evaluate the soil moisture products [26]. The formulas are expressed as follows:

$$\mathrm{RMSE} = \sqrt{E\left[\left(SM_{\mathrm{pro}} - SM_{\mathrm{insitu}}\right)^2\right]} \tag{1}$$

$$\mathrm{ubRMSE} = \sqrt{E\left\{\left[\left(SM_{\mathrm{pro}} - E\left[SM_{\mathrm{pro}}\right]\right) - \left(SM_{\mathrm{insitu}} - E\left[SM_{\mathrm{insitu}}\right]\right)\right]^2\right\}} \tag{2}$$

$$b = E\left[SM_{\mathrm{pro}}\right] - E\left[SM_{\mathrm{insitu}}\right] \tag{3}$$

$$R = \frac{\sum\limits_{i=1}^{n}\left(SM_{pro_i} - E\left[SM_{pro}\right]\right)\left(SM_{insitu_i} - E\left[SM_{insitu}\right]\right)}{\sqrt{\sum\limits_{i=1}^{n}\left(SM_{pro_i} - E\left[SM_{pro}\right]\right)^2 \cdot \sum\limits_{i=1}^{n}\left(SM_{insitu_i} - E\left[SM_{insitu}\right]\right)^2}} \tag{4}$$

where $E[\cdot]$ is the expected or linear average operator, SM_{pro} represents the passive microwave remote sensing soil moisture product estimate, and SM_{insitu} represents the ascending scale measured soil moisture. i means the data number.

3. Results

3.1. In situ *Soil Moisture Data from the Network*

All the in situ soil moisture data, separately for the AMSR2 and FY3B pixels, from the in situ observation network are shown in Figure 10. The effective period of the data is from a consecutive period of 141 days from 129th to 269th day of the year (DOY). The numbers shown in the figure represent different sensors. Since the AMSR2 pixel and the FY3B pixel mostly overlap, some sensor data were shared by both. The sensors needed replacing the battery and memory card. To ensure the in situ data were continuously obtained, the data may be collected alternately using different numbered sensors at the same in situ point. It can be seen that the soil moisture at each experimental point shows a highly distinct correlation with precipitation. The soil moisture at each point shows a sharp rise upon the increase and shows a slow downward trend when decreasing.

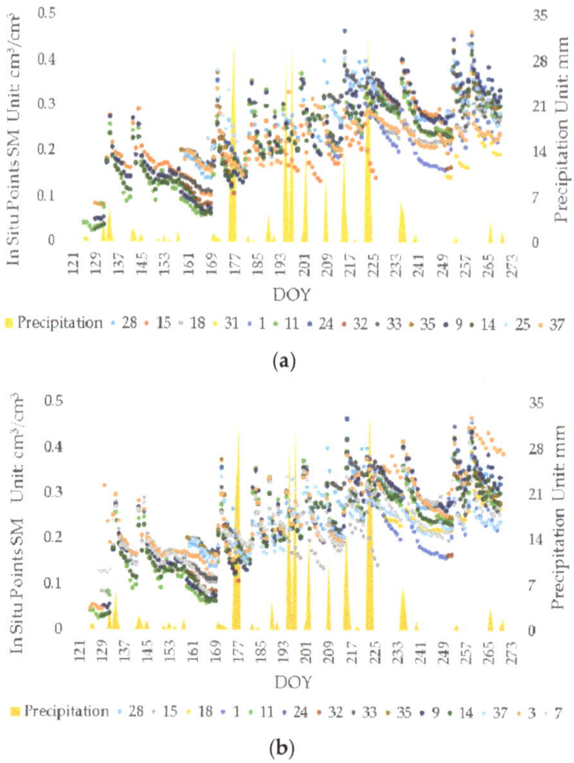

Figure 10. Measured soil moisture data from the in situ observation network and the cumulative 24-hour precipitation. Among them, (**a**) is the AMSR2 pixel case, and (**b**) is the FY3B pixel case. The numbers shown at the bottom of the figure are the sensors' numbers.

As shown in Figure 11, the in situ soil moisture data at each point were compared with the up-scaled results. It reveals that the data point from both AMSR2 and FY3B pixel are evenly distributed on both sides of the diagonal, and the direct average is very close to the result of Thiesson Polygons method. That indicates that the in situ points were evenly distributed in the pixel, and the up-scaled average value could well represent the average soil moisture of the entire pixel. There is not much difference between the results of the two methods. The results of Thiessen Polygons method were mainly used as the in situ soil moisture in the later part of this paper. The evaluation results using the direct average will be displayed in Appendix A. Figure 11g shows the comparison of the up-scaled results of AMSR2 pixel and FY3B pixel using Thiessen Polygons method, which reflects that the results are almost the same due to pixels overlapping.

In this experiment, we usually placed more than one sensor in close distance to prevent the individual sensor from failing, and most of the results of the TP approach and the direct average were also very close. To some extent, that indicated that the in situ points were evenly distributed and the spatial variation within the pixel was stable and uniform. However, there were also some large differences between the results of the TP approach and the direct average because of the big changes of the gathered points' data. Because such situations were relatively small, the impacts on the whole time period were limited, so the statistical results also would not be significantly affected. To eliminate the error caused by the artificial distribution of points, the TP results were preferentially used for calculation. The direct average results were also given as a reference in the attached table.

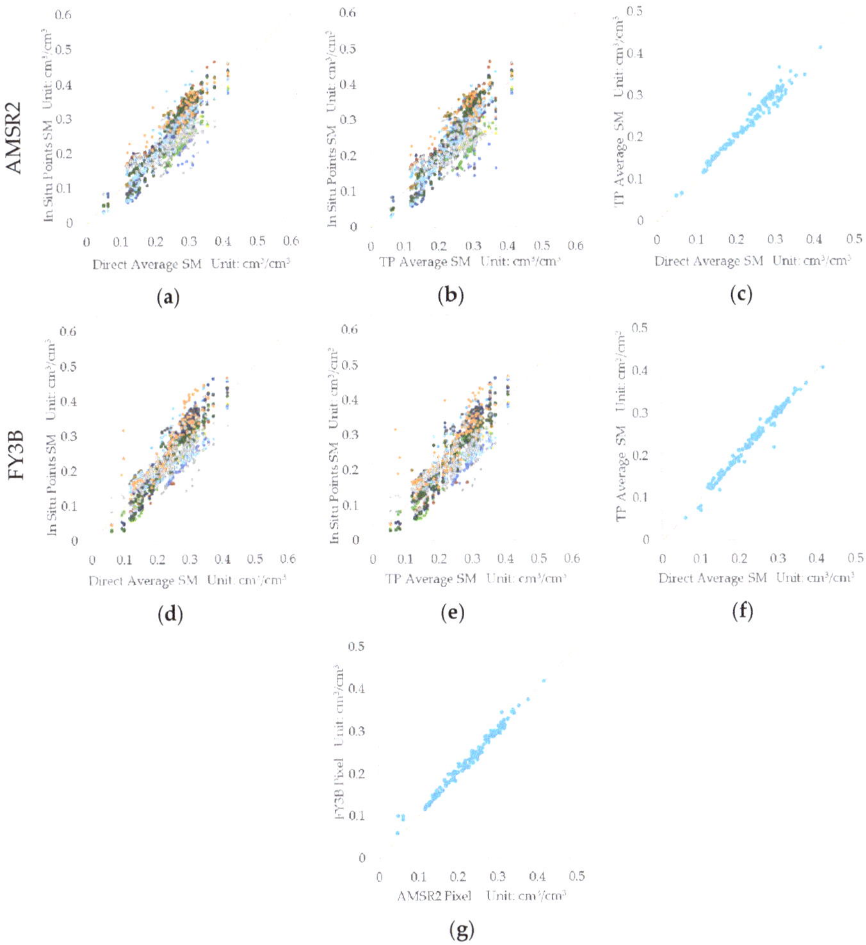

Figure 11. The comparison of the in situ point soil moisture data and the up-scaled results in the study area. (**a**,**d**) are respectively the comparison of the in situ soil moisture at each point and the direct average in the AMSR2 pixel and the FY3B pixel. (**b**,**e**) are respectively the comparison of the in situ soil moisture at each point and the result of Thiessen Polygons method in the AMSR2 pixel and the FY3B pixel. (**c**,**f**) are respectively the comparison of the direct average and the result of Thiessen Polygons method in the AMSR2 pixel and the FY3B pixel. (**g**) is the comparison of the results of Thiessen Polygons method in AMSR2 pixel and FY3B pixel.

The experimental period was 141 days, nearly four and a half months, and the seasons spanned spring, summer, and autumn when the vegetation cover and temperature changed greatly. Therefore, according to the objective environmental conditions, mainly seasonal vegetation cover, the whole experimental period was divided into two stages with the DOY of 169 as the node. The first stage was the DOY from 129th to 169th day. At this time, the study area was from spring to early summer, and the surface was mostly from bare soil to low vegetation. The second stage was the DOY from 170th to

269th day. During that period, the season was summer and ended at early autumn while the EVI was high and the land surface was densely covered by vegetation.

As shown in Figure 12, the distribution of the in situ soil moisture of the AMSR2 and FY3B pixels are totally different at different stages. It manifests that the in situ soil moisture is clearly divided into two parts at different stages. Soil moisture in both stages shows a downward trend with increasing temperature. It also can be found that there are two parallel lines separately on the upper side of the data at the first stage and on the bottom side of the data at the second stage, and there is an obvious space between the two sides. The data on the upper side correspond to the maximum value of soil moisture at a specific temperature at the first stage, and the data on the bottom side correspond to the minimum value of soil moisture at a specific temperature at the second stage. The main difference between the two stages was the EVI. At the first stage, the soil can no longer absorb water after soil moisture increased to a certain value because of its yield capacity. Given the land surface of bare soil without vegetation, the yield capacity was controlled by the soil itself and was inversely proportional to the SST. The excessive rainwater would flow away with surface runoff. At the second stage, it was summer with abundant sunshine and high temperature, when much dense vegetation was covering the surface. By virtue of the evapotranspiration, the vegetation roots had a locking effect on soil moisture, which would greatly increase the yield capacity. Therefore, it maintained the lowest value of soil moisture change, and the lowest value was also negatively correlated with the SST. This was mainly resulted from the evapotranspiration of vegetation. The high temperature strengthened the evapotranspiration, causing a decrease in surface soil moisture in the root zone. When the temperature was low, sometimes accompanied by precipitation events, the evapotranspiration became weaker, causing the surface soil moisture to rise.

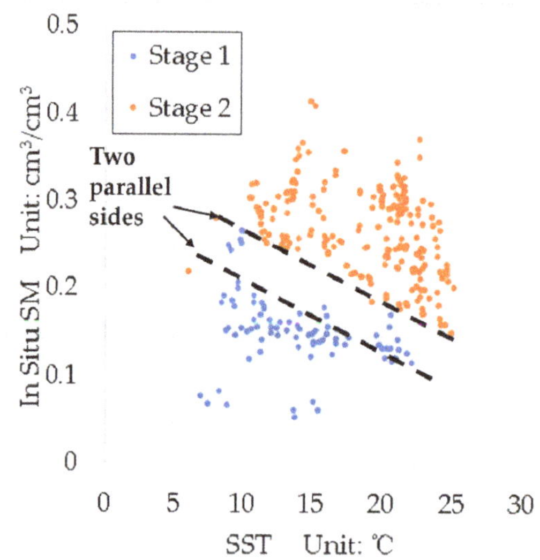

Figure 12. The relationship between the in situ soil moisture and the SST. Both of the in situ soil moisture in the AMSR2 and FY3B pixels are shown in the figure at two different stages. There are two parallel lines separately on the upper side of the data at the first stage and on the bottom side of the data at the second stage, and there is an obvious space between the two sides.

3.2. Satellite Data Evaluation and Intercomparison

Figure 13 shows the change of the JAXA, LPRM, and FY3B soil moisture products, daily precipitation, and the EVI with the DOY in the study area. It can be seen that the in situ soil moisture is sensitive to precipitation events, with which the local peaks of the soil moisture always appear.

The increasing magnitude of the soil moisture is also consistent with the amount of precipitation at the first stage. The in situ soil moisture generally shows an upward trend during the whole experimental period. The EVI remained at a low level at the first stage while it was spring, and the surface was mostly bare soil. Then the EVI increased slightly from the 161st to the 176th day when the season was late spring and early summer and the surface was covered with low vegetation, but the EVI still remains at a low level. After that, the EVI increased sharply in the DOY from 177th to 192th day when the season was summer, and the vegetation grew densely on the land surface. In the last few periods of the experimental period, the EVI decreased significantly, but it was still significantly higher than that at the first stage. The JAXA soil moisture product and LPRM soil moisture product well cover the experimental period. Among them, JAXA soil moisture products are relatively close to the in situ soil moisture at the first stage and generally less at later. The LPRM product is consistently higher than the in situ soil moisture all the period and the changes are severe. The FY3B soil moisture product are severely deficient in the experimental period, and the data are cut off from the 161th of DOY and recover from the 256th of DOY. The large amount of missing data of the FY3B product is mainly concentrated in the period with high EVI, high SST, and high precipitation in summer. Moreover, the brightness temperature data of FY3B at 10 GHz are intact and very close to the AMSR2 data. Therefore, it can be judged that the missing of the FY3B product is mainly due to the FY3B soil moisture product algorithm. The maximum of the FY3B product is 0.5 cm^3/cm^3. In previous studies, it was common for the algorithms to be saturated or even overflowed under dense vegetation cover. [42,43].

The precipitation data from the meteorological station is the cumulative amount every 3 hours. The daily precipitation data in this paper was the accumulated precipitation data from the past 24 hours since the satellite transited. The large amount of precipitation that occurred in 176 days was mainly concentrated at 3 to 6 o'clock, which was a short-term concentrated precipitation. The satellite transit time was around 17:30, with a gap of 9.5–12.5 hours. The in situ data showed that most of the in situ points barely reacted. The meteorological station is about 60km away from the study area, and the study area covered 36 km×36 km. It may be that the precipitation in the area was not obvious. On the other hand, the EVI rose sharply and the crop grew rapidly at this time. The surface temperature was also at the highest value during the experimental period. So, the evaporation and transpiration cannot be ignored. The moisture might be reduced largely before the satellite transited. Therefore, for all the analysis above, the high precipitation at 176th day did not lead to an increase in soil moisture.

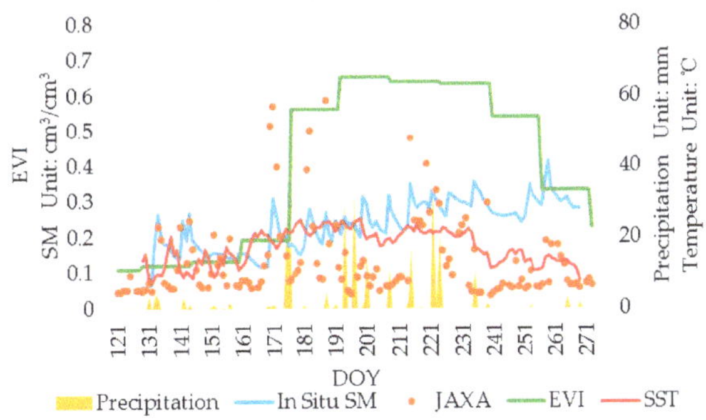

(a)

Figure 13. *Cont.*

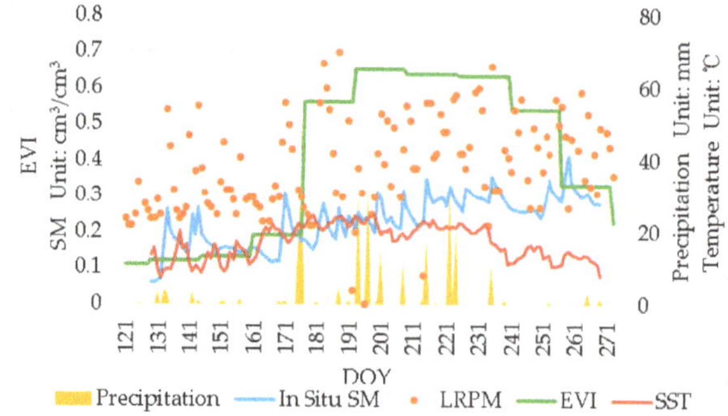

(**b**)

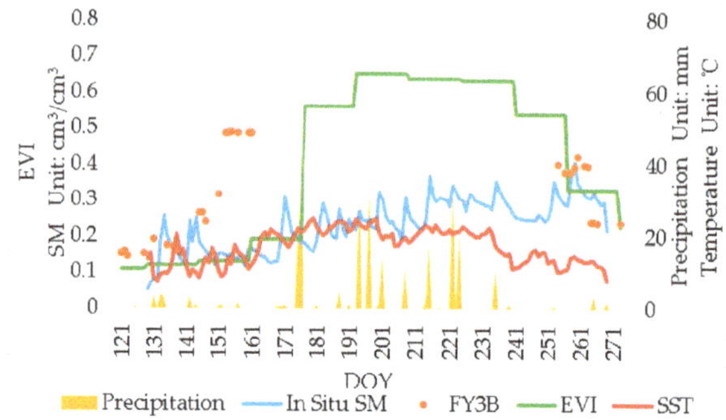

(**c**)

Figure 13. The change of soil moisture products with in situ soil moisture, daily precipitation, EVI, and surface soil temperature. Among them, (**a**) is about the JAXA soil moisture product, (**b**) is about the LPRM soil moisture product, and (**c**) is about the FY3B soil moisture product.

Figure 14 shows the comparisons among the JAXA product, the LPRM product and the in situ soil moisture in different periods, and Table 3 lists the results of the performance metrics including the RMSE, the ubRMSE, the *b*, and the *R*. The results using the direct average is displayed in Table A1 in Appendix A. It exhibits that the JAXA product generally shows an underestimation of the soil moisture with a *b* of -0.094 cm^3/cm^3. In contrast, the LPRM product generally demonstrates a large overestimation of soil moisture with a bias of 0.156 cm^3/cm^3. The JAXA's RMSE is 0.150 cm^3/cm^3 that is smaller than the LPRM's 0.191 cm^3/cm^3 but the ubRMSE of them are almost the same during the whole period. Both the JAXA product and the LPRM product are close to the in situ data at the first stage. The JAXA points are evenly distributed on both sides of the diagonal while the in situ soil moisture is low in this period. However, the LRPM product has already provided an obvious overestimation of soil moisture at this time. At the second stage, the JAXA product displays a significantly underestimation when the in situ soil moisture overall increases, and there are also some overestimated data. All the performance metrics of the two products become worse except for the LPRM's *b*. However, the LPRM still keep an overestimation of the soil moisture at a high level. The JAXA product has the best performance of error with the lowest RMSE and ubRMSE of 0.066 cm^3/cm^3 and 0.049 cm^3/cm^3, and the *b* is -0.043 cm^3/cm^3 that reaches the goal accuracy of the product (± 0.05 cm^3/cm^3). However, the LPRM product has the highest correlation coefficient of 0.654 at this time. The JAXA product range is 0 to 0.6 cm^3/cm^3, while the LPRM product is 0 to 1 cm^3/cm^3. It can be seen from the linear prediction of the data in Figure 14 that in the two periods, the LPRM product generally shows an upward trend with the increase of the in situ soil moisture.

The slope of the linear prediction is basically consistent with the range ratio of the LPRM product to the in situ soil moisture. Moreover, the LPRM product has better correlation with in situ soil moisture than the JAXA product throughout the whole experimental period. Comparing the two products, the LPRM product are generally higher than the JAXA's. They show a good linear relationship with the correlation coefficient of 0.94 at the first stage, and the change of the difference between them becomes very severe at the second stage.

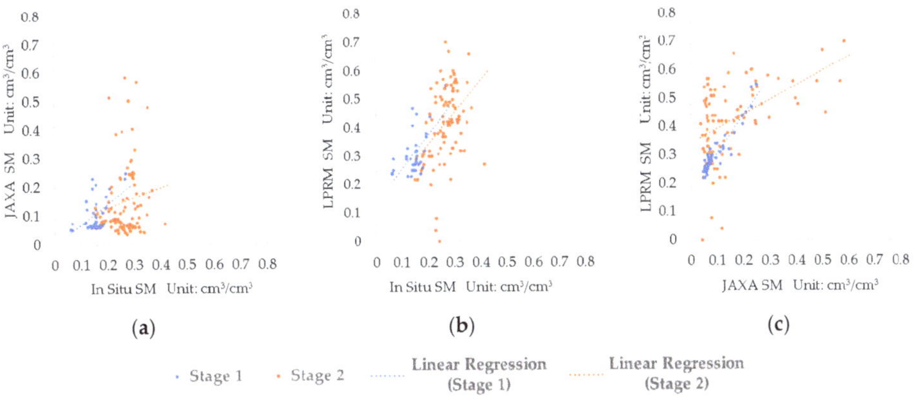

Figure 14. Comparisons between the JAXA product, the LPRM product, and the in situ soil moisture in two stages. (**a**) is the comparison between the JAXA product and the in situ soil moisture. (**b**) is the comparison between the LPRM product and the in situ soil moisture. (**c**) is the comparison between the two AMSR2 products.

Table 3. The performance metrics of the JAXA and the LPRM soil moisture products at different study periods. The best one for each performance metric is in bold.

Period	Products	RMSE (cm^3/cm^3)	ubRMSE (cm^3/cm^3)	b (cm^3/cm^3)	R
Whole Period	JAXA	0.150	0.117	−0.094	0.259
	LPRM	0.191	0.110	0.156	0.542
First Stage	JAXA	**0.066**	**0.049**	**−0.043**	0.565
	LPRM	0.177	0.063	0.166	**0.654**
Second Stage	JAXA	0.173	0.129	−0.115	0.136
	LPRM	0.196	0.124	0.152	0.403

Due to the serious lack, the FY3B soil moisture product cannot be evaluated throughout the experimental period like the JAXA and the LPRM products. Therefore, the three soil moisture products were evaluated based on the coverage of the FY3B soil moisture product. The results are shown in Table 4. The results using the direct average is displayed in Table A2 in Appendix A. It can be seen that the RMSEs of the three products are almost the same. The results of the FY3B product is the neither the best nor the worst except for the R. The LPRM still has the best R of 0.516 with a P-value of 0.0117 in the significance level. However, due to only 23 days of data in total, the FY3B product does not have a good performance during the study period.

Table 4. The comparison of the performance metrics of the three soil moisture products as the period of the FY3B product. The best one for each performance metric is in bold.

Period	Products	RMSE (cm^3/cm^3)	ubRMSE (cm^3/cm^3)	b (cm^3/cm^3)	R
as FY3B	**FY3B**	0.237	0.155	0.179	0.042
	JAXA	**0.231**	**0.085**	0.215	0.180
	LPRM	0.233	0.156	**0.174**	**0.516**

4. Discussion

The winter in the study area is long and cold, and the frost-free period of the year is only about 5 months. Only in this period, the soil moisture is guaranteed to be liquid and free of ice. Therefore, the local climatic conditions determine the effective monitoring period of soil moisture in this study. The land types and soil types in the study area are stable, most of which are agricultural land. The crops are basically corn. Although the crops in this area only can be planted once a year, the black soil here is fertile and the quality of crops is almost the best in the country. The annual climate change in the study area is basically stable, and the experimental period is representative. Therefore, it is of great significance to monitor the climatic conditions in this region and the soil moisture change during the growing season. The effect of the environment on high-quality crop production can be analyzed by studying the effects of various ecological climate changes in the study area on crop growth. The preceding results reveal obvious differences among the three soil moisture products, and the products also have highly varying performance at different periods. Since the pixels in the study area were monotonous, we analyzed the factors including vegetation, SST, and even the actual soil moisture itself that may affect the products.

4.1. The Vegetation Cover Effect

As shown in Figure 15, the EVI, representing the vegetation change is used to be compared with the differences among the products and the in situ soil moisture. With the EVI increasing, only the JAXA product underestimates more obviously. The differences grow between the product and the in situ soil moisture. The differences between the LPRM product and the in situ data almost remain as the EVI rises, but some soil moisture data are clearly underestimated. In the case of lower EVI, the differences between the LPRM product and the JAXA product are concentrated at a lower level. With the increase of the EVI, the vegetation coverage becoming dense, the range of the differences is expanded. According to Figure 13, the EVI keeps rising until declining to 0.33 at the end of the experimental period. Compared with the high EVI (0.54~0.65) in the dense vegetation cover in summer, the EVI has decreased significantly, which is much closer to the spring EVI (0.11~0.19). It manifests that the differences between the products and the in situ data do not grow smaller as the EVI drops to 0.33. In particular, the JAXA product is still similar to the situation with the high EVI in summer. The differences between the three products and the in situ data were analyzed after the recovery of the FY3B product in the later experimental period. It was found that the performance of FY3B seems to be better than the JAXA and LPRM products at this time. The differences between the FY3B product and the in situ soil moisture are completely lower than the JAXA's and basically lower than the LPRM's. In addition, the P values in the significance level were 0.0341 and 0.0001 respectively.

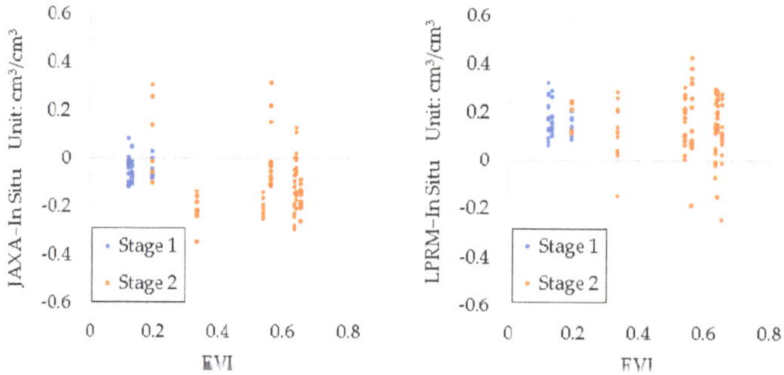

(a) (b)

Figure 15. *Cont.*

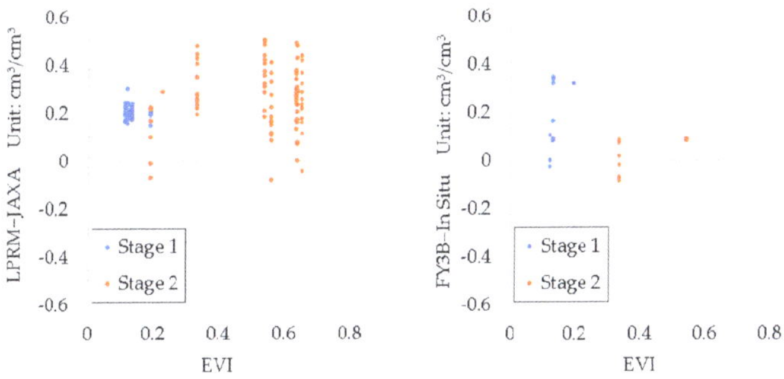

(c) (d)

Figure 15. The effect of the EVI to the soil moisture products. (**a**) is the difference between the JAXA product and the in situ soil moisture with the EVI, (**b**) is the difference between the LPRM product and the in situ soil moisture with the EVI, (**c**) is the difference between LPRM product and the JAXA product with the EVI, and (**d**) is the difference between the FY3B product and the in situ soil moisture with the EVI.

4.2. The Effect of the SST

The SST was in situ measurement like soil moisture, and both were consistent in time and space, which means that the SST is more accurate and reliable. On the other hand, the SST can directly affect the brightness temperature and emissivity of the surface soil, which will affect the retrieval results of soil moisture. Therefore, it is meaningful to study the effects of the SST on the results. The performance of the soil moisture products with the SST is shown in Figure 16. It can be seen that the JAXA product is very close to the in situ soil moisture at the first stage, and its distribution with SST is also similar to the in situ soil moisture's. At the second stage, the JAXA's distribution becomes dispersed with the SST increasing. It is different from the in situ soil moisture in Figure 12, and no obvious bevel edge appears on the lower side. At the same time, the underestimation of the JAXA product with the SST is not stable, and some data even overestimate the soil moisture. At the second stage, the JAXA product has larger gap to the in situ data overall. From the linear prediction in Figure 16a, the JAXA increases with the SST increasing. This offsets the previous underestimation to some extent. Although some of this is caused by precipitation, the LPRM product that are more sensitive to precipitation events has not shown an upward trend at this stage. Therefore, we believe that the soil moisture product of JAXA algorithm is still obviously affected by the SST at this stage. The variation trend of the LPRM product is relatively consistent during the whole experimental period. At the second stage, its distribution also has a fuzzy edge similar to Figure 12. However, the range of the LPRM product is different due to its overall overestimation. In addition, the LPRM product also shows some underestimation as the SST rises. It all occurs at the SST above 15 °C except for one time. The difference between the in situ data and the products all become more dispersed with the increasing SST at the second stage. The difference between the JAXA and the LPRM soil moisture products is stable at the first stage, but it becomes smaller as the SST increases at the second stage. This may be that the JAXA product range is 0 to 0.6 cm^3/cm^3, while the LPRM product is 0 to 1 cm^3/cm^3. When the actual soil moisture rises, the difference between the two products is enlarged. When there is no rain event accompanied by an increasing SST, both products become smaller, so that the gap between them is shrunk.

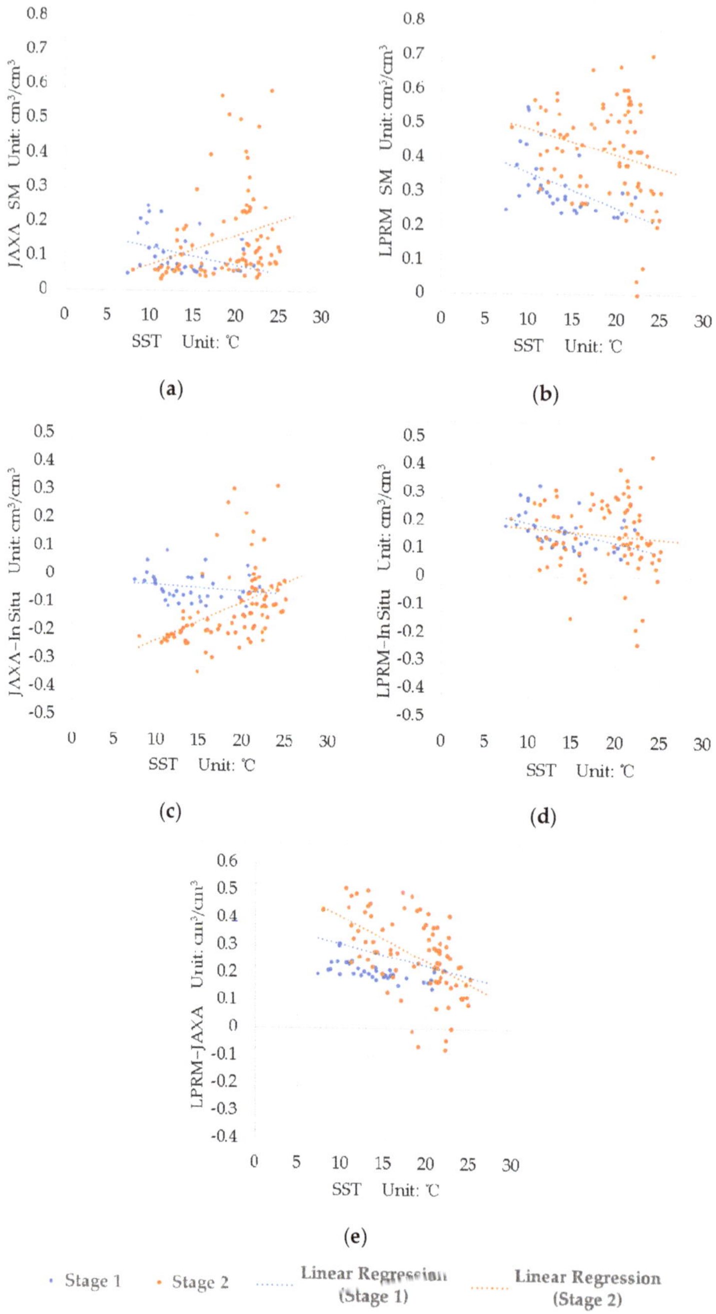

Figure 16. The effect of the SST to the soil moisture products. (**a**) shows the JAXA product with the SST, (**b**) shows the LPRM product with the SST, (**c**) shows the difference between the JAXA product and the in situ soil moisture with the SST, (**d**) shows the difference between the LPRM product and the in situ soil moisture with the SST, and (**e**) shows the difference between the LPRM and the JAXA products with the SST.

4.3. The Actual Soil Moisture Change

Both the EVI and the SST have a very significant increase from the first stage to the second stage. In addition, the performance of the JAXA product and the LPRM product at the first stage is generally better than that at the second stage. As can be seen from Figure 13, the EVI and the SST decline at the end of the experimental period, but the errors in the JAXA and the LPRM products are not improved apparently. On the other hand, although the in situ soil moisture varies locally, it has been increasing in the overall experimental period.

As shown in Figure 17, the change of the products' errors and the different between the two AMSR2 products are compared with the in situ soil moisture. It can show the impact of the actual surface soil moisture itself on the products' performance more clearly than Figure 14. It can be seen that although some data may be close or overestimated, most of the JAXA product more underestimate the soil moisture as the in situ soil moisture increases. There is a significant bevel edge on the bottom of the distribution of the difference between the JAXA product and the in situ data, which is related to the fact that the product value has been maintained at a low level and the product range is 0 to 0.6 cm^3/cm^3. However, the points of scatterplots of the LPRM product does not have such a clear edge. Its range of the variability become larger, and so is its error. However, the range of the in situ soil data is basically below 0.4 cm^3/cm^3 that is not beyond the product range. As shown in Figure 17b, the linear predictions of the differences between the LPRM product and the in situ data slightly increase with the actual soil moisture increasing during the both periods. In addition, two slopes in two periods are almost the same. One reason is that the difference between the range of the LPRM product and the field capacity becomes obvious as the actual soil moisture increasing. Previous studies have shown that the LPRM algorithm is very sensitive to the temporal variability of soil moisture, but its absolute accuracy is difficult to guarantee [31,32,44,45]. The LPRM product generally overestimates the soil moisture and is also higher than the JAXA product [24,25]. The reason may be that the soil moisture range of the LPRM algorithm is 0-1 cm^3/cm^3, but the field capacity is generally ~0.5 cm^3/cm^3 [25], and the LPRM is very sensitive to the temporal variability, so it is more likely to exceed the actual soil moisture. The LPRM product generally has a good correlation with in situ data without considering the absolute accuracy [46]. This may also explain that the overestimation bias of the LPRM product is large, but the correlation is good in this paper. The changes in the two AMSR2 products also lead to an increase and complexity in the difference between the two soil moisture products as the in situ soil moisture increases. After that, even though the EVI and the SST both fall to the level close to the first stage, the performance of the JAXA and the LPRM products are not enhanced. It is worth noting that as shown in Figure 17d, the FY3B product performs better than the JAXA and the LPRM products in the case of high soil moisture. Although the amount of the data is limited, the FY3B product seems to have certain advantages at the end of the experimental range where both the EVI and the SST decrease but the in situ soil moisture continues to rise.

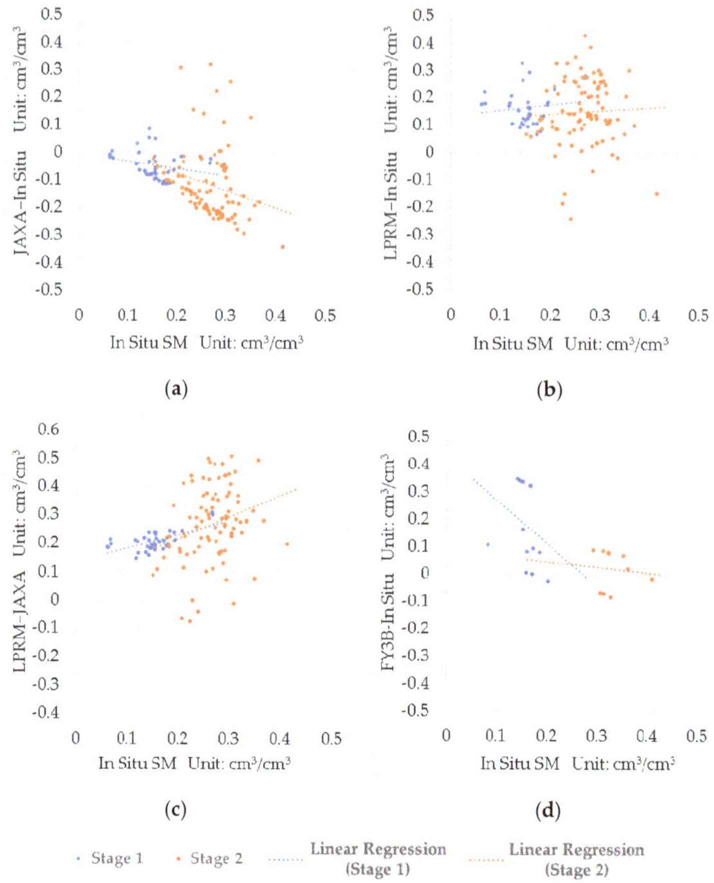

(a) (b) (c) (d)

Figure 17. The effect of the actual soil moisture to the soil moisture products. (**a**) shows the difference between the JAXA product and the in situ soil moisture with the in situ soil moisture, (**b**) shows the difference between the LPRM product and the in situ soil moisture with the in situ soil moisture, (**c**) shows the difference between the LPRM product and the JAXA product with the in situ soil moisture, and (**d**) is the difference between the FY3B product and the in situ soil moisture with the in situ soil moisture.

5. Conclusions

In this paper, an in situ soil moisture observation network in cropland on pixel scale in the northeast of China was designed considering the unique climatic characteristics of the regional area and the detection depth of the satellite sensors. The crop-growing season from May to September 2017, almost covering the whole frost-free period, was selected as the experimental period. Multiple EC-5 soil moisture sensors were arranged in a typical cropland in the northeast of China as the study area to obtain data every hour. All the sensors were calibrated according to the soil in the experimental area, so that the in situ soil moisture was consistent with the satellite products in terms of time, space, and depth. The results showed that JAXA product underestimated with a b of -0.094 cm^3/cm^3 and the LPRM product seriously overestimated the soil moisture with a b of 0.156 cm^3/cm^3 throughout the whole experimental period. The FY3B product was severely deficient in the experimental period and was all absent when the EVI was above 0.5. When it was bare soil or less vegetation cover, the JAXA product had the best performance of errors with the lowest ubRMSE at 0.049 cm^3/cm^3 and the b at

-0.043 cm^3/cm^3 that reached the goal accuracy of the product (±0.05 cm^3/cm^3), and the LPRM product had the best correlation (R = 0.654). When the EVI increased over time, all of them declined. When the EVI was low, the JAXA product errors were less affected by the SST, but the LPRM product slightly decreased with the SST increasing. When the EVI was high, the distribution of the JAXA product errors became more complicated, and some data decreased with the SST increasing, but the LPRM product did not change significantly at this time. Throughout the experimental period, the ranges and errors of the JAXA and LPRM products showed an upward trend, and only the in situ soil moisture showed a similar trend throughout the whole period. Therefore, the JAXA product error was smaller and the LPRM product correlation was better. At the end, the EVI and SST turned to decrease and the in situ soil moisture kept increasing. Although the amount of data was limited, the FY3B product seemed to have a better error performance than the JAXA and LPRM products.

Author Contributions: Conceptualization, H.F.; methodology, H.F. and T.Z.; software, H.F. and T.Z.; validation, H.F. and T.Z.; formal analysis, H.F.; investigation, H.F.; resources, C.S.; data curation, H.F.; writing—original draft preparation, H.F.; writing—review and editing, H.F. T.Z. and C.S.; visualization, H.F. and T.Z.; supervision, C.S.; project administration, C.S.; funding acquisition, C.S.

Funding: This work was supported by National Natural Science Foundation of China (NSFC) (11574113, 11374123, 11104106); Science and Technology Planning Project of Jilin Province (20180101238JC, 20170204076GX, 20180101006JC, 20190103041JH), China Postdoctoral Science Foundation (BX20180127).

Acknowledgments: The first author would like to thank Changchun Jingyuetan Remote Sensing Experiment Station, Chinese Academy of Sciences for the support for the in situ soil moisture and temperature data. The first author especially thank Prof. Xingming Zheng at the Northeast Institute of Geography and Agroecology, Chinese Academy of Sciences. Without his great contribution to the conceptualization, this article cannot be done. The first author also thank Prof. Lingjia Gu at College of Electronic Science and Engineering, Jilin University for her the guidance to this article. The first author also is grateful to Dr. Tao Jiang and Dr. Yu Bai at the Northeast Institute of Geography and Agroecology, Chinese Academy of Sciences for the cooperation in the construction of the In Situ Observation Network.

Conflicts of Interest: The authors declare no conflict of interest.

Appendix A

Table A1. The performance metrics of the JAXA and the LPRM soil moisture products at different study periods with the in situ data using direct average. The best one for each performance metric is in bold.

Period	Products	RMSE (cm^3/cm^3)	ubRMSE (cm^3/cm^3)	b (cm^3/cm^3)	R
Whole Period	JAXA	0.144	0.113	−0.090	0.321
	LPRM	0.194	0.109	0.160	0.551
First Stage	JAXA	**0.061**	**0.049**	**−0.037**	0.413
	LPRM	0.184	0.063	0.173	**0.579**
Second Stage	JAXA	0.167	0.124	−0.112	0.252
	LPRM	0.198	0.122	0.155	0.424

Table A2. The comparison of the performance metrics of the three soil moisture products as the period of the FY3B product with the in situ data using direct average. The best one for each performance metric is in bold.

Period	Products	RMSE (cm^3/cm^3)	ubRMSE (cm^3/cm^3)	b (cm^3/cm^3)	R
as FY3B	**FY3B**	0.236	0.155	0.178	0.018
	JAXA	**0.228**	**0.087**	0.211	0.144
	LPRM	0.232	0.158	**0.170**	**0.500**

References

1. Chen, F.; Avissar, R. Impact of land-surface moisture variability on local shallow convective cumulus and precipitation in large-scale models. *J. Appl. Meteorol.* **1994**, *33*, 1382–1401. [CrossRef]
2. Seneviratne, S.I.; Corti, T.; Davin, E.L.; Hirschi, M.; Jaeger, E.B.; Lehner, I.; Orlowsky, B.; Teuling, A.J.J.E.-S.R. Investigating soil moisture–climate interactions in a changing climate: A review. *Earth-Sci. Rev.* **2010**, *99*, 125–161. [CrossRef]
3. Koster, R.D.; Dirmeyer, P.A.; Guo, Z.; Bonan, G.; Chan, E.; Cox, P.; Gordon, C.; Kanae, S.; Kowalczyk, E.; Lawrence, D.J.S. Regions of strong coupling between soil moisture and precipitation. *Science* **2004**, *305*, 1138–1140. [CrossRef] [PubMed]
4. Gentine, P.; Polcher, J.; Entekhabi, D. Harmonic propagation of variability in surface energy balance within a coupled soil-vegetation-atmosphere system. *Water Resour. Res.* **2011**, *47*. [CrossRef]
5. Seneviratne, S.I.; Lüthi, D.; Litschi, M.; Schär, C. Land–atmosphere coupling and climate change in Europe. *Nature* **2006**, *443*, 205–209. [CrossRef] [PubMed]
6. Bateni, S.; Entekhabi, D. Relative efficiency of land surface energy balance components. *Water Resour. Res.* **2012**, *48*, 4. [CrossRef]
7. Kędzior, M.; Zawadzki, J. SMOS data as a source of the agricultural drought information: Case study of the Vistula catchment, Poland. *Geoderma* **2017**, *306*, 167–182. [CrossRef]
8. Fujii, H.; Koike, T.; Imaoka, K. Improvement of the AMSR-E algorithm for soil moisture estimation by introducing a fractional vegetation coverage dataset derived from MODIS data. *J. Remote Sens. Soc. Jpn.* **2009**, *29*, 282–292.
9. Owe, M.; de Jeu, R.; Walker, J.; Zukor, D.J. A methodology for surface soil moisture and vegetation optical depth retrieval using the microwave polarization difference index. *IEEE Trans. Geosci. Remote Sens.* **2001**, *39*, 1643–1654. [CrossRef]
10. Mo, T.; Choudhury, B.; Schmugge, T.; Wang, J.; Jackson, T.J. A model for microwave emission from vegetation-covered fields. *J. Geophys. Res.: Oceans.* **1982**, *87*, 11229–11237. [CrossRef]
11. Jackson, T.J. Measuring surface soil moisture using passive microwave remote sensing. *Hydrol. Process.* **1993**, *7*, 139–152. [CrossRef]
12. Holmes, T.R.H.; De Jeu, R.A.M.; Owe, M.; Dolman, A.J. Land surface temperature from Ka band (37 GHz) passive microwave observations. *J.Geophys. Res.* **2009**, *114*. [CrossRef]
13. Kerr, Y.H.; Waldteufel, P.; Wigneron, J.-P.; Delwart, S.; Cabot, F.; Boutin, J.; Escorihuela, M.-J.; Font, J.; Reul, N.; Gruhier, C.; et al. The SMOS Mission: New Tool for Monitoring Key Elements ofthe Global Water Cycle. *Proc. IEEE* **2010**, *98*, 666–687. [CrossRef]
14. Entekhabi, D.; Njoku, E.G.; O'Neill, P.E.; Kellogg, K.H.; Crow, W.T.; Edelstein, W.N.; Entin, J.K.; Goodman, S.D.; Jackson, T.J.; Johnson, J.; et al. The Soil Moisture Active Passive (SMAP) Mission. *Proc. IEEE* **2010**, *98*, 704–716. [CrossRef]
15. Imaoka, K.; Maeda, T.; Kachi, M.; Kasahara, M.; Ito, N.; Nakagawa, K. Status of AMSR2 instrument on GCOM-W1. In Proceedings of Earth Observing Missions and Sensors: Development, Implementation, and Characterization II. *Intern. Soc. Opt. Photonics* **2012**, *8528*, 852815.
16. Njoku, E.G.; Jackson, T.J.; Lakshmi, V.; Chan, T.K.; Nghiem, S.V. Soil moisture retrieval from AMSR-E. *IEEE Trans. Geosci.Remote Sens.* **2003**, *41*, 215–229. [CrossRef]
17. Sun, R.; Zhang, Y.; Wu, S.; Yang, H.; Du, J. The FY-3B/MWRI soil moisture product and its application in drought monitoring. In Proceedings of the 2014 IEEE Geoscience and Remote Sensing Symposium, Quebec City, QC, Canada, 13–18 July 2014; pp. 3296–3298.
18. Dorigo, W.; de Jeu, R.; Chung, D.; Parinussa, R.; Liu, Y.; Wagner, W.; Fernández-Prieto, D. Evaluating global trends (1988–2010) in harmonized multi-satellite surface soil moisture. *Geophys. Res. Lett.* **2012**, *39*, 18. [CrossRef]
19. Cui, C.; Xu, J.; Zeng, J.; Chen, K.-S.; Bai, X.; Lu, H.; Chen, Q.; Zhao, T. Soil Moisture Mapping from Satellites: An Intercomparison of SMAP, SMOS, FY3B, AMSR2, and ESA CCI over Two Dense Network Regions at Different Spatial Scales. *Remote Sens.* **2017**, *10*, 33. [CrossRef]
20. Wu, Q.; Liu, H.; Wang, L.; Deng, C. Evaluation of AMSR2 soil moisture products over the contiguous United States using in situ data from the International Soil Moisture Network. *Int. J.Appl. Earth Obs.Geoinfomr.* **2016**, *45*, 187–199. [CrossRef]

21. Cho, E.; Su, C.-H.; Ryu, D.; Kim, H.; Choi, M. Does AMSR2 produce better soil moisture retrievals than AMSR-E over Australia? *Remote Sens.Env.* **2017**, *188*, 95–105. [CrossRef]
22. Cho, E.; Moon, H.; Choi, M. First Assessment of the Advanced Microwave Scanning Radiometer 2 (AMSR2) Soil Moisture Contents in Northeast Asia. *J. Meteorol. Soc. Japan. Ser. II* **2015**, *93*, 117–129. [CrossRef]
23. Kim, H.; Parinussa, R.; Konings, A.G.; Wagner, W.; Cosh, M.H.; Lakshmi, V.; Zohaib, M.; Choi, M. Global-scale assessment and combination of SMAP with ASCAT (active) and AMSR2 (passive) soil moisture products. *Remote Sens. Env.* **2018**, *204*, 260–275. [CrossRef]
24. Bindlish, R.; Cosh, M.H.; Jackson, T.J.; Koike, T.; Fujii, H.; Chan, S.K.; Asanuma, J.; Berg, A.; Bosch, D.D.; Caldwell., T.; Collins, C.H.; McNairn, H.; Martínez-Fernández, J.; et al. GCOM-W AMSR2 soil moisture product validation using core validation sites. *IEEE J. Sel. Top. Appl. Earth Obs.Remote Sens.* **2018**, *11*, 209–219. [CrossRef]
25. Kim, S.; Liu, Y.Y.; Johnson, F.M.; Parinussa, R.M.; Sharma, A. A global comparison of alternate AMSR2 soil moisture products: Why do they differ? *Remote Sens. Env.* **2015**, *161*, 43–62. [CrossRef]
26. Entekhabi, D.; Reichle, R.H.; Koster, R.D.; Crow, W.T. Performance metrics for soil moisture retrievals and application requirements. *J. Hydrometeorol.* **2010**, *11*, 832–840. [CrossRef]
27. Njoku, E.G.; Ashcroft, P.; Chan, T.K.; Li, L.; Sensing, R. Global survey and statistics of radio-frequency interference in AMSR-E land observations. *IEEE Trans. Geosci. Remote Sens.* **2005**, *43*, 938–947. [CrossRef]
28. Li, L.; Njoku, E.G.; Im, E.; Chang, P.S.; Germain, K.; Sensing, R. A preliminary survey of radio-frequency interference over the US in Aqua AMSR-E data. *IEEE Trans. Geosci. Remote Sens..* **2004**, *42*, 380–390. [CrossRef]
29. Njoku, E.; Chan, S. Vegetation and surface roughness effects on AMSR-E land observations. *Remote Sens. Env.* **2006**, *100*, 190–199. [CrossRef]
30. Jackson, T.J.; Schmugge, T.J. Vegetation Effects on Microwave Emission of soils. *Remote Sens. Env.* **1991**, *36*, 203–212. [CrossRef]
31. Draper, C.S.; Walker, J.P.; Steinle, P.J.; De Jeu, R.A.; Holmes, T.R.H. An evaluation of AMSR–E derived soil moisture over Australia. *Remote Sens. Env.* **2009**, *113*, 703–710. [CrossRef]
32. de Jeu, R.A.M.; Wagner, W.; Holmes, T.R.H.; Dolman, A.J.; van de Giesen, N.C.; Friesen, J. Global Soil Moisture Patterns Observed by Space Borne Microwave Radiometers and Scatterometers. *Surv. Geophys.* **2008**, *29*, 399–420. [CrossRef]
33. Maeda, T.; Taniguchi, Y. Descriptions of GCOM-W1 AMSR2 Level 1R and Level 2 Algorithms. Available online: https://suzaku.eorc.jaxa.jp/GCOM_W/data/doc/NDX-120015A.pdf (accessed on 21 February 2019).
34. Lu, H.; Koike, T.; Fujii, H.; Ohta, T.; Tamagawa, K. Development of a physically-based soil moisture retrieval algorithm for spaceborne passive microwave radiometers and its application to AMSR-E. *J. Remote Sens. Soc. Jpn.* **2009**, *29*, 253–262.
35. Owe, M.; de Jeu, R.; Holmes, T. Multisensor historical climatology of satellite-derived global land surface moisture. *J. Geophys. Res. Earth Surf.* **2008**, *113*, F1. [CrossRef]
36. Meesters, A.G.; De Jeu, R.A.; Owe, M.; Letters, R.S. Analytical derivation of the vegetation optical depth from the microwave polarization difference index. *IEEE Geosci. Remote Sens. Lett..* **2005**, *2*, 121–123. [CrossRef]
37. Shi, J.; Jiang, L.; Zhang, L.; Chen, K.-S.; Wigneron, J.-P.; Chanzy, A.; Jackson, J.P.; Sensing, R. Physically based estimation of bare-surface soil moisture with the passive radiometers. *IEEE Trans.Geosci. Remote Sens.* **2006**, *44*, 3145–3153. [CrossRef]
38. Jackson, T.J.; Le Vine, D.M.; Hsu, A.Y.; Oldak, A.; Starks, P.J.; Swift, C.T.; Isham, J.D.; Haken, M. Soil moisture mapping at regional scales using microwave radiometry: The Southern Great Plains Hydrology Experiment. *IEEE Trans. Geosci. Remote Sens.* **1999**, *37*, 2136–2151. [CrossRef]
39. MODIS vegetation index user's guide (MOD13 Series). Available online: https://icdc.cen.uni-hamburg.de/fileadmin/user_upload/icdc_Dokumente/MODIS/MODIS_Collection6_VegetationIndex_UsersGuide_MOD13_V03_June2015.pdf (accessed on 21 February 2019).
40. Şen, Z. Average areal precipitation by percentage weighted polygon method. *J. Hydrologic Engineering.* **1998**, *3*, 69–72. [CrossRef]
41. Rhynsburger, D. Analytic delineation of Thiessen polygons. *Geogr. Anal.* **1973**, *5*, 133–144. [CrossRef]
42. Song, C.; Jia, L. A Method for Downscaling FengYun-3B Soil Moisture Based on Apparent Thermal Inertia. *Remote Sens.* **2016**, *8*, 703. [CrossRef]

43. Wang, G.; Hagan, D.F.T.; Lou, D.; Chen, T. Evaluation of soil moisture derived from FY3B microwave brightness temperature over the Tibetan Plateau. *Remote Sens. Lett.* **2016**, *7*, 817–826. [CrossRef]
44. Brocca, L.; Hasenauer, S.; Lacava, T.; Melone, F.; Moramarco, T.; Wagner, W.; Dorigo, W.; Matgen, P.; Martínez-Fernández, J.; Llorens, P.; et al. Soil moisture estimation through ASCAT and AMSR-E sensors: An intercomparison and validation study across Europe. *Remote Sens. Env.* **2011**, *115*, 3390–3408. [CrossRef]
45. Parinussa, R.M.; Holmes, T.R.H.; De Jeu, R.A.M. Soil Moisture Retrievals From the WindSat Spaceborne Polarimetric Microwave Radiometer. *IEEE Trans. Geosci. Remote Sens.* **2012**, *50*, 2683–2694. [CrossRef]
46. Parinussa, R.M.; Wang, G.; Holmes, T.R.H.; Liu, Y.Y.; Dolman, A.J.; De Jeu, R.A.M.; Jiang, T.; Zhang, P.; Shi, J. Global surface soil moisture from the Microwave Radiation Imager onboard the Fengyun-3B satellite. *Int. J Remote Sens.* **2014**, *35*, 7007–7029. [CrossRef]

remote sensing

MDPI

Letter

Regionalization of Coarse Scale Soil Moisture Products Using Fine-Scale Vegetation Indices—Prospects and Case Study

Mengyu Liang [1,*], Marion Pause [2], Nikolas Prechtel [2] and Matthias Schramm [3]

1 Department of Geographical Sciences, University of Maryland–College Park, College Park, MD 20740, USA
2 Faculty of Environmental Sciences, TU Dresden, 01062 Dresden, Germany; marion.pause@tu-dresden.de (M.P.); nikolas.prechtel@tu-dresden.de (N.P.)
3 Department of Geodesy and Geoinformation, TU Wien, 1040 Vienna, Austria; matthias.schramm@geo.tuwien.ac.at
* Correspondence: mliang77@terpmail.umd.edu; Tel.: +1-301-405-4050

Received: 31 December 2019; Accepted: 4 February 2020; Published: 7 February 2020

Abstract: Surface soil moisture (SSM) plays a critical role in many hydrological, biological and biogeochemical processes. It is relevant to farmers, scientists, and policymakers for making effective land management decisions. However, coarse spatial resolution and complex interactions of microwave radiation with surface roughness and vegetation structure present limitations within active remote sensing products to directly monitor soil moisture variations with sufficient detail. This paper discusses a strategy to use vegetation indices (VI) such as greenness, water stress, coverage, vigor, and growth dynamics, derived from Earth Observation (EO) data for an indirect characterization of SSM conditions. In this regional-scale study of a wetland environment, correlations between the coarse Advanced SCATterometer-Soil Water Index (ASCAT-SWI or SWI) product and statistical measurements of four vegetation indices from higher resolution Sentinel-2 data were analyzed. The results indicate that the mean value of Fraction of Absorbed Photosynthetically Active Radiation (FAPAR) correlates most strongly to the SWI and that the wet season vegetation traits show stronger linear relation to the SWI than during the dry season. The correlation between VIs and SWI was found to be independent of the underlying dominant vegetation classes which are not derived in real-time. Therefore, fine-scale vegetation information from optical satellite data convey the spatial heterogeneity missed by coarse synthetic aperture radar (SAR)-derived SSM products and is linked to the SSM condition underneath for regionalization purposes.

Keywords: surface soil moisture; regional scale; vegetation traits; multi-sensor approach; wetland; environmental monitoring

1. Introduction

Detecting surface soil moisture (SSM) is a key challenge in EO and of great interest in environmental monitoring at all spatial scales [1–5]. Advancements are available for SSM remote sensing and algorithms with a focus on large-scale (continental/global) applications [1–5]. Regional, intermediate-to-small-catchment scale soil-moisture monitoring can be performed without or independently of satellite remote sensing observations using techniques such as low-energy cosmic-ray neutrons and Proximal Gamma-Ray (PGR) spectroscopy [6,7]. However, it is of increasing importance for governmental agencies, scientists, and farmers to monitor SSM change in relations with climate and weather with fine spatial and temporal resolution remote sensing products [1–5], due to the fact that SSM tightly links to many hydrological, biological and biogeochemical processes at these finer scales. Thus, the mismatch between the availability and need of fine spatial resolution remote-sensing-based

SSM presents a clear gap for many environmental monitoring and applications on the one hand, and on the other hand, it highlights the importance of such information for local adaption efforts to mitigate the effects of climate change.

Due to its high significance in the Earth system, SSM has always solicited much attention in its measurement and monitoring using remote sensing techniques. SSM observations are retrieved from instruments sensing at microwave and optical/thermal infrared wavelength. Some key soil moisture products include the well-known ESA SMOS (Soil Moisture Ocean Salinity), SMAP (Soil Moisture Active Passive), and ASCAT-SWI [8–13]. However, most publicly available SSM datasets have coarse spatial resolutions ranging from 25–50 km [8–13]. A recent breakthrough is the 1 km Sentinel-1/ASCAT fusion product [13], which is currently available across Europe. Moreover, microwave radiation of soil water content is sensitive to surface roughness, as induced by a dynamic vegetation structure, for instance. As surface roughness increases or the vegetation canopy gets higher, the backscatter from differently polarized signals converges, whilst the noise level is getting higher [14,15]. Thus, in addition to the scale issue, traditional active remote-sensing-based SSM products cannot easily account for high variability in the terrain parameters and are affected by such noise when sensing landscapes with complex land cover and water patterns, for example, in the case of wetland areas.

Many efforts are currently taking place to explore the use of vegetation traits derived from passive remote sensing products for SSM monitoring. Alexandridis et al. (2016) adopted an integrated approach to derive evaporative fraction and saturated water content with the thermal infrared data from MODIS, in combination with ancillary soil and meteorology data, to produce 250m resolution soil moisture map over sites in Europe [16]. Torres-Rua et al. (2016) combined Normalized Difference Vegetation Index (NDVI), Leaf Area Index (LAI), energy balance product from Landsat 7, and weather data, and used Relevance Vector Machine (RVM) to relate these potential predictors to SSM [17]. Pause et al. (2012) combined L-band brightness temperature observations and hyperspectral vegetation indices to estimate and improve SSM patterns at the field scale [18]. Qiu et al. (2018) explored the parameterization of SAR vegetation scattering model for high-resolution SSM retrieval with VIs (NDVI, EVI, LAI) and surface roughness derived from Moderate Resolution Imaging Spectroradiometer (MODIS) and Landsat using both the Advanced Integral Equation Model (AIEM) and the Water Cloud Model (WCM) [19]. Klinke et al. (2018) used plant characteristics and temperature as indicators from Sentinel (1, 2) and Landsat archives to derive a high spatial resolution soil moisture product for wetlands in northeastern Germany [20]; same potential of coupling Sentinel 1 and 2 for soil moisture downscaling has also been examined by El Hajj et al. (2017) [21]. Dabrowska-Zielinska et al. (2018) estimated wetland SSM using Sentinel-1 data and addressed the vegetation effect on Radar backscattering change under different SSM and NDVI conditions. In her study, she pointed out that vegetation has a different influence on the backscattering of different polarizations, depending on measurements under dry (soil moisture $<$ 30 vol. %) or moist conditions (soil moisture $>$ 60 vol. %) [22]. Additionally, Samaniego et al. (2010) highlighted the issue of over parameterization and ineffectiveness in integrating spatial heterogeneity in multiscale hydrological models and proposed a multiscale parameter regionalization technique (MPR) to link the dominant process parameter with the finer resolution input data through upscaling operators such as the harmonic mean [23]. Therefore, the work discussed in this paper is based on these critical efforts and aims to address the current SSM monitoring challenges with innovative approaches.

This work mainly focuses on obtaining vegetation information from fine spatial resolution optical EO data and using this information to understand the influence of vegetation on the estimation of spatially varying SSM. The downscaling efforts allow a closer examination of SSM variations over the wetland ecosystem. Specifically, the paper addresses the research question of whether the spatial-temporal heterogeneity in vegetation traits as observed by Sentinel-2 data can be an indicator for SSM as represented by the ASCAT-SWI product. The results illustrate a link between fine-resolution vegetation traits and the soil humidity conditions in wetland environments and demonstrate the potential of using vegetation as sensors for SSM. The work also highlights the commonly used vegetation indices (VIs) and their usability in uncovering the spatial and seasonal relationships between vegetation and SSM on a regional scale. The paper first describes the current progress and gaps in SSM remote sensing, characterizes the study site selection and experiment workflow, highlights the results, provides a discussion on the results, ecological relevance, and limitation, and then concludes the study for future implications.

2. Materials and Methods

2.1. Study Areas

The study area is the Okavango Delta, located in northern Botswana between −18.23 and −18.51 °S, 21.84 and 23.81 °E, shown in Figure 1a. The size of the delta is approximately 16,000 km^2, varying between dry and wet seasons. The climate of the surrounding area is semi-arid; the annual average precipitation ranges from approximately 400 to 500 mm, and the mean annual temperature ranges from 15~20 °C. The wet season usually begins in December, peaks in January and February, and finishes by March. Water infiltrates the Okavango Delta through the Okavango River from the Angolan Plateau in the Northwest. In Figure 1a, the locations of the in-situ water level stations Mohembo (North) and Guma (South) are marked.

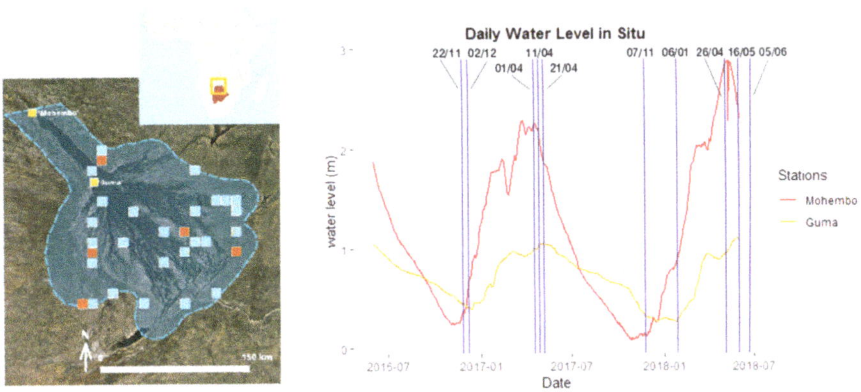

(a) (b)

Figure 1. Study area and the regional flow dynamics: (**a**) Extent of the Okavango Delta and the locations of the in-situ water stations (yellow); the locations of five experimental sites (red) and the extended sites (cyan) are indicated; (**b**) Water level records from 2016-06 to 2018-06 measured at Mohembo (red line) and Guma (yellow line) stations, and blue vertical lines indicate the dates of Sentinel-2 imagery and ASCAT-SWI data retrieval.

2.2. Data and Pre-Processing

The Sentinel-2A Level-1C products used in this research were obtained from the United States Geological Survey (USGS) EarthExplorer [24]. This cloud-free multispectral imagery corresponds to ten dates (2016-11-22, 2016-12-02, 2017-04-01, 2017-04-11, 2017-04-21, 2017-11-07, 2018-01-06, 2018-04-26, 2018-05-16, 2018-06-05), and they represent the regional dry and wet seasons (Figure 1 b). For each date, six 100 km × 100 km tiles were retrieved for the study area. The Level 1C Top-of-Atmosphere (TOA) reflectance data were first resampled to the spatial resolution of Band 2 (Table 1), and then corrected for atmosphere and cirrus in the Sen2Cor processor (version 2.8.0) distributed by the Sentinel Toolbox Exploitation Platform (STEP) with its Graph Processing Tool (GPT) [25]. The corrected results were reformatted to Level 2A Top-of-Canopy (TOC) reflectance data and subset to spatial extents of the sample sites illustrated in Figure 1a.

Table 1. Overview of primary and ancillary datasets used in the research.

Data	Spatial Resolution	Temporal Coverage	Source
Sentinel-2A Level-1C	10 m (Band 2,3,4,8) 20 m (Band 5,6,7,8a,11,12)	11/22/2016~06/05/2018	Earth Explorer
ASCAT-SWI	25 km	11/22/2016~06/05/2018	CGLS
PROBA-Vegetation (PROBA-V) Land Cover	100 m	2015	CGLS
In-situ Water Level	Site-based	06/01/2016~06/01/2018	Okavango Delta Monitoring and Forecasting

The daily ASCAT-SWI data or simply SWI data, along with the quality flag and metadata, were retrieved from Copernicus Global Land Service (CGLS) as Network Common Data Form (NetCDF) files for the same dates as Sentinel-2 data [9,10]. The spatial resolution of the SWI product matches the SSM by ASCAT [9,10,26]. On board the MetOp satellite series, ASCAT is a real aperture radar instrument, and the scatterometer radar signals can penetrate the surface, thus allowing the detection the subsurface climate feature such as soil wetness [26–28]. The instrument operates during day and night, under all weather conditions, hence the rapid global coverage. The processing of the NetCDF files was done in RStudio.

The in-situ water level records measured at two stations, Mohembo (Latitude: −18.275733, Longitude: 21.787312) and Guma (Latitude: −18.96266, Longitude: 22.373213), were retrieved from the Okavango Delta Monitoring & Forecasting service at daily resolution [29]. The missing data were interpolated with the Kriging method in RStudio. The water level records at both gauges in Figure 1b show clear seasonality. As previously discussed, the Delta region receives a low amount of precipitation; thus, the majority of the water supplied to the Delta is largely related to the Okavango River runoff measured at the two inlet stations. Hence, the wet season in the scope of this research is defined as the time of the year at which high water levels occur at both stations in April through June; dry season is when the water level is low at both gauges in November through January.

The 100m resolution Dynamic Land Cover map of Africa was also obtained from CGLS. The product was derived from the PROBA-V time series for the year 2015 over the continental Africa [30]. The discrete land cover classification and the cover fraction layer for seasonal inland water areas were used in this study to identify the dominant land cover type of each sample site and to eliminate sites with a large water extent that would pollute the SWI signals. The dominant land cover type for each site is the discrete class that has the highest percent coverage at the site.

2.3. SWI Products and Sample Sites

SSM can be directly estimated from ASCAT observations at daily temporal resolution, but profile soil moisture cannot be directly measured by remote sensing [26]. To gain insight into the moisture condition beyond the surface soil layer, a relationship between surface and profile soil moisture has to be established, and Wagner et al. (1999) developed a two-layer water balance model to describe this relationship as a function of time [26]. The ASCAT-SWI product was developed within this framework by using the moisture conditions for different characteristic time lengths to represent different depths. Furthermore, the SSM over the preceding time period, T, was summed and exponentially weighted. T determines how fast the weight becomes smaller and how strongly the SSM observations taken in the past influence the current SWI [26]. The selection of $T = 10$ *days* was found to be suitable for estimating the influence of recent SSM measures on the SWI [28]. Since the described model was designed to be independent of soil texture and does not involve any vegetation information in its calculation [9,10,26,27], a correlation analysis between the SSM product and the vegetation traits was appropriate to conduct.

A total of 30 sample sites are selected across the Delta (Figure 1). Each site's spatial extent corresponds to a 25km resolution SWI cell and is referenced by the CGLS SWI product cell number [9]. Therefore, five experimental sites were selected in a first step, based on their geographical locations in the Delta and their land cover types to serve as a proof of concept for the workflow developed. Then, 25 additional sample sites were randomly selected across the Delta and evaluated regarding their suitability to be analyzed with the same methods as the experimental sites. Two sites were excluded manually due to their high cover fraction of inland seasonal water. A total of 28 sites were analyzed in the study for all ten dates with cloud-free images. This workflow, including data preprocessing, SWI retrieval, VI calculation, and correlation analysis, cannot be applied to every SWI pixel because some SWI pixels crossing the Sentinel-2 tiles were cloud-covered or flooded. Therefore, an analysis of continuous spatial coverage is not feasible.

2.4. Vegetation Indices (VIs) Retrieval

Four VIs, NDVI, NDWI, LAI, and FAPAR, were retrieved for the six tiles from the pre-processed Sentinel-2 data over the 10 dates (Figure 1b). VIs were batch-calculated with SNAP GPT [25] using the bands listed in Table 2. NDVI measures the photosynthetic activity of vegetation and describes the vitality of vegetation on the Earth's Surface [31–33]. It is included here to correlate with SWI and analyze whether the vitality and greenness of the vegetation are related to the soil water content. The algorithm for calculating NDVI is as below:

$$NDVI = (NIR - RED)/(NIR + RED), \quad (1)$$

Table 2. Sentinel-2A Level-1C spectral bands and center wavelength used for VI retrieval.

Calculated VIs	Spectral Bands	Central Wavelength (nm)	Band Width (nm)
LAI, FAPAR	B3 Green	560	35
NDVI, LAI, FAPAR	B4 Red	665	30
LAI, FAPAR	B5 Vegetation Red Edge	705	15
LAI, FAPAR	B6 Vegetation Red Edge	740	15
LAI, FAPAR	B7 Vegetation Red Edge	783	20
NDWI, NDVI	B8 NIR	842	115
LAI, FAPAR	B8a Vegetation Red Edge	865	20
NDWI, LAI, FAPAR	B11 SWIR1	1610	90
LAI, FAPAR	B12 SWIR2	2190	180

NDWI is another important index that measures the liquid water content in the canopy that interacts with the incoming solar radiation [34]. It was included in this study to analyze whether the water content in vegetation is related to the water content in soil. NDWI generally increases as the vegetation fractions and the leaf layer increase, while NDWI is generally negative in areas with naked soil [34]. Gao suggested NDWI contains information independent of NDVI [34]. The equation followed to calculate NDWI is as follow, and Band 11/ SWIR1 is used (Table 2):

$$NDWI = (NIR - SWIR) / (NIR + SWIR), \quad (2)$$

LAI and FAPAR are both calculated with the neural networks built in the Biophysical Processor of SNAP software [25,35]. LAI is defined as half the developed area of photosynthetically active elements of the vegetation per unit horizontal ground area. It is used to determine the size of the interface for energy and mass exchange between canopy and atmosphere [36,37]. FAPAR measures the fraction of photosynthetically active radiation absorbed by the canopy, and it corresponds to the canopy's primary productivity of photosynthesis [37,38]. Both VIs were analyzed to understand the vegetation's evapotranspiration and photosynthetic primary production capacity as related to SSM. Based on SNAP algorithm descriptions, to calculate each input biophysical variable (LAI or FAPAR), the neural network is trained with a representative set of TOC reflectance and with prior information on the distribution of the input variables from the training data. After adjusting the synaptic weights and neuron bias according to a combination of tangent sigmoid and linear transfer functions, the trained neural network can then be used in operational mode for new calculation. The network takes 11 normalized input data including 8 Sentinel-2 TOC reflectance wavebands (B3, B4, B5, B6, B7, B8a, B11 and B12) and the geometry of acquisitions ($\cos(\theta_s)$, $\cos(\theta_v)$, and $\cos(\theta_\varnothing)$) to output targeted biophysical variable (LAI or FAPAR) for each pixel [35].

2.5. Statistics Retrieval

The mean value, standard deviation (SD), and coefficient of variation (CV) are calculated as statistical parameters of the VIs over each study site and each date of analysis (Figure 1) to capture the central tendency and spatial heterogeneity of the VIs. Second-order image entropy and homogeneity are used to describe the image texture and to reflect vegetation structure.

Mean (μ) of the VIs is calculated using the following formula:

$$\mu = \frac{\sum x}{n}, \tag{3}$$

In (3) x denotes the VI value at each pixel; n indicates the total number of pixels in a given image.

SD (σ) describes the variation of data values in the VIs about the mean using the formula below:

$$\sigma = \sqrt{\frac{\sum |x - \mu|^2}{n}}, \tag{4}$$

In (4), x denotes VI value at each pixel; n indicates the total number of pixels in a given image.

CV is the ratio of SD to mean, and it measures the relative variability in the dataset. It adjusts the variation for the mean so it allows comparison across data values from different datasets; in comparison with SD, it highlights the variability in data overshadowed by low SD (5).

$$CV = \frac{\sigma}{\mu}, \tag{5}$$

In addition to the standard statistics, the second-order image texture, entropy and heterogeneity, are calculated using the Grey Level Co-occurrence Matrices (GLCM). The image texture of VIs can capture gradients in vegetation structure that may be overshadowed by the discrete land cover [39,40]. GLCM are a statistical texture analysis method that describe the spatial distribution of the observed intensity pairs in respect to their relative distances [40]. Entropy and homogeneity are selected among many statistical measures derivable from GLCM to represent the orderliness and contrast group within the second-order image texture measures. Entropy mainly measures the disorderedness of the image pixel and when GLCM has the same values, the entropy is the highest (6). Homogeneity measures the closeness of the distribution of elements to the GLCM diagonal (7). The mean and SD for entropy and homogeneity are calculated to summarize the texture analysis for each VI using the "glcm" package in R. The calculations use N = 32 as the number of grey levels for all directions (0 degrees, 45 degrees, 90 degrees, and 135 degrees), and a 3 × 3 window size. This window size has the advantage of capturing the heterogeneity of pixel values over a small distance [39,40].

$$entropy = \sum_{i,j=0}^{N-1} -ln\left(P_{ij}\right)P_{ij}, \tag{6}$$

$$homogenieity = \sum_{i,j=0}^{N-1} \frac{P_{ij}}{1 + (i - j)^2}, \tag{7}$$

where i is the row number and j is the column number. P_{ij} is the probability value recorded for the cell i, j; N is the number of rows or columns.

3. Results

3.1. SWI and VIs Correlation by Season

In terms of the means of most VIs, stronger positive correlations are noted for wet season observations (Figure 2). Correlations for wet seasons are slightly higher than those for mixed observations from both seasons but generally, they are not very distinct. However, it is obvious to see the dry season observations show little positive linear correlation to SWI. In the case of LAI, a slightly negative correlation can be observed but the correlation is not significant. This indicates that mean values of four VIs are positively correlated to SWI in the wet season comparing to the dry season, meaning dry season vegetation conditions vary greatly over these sample sites.

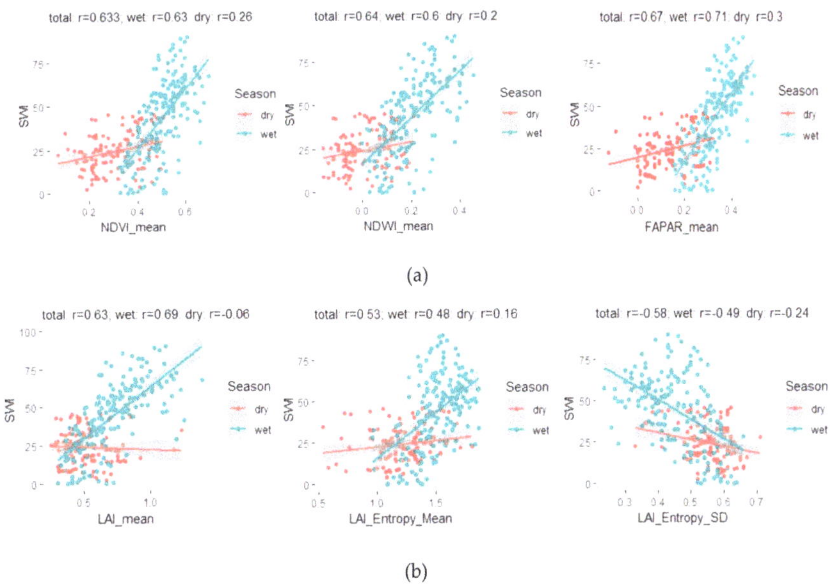

Figure 2. Selected scatterplots showing the distribution of observation by season, and the relationship between SWI and key VI statistics for the 28 samples sites across the Delta. Only the top-performing statistics (in terms of r values) are shown here. Blue indicates wet season and red for dry season. Grey lines display the confidence interval at 0.95. In each panel: (**a**) Correlating NDVI, NDWI, and FAPAR mean to SWI; (**b**) Correlating LAI mean, LAI entropy mean and SD to SWI.

In the SD plots for the four VIs, a moderately positive correlation (> 0.50) is observed for the wet seasons and weak negative correlation is observed for the dry seasons (Figure 2). This indicates a difference between the variations of vegetation conditions in different seasons and how the variation correlates to soil humidity conditions. Moreover, this demonstrates, as expected, that higher variation in activated vegetation cover during the wet season occurs in association with increased variance in soil moisture conditions as captured through the ASCAT-SWI. This is also verified by the observation that the higher the variation in vegetation's leaf surface area or the area of photosynthetic activity becomes, the higher the SWI.

CVs depicting relative variability indicate a stronger negative linear correlation, r = −0.4 for NDVI SD and r = −0.21 for LAI SD. This demonstrates a higher absolute deviation of the individual values within a cell in a high moisture saturation period, but a higher relative variability in NDVI and LAI in a lower moisture saturation period. Because NDWI and FAPAR can get 0 or negative values, a CV calculation for NDWI and FAPAR cannot be performed.

Texture information (entropy and homogeneity) also contributes to the explanation of the variance in SWI/VI relations but to a smaller extent (Figure 2, Table 3). The mean and SD values for entropy (calculated from the GLCM) indicate the level of disorder in the VI distribution, and smooth image values result in high entropy. The mean entropy of LAI shows a positive linear correlation with SWI at r = 0.53; the SDs of entropy for LAI are negatively correlated to SWI at r = −0.58. The FAPAR entropy SD (r = −0.48) for the dry season also shows a moderately negative correlation with the SWI (r = −0.48). Therefore, moderate-to-weak correlations can be found between soil moisture condition and the second-order texture measures of variables estimating vegetation's evapotranspiration and photosynthetic primary production capacity.

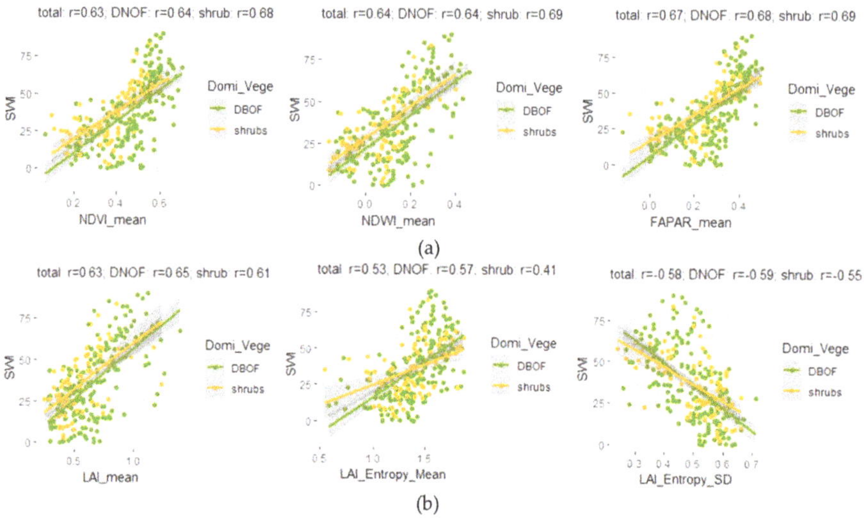

Figure 3. Selected scatterplots showing the distribution of observations by dominant land cover types, and the relationship between SWI and key VI statistics for the 28 samples sites across the Delta. Only the top-performing statistics (in terms of r values) are shown here. Green indicates observations with DBOF as dominant vegetation and yellows are for shrubs. The grey lines display the confidence interval at 0.95. In each panel: (**a**) Correlating NDVI, NDWI, and FAPAR mean to SWI; (**b**) Correlating LAI mean, LAI entropy mean and SD to SWI.

Table 3. Remaining top performing statistics (in terms of r values) describe the second-order homogeneity (mean and SD) that are not illustrated in Figures 2 and 3.

VI_Stats	Total r	Total p	Wet r	Wet p	Dry r	Dry p	DBOF r	DBOF p	Shrub r	Shrub p
NDVI_hom_SD	−0.39	1.16×10^{11}	−0.30	9.39×10^{5}	−0.35	0.000148	−0.36	1.09×10^{7}	−0.49	3.97×10^{6}
LAI_hom_mean	−0.50	2.43×10^{19}	−0.42	1.57×10^{8}	−0.17	0.075967	−0.55	5.20×10^{17}	−0.38	0.000497
FAPAR_hom_SD	−0.38	3.49×10^{11}	−0.19	0.015037	−0.35	0.000176	−0.36	2.11×10^{7}	−0.47	9.49×10^{6}

The observations for the two seasons do not form completely separate clusters. The lower range of the wet season observations intermingles with part of dry season at mid to low signals. This may be related to the contribution of other water sources, such as groundwater and precipitation, in vegetation conditions, but they are not in the scope of the wet/dry season definition in this research.

3.2. SWI and VIs Correlation by Dominant Land Cover Type

Two land cover types, deciduous broad open forest (DBOF) and shrubs, are dominating the sites based on the CGLS 2015 Africa Land Cover product [30]. Overall, the distribution of observations for both classes does not result in separate clusters (Figure 3). This indicates that these two dominant land cover types behave similarly with respect to soil moisture dynamics.

Mean plots for all four VIs show a strong positive linear correlation with the SWI ($r > 0.60$). The shrub-dominant sites indicate a strong correlation with SWI ($r > 0.68$) in terms of mean NDVI, NDWI, and FAPAR; DBOF-dominant sites also show a strong correlation. Spatial variance described by SD and CV are moderate in correlation strength with SWI for NDVI, NDWI, and FAPAR. The SD of FAPAR shows weak to no correlation with SWI. Second-order entropy in Figure 3 and second-order homogeneity in Table 3 both show weak-to-no correlation except for LAI, where a moderate correlation strength can be observed. This indicates a partial significance of second-order image texture measures in the discussion of vegetation structures relevant indices, such as LAI, over the study area.

4. Discussion

The series of correlation analyses demonstrates the possibility of using VIs to downscale SSM in the wetland environment. Seasonal differences in using vegetation proxies for soil moisture are obvious—in the wet season, vegetation information has a strong linkage to soil moisture condition, while very scattered results are obtained during the dry season. Different reactions of dominant vegetation types in respect to the soil moisture distributions exist, but they are not drastic in the Okavango study sites—at sites with shrubs as dominant vegetation, vegetation proxies performed generally well in estimating soil moisture; at sites with DBOF as dominant vegetation, a moderately strong correlation could be found as well.

The mean values for all VIs correlate moderately to strongly with SWI. The FAPAR mean is marginally stronger than the rest. An interpretation is that the vegetation's evapotranspiration and photosynthetic primary production capacity is well linked to SSM. LAI and FAPAR are closely linked biophysical variables that characterize the total canopy and the photosynthetic activity of plants., while LAI, only accounting for the amount of foliage in the plant canopy including the understory, FAPAR, reveals more of the amount of light absorbed by canopy at a given time. Based on the crop-specific empirical relationship between these two indices analyzed by Kukal and Irmak (2020), the increasing in the leaf area is accompanied with the linear increase of light absorption by the plant but the linearity diminishes at a threshold (LAI of 2-4) and this diminishing return denotes that FAPAR is a more direct proxy of vegetation's light absorption capacity than the canopy area [41]. Moreover, the vegetation's vitality and greenness, conveyed through the NDVI, and NDWI, which measures the liquid water content in vegetation, both correlate moderately to strongly with SWI.

SD and CV measure first-order spatial variation in the four VIs. They are moderately correlated with SWI (Figure 2, Figure 3). CVs are generally negatively correlated to SWI while SDs are positively correlated to SWI. This indicates that spatial variability in the VIs could be meaningful in understanding VIs' relations with SWI. The texture information, homogeneity, and entropy, correlated weakly to moderately to SWI. These second-order texture measures describe properties of the horizontal vegetation structure through the relationship comparison of directly neighboring pixels. They are mathematically more complex than the first-order statistics discussed above but can reveal particular patterns. In the wet season where foliage is abundant, textural information is more distinct since the original signals are more dynamic. Correlation strength with SWI is, therefore, stronger in comparison to the dry season. In comparing the four selected VIs, textural information for LAI is consistently more relevant in its correlation with SWI.

The stratification strategy both by season and by dominant land cover type in the study of the relationship between VIs and SWI encompasses a temporal and a spatial aspect. The observation dates were selected firstly to reflect two states of the regional flow dynamics via measured in-situ water levels and, secondly, to reflect the availability of cloud-free Sentinel-2 images for the target seasons. However, analyses of the temporal behavior are limited to a comparison of peak season variation by grouping these dates into dry and wet season data, because an evenly spaced time series could not be constructed. Regarding the dominant land cover selection, among the 28 sample sites, 20 sites have DBOF as the dominant type and eight sites have shrubs. When two competing land cover types appear within one site, the one with the higher coverage has been defined as the dominant type. The signals observed in VIs, however, do not purely stem from one uniform type of vegetation. In a fairly heterogeneous landscape as the Okavango Delta, it is unlikely to find one site with a uniform vegetation type to be used as a stereotype. Such purity would have revealed a closer correlation between vegetation status and SWI. Nevertheless, some distinction was possible in defining dominant vegetation.

The results indicated in this study align with the conclusion of Torres-Rua et al. (2016) that one single vegetation index, such as NDVI, is insufficient to describe the internal variance of SSM fully; the addition of other vegetation indices (such as LAI) and spatial heterogeneity parameters can provide an improved spatial estimation of SSM [17]. In this study, different characteristics of each investigated VI infer that the heterogeneity in the vitality and evapotranspiration of vegetation, and the photosynthetic primary production capacity over the landscape contribute to the explanation of the SSM patterns underneath. As Figure 2; Figure 3 indicate, the mean values of vegetation characteristics perform best among all statistical descriptors of vegetation conditions in correlating the most strongly with SWI. This result is consistent with Klinke et al.'s (2018) findings [20] and is responsive to the multiscale parameter regionalization (MPR) technique proposed by Samaniego et al. (2010) to downscale coarse resolution parameters with finer resolution input data through upscaling operators like the harmonic mean [23]. Additionally, Klinke et al. (2018) pointed out the long-range and intra-annual variations in SSM [20], and Dabrowska-Zielinska et al. (2018) analyzed the different contributions of vegetation in dry and wet moist conditions of soil on backscattering [22]. Both indicate the seasonal effect in vegetation–soil relations. This idea has been taken up through a dry/wet season observation stratification in this study, whereas a stronger linear correlation between VI and SWI can be found for the wet season observations. However, no obvious differences could be found with respect to dominant vegetation types based on the 2015 PROBA-V Land Cover map of Africa [30]. This may be due to similarities in the reaction of these two dominant savanna vegetation classes. Moreover, this demonstrates that real-time vegetation information demonstrates a stronger capability to estimate SSM while non-real-time vegetation information from past land cover map does not add much information to the estimation of SWI in this study. To further understand how the near real-time vegetation traits are associated with soil moisture condition, long term time series of VIs and time-lagged analysis should be used to interpret when the vegetation is stressed after some accumulation in a plant's water-limited features over time.

5. Conclusions

The remote sensing analyses in this study make use of the popular remote sensing products, ASCAT-SWI for SSM and the Sentinel-2 images for VI retrieval. They demonstrate the relationships between vegetation and soil moisture in a complex wetland environment and the potential of using vegetation proxies from multispectral data for downscaling and regionalization of soil moisture. Fine spatial resolution vegetation traits calculated from the Sentinel-2 data convey information about the internal moisture variation within each coarse SWI pixel and indicate that both the absolute VI values and the individual spatial variation in the vegetation structure relate to soil moisture conditions at the time of VI retrieval. The stratification techniques, by seasons and by dominant land cover type, also shed light on how the soil-vegetation relations change under the influence of regional flow dynamics and dominant vegetation patterns. Finally, it can be inferred that the time when vegetation flourishes are clearly reflected through high values in VI signals and are associated with stronger correlations with soil humidity; low VIs, on the other hand, indicating low vegetation vitality, have a very limited correlation with SWI. In sites with different dominant vegetation, correlations strength does not differ much, but vegetation structures could make an influential distinction. The latter needs further analysis. Future scientific efforts are needed to understand vegetation signals' delayed response to changes in soil moisture; this is hindered in this research by a limited number of cloud-free Sentinel-2 data. Long-term and high temporal resolution time series of vegetation traits, however, will have great potential in uncovering the relation between coarse-scale soil moisture and fine-scale vegetation. Therefore, the research exhibits elements in line with state-of-the-art soil monitoring in remote sensing and informs that vegetation data should be implemented in soil moisture retrieval algorithms to improve the estimation of spatially varying soil moisture. Further space missions collecting hyperspectral data and terrestrial vegetation's chlorophyll fluorescence data from the FLuorescence EXplorer (FLEX) can also be integrated into this line of studies to provide more detailed information on the linkage of vegetation traits and soil water availability.

Author Contributions: Conceptualization, M.L. and M.P.; Formal analysis, M.L.; Investigation, M.L.; Methodology, M.L.; Software, M.L.; Supervision, M.P., N.P. and M.S.; Visualization, M.L.; Writing—original draft, M.L.; Writing—review & editing, M.P., N.P. and M.S. All authors have read and agreed to the published version of the manuscript.

Acknowledgments: Thanks are extended to the technical support this work received from the Institute of Photogrammetry and Remote Sensing and Institute of Cartography at TU Dresden, as well as the support in editing the written work received the University of Maryland Graduate School Writing Center.

Conflicts of Interest: All authors declare no conflict of interest.

References

1. Peng, J.; Loew, A. Recent Advances in Soil Moisture Estimation from Remote Sensing. *Water* **2017**, *9*, 530. [CrossRef]
2. Mohanty, B.P.; Cosh, M.H.; Lakshmi, V.; Montzka, C. Soil Moisture Remote Sensing: State-of-the-Science. *Vadose Zo. J.* **2017**, *16*, 0. [CrossRef]
3. Wagner, W.; Dorigo, W.; De Jeu, R.; Fernandez, D.; Benveniste, J.; Haas, E.; Ertl, M. Fusion of Active and Passive Microwave Observations to Create an Essential Climate Variable Data Record on Soil Mmoisture. *ISPRS Ann. Photogramm. Remote Sens. Spat. Inf. Sci.* **2012**, *1*, 315–321. [CrossRef]
4. Dorigo, W.A.; Gruber, A.; De Jeu, R.A.M.; Wagner, W.; Stacke, T.; Loew, A.; Albergel, C.; Brocca, L.; Chung, D.; Parinussa, R.M.; et al. Evaluation of the ESA CCI Soil Moisture Product Using Ground-Based Observations. *Remote Sens. Environ.* **2015**, *162*, 380–395. [CrossRef]
5. Dorigo, W.; de Jeu, R.; Chung, D.; Parinussa, R.; Liu, Y.; Wagner, W.; Fernández-Prieto, D. Evaluating Global Trends (1988-2010) in Harmonized Multi-Satellite Surface Soil Moisture. *Geophys. Res. Lett.* **2012**, *39*. [CrossRef]
6. Filippucci, P.; Tarpanelli, A.; Massari, C.; Serafini, A.; Strati, V.; Alberi, M.; Raptis, K.G.C.; Mantovani, F.; Brocca, L. Soil Moisture as a Potential Variable for Tracking and Quantifying Irrigation: A Case Study with Proximal Gamma-Ray Spectroscopy Data. *Adv. Water Resour* **2020**, *136*, 103502. [CrossRef]
7. Zreda, M.; Desilets, D.; Ferré, T.P.A.; Scott, R.L. Measuring Soil Moisture Content Non-Invasively at Intermediate Spatial Scale Using Cosmic-Ray Neutrons. *Geophys. Res. Lett.* **2008**, *35*, L21402. [CrossRef]
8. Kerr, Y.H.; Waldteufel, P.; Wigneron, J.-P.; Martinuzzi, J.; Font, J.; Berger, M. Soil moisture retrieval from space: the Soil Moisture and Ocean Salinity (SMOS) mission. *IEEE Trans. Geosci. Remote Sens.* **2001**, *39*, 1729–1735. [CrossRef]
9. Bartalis, Z.; Naeimi, V.; Hasenauer, S.; Wagner, W. ASCAT Soil Moisture Report Series No. *15 ASCAT Soil Moisture Product Handbook*. **2008**.
10. Bartalis, Z.; Wagner, W.; Naeimi, V.; Hasenauer, S.; Scipal, K.; Bonekamp, H.; Figa, J.; Anderson, C. Initial Soil Moisture Retrievals from the METOP-A Advanced Scatterometer (ASCAT). *Geophys. Res. Lett.* **2007**, *34*. [CrossRef]
11. Entekhabi, D.; Njoku, E.G.; O'Neill, P.E.; Kellogg, K.H.; Crow, W.T.; Edelstein, W.N.; Entin, J.K.; Goodman, S.D.; Jackson, T.J.; Johnson, J.; et al. The Soil Moisture Active Passive (SMAP) Mission. *Proc. IEEE* **2010**, *98*, 704–716. [CrossRef]
12. Colliander, A.; Jackson, T.J.; Bindlish, R.; Chan, S.; Das, N.; Kim, S.B.; Cosh, M.H.; Dunbar, R.S.; Dang, L.; Pashaian, L.; et al. Validation of SMAP surface soil moisture products with core validation sites. *Remote Sens. Environ.* **2017**, *191*, 215–231. [CrossRef]
13. Bauer-Marschallinger, B.; Paulik, C.; Hochstöger, S.; Mistelbauer, T.; Modanesi, S.; Ciabatta, L.; Massari, C.; Brocca, L.; Wagner, W.; Bauer-Marschallinger, B.; et al. Soil Moisture from Fusion of Scatterometer and SAR: Closing the Scale Gap with Temporal Filtering. *Remote Sens.* **2018**, *10*, 1030. [CrossRef]

14. Calvet, J.-C.; Wigneron, J.-P.; Walker, J.; Karbou, F.; Chanzy, A.; Albergel, C. Sensitivity of Passive Microwave Observations to Soil Moisture and Vegetation Water Content: L-Band to W-Band. *IEEE Trans. Geosci. Remote Sens.* **2011**, *49*, 1190–1199. [CrossRef]
15. Elachi, C.; Van Zyl, J. Introduction to the Physics and Techniques of Remote Sensing. *Wiley-Interscience* **2006**.
16. Alexandridis, T.; Cherif, I.; Bilas, G.; Almeida, W.; Hartanto, I.; van Andel, S.; Araujo, A.; Alexandridis, T.K.; Cherif, I.; Bilas, G.; et al. Spatial and Temporal Distribution of Soil Moisture at the Catchment Scale Using Remotely-Sensed Energy Fluxes. *Water* **2016**, *8*, 32. [CrossRef]
17. Torres-Rua, A.; Ticlavilca, A.; Bachour, R.; McKee, M. Estimation of Surface Soil Moisture in Irrigated Lands by Assimilation of Landsat Vegetation Indices, Surface Energy Balance Products, and Relevance Vector Machines. *Water* **2016**, *8*, 167. [CrossRef]
18. Pause, M.; Zacharias, S.; Schulz, K.; Lausch, A. Near-Surface Soil Moisture Estimation by Combining Airborne L-Band Brightness Temperature Observations and Imaging Hyperspectral Data at the Field Scale. *J. Appl. Remote Sens.* **2012**, *6*, 063516.
19. Qiu, J.; Crow, W.T.; Wagner, W.; Zhao, T. Effect of Vegetation Index Choice on Soil Moisture Retrievals via the Synergistic Use of Synthetic Aperture Radar and Optical Remote Sensing. *Int. J. Appl. Earth Obs. Geoinf.* **2019**, *80*, 47–57. [CrossRef]
20. Klinke, R.; Kuechly, H.; Frick, A.; Förster, M.; Schmidt, T.; Holtgrave, A.-K.; Kleinschmit, B.; Spengler, D.; Neumann, C. Indicator-Based Soil Moisture Monitoring of Wetlands by Utilizing Sentinel and Landsat Remote Sensing Data. *PFG J. Photogramm. Remote Sens. Geoinf. Sci.* **2018**, *86*, 71–84. [CrossRef]
21. El Hajj, M.; Baghdadi, N.; Zribi, M.; Bazzi, H. Synergic Use of Sentinel-1 and Sentinel-2 Images for Operational Soil Moisture Mapping at High Spatial Resolution over Agricultural Areas. *Remote Sens.* **2017**, *9*, 1292. [CrossRef]
22. Dabrowska-Zielinska, K.; Musial, J.; Malinska, A.; Budzynska, M.; Gurdak, R.; Kiryla, W.; Bartold, M.; Grzybowski, P. Soil Moisture in the Biebrza Wetlands Retrieved from Sentinel-1 Imagery. *Remote Sens.* **2018**, *10*, 1979. [CrossRef]
23. Samaniego, L.; Kumar, R.; Attinger, S. Multiscale parameter regionalization of a grid-based hydrologic model at the mesoscale. *Water Resour. Res.* **2010**, *46*. [CrossRef]
24. USGS EROS Archive - Sentinel-2. Available online: https://www.usgs.gov/centers/eros/science/usgs-eros-archive-sentinel-2?qt-science_center_objects=0#qt-science_center_objects (accessed on 27 November 2019).
25. STEP | Science Toolbox Exploitation Platform. Available online: http://step.esa.int/main/ (accessed on 27 November 2019).
26. Wagner, W.; Lemoine, G.; Rott, H. A Method for Estimating Soil Moisture from ERS Scatterometer and Soil Data. *Remote Sens. Environ.* **1999**, *70*, 191–207. [CrossRef]
27. Bauer-Marschallinger, B.; Paulik, C. "CGLOPS-1" Algorithm Theoretical Basis Document Soil Water Index Collection 1km Version 1. *Copernic. Glob. Land Oper.* **2019**.
28. Paulik, C.; Dorigo, W.; Wagner, W.; Kidd, R. Validation of the ASCAT Soil Water Index Using in Situ Data from the International Soil Moisture Network. *Int. J. Appl. Earth Obs. Geoinf.* **2014**, *30*, 1–8. [CrossRef]
29. Okavango Delta Monitoring & Forecasting. Available online: http://okavangodata.ub.bw/ori/monitoring/water/ (accessed on 27 November 2019).
30. Buchhorn, M.; Smets, B.; Bertels, L.; Lesiv, M.; Tsendbazar, N.-E.; Herold, M.; Fritz, S. Land Cover 100m: Collection 2: Epoch 2015. *Copernic. Glob. Land Serv.* **2019**. [CrossRef]
31. Deering, D.W. Rangeland Reflectance Characteristics Measured by Aircraft and Spacecraft Sensors.
32. Tucker, C.J. Red and Photographic Infrared Linear Combinations for Monitoring Vegetation. *Remote Sens. Environ.* **1979**, *8*, 127–150. [CrossRef]
33. Rouse, J.W., Jr.; Haas, R.H.; Schell, J.A.; Deering, D.W. *Monitoring Vegetation Systems in the Great Plains with Erts*; 1974.
34. Gao, B. NDWI—A Normalized Difference Water Index for Remote Sensing of Vegetation Liquid Water from Space. *Remote Sens. Environ.* **1996**, *58*, 257–266. [CrossRef]
35. Weiss, M.; Baret, F. *S2ToolBox Level 2 Products: LAI, FAPAR, FCOVER*; INRA; Paris, France, 2016.
36. Garrigues, S.; Allard, D.; Baret, F.; Weiss, M. Influence of Landscape Spatial Heterogeneity on the Non-Linear Estimation of Leaf Area Index from Moderate Spatial Resolution Remote Sensing Data. *Remote Sens. Environ.* **2006**, *105*, 286–298. [CrossRef]

37. Prince, S.D. A Model of Regional Primary Production for Use with Coarse Resolution Satellite Data. *Int. J. Remote Sens.* **1991**, *12*, 1313–1330. [CrossRef]
38. Weiss, M.; Baret, F.; Myneni, R.; Pragnère, A.; Knyazikhin, Y.; Myneni, R.B.; Weiss, M. Investigation of a Model Inversion Technique to Estimate Canopy Biophysical Variables from Spectral and Directional Reflectance Data Investigation of a Model Inversion Technique to Estimate Canopy Biophysical Variables from Spectral and Directional Reflectance Data Investigation of a Model Inversion Technique to Estimate Canopy Biophysical Variables from Spectral and Directional Reflectance Data. *Agron. EDP Sci.* **2000**, *20*.
39. Wood, E.M.; Pidgeon, A.M.; Radeloff, V.C.; Keuler, N.S. Image Texture as a Remotely Sensed Measure of Vegetation Structure. *Remote Sens. Environ.* **2012**, No. 121. 516–526. [CrossRef]
40. Lu, D.; Batistella, M. Exploring TM Image Texture and Its Relationships with Biomass Estimation in Rondônia, Brazilian Amazon. *Acta Amaz.* **2005**, *35*, 249–257. [CrossRef]
41. Kukal, M.S.; Irmak, S. Light Interactions, Use and Efficiency in Row Crop Canopies under Optimal Growth Conditions. *Agric. For. Meteorol.* **2020**, *284*, 107887. [CrossRef]

remote sensing

MDPI

Article

Soil Moisture Retrievals by Combining Passive Microwave and Optical Data

Cheng Tong [1], Hongquan Wang [1,*], Ramata Magagi [2], Kalifa Goïta [2], Luyao Zhu [1], Mengying Yang [1] and Jinsong Deng [1]

[1] College of Environmental and Resource Sciences, Zhejiang University, Hangzhou 310029, China; 21714124@zju.edu.cn (C.T.); luyao_zhu@zju.edu.cn (L.Z.); 21914160@zju.edu.cn (M.Y.); Jsong_deng@zju.edu.cn (J.D.)

[2] Centre d'Applications et de Recherche en Télédétection, Université de Sherbrooke, QC J1K 2R1, Canada; Ramata.Magagi@USherbrooke.ca (R.M.); Kalifa.Goita@Usherbrooke.ca (K.G.)

* Correspondence: hongquan_wang@zju.edu.cn

Received: 16 August 2020; Accepted: 25 September 2020; Published: 28 September 2020

Abstract: This paper aims to retrieve the temporal dynamics of soil moisture from 2015 to 2019 over an agricultural site in Southeast Australia using the Soil Moisture Active Passive (SMAP) brightness temperature. To meet this objective, two machine learning approaches, Random Forest (RF), Support Vector Machine (SVM), as well as a statistical Ordinary Least Squares (OLS) model were established, with the auxiliary data including the 16-day composite MODIS NDVI (MOD13Q1) and Surface Temperature (ST). The entire data were divided into two parts corresponding to ascending (6:00 p.m. local time) and descending (6:00 a.m. local time) orbits of SMAP overpasses. Thus, the three models were trained using the descending data acquired during the five years (2015 to 2019), and validated using the ascending product of the same period. Consequently, three different temporal variations of the soil moisture were obtained based on the three models. To evaluate their accuracies, the retrieved soil moisture was compared against the SMAP level-2 soil moisture product, as well as to in-situ ground station data. The comparative results show that the soil moisture obtained using the OLS, RF and SVM algorithms are highly correlated to the SMAP level-2 product, with high coefficients of determination (R^2_{OLS} = 0.981, R^2_{SVM} = 0.943, R^2_{RF} = 0.983) and low RMSE ($RMSE_{OLS}$ = 0.016 cm^3/cm^3, $RMSE_{SVM}$ = 0.047 cm^3/cm^3, $RMSE_{RF}$ = 0.016 cm^3/cm^3). Meanwhile, the estimated soil moistures agree with in-situ station data across different years (R^2_{OLS} = 0.376~0.85, R^2_{SVM} = 0.376~0.814, R^2_{RF} = 0.39~0.854; $RMSE_{OLS}$ = 0.049~0.105 cm^3/cm^3, $RMSE_{SVM}$ = 0.073~0.1 cm^3/cm^3, $RMSE_{RF}$ = 0.047~0.102 cm^3/cm^3), but an overestimation issue is observed for high vegetation conditions. The RF algorithm outperformed the SVM and OLS, in terms of the agreement with the ground measurements. This study suggests an alternative soil moisture retrieval scheme, in complementary to the SMAP baseline algorithm, for a fast soil moisture retrieval.

Keywords: soil moisture; SMAP; random forest; support vector machine; ordinary least square regression

---

1. Introduction

Soil moisture (SM) influences water, energy and biogeochemical cycles [1], and is a key parameter in meteorological, hydrological, ecological and agricultural systems [2–6]. Traditionally, the soil moisture is measured over fields at some measuring points or ground stations. Although the accuracy of the measured values is high, the point-based soil moisture data cannot reflect the soil moisture spatial patterns at a large scale, due to the limited number of measurements. In addition, the point measurements of SM are time-consuming, labor-intensive and affected by extreme weather [7,8]. With the advantages of timely and objective observations, remote sensing technology was widely used to monitor surface information such as land use, land surface temperature and vegetation cover [9–12]. In comparison with optical remote sensing, microwave has the penetrating ability, allowing to observe the soil surface under moderate vegetation cover for all-weather and day/night conditions. Thus, it has become an effective method to estimate soil moisture [13–16].

Microwave remote sensing consists of active and passive options, and both can perform the tasks of soil moisture retrievals: (i) passive sensors measure the surface emissivity with high temporal resolution but at coarse spatial resolution; (ii) active sensor measures the backscattering power at high spatial resolution but it is limited by low temporal resolution, except the measurements provided by radar constellations such as the Sentinel-1 and Radarsat constellation missions [17,18]. Considering their individual advantages and limitations, the authors in [19–21] focused on synergistic studies between active-passive microwave observations for soil moisture estimation. For instance, Bai, et al. [22] used the synergy of Soil Moisture Active and Passive (SMAP) radar and radiometer observations at the same spatial scale into a discrete radiative transfer model, Tor Vergata (TVG) model, to gain insights into microwave scattering and emissions mechanisms over grasslands. The SMAP satellite launched in January 2015 consists of L-band (1.26 GHz) radar with resolution of 3 km and radiometer (1.41 GHz) with resolution of 36 km. Unfortunately, the radar sensor stopped working in July 2015. Since then, the SMAP passive radiometer operates alone, providing only passive brightness temperature at a coarse resolution [23,24]. SMAP mission provides a set of soil moisture products at different spatial scales through inversion of physical radiative transfer models [25]. The zeroth-order solution of the radiative transfer equation, known as τ-ω model was used to account for the vegetation effect on the brightness temperature [26,27]. In addition to brightness temperature, a number of ancillary data concerning the vegetation and soil characteristics such as effective soil temperature and vegetation water content were also required as inputs to generate the soil moisture products [28]. The Single Channel Algorithm (SCA) based on brightness temperature at V-polarization was considered as a baseline algorithm, and the Dual Channel Algorithm (DCA) was also proposed to achieve better retrieval performance. Compared to the SMAP SCA and DCA algorithms which used the NDVI climatology to account for vegetation contribution in the brightness temperature, Ebtehaj and Bras [29] proposed a multi-channel retrieval algorithm that considers the soil types and vegetation density as a priori information to constrain the temporal changes of vegetation characteristics. This algorithm allows soil moisture retrieval at higher spatial resolution than the original radiometer data.

With the developments of artificial intelligence in recent decades, machine learning methods such as Artificial Neutral Network (ANN), Support Vector Machine (SVM) and Random Forest (RF) provide new ideas to retrieve soil moisture from satellite data [30,31]. Compared to traditional physical models [25,32], machine learning methods avoid complex physical relationships, although they lack the interpretability of the retrieval results due to their black-box nature. In the machine learning methods, the estimators are trained using one portion of the total data to optimize the nonlinear relationships between satellite observations and soil moisture, followed by the validation process performed using the other portion of the data.

To retrieve the soil moisture, Lu, et al. [33] used a recurrent autoregressive neural network algorithm with AMSR2 (Advanced Microwave Scanning Radiometer) and SMOS (Soil Moisture and Ocean Salinity) data, daily NDVI, Land Surface Temperature, precipitation and DEM as trained datasets. When compared with in-situ soil moisture measurements, the retrieved results show a higher correlation coefficient (R) and lower Root Mean Square Error (RMSE) than other satellite soil moisture products such as AMSR-E. Yao, et al. [34] used a back-propagation neural network (BPNN) method to derive global long-term soil moisture series from AMSR-E/AMSR2 brightness temperature. The retrieved results agree well with SMOS soil moisture products as well as the ground station measurements. This indicates that the BPNN method can capture the surface soil moisture in terms of absolute values and temporal variations. Qu, et al. [35] applied the RF model to AMSR-E/AMSR2 brightness temperature and auxiliary data such as latitude, longitude, DEM, Day of Year (DOY) and land classification data, in order to estimate soil moisture from 2010 to 2015 in the Qinghai-Tibet plateau.

During the unfrozen seasons, the retrieved soil moisture correlated well (R = 0.75, RMSE = 0.06 m^3/m^3) with the in-situ soil moisture networks. In addition, the performance of the trained RF estimator was evaluated against the SMAP Single Channel Algorithm at V polarization (SCA-V), indicating a high reliability of the RF model. Kolassa, et al. [36] developed an ANN-based retrieval algorithm to estimate global surface soil moisture from SMAP brightness temperatures. Compared with ground validation data, the ANN retrievals have a significantly higher performance than the NASA Goddard Earth Observing System Model version 5 (GEOS-5) land modeling system. However, the accuracy of the ANN derived soil moisture is less than that of the SMAP Level-2 product, probably due to the inappropriate target soil moisture during the ANN training process. Senyurek, et al. [37] adopted a machine learning framework for the soil moisture retrieval using NASA's Cyclone GNSS observations. Three widely-used machine learning approaches, namely ANN, RF, and SVM were tested and validated. The results reveal that the machine learning algorithms particularly the RF can be applied in soil moisture monitoring over the agricultural areas. Furthermore, to obtain finer resolution soil moisture information, Park, et al. [38] used MODIS products to downscale the 25 km AMSR2 soil moisture products to 1km via statistical ordinary least squares and RF methods.

Previous studies on soil moisture retrievals were usually based on a single method such as physical models, empirical statistical models or machine learning approaches. However, each method has distinct advantages and limitations: physical models follow complex physical laws and have high universality, but involve many physical parameters which increase the complexity; empirical statistical model is based on fitting a large set of data, but with poor applicability; machine learning approaches have high retrieval accuracy, but need a large number of training samples. Therefore, it is thus necessary to compare their performances under different soil and vegetation conditions [31,39]. In addition, the data assimilation methods were also considered for soil moisture estimation [40], but they require many variables such as atmospheric parameters (e.g., temperature, humidity), soil parameters (e.g., temperature, texture, albedo), precipitation, wind speed, elevation and land surface process models [40,41]. Thus, the number of input parameters for data assimilation approaches was larger than the statistical or machine learning approaches. Within this context, the motivation of the current paper is to identify an alternative statistical or machine learning based soil moisture retrieval algorithm that is capable to provide soil moisture information when the SMAP physical algorithms fail due to the strict constraints in surface roughness and vegetation conditions. It provides the potential to fill the data gaps when the SMAP soil moisture products are not available. To realize this objective, our study proposes to retrieve the soil moisture from SMAP brightness temperature using two representative machine learning methods, SVM [42] and RF [43], as well as an Ordinary Least Square (OLS) algorithm. These three algorithms were selected, because they were widely used in the geophysical parameter retrievals from remote sensing data, and can provide reasonable retrieval accuracies [37,38]. Then, the retrieved soil moisture from the three methods was compared to the SMAP L2 soil moisture product obtained using physical radiative transfer models, and validated against the ground measured soil moisture.

2. Study Area and Datasets

2.1. Study Area

The study site is located in the western plains of the Murrumbidgee catchment near the town of Yanco, Australia. The Yanco hydrological monitoring network data available over this site will be used to validate our soil moisture retrieval algorithms. The Yanco network contains 37 soil moisture stations distributed over a 60 × 60 km area. The soil texture was mainly composed of 11% clay, 83% sand and 6% silt. However, as shown in Figure 1, we selected only 15 stations that are distributed within a single SMAP pixel of 36 × 36 km. This study site belongs to a semi-arid agricultural and grazing area, and is dominated by annual crops including rice, corn, soybeans, wheat, barley, oats, and canola [39].

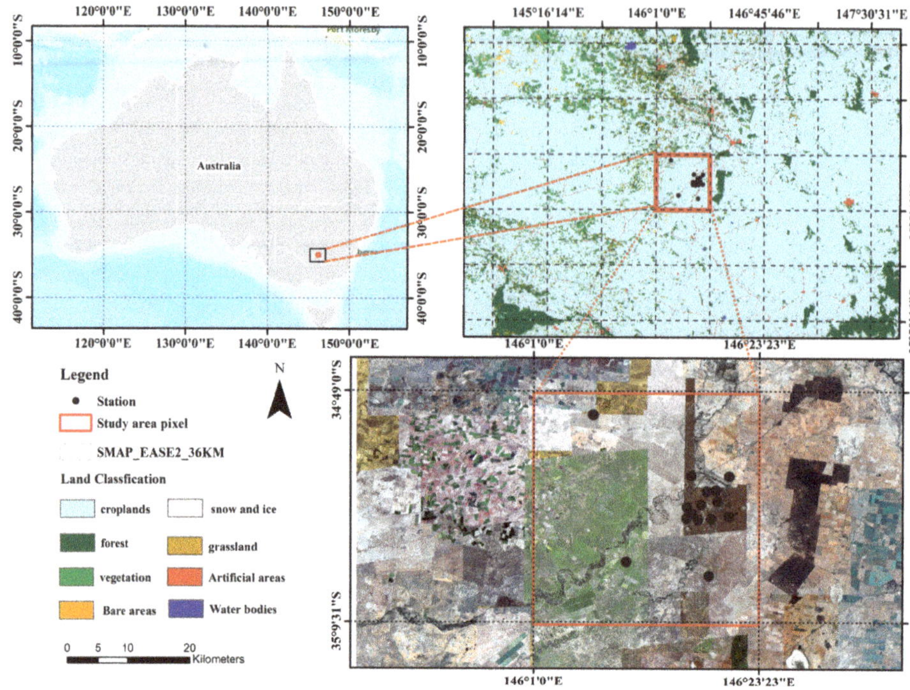

Figure 1. Study area and Yanco soil moisture ground stations.

2.2. Datasets

2.2.1. SMAP Data

The SMAP mission provides L-band microwave brightness temperature data and a series of application products. For instance, the SMAP Level-1C (L1C) data are the calibrated, geo-located and time-ordered brightness temperatures. The SMAP L2 products (version 6) include the global soil moisture and soil surface temperature at a spatial resolution of 36 km. These products were acquired from the National Snow and Ice Data Center (http://nsidc.org/data/smap/smap-data.html).

In Figure 2a,b, the SMAP L1C brightness temperature (descending orbits) at horizontal polarization TB_H and vertical polarization TB_V show seasonal and annual dynamics from 2015 to 2019. According to τ-ω model, the brightness temperature over vegetated soil was impacted by three emission processes: (i) vegetation direct upwelling emission; (ii) vegetation down-welling emission reflected by underlying soils and then attenuated by vegetation itself; (iii) soil emission attenuated by vegetation. Hence, the variation of brightness temperature is highly related to changes in vegetation and soil moisture.

However, compared to the radiative transfer model algorithms used to provide the SMAP soil moisture products, this study contributes to alternative machine learning algorithms to retrieve the soil moisture from the brightness temperature, surface temperature and vegetation feature such as the remotely sensed NDVI.

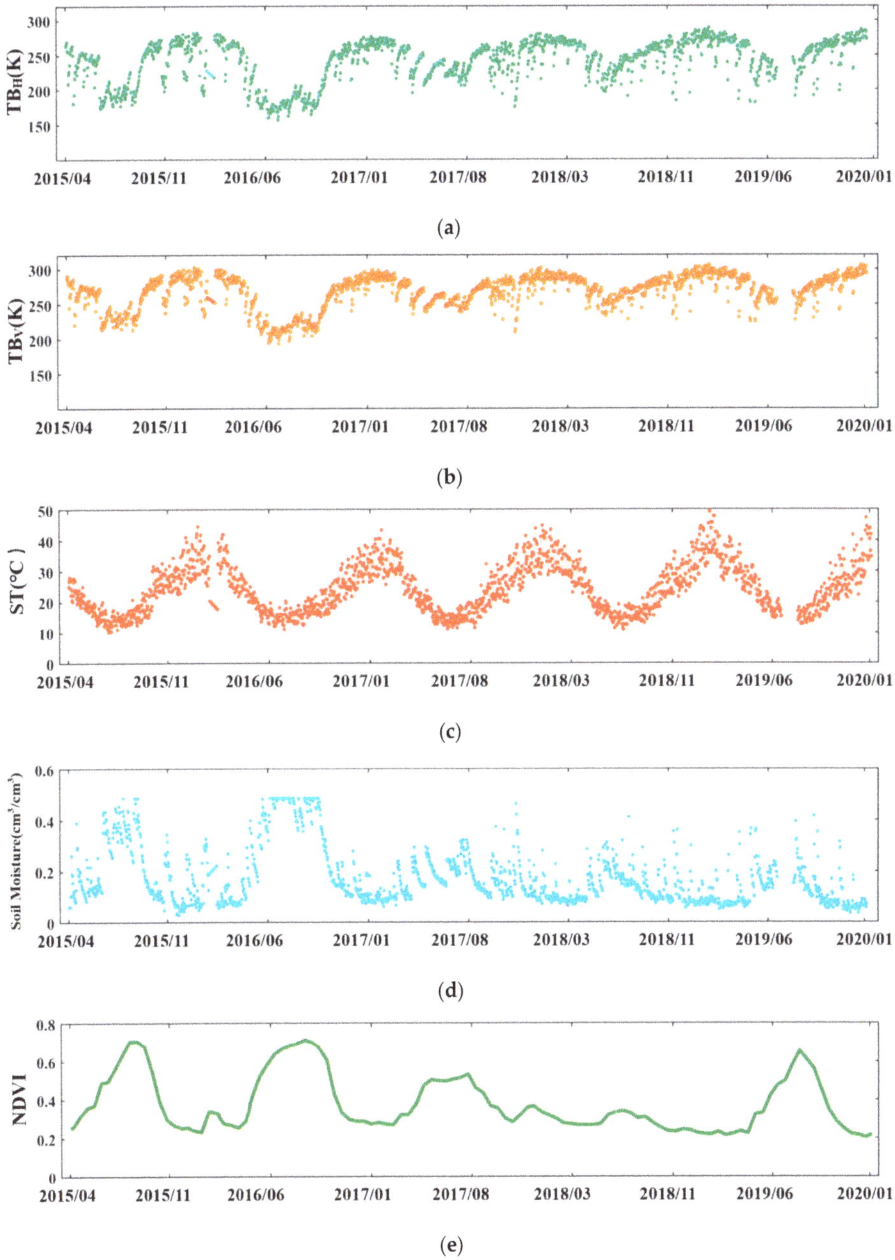

Figure 2. Temporal variations of (**a**) Soil Moisture Active Passive (SMAP) brightness temperature TB_H; (**b**) TB_V; (**c**) in situ surface temperature; (**d**) in situ soil moisture and; (**e**) MODIS NDVI composite.

2.2.2. In-Situ Soil Moisture and Surface Temperature

In-situ soil moisture, surface temperature and precipitation data were obtained from OzNet hydrological monitoring network (www.oznet.org.au) [44]. This network provides long time series of ground measurements at 20 min intervals from 2001 to the present, including the soil moisture at 0~5,

0~30, 30~60, and 60~80 cm depths, the surface temperature at 2.5 and 15 cm depths, and the precipitation measured using a rain gauge with a precision of 0.2 mm. Over each station, the soil moisture was measured using vertically installed 30 cm Campbell Scientific water content reflectometers, followed by a verification using the Time Domain Reflectometer (TDR) [44]. The accuracy of soil moisture in-situ measurement is about 0.03 m^3/m^3 across the stations. Given the penetration depth of the L-band microwave, this study used the in-situ soil moisture and surface temperature measurements at the top soil layer (0~5 cm and 0~2.5 cm, respectively) for the model evaluation and validation.

The SMAP satellite applied sun-synchronous high-inclination orbit, overpassing the test site at 6 a.m. and 6 p.m. for descending and ascending acquisitions, respectively. In order to match the SMAP acquisitions, the in-situ soil moisture and surface temperature data collected 3 h before and after the SMAP satellite overpasses were selected. However, the in-situ data are based on point measurements, reflecting the soil attribute for that point. Due to the spatial heterogeneity and scaling effects, it is still challenging to match the point-based in-situ measurements to the large-scale satellite product [45]. In this study, to lessen the spatial scale mismatch between the in-situ measurements and SMAP pixels, the average value of all the ground stations located within the selected pixel is calculated to represent the soil moisture for that pixel [46,47]. Furthermore, the current study used the in-situ soil temperature, since we assumed it was closer to the effective temperature adopted by the radiative transfer models. However, in case that the ground measured temperature is not available, other land surface temperature products simulated using land surface models such as the GLDAS Noah Model or the products retrieved from satellite data such as the MODIS can be considered to implement the proposed three algorithms.

Figure 2c,d shows the evolution of daily surface temperature and soil moisture from 2015 to 2019, respectively. The surface temperature shows significant annual and seasonal variation with the highest austral temperature in February and the lowest in August for each year. The study area belongs to a semi-arid agricultural and grazing area, where the increased temperature in austral summer (from December to the next year February) results in high evaporation, leading to decreasing soil moisture. Consequently, the soil moisture is lower from December to February for each year.

2.2.3. MODIS NDVI Composite

The MODIS-Terra MOD13Q1 is a level-3 grid data in a Sinusoidal projection mode at a 250 m spatial resolution and a 16-day temporal resolution. The NDVI products were generated by selecting the best available pixels from all the acquisitions during the 16-day period [48]. This paper employed the 16-day composite NDVI products to characterize the vegetation dynamics from 2015 to 2019. The MODIS data were re-projected to WGS84 using the MODIS Re-Projection Tool (MRT). To account for the difference in the spatial resolution between the SMAP and MODIS sensors, the MODIS NDVI data were aggregated into the SMAP pixel to represent the overall vegetation characteristics. Figure 2e shows the temporal variation of the aggregated NDVI for the studied pixel (Figure 1). As the study area is covered by agricultural crops, the NDVI behaviors are more induced by the phenological rhythms of the crop growths. In agreement with He, et al. [49], the NDVI reached the highest value in August of each year, revealing flourishing crop growth stages which include the flowering and anthesis. In contrast, the NDVI decreases for the ripe and the harvest periods.

3. Methodology

We developed a scheme to retrieve soil moisture from three different algorithms including Support Vector Machine (SVM), Random Forest (RF) and Ordinary Least Squares (OLS), as shown in Figure 3. Indeed, our study was motivated by the previous researches on the reconstruction of soil moisture time series. For instance, Qu, et al. [33] rebuilt a time series of soil moisture by applying the RF algorithm into the AMSR-E and AMSR2 brightness temperature. In the training process of the RF algorithm, the SMAP soil moisture was considered as the target output. In Yao, et al. [34], a global long time series of soil moisture was developed using a backpropagation artificial neural network.

They considered AMSR-E and AMSR2 brightness temperature as input variables, while the SMOS Level 3 soil moisture product as the target output. Following the above work, the current paper is to establish an alternative soil moisture retrieval algorithm when the SMAP physical algorithms fail due the roughness and vegetation conditions. In the retrieval models, the input features included SMAP brightness temperature (TB_H, TB_V), in situ Surface Temperature (ST) and MODIS NDVI product, and the output feature is the soil moisture from SMAP L2 product. By fitting the training data to the three proposed models, linear or non-linear relationships were developed between the selected input features and the SMAP soil moisture. After completing the training process, the established models were used to estimate the soil moisture from the testing data. Finally, the performances of the different models were compared against the ground station measurements and SMAP L2 product.

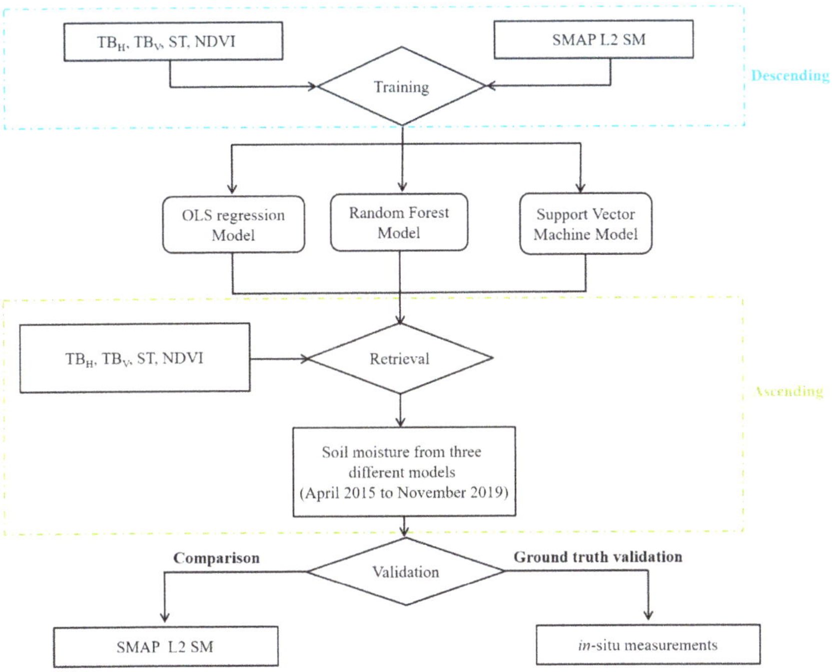

Figure 3. Diagram of soil moisture retrieval method.

3.1. Input Feature Selection Strategy

SMAP mission used the radiative transfer model τ-ω to retrieve the soil moisture from the brightness temperatures along with other ancillary data such as surface roughness, soil temperature, land cover and vegetation parameters. Thus, vegetation covers and to a lesser extent surface roughness are two key variables affecting the SMAP soil moisture retrieval algorithms. Previous studies show that the brightness temperature has a strong correlation with surface roughness, and also that the vegetation cover imposes a significant impact on surface emissivity [50]. For this reason, to select the ancillary input variables for our retrieval models which are based on the SMAP observations, the surface roughness is considered, at first. However, as our study is limited within a single SMAP pixel, the surface roughness is considered relatively stable during the multi-temporal radiometer observations. In addition, TB is less sensitive to surface roughness at coarse than high spatial resolution [51]. Consequently, the surface roughness acts like a constant in the proposed retrieval algorithms so that it isn't necessary to be included in the input features. As the brightness temperature is a product between soil effective temperature and land surface microwave emissivity, the surface temperature is chosen as an input variable [25].

As for the vegetation characteristics, the vegetation water content (VWC) is usually used in the radiative transfer models. However, it is labor-consuming to directly measure the VWC. Previous studies reveal that NDVI can be transformed into VWC through empirical equations for different vegetation types [52,53]. As our study is based on the machine learning approaches, we assumed that the potential empirical relationships can be learned in the training process. Thus, we used the NDVI instead of VWC. As a vegetation descriptor, NDVI has the following advantages: (1) Over a large area such as the SMAP pixel, the NDVI is sensitive to abrupt accidents which changed external environment such as drought, flood and human activities. (2) It is associated with the overall vegetation coverage of the study area. (3) For long time series, the NDVI displays seasonal and inter-annual variations [54,55]. To develop the machine learning based soil moisture retrieval algorithms, daily NDVI was required. As the vegetation growth is a gradual process, the temporal NDVI data possesses significant seasonal patterns with respect to the vegetation phenology. Consequently, considering the temporal periodicity and stability of MODIS NDVI [56,57], a linear interpolation was applied to these 16-day composite products to provide daily NDVI, which can be consistent with the SMAP overpass. To derive the NDVI for a given day corresponding to the SMAP acquisition, two NDVI products before and after that day were used to realize the linear interpolation.

3.2. Model Development and Retrieval

With the selected observables, this study develops three retrieval models (OLS, SVM and RF) applied to the time series of SMAP data and MODIS products. In order to increase the dataset and to retrieve the soil moisture during the entire period 2015–2019, the training and validation processes make use of the descending and ascending SMAP measurements, respectively [58]. For soil moisture retrievals, some studies have shown the better quality of SMAP descending product at 6 a.m. than that of the ascending at 6 p.m. owing to more uniform soil/vegetation temperature and lower Faraday rotation in the morning [59,60]. Nevertheless, although the heterogeneity in soil and vegetation characteristics for the ascending orbit at 6 p.m. is complex, it was reported that the retrieved soil moisture products from ascending orbit are only slightly worse than those from the descending orbit [61,62]. In the current study, the five years SMAP descending product and the auxiliary data for the same period are selected as the training set, and the ascending data as the testing set to retrieve the 2015–2019 temporal soil moisture profile from the three algorithms. The retrieved soil moisture was then validated against the ground measurements. In addition, the SMAP L2 products that were used in training the algorithms, were also analyzed with respect to the in-situ measurements to identify the potential factors influencing the retrieval accuracy.

A. OLS Model

As a linear regression model, OLS aims to estimate the optimal coefficients for different observables to predict the soil moisture by minimizing the sum of squared errors [63]. The linearity between the four selected input variables (TB_H, TB_V, ST, NDVI) and the soil moisture was evaluated to determine the formula. For instance, Figure 4a shows the relationship between the SMAP brightness temperature and the soil moisture, indicating a high negative correlation ($R = -0.97$, $p < 0.001$). On the other hand, in Figure 4b, the surface temperature presents a moderate positive relationship ($R = 0.72$, $p < 0.001$) with the SMAP brightness temperature. These findings are in accordance with the formulation of linear regression for estimating soil moisture from NDVI, ST and other auxiliary data in Park, et al. [38]. Furthermore, the studies in [64,65] also reported that NDVI and ST are strongly correlated with soil moisture and can be used to retrieve soil moisture.

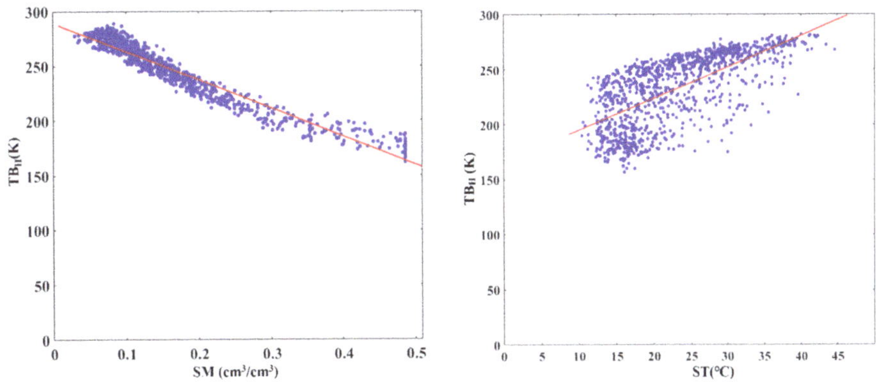

(**a**) (**b**)

Figure 4. Relationship of brightness temperature with (**a**) soil moisture (**b**) land surface temperature.

Thus, a linear formulation for predicting soil moisture is assumed as:

$$mv = b + a_1 NDVI + a_2 ST + a_3 TB_H + a_4 TB_V \quad (1)$$

where a_1, a_2, a_3, a_4 are the regression coefficients, and b represents an intercept. When the determined coefficient is positive, it means that the input variable positively influences the soil moisture values. Otherwise, the input variable negatively impacts the soil moisture values.

B. SVM Model

The non-parametric kernel-based SVM model is one of the most popular machine learning algorithms, and has been used in linear or nonlinear classification, regression and prediction [66]. The basic principle of SVM is to find optimal separating hyperplanes that divide training samples into different classes. Meanwhile, the SVM used kernel functions to transform data into a higher-dimensional space where the data can be separated via a hyperplane with maximum width. The accuracy of SVM depends on the selection of kernel functions, and the distance between the hyperplanes and training samples. Our study will evaluate the linear, polynomial and radial kernel functions used in the SVM algorithm to retrieve the soil moisture from the SMAP brightness temperature.

C. RF Model

The ensemble-learning RF algorithm has been widely applied in classification [67,68] and regression [43,69,70]. The RF consists of numerous decision trees, where each tree is built using a random subset of independent variables. Each tree casts a unit vote for the most popular class to classify the input vector [71]. Compared with other machine learning methods, RF has the advantages of high sensitivity, high precision and fast training speed. It reduced the issue of overfitting, and is especially suitable for processing high-dimensional data. In addition, RF model provides the relative importance of each variable, by measuring the increased mean square error (MSE) when that variable is changed [72]. The increased MSE represents the effect of a variable change on the prediction accuracy of the RF model: the larger the MSE value, the more important the corresponding input variable [38]. Thus, this algorithm is helpful for determining the important remote sensing variables for the soil moisture retrievals.

3.3. Statistic Metrics for Validation

To evaluate the accuracy of the developed soil moisture retrieval models, four statistical metrics including the correlation coefficient (R), root mean square error (RMSE), mean absolute percentage error (MAPE) and bias between the retrieved and measured soil moisture were computed [73]:

$$R = \frac{\sum(x_i - \overline{x})(y_i - \overline{y})}{\sqrt{\sum(x_i - \overline{x})^2 \sum(y_i - \overline{y})^2}} \quad (2)$$

$$RMSE = \sqrt{\frac{\sum(x_i - y_i)^2}{N}} \quad (3)$$

$$MAPE = \frac{1}{N}\sum\left|\frac{x_i - y_i}{y_i}\right| \cdot 100\% \quad (4)$$

$$bias = \frac{1}{N}\sum(x_i - y_i) \quad (5)$$

where x_i, y_i represents the retrieved and referenced soil moisture for validation; $\overline{x}$ and $\overline{y}$ indicate their mean values, and N is the number of matching data for comparison.

4. Results and Discussion

This section discusses the training and testing retrieval results obtained from the proposed OLS, SVM and RF algorithms applied to the SMAP brightness temperature, MODIS NDVI data, and in situ soil temperature. The retrieved soil moisture was compared to SMAP L2 products, and validated using ground measurements.

4.1. Training Results

A. OLS Model

After fitting the OLS model to the training dataset, the optimal coefficients were determined by considering the minimum sum of errors. The resulting formula to retrieve soil moisture from 2015 to 2019 is given as:

$$mv = b + a_1 NDVI + a_2 ST + a_3 TB_H + a_4 TB_V \quad (6)$$

In accordance with Equation (6), the fitted positive coefficients for NDVI and ST indicate a positive relationship with the soil moisture, while the negative coefficients for TB_H and TB_V reveal a negative relationship, as expected. The R and RMSE of the formulation is 0.99 and 0.018 cm^3/cm^3 (p-value < 0.0001) respectively.

B. SVM Model

The SVM model uses different kernel functions for classification and regression. Table 1 compares the performances of the linear, polynomial and radial based functions in the SVM model. The linear kernel function achieves the best results with the highest correlation coefficient (0.93) and the lowest RMSE (0.047 cm^3/cm^3), and the corresponding cost value and gamma value are 1 and 0.25, respectively.

Table 1. Performances of different kernel functions in the Support Vector Machine (SVM) algorithm.

Kernel Function	Linear	Polynomial	Radial Basis
R^2	0.93	0.88	0.45
RMSE	0.047	0.047	0.11

C. RF Model

Once the training process is completed, the RF estimator with the importance rank of different input observables is obtained. Figure 5 shows the importance order of the four input variables in the RF algorithm. It indicates that the TB_H and TB_V play the most significant role in retrieving soil moisture, but the auxiliary NDVI and ST also have significant contributions.

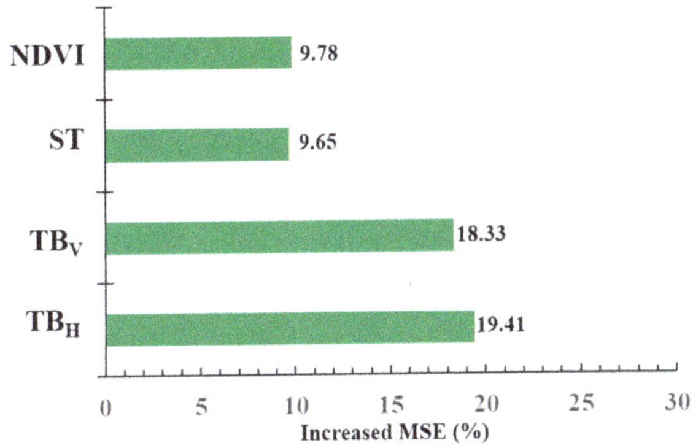

Figure 5. Relative importance of input variables for soil moisture retrieval using Random Forest (RF) algorithm.

4.2. Validation of Testing Results

The quality of the three retrieval algorithms is quantitatively evaluated by comparing the testing soil moisture results with the SMAP L2 soil moisture product and with the ground measurements from April 2015 to December 2019. The metrics used are R, RMSE, MAPE and bias values.

4.2.1. Comparison with SMAP L2 Product

Figure 6 shows linear regressions between the retrieved soil moisture from the three proposed algorithms and the SMAP L2 product. The red and black lines indicate the best-fitted curve and the nonbiased 1:1 line, respectively.

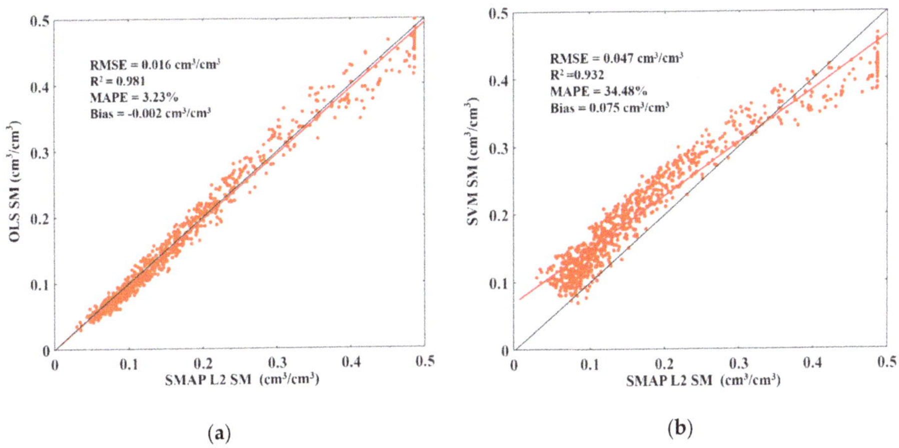

Figure 6. *Cont.*

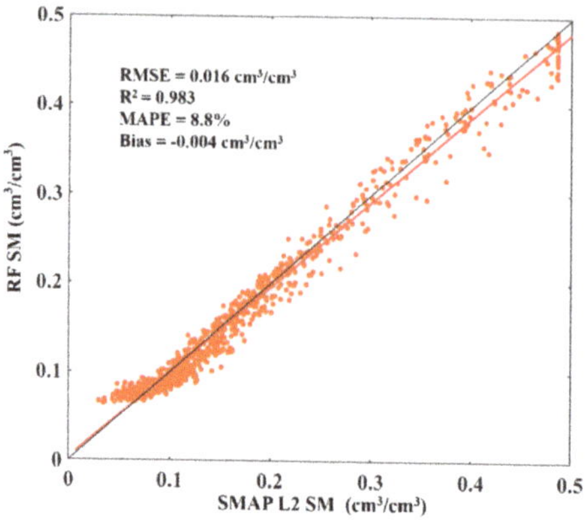

(c)

Figure 6. Comparison of the retrieved soil moisture using (**a**) Ordinary Least Squares (OLS) (**b**) SVM and (**c**) RF algorithms with the SMAP L2 soil moisture product.

A. OLS Model

As can be seen in Figure 6a, the retrieved soil moisture from OLS model presents strong correlation (R^2 = 0.981) and low RMSE (0.016 cm^3/cm^3) with the SMAP L2 soil moisture product. Thus, the developed empirical OLS formula seems promising to briefly describe the physical relationships between soil moisture and the four input observables. However, for the small study area corresponding to only one SMAP pixel (Figure 1), land use and vegetation cover types are relatively single and not fully representative. Consequently, the obtained simple linearly mathematical formula may only be applicable to the environments with similar soil and vegetation characteristics as our study site.

B. SVM Model

In Figure 6b, the SVM retrieved soil moisture was related to the SMAP L2 product with a high correlation of R^2 = 0.932 and RMSE of 0.047 cm^3/cm^3. However, an overestimation occurs for soil moisture less than 0.3 cm^3/cm^3; beyond this threshold value, the retrieved soil moisture is underestimated compared to SMAP L2 data. Consequently, the obtained MAPE (34.48%) and bias (0.075 cm^3/cm^3) using the SVM algorithm are higher than the other two approaches. The performances of the SVM is therefore the worst among the three proposed algorithms in this study.

C. RF Model

Similarly, Figure 6c shows the comparison between the RF retrieved soil moisture and SMAP L2 product, indicating a correlation of R^2 = 0.983 and a RMSE of 0.016 cm^3/cm^3. The RF model is likely suitable to retrieve the soil moisture from the SMAP brightness temperature in an efficient and simple way. This is because RF comprises numerous decision trees and adopts the average values of multiple decision trees, so that the overfitting of single linear regression is relieved. This algorithm seems powerful in the practically near-real-time retrieval of soil moisture from the SMAP instantaneous observations [35].

4.2.2. Comparison with In-Situ Measurements

To further quantify the performances of the proposed three soil moisture retrieval algorithms, we compare the retrieved soil moisture to the in-situ data measured from the Yanco ground stations. For each algorithm, Figure 7 shows the temporal variation of the retrieved soil moisture with respect to the in-situ measurements and daily precipitation from April 2015 to November 2019. The statistical metrics between the retrieved and measured soil moisture are presented in Figure 8. In addition, the SMAP L2 product was also analyzed in terms of the in-situ measurements to identify the potential factors influencing its retrieval accuracy.

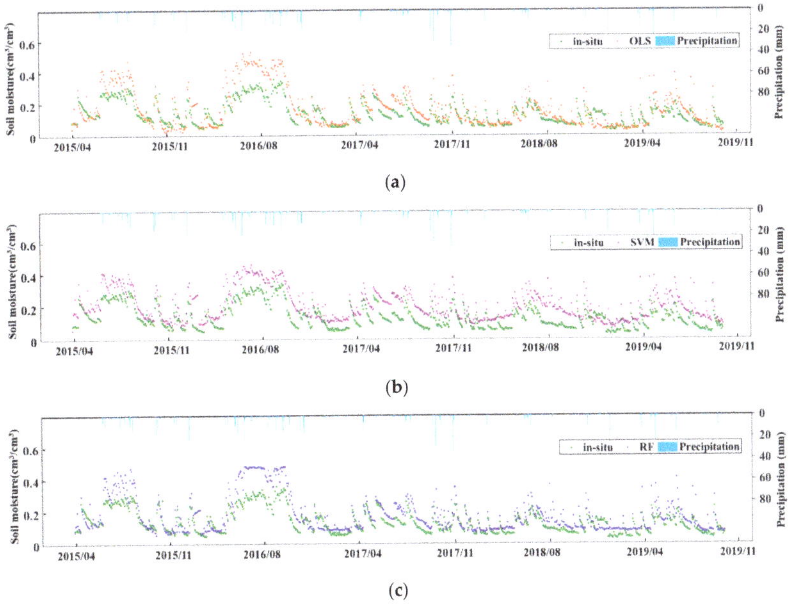

Figure 7. Temporal variation (from April 2015 to November in 2019) of the retrieved soil moisture using (**a**) OLS (**b**) SVM (**c**) RF along with the in-situ soil moisture and precipitation.

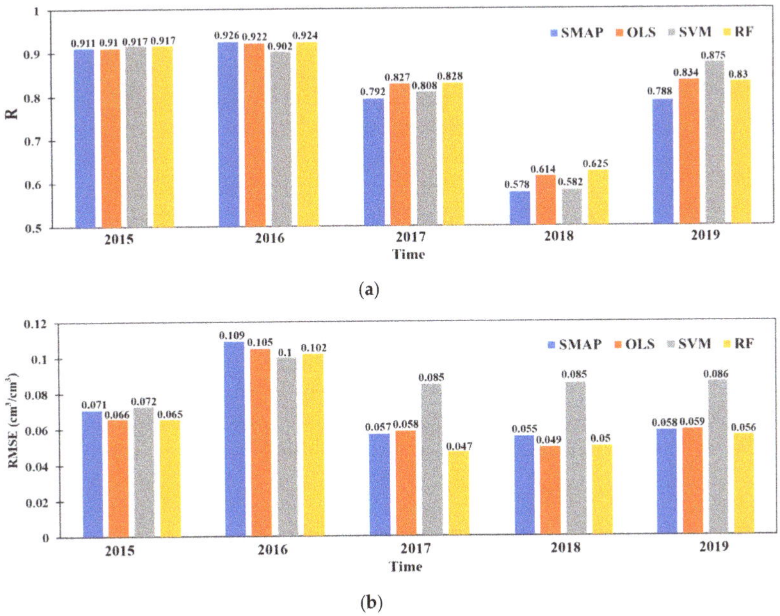

Figure 8. Annual evaluation of the retrieved soil moisture and SMAP L2 product using in-situ measurements from 2015 to 2019: (**a**) correlation R values; (**b**) RMSE values.

A. Performances of the Three Proposed Algorithms

In Figure 7, the retrieved soil moisture using the three developed models and the SMAP L2 SM product capture the temporal dynamic of in-situ measurements. However, the SVM algorithm (Figure 7a) overestimated the soil moisture during the entire period from 2015 to 2019. In contrast, for the RF (Figure 7b) and OLS (Figure 7c) derived soil moisture, an overestimation is mainly observed for specific times of the year. For example, from June to August of every year, the significant overestimation in the RF and OLS algorithms may be due to the high vegetation conditions as discussed in the next section. Furthermore, the retrieved soil moisture is related to the precipitation. The peak of soil moisture followed a significant rainfall amount but with a certain temporal lag.

Generally, the retrieved soil moisture correlates well with in-situ measurements, with encouraging metrics which are however weaker for 2018 (Figure 8). The validation against the in-situ data shows different temporal patterns. In 2015, the RF retrieved soil moisture presents the highest R = 0.917 and the lowest RMSE = 0.065 cm^3/cm^3. In 2016, the SMAP L2 product outperforms the soil moisture retrievals from the proposed three algorithms. In 2017, RF retrieval obtains again the highest R = 0.828, while in 2018 OLS and RF retrieved soil moisture perform better than SVM as well as the SMAP L2 product. In 2019, SVM retrieved soil moisture provides the highest R, but with the highest RMSE, and RF results keep the lowest RMSE.

For the entire period from 2015 to 2019, Table 2 summarizes the correlation coefficient, RMSE and Bias between the in-situ soil moisture measurements and the retrieved soil moisture using the three algorithms. The SMAP L2 soil moisture product was also included in the comparison with ground measurements. Compared to the SVM, the RF and OLS obtained retrieved soil moisture with a high resemblance to the SMAP L2 product.

Table 2. Comparison of model performances during the whole study period from 2015 to 2019.

	OLS	SVM	RF	SMAP
R	0.886	0.874	0.889	0.879
RMSE (cm^3/cm^3)	0.074	0.088	0.072	0.074
Bias (cm^3/cm^3)	0.04	0.074	0.044	0.042

B. Annual Performances of SMAP L2 Product

Figure 9 presents the temporal variation of the SMAP L2 soil moisture along with the NDVI variation from 2015 to 2019. For each year, during the period of high NDVI values, the SMAP algorithm overestimates the soil moisture. In 2015 (Figure 9a) and 2016 (Figure 9b), when the NDVI values were above 0.5 during the vegetation rapid growth season, a series of abnormal retrieved results appeared. In contrast, for the year 2017 (Figure 9a), the changes in NDVI are less pronounced and only a few NDVI values were greater than 0.5. Consequently, the issue of significant overestimation is not presented. In 2018, the NDVI fluctuated slightly with lower values than the past three years (2015~2017). Indeed, the drought of that year directly affected the normal growth of vegetation, resulting thus to low NDVI values. Meanwhile, the correlation (R = 0.6) between SMAP L2 product and in-situ soil moisture is the lowest compared with other years, due to the less variability in soil moisture as a result of the drought. In 2019, the drought ended, the NDVI value returned to normal; the performance of SMAP product also reached the former level (correlation coefficient R restored around 0.8). The study of Ma, et al. [74] has demonstrated the SMAP products perform well under moderate or dense vegetation conditions, whereas in the areas with sparse vegetation conditions, the SMAP products present relatively poor skills with lower time-series correlations. Our results are consistent with former research.

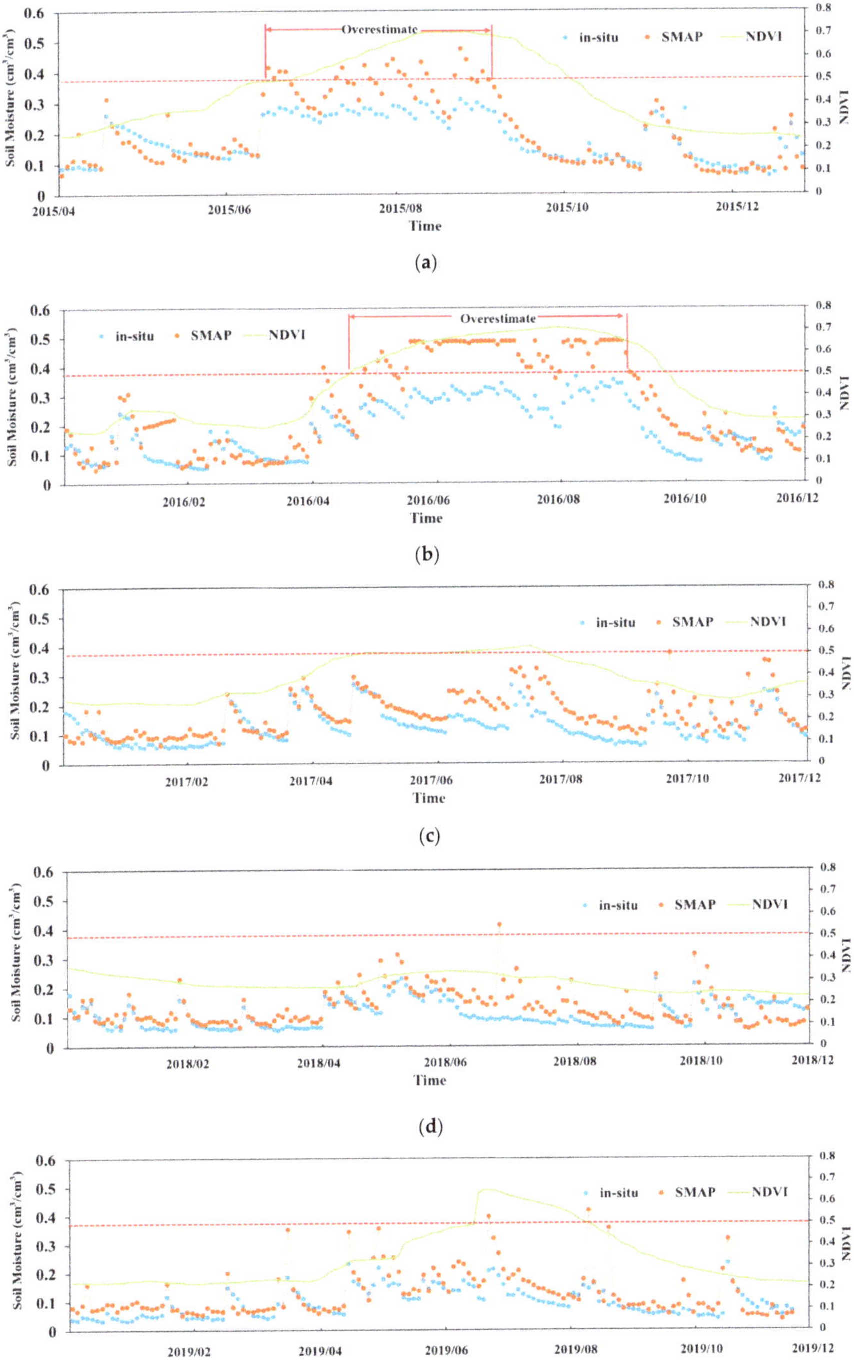

Figure 9. Temporal variation of SMAP L2 soil moisture along with in-situ measurements and NDVI in (**a**) 2015 (**b**) 2016 (**c**) 2017(**d**) 2018 (**e**) 2019.

4.3. Discussion

In this study, we explored the added values of the combination of passive microwave and optical observations to retrieve soil moisture over the vegetated area. Soil moisture is not only related to these four variables, but it is also relevant to other variables such as elevation, soil texture, precipitation and land cover [16,75]. This study selects the most important physical variables similar to those used in the physical models, and obtains comparable results. Compared to the traditional physical models, the machine learning based retrieval algorithms avoid the complex formulation, and enhance the soil moisture retrieval efficiency.

Once completed, the validation process against ground measured data pointed out that the developed algorithms could provide soil moisture retrieval with consistent temporal behaviors. In recent research, Ma, et al. [74] assess several satellite soil moisture products including SMAP, SMOS, AMSR2 and ESA CCI, using global ground-based observations from dense and sparse networks. The results show that SMAP product has the capacity of capturing temporal trends of ground soil moisture. In our study, the soil moisture was retrieved using alternative machine learning algorithms instead of the radiative transfer models used in SMAP algorithms. The proposed machine learning algorithms provided retrieval results that are well correlated with the SMAP L2 product. Therefore, when the SMAP L2 soil moisture product is not available due to the limitation of the physical algorithms, the proposed machine learning approaches such as RF can cover the temporal gap using only the brightness temperature, surface temperature and NDVI as observables.

The validation of the results indicates that the retrieved soil moisture matched the temporal variation of SMAP soil moisture product, with a considerable overestimation that occurs around June to August due to the dense vegetation during this period. Indeed, when the NDVI reaches a certain high value, the flourished vegetation weakens the microwave penetration as well as the emitted signals from the soil layer, which in turn negatively impacts the accuracy of the soil moisture retrieval. On another hand, the empirical formula in Equation (6) indicated that vegetation had a positive correlation with soil moisture, and thus high NDVI value may be related to high soil moisture, justifying the observed overestimation. In further SMAP algorithm refinement, the issues of the overestimation caused by high NDVI are supposed to be taken into consideration [74].

For the SVM algorithm, we compared three kernel functions, and the linear one obtained the highest retrieval accuracy. From our knowledge, this is due to the high linearity between the SMAP brightness temperature and the soil moisture (Figure 4). In contrast, the radial basis function is more suitable for nonlinear and high dimensional relationships, and produced thus degraded performance than the linear kernel function. Furthermore, to train a reliable SVM estimator, a larger number of training data is required. However, in the current study, the number of samples is not sufficient. As for the RF model, the performance depends on the number and structure of decision trees. In the current study, the best performance was obtained with about 100 trees. Compared to SVM, the RF algorithm requires less number of training data [43], resulting thus a better performance than the SVM, given the limited number of the available training samples. Compared with the two machine-learning algorithms (SVM, RF), the empirical OLS model reflects the physical relationships between input observables and soil moisture, and provide a simple way to retrieve soil moisture. Surprisingly, the OLS obtained better performance than the SVM, which may also attribute to the significant linear relationship between TB and soil moisture. However, the regression coefficients need to be adjusted for any use with a different dataset or over different regions. Indeed, the machine learning methods can be also categorized as empirical statistical models, although they show limitations in analyzing the retrieval mechanisms due to the encapsulation in the black-box learning process. In the current study, the RF algorithm provides an overall better result than the SVM model which is impacted by a significant overestimation. This is because the RF algorithm involves the average of the estimates from multiple decision trees [35], which may decrease the variation in the predictions.

Furthermore, for a given area, the SMAP L2 soil moisture products are not always available for training the algorithms. In this case, two solutions were suggested. First, if the SMAP L2 products were missing only for a short period, we can train the algorithms using the SMAP data beyond that missing time. Furthermore, the gap-filling algorithms [76] can be also considered to reconstruct the missing pixels in the SMAP baseline products, before training the proposed algorithms. Second, in the extreme condition that the SMAP L2 products were not available for almost all the observation time, we may consider other similar soil moisture products such as the European Space Agency (ESA) Climate Change Initiative (CCI) as the target output in the training process. According to Ma, et al. [74], the ESA CCI soil moisture is an appropriate complementarity to the SMAP L2 products across different climate conditions. However, additional bias may be introduced due to the uncertainty in the alternative soil moisture products, and the different configurations among diverse sensors.

5. Conclusions

This study proposed multiple models including Ordinary Least Squares, Random Forest and Support Vector Model to retrieve the temporal dynamics of soil moisture from SMAP brightness temperature, surface temperature and auxiliary MODIS data. The brightness temperature (TB_H, TB_V), NDVI and surface temperature were considered as input observables, while the SMAP L2 soil moisture product was considered as the output target variable. The linear or nonlinear relationships between the input and output features were obtained via the training process using a portion of the dataset. Then the obtained estimators were used to estimate the soil moisture from SMAP brightness temperature with the auxiliary vegetation feature provided by the NDVI of MODIS.

To evaluate the accuracy of the proposed three algorithms, we compared the retrieved soil moisture with the SMAP L2 product, based on the second portion of the dataset. The results indicate that the retrieved soil moisture agrees well with SMAP L2 soil moisture products. Both the RF and OLS retrieval results obtained high correlation coefficients ($R^2{}_{OLS} = 0.981$, $R^2{}_{RF} = 0.983$) and low RMSE ($RMSE_{OLS} = 0.016\ m^3/m^3$, $RMSE_{RF} = 0.016\ m^3/m^3$) with the SMAP L2 product. In contrast, the overall performance of the SVM was relatively weak.

The retrieved soil moisture was also compared to in-situ measurements. Good agreement is observed between the temporal profiles of the retrieved and in-situ soil moisture; an overestimation was also found for high vegetation conditions. Through a comprehensive comparison of the three methods, RF and OLS algorithms outperformed the SVM for soil moisture retrieval from the SMAP observations. However, the proposed algorithms were only evaluated over a limited area. In the perspective, the potential applications of the proposed algorithms along with the deep learning approaches [77,78] will also be investigated for soil moisture retrievals over other areas with diverse roughness and vegetation conditions, particularly at regional and global scales. Furthermore, other vegetation parameters such as LAI, vegetation water content and different kinds of vegetation index will be exploited to characterize the vegetation influence on soil moisture retrievals from brightness temperature.

Author Contributions: Conceptualization, H.W. and C.T.; methodology, H.W.; software, C.T., L.Z. and M.Y.; validation, H.W., R.M., K.G. and J.D.; formal analysis, C.T.; investigation, C.T., L.Z. and M.Y.; resources, H.W.; writing—original draft preparation, C.T.; writing—review and editing, H.W., R.M., K.G. and J.D.; supervision, H.W., R.M., K.G. and J.D.; funding acquisition, H.W. All authors have read and agreed to the published version of the manuscript.

Funding: This study was supported by the National Natural Science Foundation of China (No. 41801232) and the Fundamental Research Funds for the Central Universities (No. 2018QNA6011).

Acknowledgments: The authors thank the OzNet hydrological monitoring network for providing the soil moisture station data, and the anonymous reviewers for improving the paper.

Conflicts of Interest: The authors declare no conflict of interest.

References

1. Seneviratne, S.; Corti, T.; Davin, E.; Hirschi, M.; Jaeger, E.; Lehner, I.; Orlowsky, B.; Teuling, A. Investigating soil moisture-climate interactions in a changing climate: A review. *Earth-Sci. Rev.* **2010**, *99*, 125–161. [CrossRef]
2. Zveryaev, I.; Arkhipkin, A.V. Leading modes of interannual soil moisture variability in European Russia and their relation to regional climate during the summer season. *Clim. Dyn.* **2019**, *53*, 3007–3022. [CrossRef]
3. Albertson, J.; Parlange, M. Natural integration of scalar fluxes from complex terrain. *Adv. Water Resour.* **1999**, *23*, 239–252. [CrossRef]
4. Koster, R.D.; Dirmeyer, P.; Guo, Z.; Bonan, G.; Chan, E.; Cox, P.; Gordon, C.T.; Kanae, S.; Kowalczyk, E.; Lawrence, D.; et al. Regions of Strong Coupling between Soil Moisture and Precipitation. *Science* **2004**, *305*, 1138–1140. [CrossRef] [PubMed]
5. Fang, B.; Lakshmi, V. Soil moisture at watershed scale: Remote sensing techniques. *J. Hydrol.* **2014**, *516*, 258–272. [CrossRef]
6. Zhang, L.; Meng, Q.; Yao, S.; Wang, Q.; Zeng, J.; Zhao, S.; Ma, J. Soil Moisture Retrieval from the Chinese GF-3 Satellite and Optical Data over Agricultural Fields. *Sensors* **2018**, *18*, 2675. [CrossRef]
7. Shin, Y.; Mohanty, B. Development of a deterministic downscaling algorithm for remote sensing soil moisture footprint using soil and vegetation classifications. *Water Resour. Res.* **2013**, *49*, 6208–6228. [CrossRef]
8. Owe, M.; de Jeu, R.; Walker, J. A Methodology for surface soil moisture and vegetation optical depth retrieval using the microwave polarization difference index. *IEEE Trans. Geosci. Remote Sens.* **2001**, *39*, 1643–1654. [CrossRef]
9. Deng, J.; Wang, K.; Hong, Y.; Qi, J. Spatio-temporal dynamics and evolution of land use change and landscape pattern in response to rapid urbanization. *Landsc. Urban Plan.* **2009**, *92*, 187–198. [CrossRef]
10. Phan, T.-N.; Kappas, M.; Thoi Nguyen, K.; Tran, T.; Vinh Tran, Q.; Rafiei Emam, A. Evaluation of MODIS land surface temperature products for daily air surface temperature estimation in northwest Vietnam. *Int. J. Remote Sens.* **2019**, *40*, 5544–5562. [CrossRef]
11. Carlson, T.; Ripley, D.A. On the Relation between NDVI, Fractional Vegetation Cover, and Leaf Area Index. *Remote Sens. Environ.* **1997**, *62*, 241–252. [CrossRef]
12. Zhuo, L.; Dai, Q.; Han, D.; Chen, N.; Zhao, B.; Berti, M. Evaluation of Remotely Sensed Soil Moisture for Landslide Hazard Assessment. *IEEE J. Sel. Top. Appl. Earth Obs. Remote Sens.* **2019**, *12*, 162–173. [CrossRef]
13. Sabaghy, S.; Walker, J.; Renzullo, L.; Jackson, T.J. Spatially enhanced passive microwave derived soil moisture: Capabilities and opportunities. *Remote Sens. Environ.* **2018**, *209*, 551–580. [CrossRef]
14. Mohanty, B.; Cosh, M.; Lakshmi, V.; Montzka, C. Soil Moisture Remote Sensing: State-of-the-Science. *Vadose Zone J.* **2017**, *16*, 1–9. [CrossRef]
15. Zhang, X.; Chen, B.; Zhao, H.; Li, T.; Chen, Q. Physical-based soil moisture retrieval method over bare agricultural areas by means of multi sensor SAR data. *Int. J. Remote Sens.* **2018**, *39*, 3870–3890. [CrossRef]
16. Mohamed, E.S.; Ali, A.; El-Shirbeny, M.; Abutaleb, K.; Shaddad, S.M. Mapping soil moisture and their correlation with crop pattern using remotely sensed data in arid region. *Egypt. J. Remote Sens. Space Sci.* **2019**. [CrossRef]
17. Montzka, C.; Rötzer, K.; Bogena, H.; Sanchez, N.; Vereecken, H. A new soil moisture downscaling approach for SMAP, SMOS, and ASCAT by predicting sub-grid variability. *Remote Sens.* **2018**, *10*, 427. [CrossRef]
18. Amazirh, A.; Merlin, O.; Er-Raki, S.; Gao, Q.; Rivalland, V.; Malbeteau, Y.; Khabba, S.; Escorihuela, M.J. Retrieving surface soil moisture at high spatio-temporal resolution from a synergy between Sentinel-1 radar and Landsat thermal data: A study case over bare soil. *Remote Sens. Environ.* **2018**, *211*, 321–337. [CrossRef]
19. Du, Y.; Ulaby, F.T.; Craig Dobson, M. Sensitivity to soil moisture by active and passive microwave sensors. *IEEE Trans. Geosci. Remote Sens.* **2000**, *38*, 105–114. [CrossRef]
20. Kolassa, J.; Reichle, R.H.; Draper, C. Merging active and passive microwave observations in soil moisture data assimilation. *Remote Sens. Environ.* **2017**, *191*, 117–130. [CrossRef]
21. Kim, H.; Parinussa, R.; Konings, A.G.; Wagner, W.; Choi, M.; Zohaib, M. Global-scale assessment and combination of SMAP with ASCAT (active) and AMSR2 (passive) soil moisture products. *Remote Sens. Environ.* **2018**, *204*, 260–275. [CrossRef]
22. Bai, X.; Zeng, J.; Chen, K.; Li, Z.; Zeng, Y.; Wen, J.; Wang, X.; Dong, X.; Su, Z. Parameter Optimization of a Discrete Scattering Model by Integration of Global Sensitivity Analysis Using SMAP Active and Passive Observations. *IEEE Trans. Geosci. Remote Sens.* **2019**, *57*, 1084–1099. [CrossRef]

23. Entekhabi, D.; Njoku, E.G.; Neill, P.E.O.; Kellogg, K.H.; Crow, W.T.; Edelstein, W.N.; Entin, J.K.; Goodman, S.D.; Jackson, T.J.; Johnson, J.; et al. The Soil Moisture Active Passive (SMAP) Mission. *Proc. IEEE* **2010**, *98*, 704–716. [CrossRef]
24. Das, N.N.; Entekhabi, D.; Njoku, E.G. An Algorithm for Merging SMAP Radiometer and Radar Data for High-Resolution Soil-Moisture Retrieval. *IEEE Trans. Geosci. Remote Sens.* **2011**, *49*, 1504–1512. [CrossRef]
25. Karthikeyan, L.; Pan, M.; Wanders, N.; Kumar, D.N.; Wood, E.F. Four decades of microwave satellite soil moisture observations: Part 1. A review of retrieval algorithms. *Adv. Water Resour.* **2017**, *109*, 106–120. [CrossRef]
26. Zheng, D.; Wang, X.; Van Der Velde, R.; Ferrazzoli, P.; Wen, J.; Wang, Z.; Schwank, M.; Colliander, A.; Bindlish, R.; Su, B. Impact of surface roughness, vegetation opacity and soil permittivity on L-band microwave emission and soil moisture retrieval in the third pole environment. *Remote Sens. Environ.* **2018**, *209*, 633–647. [CrossRef]
27. Ebrahimi-Khusfi, M.; Alavipanah, S.K.; Hamzeh, S.; Amiraslani, F.; Neysani Samany, N.; Wigneron, J.-P. Comparison of soil moisture retrieval algorithms based on the synergy between SMAP and SMOS-IC. *Int. J. Appl. Earth Obs. Geoinf.* **2018**, *67*, 148–160. [CrossRef]
28. O'Neill, P.E.; Chan, S.; Njoku, E.G.; Jackson, T.; Bindlish, R. *SMAP L3 Radiometer Global Daily 36 km EASE-Grid Soil Moisture, Version 6*; NASA National Snow and Ice Data Center Distributed Active Archive Center: Boulder, CO, USA, 2019.
29. Ebtehaj, A.; Bras, R.L. A physically constrained inversion for high-resolution passive microwave retrieval of soil moisture and vegetation water content in L-band. *Remote Sens. Environ.* **2019**, *233*, 111346. [CrossRef]
30. Ali, I.; Greifeneder, F.; Stamenkovic, J.; Neumann, M.; Notarnicola, C. Review of Machine Learning Approaches for Biomass and Soil Moisture Retrievals from Remote Sensing Data. *Remote Sens.* **2015**, *7*, 16398–16421. [CrossRef]
31. Ahmad, S.; Kalra, A.; Stephen, H. Estimating soil moisture using remote sensing data: A machine learning approach. *Adv. Water Resour.* **2010**, *33*, 69–80. [CrossRef]
32. Xing, M.; He, B.; Ni, X.; Wang, J.; An, G.; Shang, J.; Huang, X. Retrieving Surface Soil Moisture over Wheat and Soybean Fields during Growing Season Using Modified Water Cloud Model from Radarsat-2 SAR Data. *Remote Sens.* **2019**, *11*, 1956. [CrossRef]
33. Lu, Z.; Chai, L.; Ye, Q.; Zhang, T. Reconstruction of time-series soil moisture from AMSR2 and SMOS data by using recurrent nonlinear autoregressive neural networks. In Proceedings of the IEEE International Geoscience and Remote Sensing Symposium (IGARSS), Milan, Italy, 26–31 July 2015.
34. Yao, P.; Shi, J.; Zhao, T.; Lu, H.; Al-Yaari, A. Rebuilding Long Time Series Global Soil Moisture Products Using the Neural Network Adopting the Microwave Vegetation Index. *Remote Sens.* **2017**, *9*, 35. [CrossRef]
35. Qu, Y.; Zhu, Z.; Chai, L.; Liu, S.; Montzka, C.; Liu, J.; Yang, X.; Lu, Z.; Jin, R.; Li, X.; et al. Rebuilding a Microwave Soil Moisture Product Using Random Forest Adopting AMSR-E/AMSR2 Brightness Temperature and SMAP over the Qinghai–Tibet Plateau, China. *Remote Sens.* **2019**, *11*, 683. [CrossRef]
36. Kolassa, J.; Reichle, R.H.; Liu, Q.; Alemohammad, H.; Gentine, P.; Aida, K.; Asanuma, J.; Bircher, S.; Caldwell, T.; Colliander, A.; et al. Estimating surface soil moisture from SMAP observations using a Neural Network technique. *Remote Sens. Environ.* **2017**, *204*, 43–59. [CrossRef]
37. Senyurek, V.; Lei, F.; Boyd, D.; Kurum, M.; Gurbuz, A.C.; Moorhead, R. Machine Learning-Based CYGNSS Soil Moisture Estimates over ISMN sites in CONUS. *Remote Sens.* **2020**, *12*, 1168. [CrossRef]
38. Park, S.; Im, J.; Park, S.; Rhee, J. AMSR2 soil moisture downscaling using multisensor products through machine learning approach. In Proceedings of the IEEE International Geoscience and Remote Sensing Symposium (IGARSS), Milan, Italy, 26–31 July 2015.
39. Gruber, A.; Paloscia, S.; Santi, E.; Notarnicola, C.; Pasolli, L.; Smolander, T.; Pulliainen, J.; Mittelbach, H.; Dorigo, W.; Wagner, W. Performance inter-comparison of soil moisture retrieval models for the MetOp-A ASCAT instrument. In Proceedings of the IEEE International Geoscience and Remote Sensing Symposium (IGARSS), Quebec City, QC, Canada, 13–18 July 2014.
40. Galantowicz, J.F.; Entekhabi, D.; Njoku, E.G. Tests of sequential data assimilation for retrieving profile soil moisture and temperature from observed L-band radiobrightness. *IEEE Trans. Geosci. Remote Sens.* **1999**, *37*, 1860–1870. [CrossRef]
41. Dumedah, G.; Walker, J. Evaluation of Model Parameter Convergence when Using Data Assimilation for Soil Moisture Estimation. *J. Hydrometeorol.* **2014**, *15*, 359–375. [CrossRef]

42. Pasolli, L.; Notarnicola, C.; Bertoldi, G.; Bruzzone, L.; Remelgado, R.; Greifeneder, F.; Niedrist, G.; Chiesa, S.D.; Tappeiner, U.; Zebisch, M. Estimation of Soil Moisture in Mountain Areas Using SVR Technique Applied to Multiscale Active Radar Images at C-Band. *IEEE J. Sel. Top. Appl. Earth Obs. Remote Sens.* **2015**, *8*, 262–283. [CrossRef]
43. Wang, H.; Magagi, R.; Goïta, K.; Trudel, M.; McNairn, H.; Powers, J. Crop phenology retrieval via polarimetric SAR decomposition and Random Forest algorithm. *Remote Sens. Environ.* **2019**, *231*, 111234. [CrossRef]
44. Smith, A.; Walker, J.; Western, A.; Young, R.; Ellett, K.; Pipunic, R.; Grayson, R.; Siriwardena, L.; Chiew, F.; Richter, H. The Murrumbidgee Soil Moisture Monitoring Network data set. *Water Resour. Res.* **2012**, *48*, 7701. [CrossRef]
45. Wu, X.; Xiao, Q.; Wen, J.; You, D.; Hueni, A. Advances in quantitative remote sensing product validation: Overview and current status. *Earth-Sci. Rev.* **2019**, *196*, 102875. [CrossRef]
46. Jackson, T.J.; Cosh, M.; Bindlish, R.; Starks, P.; Bosch, D.; Seyfried, M.; Goodrich, D.; Susan Moran, M.; Du, J. Validation of Advanced Microwave Scanning Radiometer Soil Moisture Products. *IEEE Trans. Geosci. Remote Sens.* **2011**, *48*, 4256–4272. [CrossRef]
47. Zeng, J.; Li, Z.; Quan, C.; Bi, H.; Qiu, J.; Zou, P. Evaluation of remotely sensed and reanalysis soil moisture products over the Tibetan Plateau using in-situ observations. *Remote Sens. Environ.* **2015**, *163*, 91–110. [CrossRef]
48. Didan, K.; Munoz, A.B.; Solano, R.; Huete, A. *MODIS Vegetation Index User's Guide (MOD13 Series)*; Vegetation Index and Phenology Lab, University of Arizona: Tucson, AZ, USA, 2015.
49. He, L.; Hong, Y.; Wu, X.; Ye, N.; Walker, J.P.; Chen, X. Investigation of SMAP Active–Passive Downscaling Algorithms Using Combined Sentinel-1 SAR and SMAP Radiometer Data. *IEEE Trans. Geosci. Remote Sens.* **2018**, *56*, 4906–4918. [CrossRef]
50. Schmugge, T.; Neill, P.E.O.; Wang, J.R. Passive Microwave Soil Moisture Research. *IEEE Trans. Geosci. Remote Sens.* **1986**, *GE-24*, 12–22. [CrossRef]
51. Gherboudj, I.; Magagi, R.; Goita, K.; Berg, A.; Toth, B.; Walker, A. Validation of SMOS Data over Agricultural and Boreal Forest Areas in Canada. *IEEE Trans. Geosci. Remote Sens.* **2012**, *50*, 1623–1635. [CrossRef]
52. Cheng, Y.-B.; Zarco-Tejada, P.J.; Riaño, D.; Rueda, C.A.; Ustin, S.L. Estimating vegetation water content with hyperspectral data for different canopy scenarios: Relationships between AVIRIS and MODIS indexes. *Remote Sens. Environ.* **2006**, *105*, 354–366. [CrossRef]
53. Gao, Y.; Walker, J.P.; Allahmoradi, M.; Monerris, A.; Ryu, D.; Jackson, T.J. Optical Sensing of Vegetation Water Content: A Synthesis Study. *IEEE J. Sel. Top. Appl. Earth Obs. Remote Sens.* **2015**, *8*, 1456–1464. [CrossRef]
54. Jin, Z.; Xu, B. A novel compound smoother—RMMEH to reconstruct MODIS NDVI time series. *IEEE Geosci. Remote Sens. Lett.* **2013**, *10*, 942–946. [CrossRef]
55. Beck, P.S.A.; Atzberger, C.; Høgda, K.A.; Johansen, B.; Skidmore, A.K. Improved monitoring of vegetation dynamics at very high latitudes: A new method using MODIS NDVI. *Remote Sens. Environ.* **2006**, *100*, 321–334. [CrossRef]
56. Xu, L.; Li, B.; Yuan, Y.; Gao, X.; Zhang, T. A Temporal-Spatial Iteration Method to Reconstruct NDVI Time Series Datasets. *Remote Sens.* **2015**, *7*, 8906–8924. [CrossRef]
57. Hu, Z.; Zhengwei, Y.; Liping, D.; Lin, L.; Haihong, Z. Crop phenology date estimation based on NDVI derived from the reconstructed MODIS daily surface reflectance data. In Proceedings of the 17th International Conference on Geoinformatics, Fairfax, VA, USA, 12–14 August 2009; pp. 1–6.
58. Chan, S.K.; Bindlish, R.; Neill, P.E.O.; Njoku, E.; Jackson, T.; Colliander, A.; Chen, F.; Burgin, M.; Dunbar, S.; Piepmeier, J.; et al. Assessment of the SMAP Passive Soil Moisture Product. *IEEE Trans. Geosci. Remote Sens.* **2016**, *54*, 4994–5007. [CrossRef]
59. de Jeu, R.; Wagner, W.; Holmes, T.; Dolman, H.; van de Giesen, N.; Friesen, J. Global Soil Moisture Patterns Observed by Space Borne Microwave Radiometers and Scatterometers. *Surv. Geophys.* **2008**, *29*, 399–420. [CrossRef]
60. Burgin, M.; Colliander, A.; Njoku, E.; Chan, S.; Cabot, F.; Kerr, Y.; Bindlish, R.; Jackson, T.; Entekhabi, D.; Yueh, S. A Comparative Study of the SMAP Passive Soil Moisture Product with Existing Satellite-Based Soil Moisture Products. *IEEE Trans. Geosci. Remote Sens.* **2017**, *55*, 2959–2971. [CrossRef] [PubMed]
61. Jackson, T.; O'Neill, P.; Chan, S.; Bindlish, R.; Colliander, A.; Chen, F.; Burgin, M.; Dunbar, S.; Piepmeier, J.; Cosh, M.; et al. *Calibration and Validation for the L2/3_SM_P Version 4 and L2/3_SM_P_E Version 1 Data Products*; Jet Propulsion Laboratory: Pasadena, CA, USA, 2016.

62. Jackson, T.; O'Neill, P.; Chan, S.; Bindlish, R.; Colliander, A.; Chen, F.; Burgin, M.; Dunbar, S.; Piepmeier, J.; Cosh, M.; et al. *Calibration and Validation for the L2/3_SM_P Version 5 and L2/3_SM_P_E Version 2 Data Products*; Jet Propulsion Laboratory: Pasadena, CA, USA, 2018.
63. Ouadhi, K.; Ourbih-Tari, M. Monte Carlo simulation of ordinary least squares estimator through linear regression adaptive refined descriptive sampling algorithm. *Commun. Stat. Theory Methods* **2018**, *48*, 865–875. [CrossRef]
64. Park, J.-Y.; Ahn, S.-R.; Hwang, S.J.; Jang, C.; Park, G.-A.; Kim, S. Evaluation of MODIS NDVI and LST for indicating soil moisture of forest areas based on SWAT modeling. *Paddy Water Environ.* **2014**, *12*, 77–88. [CrossRef]
65. Engstrom, R.; Hope, A.; Kwon, H.; Stow, D. The Relationship between Soil Moisture and NDVI near Barrow, Alaska. *Phys. Geogr.* **2013**, *29*, 38–53. [CrossRef]
66. Cao, X.; Chen, J.; Imura, H.; Higashi, O. A SVM-based method to extract urban areas from DMSP-OLS and SPOT VGT data. *Remote Sens. Environ.* **2009**, *113*, 2205–2209. [CrossRef]
67. Zhang, Y.; Li, Q.; Huang, H.; Wu, W.; Du, X.; Wang, H. The Combined Use of Remote Sensing and Social Sensing Data in Fine-Grained Urban Land Use Mapping: A Case Study in Beijing, China. *Remote Sens.* **2017**, *9*, 865. [CrossRef]
68. Rodriguez-Galiano, V.; Chica-Olmo, M.; Abarca-Hernández, F.; Atkinson, P.; Chockalingam, J. Random Forest classification of Mediterranean land cover using multi-seasonal imagery and multi-seasonal texture. *Remote Sens. Environ.* **2012**, *121*, 93–107. [CrossRef]
69. Esteban, J.; McRoberts, R.; Fernández-Landa, A.; Tomé, J.; Næsset, E. Estimating Forest Volume and Biomass and Their Changes Using Random Forests and Remotely Sensed Data. *Remote Sens.* **2019**, *11*, 1944. [CrossRef]
70. Wang, Y.; Qi, Y.; Chen, Y.; Xie, F. Prediction of soil organic matter based on multi-resolution remote sensing data and random forest algorithm. *Acta Pedol. Sin.* **2016**, *53*, 342–354. [CrossRef]
71. Río, S.; López, V.; Benítez, J.; Herrera, F. On the use of MapReduce for Imbalanced Big Data using Random Forest. *Inf. Sci.* **2014**, *285*, 112–137. [CrossRef]
72. Chehreh Chelgani, S.; Matin, S.S.; Makaremi, S. Modeling of free swelling index based on variable importance measurements of parent coal properties by random forest method. *Measurement* **2016**, *94*, 416–422. [CrossRef]
73. Colliander, A.; Jackson, T.J.; Bindlish, R.; Chan, S.; Das, N.; Kim, S.B.; Cosh, M.H.; Dunbar, R.S.; Dang, L.; Pashaian, L.; et al. Validation of SMAP surface soil moisture products with core validation sites. *Remote Sens. Environ.* **2017**, *191*, 215–231. [CrossRef]
74. Ma, H.; Zeng, J.; Chen, N.; Zhang, X.; Cosh, M.H.; Wang, W. Satellite surface soil moisture from SMAP, SMOS, AMSR2 and ESA CCI: A comprehensive assessment using global ground-based observations. *Remote Sens. Environ.* **2019**, *231*, 111215. [CrossRef]
75. Hassan, A.M.; Belal, A.A.; Hassan, M.A.; Farag, F.M.; Mohamed, E.S. Potential of thermal remote sensing techniques in monitoring waterlogged area based on surface soil moisture retrieval. *J. Afr. Earth Sci.* **2019**, *155*, 64–74. [CrossRef]
76. Mao, H.; Kathuria, D.; Duffield, N.; Mohanty, B.P. Gap Filling of High-Resolution Soil Moisture for SMAP/Sentinel-1: A Two-Layer Machine Learning-Based Framework. *Water Resour. Res.* **2019**, *55*, 6986–7009. [CrossRef]
77. Chi, J.; Kim, H.-C.; Lee, S.; Crawford, M.M. Deep learning based retrieval algorithm for Arctic sea ice concentration from AMSR2 passive microwave and MODIS optical data. *Remote Sens. Environ.* **2019**, *231*, 111204. [CrossRef]
78. Fang, K.; Pan, M.; Shen, C. The Value of SMAP for Long-Term Soil Moisture Estimation with the Help of Deep Learning. *IEEE Trans. Geosci. Remote Sens.* **2019**, *57*, 2221–2233. [CrossRef]

remote sensing

MDPI

Article

Monitoring Residual Soil Moisture and Its Association to the Long-Term Variability of Rainfall over the Upper Blue Nile Basin in Ethiopia

Getachew Ayehu [1,2,*], Tsegaye Tadesse [3] and Berhan Gessesse [1]

1 Remote Sensing Research and Development Department, EORC, Ethiopian Space Science and Technology Institute, Addis Ababa P.O. Box 33679, Ethiopia; berhang@essti.gov.et

2 Institute of Land Administration, Bahir Dar University, Bahir Dar P.O. Box 79, Ethiopia

3 National Drought Mitigation Center, University of Nebraska-Lincoln, Lincoln, NE 68583, USA; ttadesse2@unl.edu

* Correspondence: getachewt@essti.gov.et; Tel.: +251-911-33-6972

Received: 30 March 2020; Accepted: 29 June 2020; Published: 3 July 2020

Abstract: Monitoring soil moisture and its association with rainfall variability is important to comprehend the hydrological processes and to set proper agricultural water use management to maximize crop growth and productivity. In this study, the European Space Agency's Climate Change Initiative (ESA CCI) soil moisture product was applied to assess the dynamics of residual soil moisture in autumn (September to November) and its response to the long-term variability of rainfall in the Upper Blue Nile Basin (UBNB) of Ethiopia from 1992 to 2017. The basin was found to have autumn soil moisture (ASM) ranging from 0.09–0.38 m^3/m^3, with an average of 0.26 m^3/m^3. The ASM time series resulted in the coefficient of variation (CV) ranging from 2.8%–28% and classified as low-to-medium variability. In general, the monotonic trend analysis for ASM revealed that the UBNB had experienced a wetting trend for the past 26 years (1992–2017) at a rate of 0.00024 m^3/m^3 per year. A significant wetting trend ranging from 0.001 to 0.006 m^3/m^3 per year for the autumn season was found. This trend was mainly showed across the northwest region of the basin and covers about 18% of the total basin area. The spatial patterns and variability of rainfall and ASM were also found to be similar, which implies the strong relationship between rainfall and soil moisture in autumn. The spring and autumn season rainfall explained a considerable portion of ASM in the basin. The analyses also signified that the rainfall amount and distribution impacted by the topography and land cover classes of the basin showed a significant influence on the characteristics of the ASM. Further, the result verified that the behavior of ASM could be controlled by the loss of soil moisture through evapotranspiration and the gain from rainfall, although changes in rainfall were found to be the primary driver of ASM variability over the UBNB.

Keywords: ESA CCI; residual soil moisture; evapotranspiration; trend; rainfall variability; CHIRPS

1. Introduction

Soil moisture is an essential parameter to understand various processes in agriculture, hydrology, and climate [1]. In agriculture, the spatio-temporal distribution of soil moisture determines the success of crop production because the growth and productivity of crops highly depends on the sufficient amount and timing of available moisture. Notably, in developing countries such as Ethiopia, where the livelihood and economy of the country are highly dependent on rainfed agriculture [2–4], water is a distinctly scarce and valuable resource. The Upper Blue Nile Basin (UBNB) in Ethiopia receives an adequate amount of annual rainfall (> 2000 mm), the majority of the rainfall occurs during the summer growing season [5,6]. Agriculture in the UBNB is dominated by smallholder farmers, who are unable to produce enough amount of food by a single harvest during the main rainy season to sustain their livelihoods [7,8]. However, following the harvest of main season cropping in the UBNB, a certain amount of soil water, as residual soil moisture, is left in the soil that can last up to several months [9,10]. Residual soil moisture in this study was defined as the amount of water left in the soil following the physiological maturity or harvesting of main season cropping. Given the low irrigation facilities [11–13] and meager crop production during the main season [3,14], additional cropping in the off-season using residual soil moisture could be an alternative option to increase food and feed production in the basin.

The extent of residual soil moisture available varies temporally with hydroclimate conditions, and spatially depending on biophysical factors such as topography and soil properties of the basin [15,16]. Different works of literature signify that the dynamics of soil moisture mainly instigates from local rainfall [17,18]. Some other scholars argued that the spatio-temporal dynamics of soil moisture depends on the temporal variability of both rainfall and evapotranspiration [19,20]. For example, Cheng et al. [21] revealed that the amount of soil moisture available to crops mainly depends on rainfall, and the spatial distribution of the soil moisture trend looks like that of rainfall. Similarly, Robinson et al. [22] and He et al. [23] indicated that the spatio-temporal dynamics of soil moisture is affected by rainfall variability to a more significant extent. Again, the result obtained by Jia et al. [24] suggested that compared to other climatic variables rainfall is the primary factor responsible for the trends and variability of soil moisture. Indeed, rainfall from different season has various impact on the amount and spatio-temporal distribution of soil moisture over various seasons [25]. Yang et al. [25] reported that both the current (i.e., spring rainfall) and previous rainfall (i.e., rainfall in summer, autumn, and winter remaining in the soil by its own "memory") influenced the variability of spring soil moisture with a difference in magnitude. As a result, knowledge of soil moisture and its link with rainfall is fundamental to better understand and predict the hydrological process of the basin.

On the other hand, due to a lack of continuously measured soil moisture datasets the soil moisture conditions and its association with rainfall variability poorly understood in the UBN B. Soil moisture information is not available or not being measured regularly like other climate variables such as rainfall and temperature in the basin. Such a study indeed requires a large-scale soil moisture dataset, although obtaining soil moisture observation at this scale is often a challenge [26]. In this connection, it is hard to find studies that characterize the spatial-temporal dynamics of soil moisture and its linkage with local hydro-climate conditions in Ethiopia. However, in other places a great effort was undertaken to understand soil moisture dynamics [27–29] and its relationship with other climate variables (e.g., rainfall, temperature, evapotranspiration, and solar radiation) [19,30,31] at a range of field to a global scale, although many of the previous studies depend on point-based in-situ measurements, which are commonly inadequate to perform soil moisture monitoring on larger scales [21].

In recent years, a broad range of regional and global soil moisture products with reasonable temporal and spatial resolution from different sources, such as land surface modeling [32], remote sensing [33], and data assimilation techniques [34,35] are now available. The dependability of model-based products is controlled by the feature of meteorological forcing data [36,37]. Consequently, recent studies of soil moisture and its concurred response to climate variation over higher spatial and temporal scales are largely based on remote sensing [38,39] and data assimilations techniques [21,40]. The latest version of the soil moisture product released by the European Space Agency's Climate Change Initiative (ESA CCI), which merged active and passive microwave observations, provides relatively consistent and reliable soil moisture information worldwide over the period of 1978 to 2018 [38,41,42]. The performance of ESA CCI has been extensively validated in different regions against in-situ network observations. Dorigo et al. [43] provided comprehensive reviews of these studies. Particularly, McNally et al. [44] evaluated ESA CCI soil moisture over East Africa (including Ethiopia) comparing with the Normalized Difference Vegetation Index (NDVI) and modeled soil moisture products. The validation of the dataset has been proven to be useful in a large number of applications such as long-term soil moisture trend analysis [24,45,46] and drought monitoring [47].

Although earlier studies conducted on the variability of soil moisture provided pertinent information, the long-term trend and variability of soil moisture in the off-season were not adequately addressed. It is also unclear to what extent rainfall variability can control soil moisture dynamics in the off-season. Thus, investigating the dynamics of soil moisture in the off-season and its response to the long-term variability of rainfall is crucial to set proper agricultural water use management in terms of maximizing crop growth and productivity [27,48]. The main objective of this study is to leverage the readily available ESA CCI soil moisture product in order to investigate the dynamics of autumn (September to November) soil moisture and its linked response to the long-term variability of rainfall over the UBNB in Ethiopia. The change in evapotranspiration (ET) might considerably affect soil moisture characteristics; therefore, it is also essential to understand how the mutual effects of ET and rainfall may bring about changes in autumn soil moisture over the UBNB. This study answers the following three questions: What is the characteristic of the autumn soil moisture inter-annual variability and trend over the UBNB from 1992 to 2017? What is the response of autumn soil moisture to the long-term effects of seasonal and annual rainfall? How does soil water loss via ET and gain through rainfall at a seasonal and annual scale determine the availability of autumn soil moisture in the UBNB?

Our paper is organized into five sections. Section 2 presents the materials and methods used in proposed study. Section 3 describes the results of the study. Major findings of the study are discussed in Section 4. Finally, Section 5 addressed the main conclusions.

2. Materials and Methods

2.1. Descriptions of the Study Area

The Upper Blue Nile Basin (UBNB) is located in the northwestern part of Ethiopia (Figure 1) and contributes the major share of the Nile River's annual water flow [5,49]. The basin is described by rugged topography with elevation ranging from 490 to 4239 m a.s.l. (Figure 1). The annual climate of the basin can be divided into two (i.e., rainy and dry) seasons. The rainy season can be split into a short rainy season from February to May and a main rainy season from June to September. The dry season occurs between October and January. The mean annual temperature in the study site is about 20.4 °C. The basin receives an annual rainfall up to 2200 mm, which mostly occurs during the wet season (June to September) and is known locally as "Kiremt" [50]. However, the basin is known by considerable spatial and temporal variations in rainfall [5,51], which makes the hydrological process of the basin very complex. Despite a range of land-use systems occurring, the livelihoods of the majority of the populations in the basin are greatly reliant on rainfed agriculture.

2.2. Datasets

2.2.1. ESA CCI Microwave Soil Moisture

The ESA CCI soil moisture product is provided by the European Space Agency. The ESA CCI dataset is a merged multi-satellite microwave soil moisture product which combines observation from active and passive sensing microwave sensing systems [38,41]. To combine these products, the datasets were first rescaled using the Global Land Data Assimilation System (GLDAS) data as a standard reference, which has a spatial resolution of 0.25° at a daily basis and represents soil moisture layers up to 10 cm [52]. The ESA CCI soil moisture dataset provides surface soil moisture information in a volumetric unit (m^3m^{-3}) and was available daily from 1978 to 2018 (v04.4) with a spatial resolution of 0.25° [41]. Besides the variation in the feature of individual data sources, the reliability of ESA CCI soil moisture products could also be affected by the adapted merging methods for combining observations from different mission and retrieval algorithms [45]. However, the merged products are superior to either the passive or active alone [41]. In this study, the latest version (v04.4) of ESA CCI soil moisture data was used for long-term trend and variability analysis of residual soil moisture in the UBNB.

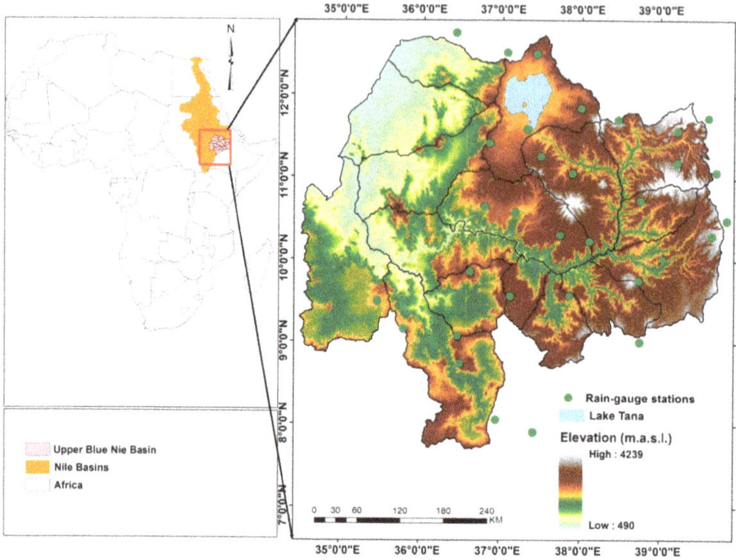

Figure 1. Location of the Upper Blue Nile Basin (UBNB) (Imagery source for UBNB: Shuttle Radar Topography Mission (SRTM) Global elevation data). The green bold point shows the distributions of rain gauge stations used to validate the performance of the Climate Hazards Group InfraRed Precipitation with Station (CHIRPS) satellite rainfall datasets.

2.2.2. FLDAS Noah Land Surface Model

The Famine Early Warning Systems Network (FEWS NET) Land Data Assimilation System (FLDAS) dataset contains a series of land surface parameters simulated from the Noah 3.6.1 model. The system generates ensembles of soil moisture, evapotranspiration (ET), and other variables based on multiple meteorological inputs or land surface models. In this study, FLDAS Noah land surface global data (Model L4) were used to extract soil moisture data (at a depth of 0–10, 10–40, and 40–100 cm depths) and ET datasets. The simulation was forced by a combination of the Modern-Era Retrospective analysis for Research and Applications version 2 (MERRA-2) data and Climate Hazards Group InfraRed Precipitation with Station (CHIRPS) 6-hourly rainfall data that was downscaled using the NASA Land Data Toolkit. The dataset provides soil moisture in volumetric units ($m^3 m^{-3}$), and ET in kilograms per meter squared per second (kg m^{-2} s^{-1}) and is available from 1982 to present at monthly intervals with a spatial resolution of 0.10°. The dataset was obtained from the NASA Goddard Earth Science Data and Information Services Center (GES DISC) website (https://disc.gsfc.nasa.gov/). The atmospheric and land modeling communities widely use the FLDAS land surface model, and therefore, model parameters are well tested [44]. Several researchers over East Africa e.g., [53,54], used the model for various hydrological functions.

2.2.3. CHIRPS Rainfall Product

The CHIRPS-v2 rainfall estimate was used as a source of rainfall data in the UBNB, Ethiopia. CHIRPS is a quasi-global (500 S–500N) product provided from 1981 to near present at 0.050 spatial resolutions (~5.3 km) and at daily, pentadal (5-day), dekadal (10-day), and monthly temporal resolution (Funk et al., 2015). The dataset was created by the U.S. Geological Survey (USGS) and the Climate Hazards Group (CHG) at the University of California [55]. It integrates both ground and satellite observations to grant a global rainfall estimate with reasonably low latency, high resolution, low bias and a long period of record. A validation study made over the UBNB revealed the great skill and immense potential of CHIRPS-v2. It can be used for various operational applications such as hydro-climate studies in areas where the gauge stations are very sparse and unevenly distributed [56]. Figure A1 (Appendix A) provides a summary of the validation results [56].

2.2.4. ESA CCI Land Cover

The ESA has derived a high-quality historical land cover (LC) dataset as part of their CCI program [57]. The ESA CCI LC dataset is available from 1992 to 2015 at 300m spatial resolution and with a one-year temporal interval [58]. This product was created based on ESA's Glob Cover products using the Glob Cover unsupervised classification chain. The dataset is generated using multiple sensors e.g., Systeme Probatoire d'Observation de la Terre Vegetation (SPOT-VGT), Advanced Very High-Resolution Radiometer (AVHRR), and PROBA-V with an overall accuracy of about 71% [59]. The product provides 37 LC types (Table A1, Appendix B) based on FAO LC classification systems [58]. Interested readers are referred to Plummer et al. [60]. The accuracy of the ESA CCI LC product was evaluated at a global scale [61] over Africa [62] and China [63], which can give valuable insights for specific applications. For example, Guidigan et al. [64] and Li et al. [65] used the ESA CCI LC product to study the land use and land cover changes in Benin and over the globe, respectively.

2.3. Methods

Since the spatio-temporal coverage of the ESA CCI soil moisture dataset is generally poor in Ethiopia before 1992, the datasets from 1992 to 2017 were selected as the study period to maintain coincident temporal coverage in the study area. Then, daily values of ESA CCI soil moisture were converted to monthly time series and averaged for September to November to construct the annual mean soil moisture for the autumn season. The same approach was applied to FLDAS NOAH soil moisture dataset. The months of September indeed belong to the wettest period in the study area; however, during this month, main season cropping reaches to the stage of physiological maturity and crops will have limited moisture intake. As a result, dry season farming may start as of September for efficient utilization of the residual soil moisture. The original unit of monthly ET is kg $m^{-2}s^{-1}$, which was first converted to millimeter (mm) and aggregated for each season: December to February (winter), March to May (spring), June to August (summer), and September to November (autumn). The annual series was calculated from the sum of the seasonal estimates. The same method was applied to CHIRPS rainfall values to calculate the seasonal and annual time series for the period of 1992 to 2017. For the FLDAS NOAH and CHIRPS dataset, average values of their estimates were calculated at a 0.25° grid to keep the spatial resolution consistent with ESA CCI soil moisture dataset. The datasets used in this study including their temporal and spatial resolutions are summarized in Table 1.

Table 1. Summary of the products used in this study.

Product	Data Type	Spatial Cover	Temporal Cover	Spatial Reso	Temporal Reso	Sensor
ESA CCI	Soil moisture (m^3 m^{-3})	Global	1978–2018	0.25°	Daily	SMMR, SSM/I, TMI, AMSR-E, ASCAT, ERS
FLDAS Noah LSM	Soil moisture (m^3 m^{-3}) ET (kg $m^{-2}s^{-1}$)	Global	1982–present	0.10°	Monthly	MERRA-2, CHIRPS
CHIRPS v2	Rainfall	Global	1981–present	0.05°	Daily, Pentadal, Dekadal, Monthly	NOAA TIR satellite, TRMM
ESA CCI	Land cover	Global	1992–2015	300 m	1 year	MERIS; FR/RR SPOT-VGT; AVHRR; PROBA-V
DEM	Elevation	Global	-	30 m	-	SRTM

2.3.1. Variability and Trend Analysis

Variability analysis involves the generation of the long-term mean (LTM), calculating the coefficient of variation (CV), and an anomaly. The LTM for autumn soil moisture (ASM) and rainfall (both at a seasonal and annual time scale) were calculated from 1992 to 2017. The analyses were done using "raster" packages in the R platform [66]. The CV is calculated to evaluate the spatio-temporal variation for the time series and is computed using Equation (1):

$$CV = \frac{\sigma}{\mu} \times 100 \tag{1}$$

where CV is the coefficient of variation; σ is the standard deviation and μ is the mean value. The standardized anomalies of soil moisture and rainfall have been calculated to indicate the departures of each year total from the LTM, and were computed as follows (Equation (2)):

$$Z_a = \frac{\left(X_i - \overline{X}_i\right)}{s} \tag{2}$$

where Z_a is standard anomaly; X_i is annual value of a particular year; $\overline{X}_i$ is long-term mean (LTM) annual values, and s is the standard deviation over a period of observations (1992 to 2017, in our case). A negative value of Z_a represents periods of below-normal value, while positive value indicates above-normal values. To further assess the variability of soil moisture across the different land cover classes and for three different decades of the UBNB, soil moisture values from 1992 to 2000 were extracted using a ESA CCI land cover map of 1995, likewise from 2001 to 2010 (2005) and 2011 to 2017 (2015), respectively. Figure 2b presents the land cover map of the UBNB for 2015.

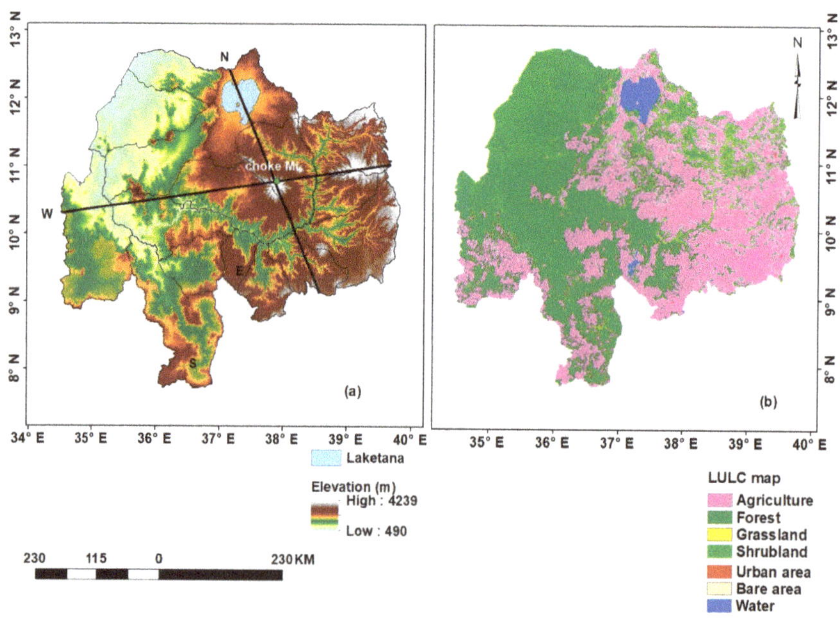

Figure 2. (**a**) Digital elevation model (DEM) with sub-basins of the Upper Blue Nile Basin (two transects from north to south-east (NS) and from west to east (WE)), and the green point shows the location of the peak "Choke" Mountain; (**b**) The land use land cover map of the UBNB for 2015 extracted from the European Space Agency's Climate Change Initiative (ESA CCI) land cover map.

In this study, the trend analysis was done using the Mann–Kendall (MK) trend test [67,68]. According to Mann, the null hypothesis H_0 states that a data series is serially independent and identically distributed with no monotonic trend. The alternative hypothesis H_1 is that the data follows a monotonic trend. In a two-sided test for a trend at a significance level of α, H_0 should be rejected, and H_1 is accepted if $|Z| > z_{\alpha/2}$, where $F_N(z_{\alpha/2})$ is the standard normal cumulative distribution function and Z is the test statistic used to identify the direction of the trend and its significance (Equation (3) to Equation (5)).

$$Z = \begin{cases} \frac{S-1}{\sqrt{var(s)}} if S > 0 \\ 0 \quad if S = 0 \\ \frac{S+1}{\sqrt{var(s)}} if S < 0 \end{cases} \tag{3}$$

where S is the MK test statistic, which measures the trend in the data and is defined as:

$$S = \sum_{i=1}^{n-1} \sum_{j=i+1}^{n} sgn(x_j - x_i) \tag{4}$$

where x is the sequential data values, and n is the length of the dataset.

$$sgn(\theta) = \begin{cases} 1 & if \theta > 0 \\ 0 & if \theta = 0 \\ -1 & if \theta < 0 \end{cases} \tag{5}$$

Kendall indicated that the distribution of S may be well approximated by a normal distribution with mean zero and variance (Equations (6) and (7)) under the assumption of no trend:

$$E(S) = 0 \tag{6}$$

$$var(S) = \left[\frac{n(n-1)(2n+5) - \sum_{i=1}^{m} t_i\,(t_i-1)(2t_i+5)}{18}\right] \tag{7}$$

where m is the number of tied groups and t_i is the size of the ith tied group.

Since the MK test does not provide the magnitude of the trend, the Theil-Sen Slope Estimator [69] in Equation (8) has been used to estimate the magnitude of the trend (β).

$$\beta = Median = \left[\frac{T_j - T_i}{j-i}\right], \forall i < j \tag{8}$$

where β is the slope between data point T_j and T_i measured at times j and i, respectively. In this paper, the trend analysis has been carried out using "ZYP" R package [70], developed based on Yue et al.'s [71] trend-free pre-whitening method. The trend result has been evaluated at the significance level of $\alpha = 0.05$. This implies that the null hypothesis is rejected when $|Z| > 1.96$ in Equation (3), thus, $Z > 1.96$ indicates a significant increasing trend and $Z < -1.96$ indicates a significant decreasing trend.

2.3.2. The Relationship Between Soil Moisture and Rainfall in Autumn

To understand the magnitude of the association between autumn soil moisture (ASM) and rainfall grid-level, a Pearson correlation was performed using the "corLocal" function in R.

The Dgital Elevation Model (DEM) and land cover map given in Figure 2a,b obtained from the Shuttle Radar Topography Mission (SRTM) and ESA CCI land cover, respectively, were used to gauge the impact of topography and land cover classes on the correlation between ASM and rainfall in autumn. To investigate the effect of topography, first two transects that cross both incised river gorges and the pick mountains at the center of the basin that runs from the north to south-east (NS) and from west to east (WE) directions were developed (Figure 2a). Then, we assessed the correlation between ASM and annual rainfall along the two transects. The anomalies of rainfall and ASM were also correlated to analyze the effect of rainfall on ASM variability. Further, the effect of the relationship between rainfall and evapotranspiration to ASM change over the UBNB was investigated.

3. Results

3.1. *Variability and Trends of Soil Moisture*

Figure 3 presents the spatial pattern of multiyear (1992 to 2017) mean autumn (average of September to November) soil moisture and coefficient of variation (CV, %), while Figure 4 provides the relation between the CV and long-term mean ASM in the UBNB. Figure 5 depicts the annual soil moisture anomalies for the autumn season. In addition, Table 2 gives the spatial mean and CV of soil moisture for three major land cover classes of the UBNB over the three decades.

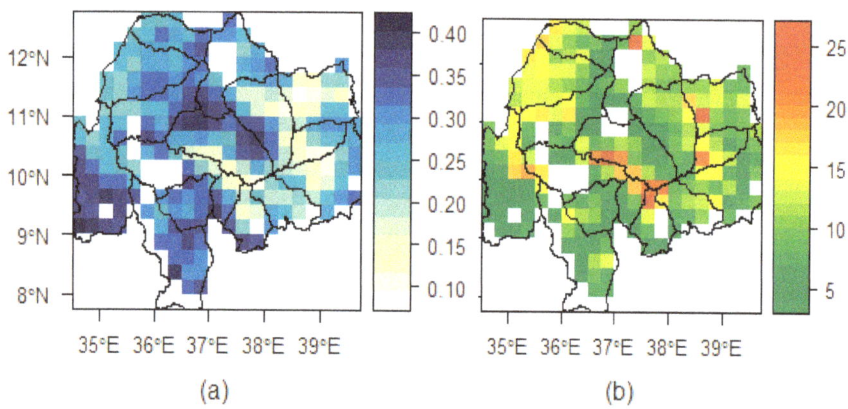

Figure 3. Spatial patterns of (**a**) mean autumn (September to November average) soil moisture (ASM) (m^3/m^3) and (**b**) coefficient of variation (CV, %) over the UBNB in Ethiopia for the period of 1992 to 2017. Note that the white regions (open pixels) show the pixels with missing values in the ESA CCI soil moisture dataset.

The multiyear mean soil moisture analysis showed a general overview of basin soil moisture distribution in the autumn season (Figure 3a). The result reveals that ASM in the basin varied considerably (0.09 to 0.38 m^3/m^3) with an average of 0.26 m^3/m^3 (Figure 3a). It is worth noting that the spatial patterns of mean ASM showed local characteristics and regions with remarkably high values are mainly located in the central, south-western and southern tips of the basin, while the lowest values majorly occur over the eastern parts (Figure 3a). On the other hand, the CV ranged from 2.8% to 28% (Figure 3b), which represents low-to-moderate variability over the study basin [72]. The relatively high variability of ASM generally occurred over lower mean soil moisture regions, while lower variability appeared over high soil moisture locations of the UBNB. To further investigate how changes in mean autumn soil moisture affect the CV over the basin, we plotted the relationship between mean soil moisture and CV for the period of 1992 to 2017 (Figure 4). The result indicated significant negative correlation ($r = -0.6$, $p < 0.01$) between mean soil moisture and soil moisture variability (expressed by the CV) over the UBNB, and the CV typically reduced with an increase in mean soil moisture.

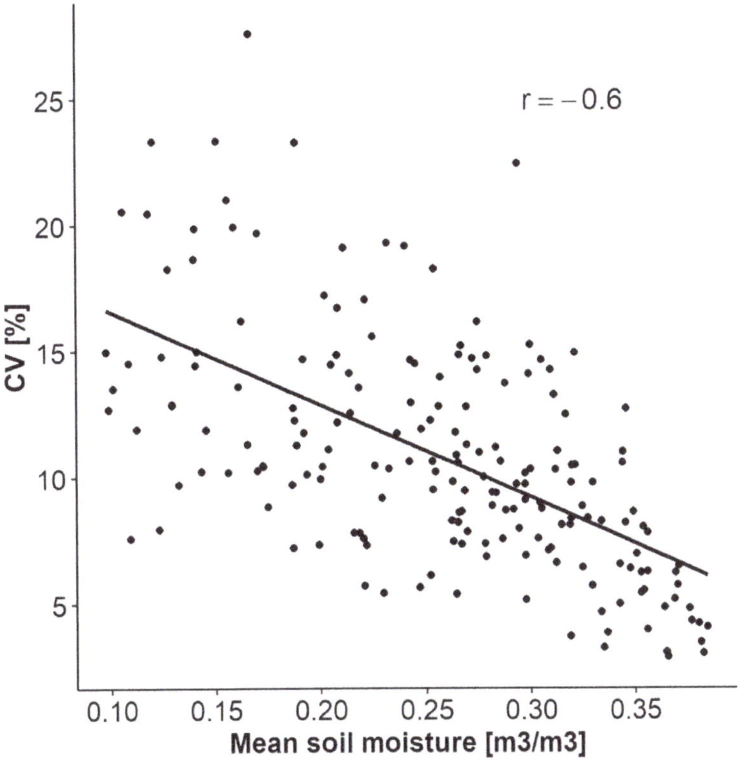

Figure 4. The relation between the coefficients of variation (CV, %) and long-term mean ASM in the UBNB.

In addition, Table 2 shows ASM variability for the dominant land cover classes of the basin arranged for three decades. Variations in the mean soil moisture content were found across the three land cover classes. Looking to the mean soil moisture values for all three decades, the shrub land was drier than agricultural and forest lands. As shown in Table 2, the analysis did not show much difference in the mean values of soil moisture over the agricultural and forest land covers. However, the ASM temporal variability exhibited a considerable variation across the different land cover classes of the basin. Accordingly, the highest variability was observed over forestland even for nearly similar or higher mean values to agricultural and shrub lands.

For example, during the second decade (2001 to 2010), agricultural, forest, and shrub lands resulted in a mean soil moisture of 0.24, 0.22, and 0.21 m^3/m^3 and CV of 11.20%, 27.07%, and 12.88%, and for the third decade (2010 to 2017), they resulted in a mean soil moisture of 0.27, 0.29, and 0.25 m^3/m^3 and CV of 3.90%, 8.28%, and 7.04%, respectively.

Annual soil moisture anomalies for the autumn season in Figure 5 showed that there are variations in the amount and distributions of ASM at different years and parts of the basin. Consequently, periods wetter than an average soil moisture year were mainly observed in 1997, 1998, 2016, and 2017, while the most critical periods that were drier than an average year were prominently noted in 1995, 1996, 2002, 2003, 2004, 2005, 2009, and 2015.

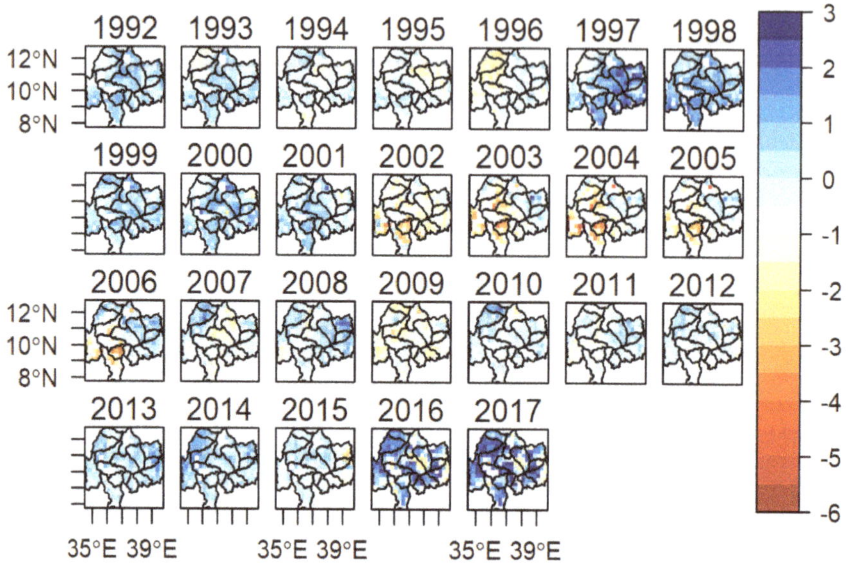

Figure 5. Standard soil moisture anomalies for autumn season in the UBNB indicating the magnitude of departure from long-term mean ASM over a period of 26 years of observation (1992 to 2017). A negative value represents periods drier than normal years, while a positive value indicates wetter than normal soil moisture periods.

Table 2. Decadal variability of autumn soil moisture (ASM) across the major land use land cover classes in the UBNB, Ethiopia. A lookup table for the land use land cover classification system is provided in Appendix B.

LULC Type	ESA CCI LCM Code	1992 to 2000		2001 to 2010		2010 to 2017	
		Mean	CV (%)	Mean	CV (%)	Mean	CV (%)
Agriculture	10,11,12,20,30,40	0.28	5.27	0.24	11.20	0.27	3.90
Forest	50,60,61,62,70,72,80,81,82,90,100,160,170	0.28	5.01	0.22	27.07	0.29	8.28
Shrub	120,121,122	0.24	5.93	0.21	12.88	0.25	7.04

The ASM for the other years showed a small deviation (negative or positive anomaly) from the long-term mean, confirming that the basin has near-average autumn soil moisture during these periods.

In addition, the trend analysis was undertaken to explore the spatial consistency of ASM over the study periods from 1992 to 2017. The soil moisture monotonic trend and pixels having $p < 0.05$ values are depicted in Figure 6. The results of MK trend analysis for mean autumn (average of September to November) soil moisture showed different spatial-temporal trends in the UBNB. The trend ranges from -0.0094 to 0.0055 m^3/m^3 per year for the autumn season, while an average over the entire basin indicated a wetting trend for the past 26 years (1992 to 2017) at a rate of 0.00024 m^3/m^3 per year (Figure 6a). Considering the spatial coverage of the trends, both the drying and wetting trends proportionally cover the study area, with relatively high magnitudes for the wetting trends. The autumn time series reveals a significant wetting trend (ranging from 0.001 to 0.006 m^3/m^3 per year for the autumn season) that primarily occurred over the northwest region of the basin and covers about 18% of the total basin (Figure 6b). A significant drying trend has also been noted over the southeast of the UBNB and only covers about 1.5% of the total basin area (Figure 6b).

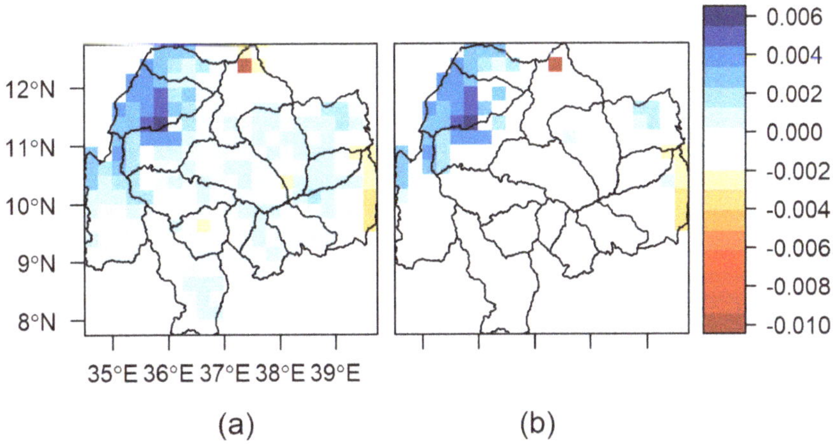

Figure 6. This figure shows (**a**) the monotonic trends (m^3/m^3 $year^{-1}$ for ASM) and (**b**) pixel having a significant ($p < 0.05$) increase or decrease trend of residual soil moisture for the period of 1992 to 2017. The positive and negative values indicate the wetting and drying trends of soil moisture in the UBNB, respectively.

3.2. Variability and Trends of Rainfall

Figures 7 and 8 present the long-term annual mean and anomalies of rainfall for the period from 1992 to 2017, respectively. The UBNB receives an annual rainfall > 2000 mm and its distribution has local characteristics, and the maxima and strong rainfall gradients are oriented along the central and southern tip of the basin (Figure 7). The lowest total annual rainfall occurred over the eastern and western margins of the basin. The wettest periods (e.g., 1996, 1998, 2000, 2006, 2008, and 2017) and driest periods (e.g., 1992, 1994, 1995, 2002, 2009, and 2015) of the UBNB were well captured in the analysis using the CHIRPS satellite rainfall product (Figure 8). Figures 9 and 10 provide the rainfall trends and masks of their respective pixel values, having p values <0.05 for the study period, respectively.

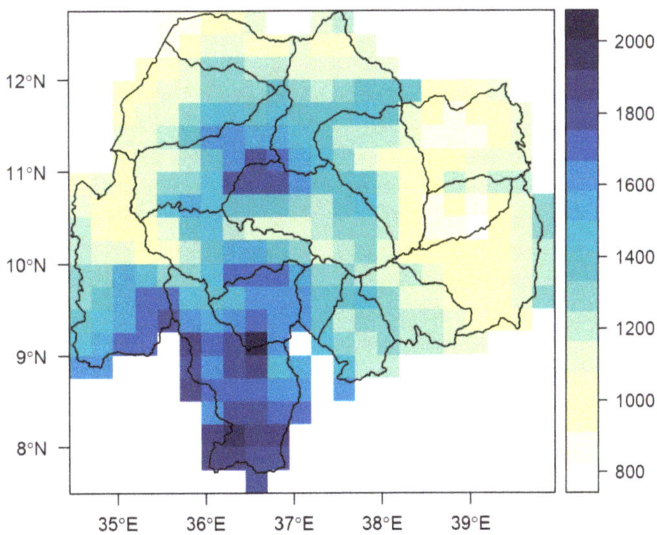

Figure 7. The spatial distribution of mean annual rainfall (mm) over the Upper Blue Nile Basin (UBNB) in Ethiopia (1992 to 2017).

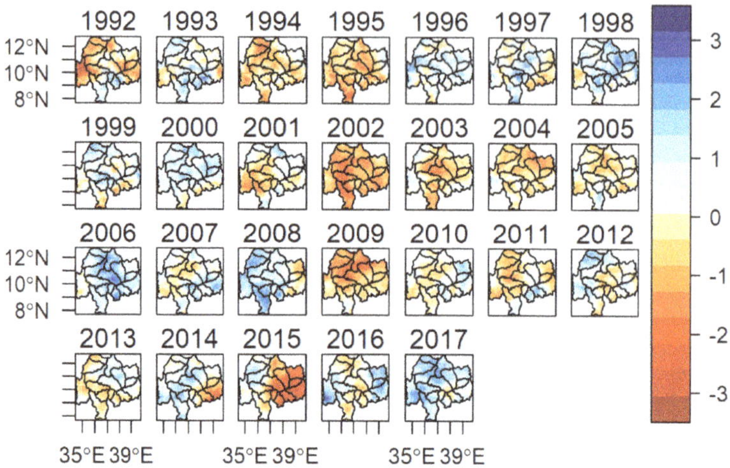

Figure 8. Standard rainfall anomalies for annual rainfall in the UBNB indicating the magnitude of departure from long-term mean rainfall over a period of 26 years observation (1992–2017). A negative value represents periods of below-normal rains (drought), while positive values indicate above-normal rains (with the possible risk of flood).

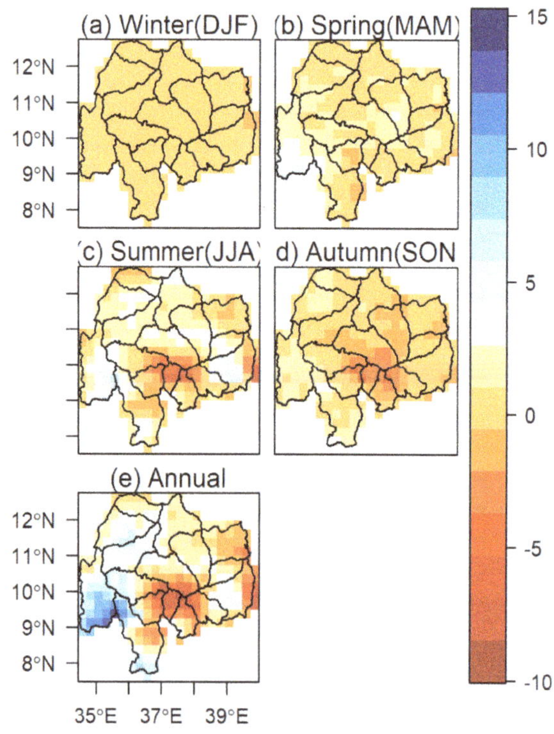

Figure 9. Monotonic trends for seasonal (mm/season) rainfall: (**a**) winter, (**b**) spring, (**c**) summer, (**d**) autumn, and (**e**) annually (mm/year) in the UBNB in Ethiopia for the period of 1992 to 2017. The annual rainfall is characterized by four distinct seasons: winter (December, January, February), spring (March, April, May), summer (June, July, August), and autumn (September, October, November).

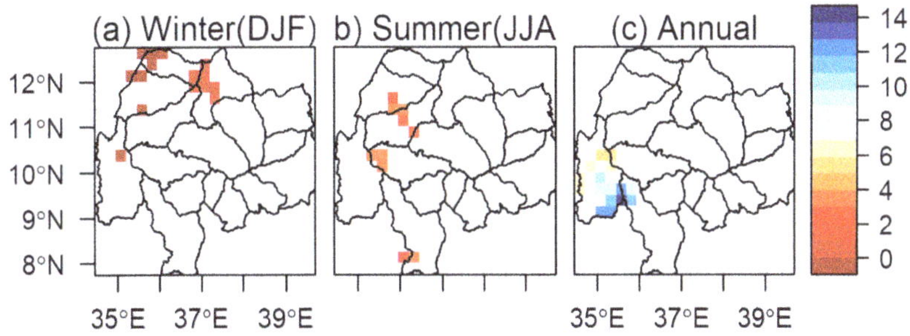

Figure 10. Zones with significant ($p < 0.05$) increasing or decreasing trends for (**a**) winter, (**b**) summer, and (**c**) annually in the UBNB in Ethiopia for the period of 1992 to 2017. No significant trends were found over the spring and autumn seasons.

When the seasonal rainfall trends were averaged over the entire basin, an increasing trend was observed in the spring, summer, and autumn rainfall with the rate of change 0.936 mm year^{-1}, 1.027 mm year^{-1}, and 0.071 mm year^{-1}, respectively (Figure 9b–d). However, the average rainfall trend during the winter season indicated a decreasing trend with the rate of change −0.0014 mm year^{-1} (Figure 9a). Although it covers a small portion of the study basin, statistically significant increasing trends at $p < 0.05$ were identified during the winter (with the rate of change up to 0.313 mm year^{-1}) and summer (up to 3.714 mm year^{-1}) rainfall seasons (Figure 10a,b). The significant trends of rainfall during the winter and summer season are majorly marked in the north and west of the UBNB, respectively (Figure 10a,b). Figure 10 also revealed that there was no significant trend in spring and autumn rainfall, while the annual rainfall trend average over the entire basin showed an increasing trend with the rate of change 1.819 mm year^{-1} (Figure 9e), with a significant increasing trend reaching up to 13.714 mm year^{-1} located in the southwest of the basin (Figure 10c).

3.3. Correlation Coefficient Between Autumn Soil Moisture and Rainfall

Figure 11 illustrates the pixel-level correlation computed between soil moisture for the autumn season and rainfall over the UBNB from 1992 to 2017. The correlation coefficient (r) indicates how much ASM can be explained by rainfall in the UBNB. In general, the soil moisture for the autumn season exhibits a positive correlation with spring, autumn, and annual rainfall (Figure 11b,d,e), and a negative correlation with winter and summer rainfall over most parts of the UBNB (Figure 11a,c).

Notably, one may expect an increase in ASM due to the predecessor wet summer rainfall in comparison to the relatively low rainfall during the spring and autumn seasons. However, this may not sometimes be proportional to the amount of rainfall received because of the surface runoff, intense evapotranspiration, etc., which usually happens during the wet periods [73]. The ASM conferred a strong positive correlation ($r > 0.5$) with the spring, autumn, and annual rainfall, which covers 6%, 35%, and 15% of the total basin area, respectively. Distinctly, the correlation between ASM and wet summer rainfall was negative over a significant portion of the basin, although less than 3% of the total basin area showed a statistically strong correlation ($r < -0.5$). The ASM and winter rainfall have also shown both a negative and positive correlation, with a significant correlation ($r > 0.4$) over a considerable portion of the basin. Overall, the correlation analysis indicates that ASM was strongly correlated to rainfall during the autumn and spring seasons than that of the summer season.

Generally, the correlation between ASM and rainfall reduced from the west to east region (or from low- to high-elevation areas) of the basin (Figure 11). Since the high variation of terrain elevation characterizes the UBNB, the spatial correlation between soil moisture and rainfall could be affected by a change in topography. Consequently, the study further assessed the effect of elevation variation on the correlation between autumn soil moisture and annual rainfall. Two transects that run from the north to south-east (NS) and from west to east (WE) directions were developed (Figure 2a). Then, the correlation between ASM and annual rainfall was assessed along the two transects (Figure 12). The result from both NE (Figure 12a) and WE (Figure 12b) indicates that the correlation between ASM and annual rainfall is affected by the variation in elevation as the correlation reduces as the elevation increases. The lowest correlation between soil moisture and rainfall has occurred over peak areas of the basin.

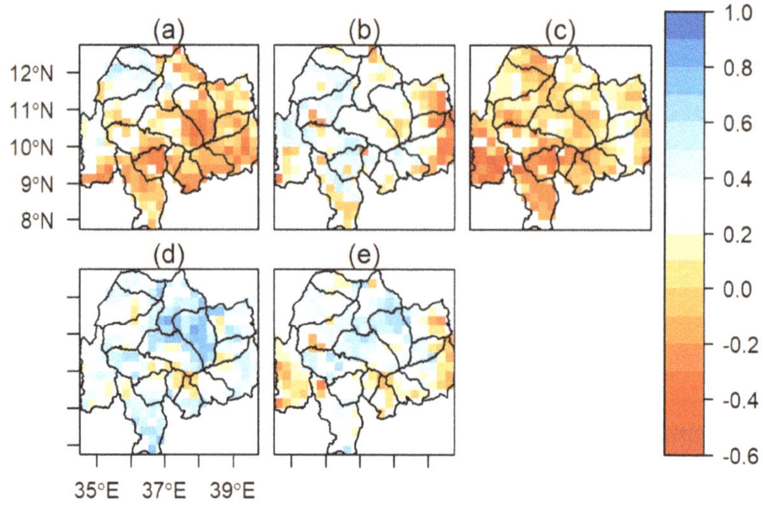

Figure 11. Pearson correlation (r ranging from −1 to 1) between ASM and rainfall for (**a**) winter, (**b**) spring, (**c**) summer, (**d**) autumn season, and (**e**) annually (1992 to 2017). The correlation coefficient ($r \geq |0.39|$) is significant at $p < 0.05$.

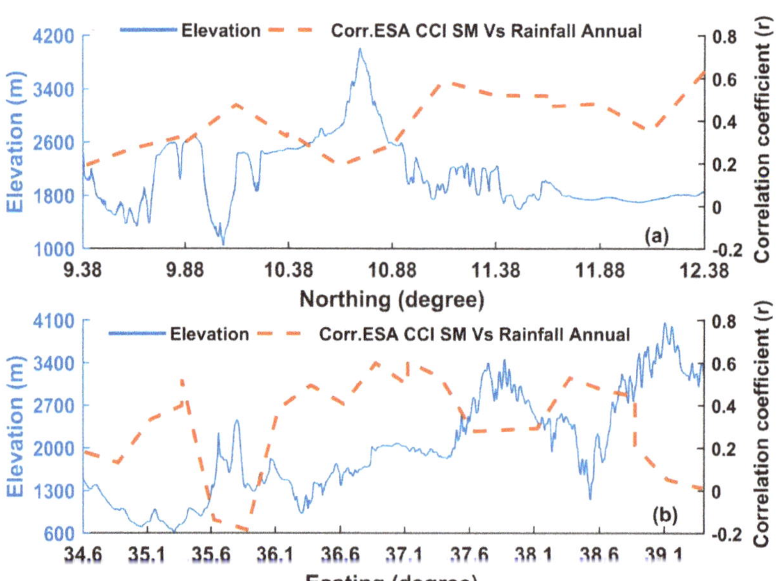

Figure 12. The correlation between autumn soil moisture (ASM) and annual rainfall along transects from (**a**) from north to south-east and (**b**) from west to east (WE) (1992 to 2017).

The degrees to which land cover types affect the correlation between ASM and rainfall could vary with vegetation type, density, and seasons. Table 3 shows the results of the spatial correlation between ASM and rainfall over the dominant land cover classes of the basin. The correlation between ASM and rainfall respond distinctively across the different seasons and land cover classes. For example, the highest correlation between ASM and spring rainfall was observed over the forestland (r = 0.51); however, autumn (r = 0.19) and summer (r = −0.28) rainfall showed their lowest positive and highest negative correlation in forestland, respectively.

Table 3. The correlation between ASM and rainfall over the major land use land cover classes of the UBNB.

LULC Type	ESA CCI LCM Code	Rainfall				
		Winter	Spring	Summer	Autumn	Annual
Agriculture	10,11,12,20,30,40	−0.16	0.37	−0.04	0.40 *	0.35
Forest	50,60,61,62,70,72,80,81,82,90,100,160,170	−0.12	0.51 *	−0.28	0.19	0.26
Shrub	120,121,122	−0.10	0.46 *	0.12	0.40 *	0.46 *

* significant at $p < 0.05$.

The highest correlation for the annual (r = 0.46) and positive correlation for summer (r = 0.12) rainfall has been indicated over the shrub land. The ASM and rainfall in autumn have shown the same correlation values for both shrub land and agricultural land. Over the agricultural field, the correlation between soil moisture and rainfall resulted in an optimal correlation in the autumn (r = 0.40), annual (r = 0.37), and spring (r = 0.37) seasons. The correlation in winter was negative for all three land cover classes with a better correlation over the agricultural land. However, the correlation between summer rainfall and ASM over the agricultural field was insignificant.

3.4. *Impact of Rainfall Variability on Soil Moisture Dynamics*

To examine how the variability of rainfall affects ASM in the UBNB, the correlation between ASM and rainfall anomalies was analyzed. Their correlation values at the seasonal and annual scale were averaged over the entire basin, as shown in Table 4. Moreover, Figures 13 and 14 show the temporal (time series) variation and the strength and sign of the correlation between ASM and rainfall anomalies.

Table 4. The correlation between ASM anomalies at different depths and rainfall anomalies at the seasonal and annual scale from 1992–2017.

Soil Moisture	Rainfall				
	Winter	Spring	Summer	Autumn	Annual
ESA CCI SM	0.12	0.46 *	0.10	0.64 *	0.64 *
FLDAS NOAH (0–10 cm)	−0.07	0.36	0.07	0.91 *	0.69 *
FLDAS NOAH (0–40 cm)	−0.08	0.35	0.10	0.91 *	0.71 *
FLDAS NOAH (0–100 cm)	−0.08	0.35	0.12	0.91 *	0.73 *

* significant at $p < 0.05$.

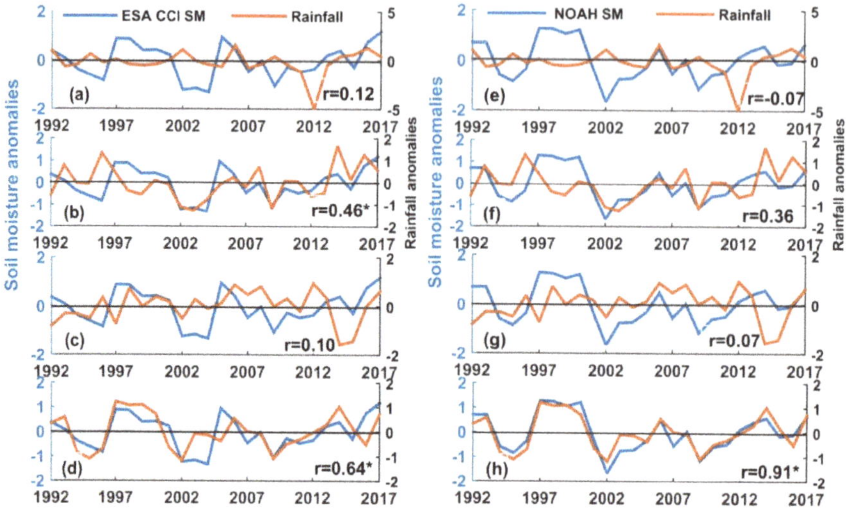

Figure 13. Time series of the autumn soil moisture (ASM) anomalies (estimated using ESA CCI) and rainfall anomalies in the preceding (**a**) winter: December, January, and February (DJF), (**b**) spring: March, April, and May (MAM), (**c**) summer: June, July, and August (JJA), and (**d**) autumn: September, October, and November (SON) averaged over the entire basin. Figure (**e–h**) are the same as Figure (**a–d**), but ASM anomalies were calculated using FLDAS NOAH at 0 to 10 cm depth. The correlation coefficients (r) between two datasets are also shown; the asterisk represents the values significant at $p < 0.05$.

The use of temporal anomalies could reduce the impacts of static properties (i.e., topography and soil properties) on soil moisture dynamics and thus are assumed to primarily reflect the impact of climate variables [74]. The correlation analysis was undertaken using soil moisture derived from ESA CCI, which denotes the top few centimeters of the soil, and FLDAS NOAH soil moisture that represents soil moisture at 0–10, 0–40, and 0–100 cm depths. A weighted average method was used to calculate soil moisture at 0–40 and 0–100 cm depths. The correlation between ASM and rainfall in the autumn or the preceding three seasons was positive (Figure 13a–d, Table 4), although the contribution of the wet summer rainfall to ASM was insignificant (relatively low) in the UBNB. On the other hand, the correlation between ASM and annual rainfall anomalies was significant and revealed that the total amount of rainfall in each year has a considerable impact on the change in the ASM (Figure 14a, Table 4). The same result has been observed for FLDAS soil moisture at different depths (Figure 13e–h, Table 4), except for winter rainfall, which was negatively correlated with ASM. The ASM derived from FLDAS NOAH showed that the contributions of spring rainfall decreased, while the contribution of the current rainfall in autumn was substantial, and its correlation with the ASM was robust (r = 0.91) at all depths of the soil. In general, the result indicated the significant contribution of rainfall to ASM at different depths, although the magnitude of contribution was large for sub-surface soil moisture.

In addition, the wetting trends of ASM derived from ESA CCI were mainly associated with the increasing trend for spring, summer and autumn rainfall, which was generally exhibited in the northwest and southwest of the basin (Figures 6 and 9). The spatial similarity in the trends of ASM with autumn rainfall was generally higher than the other temporal scales, although some noticeable differences existed between trends of rainfall and ASM. For instance, ASM showed a statistically significant increasing trend that was observed over the northwest of the basin where an increase in rainfall trend was not significant.

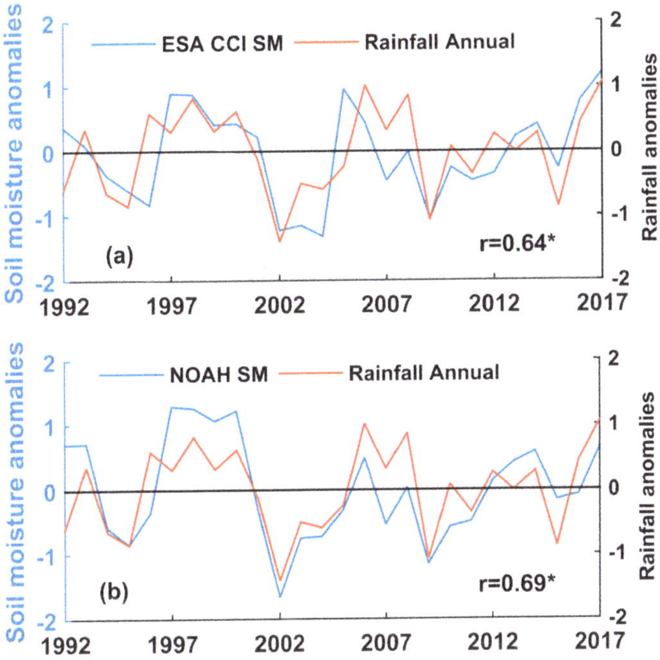

Figure 14. Time series of annual rainfall anomalies and ASM anomalies estimated using (**a**) ESA CCI and (**b**) FLDAS NOAH at 0 to 10cm depth averaged over the entire UBNB. The correlation coefficient (r) between the two datasets is also shown; the asterisk represents the value significant at $p < 0.05$.

3.5. The Effect of the Relationship Between Rainfall and ET on Soil Moisture

Besides rainfall, evapotranspiration (ET), which is the sum of soil evaporation and vegetation transpiration, is a crucial climate variable influencing the distribution and availability of soil moisture [75]. It is essential to determine the effect of the quantitative relationship between ET and rainfall on the temporal patterns of ASM in the UBNB. The ratio of ET to rainfall (RF) was calculated to determine their effect on the wetness and dryness of ASM derived from ESA CCI in the UBNB (Figures 15 and 16).

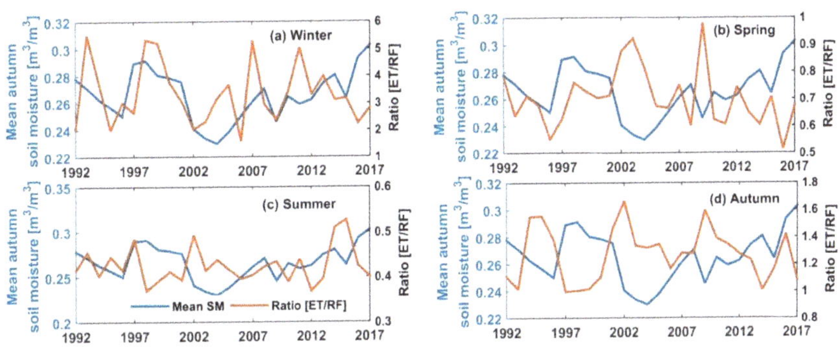

Figure 15. The effects of (**a**) winter, (**b**) spring, (**c**) summer, and (**d**) autumn rainfall (RF) and evapotranspiration (ET) (expressed as ET/RF ratio) on the temporal distribution of ASM in the UBNB from 1992 to 2017.

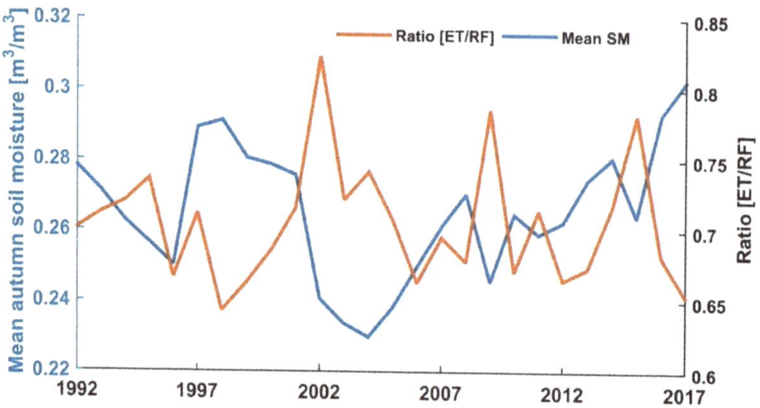

Figure 16. The effects of annual rainfall (RF) and evapotranspiration (ET) (expressed as ET/RF ratio) on the temporal distribution of autumn soil moisture over the UBNB from 1992 to 2017. We can see the highest ET/RF values over the known drought periods of 2002, 2009, and 2015 in the UBNB, as well as in Ethiopia in general [76].

In general, from Figures 15 and 16 we can see that the amount of mean soil moisture decreases with an increase in ET/RF ratio. Over the winter season (Figure 15a), ET has an average rate of 83.98 mm year^{-1} and stays low since water availability is reduced due to the absence or significantly low rainfall (an average of 27.56 mm year-1) events in winter. Consequently, ASM has a positive correlation (r = 0.22) with ET/RF ratio. The significant correlation between ASM and winter ET (r = 0.41, $p < 0.05$) to ET/RF ratio indicates that ET explains the soil moisture variability better than rainfall in winter. The winter rainfall has a very low correlation (r = −0.13) with ASM as expected. In spring (Figure 15b), the rainfall (with an average of 219.70 mm year^{-1}) and ET (mean rate of 150.01 mm year^{-1}) have a strong positive correlation (r = 0.85) and ET increases with rainfall. Both the rainfall and ET have a significant positive correlation with ASM, with r = 0.49 ($p < 0.05$) and r = 0.46 ($p < 0.05$), respectively. However, rainfall increases more than ET; thus, ASM has a negative correlation (r = −0.47) with ET/RF ratio over the spring season. In summer (Figure 15c), rainfall exceeds (mean rainfall 764.97 mm year^{-1}) ET (average rate of 323.09 mm year^{-1}), but ASM has a small negative correlation with ET/RF ratio (r = −0.11). The summer rainfall has no correlation with autumn soil moisture (r = 0.016), and no apparent contribution to the amount and variability of ASM was observed during the summer season. Instead, summer ET showed a negative correlation (r = −0.24) with ASM over the UBNB. In autumn (Figure 15d), ET (with the mean rate of 351.09 mm year^{-1}) exceeds that of rainfall (an average of 282.15 mm year^{-1}) and ASM has a strong negative correlation (r = −0.57, $p < 0.05$) with the ET/RF ratio. However, the ASM showed a strong positive correlation (r = 0.56, $p < 0.05$) with autumn rainfall compared to the low negative correlation (r =−0.27) with autumn ET. Over the annual analysis (Figure 16), although both the mean annual rainfall (1294.63 mm year^{-1}) and mean ET (909.73 mm year^{-1}) have a positive correlation with autumn soil moisture, r = 0.57($p < 0.05$) and r = 0.24, respectively, the rainfall exceeds that of ET, and ASM has a significant negative correlation (r = −0.59, $p < 0.05$) with the annual ET/RF ratio.

4. Discussion

Soil moisture is a vital component of agriculture and hydrology. Its spatial-temporal change can considerably affect the socioeconomics of agrarian countries such as Ethiopia. Understanding the soil moisture dynamics and its associated response to climate variables (such as rainfall) is vital for proper water resource planning. This study investigated the spatial and temporal characteristics of residual soil moisture in autumn (September to November) over the UBNB in Ethiopia. The result indicated that the variability of soil moisture is affected by the extent of moisture available in the soil and found that dry soil regions vary more than that of wet soil regions (Figure 4). Indeed, this result is compatible with the findings of different scholars who found an increase in soil moisture spatial variability with decreasing mean soil moisture values [77,78]. Similarly, a comparable result for rainfall variability and mean rainfall over the UBNB has been found (Figure A2, Appendix A), which indicates the direct impact of rainfall on the soil water content of the basin.

Further, the analysis showed that the moisture status of the soil was dependent on the land cover classes of the UBNB (Table 2). The different land cover classes of the basin could determine the distribution and availability of soil moisture through controlling the soil infiltration process and the rate of evapotranspiration [78]. In the study, the shrub lands show a drier condition than agricultural and forest land covers. Different scholars e.g., [78,79] found a similar result over shrub lands in comparison to other land cover classes. The drier soil moisture conditions beneath the shrub land could be associated with the higher density of plant roots for the shrubs than the other land cover classes, which eventually leads to a more significant loss of soil water through transpiration [78]. The agricultural and forest land covers of the basin showed a relatively wetter soil condition. Similarly, Feng et al. [79] have reported wetter soil conditions for cultivated lands. The tremendous infiltration rate of cultivated soil just after rainfall events and little soil water loss via transpiration from crops with lower leaf area index could contribute to the relatively wetter soil conditions over the agricultural fields. The wetter condition over the forest land could partly be explained by the maximum porosity of the soil in forest areas, less direct soil water evaporation under a higher coverage of plants, and little loss of water through transpiration at the surface soil for deep-rooted forests [28]. The relatively high soil moisture variability over forest areas could be explained by the diversity of trees within the forest, which may result in low soil moisture values around the trees and higher values in the interspaces between trees [28]. The high soil moisture variation over forest lands could also be due to the reduced quality and uncertainty of the ESA CCI soil moisture product over vegetated areas [80]. The annual anomalies for ASM are broadly comparable with wet and drought years of the basin identified by different scholars [76] and anomalously for dry and wet rainfall periods of the UBNB (Figure 8). These results suggest that the broad patterns exhibited by ESA CCI soil moisture data are likely reliable. The result could further indicate that the spatial and temporal distribution of rainfall is one of the significant factors which governs the variability of ASM, among the other climate variables.

Further, it is noted that the inter-annual variability of rainfall and ASM are very much associated. The annual soil moisture for the autumn season indicates a positive correlation with spring, autumn, and annual rainfall (Figure 11). Again, the changes in ASM anomalies at different soil depths also correspond to changes in the variability of rainfall in autumn and the previous spring seasons and showed a significant correlation ($r = 0.35$ to 0.91) (Figure 13, Table 4). The same result was also observed using the annual rainfall presented in Figure 14 and Table 4. The result implies that the impact of the current (autumn) and previous spring rainfall on ASM is positive and anomalies in ASM mainly originate from the variability of rainfall during the previous spring and autumn season as well as to the annual total. Although the intensity of rainfall is low in spring compared to autumn and the summer seasons, the spring rainfall could infiltrate into the soil layer and thus has a significant contribution to the amount of soil water left in the autumn season. Certainly, rainfall from the autumn season is the dominant contributor to the annual variability of ASM over the UBNB. Cai et al. [40] reported that rainfall and soil moisture have a high degree of correlation in autumn over eastern China. Likewise, Longobardi [81] indicated that the volume of rainfall occurs at the end of the wet season perhaps determines the amount and distributions of soil moisture at the beginning of the dry season. On the other hand, the contribution of summer rainfall to ASM variability is very insignificant (Figures 11 and 13, Table 4) and generally has a weak correlation with ASM compared to spring and autumn rainfall over the UBNB. This could be due to the increased rate of soil moisture depletion, reduced infiltration rate [82], and loss of incoming rain through surface runoff [73] over the summer growing season. According to Yang et al. [83], rainfall storage occurred in April and May (months in spring) because of soil water consumed over the wet months from June to mid-September (summer season) and recovery of soil water from late September to October (months of the autumn season). Consequently, the contribution of summer rainfall to soil moisture storage is limited; therefore, this could partly be the reason for the low correlation between ASM and summer rainfall in the UBNB.

The findings of this study indicated that soil moisture response to rainfall variability is considerably controlled by the variation in topography and land use types of the UBNB (Figure 12, Table 3). The correlation between ASM and rainfall decreases with an increase in elevation because locations at a relatively higher elevation could have less available soil moisture due to gravity and exposure to sunlight warming, which may result in water drains downhill and higher evaporation rates, respectively [78]. Similar findings have been reported by Crave and Gascuel-Odoux [84]. Furthermore, the difference in the correlation between rainfall and ASM at different land-use types and seasons could be explained by the soil moisture build-up process over various land cover classes. In the shrub and agricultural lands, the build-up of soil moisture is relatively fast, and soil moisture peak is attained rapidly in comparison to the slow rate build-up process in forest land [85].

Figures 15 and 16, in general, imply that a significant portion of rainfall received by the UBNB returns to the atmosphere via evapotranspiration (ET). Accordingly, the amount of ASM over the periods of a higher ET/RF ratio in general reduced, while the magnitude of mean soil moisture peak corresponds to low ET/RF ratio (Figures 15 and 16). In each season the effects of rainfall and ET on ASM are varied. The highest ET/RF ratio in winter shows the dominance of ET over rainfall due to the absence or low amount of rainfall over this season. The result implies that in winter, ET could explain a considerable share of autumn soil moisture variance than that of rainfall. In spring, rainfall increases more than ET and produces a slightly higher rainfall than ET demands, suggesting that both the previous spring rainfall and ET have considerable contributions to the variability of autumn soil moisture in the UBNB, with the highest magnitude from spring rainfall. Over the wet summer season, despite rainfall exceeding that of ET, its contribution to ASM is very insignificant. Again, in autumn ET slightly exceeds rainfall and ASM has a strong negative correlation ($r = -0.57$, $p < 0.05$) with ET/RF ratio, which indicates that ET could explain the majority of autumn soil moisture variability in comparison to rainfall in autumn. However, the ASM has a strong positive correlation ($r = 0.56$, $p < 0.05$) with autumn rainfall compared to the low negative correlation ($r = -0.27$) with autumn ET, which may indicate the positive contributions of autumn rainfall to soil moisture rather than the loss of soil water via ET. This is because during autumn, crops and other vegetation canopies shade more and more of the ground area and may lead to less evaporation of water from the soil and thus a considerable portion of ET could come from the root zone via plant transpiration. Although other climate variables, including surface solar radiation, relative humidity, and wind speed might affect the inter-annual variability of residual soil moisture, we assume these effects to be small.

5. Conclusions

Understanding the availability and dynamics of residual soil moisture over the rainfed agricultural system, characterized by low crop production, is imperative for supplementary food and feed production in the off-season. In this study, we applied ESA CCI soil moisture products from 1992 to 2017 to assess the long-term trend and dynamics of residual soil moisture in the autumn (September to November) season and its linked response to the long-term variability of rainfall in the UBNB of Ethiopia. Besides, the mutual effect of rainfall and evapotranspiration on the variability of autumn soil moisture was analyzed. The basin was found to have soil moisture ranging from 0.09–0.38 m^3/m^3, with an average of 0.26 m^3/m^3 in autumn. The ASM time series resulted in the CV ranging from 2.8–28% and was classified with a low-to-medium variability. Moreover, ASM variability showed a strong relationship with the mean ASM, and the highest inter-annual variability occurred over low mean ASM areas of the basin. The mean and variability of ASM changes with the different land-cover classes of the basin. In general, the MK monotonic trend analysis for ASM revealed that the UBNB had experienced a wetting trend for the past 26 years (1992–2017) at a rate of 0.00024 m^3/m^3 per year. Besides, the study provided strong evidence that the previous spring and current autumn rainfall could explain a considerable portion of ASM in the basin. Furthermore, the result indicates that the rainfall degree of influence on the characteristics of ASM could be induced by topography and dominant land cover classes of the study basin. In addition, the behavior of the mean ASM could be determined by the effect of soil water loss through evapotranspiration.

To conclude, the ESA CCI soil moisture product provides valuable insights into the spatial and temporal characteristics of autumn soil moisture over the UBNB. Thus, ESA CCI soil moisture estimates could be used as an alternative data to monitor the extent and dynamics of soil moisture over the data scarce regions such as the UBNB. However, high-resolution soil moisture datasets are still crucial for better understanding the characteristics of soil moisture in a complex topographic area of the UBNB. The information provided in this study could provide pertinent information to comprehend the soil moisture status of the basin in the off-season and its potential to support additional short or medium cycle cropping. Further, the link between soil moisture and rainfall presented in this study could play an important role to predict the extent and conditions of residual soil moisture in advance. Future studies should include the contributions of land use and land cover change in the extent and dynamics of soil moisture in the basin.

Author Contributions: G.A., T.T. and B.G. envisioned and designed the research; G.A. performed the data collection; G.A., T.T. and B.G. analyzed the results; G.A. wrote the original manuscript and T.T., and B.G. reviewed the manuscript. All authors have read and agreed to the published version of the manuscript.

Funding: This research was funded by Geospatial Data and Technology Center of Bahir Dar University (Grant No. BDU/RCS/GDTC/2009-04) and Entoto Observatory and Research Center postgraduate research fund.

Acknowledgments: The authors would like to thank the European Space Agency (ESA), NASA, and USGS for providing satellite and model products.

Conflicts of Interest: The authors declare no conflict of interest.

Appendix A

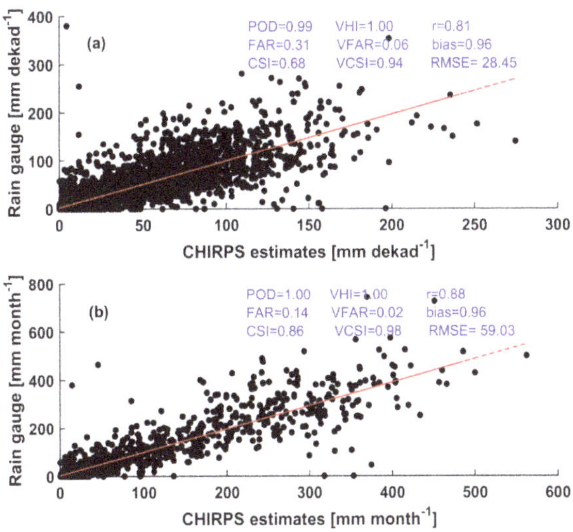

Figure A1. Scatter plot between rain gauge observations and CHIRPS rainfall estimates at dekadal (**a**) and monthly (**b**) temporal scale over the Upper Blue Nile basin for the period of 2000–2015. Probability of detection (POD), false alarm ratio (FAR), critical success index (CSI), volumetric hit index (VHI), volumetric false alarm ratio (VFAR), volumetric critical success index (VCSI), correlation coefficient (r), bias, and the root mean square error (RMSE). The perfect score for POD, CSI, VHI, VCSI and bias is 1, while 0 is the perfect score for FAR and VFAR. The RMSE values are presented in millimeters (mm) (See Ayehu et al., 2018 for the details of the analysis).

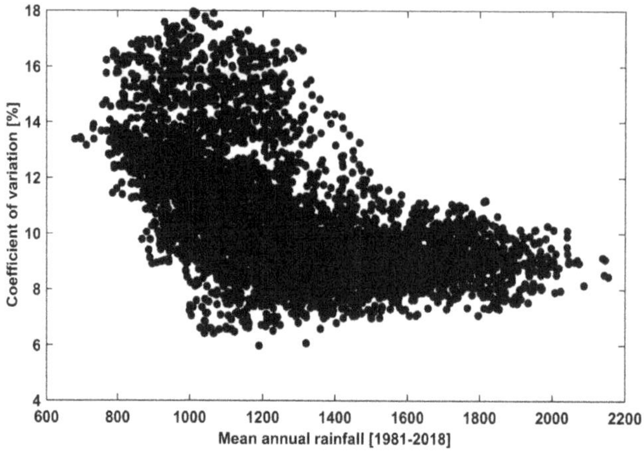

Figure A2. The relationship between the coefficients of variation (%) and long-term mean annual rainfall (mm). The CV increases with a decrease in mean annual rainfall with the coefficient of correlation (r) equal to −0.57.

Appendix B

Table A1. A look up table for reclassification of land cover types (adapted from Land Cover CCI: product user guide, Version 2).

Classes Considered in this Study	Land Types used in the CCI–LC Maps	
Agriculture	10,11,12	Rainfed cropland
	20	Irrigated cropland
	30	Mosaic cropland (> 50%) / natural vegetation (tree, shrub, herbaceous cover) (< 50%)
Forest	40	Mosaic natural vegetation (tree, shrub, herbaceous cover) (> 50%) / cropland (< 50%)
	50	Tree cover, broadleaved, evergreen, closed to open (> 15%)
	60,61,62	Tree cover, broadleaved, deciduous, closed to open (> 15%)
	70,71,72	Tree cover, needleleaved, evergreen, closed to open (> 15%)
	90	Tree cover, mixed leaf type (broadleaved and needleleaved)
	100	Mosaic tree and shrub (> 50%) / herbaceous cover (< 50%)
	160	Tree cover, flooded, fresh or brakish water
	170	Tree cover, flooded, saline water
Grassland	110	Mosaic herbaceous cover (> 50%) / tree and shrub (< 50%)
	130	Grassland
Wetland	180	Shrub or herbaceous cover, flooded, fresh-saline or brakish water
Urban	190	Urban
Shrubland	120, 121, 122	Shrubland
Bare area	200, 201, 202	Bare areas
Water	210	Water

References

1. Lin, H.; Bouma, J.; Wilding, L.; Richardson, J.; Kutilek, M.; Nielsen, D. Advances in hydropedology. *Adv. Agron.* **2005**, *85*, 1–89.
2. Awulachew, S.B.; Yilma, A.; Loulseged, D.; Loiskandl, W.; Ayana, M.; Alamirew, T. *Water Resources and Irrigation Development in Ethiopia. Colombo*; International Water Management Institute: Battaramulla, Sri Lanka, 2007; p. 78.
3. Cheung, W.H.; Senay, G.B.; Singh, A. Trends and spatial distribution of annual and seasonal rainfall in Ethiopia. *Int. J. Climatol.* **2008**, *28*, 1723–1734. [CrossRef]
4. MOA (Ministry of Agriculture). *Natural Resources Management Directorates. Small-Scale Irrigation Situation Analysis and Capacity Needs Assessment*; Ministry of Agriculture: Addis Ababa, Ethiopia, 2011.
5. Conway, D. The climate and Hydrology of the Upper Blue Nile River. *Geogr. J.* **2000**, *166*, 49–62. [CrossRef]
6. Engida, N.A.; Esteves, M. Characterization and disaggregation of daily rainfall in the upper Blue Nile Basinin Ethiopia. *J. Hydrol.* **2011**, *399*, 226–234. [CrossRef]
7. Food and Agricultural Organization (FAO). *Ethiopia Country Programming Framework*; Office of the FAO Representative to Ethiopia; FAO: Addis Ababa, Ethiopia, 2014.
8. Central Statistical Agency (CSA). *Report on the Year 2000 Welfare Monitoring Survey*; Central Statistical Authority; CSA: Addis Ababa, Ethiopia, 2001.
9. Pielke, R.A.; Liston, G.E.; Eastman, J.L.; Lu, L.; Coughenour, M. Seasonal weather prediction as an initial value problem. *J. Geophys. Res. Atmos.* **1999**, *104*, 19463–19479. [CrossRef]
10. Vinnikov, K.Y.; Robock, A.; Speranskaya, N.A.; Schlosser, C.A. Scales of temporal and spatial variability of mid-latitude soil moisture. *J. Geophys. Res. Atmos.* **1996**, *101*, 7163–7174. [CrossRef]
11. World Bank. *Ethiopia-Managing Water Resources to Maximize Sustainable Growth: A World Bank Water Resources Assistance Strategy for Ethiopia*; Report No. 36000-ET; The World Bank Agriculture and Rural Development Department: Washington, DC, USA, 2006.
12. Awulachew, S.B.; Merry, J.; Kamara, A.B.; van Koppen, B.; de Vries, F.P.; Boelee, E.; Makombe, G. *Experiences and Opportunities for Promoting Small-Scale/Micro Irrigation and Rainwater Harvesting for Food Security in Ethiopia*; International Water Management Institute (IWMI): Colombo, Sri Lanka, 2005; p. 91.
13. Makombe, G.; Namara, R.; Hagos, F.; Awulachew, S.B.; Ayana, M.; Bossio, D. *A Comparative Analysis of the Technical Efficiency of Rainfed and Smallholder Irrigation in Ethiopia*; International Water Management Institute: Colombo, Sri Lanka, 2011; p. 37.
14. Bekele, Y.; Nata, T.; Bheemalingswara, K. Preliminary Study on the Impact of Water Quality and Irrigation Practices on Soil Salinity and Crop Production, Gergera Watershed, Atsbi-Wonberta, Tigray, Northern Ethiopia. *MEJS* **2012**, *4*, 29–46. [CrossRef]
15. Vanderlinden, K.; Vereecken, H.; Hardelauf, H.; Herbst, M.; Martinez, G.; Cosh, M.H.; Pachepsky, Y. Temporal stability of soil water contents: A review of data and analyses. *Vadose Zone J.* **2012**, *11*, 19. [CrossRef]
16. Bosson, E.; Sabel, U.; Gustafsson, L.G.; Sassner, M.; Destouni, G. Influences of shifts in climate, landscape, and permafrost on terrestrial hydrology. *J. Geophys. Res. Atmos.* **2012**, *117*, D05120. [CrossRef]
17. Sun, F.; Roderick, M.L.; Farquhar, G.D. Changes in the variability of global land precipitation. *Geophys. Res. Lett.* **2012**, *39*, L19402. [CrossRef]
18. Koster, D.; Dirmeyer, P.; Guo, Z. Regions of strong coupling between soil moisture and precipitation. *Science* **2004**, *305*, 1138–1140. [CrossRef]
19. Wu, W.; Geller, A.M.; Dickinson, E.R. The Response of Soil Moisture to Long-Term Variability of Precipitation. *J. Hydrometeorol.* **2002**, *3*, 604–613. [CrossRef]
20. Jia, D.; Wen, J.; Zhang, T.; Xi, J. Responses of soil moisture and thermal conductivity to precipitation in the mesa of the Loess Plateau. *Environ. Earth Sci.* **2016**, *75*, 395. [CrossRef]
21. Cheng, S.; Guan, X.; Huang, J.; Ji, F.; Guo, R. Long-term trend and variability of soil moisture over East Asia. *J. Geophys. Res. Atmos.* **2015**, *120*, 8658–8670. [CrossRef]
22. Robinson, D.A.; Campbell, C.S.; Hopmans, J.W.; Hornbuckle, B.K.; Jones, S.B.; Knight, R.; Ogden, F.L.; Selker, J.S.; Wendroth, O. Soil moisture measurement for ecological and hydrological watershed-scale observatories: A review. *Vadose Zone J.* **2008**, *7*, 358–389. [CrossRef]

23. He, Z.; Zhao, W.; Liu, H.; Chang, X. The response of soil moisture to rainfall event size in subalpine grassland and meadows in a semi-arid mountain range: A case study in northwestern China's Qilian Mountains. *J. Hydrol.* **2012**, *420*, 183–190. [CrossRef]
24. Jia, B.; Liu, J.; Xie, Z.; Shi, C. Interannual Variations and Trends in Remotely Sensed and Modeled Soil Moisture in China. *J. Hydrometeorol.* **2018**, *19*, 831–847. [CrossRef]
25. Yang, K.; Wang, C.; Bao, H. Contribution of soil moisture variability to summer precipitation in the Northern Hemisphere. *J. Geophys. Res. Atmos.* **2016**, *12*, 108–124. [CrossRef]
26. Seneviratne, S.I.; Corti, T.; Davin, E.L.; Hirschi, M.; Jaeger, E.B.; Lehner, I.; Orlowsky, B.; Teuling, A.J. Investigating soil moisture-climate interactions in a changing climate: A review. *Earth Sci. Rev.* **2010**, *99*, 125–161. [CrossRef]
27. Vereecken, H.; Huisman, A.J.; Pachepsky, Y.; Montzka, C.; van der Kruk, J.; Bogena, H.; Weihermüller, L.; Herbst, M.; Martinez, G.; Vanderborght, J. On the spatio-temporal dynamics of soil moisture at the field scale. *J. Hydrol.* **2014**, *516*, 76–96. [CrossRef]
28. Korres, W.; Reichenau, G.T.; Fiener, P.; Koyama, N.C.; Bogena, R.H.; Cornelissen, T.; Baatz, R.; Herbst, M.; Diekkrüger, B.; Vereecken, H.; et al. Spatio-temporal soil moisture patterns – A meta-analysis using plot to catchment scale data. *J. Hydrol.* **2015**, *520*, 326–341. [CrossRef]
29. Illston, B.; Basara, B.J.; Crawford, C.K. Seasonal to interannual variations of soil moisture measured in Oklahoma. *Int. J. Climatol.* **2004**, *24*, 1883–1896. [CrossRef]
30. Entin, J.K.; Robock, A.; Vinnikov, K.Y.; Hollinger, S.E.; Liu, S.X.; Namkhai, A. Temporal and spatial scales of observed soil moisture variations in the extratropics. *J. Geophys. Res. Atmos.* **2000**, *105*, 11865–11877. [CrossRef]
31. Li, B.; Wang, L.; Kaseke, K.F.; Li, L.; Seely, M.K. The Impact of Rainfall on Soil Moisture Dynamics in a Foggy Desert. *PLoS ONE* **2016**, *11*, e0164982. [CrossRef] [PubMed]
32. Wang, A.; Lettenmaier, D.; Sheffield, J. Soil moisture drought in China, 1950–2006. *J. Clim.* **2011**, *24*, 3257–3271. [CrossRef]
33. Njoku, E.; Jackson, T.; Lakshmi, V.; Chan, T.; Nghiem, S. Soil moisture retrieval from AMSR-E. *IEEE Trans. Geosci. Remote Sens.* **2003**, *41*, 215–229. [CrossRef]
34. Albergel, C.; de Rosnay, P.; Gruhier, C.; Muñoz-Sabater, J.; Hasenauer, S.; Isaksen, L.; Kerr, Y.; Wagner, W. Evaluation of remotely sensed and modelled soil moisture products using global ground-based in situ observations. *Remote Sens. Environ.* **2012**, *118*, 215–226. [CrossRef]
35. de Rosnay, P.; Drusch, M.; Vasiljevic, D.; Balsamo, G.; Albergel, C.; Isaksen, L. A simplified extended Kalman filter for the global operational soil moisture analysis at ECMWF. *Quart. J. R. Meteorol. Soc.* **2013**, *139*, 1199–1213. [CrossRef]
36. Wang, A.; Zeng, X. Sensitivities of terrestrial water cycle simulations to the variations of precipitation and air temperature in China. *J. Geophys. Res.* **2011**, *116*, D02107. [CrossRef]
37. Hain, C.; Crow, W.; Mecikalski, J.; Anderson, M.; Holmes, T. Anintercomparison of available soil moisture estimates from thermal infrared and passive microwave remote sensing and land surface modeling. *J. Geophys. Res.* **2011**, *116*, 1158–1162. [CrossRef]
38. Wagner, W.; Dorigo, W.; de Jeu, R.; Fernandez, D.; Benveniste, J.; Haas, E.; Ertl, M. Fusion of active and passive microwave observations to create an essential climate variable data record on soil moisture. *ISPRS Ann. Photogramm. Remote Sens. Spat. Inf. Sci.* **2012**, *7*, 315–321.
39. Pangaluru, K.; Velicogna, I.; Geruo, A.; Mohajerani, Y.; Ciracì, E.; Charakola, S.; Basha, G.; Rao, S.V.B. Soil Moisture Variability in India: Relationship of Land Surface–Atmosphere Fields Using Maximum Covariance Analysis. *Remote Sens.* **2019**, *11*, 335. [CrossRef]
40. Cai, J.; Zhang, Y.; Li, Y.; Liang, X.S.; Jiang, T. Analyzing the Characteristics of Soil Moisture Using GLDAS Data: A Case Study in Eastern China. *Appl. Sci.* **2017**, *7*, 566. [CrossRef]
41. Liu, Y.Y.; Parinussa, R.M.; Dorigo, W.A.; De Jeu, R.A.M.; Wagner, W.; van Dijk, A.I.J.M.; McCabe, M.F.; Evans, J.P. Developing an improved soil moisture dataset by blending passive and active microwave satellite-based retrievals. *Hydrol. Earth Syst. Sci.* **2011**, *15*, 425–436. [CrossRef]
42. Gruber, A.; Scanlon, T.; van der Schalie, R.; Wagner, W.; Dorigo, W. Evolution of the ESA CCI Soil Moisture climate data records and their underlying merging methodology. *Earth Syst. Sci. Data* **2019**, *11*, 717–739. [CrossRef]

43. Dorigo, W.; Gruber, A.; De Jeu, R.; Wagner, W.; Stacke, T.; Loew, A.; Albergel, C.; Brocca, L.; Chung, D.; Parinussa, R.; et al. Evaluation of the ESA CCI soil moisture product using ground-based observations. *Remote Sens. Environ.* **2015**, *162*, 380–395. [CrossRef]
44. McNally, A.; Shukla, S.; Arsenault, R.K.; Wang, S.; Peters-Lidard, D.C.; Verdin, P.J. Evaluating ESA CCI soil moisture in East Africa. *Int. J. Appl. Earth Obs. Geoinf.* **2016**, *48*, 96–109. [CrossRef]
45. Qiu, J.; Gao, Q.; Wang, S.; Su, Z. Comparison of temporal trends from multiple soil moisture data sets and precipitation: The implication of irrigation on regional soil moisture trend. *Int. J. Appl. Earth Observ. Geoinf.* **2016**, *48*, 17–27. [CrossRef]
46. Chen, X.; Su, Y.; Liao, J.; Shang, J.; Dong, T.; Wang, C.; Liu, W.; Zhou, G.; Liu, L. Detecting significant decreasing trends of land surface soil moisture in eastern China during the past three decades (1979–2010). *J. Geophys. Res. Atmos.* **2016**, *121*, 5177–5192. [CrossRef]
47. Yuan, X.; Ma, Z.; Pan, M.; Shi, C. Microwave remote sensing of short-term droughts during crop growing seasons. *Geophys. Res. Lett.* **2015**, *42*, 4394–4401. [CrossRef]
48. Dobriyal, P.; Qureshi, A.; Badola, R. A review of the methods available for estimating soil moisture and its implications for water resource management. *J. Hydrol.* **2012**, *458*, 110–117. [CrossRef]
49. Degefu, G.T. *The Nile Historical Legal and Developmental Perspectives*; Trafford Publishing: Victoria, BC, Canada, 2003.
50. Kim, U.; Kaluarachchi, J.; Smakhtin, V. Generation of monthly precipitation under climate change for the upper Blue Nile River Basin, Ethiopia. *JAWRA J. Am. Water Resour. Assoc.* **2008**, *44*, 1231–1247. [CrossRef]
51. Taye, M.; Willems, P. Temporal variability of hydro-climatic extremes in the Blue Nile basin. *Water Resour. Res.* **2012**, *48*, 1–13. [CrossRef]
52. ESA: ESA Climate Change Initiative Phase II—Soil Moisture. Product Specification Document (PSD) D1.2.1 Version 4.2. 2018. Available online: https://www.esa-soilmoisture-cci.org/ (accessed on 20 August 2019).
53. Anderson, W.; Zaitchik, B.; Hain, C.; Anderson, M.; Yilmaz, M.; Mecikalski, J.; Schultz, L. Towards an integrated soil moisture drought monitor for East Africa. *Hydrol. Earth Syst. Sci.* **2012**, *16*, 2893–2913. [CrossRef]
54. Ayehu, G.; Tadesse, T.; Gessesse, B.; Yigrem, Y. Soil Moisture Monitoring Using Remote Sensing Data and a Stepwise-Cluster Prediction Model: The Case of Upper Blue Nile Basin, Ethiopia. *Remote Sens.* **2019**, *11*, 125. [CrossRef]
55. Funk, C.; Peterson, P.; Landsfeld, M.; Pedreros, D.; Verdin, J.; Shukla, S.; Husak, G.; Rowland, J.; Harrison, L.; Hoell, A.; et al. The climate hazards infrared precipitation with stations—A new environmental record for monitoring extremes. *Sci. Data* **2015**, *2*, 1–21. [CrossRef] [PubMed]
56. Ayehu, G.; Tadesse, T.; Gessesse, B.; Dinku, T. Validation of new satellite rainfall products over the Upper Blue Nile basin, Ethiopia. *Atmos. Meas. Tech.* **2018**, *11*, 1921–1936. [CrossRef]
57. Bojinski, S.; Verstraete, M.; Peterson, T.C.; Richter, C.; Simmons, A.; Zemp, M. The concept of essential climate Variables in support of climate research, applications, and policy. *Bull. Am. Meteorol. Soc.* **2014**, *95*, 1431–1443. [CrossRef]
58. ESA. Land Cover CCI Product User Guide Version 2.0. 2017. Available online: http://maps.elie.ucl.ac.be/CCI/viewer/download/ESACCI-LC-Ph2-PUGv2_2.0.pdf (accessed on 31 July 2019).
59. Defourny, P.; Schouten, L.; Bartalev, S.; Bontemps, S.; Caccetta, P.; De Wit, A.J.W.; Di Bella, C.; Gérard, B.; Giri, C.; Gond, V.; et al. Accuracy Assessment of a 300 m Global Land Cover Map: The GlobCover Experience. 2008. Available online: http://www.un-spider.org/space-application/space-application-matrix/accuracy-assessment-300-m-global-land-cover-map-globcover (accessed on 31 July 2019).
60. Plummer, S.; Lecomte, P.; Doherty, M. The ESA climate change initiative (CCI): A European contribution to the generation of the global climate observing system. *Remote Sens. Environ.* **2017**, *203*, 2–8. [CrossRef]
61. Bontemps, S.; Defourny, P.; van Bogaert, E.; Kalogirou, V.; Arino, O. Glob Cover 2009: Products Description and Validation Report. 2011. Available online: http://due.esrin.esa.int/files/GLOBCOVER2009_Validation_Report_2.2.pdf (accessed on 31 July 2019).
62. Tsendbazar, N.E.; de Bruin, S.; Fritz, S.; Herold, M. Spatial accuracy assessment and integration of global land cover datasets. *Remote Sens.* **2015**, *7*, 15804–15821. [CrossRef]
63. Yang, Y.; Xiao, P.; Feng, X.; Li, H. Accuracy assessment of seven global land cover datasets over China. *ISPRS J. Photogramm. Remote Sens.* **2017**, *125*, 156–173. [CrossRef]

64. Guidigan, L.M.; Sanou, L.C.; Ragatoa, S.D.; Fafa, O.C.; Mishra, N.V. Assessing Land Use/Land Cover Dynamic and Its Impact in Benin Republic Using Land Change Model and CCI-LC Products. *Earth Syst. Environ.* **2019**, *3*, 127–137. [CrossRef]
65. Li, W.; MacBean, N.; Ciais, P.; Defourny, P.; Lamarche, C.; Bontemps, S.; Houghton, A.R.; Peng, S. Gross and net land cover changes in the main plant functional types derived from the annual ESA CCI land cover maps (1992–2015). *Earth Syst. Sci. Data* **2018**, *10*, 219–234. [CrossRef]
66. Hijmans, R.J. Geographic Data Analysis and Modeling "Raster" Package in R. 2018. Available online: https://rdrr.io/github/rspatial/raster/ (accessed on 2 July 2020).
67. Mann, H.B. Non-parametric tests against trend. *Econometrica* **1945**, *13*, 245–259. [CrossRef]
68. Kendall, M.G. *Rank Correlation Measures*; Charles Griffin: London, UK, 1975.
69. Sen, P.K. Estimates of the regression coefficient based on Kendall's tau. *J. Am. Stat. Assoc.* **1968**, *63*, 1379–1389. [CrossRef]
70. Bronaugh, D.; Werner, A. "ZYP (Zhang+Yue_pilon)" Prewhitened nonlinear trend analysis package in R. 2019. Available online: https://cran.r-project.org/web/packages/zyp/index.html (accessed on 2 July 2020).
71. Yue, S.; Pilon, P.; Phinney, B.; Cavadias, G. The influence of autocorrelation on the ability to detect trend in hydrological series. *Hydrol. Process.* **2002**, *16*, 1807–1829. [CrossRef]
72. Hare, W. *Assessment of Knowledge on Impacts of Climate Change, Contribution to the Specification of Art, 2 of the UNFCCC*; WBGU: Berlin, Germany, 2003.
73. Usowicz, B.; Lipiec, J.; Lukowski, M. Evaluation of Soil Moisture Variability in Poland from SMOS Satellite Observations. *Remote Sens.* **2019**, *11*, 1280. [CrossRef]
74. Brocca, L.; Zucco, G.; Mittelbach, H.; Moramarco, T.; Seneviratne, S.I. Absolute versus temporal anomaly and percent of saturation soil moisture spatial variability for six networks worldwide. *Water Resour. Res.* **2014**, *50*, 5560–5576. [CrossRef]
75. Liu, L.; Zhang, R.; Zuo, Z. The Relationship between Soil Moisture and LAI in Different Types of Soil in Central Eastern China. *J. Hydrometeorol.* **2016**, *17*, 2733–2742. [CrossRef]
76. Tilahun, K. Analysis of rainfall climate and evapo-transpiration in arid and semi-arid regions of Ethiopia using data over the last half a century. *J. Arid Environ.* **2006**, *64*, 474–487. [CrossRef]
77. Hu, W.; Shao, M.G.; Han, F.P.; Reichardt, K. Spatio-temporal variability behavior of land surface soil water content in shrub- and grass-land. *Geoderma* **2011**, *162*, 260–272. [CrossRef]
78. Qiu, Y.; Fu, B.; Wang, J.; Chen, L. Spatial variability of soil moisture content and its relation to environmental indices in a semi-arid gully catchment of the Loess Plateau, China. *J. Arid Environ.* **2001**, *49*, 723–750. [CrossRef]
79. Feng, Q.; Zhao, W.; Qiu, Y.; Zhao, M.; Zhong, L. Spatial Heterogeneity of Soil Moisture and the Scale Variability of Its Influencing Factors: A Case Study in the Loess Plateau of China. *Water* **2013**, *5*, 1226–1242. [CrossRef]
80. Dorigo, W.; Wagner, W.; Calvet, J.-C.; Albrecht, F.; Balsamo, G.; Brocca, L.; Chung, D.; Ertl, M.; Forkel, M.; Gruber, A.; et al. ESA CCI Soil Moisture for improved Earth system understanding: State-of-the art and future directions. *Remote Sens. Environ.* **2017**, *203*, 185–215. [CrossRef]
81. Longobardi, A. Observing soil moisture temporal variability under fluctuating climatic conditions. *Hydrol. Earth Syst. Sci. Discuss.* **2008**, *5*, 935–969. [CrossRef]
82. Bayabil, H.K. *Modeling Rainfall-runoff Relationship and Assessing Impacts of Soil Conservation Research Program Intervention on Soil Physical and Chemical Properties at Maybar Research Unit*; Cornell University: Wollo, Ethiopia, 2009.
83. Yang, T.; Ala, M.; Zhang, Y.; Wu, J.; Wang, A.; Guan, D. Characteristics of soil moisture under different vegetation coverage in Horqin Sandy Land, northern China. *PLoS ONE* **2018**, *13*, e0198805. [CrossRef] [PubMed]
84. Crave, A.; Gascuel-Odux, C. The influence of topography on the time and space distribution of soil surface water content. *Hydrol. Process.* **1997**, *11*, 203–210. [CrossRef]
85. Venkatesh, B.; Lakshman, N.; Purandara, B.K.; Reddy, V.B. Analysis of observed soil moisture patterns under different land covers in Western Ghats, India. *J. Hydrol.* **2011**, *397*, 281–294. [CrossRef]

remote sensing

MDPI

Article

Estimation of Soil Moisture Applying Modified Dubois Model to Sentinel-1; A Regional Study from Central India

Abhilash Singh [1], Kumar Gaurav [1,*], Ganesh Kumar Meena [1] and Shashi Kumar [2]

[1] Department of Earth and Environment Sciences, Indian Institute of Science Education and Research, Bhopal 462066, Madhya Pradesh, India; sabhilash@iiserb.ac.in (A.S.); gkmeena@iiserb.ac.in (G.K.M.)

[2] Department of Photogrammetry and Remote Sensing, Indian Institute of Remote Sensing, ISRO, Dehradun 248001, Uttarakhand, India; shashi@iirs.gov.in

* Correspondence: kgaurav@iiserb.ac.in; Tel.: +91-755-669-1383

Received: 11 May 2020; Accepted: 3 June 2020; Published: 15 July 2020

Abstract: Surface soil moisture has a wide application in climate change, agronomy, water resources, and in many other domain of science and engineering. Measurement of soil moisture at high spatial and temporal resolution at regional and global scale is needed for the prediction of flood, drought, planning and management of agricultural productivity to ensure food security. Recent advancement in microwave remote sensing, especially after the launch of Sentinel operational satellites has enabled the scientific community to estimate soil moisture at higher spatial and temporal resolution with greater accuracy. This study evaluates the potential of Sentinel-1A satellite images to estimate soil moisture in a semi-arid region. Exactly at the time when satellite passes over the study area, we have collected soil samples at 37 different locations and measured the soil moisture from 5 cm below the ground surface using ML3 theta probe. We processed the soil samples in laboratory to obtain volumetric soil moisture using the oven dry method. We found soil moisture measured from calibrated theta probe and oven dry method are in good agreement with Root Mean Square Error (RMSE) 0.025 m^3/m^3 and coefficient of determination (R^2) 0.85. We then processed Sentinel-1A images and applied modified Dubois model to calculate relative permittivity of the soil from the backscatter values (σ°). The volumetric soil moisture at each pixel is then calculated by applying the universal Topp's model. Finally, we masked the pixels whose Normalised Difference Vegetation Index (NDVI) value is greater than 0.4 to generate soil moisture map as per the Dubois NDVI criterion. Our modelled soil moisture accord with the measured values with RMSE = 0.035 and $R^2 = 0.75$. We found a small bias in the modelled soil moisture (0.02 m^3/m^3). However, this has reduced significantly (0.001 m^3/m^3) after applying a bias correction based on Cumulative Distribution Function (CDF) matching. Our approach provides a first-order estimate of soil moisture from Sentinel-1A images in sparsely vegetated agricultural land.

Keywords: soil moisture; theta probe; Sentinel-1A; NDVI; modified Dubois model

1. Introduction

Soil moisture is a temporary storage of water in soil pores that controls various processes occurring at the air–soil interface [1–5]. Quantification of soil moisture is required on a regular basis for predicting flood, drought, agricultural productivity, hydrological modelling and climate studies [6–11]. Based on the specific application, soil moisture is needed at different spatial and temporal scales. At a local scale, it can be measured in the field using Time Domain Reflectometry (TDR) or gravimetric methods. These measurement techniques provide a more accurate estimate of soil moisture, but they are tedious and time-consuming. This limits the use of in-situ measurement techniques to measure soil moisture on global or regional scales.

An alternative to in-situ measurements, surface soil moisture, can be modelled from remote sensing images. Active microwave remote sensing, specifically Synthetic Aperture Radar (SAR) imaging has emerged as an effective tool to estimate surface soil moisture. The SAR sensors transmit microwave electromagnetic pulses and record the backscattered energy from the earth's surface. The microwave pulses have high sensitivity towards the dielectric properties of the target and surface roughness [12]. At a given incidence angle, when a SAR signal interacts with the soil-water mixture, a permittivity (ϵ) gradient exists between the dry soil ($\epsilon = 2$) and water ($\epsilon = 80$), that reflects in the intensity of radar backscatter [13–16].

To retrieve the relative soil permittivity and surface roughness component from the SAR backscatter, various empirical, semi-empirical and theoretical models have been proposed [17–24]. These models are developed for the quad polarised SAR images and are mainly applicable on barren land. However, the Water Cloud and Dubois models have been successfully used to estimate soil moisture over barren and sparsely vegetated land [25–28].

In a study, Zribi et al. [29] have applied Water Cloud Model (WCM) on L-band PALSAR/ALOS-2 satellite data to estimate soil moisture in a tropical agricultural area under dense vegetation cover conditions. Yang et al. [30] have used the fully polarimetric C-band Radarsat-2 SAR data for soil moisture mapping in Juyanze Basin, China. They concluded that increasing the number of polarimetric parameters at C-band can provide a more robust estimate of surface soil moisture. El Hajj et al. [31] and Bousbih et al. [32] have shown that the synergic use of radar (Sentinel-1) and optical (Sentinel-2) data can be utilised to estimate soil moisture at higher spatial resolution at field scale. They have used a neural network (NN) model to estimate soil moisture by the inversion of radar signals. In doing so, they have generated a synthetic database of the backscattering coefficient in the VV and HH polarisation for a range of soil moisture, surface roughness and NDVI values by the parameterisation of coupled WCM and modified Integral Equation Model (IEM). They concluded that their approach can be used to estimate soil moisture in agricultural plots having NDVI less than 0.75. Further, Qiu et al. [33] used the similar WCM and IEM coupled model to evaluate the impact of different vegetation indices (NDVI, Enhanced Vegetation Index, and Leaf Area Index) in the estimation of soil moisture. They reported the accuracy of estimated soil moisture is independent of the choice of specific vegetation indices. Hachani et al. [34] used sentinel-1 images to estimate soil moisture in an arid climate in Tunisia. They have developed an artificial neural network (ANN) using the training samples obtained by combining satellite measurements and the simulated backscatter values using the Integral Equation Model (IEM). They claimed their model is almost site independent and able to simulate soil moisture content with limited or no ancillary information (i.e., DEM, local Incidence angle, NDVI). More recently, Ezzahar et al. [35] have used Support Vector Machine (SVM), IEM and Oh models to estimate soil moisture over bare agricultural soil in the Tensfit basin of Morocco from sentinel-1 satellite images.

Hosseini and McNairn [36] applied WCM-Ulaby model on C-band (Radarsat-2) and L-band (UAVSAR) SAR data to estimate soil moisture and biomass in wheat fields in western Canada. Some authors [37,38] have used modified Dubois model and dual polarised (HH and HV) RISAT-1, C-band data to estimate soil moisture of the Bhal region in Gujrat, India. They found promising results with good correlation during the initial period of the crop. However, the accuracy decreases greatly at the locations of dense canopy cover and higher NDVI values in the study area.

After the launch of Sentinel-1A satellite mission in April 2014, the global coverage of SAR data is easily available at higher spatial and temporal resolution, and is being widely used for soil moisture estimation [39]. Sentinel-1A satellite sensor operates in C-band at frequency 5.405 GHz and acquire information about the earth's surface in selectable single (HH or VV) and dual polarisation (HH + HV, VV + VH) [40]. This data is freely available and is widely used in various applications, including soil moisture estimation [34,38,41–43].

This study uses dual polarised Sentinel-1A satellite images to estimate soil moisture in a semi-arid region in central India. We used modified Dubois model to calculate the relative soil permittivity from

SAR backscatter values. Eventually, we input the relative soil permittivity in universal Topp's model to obtain volumetric soil moisture in barren or sparsely vegetated farmlands.

2. Study Area

This study is conducted in Bhopal district of Madhya Pradesh in central India (Figure 1). Bhopal is divided into two administrative blocks, Berasia in the north and Phanda in the south having a surface area of about 1424 km^2 and 1348 km^2 respectively (Figure 1). Climatically, Bhopal lies in a semiarid zone and is typically covered by agriculture land (64.5%), barren land (7.3%), forest (13%), and water bodies (4.6%) [44]. The average elevation in the study area varies between 450 and 550 m from the mean sea level with the gently undulated landscape. The average air temperature ranges between 6 °C and 41 °C. About 75% of the study area is covered by black cotton soil formed due to the weathering of basaltic rocks. The remaining 25% is covered with yellowish-red, mixed soils [45,46].

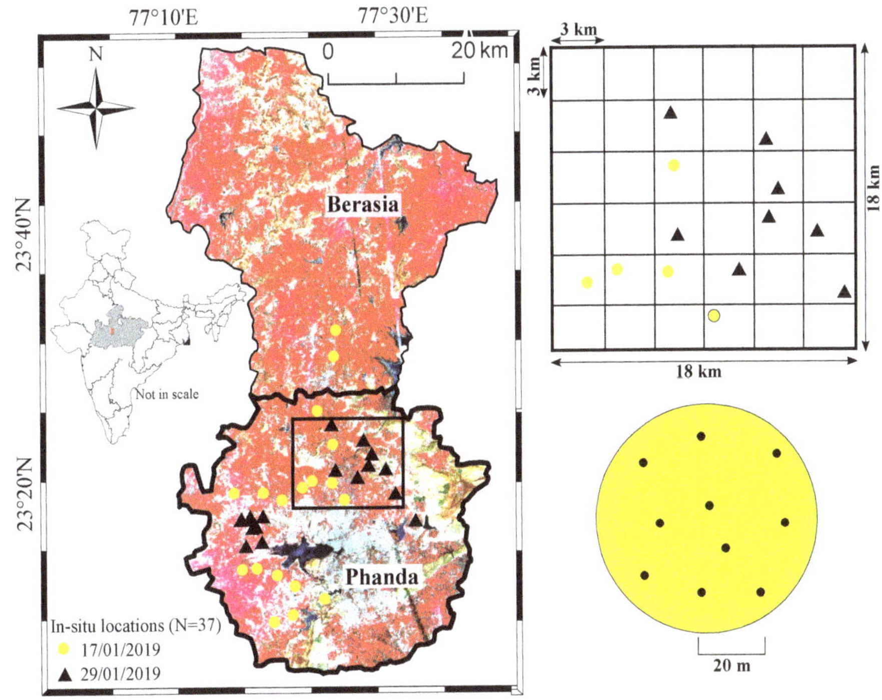

Figure 1. Landsat-8 image in False Color Composite (FCC) shows the administrative blocks (Phanda and Berasia) of Bhopal district, Madhya Pradesh. Circles in Yellow and triangles in black are the locations of soil moisture measurement in the field. Grid on the top right illustrates the random sampling strategy to collect soil moisture.

In this study, we have considered Phanda block as a test site to estimate soil moisture. About 44% of the area of Phanda is cultivable and used for agricultural purposes. The agricultural practice in the region largely depends on the Indian summer monsoon in June, July, August, and September (JJAS), where it receives about 92% of total rainfall.

In recent years, the frequency of droughts in central India has increased, which has adversely impacted the agricultural productivity [47]. To obtain the frequency of drought events in the study area, we have calculated Standardized Precipitation Index (SPI) for the monsoon period (JJAS) from 1990 to 2018 using gridded ($0.25^0 \times 0.25^0$) rainfall data obtained from the Indian Meteorological

Department [48,49]. SPI is used to characterise meteorological drought. For example, SPI values in a range between (−0.99 to 0.99) is considered normal, whereas SPI values less than −1 and greater than 1 are considered to be dry and wet period respectively [50]. The SPI values of Phanda block (Figure 2), clearly suggests that the frequency of drought events has increased in last two decades. In total, six drought events (2002, 2004, 2010, 2014, 2015, and 2017) have occurred between the years 2000 to 2018 (Figure 2).

In this scenario, monitoring soil moisture at higher spatial and temporal resolution has become important in planning and management of agricultural productivity, water resources and ensuring food security. The landuse, geology, and soil types of the Phanda block makes it an ideal field site to study the soil moisture.

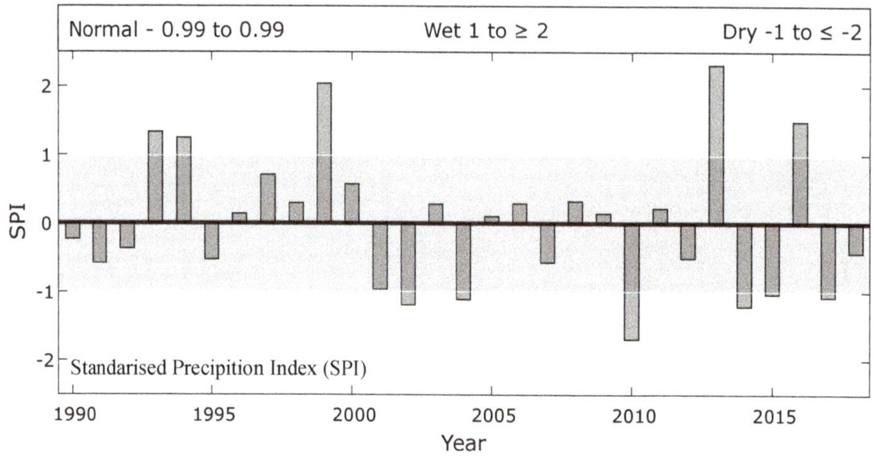

Figure 2. Areal averaged Standardized precipitation Index (SPI) for Indian summer monsoon (June, July, August, September) from 1990 to 2018. Shaded region (SPI −0.99 to 0.99) shows the normal precipitation condition. SPI in range ($-1 \leq -2$) and ($1 \geq 2$) suggests drought and wet condition respectively.

3. Material and Methods

3.1. Satellite Data

This study uses microwave and optical satellite images (Table 1) to estimate soil moisture. We have downloaded publicly available Sentinel-1A images of two consecutive pass, i.e., 17 and 29 January 2019 from European Space Agency (https://scihub.copernicus.eu/). Sentinel-1A, C-band SAR records the backscatter signals day and night, independent of the illumination and weather conditions. It acquires microwave images in four exclusive modes; Stripmap (SM), Interferometric Wide swath (IW), Extra Wide swath (EW), Wave (WV). It can capture images (in terms of polarization) using same set of transmitted pulses by using its antenna. Depending upon the acquisition mode, Sentinel-1 can acquires images in dual polarisation modes, Vertical–Vertical (VV) and Vertical–Horizontal (VH), or in single polarisation (HH or VV) at 10 m× 10 m cell size with a swath 250 km. At this swath, the incidence angle ranges between 29° to 46° for near and far range respectively. It has a temporal resolution of 12 days (in combination with Sentinel-1B the temporal resolution is 6 days). Microwave signals at C-band can penetrate up to 5 cm deep below the soil surface [51,52]. The Sentinel-1A level-1 data is categorised into two product types: Ground Range Detected (GRD) and Single Look Complex (SLC). Wide range of applications requires Sentinel-1A GRD product with standard corrections. After applying standard corrections, the Sentinel-1A GRD images will have square pixels (consisting of amplitude) with reduced speckle [53–57]. We have also downloaded Landsat 8 images from the United States Geological Survey (https://earthexplorer.usgs.gov/) pertaining to the date closest to the Sentinel-1A images. The Landsat 8 satellite mission carries a two-sensor payload, the Operational Land Imager (OLI) and the Thermal Infrared Sensor (TIRS). The OLI sensor consists of nine spectral bands (0.43–1.38 μm) and acquire images at 16 days revisit time with the spatial resolution 15 m for the panchromatic band (0.50–0.68 μm) and 30 m for the remaining bands [58].

Table 1. Details of the Sentinel-1 and Landsat 8 images used in this study.

		Sentinel-1A		
Date	Polarisations	Incidence angle (°)	Pixel size (m × m)	Direction
17/01/2019	VH + VV	38.4	10 × 10	NE
29/01/2019	VH + VV	38.5	10 × 10	NE
		Landsat 8		
	Row/Path	Band	Wavelength (μm)	Resolution (m)
15/01/2019	44/145	4, 5	0.64–0.67, 0.85–0.88	30
22/01/2019	43/146	4, 5	0.64–0.67, 0.85–0.88	30

3.2. Field Measurement

We have used ML3 theta probe sensor and gravimetric method to measure soil moisture in the field. Theta probe works on the principle of TDR, and it measures the bulk soil relative permittivity (ϵ) at a frequency 100 MHz. This bulk relative permittivity is then converted into volumetric soil moisture by applying a soil specific calibration of TDR. We have calibrated the theta probe for three different dominant soil types of our study area. A detailed procedure of calibration is mentioned in the Appendix A. To measure soil moisture, we have conducted a field campaign on 17 and 29 January 2019 in Bhopal, Madhya Pradesh. These dates coincide with the Sentinel-1A pass over Bhopal at 5:50 a.m. (IST). At the time of satellite passes, we have measured the soil moisture at 37 locations.

To conduct measurements, we overlay a square grid of 3 km × 3 km on our study area [59]. We then randomly select few grids to measure the soil moisture. At the centre of each grid, we measured the soil moisture by inserting the metal rod of the theta probe at 5 cm below the ground surface and record the soil moisture value (m^3/m^3) in the data logger and acquired their locations using a Garmin-64S handheld GPS (Figure 3a). We repeated this procedure at least in 8 to 10 different locations within the grid and finally averaged them to get soil moisture (Figure 1). Simultaneously at each location, we have collected about 100 g of soil samples at 5 cm below the ground surface using tubular samplers (Figure 3b). We used these samples to measure soil moisture by oven drying method (Figure 3c). At each of the measurement location, we have also taken observations on land use, soil type, vegetation height and weather parameters (temperature, precipitation, etc.). Table 2 reports the detailed characteristics and location of our sampling sites.

Figure 3. Measurement of soil moisture using (**a**) ML3 sensor and (**b**) Collection of soil samples in the field. Photograph (**c**) Shows the sample processing in the laboratory for oven drying to calculate soil moisture.

Table 2. Details of the ground condition and soil moisture measured in the field using ML3 theta probe.

Date	Site ID	Latitute	Longitude	Elevation (msl)	Land Use	Sampling Depth (cm)	Vegetation Height (cm)	$\overline{SM}$ (m^3/m^3) (TDR)
17-01-2019	1	23.22913	77.26613	516	Agriculture land (Chickpeas)	5	10–15	0.291 ± 0.053
17-01-2019	2	23.22978	77.28679	517	Agriculture land (Chickpeas)	5	15–20	0.170 ± 0.028
17-01-2019	3	23.22081	77.31552	515	Agriculture land (Chickpeas)	5	0–5	0.167 ± 0.020
17-01-2019	4	23.20669	77.34076	538	Barren land	5	0	0.043 ± 0.017
17-01-2019	5	23.18931	77.38358	507	Barren land	5	0	0.109 ± 0.032
17-01-2019	6	23.16962	77.33730	532	Agriculture land (Mustard)	5	20–30	0.185 ± 0.025
17-01-2019	7	23.16144	77.31003	546	Barren land	5	0	0.111 ± 0.014
17-01-2019	8	23.31624	77.41531	495	Agriculture land (Wheat)	5	15–20	0.266 ± 0.052
17-01-2019	9	23.33788	77.39777	497	Agriculture land (Wheat)	5	30–35	0.175 ± 0.022
17-01-2019	10	23.33280	77.35484	494	Agriculture land (Coriander)	5	10–15	0.275 ± 0.046
17-01-2019	11	23.31749	77.32511	503	Agriculture land (Mixed)	5	20–30	0.198 ± 0.032
17-01-2019	12	23.32754	77.25596	529	Agriculture land (Wheat)	5	20–30	0.255 ± 0.051
17-01-2019	13	23.32683	77.29851	510	Agriculture land (Chickpeas)	5	10–15	0.254 ± 0.037
17-01-2019	14	23.34099	77.36872	502	Agriculture land (Wheat)	5	30–40	0.098 ± 0.015
17-01-2019	15	23.38717	77.39961	488	Agriculture land (Daikon)	5	0	0.270 ± 0.042
17-01-2019	16	23.43098	77.37777	487	Agriculture land (Chickpeas)	5	15–20	0.218 ± 0.038
17-01-2019	17	23.50044	77.40357	476	Agriculture land (Wheat)	5	20–25	0.285 ± 0.052
17-01-2019	18	23.53360	77.40742	478	Agriculture land (Chickpeas)	5	10-15	0.186 ± 0.021
29-01-2019	19	23.41329	77.39910	484	Agriculture land (Wheat)	5	30–40	0.189 ± 0.025
29-01-2019	20	23.35487	77.40400	483	Agriculture land (Wheat)	5	40–50	0.174 ± 0.024
29-01-2019	21	23.34471	77.43551	490	Agriculture land (Wheat)	5	20–25	0.175 ± 0.024
29-01-2019	22	23.35516	77.47569	512	Agriculture land (Wheat)	5	10–15	0.270 ± 0.060
29-01-2019	23	23.37469	77.45610	492	Agriculture land (Wheat)	5	40–50	0.129 ± 0.018
29-01-2019	24	23.39272	77.44480	478	Agriculture land (Wheat)	5	20–25	0.069 ± 0.013
29-01-2019	25	23.36028	77.45131	501	Agriculture land (Mixed)	5	10–15	0.262 ± 0.033
29-01-2019	26	23.32369	77.48873	518	Agriculture land (Wheat)	5	30–40	0.099 ± 0.024
29-01-2019	27	23.28805	77.51691	491	Agriculture land (Chickpeas)	5	15–20	0.267 ± 0.041
29-01-2019	28	23.29685	77.27975	516	Barren land	5	0	0.221 ± 0.042
29-01-2019	29	23.29730	77.99768	519	Agriculture land (Chickpeas)	5	20–25	0.180 ± 0.027
29-01-2019	30	23.29316	77.26544	528	Agriculture land (Wheat)	5	15–20	0.267 ± 0.042
29-01-2019	31	23.29380	77.28096	520	Agriculture land (Wheat)	5	10–15	0.159 ± 0.027
29-01-2019	32	23.28579	77.28620	523	Agriculture land (Mixed)	5	5–10	0.234 ± 0.046
29-01-2019	33	23.28255	77.28209	519	Agriculture land (Wheat)	5	40–50	0.117 ± 0.026
29-01-2019	34	23.28405	77.28794	520	Agriculture land (Chickpeas)	5	10–15	0.153 ± 0.028
29-01-2019	35	23.29396	77.26690	526	Agriculture land (Wheat)	5	40–50	0.276 ± 0.032
29-01-2019	36	23.25948	77.27253	511	Barren land	5	0	0.102 ± 0.034
29-01-2019	37	23.26410	77.29583	515	Agriculture land (Chickpeas)	5	10–15	0.124 ± 0.029

Note: Temperature range (at 2 m): (18.4–19.1) °C; Earth surface temperature: (13.7–18.1) °C; Precipitation: Nil (Source: Weather station Bhopal).

3.3. Data Processing

3.3.1. Soil Samples

We processed the soil samples in laboratory to measure their moisture content. We took about 100 g each of soil samples from a grid and placed them in a separate beaker (Figure 3c). We placed the samples in an electric oven at 105 °C for about 24 h [60,61]. Once the samples are completely dried, we weighed them. The difference between initial and dry weight provides the amount of moisture present in the soil samples. Eventually, we divide the weight of moisture content with the dry weight of soil samples. This quantity is the gravimetric soil moisture (m_g) and expressed in [kg/kg].

Since theta probe measures the volumetric soil moisture (m_v), we need to convert our laboratory measurement into comparable metric for any meaningful comparison. We multiply the gravimetric soil moisture (m_g) to the density ratio of soil (ρ_{soil}) and water (ρ_{water}) to compute the volumetric soil moisture (m_v) in [m^3/m^3] according to;

$$m_v = m_g \cdot \frac{\rho_{soil}}{\rho_{water}} \tag{1}$$

3.3.2. Satellite Images

We have used Sentinel Application Platform (SNAP) v6.0 to process the Sentinel-1A images. The processing involves four major steps, radiometric calibration, multilook, speckle noise reduction using refined Lee filter, and terrain correction. Terrain or geometric calibration uses the Shuttle Radar Topography Mission (SRTM) digital elevation model of spatial resolution 30 m. The resulting image pixels contain the true backscatter (σ°) values on a linear scale. Finally, we convert the backscatter values into decibel scale (σ_{dB}) according to, $\sigma^\circ_{dB} = 10 * log_{10}(\sigma^\circ)$. We have also processed Landsat-8 images to compute NDVI. It helps to measure the intensity of vegetation cover in terms of vegetation density and vegetation height. We take the ratio of the difference between band 5 (Near Infrared) and band 4 (Red) to the sum of band 5 and band 4 of Landsat-8 images. The NDVI is required to specify the validity range (NDVI $\leq$ 0.4) of modified Dubois model for soil moisture.

3.4. Soil Moisture Modelling

We have used the backscatter values of Sentinel-1A images and incidence angle in the modified Dubois model to calculate the relative soil permittivity. Finally, we used the relative permittivity in universal Topp's model to compute volumetric soil moisture.

3.4.1. Radar Backscattering Model

Dubois et al. [19] developed an empirical model to calculate the relative soil permittivity from quad polarised SAR images. Initially, this model was developed for L-, C- and X-band data obtained from scatterometer and later applied on airborne images as well. The model structure is developed on a strong physical reasoning; however, some of the unknown coefficients are obtained by fitting to the experimental data. The backscattering coefficient for HH and VV polarisation is given by Equations (2) and (3).

$$\sigma^\circ_{HH} = 10^{-2.75}\left(\frac{\cos^{1.5}\theta}{\sin^5\theta}\right)10^{0.028\epsilon\tan\theta}(k.s.\sin\theta)^{1.4}\lambda^{0.7} \tag{2}$$

$$\sigma^\circ_{VV} = 10^{-2.35}\left(\frac{\cos^3\theta}{\sin^3\theta}\right)10^{0.046\epsilon\tan\theta}(k.s.\sin\theta)^{1.1}\lambda^{0.7} \tag{3}$$

where θ is the incidence angle, ϵ is the relative soil permittivity, s is the surface roughness (cm), $k = (2\pi/\lambda)$ is the wavenumber, and λ is the SAR wavelength. These parameters can be grouped into two; sensor parameters (θ and λ) and target parameters (ϵ and s). In Equations (2) and (3), target parameters are unknown. These equations can be inverted to compute the relative soil permittivity and surface roughness parameters.

Dubois model is applicable for the SAR images acquired at incidence angle θ between 30° to 65° and in the frequency range between 1.5–11 GHz. Further, this model is valid on the region having sparsely vegetation or barren land. The performance of the Dubois model is maximum where NDVI of the image pixels is less than 0.4 (Dubois NDVI criterion) or in the region on SAR images where the ratio of cross-polarised $\sigma^{\circ}_{HV}/\sigma^{\circ}_{VV}$ is less than -11 dB [19].

Dubois model was initially developed for quad polarised SAR images; it can not be directly applied to dual polarised data. In a study Rao et al. [41] modified the Dubois model by incorporating in-situ measurements and field conditions and used the Equation (2) to calculate the soil moisture from the SAR images acquired in HH polarisation.

To derive the unknown parameter (surface roughness) of Equation (3), we obtain the parameter from regression model proposed by Srivastava et al. [62]. Once surface roughness is known, Equation (2) or (3) can be solved to compute the other unknown, the relative soil permittivity. A flowchart in Figure 4 illustrates the detailed methodology used for in-situ data acquisition, image processing and modelling to estimate soil moisture.

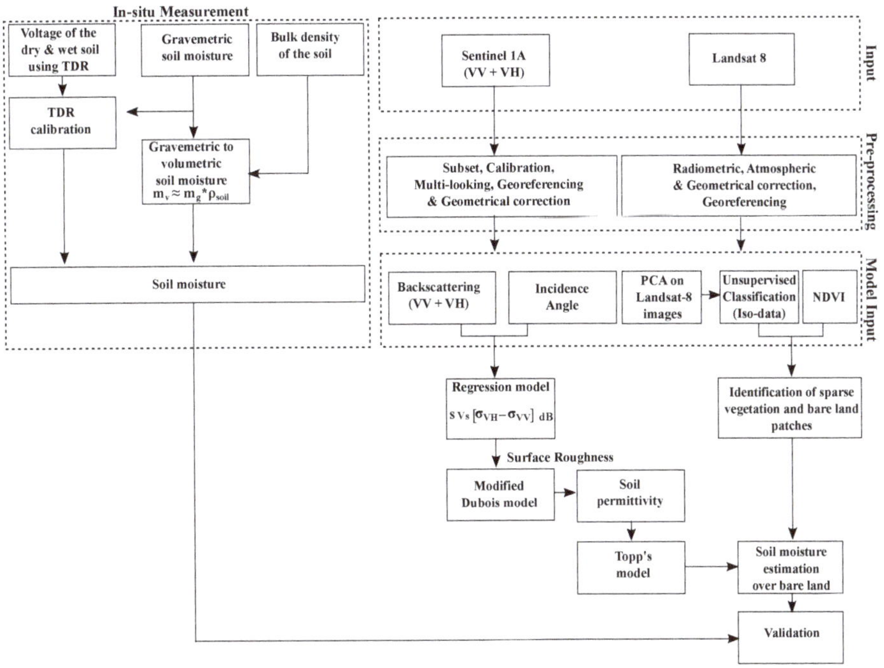

Figure 4. Flow chart illustrates the methodology used for the estimation of soil moisture from Sentinel-1 images.

3.4.2. Estimation of Soil Moisture

To estimate soil moisture, we used Topp's model [63]. It takes the relative soil permittivity derived from Sentinel-1A image as an input to estimate volumetric soil moisture content according to

Equation (4). This model does not require a priori knowledge of soil properties (texture, grain size) and is proven to be a robust approach for the estimation of soil moisture [64].

$$m_v = -5.3 \times 10^{-2} + 2.92 \times 10^{-2}\epsilon - 5.5 \times 10^{-4}\epsilon^2 + 4.3 \times 10^{-6}\epsilon^3 \quad (4)$$

4. Results

4.1. Performance of Calibrated Theta Probe

We compared the soil moisture measured from the theta probe and oven dry method. We observed no obvious difference, despite a mild scatter, all data points seem to gather around a single line (Figure 5). This suggests a good agreement between (R^2 = 0.84 with RMSE = 0.025 m^3/m^3) both the methods, provided theta probe is correctly calibrated for the specific soil types in the study area. Hereafter we will use the measurements of theta probe for further analysis.

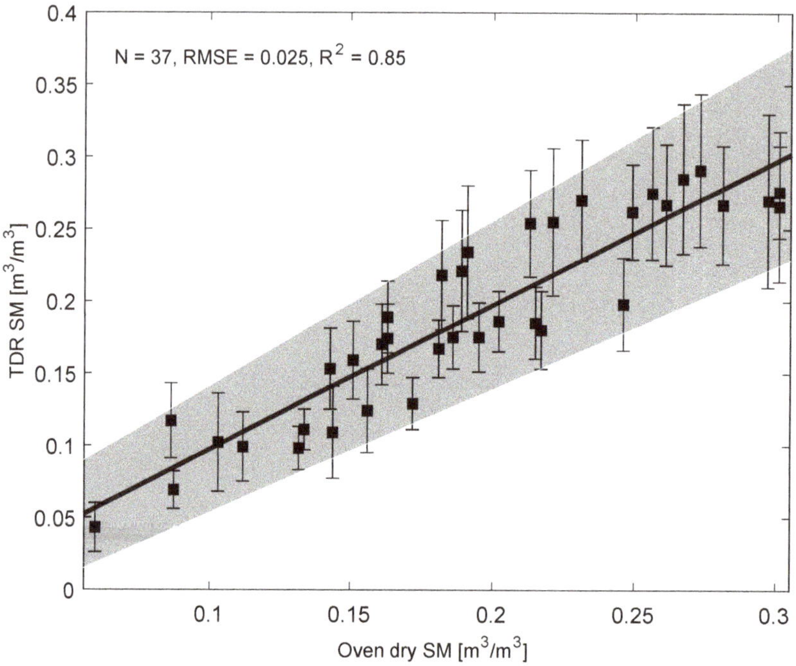

Figure 5. Soil moisture measured using theta probe as a functions of gravimetric method. Shaded gray region is the 95% confidence level of the regression curve.

4.2. Soil Moisture from Sentinel-1A

Once we calculated the soil moisture from Sentinel-1A images, we evaluated their accuracy with the in-situ measurement. We first identified the valid region for modified Dubois model using the Dubois NDVI criterion. For the locations where ground measurement is valid, we extracted the corresponding pixels from the modelled soil moisture maps. For both the dates, we observed a good agreement (R^2) between the measured and modelled soil moisture (Figure 6). However, at some locations, the difference between the modelled and measured values of soil moisture is comparatively large. This is probably due to the spatial scale mismatch, heterogeneous field conditions, measurement uncertainty, and model bias. When in-situ measurement is compared with the satellite derived soil moisture, a representative error arises due to the difference in the spatial scale of in-situ measurement and satellite observation. The spatial scale mismatch becomes more prominent when the land surface is heterogeneous. This limits the competency to compare the point measurement with the satellite [65]. The error in the retrieval of soil moisture increases with the heterogeneity in land surface [66]. Further, the measurement uncertainty is mainly due to the error associated with in-situ measurement. For example, error resulting from the conversion of dielectric constant to soil moisture in TDR, presence of organic matter in the soil, overestimation of drying condition in oven and difference in the sampling depth of the in-situ and satellite measurements [67].

To reduce the measurement uncertainty, we have calibrated the TDR, removed the organic matter from soil the soil samples, and collected soil samples in the field according to the simulated penetration depth of the C-band SAR signal in the ground [51]. Model error (or bias) is mainly due to the assumptions followed by its use.

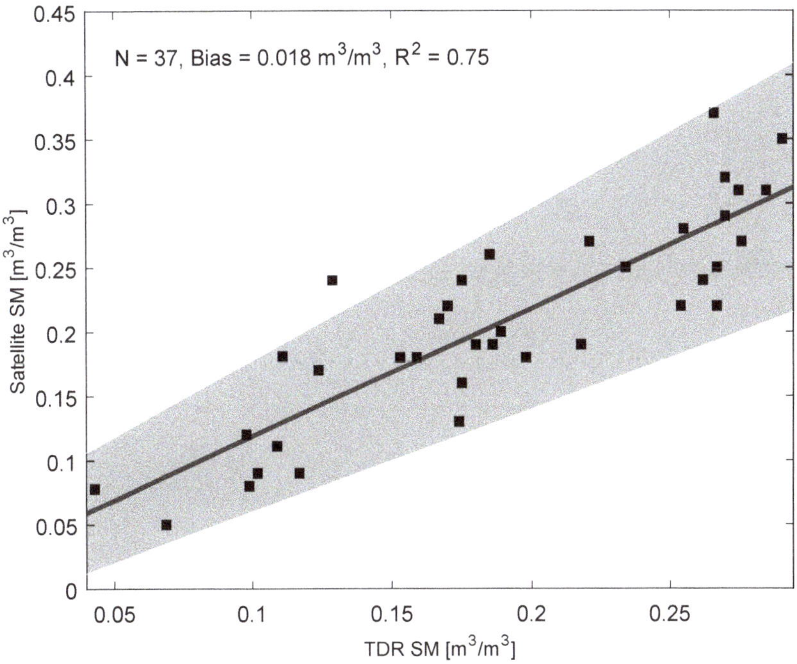

Figure 6. Regression curve between satellite derived and in-situ (TDR) measured soil moisture. Shaded gray region is the 95% confidence level of the regression curve.

To overcome the systematic errors (instrument calibration and model error), we performed a bias correction by applying the CDF matching approach [68]. This is one of the widely used statistical methods to minimise the bias in satellite-derived soil moisture [69–71]. In principle, we adjusted the satellite-derived soil moisture according to the CDF of in-situ (TDR) soil moisture to minimise their difference (Figure 7).

Applying the CDF correction on our data, biases in the soil moisture derived from Sentinel-1A has reduced from 0.02 m^3/m^3 to 0.001 m^3/m^3. Finally, we used the bias corrected values to generate soil moisture maps for 17 and 29 January 2019 (Figure 8) of the study area.

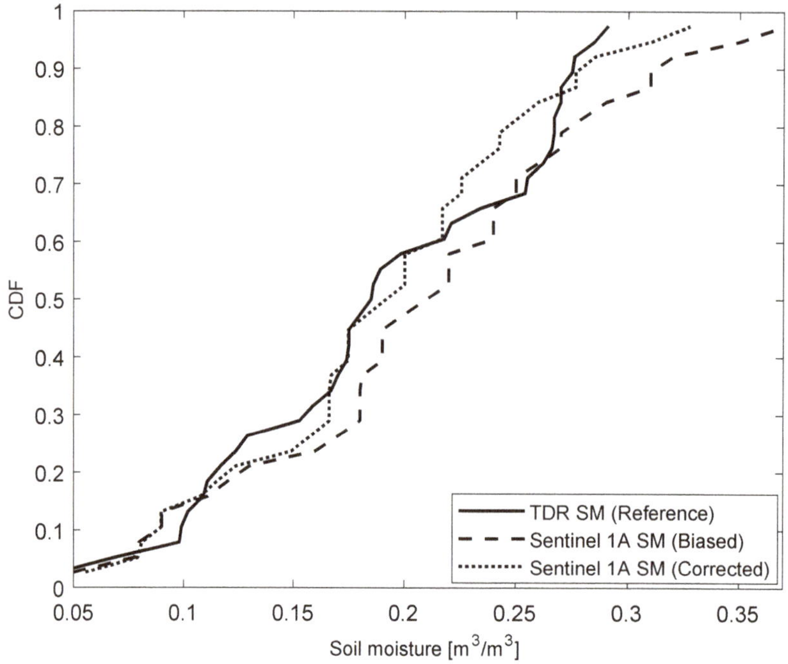

Figure 7. Bias correction using the CDF matching technique. Solid black line is the CDF of in-situ measured soil moisture using TDR, dashed and dotted lines in black are the CDF of biased and corrected soil moisture estimated from sentinel-1A respectively.

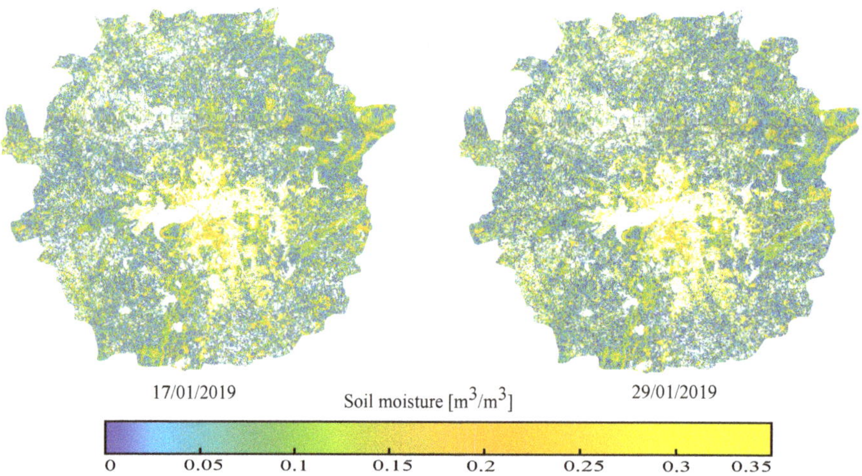

Figure 8. Spatial distribution of soil moisture estimated from Sentinel-1. Pixels shown in white correspond to the region where modified Dubois model is not valid.

5. Discussion

This study uses modified Dubois model to estimate soil moisture from dual polarised (VV and VH) Sentinel-1A images. Our result suggests that the model derived soil moisture accord well (R^2 = 0.75

and RMSE = 0.035 m^3/m^3) with the in-situ measurement (Figure 6). This is consistent with the range of RMSE (0.03–0.04) m^3/m^3 reported by other researchers [72].

The modelled soil moisture is subject to bias due to various geometric, atmospheric and modelling errors. The magnitude of these uncertainties depends on various factors such as choice of a backscatter model, frequency of SAR images, ground condition and vegetation types. Several methods such as mean based (linear, local intensity and variance scaling) and the distribution-based (quantile and CDF matching) have been developed to correct the bias from the modelled soil moisture values [68,73–77].

We have used the CDF matching approach to minimize the bias from our modelled soil moisture. In doing so, we have adjusted the CDF of biased values according to the CDF of reference data. This has significantly reduced the bias from our modelled soil moisture. Figure 9, shows the relative soil moisture values estimated from Sentinel-1A images before and after the CDF correction.

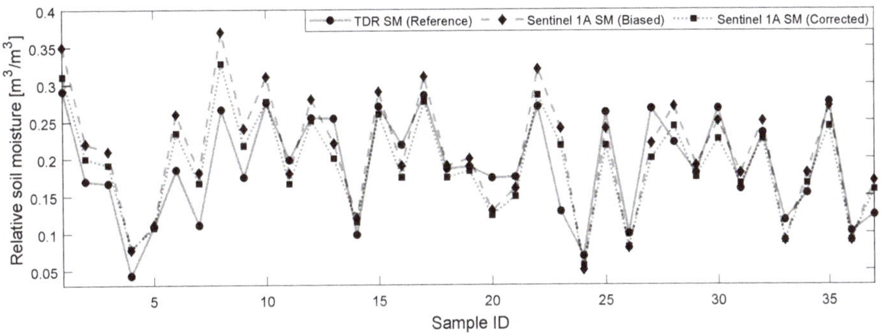

Figure 9. Relative soil moisture plotted at their corresponding sample location ID. Difference between the measured (solid line) and modelled (dotted lines) soil moisture is shown before and after the bias correction.

We observed, at some locations the modelled soil moisture is overestimated or underestimated with respect to the measured values. The underestimation is probably related to low backscattering coefficients from relatively smooth surfaces. Similarly, the overestimation of soil moisture could be associated with the high surface roughness resulting in high backscattering coefficients.

Though Sentinel-1A images provide a first-order estimation of soil moisture, modelling soil moisture from satellite images has substantial limitations and challenges. Most of the backscatter models used for the estimation of the relative soil permittivity from SAR backscatter are only valid for a specific range of soil moisture. For example, the modified Dubois model is valid for soil moisture in a range between 0 and 0.35 m^3/m^3. Microwave SAR images with appropriate backscatter models can estimate soil moisture only upto a few cms below the soil surface. In our case, we have estimated the soil moisture from the top about 5 cm below the ground surface. Further, many of the existing backscatter models are applicable for specific landuse and landcover classes. For example, Oh and water cloud models are applicable for barren and vegetated lands, respectively. The modified Dubois model can be applied on both barren and sparsely vegetated (NDVI $\leq$ 0.4) land. In summary, modified Dubois model performs well over the semi-arid region on agriculture and barren land. The performance is further improved when the bias correction method is used.

Moreover, estimation of soil moisture from C-band SAR is sensitive to various parameters such as; the incidence angle, polarisation, vegetation height and vegetation indices, i.e., Leaf Area Index (LAI). Ulaby [13] observed that in a typical agricultural field (LAI > 0.5), the effect of vegetation is more prominent in the radar backscatter compared to other target properties.

As a consequence, when the vegetation effect starts dominating, it complicates the soil moisture retrieval. The backscatter values results from the combined signature of vegetation and underlying soil water [78]. In such condition we can not ignore the attenuation due to vegetation. To minimise the effect of vegetation, scattering model such as the WCM can be implemented [79–83].

6. Conclusions

We have shown that the modified Dubois model provides a good estimate (R^2 = 0.75) of soil moisture in a region of heterogeneous land cover. The modelled soil moisture is subject to error due to the model (systematic) and sampling (random) errors. We have shown that the systematic error (also called systematic bias) can be largely minimised by the CDF matching technique. Whereas random errors can be reduced by increasing the number of sample size.

We found that the VV polarisation of Sentinel-1A is suitable for soil moisture monitoring. This is mainly because the VV polarisation is more sensitive to the soil contribution. In contrast, VH polarisation is more sensitive to the volume scattering, and it describes the vegetation contribution more effectively. Since our model is only for VV or HH, we have used the VV polarisation based on the polarisation of the Sentinel-1A data.

Further, the potential of VV and HH polarisation of Sentinel-1A may be evaluated, especially for the models (i.e., WCM) that can estimate soil moisture in diverse vegetation condition. Finally, our first-order analysis calls for a more detailed study on soil moisture modelling from Sentinel-1A images in diverse soil, landuse and landcover conditions.

This study is a step towards monitoring the surface soil moisture at higher spatial and temporal resolution from remote sensing data. Our methodology can be used to predict and monitor meteorological droughts, agricultural productivity and managing water resources in the region.

Author Contributions: Conceptualization, A.S. and K.G.; methodology, A.S., G.K.M. and K.G.; software, A.S. and K.G.; validation, G.K.M., A.S. and K.G.; formal analysis, A.S. and G.K.M.; investigation, K.G. and A.S.; resources, K.G. and S.K.; data curation, G.K.M., A.S. and K.G.; writing—original draft preparation, A.S. and K.G.; writing—review and editing, K.G. and A.S.; visualization, K.G. and S.K.; supervision and project administration, K.G.; funding acquisition, K.G. and S.K. All authors have read and agreed to the published version of the manuscript.

Funding: This research was funded by the Space Applications Centre (SAC-ISRO) under NASA-ISRO Synthetic Aperture Radar (NISAR) mission through grant Hyd-01.

Acknowledgments: We would like to acknowledge IISER Bhopal for providing institutional support. Abhilash Singh PhD is supported by the Department of Science and Technology (DST), Government of India through DST-INSPIRE fellowship. We gratefully acknowledge Prof. S.K. Tandon for fruitful suggestions. We thanks to the editor and all the three anonymous reviewers for providing helpful comments and suggestions.

Conflicts of Interest: The authors declare no conflict of interest.

Appendix A

Theta probe (ML3 sensor) requires an intensive soil specific and a sensor specific calibration before being used for data acquisition in the field. The soil specific calibration explores a functional relationship between the relative soil permittivity and soil moisture. In contrast, the sensor specific calibration explores the functional relationship between the relative soil permittivity and ML3 output (volts). Both the soil and sensor specific calibration equations are combined to finally convert the sensor output directly into soil moisture. The soil specific calibration allows to find the two constants (a_0, a_1) of Equation (A1).

$$\sqrt{\epsilon} = a_0 + a_1 \cdot m_v \tag{A1}$$

For sensor calibration, ML3 sensor measures the bulk relative soil permittivity using the empirical relation given by Equation (A2);

$$\sqrt{\epsilon} = 1.0 + 6.175 \cdot V + 6.303 \cdot V^2 - 73.578 \cdot V^3 + 183.44 \cdot V^4 - 184.78 \cdot V^5 + 68.017 \cdot V^6 \tag{A2}$$

where, V is the voltage in Volt. Combining soil and sensor specific calibration equation reads (Equation (A3));

$$m_v = \frac{\sqrt{\epsilon} - a}{a_1} \tag{A3}$$

Before the field campaign, we have calibrated the theta probe at three different locations in the study area (Figure A1).

Figure A1. Soil samples used from three different fields for the calibration of MLT-3 theta probe. Samples from S1 and S3 corresponds to agricultural lands composed of black cotton soil and rich in organic matters. S2 consists of black soil with granules rubble (about 20–25%) formed due to the subsequent weathering of Deccan basalt.

The corresponding values of the calibration constant is given Table A1.

Table A1. Field derived calibration constants for MLT-3 theta probe.

Measuring Unit	Site 1 (S1)	Site 2 (S2)	Site 3 (S3)
L_s (cm^3)	563	563	563
W_w (gm)	491	505	504
V_w (mV)	342	333	350
W_0 (gm)	410	415	421
V_0 (mV)	101	72	90
a_0	1.6	1.5	1.6
a_1	7.1	7.4	7.6

To calculate a_0 and a_1, we have measured two voltage from Equation (A2). V_0 and V_w correspond to the dry and wet samples respectively. We have also measured wet weight (W_w) and dry weight (W_0) of the samples.

Calculation of a_0: For dry soil (i.e., m_v=0), Equation (A1), is reduced to $\sqrt{\epsilon_0} = a_0$. Substituting the value of V_0 in Equation (A2), we have calculated the value of $\sqrt{\epsilon_0}$. Eventually we calculated a_0 according to;

$$a_0 = \sqrt{\epsilon_0} \tag{A4}$$

Calculation of a_1: For wet soil, we calculated the soil moisture as;

$$m_v = \frac{W_w - W_0}{L_s} \tag{A5}$$

where, L_s is the volume occupied by the sample in the beaker. By substituting the value of V_m in Equation (A2), we have calculated the values of $\sqrt{\epsilon_m}$. Finally a_1 is calculated according to;

$$a_1 = \frac{\sqrt{\epsilon_w} - \sqrt{\epsilon_0}}{m_v} \tag{A6}$$

References

1. Kite, G.; Pietroniro, A. Remote sensing applications in hydrological modelling. *Hydrol. Sci. J.* **1996**, *41*, 563–591. [CrossRef]
2. Jackson, T.J. Measuring surface soil moisture using passive microwave remote sensing. *Hydrol. Process.* **1993**, *7*, 139–152. [CrossRef]
3. Entekhabi, D. Retrieval of Soil Moisture Profile by Combined Remote Sensing and Modeling. 1993. Available online: https://trs.jpl.nasa.gov/handle/2014/34543 (accessed on 11 May 2020).
4. Jackson, T.; Hawley, M.; O'neill, P. Preplanting soil moisture using passive microwave sensor. *JAWRA J. Am. Water Resour. Assoc.* **1987**, *23*, 11–19. [CrossRef]
5. Martin, M.E.; Aber, J.D. High spectral resolution remote sensing of forest canopy lignin, nitrogen, and ecosystem processes. *Ecol. Appl.* **1997**, *7*, 431–443. [CrossRef]
6. Sivapalan, M.; Takeuchi, K.; Franks, S.; Gupta, V.; Karambiri, H.; Lakshmi, V.; Liang, X.; McDonnell, J.; Mendiondo, E.; O'connell, P. IAHS Decade on Predictions in Ungauged Basins (PUB), 2003–2012: Shaping an exciting future for the hydrological sciences. *Hydrol. Sci. J.* **2003**, *48*, 857–880. [CrossRef]
7. Field, C.B.; Randerson, J.T.; Malmström, C.M. Global net primary production: Combining ecology and remote sensing. *Remote Sens. Environ.* **1995**, *51*, 74–88. [CrossRef]
8. Seneviratne, S.I.; Corti, T.; Davin, E.L.; Hirschi, M.; Jaeger, E.B.; Lehner, I.; Orlowsky, B.; Teuling, A.J. Investigating soil moisture–climate interactions in a changing climate: A review. *Earth-Sci. Rev.* **2010**, *99*, 125–161. [CrossRef]
9. Kerr, J.T.; Ostrovsky, M. From space to species: Ecological applications for remote sensing. *Trends Ecol. Evol.* **2003**, *18*, 299–305. [CrossRef]
10. Mulla, D.J. Twenty five years of remote sensing in precision agriculture: Key advances and remaining knowledge gaps. *Biosyst. Eng.* **2013**, *114*, 358–371. [CrossRef]
11. Walker, J.P. Estimating Soil Moisture Profile Dynamics from Near-Surface Soil Moisture Measurements and Standard Meteorological Data. Ph.D. Thesis, University of Newcastle, Callaghan, Australia, 1999.
12. Narvekar, P.S.; Entekhabi, D.; Kim, S.; Njoku, E.G. Soil Moisture Retrieval Using L-Band Radar Observations. *IEEE Trans. Geosci. Remote Sens.* **2015**, *53*, 3492–3506. [CrossRef]
13. Ulaby, F.T. Microwave remote sensing active and passive. In *Rader Remote Sensing and Surface Scattering and Emission Theory*; NASA Technical Report; Kansas University: Lawrence, KS, USA, 1982; pp. 848–902.
14. Altese, E.; Bolognani, O.; Mancini, M.; Troch, P.A. Retrieving soil moisture over bare soil from ERS 1 synthetic aperture radar data: Sensitivity analysis based on a theoretical surface scattering model and field data. *Water Resour. Res.* **1996**, *32*, 653–661. [CrossRef]
15. Verhoest, N.E.; Lievens, H.; Wagner, W.; Álvarez-Mozos, J.; Moran, M.S.; Mattia, F. On the soil roughness parameterization problem in soil moisture retrieval of bare surfaces from synthetic aperture radar. *Sensors* **2008**, *8*, 4213–4248. [CrossRef]
16. Barrett, B.W.; Dwyer, E.; Whelan, P. Soil moisture retrieval from active spaceborne microwave observations: An evaluation of current techniques. *Remote Sens.* **2009**, *1*, 210–242. [CrossRef]
17. Oh, Y.; Sarabandi, K.; Ulaby, F.T. An empirical model and an inversion technique for radar scattering from bare soil surfaces. *IEEE Trans. Geosci. Remote Sens.* **1992**, *30*, 370–381. [CrossRef]
18. Oh, Y.; Sarabandi, K.; Ulaby, F.T. An inversion algorithm for retrieving soil moisture and surface roughness from polarimetric radar observation. In Proceedings of the IEEE International Geoscience and Remote Sensing Symposium, Pasadena, CA, USA, 8–12 August 1994; Volume 3, pp. 1582–1584.
19. Dubois, P.C.; Van Zyl, J.; Engman, T. Measuring soil moisture with imaging radars. *IEEE Trans. Geosci. Remote Sens.* **1995**, *33*, 915–926. [CrossRef]
20. Oh, Y.; Sarabandi, K.; Ulaby, F.T. Semi-empirical model of the ensemble-averaged differential Mueller matrix for microwave backscattering from bare soil surfaces. *IEEE Trans. Geosci. Remote Sens.* **2002**, *40*, 1348–1355. [CrossRef]
21. Oh, Y. Quantitative retrieval of soil moisture content and surface roughness from multipolarized radar observations of bare soil surfaces. *IEEE Trans. Geosci. Remote Sens.* **2004**, *42*, 596–601. [CrossRef]
22. Attema, E.; Ulaby, F.T. Vegetation modeled as a water cloud. *Radio Sci.* **1978**, *13*, 357–364. [CrossRef]
23. Fung, A.K.; Li, Z.; Chen, K.S. Backscattering from a randomly rough dielectric surface. *IEEE Trans. Geosci. Remote Sens.* **1992**, *30*, 356–369. [CrossRef]

24. Fung, A.K.; Cheng, K.S. *Microwave Scattering and Emission Models and Their Applications*; Artech House Inc.: Boston, MA, USA, 1994.
25. Baghdadi, N.; Cresson, R.; El Hajj, M.; Ludwig, R.; La Jeunesse, I. Soil parameters estimation over bare agriculture areas from C-band polarimetric SAR data using neural networks. *Hydrol. Earth Syst. Sci. Discuss.* **2012**, *9*, 2897–2933. [CrossRef]
26. El Hajj, M.; Baghdadi, N.; Zribi, M.; Belaud, G.; Cheviron, B.; Courault, D.; Charron, F. Soil moisture retrieval over irrigated grassland using X-band SAR data. *Remote Sens. Environ.* **2016**, *176*, 202–218. [CrossRef]
27. Dabrowska-Zielinska, K.; Musial, J.; Malinska, A.; Budzynska, M.; Gurdak, R.; Kiryla, W.; Bartold, M.; Grzybowski, P. Soil moisture in the Biebrza Wetlands retrieved from Sentinel-1 imagery. *Remote Sens.* **2018**, *10*, 1979. [CrossRef]
28. Wagner, W.; Pathe, C.; Doubkova, M.; Sabel, D.; Bartsch, A.; Hasenauer, S.; Blöschl, G.; Scipal, K.; Martínez-Fernández, J.; Löw, A. Temporal stability of soil moisture and radar backscatter observed by the Advanced Synthetic Aperture Radar (ASAR). *Sensors* **2008**, *8*, 1174–1197. [CrossRef] [PubMed]
29. Zribi, M.; Muddu, S.; Bousbih, S.; Al Bitar, A.; Tomer, S.K.; Baghdadi, N.; Bandyopadhyay, S. Analysis of L-band SAR data for soil moisture estimations over agricultural areas in the tropics. *Remote Sens.* **2019**, *11*, 1122. [CrossRef]
30. Yang, L.; Feng, X.; Liu, F.; Liu, J.; Sun, X. Potential of soil moisture estimation using C-band polarimetric SAR data in arid regions. *Int. J. Remote Sens.* **2019**, *40*, 2138–2150. [CrossRef]
31. El Hajj, M.; Baghdadi, N.; Zribi, M.; Bazzi, H. Synergic use of Sentinel-1 and Sentinel-2 images for operational soil moisture mapping at high spatial resolution over agricultural areas. *Remote Sens.* **2017**, *9*, 1292. [CrossRef]
32. Bousbih, S.; Zribi, M.; El Hajj, M.; Baghdadi, N.; Lili-Chabaane, Z.; Gao, Q.; Fanise, P. Soil moisture and irrigation mapping in A semi-arid region, based on the synergetic use of Sentinel-1 and Sentinel-2 data. *Remote Sens.* **2018**, *10*, 1953. [CrossRef]
33. Qiu, J.; Crow, W.T.; Wagner, W.; Zhao, T. Effect of vegetation index choice on soil moisture retrievals via the synergistic use of synthetic aperture radar and optical remote sensing. *Int. J. Appl. Earth Obs. Geoinf.* **2019**, *80*, 47–57. [CrossRef]
34. Hachani, A.; Ouessar, M.; Paloscia, S.; Santi, E.; Pettinato, S. Soil moisture retrieval from Sentinel-1 acquisitions in an arid environment in Tunisia: Application of Artificial Neural Networks techniques. *Int. J. Remote Sens.* **2019**, *40*, 9159–9180. [CrossRef]
35. Ezzahar, J.; Ouaadi, N.; Zribi, M.; Elfarkh, J.; Aouade, G.; Khabba, S.; Er-Raki, S.; Chehbouni, A.; Jarlan, L. Evaluation of Backscattering Models and Support Vector Machine for the Retrieval of Bare Soil Moisture from Sentinel-1 Data. *Remote Sens.* **2020**, *12*, 72. [CrossRef]
36. Hosseini, M.; McNairn, H. Using multi-polarization C-and L-band synthetic aperture radar to estimate biomass and soil moisture of wheat fields. *Int. J. Appl. Earth Obs. Geoinf.* **2017**, *58*, 50–64. [CrossRef]
37. Dave, R.; Kumar, G.; Kr. Pandey, D.; Khan, A.; Bhattacharya, B. Evaluation of modified Dubois model for estimating surface soil moisture using dual polarization RISAT-1 C-band SAR data. *Geocarto Int.* **2019**, 1–11. doi:10.1080/10106049.2019.1655801. [CrossRef]
38. Alexakis, D.D.; Mexis, F.D.K.; Vozinaki, A.E.K.; Daliakopoulos, I.N.; Tsanis, I.K. Soil moisture content estimation based on Sentinel-1 and auxiliary earth observation products. A hydrological approach. *Sensors* **2017**, *17*, 1455. [CrossRef]
39. Wagner, W.; Sabel, D.; Doubkova, M.; Bartsch, A.; Pathe, C. The potential of Sentinel-1 for monitoring soil moisture with a high spatial resolution at global scale. In Proceedings of the Earth Observation and Water Cycle Science, Frascati, Italy, 18–20 November 2009.
40. Bauer-Marschallinger, B.; Freeman, V.; Cao, S.; Paulik, C.; Schaufler, S.; Stachl, T.; Modanesi, S.; Massari, C.; Ciabatta, L.; Brocca, L.; et al. Toward global soil moisture monitoring with Sentinel-1: Harnessing assets and overcoming obstacles. *IEEE Trans. Geosci. Remote Sens.* **2018**, *57*, 520–539. [CrossRef]
41. Rao, S.S.; Das, S.; Nagaraju, M.; Venugopal, M.; Rajankar, P.; Laghate, P.; Reddy, M.S.; Joshi, A.; Sharma, J. Modified Dubois model for estimating soil moisture with dual polarized SAR data. *J. Indian Soc. Remote Sens.* **2013**, *41*, 865–872.
42. Thanabalan, P.; Vidhya, R. A synergistic approach for soil moisture estimation using modified Dubois model with dual-polarized SAR and optical satellite data. *Proc. SPIE* **2016**, *9877*. [CrossRef]

43. Mirsoleimani, H.R.; Sahebi, M.R.; Baghdadi, N.; El Hajj, M. Bare soil surface moisture retrieval from sentinel-1 SAR data based on the calibrated IEM and Dubois models using neural networks. *Sensors* **2019**, *19*, 3209. [CrossRef]
44. NRSC. District and Category Wise Distribution of Land Use and Land Cover in Madhya Pradesh (2015-16). Available online: https://bhuvan-app1.nrsc.gov.in/2dresources/thematic/LULC503/MAP/MP.pdf (accessed on 25 May 2020).
45. Ministry of Water Resources, Central Ground Water Board, North Central Region, Bhopal. District Ground Water Information Booklet. Available online: http://cgwb.gov.in/District_Profile/MP/Bhopal.pdf (accessed on 25 May 2020).
46. Central Ground Water Board, Ministry of Water Resources, River Development and Ganga Rejuvenation, Government of India. Aquifer Mapping Report Phanda Block, Bhopal District, Madhya Pradesh. Available online: http://cgwb.gov.in/AQM/NAQUIM_REPORT/MP/PhandaBlockBhopal.pdf (accessed on 25 May 2020).
47. Mishra, V.; Shah, R.; Garg, A. *Climate Change in Madhya Pradesh: Indicators, Impacts and Adaptation*; Indian Institute of Management Ahmadabad: Ahmedabad, India, 2016.
48. Pai, D.; Sridhar, L.; Rajeevan, M.; Sreejith, O.; Satbhai, N.; Mukhopadhyay, B. Development of a new high spatial resolution (0.25 × 0.25) long period (1901–2010) daily gridded rainfall data set over India and its comparison with existing data sets over the region. *Mausam* **2014**, *65*, 1–18.
49. Indian Meteorological Department. Rainfall Data. Available online: http://www.imdpune.gov.in/Clim_Pred_LRF_New/Grided_Data_Download.html (accessed on 25 May 2020).
50. Hayes, M.J.; Svoboda, M.D.; Wiihite, D.A.; Vanyarkho, O.V. Monitoring the 1996 drought using the standardized precipitation index. *Bull. Am. Meteorol. Soc.* **1999**, *80*, 429–438. [CrossRef]
51. Singh, A.; Meena, G.K.; Kumar, S.; Gaurav, K. Analysis of the effect of incidence angle and moisture content on the penetration depth of L-and S-band SAR signals into the ground surface. *ISPRS Ann. Photogramm. Remote Sens. Spat. Inf. Sci.* **2018**, *4*, 197–202. [CrossRef]
52. Singh, A.; Meena, G.K.; Kumar, S.; Gaurav, K. Evaluation of the Penetration Depth of L-and S-Band (NISAR mission) Microwave SAR Signals into Ground. In Proceedings of the 2019 URSI Asia-Pacific Radio Science Conference (AP-RASC), New Delhi, India, 9–15 March 2019.
53. Potin, P.; Rosich, B.; Grimont, P.; Miranda, N.; Shurmer, I.; O'Connell, A.; Torres, R.; Krassenburg, M. Sentinel-1 mission status. In Proceedings of the EUSAR 2016: 11th European Conference on Synthetic Aperture Radar, Hamburg, Germany, 6–9 June 2016; pp. 1–6.
54. Filipponi, F. Sentinel-1 GRD Preprocessing Workflow. 3rd International Electronic Conference on Remote Sensing. *Proceedings* **2019**, *18*, 11. [CrossRef]
55. Frison, P.L.; Fruneau, B.; Kmiha, S.; Soudani, K.; Dufrêne, E.; Le Toan, T.; Koleck, T.; Villard, L.; Mougin, E.; Rudant, J.P. Potential of Sentinel-1 data for monitoring temperate mixed forest phenology. *Remote Sens.* **2018**, *10*, 2049. [CrossRef]
56. Gao, Q.; Zribi, M.; Escorihuela, M.J.; Baghdadi, N. Synergetic use of Sentinel-1 and Sentinel-2 data for soil moisture mapping at 100 m resolution. *Sensors* **2017**, *17*, 1966. [CrossRef]
57. Martinis, S.; Plank, S.; Ćwik, K. The use of Sentinel-1 time-series data to improve flood monitoring in arid areas. *Remote Sens.* **2018**, *10*, 583. [CrossRef]
58. Roy, D.P.; Wulder, M.A.; Loveland, T.R.; Woodcock, C.; Allen, R.G.; Anderson, M.C.; Helder, D.; Irons, J.R.; Johnson, D.M.; Kennedy, R.; et al. Landsat-8: Science and product vision for terrestrial global change research. *Remote Sens. Environ.* **2014**, *145*, 154–172. [CrossRef]
59. Guo, M. Soil sampling and methods of analysis. *J. Environ. Qual.* **2009**, *38*, 375. [CrossRef]
60. Gardner, C.M.; Robinson, D.A.; Blyth, K.; Cooper, J.D. Soil water content. In *Soil and Environmental Analysis: Physical Methods*; Smith, K.A., Mullins, C.E., Eds.; Marcel Dekker: New York, NY, USA, 2001; pp. 1–64.
61. Reynolds, S. The gravimetric method of soil moisture determination Part I. A study of equipment, and methodological problems. *J. Hydrol.* **1970**, *11*, 258–273. [CrossRef]
62. Srivastava, H.; Patel, P.; Navalgund, R.; Sharma, Y. Retrieval of surface roughness using multi polarized Envisat-1 ASAR data. *Geocarto Int.* **2008**, *23*, 67–77. [CrossRef]
63. Topp, G.C.; Davis, J.; Annan, A.P. Electromagnetic determination of soil water content: Measurements in coaxial transmission lines. *Water Resour. Res.* **1980**, *16*, 574–582. [CrossRef]

64. Song, K.; Zhou, X.; Fan, Y. Empirically adopted IEM for retrieval of soil moisture from radar backscattering coefficients. *IEEE Trans. Geosci. Remote Sens.* **2009**, *47*, 1662–1672. [CrossRef]
65. Talone, M.; Camps, A.; Monerris, A.; Vall-llossera, M.; Ferrazzoli, P.; Piles, M. Surface topography and mixed-pixel effects on the simulated L-band brightness temperatures. *IEEE Trans. Geosci. Remote Sens.* **2007**, *45*, 1996–2003. [CrossRef]
66. Zhang, T.; Zhang, L.; Jiang, L.; Zhao, S.; Zhao, T.; Li, Y. Effects of spatial distribution of soil parameters on soil moisture retrieval from passive microwave remote sensing. *Sci. China Earth Sci.* **2012**, *55*, 1313–1322. [CrossRef]
67. Lee, J.H.; Zhao, C.; Kerr, Y. Stochastic bias correction and uncertainty estimation of satellite-retrieved soil moisture products. *Remote Sens.* **2017**, *9*, 847. [CrossRef]
68. Michelangeli, P.A.; Vrac, M.; Loukos, H. Probabilistic downscaling approaches: Application to wind cumulative distribution functions. *Geophys. Res. Lett.* **2009**, *36*, 1–6. [CrossRef]
69. Reichle, R.H. Data assimilation methods in the Earth sciences. *Adv. Water Resour.* **2008**, *31*, 1411–1418. [CrossRef]
70. Reichle, R.H.; Koster, R.D. Bias reduction in short records of satellite soil moisture. *Geophys. Res. Lett.* **2004**, *31*, 1–4. [CrossRef]
71. Draper, C.; Mahfouf, J.F.; Walker, J.P. An EKF assimilation of AMSR-E soil moisture into the ISBA land surface scheme. *J. Geophys. Res. Atmos.* **2009**, *114*, 1–13. [CrossRef]
72. Hosseini, R.; Newlands, N.K.; Dean, C.B.; Takemura, A. Statistical modeling of soil moisture, integrating satellite remote-sensing (SAR) and ground-based data. *Remote Sens.* **2015**, *7*, 2752–2780. [CrossRef]
73. Kumar, S.; Peters-Lidard, C.; Santanello, J.; Reichle, R.; Draper, C.; Koster, R.; Nearing, G.; Jasinski, M. Evaluating the utility of satellite soil moisture retrievals over irrigated areas and the ability of land data assimilation methods to correct for unmodeled processes. *Hydrol. Earth Syst. Sci.* **2015**, *19*, 4463–4478. [CrossRef]
74. Leander, R.; Buishand, T.A. Resampling of regional climate model output for the simulation of extreme river flows. *J. Hydrol.* **2007**, *332*, 487–496. [CrossRef]
75. Schmidli, J.; Frei, C.; Vidale, P.L. Downscaling from GCM precipitation: A benchmark for dynamical and statistical downscaling methods. *Int. J. Climatol.* **2006**, *26*, 679–689. [CrossRef]
76. Lenderink, G.; Buishand, A.; Deursen, W.V. Estimates of future discharges of the river Rhine using two scenario methodologies: Direct versus delta approach. *Hydrol. Earth Syst. Sci.* **2007**, *11*, 1145–1159. [CrossRef]
77. Panofsky, H.A.; Brier, G.W. *Some Applications of Statistics to Meteorology*; Pennsylvania State University: Pennsylvania, PA, USA, 1958.
78. He, B.; Xing, M.; Bai, X. A synergistic methodology for soil moisture estimation in an alpine prairie using radar and optical satellite data. *Remote Sens.* **2014**, *6*, 10966–10985. [CrossRef]
79. Bindlish, R.; Barros, A.P. Parameterization of vegetation backscatter in radar-based, soil moisture estimation. *Remote Sens. Environ.* **2001**, *76*, 130–137. [CrossRef]
80. Gherboudj, I.; Magagi, R.; Berg, A.A.; Toth, B. Soil moisture retrieval over agricultural fields from multi-polarized and multi-angular RADARSAT-2 SAR data. *Remote Sens. Environ.* **2011**, *115*, 33–43. [CrossRef]
81. Du, J.; Shi, J.; Sun, R. The development of HJ SAR soil moisture retrieval algorithm. *Int. J. Remote Sens.* **2010**, *31*, 3691–3705. [CrossRef]
82. Lievens, H.; Verhoest, N.E. On the retrieval of soil moisture in wheat fields from L-band SAR based on water cloud modeling, the IEM, and effective roughness parameters. *IEEE Geosci. Remote Sens. Lett.* **2011**, *8*, 740–744. [CrossRef]
83. De Roo, R.D.; Du, Y.; Ulaby, F.T.; Dobson, M.C. A semi-empirical backscattering model at L-band and C-band for a soybean canopy with soil moisture inversion. *IEEE Trans. Geosci. Remote Sens.* **2001**, *39*, 864–872. [CrossRef]

remote sensing

MDPI

Article

Evaluation of Different Radiative Transfer Models for Microwave Backscatter Estimation of Wheat Fields

Thomas Weiß *, Thomas Ramsauer , Alexander Löw and Philip Marzahn

Department of Geography, Ludwig-Maximilians-Universität München, Luisenstraße 37, 80333 Munich, Germany; t.ramsauer@iggf.geo.uni-muenchen.de (T.R.); alexander.loew@lmu.de (A.L.); p.marzahn@iggf.geo.uni-muenchen.de (P.M.)

* Correspondence: weiss.thomas@lmu.de

Received: 28 July 2020; Accepted: 10 September 2020; Published: 17 September 2020

Abstract: This study aimed to analyze existing microwave surface (Oh, Dubois, Water Cloud Model "WCM", Integral Equation Model "IEM") and canopy (Water Cloud Model "WCM", Single Scattering Radiative Transfer "SSRT") Radiative Transfer (RT) models and assess advantages and disadvantages of different model combinations in terms of VV polarized radar backscatter simulation of wheat fields. The models are driven with field measurements acquired in 2017 at a test site near Munich, Germany. As vegetation descriptor for the canopy models Leaf Area Index (LAI) was used. The effect of empirical model parameters is evaluated in two different ways: (a) empirical model parameters are set as static throughout the whole time series of one growing season and (b) empirical model parameters describing the backscatter attenuation by the canopy are treated as non-static in time. The model results are compared to a dense Sentinel-1 C-band time series with observations every 1.5 days. The utilized Sentinel-1 time series comprises images acquired with different satellite acquisition geometries (different incidence and azimuth angles), which allows us to evaluate the model performance for different acquisition geometries. Results show that total LAI as vegetation descriptor in combination with static empirical parameters fit Sentinel-1 radar backscatter of wheat fields only sufficient within the first half of the vegetation period. With the saturation of LAI and/or canopy height of the wheat fields, the observed increase in Sentinel-1 radar backscatter cannot be modeled. Probable cause are effects of changes within the grains (both structure and water content per leaf area) and their influence on the backscatter. However, model results with LAI and non-static empirical parameters fit the Sentinel-1 data well for the entire vegetation period. Limitations regarding different satellite acquisition geometries become apparent for the second half of the vegetation period. The observed overall increase in backscatter can be modeled, but a trend mismatch between modeled and observed backscatter values of adjacent time points with different acquisition geometries is observed.

Keywords: Oh; Dubois; IEM; WCM; SSRT; SAR; soil moisture; LAI; wheat; Sentinel-1

1. Introduction

Soil moisture plays an important role in land surface processes, such as water and energy fluxes. Therefore, soil moisture is a key variable in scientific fields like climatology, hydrology, meteorology, or agriculture [1,2]. In recent decades, microwave data has proven to be a suitable tool for long-term soil moisture derivation of large areas and different land cover types [3–9]. The retrieved soil moisture information is widely used in applications, like climate modeling, precision farming, water management, flood forecast, and drought monitoring [10–14]. With different available Synthetic Aperture Radar (SAR) data from different sensors and for different usage in terms of absolute accuracy and spatial scale, various soil moisture retrieval approaches, like change detection, microwave data fusion (active and passive), differential Synthetic Aperture Radar (SAR) interferometry, or

SAR polarimetry, are available [15]. Furthermore, land surface parameters, like soil moisture, can be also derived by using Radiative Transfer (RT) models. Starting in 1974 [16], with the first publication examining radar response and soil moisture [17], hundreds of different studies developing and/or analyzing new or existing RT models have been conducted. RT models try to simulate the interaction of the radar wave with the soil and the vegetation to derive different soil and vegetation parameters [18]. Complexities in RT models for surface backscatter calculations range from simple empirical regression-based models [18–21] and different empirical models based on the Water Cloud approach (WCM surface part) [22–25], over semi-empirical models from Oh (Oh92, Oh04) [26,27] or Dubois (Dubois95), [28] to physical-based models, like the Integral Equation Model (IEM) in its original form [29] or refined versions [30–32]. Common RT models for canopy backscatter calculations range from empirical models, like Water Cloud Model (WCM canopy part) [22], to more sophisticated and multi-layered models, like the Michigan Canopy Scattering Model (MIMICS) [33], Tor Vergata model [34], Single Scattering Radiative Transfer (SSRT) models described by De Roo [35] or Ulaby [17], or a first order scattering model from Quast [36,37]. Despite the large numbers of existing models, there is still the need of an algorithm generating soil moisture maps with acceptable accuracy of 3–4% [17].

So far, several studies have been carried out to test and compare pure surface RT models [15,38–42]. Research analyzing radar backscatter calculations and soil moisture retrieval approaches of combined surface and canopy RT models has been performed. For these studies different test sites, land cover types, and vegetation descriptors were used [25,43–55]. Investigations on how different vegetation descriptors, like Leaf Area Index (LAI), Vegetation Water Content (VWC), Leaf Water Area Index (LWAI), normalized Plant Water Index (PWI), or Normalized Different Water Index (NDWI), affect soil moisture retrievals have been carried out, as well [25,48,53,54]. In this context, synergistic retrieval approaches by using vegetation descriptors derived from optical sensors as input data for microwave RT models have been published more and more in recent years [45,52,56–61].

Despite the existing analyses, a study testing and comparing different surface and canopy RT model combinations with focus on the interaction between surface and canopy part and advantages or disadvantages of the model combinations is missing. The launch of Sentinel-1A/B, and, therefore, the availability of free SAR data with high temporal and spatial coverage, constitutes a suitable basis for such an analysis. Investigations of the usage of dense Sentinel-1 time series with observations up to every 1.5 days in terms of future synergistic retrieval approaches of SAR and optical data are needed. The models and the knowledge gained from this paper shall be used within a newly developed platform called MULTIPLY, which combines data from different optical and microwave satellites by using state-of the-art RT models within a data assimilation framework to consistently acquire and interpret different land surface parameters.

This study was performed on time series data with high temporal and spatial (field scale) resolution. The surface RT models WCM, Oh92, Oh04, Dubois95, and IEM were coupled with canopy models WCM or SSRT. With these model combinations, VV polarized backscatter values for an entire vegetation period of different wheat fields was calculated. The different input variables for the model combinations, such as soil moisture, canopy height, LAI, or soil properties, were provided by field measurements. For other parameters, like surface roughness or single scattering albedo, suitable literature values were chosen. LAI was used as vegetation descriptor because of available field measurements and its straight forward derivation from optical sensors [62–64]. The remaining empirical parameters were calibrated by comparing modeled backscatter with Sentinel-1 backscatter values. In summary, this paper aimed to

- investigate different RT model combinations and asses their advantages and disadvantages;
- evaluate if different radar acquisition geometries are modeled adequately with the used RT models; and
- serve as preliminary work for future synergistic retrieval approaches of SAR and optical sensors with a focus on high spatial and temporal resolutions.

Section 2 presents the used dataset. Section 3 summarizes the used RT models. In Section 4, calibration and validation results are shown and discussed. Finally, the main conclusions are drawn in Section 5.

2. Datasets

2.1. Study Area

The study area, Munich North Isar (MNI), is located in southern Germany (Bavaria), near Munich (48°13′N–48°20′N, 11°39′E–11°45′E, Figure 1). Since 2014, almost every year from spring until autumn, different field campaigns targeting agricultural purposes were carried out [65–68]. From March to September 2017, an intensive field campaign focusing on maize and wheat fields for validation of soil and vegetation parameter retrievals from Sentinel-1, Sentinel-2 and the future EnMAP satellite was conducted. MNI is characterized by intensive agriculture with wheat, maize, and grassland as main crop types. In close vicinity (<10 km) to the test site two meteorological stations, Freising (470 m a.s.l.) and Eichenried (475 m a.s.l.) managed by the Bavarian State Research Institute (LFL) and one meteorological station Munich-airport (446 m a.s.l.) managed by the German Meteorological Service (DWD) are situated. The annual measured mean temperature for 2017 ranges between 9 °C (Freising) and 9.3 °C (Eichenried). The average annual precipitation for 2017 reached 753 mm (Munich-airport) to 853 mm (Eichenried). The used data for this study includes field campaign data of wheat fields from 2017 (Section 2.2) and Sentinel-1 satellite data (Section 2.3).

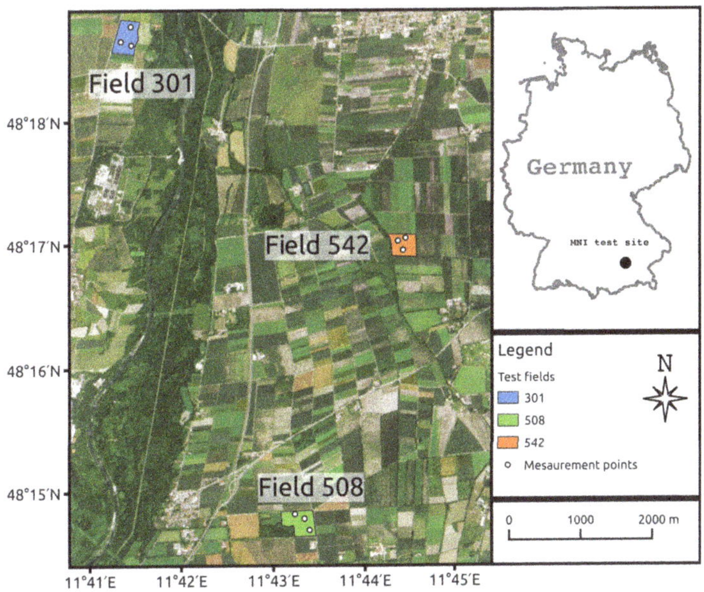

Figure 1. Overview of study area Munich-North-Isar (MNI) located in southern Germany (Bavaria). Three wheat test fields—508 (green), 542 (orange), and 301 (blue)—with three measurement points each of the field campaign in 2017 are highlighted. Reference system: WGS84 (EPSG:4326)

2.2. Field Data

During the MNI field campaign of 2017, weekly field measurements of different biophysical parameters (Table 1) were conducted. The total LAI was measured with a LI-COR Biosciences LAI-2200C device (LI-COR Biosciences Inc., Lincoln, NE, USA) as an average of 14 measurements from the same area. The measurements were taken within each test field at three different locations (Figure 1). The accuracy of LAI in terms of mean standard deviation of repeated measurements ranges within the fields between 0.45 and 0.52. The monitoring period started end of March and ended shortly before the fields were almost simultaneously harvested in mid of July. Additionally, Decagon TM5 soil moisture sensors using the capacity method were installed permanently within the first five centimeters of the soil surface. Soil moisture changes were monitored with a time interval of 10 min. Information about the soil was provided by earlier campaigns when soil samples were taken from the fields, and the soil properties were analyzed in the laboratory (Table 2). The soils bulk density with 1.45 ± 0.13 g/cm^3, and the clay content with 7.38 ± 1.8%, show no high variability between the different fields. The sand content, on the other hand, shows higher variability, with 24.08 ± 10.46%.

Table 1. Acquisition time, time interval, and range of dynamic in-situ measurements.

Variable	Acquisition Time	Time Interval	Range
Canopy height [cm]	03/24–07/17/2017	weekly	7–105
LAI	03/24–07/17/2017	weekly	0.35–6.25
Soil moisture [m^3/m^3]	03/24–07/17/2017	continues	0.09–0.38

Table 2. Laboratory results for sand, clay, and bulk content of soil surface samples.

Variable	Time Interval	Mean	Std
Soil sand content [%]	once (several locations)	24.08	10.46
Soil clay content [%]	once (several locations)	7.38	1.80
Bulk density [g/cm^3]	once (several locations)	1.45	0.13

2.3. Satellite Data

For this study, C-band Sentinel-1 SLC data of Sentinel-1A/B was used. The Sentinel data was pre-processed with ESA's SNAP Toolbox Version 7.0.3. An overview of all applied pre-processing steps is given in Figure 2. For the geometric correction, SRTM data with 1 arc-second resolution was chosen as digital elevation model input. Afterward, the radiometric correction method of Kellndorfer et al. [69] was applied. In a second pre-processing step, a multi-temporal Lee-sigma filter was used for speckle reduction. The temporal filter was applied on each image with information of 6 other images (three before the target and three after) with a spatial window size of 5 × 5 pixels, a sigma of 0.9, and a target window size of 3 × 3 pixels. For the period of the field campaign in 2017 (March to July), in total, 78 Sentinel images covering the study area are available. Considering images with different orbit directions (ascending and descending) and different incidence angles (ranges from 35° to 45°) a revisit time of 1.5 days was archived. The spatial resolution of the processed data was 10 × 10 m. A more detailed overview of the used Sentinel-1 dataset and the image properties is given in Table 3. The primary acquisition mode of Sentinel-1 provides data with polarization VV and VH. For our study, the focus was set on polarization VV due to findings that, for retrieving soil moisture, the usage of VH alone or in addition to VV is not suitable for well-developed vegetation [45,70].

Table 3. Available Sentinel-1A/B satellite data for MNI field campaign period in 2017 (03/23–07/17/2017).

Asc./Desc.	Incidence Angle [°]	Relative Orbit	Amount	Revisit Time [Days]	Acquisition Time
Ascending	36	44	19	6	4:58 p.m.
	45	117	19	6	5:06 p.m.
Descending	43	95	20	6	5:17 a.m.
	35	168	20	6	5:25 a.m.

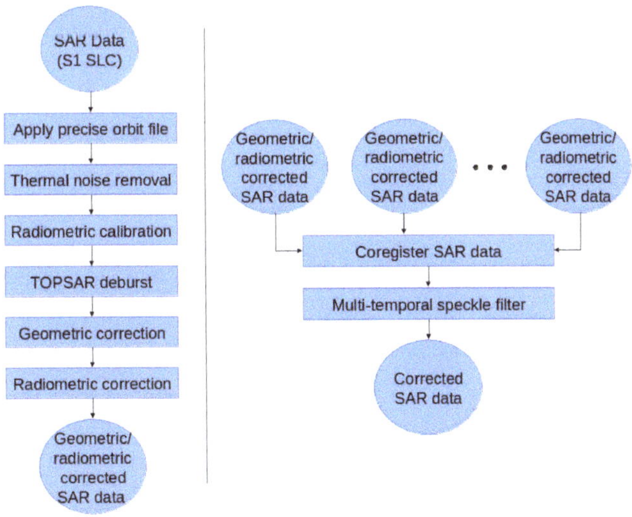

Figure 2. Schematic overview of SNAP pre-processing steps to retrieve geometric and radiometric corrected images from Sentinel-1 SLC data (**left**). Pre-processing steps for speckle reduction using a multi-temporal speckle filter (**right**).

3. Microwave Radiative Transfer Models

3.1. Surface RT Models

3.1.1. Empirical Water Cloud Model (WCM Surface Part)

The WCM, often referred to as the tau-omega model, was developed by Attema and Ulaby in 1978 [22]. For a given polarization pq ($pq = HH$, VV, or HV), the surface contribution $\sigma^0_{s_{pq}}$ of WCM to the backscattered radar signal in dB scale is defined as

$$\sigma^0_{s_{pq}} = C_{pq} + D_{pq} \cdot mv, \tag{1}$$

with empirical fitted soil parameters C_{pq} and D_{pq} and soil moisture content mv. C_{pq} is an empirical calibration constant, whereas D_{pq}, as a calibration factor, indicates the sensitivity of soil moisture on the received radar signal. The WCM surface part is a purely empirical model thus no additional information about surface roughness or incidence angle is needed. Therefore, the empirical parameters have to be calibrated for each test site separately.

3.1.2. Semi-Empirical Oh Model 1992 (Oh92)

In 1992, Oh et al. [26] developed an approach for the retrieval of soil moisture and soil surface roughness by empirical determined functions. Based on scatterometer measurements and various ground measurements of soil moisture and surface roughness, two functions for the co- ($p = \frac{\sigma^0_{s_{HH}}}{\sigma^0_{s_{VV}}}$) and cross-polarized ($q = \frac{\sigma^0_{s_{HV}}}{\sigma^0_{s_{VV}}}$) backscatter ratios were fitted. Consequently, p and q are defined as

$$p = \frac{\sigma^0_{s_{HH}}}{\sigma^0_{s_{VV}}} = \left[1 - \left(\frac{2\theta}{\pi}\right)^{\frac{1}{3R_0}} \cdot e^{-ks}\right]^2 \tag{2}$$

and

$$q = \frac{\sigma^0_{s_{HV}}}{\sigma^0_{s_{VV}}} = 0.23\,\sqrt{R_0}\,(1 - e^{-ks}), \tag{3}$$

with θ as local incidence angle, k as radar wave number ($k = 2\pi/\lambda$), where λ is the wavelength, and s as rms height. R_0 is the Fresnel reflectivity coefficient at nadir given by

$$R_0 = \left|\frac{1 - \sqrt{\epsilon_r}}{1 + \sqrt{\epsilon_r}}\right|^2, \tag{4}$$

where ϵ_r is the relative dielectric constant. The VV polarized backscatter coefficient $\sigma^0_{s_{VV}}$ is further defined as

$$\sigma^0_{s_{VV}} = 0.7\left[1 - e^{-0{,}65(ks)^{1.8}}\right]\frac{cos^3\theta}{\sqrt{p}}\left[R_v(\theta) + R_h(\theta)\right], \tag{5}$$

with the Fresnel coefficients for horizontal R_h and vertical R_v polarization

$$R_h = \frac{\mu_r cos\theta - \sqrt{\mu_r\epsilon_r - sin^2\theta}}{\mu_r cos\theta + \sqrt{\mu_r\epsilon_r - sin^2\theta}}, \tag{6}$$

$$R_v = \frac{\epsilon_r cos\theta - \sqrt{\mu_r\epsilon_r - sin^2\theta}}{\epsilon_r cos\theta + \sqrt{\mu_r\epsilon_r - sin^2\theta}}, \tag{7}$$

where μ_r is the relative permittivity. Furthermore, the backscatter coefficients $\sigma^0_{s_{HH}}$ and $\sigma^0_{s_{HV}}$ are given with respect to $\sigma^0_{s_{VV}}$, p, and q by

$$\sigma^0_{s_{HH}} = p\,\sigma^0_{s_{VV}}, \tag{8}$$

$$\sigma^0_{s_{HV}} = q\,\sigma^0_{s_{VV}}. \tag{9}$$

The model in its original form can be applied for the retrieval of soil moisture or soil surface roughness for bare soil conditions at several frequencies (X- to L-Band) and a broad range of incidence angles (10–70°). Because it is a semi-empirical model, the validity range of the model in terms of soil moisture and soil surface roughness is defined with $0.1 < ks < 6$, and 9 Vol.% $< mv <$ 31 Vol.%.

3.1.3. Semi-Empirical Oh Model 2004 (Oh04)

In 2004, Oh [27] revised and simplified his original approach to use only soil moisture (mv) as an independent variable rather than R_0 and ϵ_r (Section 3.1.2). Thus, if using mv as input variable, no additional information about the soil properties (bulk density, sand and clay content) is needed. The model is defined by

$$p = \frac{\sigma^0_{s_{HH}}}{\sigma^0_{s_{VV}}} = 1 - \left(\frac{2\theta}{\pi}\right)^{0.35\,mv^{-0.63}} \cdot e^{-0.4\,(ks)^{1.4}}, \tag{10}$$

$$q = \frac{\sigma^0_{s_{HV}}}{\sigma^0_{s_{VV}}} = 0.095\,(0.13 + sin^{1.5}\theta)^{1.4}\,[1 - e^{-1.3\,(ks)^{0.9}}], \tag{11}$$

$$\sigma^0_{s_{HV}} = 0.11\,mv^{0.7}\,(cos\theta)^{2.2}\,[1 - e^{-0.32\,(ks)^{1.8}}]. \tag{12}$$

Oh04 is optimized for bare soils with $0.13 < ks < 6.98$, 4 Vol.% $< mv <$ 29.1 Vol.% and $10° < \theta < 70°$.

3.1.4. Semi-Empirical Dubois Model (Dubois95)

Dubois et al. [28] proposed, in 1995, an empirical approach to determine backscatter values for HH and VV polarizations based on soil moisture, soil surface roughness, and system parameters, like local incidence angle, wavelength, and frequency. Two non-linear equations were fitted to measured backscatter values obtained by a scatterometer for a broad range of frequencies ranging from 2.5 GHz to 11 GHz and incidence angles ranging from 30° to 60°. The backscatter values can be calculated by

$$\sigma^0_{s_{HH}} = 10^{-2.75} \frac{cos^{1.5}\theta}{sin^5\theta} 10^{\,0.028\,\epsilon_r\,tan\theta} (ks \cdot sin\theta)^{1.4} \lambda^{0.7}, \tag{13}$$

$$\sigma^0_{s_{VV}} = 10^{-2.37} \frac{cos^3\theta}{sin^3\theta} 10^{\,0.046\,\epsilon_r\,tan\theta} (ks \cdot sin\theta)^{1.1} \lambda^{0.7}. \tag{14}$$

The Dubois model was optimized for bare soil conditions and has a validity range for soil moisture of $mv \leq 35$ Vol.% and soil surface roughness of $ks \leq 2.5$.

3.1.5. Physical Integral Equation Model (IEM)

The IEM is a theoretical backscattering model and was developed by Fung et al. [29] in 1992. Since then, Fung and colleagues extended the IEM to bistatic scattering [71]. The general co-polarized backscatter coefficient σ^0_{pp} for $pp = VV$ or HH is defined as

$$\sigma^0_{s_{pp}} = \frac{k^2}{4\pi} e^{k^2s^2cos^2\theta} \sum_{n=1}^{\infty} |I^n_{pp}|^2 \frac{W^n\,(2ksin\theta, 0)}{n!}, \tag{15}$$

where I^n_{pp} is

$$I^n_{pp} = (2\,ks\,cos\theta)^n f_{pp} e^{-k^2s^2cos^2\theta} + (ks\,cos\theta)^n\,F_{pp}, \tag{16}$$

with W^n as the Fourier transform of the nth power of the surface correlation function $p(x,y)$. Furthermore, the backscatter at cross polarization $\sigma^0_{s_{HV}}$ is calculated as

$$\begin{aligned}\sigma^0_{s_{HV}} = {} & \frac{k^2}{16\pi} e^{-2k^2s^2cos^2\theta} \sum_{n=1}^{\infty}\sum_{m=1}^{\infty} \frac{(k^2s^2cos^2\theta)^{n+m}}{n!m!} \\ & \int [|F_{HV}(u,v)|^2 + F_{HV}(u,v)F^*_{HV}(-u,-v)]\; W^n(u - k\,sin\theta, v)\; W^m(u + k\,sin\theta)\; dudv.\end{aligned} \tag{17}$$

The Kirchhoff coefficients f_{HH}, f_{VV} and complementary field coefficients F_{HH}, F_{VV}, F_{HV} are given as

$$f_{HH} = \frac{2R_h}{cos\theta}, \tag{18}$$

$$f_{VV} = \frac{2R_v}{cos\theta}, \tag{19}$$

$$F_{HH} = 2\frac{sin^2\theta}{cos\theta}[4R_h - (1 - \frac{1}{\epsilon_r}(1 + R_h)^2)], \tag{20}$$

$$F_{VV} = 2\frac{sin^2\theta}{cos\theta}[(1 - \frac{\epsilon cos^2\theta}{\mu_r\epsilon_r - sin^2\theta})(1 - R_v)^2 + (1 - \frac{1}{\epsilon_r})(1 + R_v)^2], \tag{21}$$

$$F_{HV}(u,c) = \frac{Uv}{kcos\theta}[\frac{8R^2}{\sqrt{k^2 - u^2 - v^2}} + \frac{-2 + 6R^2 + \frac{(1+R)^2}{\epsilon_r} + \epsilon_r(1 - R)^2}{\sqrt{\epsilon_r k^2 - u^2 - v^2}}], \tag{22}$$

with Fresnel coefficient at horizontal R_h (Equation (6)) and vertical R_v (Equation (7)) polarization thus R is described by

$$R = \frac{R_v - R_h}{2}. \tag{23}$$

The Fourier transform of the nth power of the surface correlation coefficient $W^n(a,b)$ is calculated by

$$W^n(a,b) = \frac{1}{2\pi} \int\int p^n(x,y) e^{-i(ax+by)} dxdy. \tag{24}$$

The distribution of surface correlation function $p(x, y)$ can be described for low surface roughness as exponential and high surface roughness values as Gaussian by

$$p(x,y) = e^{-(\frac{|x|+|y|}{L})} \ (exponential), \tag{25}$$

$$p(x,y) = e^{-(\frac{x^2+y^2}{L^2})} \ (Gaussian), \tag{26}$$

with L as correlation length.

3.2. Surface and Canopy RT Models

3.2.1. Empirical Water Cloud Model (WCM)

The WCM [22] with respect to surface $\sigma^0_{s_{pq}}$ and canopy $\sigma^0_{c_{pq}}$ contribution, as well as two-way attenuation T^2_{pq}, is defined as

$$\sigma^0_{pq} = \sigma^0_{c_{pq}} + T^2 \sigma^0_{s_{pq}}, \tag{27}$$

whereas the canopy part $\sigma^0_{c_{pq}}$ in linear scale and the two-way attenuation T^2_{pq} are written as

$$\sigma^0_{c_{pq}} = A_{pq} \ V_1 \ cos\theta \ (1 - T^2_{pq}), \tag{28}$$

$$T^2_{pq} = e^{-2 \ B_{pq} \ V_2 \ sec\theta}, \tag{29}$$

where θ is the local incidence angle, $V1$ and $V2$ are empirical vegetation descriptors, and A_{pq}, as well as B_{pq}, are fitted parameters of the model which depend on the vegetation properties and the radar configuration. For $\sigma^0_{s_{pq}}$ in Equation (27), each of the described surface models in Section 3.1 can be used.

3.2.2. Semi-Empirical Single Scattering Radiative Transfer (SSRT) Model

The SSRT model used by De Roo [35] and Ulaby [17] is a semi-empirical first-order scattering model. The model is defined as

$$\sigma^0_{pq} = \sigma^0_{g_{pq}} + \sigma^0_{c_{pq}} + \sigma^0_{cgt_{pq}} + \sigma^0_{gcg_{pq}}, \tag{30}$$

where

$$\sigma^0_{g_{pq}} = T_p T_q \ \sigma^0_{s_{pq}}, \tag{31}$$

with T_p and T_q as attenuation of the canopy for different polarizations and $\sigma^0_{s_{pq}}$ describing the pure surface scattering mechanism. Similar to the definition of WCM in Section 3.2.1, all surface models described in Section 3.1 can be used for calculating the surface contribution $\sigma^0_{s_{pq}}$ within SSRT. Furthermore, the p polarized one way transmittivity of the canopy T_p is defined as

$$T_p = e^{-\tau_p}, \tag{32}$$

with τ_p as the p polarized attenuation of the canopy given by

$$\tau_p = k^p_e \ H \ sec\theta, \tag{33}$$

whereas H represents the canopy height. The extinction coefficient k^p_e which accounts for the absorption and scattering losses of the electromagnetic wave through the canopy is defined as

$$k^p_e = k^p_a + k^p_s. \tag{34}$$

In general, a canopy consists of leaves, stalks, and branches with different shapes and orientations, which do not show a linear distribution in the vertical. However, in the applied SSRT, it is assumed that k_e^p, k_a^p, and k_s^p follow a uniform distribution in the vertical as a function of z within the canopy layer. In addition to the extinction coefficient, the scattering part k_s^p of k_e^p can be derived by

$$k_s^p = k_e^p\ \omega, \tag{35}$$

where ω represents the single scattering albedo. For the direct backscattering contribution of the canopy $\sigma^0_{c_{pq}}$, Attema and Ulaby's [22] water cloud approach of identical scatterers, which are uniformly distributed within the volume, is used. Thus, multiple scattering effects are ignored. As a consequence, the volume backscattering coefficient $\sigma^{back}_{V_{pq}}$ of the vegetation medium is defined as

$$\sigma^{back}_{V_{pq}} = N_v\ \sigma^{back}_{pq}, \tag{36}$$

with N_v as the number of scattering particles per unit volume and σ^{back}_{pq} as the pq polarized backscattering cross section of a single particle. Finally, the pq polarized canopy backscattering coefficient $\sigma^0_{c_{pq}}$ within Equation (30) can be obtained from

$$\sigma^0_{c_{pq}} = \frac{\sigma^{back}_{V_{pq}}\ cos\theta}{k_e^p + k_e^q}\ (1 - T_p T_q). \tag{37}$$

Furthermore, ground/canopy ($\sigma^0_{gc_{pq}}$) and canopy/ground ($\sigma^0_{cg_{pq}}$) scattering contributions are defined as

$$\sigma^0_{gc_{pq}} = \sigma^{bist}_{v_{pq}}\ H\ R_q\ T_p T_q, \tag{38}$$

$$\sigma^0_{cg_{pq}} = \sigma^{bist}_{v_{pq}}\ H\ R_p\ T_p T_q, \tag{39}$$

where H is the canopy height, $\sigma^{bist}_{v_{pq}}$ is the bi-static scattering cross section of a single leaf or stalk, and R_p describes the p polarized Fresnel reflectivity (Equations (6) and (7)). Thus, the total canopy ground contribution $\sigma^0_{cgt_{pq}}$ within Equation (30) as the sum of $\sigma^0_{gc_{pq}}$ and $\sigma^0_{cg_{pq}}$ can be written as

$$\sigma^0_{cgt_{pq}} = \sigma^{bist}_{v_{pq}}\ H\ [R_p + R_q]\ T_p T_q. \tag{40}$$

Furthermore, ground canopy ground contribution ($\sigma^0_{gcg_{pq}}$) within Equation (30) is defined as

$$\sigma^0_{gcg_{pq}} = \frac{\sigma^{back}_{V_{pq}}\ cos\theta}{k_e^p + k_e^q}\ (R_p R_q - T_p T_q). \tag{41}$$

3.3. Practical Considerations

Each described model requires a different set of input parameters. A summary for the different RT models is given in Table 4. For some parameters, field measurements (Section 2.2) or literature values (s and ω, Table 5) are used, and other site dependent parameters have to be fitted. The analyzed wheat fields were sown in autumn of 2016. By the starting point of the observation period (end of March 2017), the soil surface was already smoothed out by rain and the ground was covered by wheat plants (height > 10 cm). Marzahn et al. [72] showed that, for wheat fields with the above mentioned state, only minor changes in terms of roughness throughout the vegetation period are observable. Previous studies regarding periodic features and roughness changes [73,74] found out that changes in surface roughness due to soil rows as periodic feature within wheat fields are essential if the viewing angle is nearly perpendicular to the row orientation, but, for other viewing angles, the changes within surface roughness are negligible. The viewing angle and the row orientation of the wheat fields in our study are always <75°. Therefore, changes due to periodic soil rows are assumed to be negligible. Typical roughness measurements of various winter wheat fields suggest rms values between 1.0 and

1.3 [72,73,75–78]. With the assumption of only minor roughness changes throughout the vegetation period, a literature value for surface height s of 1.2 cm was chosen. The value of 1.2 cm was used due to former field campaign measurements of wheat fields in Germany [72,73]. For single scattering albedo ω, a common literature value of 0.03 [79] was set. The conversion from soil moisture field measurements to the required model input of dielectric constant ϵ for models Oh92, Dubois95, and IEM was performed by using a dielectric mixing model for soils after Dobson et al. [80]. The required soil information about sand and clay content, bulk density (Table 2), and soil moisture (Table 1) were provided by field measurements and laboratory results. Additionally, for some parameters, in their original form, adjustments were made within this study. In particular, DeRoo et al. [35] used for parameterizing of the extinction coefficient k_e^p a combination of an empirical parameter, vegetation water mass and vegetation height. In our study, the vegetation water mass, and the vegetation height were replaced by total LAI. Therefore, k_e^p is defined by

$$k_e^p = coef * \sqrt{LAI}, \tag{42}$$

with coef as an empirical parameter. To reduce the required parameters for model IEM, a well-established approach of Baghdadi et al. [31,45] was used. The correlation length L was replaced by a fitted parameter $Lopt$ which is dependent on s, θ, and the polarization. $Lopt$ for C-band VV polarization data and the Gaussian correlation function after Baghdadi et al. [45] is defined as

$$Lopt(s, \theta, VV) = 1.281 + 0.134(sin0.19\theta)^{-1.59s}. \tag{43}$$

The refined version of Baghdadi is hereinafter referred to as IEM_B. A schematic illustration of the RT model calibration approach is shown in Figure 3. All RT model combinations (surface + canopy) are driven by field measurements and the required empirical parameters (Table 4). The fitting of the empirical parameters was carried out by minimizing the sum of the squared error between modeled and measured (Sentinel-1) radar backscatter values. For the measured Sentinel-1 backscatter value of each measurement point shown in Figure 1, the mean backscatter of 5 × 5 pixel (50 × 50 m) around the measurement location was chosen. In a first fitting approach, all parameters shown in Table 4 were defined to be static for the entire vegetation period. In a second fitting approach, WCM parameters C, D, and A were set to be static; therefore, the mean (Table 5) of all model results of the static approach was used. The used values for the different parameters for validation are shown in Table 5. Additionally, the attenuation of the backscatter through the canopy was defined to be variable throughout the time series. More specifically, parameters $coef$ within SSRT and B within WCM were fitted for each time step individually by taking three observations before and after into account. By making only $coef$ or B variable, changes within the results can then clearly related to changes of the attenuation of the radar backscatter signal by the canopy. Field measurements used as model input parameters show multidimensionally unstructured inter- and intra-field correlations. Therefore, measurement point independency is assumed, and a validation of the parametrized RT models is performed using a leave-one-out cross-validation approach. Hereby, the parameter mean of the calibration results of eight measurement points is validated with the remaining measurement point.

Table 4. Overview of differences of surface models Water Cloud Model (WCM surface), Oh models Oh92 and Oh04, Dubois95, and Integral Equation Model refined version of Baghdadi (IEM_B), as well as canopy models Single Scattering Radiative Transfer (SSRT) and WCM canopy, in terms of type, validity range, site dependency, required input parameters, and polarization. Separation of used input parameter of the analyzed Radiative Transfer (RT) models in fitted parameters and parameters where field measurements or literature values were used as input data.

	Type	Validity Range	Site Dependent	Required Parameters		Pol.
				Fitted	Field Measurements or Literature Values	
WCM surface	empi.	$10° < \theta < 70°$	Yes	C, D	θ, mv	HH, VV VH
Oh92	semi-empi.	$10° < \theta < 70°$ $0.1 < ks < 6$ $9 < mv < 31 Vol.\%$	No		s, k, θ, ϵ $(clay, sand, bulk, mv)$	HH, VV VH
Oh04	semi-empi.	$10° < \theta < 70°$ $0.13 < ks < 6.98$ $4 < mv < 29.1 Vol.\%$	No		s, k, θ, mv	HH, VV VH
Dubois95	semi-empi.	$30° < \theta < 60°$ $ks \leq 2.5$ $mv \leq 35 Vol.\%$	No		s, k, θ, ϵ $(clay, sand, bulk, mv)$	HH, VV
IEM_B	theoretical	$10° < \theta < 70°$ $ks \leq 3$	No		$s, k, l, \theta, \epsilon$ $(clay, sand, bulk, mv)$	HH, VV VH
SSRT	semi-empi.		Yes	$ke\ (coef)$	H, LAI, θ, ω	HH, VV VH
WCM canopy	empi.		Yes	A, B	$V1\ (LAI), V2\ (LAI), \theta$	HH, VV VH

Table 5. Used model input parameters for validation.

k [cm^{-1}]	s [cm]	ω	C [dB]	D [dB]	A
1.13	1.2	0.03	−14.61	12.88	0.0029

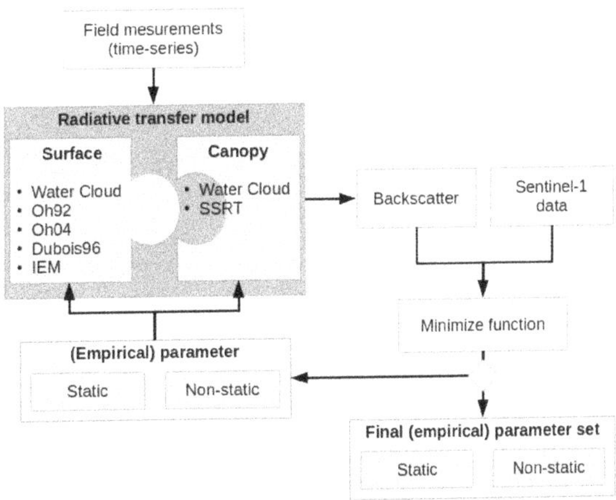

Figure 3. Schematic illustration of RT model calibration approach.

3.4. Differences Between Applied Models

WCM is a purely empirical model and depends, therefore, only on the calibrated empirical coefficients. It is applicable under almost all surface/vegetation conditions, but it has to be calibrated for different test sites separately. The transferability to other test sites and or other surface/vegetation conditions is not possible. Surface models, like Oh92, Oh04, Dubois95, and IEM, were originally only developed for bare soil and or sparse vegetation conditions. Nevertheless, different studies replaced the surface component within WCM and SSRT with Oh [17,35,43,47], Dubois95 [51,54,55], or IEM/IEM_B [17,45,53] models. One key advantage of the semi-empirical-based (Oh92, Oh04, Dubois95) or theoretical-based (IEM/IEM_B) surface models in comparison to WCM is their better transferability to other test sites and surface/vegetation conditions. Models, like Oh92, Oh04, or Dubois95, are based on a hybrid construction with experimental data guided by trends predicted by theoretical models [17]. Theoretical models, like IEM/IEM_B, on the other hand, have a theoretical foundation, whereas, for used mathematical approximations within the model, various assumptions to retrieve a analytical solution are made [17]. The Oh model of 1992 was developed based on a single experiment with information about only four different soil surfaces [26]. For the model version of 2004, Oh used information of approximately 40 bare soil fields conducted over seven experiments [27]. Furthermore, the usage of ϵ and R in model Oh92 was refined to the usage of mv in model Oh04. Model Dubois95 is the only model which was developed only for co-polarized backscatter data (HH or VV), whereas the other models can calculate co- and cross-polarized backscatter values (HH, VH, VV). Differences between vegetation models WCM and SSRT exists in the form of their simplicity. WCM calculates only the volume backscattering component, whereas, within SSRT, additional backscatter components, such as plant-ground and ground-plant interaction, as well as ground-plant-ground scattering contributions, are considered. In our implementation, the vegetation descriptor of WCM is LAI, whereas the vegetation descriptor of SSRT consists of LAI and canopy height. In general, the computational time and the required additional input parameters are increasing from empirical to semi-empirical and theoretical models. A summary of the different models with information about type, validity range, site dependency, required parameters, and used polarization are given in Table 4.

4. Results and Discussion

4.1. Model Calibration Results

4.1.1. Static Empirical Parameters

In a first calibration approach, the empirical parameters for the different models were treated as static throughout the entire time series. The modeled backscatter was then compared to the measured backscatter from Sentinel-1. Table 6 shows the mean of RMSE and R^2 of all analyzed sample points (Figure 1) for different surface and canopy model combinations. The retrieved RMSE of the calibration results ranges from 1.92 to 2.25 dB with R^2 of 0.08 to 0.34, respectively. A more detailed picture regarding differences between modeled and Sentinel-1 backscatter during the time series is shown for field 508 in Figure 4. While, for the first half of the vegetation period, all model combinations show a relatively good fit to Sentinel-1 backscatter data, in the second half, deviations are obvious. Furthermore, from the beginning of June where LAI reaches its saturation point and the maximum plant height is almost reached, no significant change over time within the modeled backscatter can be observed. The analyzed soil models (different colors) show small differences, whereas a clear separation between analyzed canopy models (solid vs dashed lines), especially for later vegetation stages, is noticeable. As described in Section 2.3, Sentinel-1 data of four different overpasses, and therefore with different incidence and azimuth angles were used for this analysis. In Figure 4, every fourth point (same icon) of the Sentinel-1 backscatter time series represents the same satellite acquisition geometry (same incidence and azimuth angle). The incidence angle is implemented within the used RT models, whereas the models do not account for difference azimuth angles. The black line (Sentinel-1 backscatter) in Figure 4 shows that the observed backscatter values differ with varying incidence angles and changes in soil moisture. The model predictions (different colors) illustrates that the models can account for varying incidence angles and changes in soil moisture only until the end of May. Furthermore, the good correlation of modeled and Sentinel-1 backscatter values suggests that the effect of different azimuth angles on backscatter values seems to be negligible until the end of May. The main change within the wheat fields in June and July in terms of phenology are the flowering, the development of the fruit, and, later on, the ripening [75,81]. For these phenology stages, the increase in backscatter is caused by higher sensitivity of the radar signal to the ground contribution due to water loss within the vegetation [81]. Mattia et al. [76] identified the heading period as turning point where the sensitivity of the radar backscatter to above-ground biomass decreases, whereas the sensitivity to soil surface increases. The temporal evolution of modeled ground contribution to the total backscatter for different model combinations (different colors) with the observed Sentinel-1 total backscatter (black line) is shown in Figure 5 (top part). Until the increase in canopy height at the beginning of May (Table 4), the modeled ground scattering part seems to be the main contributor to the total backscatter, whereas the canopy part is negligible. With increase, especially in canopy height, the ground contribution drops significantly. The expected decrease in backscatter of the ground contribution due to a bigger canopy layer (increase of canopy height and LAI) can be more clearly observed for SSRT than WCM. The differences between SSRT and WCM might be found by using different canopy descriptors (SSRT: LAI and canopy height; WCM: LAI). Differences in ground contributions between Dubois95 and the other surface models are related to differences in the modeled attenuation through the canopy T (Figure 4, bottom part). T is regulating the contribution intensity of the ground and canopy part for the total backscatter calculation. The temporal evolution of T with a value range from 0 (dominant canopy contribution) to 1 (dominant ground contribution) is shown in Figure 5 (bottom part). All model combinations show a similar temporal shape with slightly higher values for model Dubois95. T decreases from April (dominant ground contribution) to mid-May (dominant canopy contribution) and stays at its minimum after mid-May. The expected increase [76,81] of the modeled ground contribution due to higher ground sensitivity in June and July (phenology: flowering, development of the fruit, and ripening) is not observed within the modeled data. In our case, T is mainly driven by static empirical parameter B (WCM) and $coef$ (SSRT), as well as non-static vegetation descriptors LAI and/or canopy height. Therefore, with almost no changes in LAI and canopy height in June and July (Figure 4, middle part), the two-way attenuation T stays near zero, which indicates a dominant canopy model contribution to the total backscatter calculation. By definition, LAI is defined as the one-sided leaf area per measured ground unit [82]. By the time wheat plants reach their maximum height, the leaves are fully developed. Changes within the wheat plants, especially during vegetation stages of flowering, fruit development, or ripening (Figure 4), are based mainly on increasing biomass within grains and stems, as well as changes of the vegetation water content. However, biomass changes in grains and stems, as well as vegetation water content loss, especially during the ripening stage, are not reflected within the LAI. Therefore, almost no information about the increased biomass and the water loss due to ripening of the plants is given within this model configuration. Plant moisture reduction affects the attenuation of the radar signal by the canopy in a way that the canopy is more transparent for the radar wave [83]. Therefore, the sensitivity of the radar signal to the canopy should decrease, whereas the sensitivity to the surface increases. The almost non-existent deviation between Sentinel-1 and modeled backscatter in early vegetation stages suggests that the interaction between surface and canopy model, and therefore the attenuation of the backscatter signal by the canopy, described by static empirical parameters and LAI, can be modeled sufficiently only in early vegetation stages. For good backscatter model results during later vegetation stages, the backscatter changes due to water loss within the plants have to be considered. The effects of utilizing non-static empirical parameters to account for these shortcomings are discussed in the next sections.

Table 6. Calibration results of different model combinations with static empirical parameters. Mean RMSE, ubRMSE, and R^2 of all analyzed field measurement points.

Model	Calibration	
Surface + Canopy	RMSE [dB]	R^2
Oh92 + SSRT	2.11	0.20
Oh92 + WCM	1.97	0.26
Oh04 + SSRT	2.02	0.18
Oh04 + WCM	1.92	0.23
Dubois95 + SSRT	2.09	0.08
Dubois95 + WCM	2.03	0.08
WCM + SSRT	2.25	0.22
WCM + WCM	2.08	0.34
IEM_B + SSRT	2.24	0.15
IEM_B + WCM	2.13	0.24

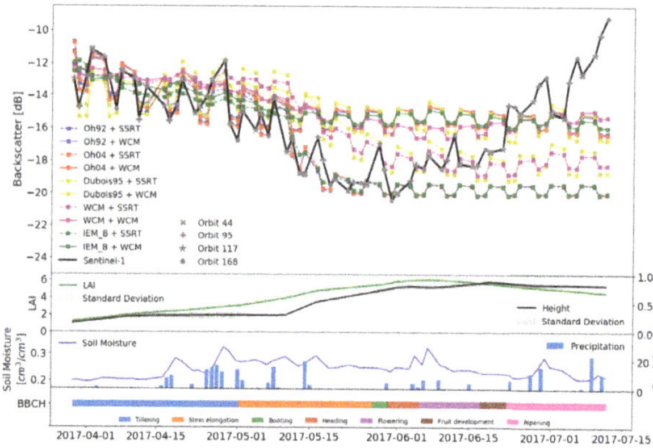

Figure 4. Measured and modeled data (static parameters) of wheat field 508 for vegetation period 2017. VV-polarized backscatter comparison of different model combinations (surface + canopy) with static empirical parameters and Sentinel-1 data. Different icons represent different acquisition geometries of Sentinel-1 (**top**). Field measurements of Leaf Area Index (LAI), canopy height, and soil moisture, as well as precipitation data from meteorological station Freising (**middle**). Observed vegetation phenology according to BBCH scale [84] (**bottom**).

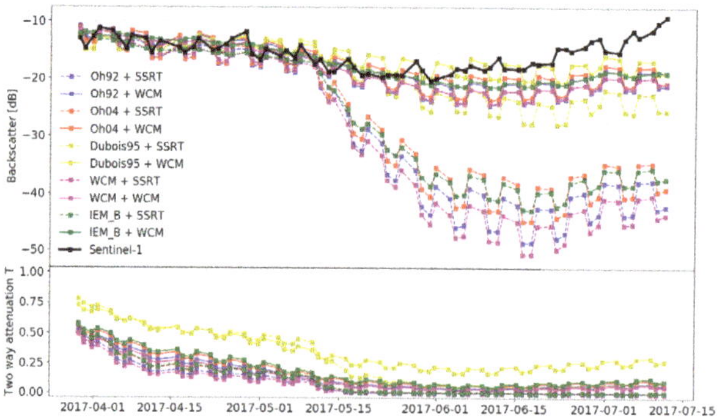

Figure 5. Model results (wheat field 508, static parameter) of ground contribution to total backscatter for different model combinations (different colors) with Sentinel-1 VV polarized total backscatter (black line) as reference (**top**). Temporal evolution of model component two way attenuation by the canopy *T* (**bottom**).

4.1.2. Non-Static Empirical Parameters

Due to the results in Section 4.1.1, which showed that LAI and height cannot account for the observed backscatter changes in June and July (increase of Sentinel-1 backscatter), a second calibration approach with non-static empirical parameter where chosen. As already mentioned, the observed increase in backscatter is caused by a higher sensitivity of the radar signal to the ground contribution. The attenuation of the backscatter T, more precisely B, for WCM and $coef$ for SSRT were identified as the main drivers for the increase or decrease of the ground contribution; therefore, a non-static approach for these parameters was tried. The other empirical parameters of WCM (surface: C, D; canopy: A) were set as static values (mean of the retrieved static values during the calibration approach shown in Section 4.1.1) to clearly relate the observed changes to the attenuation of the backscatter. The modeled backscatter was compared to Sentinel-1 observations and the statistics for different model combinations are shown in form of mean RMSE and R^2 of all analyzed field measurement points in Table 7. The retrieved RMSE ranges from 1.13 to 1.60dB with R^2 of 0.45 to 0.82, respectively. By comparing different model combinations, almost no differences between the two analyzed canopy models can be observed. A different picture is shown by comparing surface models only. WCM seems to outperform all others, whereas differences between WCM and Oh92, Oh04, and IEM_B are smaller than differences of these models to Dubois95. Similar to the approach with static empirical parameters, the evolution over time of modeled and Sentinel-1 backscatter is shown in Figure 6 for one measurement point of field 508. Like the results in Section 4.1.1, modeled backscatter results of the first half of the vegetation period fit well to the observed Sentinel-1 backscatter data. Contrary to the static parameter approach, the second half of the vegetation period shows high correlations between Sentinel-1 and modeled backscatter. Unlike the results of the static approach (Figure 5, top part), the expected increase of the ground contribution at the end of the vegetation period can be observed in Figure 7 (top part). The increase of the ground contribution to the total backscatter is also reflected within the two-way attenuation by the canopy T shown in Figure 7 (bottom part). Compared to the static approach (Figure 5, bottom part), the values of T from April to the beginning of June are very similar for all model combinations, except Dubois95. The temporal changes of non-static parameters B and $coef$ are shown for the validation results in Figures 8 and 9 and will be further discussed in the validation Section 4.2. Besides the modeled backscatter increase for the second half of the vegetation period, different trends between Sentinel-1 and modeled backscatter for individual time steps are observed. This mismatch might occur for several reasons. As already mentioned in Section 4.1.1, Sentinel-1 data with different incidence and azimuth angles were used. A closer look at every fourth modeled point (same incidence and azimuth angle) in Figure 6 (same icon) shows a steady increase of modeled backscatter values at the end of the vegetation period. Therefore, an overall trend of an increasing backscatter can be seen for Sentinel-1 and modeled data. Due to non contradictory trends during the first half of the vegetation period, it seems that the influence on backscatter due to incidence and azimuth angles increases for the second half of the vegetation period. Thus, higher canopy heights and LAI values may also increase the impact of different incidence and azimuth angles on the backscatter behavior. In addition, the acquisition time (Table 3) of the Sentinel-1 images might play another role for different attenuation effects by the canopy [81]. The acquisition time of the satellite differs for different overpasses. For the MNI test site, Sentinel-1 data was acquired during early morning or late afternoon (Table 3). This might lead to differences in the observations due to dew [85,86] or different plant alignments towards the sun which, in theory, leads to different backscatter attenuations through the canopy [87–89], which are not accounted for within the models. Despite the different trends during the temporal evolution of modeled and Sentinel-1 backscatter for the second half of the vegetation period, the observed overall increase in backscatter at the end of the vegetation period can be modeled well.

Table 7. Calibration and validation results of different model combinations with non-static empirical parameters. Mean RMSE, ubRMSE, and R^2 of all analyzed field measurement points.

Model Surface + Canopy	Calibration RMSE [dB]	 R^2	Validation RMSE [dB]	 ubRMSE [dB]	 R^2
Oh92 + SSRT	1.24	0.73	2.82	2.10	0.59
Oh92 + WCM	1.22	0.73	2.75	2.21	0.57
Oh04 + SSRT	1.33	0.69	2.87	2.14	0.57
Oh04 + WCM	1.32	0.68	2.81	2.22	0.57
Dubois95 + SSRT	1.55	0.49	3.06	2.11	0.49
Dubois95 + WCM	1.60	0.45	3.06	2.18	0.48
WCM + SSRT	1.16	0.82	2.65	1.93	0.63
WCM + WCM	1.13	0.81	2.57	2.08	0.60
IEM_B + SSRT	1.32	0.78	2.62	1.82	0.64
IEM_B + WCM	1.34	0.77	2.54	1.92	0.62

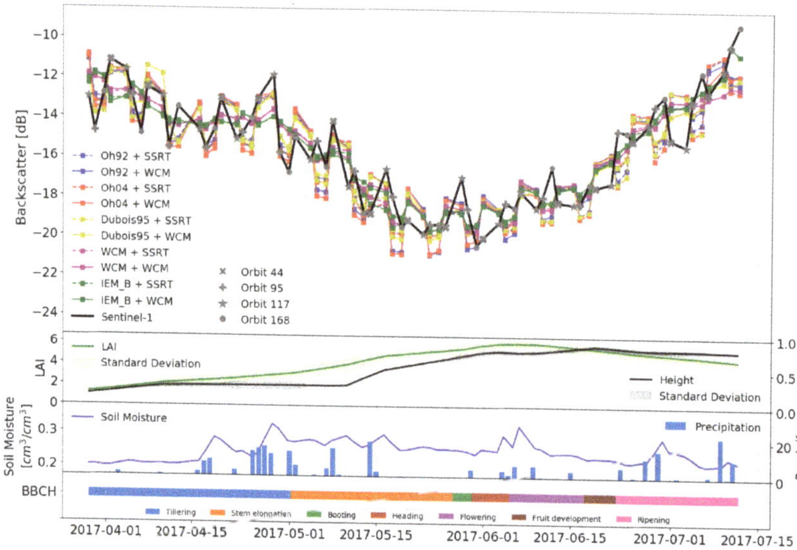

Figure 6. Measured and modeled data (non-static parameters) of wheat field 508 for vegetation period 2017. VV-polarized backscatter comparison of different model combinations (surface + canopy) with non-static empirical parameters and Sentinel-1 data. Different icons represent different acquisition geometries of Sentinel-1 (**top**). Field measurements of LAI, canopy height, and soil moisture, as well as precipitation data from meteorological station Freising (**middle**). Observed vegetation phenology according to BBCH scale [84] (**bottom**).

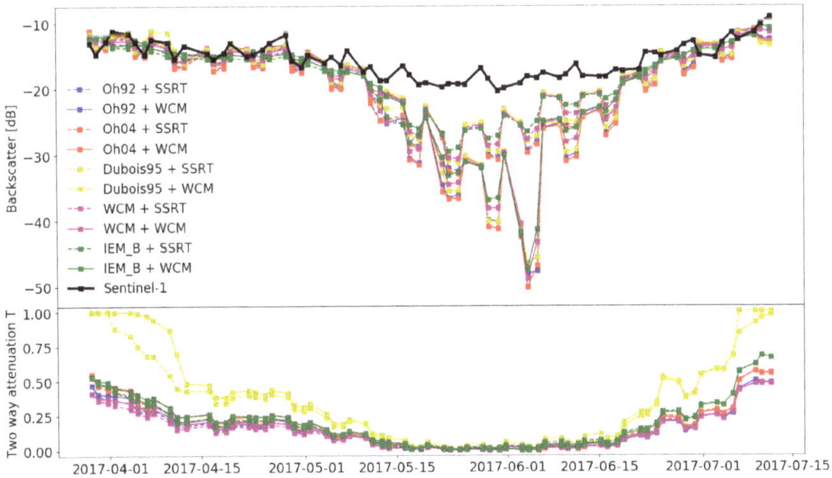

Figure 7. Model results (wheat field 508, non-static parameter) of ground contribution to total backscatter for different model combinations (different colors) with Sentinel-1 VV polarized total backscatter (black line) as reference (**top**). Temporal evolution of model component two way attenuation by the canopy T (**bottom**).

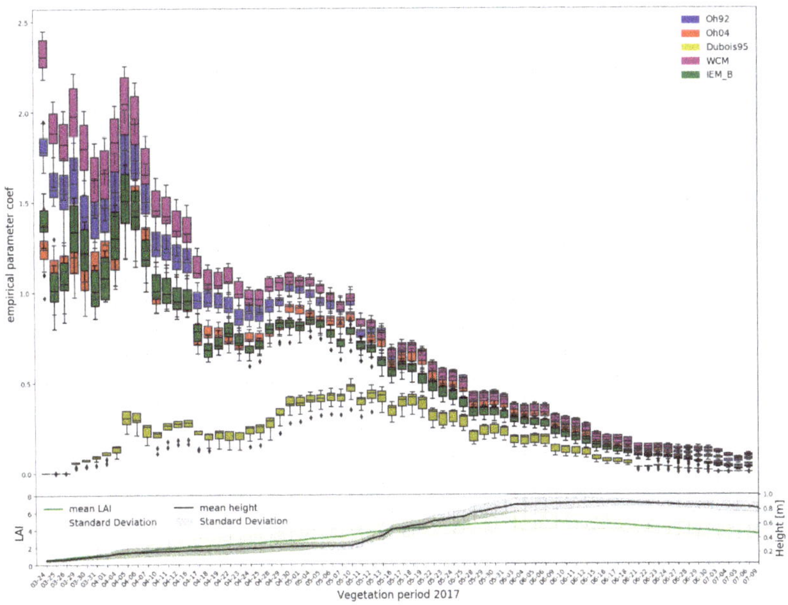

Figure 8. Evolution over time of non-static parameter $coef$ separated by different surface models (colors) for each Sentinel-1 acquisition date during vegetation period 2017. The box plots show the range of $coef$ used during the validation approach for different fields point (**top**). Mean and standard deviation of all measurement points for LAI and canopy height (**bottom**).

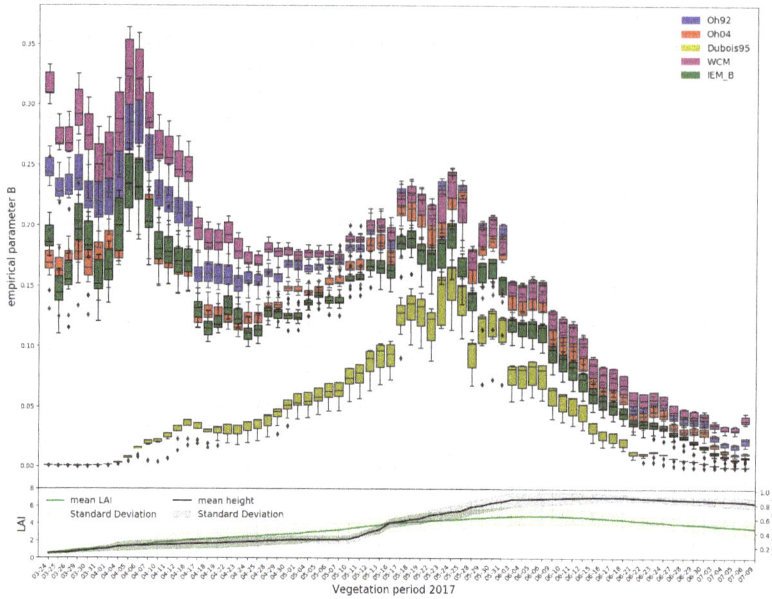

Figure 9. Evolution over time of non-static parameter *B* separated by different surface models (colors) for each Sentinel-1 acquisition date during vegetation period 2017. The box plots show the range of *B* used during the validation approach for different fields point (**top**). Mean and standard deviation of all measurement points for LAI and canopy height (**bottom**).

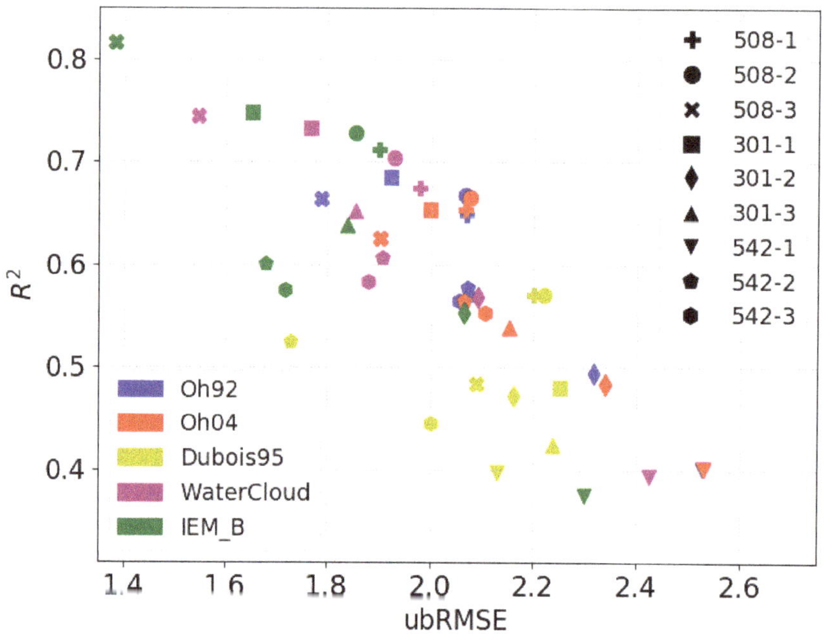

Figure 10. Scatter plot showing statistical validation results (correlation coefficient R^2 and ubRMSE) for canopy model SSRT separated by different surface models (colors) and different field points (icons).

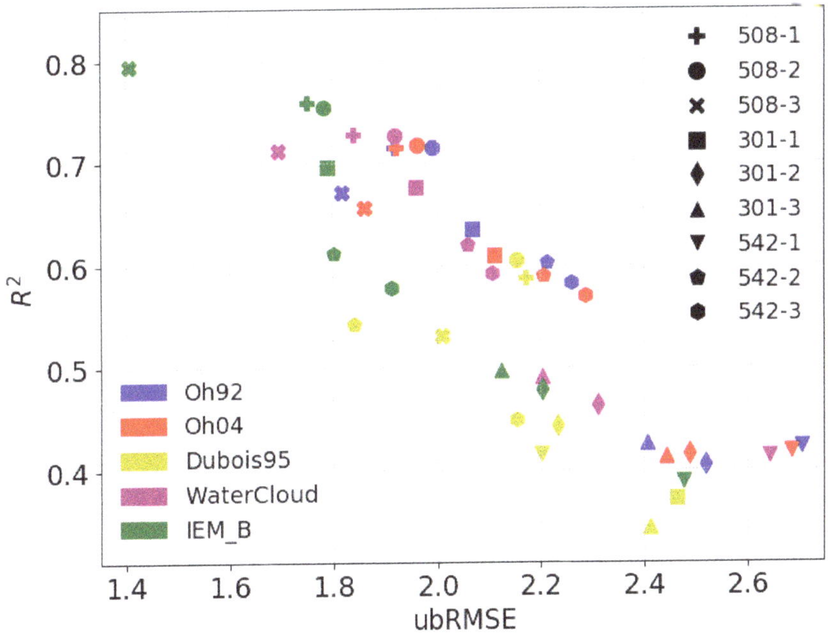

Figure 11. Scatter plot showing statistical validation results (correlation coefficient R^2 and ubRMSE) for canopy model WCM separated by different surface models (colors) and different field points (icons).

4.2. Model Validation Results

The validation for the non-static approach was performed by a leave-one-out cross-validation method. Means of RMSE, ubRMSE, and R^2 are shown in Table 7. The results for the non-static approach show ubRMSE values between 1.92 and 2.22dB. All validation (ubRMSE) results are in a range of 0.65 to 0.95dB poorer than the calibration (RMSE) results. Furthermore, for the validation results R^2 ranges between 0.49 and 0.64, whereas the poorest results are archived with surface model Dubois95. A comparison of ubRMSE for canopy model SSRT and WCM yields for all surface models slightly better results for canopy model SSRT. A more detailed overview of the validation results separated by models and measurement points is given in Figure 10 for canopy model SSRT and in Figure 11 for canopy model WCM. Comparing different fields, as well as field measurement points, a diverse picture is drawn. In general, field 508, especially 508-1, shows the best results, with R^2 values mainly higher than 0.6 and ubRMSE lower than 2.2. Field 542, and especially 542-1, show the worst results, with R^2 around 0.4 and ubRMSE higher than 2.2. Comparing different surface model combinations, IEM_B and WCM show, for both canopy models, slightly better results than the others. It is noticeable that the results of Dubois95 are, in general, poorer than the results of the other surface models, with the exception of point 542-1 (measurement point with the poorest overall results) and partly 301-2. Differences between Oh92 and Oh04 for the different measurement points are present but, compared to the other models, very low. A closer look into each model combination, and especially at the used non-static empirical parameters $coef$ and B, and thus at the change of the attenuation of the canopy in time, are given with Figures 8 and 9. The figures show the development of $coef$ and B over the vegetation period and the parameter calibration spread, which is defined by the used leave-one-out cross-validation results. Comparing different surface models, the same evolution over time (except Dubois95 for the first half of the vegetation period), with some differences in the absolute values, can be observed. The low values of $coef$ and B for model Dubois95 in the first half of the vegetation period can explain the differences in absolute values of T (Figure 7, bottom part) between Dubois95 and the other model combinations. A comparison of canopy model SSRT (parameter $coef$) and WCM (parameter B) show, for all surface models (except Dubois95), for parameter $coef$, a almost steady decrease from the beginning to the end of the vegetation period and a similar evolution for parameter B, except for a short period of increasing values to a relative maximum from mid-May to the end of May. The differences in shape between $coef$ and B might be explained due to different model input data. SSRT uses the canopy height and LAI for the description of the canopy, whereas WCM only uses LAI. The increase of parameter B correlates very well with the measured increase of the canopy height. This suggests that parameter B compensates for possible shortcomings of WCM due to missing information about the canopy height. Furthermore, not only the absolute values but also the spread of $coef$ and B for different models are higher at the beginning of the vegetation period and strongly decrease over time. This means that $coef$ and B show higher differences between fields and field measurement points at the beginning of the vegetation period. This is in line with observations made during the field campaign, where higher differences within one field and between fields could be observed more easily at early vegetation stages. At the end of the vegetation period, such differences were not detectable anymore. Another indicator of low field differences regarding the ripening stage of the wheat plants was the almost simultaneous harvesting date. Another factor to be considered if looking at higher variability of $coef$ and B at the beginning of the vegetation season is the model implementation itself. B and $coef$ are influencing the attenuation of the backscatter by the canopy and therefore how strong each model compartment (surface or canopy) contributes to the total backscatter calculation. At the beginning of the growing season, LAI and canopy height are low; therefore, notable differences within $coef$ and B might have not such a high impact on total backscatter predictions. Differences in the range of $coef$ (0 to 2.5) and B (from 0 to 0.35) are most likely based on differences between the model definitions and the required input parameters.

5. Conclusions

Modeled backscatter results of wheat field time series data using different RT model combinations (surface: Oh92, Oh04, Dubois95, IEM_B, WCM; canopy: WCM, SSRT) were compared to observed C-band data from Sentinel-1. Differences between the models were analyzed. The used dataset was acquired by an intense field campaign throughout one vegetation period in 2017. The analysis focused on coupled performance of surface and canopy models and especially on how changes of backscatter attenuation through the canopy influence the total backscatter calculation for different vegetation stages. The two novelties of this study are the evaluation of different combinations of widely used surface and canopy RT models on one test site and the analysis over time of empirical model parameters $coef$ (SSRT) and B (WCM) describing the backscatter attenuation through the canopy T.

Results show that, for total LAI, as vegetation descriptor, a static parameter influencing the backscatter attenuation through the canopy is suitable for the first half but not for the second half of the vegetation period. By using a non-static parameter approach, the backscatter increase at the end of the vegetation period can be modeled. The static calibration performance results in the form of RMSE improved from 1.92–2.25dB to 1.13–1.60dB for the non-static approach. The validation accuracy for the non-static parameter approach was evaluated with ubRMSE and ranges for all model combinations between 1.82 and 2.22dB. Validation results with SSRT as canopy model show better results in combination with all surface models when compared to respective combinations using WCM for the canopy part. Furthermore, it has been shown that the modeled backscatter results highly depend on the non-static empirical parameter. The evolution of the empirical parameter is similar for all surface models except Dubois95. At the beginning of the vegetation period, high values are observed, which decrease during the vegetation season to a minimum shortly before harvesting. Differences between the canopy model SSRT and WCM are noticeable in the form of higher variability (more outliers) of the empirical parameter from canopy model WCM. Furthermore, the empirical parameter for canopy model WCM has a relative maximum at the end of June. This increase of WCM's empirical parameter B can most likely be explained by not including information about the canopy height. Overall, the results of this study indicate that more complex models, like IEM_B as surface and SSRT as canopy model, provide the best results in our setup regarding ubRMSE and R^2. It should be mentioned that the disadvantage of more complex models, by requiring more input parameters, were set to a minimum because of model parameter reduction by model adjustments and the use of literature values for rms height and single scattering albedo. Therefore, based on this study, we suggest using surface model IEM_B in combination with canopy model SSRT.

To accomplish a very dense time series of satellite acquisitions with a revisit time of 1.5 days, Sentinel-1 images with different incidence and azimuth angles were used. The models can account for backscatter changes due to different acquisition geometries only during the first half of the vegetation period. During the second half, a trend mismatch between Sentinel-1 and modeled backscatter is apparent in all model results. Therefore, it has to be stated that the used models in this study are only partially able to handle differences due to changes in radar acquisition geometries.

To take full advantage of dense time series provided, e.g., by incorporating different sensors on top of varying radar acquisition geometries, extended research on the suitability of certain RT models is required. Respectively, arising radiometric differences added to the disparity in radar acquisition geometries must be investigated, for that matter. To advance synergistic retrieval methods of SAR and optical data for soil moisture estimation, further research on incorporating non-static empirical parameter sets retrieved from optical sensors, to account for vegetation phenological states, is needed.

Author Contributions: Conceptualization, T.W. and A.L.; Methodology and Formal Analysis, T.W.; Software, T.W. and A.L.; Writing-Original Draft Preparation, T.W., Writing-Review & Editing, T.W., T.R. and P.M.; Resources, T.W. and T.R.; Visualization, T.W.; Supervision, A.L and P.M.; Project administration, A.L. and P.M.; Funding acquisition, A.L. All authors have read and agreed to the published version of the manuscript.

Funding: The project leading to this application has received funding from the European Union's Horizon 2020 research and innovation program under Grant Agreement No. 687320.

Acknowledgments: Thomas Weiß is truly thankful for the inspiration, ideas, and support that Alexander Löw († 2. July 2017) has given him.

Conflicts of Interest: The authors declare no conflict of interest.

References

1. Hollmann, R.; Merchant, C.J.; Saunders, R.; Downy, C.; Buchwitz, M.; Cazenave, A.; Chuvieco, E.; Defourny, P.; de Leeuw, G.; Forsberg, R.; et al. The ESA Climate Change Initiative: Satellite Data Records for Essential Climate Variables. *Bull. Am. Meteorol. Soc.* **2013**, *94*, 1541–1552. [CrossRef]
2. Brocca, L.; Ciabatta, L.; Massari, C.; Camici, S.; Tarpanelli, A. Soil Moisture for Hydrological Applications: Open Questions and New Opportunities. *Water* **2017**, *9*, 140. [CrossRef]
3. Hochschild, V.; Thiel, C.; Grünler, S. Derivation of Land Cover and Soil Moisture from Airborne Polarimetric SAR-Data. *Sao/NASA Astrophys. Data Syst.* **2001**, *475*, 355–358.
4. Zribi, M.; Muddu, S.; Bousbih, S.; Al Bitar, A.; Tomer, S.K.; Baghdadi, N.; Bandyopadhyay, S. Analysis of L-Band SAR Data for Soil Moisture Estimations over Agricultural Areas in the Tropics. *Remote Sens.* **2019**, *11*, 1122. [CrossRef]
5. Oldak, A.; Jackson, T.J.; Starks, P.; Elliott, R. Mapping Near-Surface Soil Moisture on Regional Scale Using ERS-2 SAR Data. *Int. J. Remote Sens.* **2003**, *24*, 4579–4598. [CrossRef]
6. Baghdadi, N.; Aubert, M.; Cerdan, O.; Franchistéguy, L.; Viel, C.; Martin, E.; Zribi, M.; Desprats, J.F. Operational Mapping of Soil Moisture Using Synthetic Aperture Radar Data: Application to the Touch Basin (France). *Sensors* **2007**, *7*, 2458–2483. [CrossRef]
7. Kornelsen, K.C.; Coulibaly, P. Advances in Soil Moisture Retrieval from Synthetic Aperture Radar and Hydrological Applications. *J. Hydrol.* **2013**, *476*, 460–489. [CrossRef]
8. Engman, E. The Potential of SAR in Hydrology. In Proceedings of the IGARSS '94-1994 IEEE International Geoscience and Remote Sensing Symposium, Pasadena, CA, USA, 8–12 August **1994**; Volume 1, pp. 283–285. [CrossRef]

9. Le Hegarat-Mascle, S.; Zribi, M.; Alem, F.; Weisse, A.; Loumagne, C. Soil Moisture Estimation from ERS/SAR Data: Toward an Operational Methodology. *IEEE Trans. Geosci. Remote Sens.* **2002**, *40*, 2647–2658. [CrossRef]
10. Dorigo, W.; Wagner, W.; Albergel, C.; Albrecht, F.; Balsamo, G.; Brocca, L.; Chung, D.; Ertl, M.; Forkel, M.; Gruber, A.; et al. ESA CCI Soil Moisture for Improved Earth System Understanding: State-of-the Art and Future Directions. *Remote Sens. Environ.* **2017**, *203*, 185–215. [CrossRef]
11. Amitrano, D.; Martino, G.D.; Iodice, A.; Riccio, D.; Ruello, G.; Ciervo, F.; Papa, M.N.; Koussoubé, Y. Effectiveness of High-Resolution SAR for Water Resource Management in Low-Income Semi-Arid Countries. *Int. J. Remote Sens.* **2014**, *35*, 70–88. [CrossRef]
12. Holtgrave, A.K.; Förster, M.; Greifeneder, F.; Notarnicola, C.; Kleinschmit, B. Estimation of Soil Moisture in Vegetation-Covered Floodplains with Sentinel-1 SAR Data Using Support Vector Regression. *PFG J. Photogramm. Remote. Sens. Geoinf. Sci.* **2018**, *86*, 85–101. [CrossRef]
13. Urban, M.; Berger, C.; Mudau, T.E.; Heckel, K.; Truckenbrodt, J.; Onyango Odipo, V.; Smit, I.P.J.; Schmullius, C. Surface Moisture and Vegetation Cover Analysis for Drought Monitoring in the Southern Kruger National Park Using Sentinel-1, Sentinel-2, and Landsat-8. *Remote Sens.* **2018**, *10*, 1482. [CrossRef]
14. Bousbih, S.; Zribi, M.; El Hajj, M.; Baghdadi, N.; Lili-Chabaane, Z.; Gao, Q.; Fanise, P. Soil Moisture and Irrigation Mapping in A Semi-Arid Region, Based on the Synergetic Use of Sentinel-1 and Sentinel-2 Data. *Remote Sens.* **2018**, *10*, 1953. [CrossRef]
15. Barrett, B.W.; Dwyer, E.; Whelan, P. Soil Moisture Retrieval from Active Spaceborne Microwave Observations: An Evaluation of Current Techniques. *Remote Sens.* **2009**, *1*, 210–242. [CrossRef]
16. Ulaby, F. Radar Measurement of Soil Moisture Content. *IEEE Trans. Antennas Propag.* **1974**, *22*, 257–265. [CrossRef]
17. Ulaby, F.; Long, D.; Blackwell, W.; Elachi, C.; Fung, A.; Ruf, C.; Sarabandi, K.; Zyl, J.; Zebker, H. *Microwave Radar and Radiometric Remote Sensing*; University of Michigan Press: Ann Arbor, MI, USA, **2014**.
18. Baghdadi, N.; King, C.; Bourguignon, A.; Remond, A. Potential of ERS and Radarsat Data for Surface Roughness Monitoring over Bare Agricultural Fields: Application to Catchments in Northern France. *Int. J. Remote Sens.* **2002**, *23*, 3427–3442. [CrossRef]
19. Holah, N.; Baghdadi, N.; Zribi, M.; Bruand, A.; King, C. Potential of ASAR/ENVISAT for the Characterization of Soil Surface Parameters over Bare Agricultural Fields. *Remote Sens. Environ.* **2005**, *96*, 78–86. [CrossRef]
20. Wickel, A.J.; Jackson, T.J.; Wood, E.F. Multitemporal Monitoring of Soil Moisture with RADARSAT SAR during the 1997 Southern Great Plains Hydrology Experiment. *Int. J. Remote Sens.* **2001**, *22*, 1571–1583. [CrossRef]
21. Mathieu, R.; Sbih, M.; Viau, A.A.; Anctil, F.; Parent, L.E.; Boisvert, J. Relationships between Radarsat SAR Data and Surface Moisture Content of Agricultural Organic Soils. *Int. J. Remote Sens.* **2003**, *24*, 5265–5281. [CrossRef]
22. Attema, E.P.W.; Ulaby, F.T. Vegetation Modeled as a Water Cloud. *Radio Sci.* **1978**, *13*, 357–364. [CrossRef]
23. Prévot, L.; Champion, I.; Guyot, G. Estimating Surface Soil Moisture and Leaf Area Index of a Wheat Canopy Using a Dual-Frequency (C and X Bands) Scatterometer. *Remote Sens. Environ.* **1993**, *46*, 331–339. [CrossRef]
24. Champion, I.; Prevot, L.; Guyot, G. Generalized Semi-Empirical Modelling of Wheat Radar Response. *Int. J. Remote Sens.* **2000**, *21*, 1945–1951. [CrossRef]
25. Dabrowska-Zielinska, K.; Inoue, Y.; Kowalik, W.; Gruszczynska, M. Inferring the Effect of Plant and Soil Variables on C- and L-Band SAR Backscatter over Agricultural Fields, Based on Model Analysis. *Adv. Space Res.* **2007**, *39*, 139–148. [CrossRef]
26. Oh, Y.; Sarabandi, K.; Ulaby, F. An Empirical Model and an Inversion Technique for Radar Scattering from Bare Soil Surfaces. *IEEE Trans. Geosci. Remote Sens.* **1992**, *30*, 370–381. [CrossRef]
27. Oh, Y. Quantitative Retrieval of Soil Moisture Content and Surface Roughness from Multipolarized Radar Observations of Bare Soil Surfaces. *IEEE Trans. Geosci. Remote Sens.* **2004**, *42*, 596–601. [CrossRef]
28. Dubois, P.; van Zyl, J.; Engman, T. Measuring Soil Moisture with Imaging Radars. *IEEE Trans. Geosci. Remote Sens.* **1995**, *33*, 915–926. [CrossRef]
29. Fung, A.; Li, Z.; Chen, K. Backscattering from a Randomly Rough Dielectric Surface. *IEEE Trans. Geosci. Remote Sens.* **1992**, *30*, 356–369. [CrossRef]

30. Baghdadi, N.; King, C.; Chanzy, A.; Wigneron, J.P. An Empirical Calibration of the Integral Equation Model Based on SAR Data, Soil Moisture and Surface Roughness Measurement over Bare Soils. *Int. J. Remote Sens.* **2002**, *23*, 4325–4340. [CrossRef]
31. Baghdadi, N.; Abou Chaaya, J.; Zribi, M. Semiempirical Calibration of the Integral Equation Model for SAR Data in C-Band and Cross Polarization Using Radar Images and Field Measurements. *IEEE Geosci. Remote Sens. Lett.* **2011**, *8*, 14–18. [CrossRef]
32. Chen, K.; Wu, T.D.; Tsang, L.; Li, Q.; Shi, J.; Fung, A. Emission of Rough Surfaces Calculated by the Integral Equation Method with Comparison to Three-Dimensional Moment Method Simulations. *IEEE Trans. Geosci. Remote Sens.* **2003**, *41*, 90–101. [CrossRef]
33. Ulaby, F.T.; Sarabandi, K.; McDonald, K.; Whitt, M.; Dobson, M.C. Michigan Microwave Canopy Scattering Model. *Int. J. Remote Sens.* **1990**, *11*, 1223–1253. [CrossRef]
34. Ferrazzoli, P.; Guerriero, L. Radar Sensitivity to Tree Geometry and Woody Volume: A Model Analysis. *IEEE Trans. Geosci. Remote Sens.* **1995**, *33*, 360–371. [CrossRef]
35. De Roo, R.; Du, Y.; Ulaby, F.; Dobson, M. A Semi-Empirical Backscattering Model at L-Band and C-Band for a Soybean Canopy with Soil Moisture Inversion. *IEEE Trans. Geosci. Remote Sens.* **2001**, *39*, 864–872. [CrossRef]
36. Quast, R.; Wagner, W. Analytical Solution for First-Order Scattering in Bistatic Radiative Transfer Interaction Problems of Layered Media. *Appl. Opt.* **2016**, *55*, 5379–5386. [CrossRef] [PubMed]
37. Quast, R.; Albergel, C.; Calvet, J.C.; Wagner, W. A Generic First-Order Radiative Transfer Modelling Approach for the Inversion of Soil and Vegetation Parameters from Scatterometer Observations. *Remote Sens.* **2019**, *11*, 285. [CrossRef]
38. Zribi, M.; Taconet, O.; Le Hégarat-Mascle, S.; Vidal-Madjar, D.; Emblanch, C.; Loumagne, C.; Normand, M. Backscattering Behavior and Simulation Comparison over Bare Soils Using SIR-C/X-SAR and ERASME 1994 Data over Orgeval. *Remote Sens. Environ.* **1997**, *59*, 256–266. [CrossRef]
39. Baghdadi, N.; Zribi, M. Evaluation of Radar Backscatter Models IEM, OH and Dubois Using Experimental Observations. *Int. J. Remote Sens.* **2006**, *27*, 3831–3852. [CrossRef]
40. Khabazan, S.; Motagh, M.; Hosseini, M. Evaluation of Radar Backscattering Models IEM, OH, and Dubois Using L and C-Bands SAR Data over Different Vegetation Canopy Covers and Soil Depths. In *ISPRS-International Archives of the Photogrammetry, Remote Sensing and Spatial Information Sciences*; Copernicus GmbH: Göttingen, Germany, **2013**; Volume XL-1-W3, pp. 225–230. [CrossRef]
41. Choker, M.; Baghdadi, N.; Zribi, M.; El Hajj, M.; Paloscia, S.; Verhoest, N.E.C.; Lievens, H.; Mattia, F. Evaluation of the Oh, Dubois and IEM Backscatter Models Using a Large Dataset of SAR Data and Experimental Soil Measurements. *Water* **2017**, *9*, 38. [CrossRef]
42. Ezzahar, J.; Ouaadi, N.; Zribi, M.; Elfarkh, J.; Aouade, G.; Khabba, S.; Er-Raki, S.; Chehbouni, A.; Jarlan, L. Evaluation of Backscattering Models and Support Vector Machine for the Retrieval of Bare Soil Moisture from Sentinel-1 Data. *Remote Sens.* **2020**, *12*, 72. [CrossRef]
43. Kweon, S.K.; Oh, Y. A Modified Water-Cloud Model With Leaf Angle Parameters for Microwave Backscattering From Agricultural Fields. *IEEE Trans. Geosci. Remote Sens.* **2015**, *53*, 2802–2809. [CrossRef]
44. Pierdicca, N.; Pulvirenti, L.; Pace, G. A Prototype Software Package to Retrieve Soil Moisture From Sentinel-1 Data by Using a Bayesian Multitemporal Algorithm. *IEEE J. Sel. Top. Appl. Earth Obs. Remote Sens.* **2014**, *7*, 153–166. [CrossRef]
45. Baghdadi, N.; El Hajj, M.; Zribi, M.; Bousbih, S. Calibration of the Water Cloud Model at C-Band for Winter Crop Fields and Grasslands. *Remote Sens.* **2017**, *9*, 969. [CrossRef]
46. Graham, A.J.; Harris, R. Estimating Crop and Waveband Specific Water Cloud Model Parameters Using a Theoretical Backscatter Model. *Int. J. Remote Sens.* **2002**, *23*, 5129–5133. [CrossRef]
47. Liu, C.; Shi, J. Estimation of Vegetation Parameters of Water Cloud Model for Global Soil Moisture Retrieval Using Time-Series L-Band Aquarius Observations. *IEEE J. Sel. Top. Appl. Earth Obs. Remote Sens.* **2016**, *9*, 5621–5633. [CrossRef]
48. Kumar, K.; Prasad, K.S.H.; Arora, M.K. Estimation of Water Cloud Model Vegetation Parameters Using a Genetic Algorithm. *Hydrol. Sci. J.* **2012**, *57*, 776–789. [CrossRef]
49. Kumar, K.; Rao, H.P.S.; Arora, M.K. Study of Water Cloud Model Vegetation Descriptors in Estimating Soil Moisture in Solani Catchment. *Hydrol. Process.* **2015**, *29*, 2137–2148. [CrossRef]

50. Song, X.; Ma, J.; Li, X.; Leng, P.; Zhou, F.; Li, S. First Results of Estimating Surface Soil Moisture in the Vegetated Areas Using ASAR and Hyperion Data: The Chinese Heihe River Basin Case Study. *Remote Sens.* **2014**, *6*, 12055–12069. [CrossRef]
51. van Emmerik, T.; Steele-Dunne, S.C.; Judge, J.; van de Giesen, N. Impact of Diurnal Variation in Vegetation Water Content on Radar Backscatter From Maize During Water Stress. *IEEE Trans. Geosci. Remote Sens.* **2015**, *53*, 3855–3869. [CrossRef]
52. Xu, C.; Qu, J.J.; Hao, X.; Wu, D. Monitoring Surface Soil Moisture Content over the Vegetated Area by Integrating Optical and SAR Satellite Observations in the Permafrost Region of Tibetan Plateau. *Remote Sens.* **2020**, *12*, 183. [CrossRef]
53. Lievens, H.; Verhoest, N.E.C. On the Retrieval of Soil Moisture in Wheat Fields From L-Band SAR Based on Water Cloud Modeling, the IEM, and Effective Roughness Parameters. *IEEE Geosci. Remote Sens. Lett.* **2011**, *8*, 740–744. [CrossRef]
54. Bai, X.; He, B. Potential of Dubois Model for Soil Moisture Retrieval in Prairie Areas Using SAR and Optical Data. *Int. J. Remote Sens.* **2015**, *36*, 5737–5753. [CrossRef]
55. Xing, M.; He, B.; Ni, X.; Wang, J.; An, G.; Shang, J.; Huang, X. Retrieving Surface Soil Moisture over Wheat and Soybean Fields during Growing Season Using Modified Water Cloud Model from Radarsat-2 SAR Data. *Remote Sens.* **2019**, *11*, 1956. [CrossRef]
56. He, B.; Xing, M.; Bai, X. A Synergistic Methodology for Soil Moisture Estimation in an Alpine Prairie Using Radar and Optical Satellite Data. *Remote Sens.* **2014**, *6*, 10966–10985. [CrossRef]
57. Qiu, J.; Crow, W.T.; Wagner, W.; Zhao, T. Effect of Vegetation Index Choice on Soil Moisture Retrievals via the Synergistic Use of Synthetic Aperture Radar and Optical Remote Sensing. *Int. J. Appl. Earth Obs. Geoinf.* **2019**, *80*, 47–57. [CrossRef]
58. Tao, L.; Wang, G.; Chen, W.; Chen, X.; Li, J.; Cai, Q. Soil Moisture Retrieval From SAR and Optical Data Using a Combined Model. *IEEE J. Sel. Top. Appl. Earth Obs. Remote Sens.* **2019**, *12*, 637–647. [CrossRef]
59. Zhao, X.; Huang, N.; Niu, Z.; Raghavan, V.; Song, X. Soil Moisture Retrieval in Farmland Using C-Band SAR and Optical Data. *Spat. Inf. Res.* **2017**, *25*, 431–438. [CrossRef]
60. Attarzadeh, R.; Amini, J.; Notarnicola, C.; Greifeneder, F. Synergetic Use of Sentinel-1 and Sentinel-2 Data for Soil Moisture Mapping at Plot Scale. *Remote Sens.* **2018**, *10*, 1285. [CrossRef]
61. El Hajj, M.; Baghdadi, N.; Zribi, M.; Bazzi, H. Synergic Use of Sentinel-1 and Sentinel-2 Images for Operational Soil Moisture Mapping at High Spatial Resolution over Agricultural Areas. *Remote Sens.* **2017**, *9*, 1292. [CrossRef]
62. Gao, F.; Morisette, J.T.; Wolfe, R.E.; Ederer, G.; Pedelty, J.; Masuoka, E.; Myneni, R.; Tan, B.; Nightingale, J. An Algorithm to Produce Temporally and Spatially Continuous MODIS-LAI Time Series. *IEEE Geosci. Remote Sens. Lett.* **2008**, *5*, 60–64. [CrossRef]
63. Dhakar, R.; Sehgal, V.K.; Chakraborty, D.; Sahoo, R.N.; Mukherjee, J. Field Scale Wheat LAI Retrieval from Multispectral Sentinel 2A-MSI and LandSat 8-OLI Imagery: Effect of Atmospheric Correction, Image Resolutions and Inversion Techniques. *Geocarto Int.* **2019**, 1–21. [CrossRef]
64. Xie, Q.; Dash, J.; Huete, A.; Jiang, A.; Yin, G.; Ding, Y.; Peng, D.; Hall, C.C.; Brown, L.; Shi, Y.; et al. Retrieval of Crop Biophysical Parameters from Sentinel-2 Remote Sensing Imagery. *Int. J. Appl. Earth Obs. Geoinf.* **2019**, *80*, 187–195. [CrossRef]
65. Berger, K.; Atzberger, C.; Danner, M.; Wocher, M.; Mauser, W.; Hank, T. Modellbasierte Selektion Hyperspektraler EnMAP Kanäle Zur Optimalen Invertierung von Strahlungstransfermodellen Für Landwirtschaftliche Kulturen. *PFG J. Photogramm. Remote Sens. Geoinf. Sci.* **2018**, *86*, 263–272. [CrossRef]
66. Danner, M.; Berger, K.; Wocher, M.; Mauser, W.; Hank, T. Retrieval of Biophysical Crop Variables from Multi-Angular Canopy Spectroscopy. *Remote Sens.* **2017**, *9*, 726. [CrossRef]
67. Danner, M.; Berger, K.; Wocher, M.; Mauser, W.; Hank, T. Fitted PROSAIL Parameterization of Leaf Inclinations, Water Content and Brown Pigment Content for Winter Wheat and Maize Canopies. *Remote Sens.* **2019**, *11*, 1150. [CrossRef]
68. Wocher, M.; Berger, K.; Danner, M.; Mauser, W.; Hank, T. Physically-Based Retrieval of Canopy Equivalent Water Thickness Using Hyperspectral Data. *Remote Sens.* **2018**, *10*, 1924. [CrossRef]
69. Kellndorfer, J.; Pierce, L.; Dobson, M.; Ulaby, F. Toward Consistent Regional-to-Global-Scale Vegetation Characterization Using Orbital SAR Systems. *IEEE Trans. Geosci. Remote Sens.* **1998**, *36*, 1396–1411. [CrossRef]

70. El Hajj, M.; Baghdadi, N.; Zribi, M.; Belaud, G.; Cheviron, B.; Courault, D.; Charron, F. Soil Moisture Retrieval over Irrigated Grassland Using X-Band SAR Data. *Remote Sens. Environ.* **2016**, *176*, 202–218. [CrossRef]
71. Fung, A.; Liu, W.; Chen, K.; Tsay, M. An Improved Iem Model for Bistatic Scattering From Rough Surfaces. *J. Electromagn. Waves Appl.* **2002**, *16*, 689–702. [CrossRef]
72. Marzahn, P.; Ludwig, R. Using Multi-Dimensional Microwave Remote Sensing Information for the Retrieval of Soil Surface Roughness. *Int. Arch. Photogramm. Remote Sens. Spat. Inf. Sci.* **2016**, *41*, 1257–1262 [CrossRef]
73. Marzahn, P.; Rieke-Zapp, D.; Ludwig, R. Assessment of Soil Surface Roughness Statistics for Microwave Remote Sensing Applications Using a Simple Photogrammetric Acquisition System. *ISPRS J. Photogramm. Remote Sens.* **2012**, *72*, 80–89. [CrossRef]
74. Wegmüller, U.; Santoro, M.; Mattia, F.; Balenzano, A.; Satalino, G.; Marzahn, P.; Fischer, G.; Ludwig, R.; Floury, N. Progress in the Understanding of Narrow Directional Microwave Scattering of Agricultural Fields. *Remote Sens. Environ.* **2011**, *115*, 2423–2433. [CrossRef]
75. Jia, M.; Tong, L.; Zhang, Y.; Chen, Y. Multitemporal Radar Backscattering Measurement of Wheat Fields Using Multifrequency (L, S, C, and X) and Full-Polarization. *Radio Sci.* **2013**, *48*, 471–481. [CrossRef]
76. Mattia, F.; Le Toan, T.; Picard, G.; Posa, F.; D'Alessio, A.; Notarnicola, C.; Gatti, A.; Rinaldi, M.; Satalino, G.; Pasquariello, G. Multitemporal C-Band Radar Measurements on Wheat Fields. *IEEE Trans. Geosci. Remote Sens.* **2003**, *41*, 1551–1560. [CrossRef]
77. Wigneron, J.P.; Kerr, Y.; Waldteufel, P.; Saleh, K.; Escorihuela, M.J.; Richaume, P.; Ferrazzoli, P.; de Rosnay, P.; Gurney, R.; Calvet, J.C.; et al. L-Band Microwave Emission of the Biosphere (L-MEB) Model: Description and Calibration against Experimental Data Sets over Crop Fields. *Remote Sens. Environ.* **2007**, *107*, 639–655. [CrossRef]
78. Davidson, M.; Toan, T.L.; Mattia, F.; Satalino, G.; Manninen, T.; Borgeaud, M. On the Characterization of Agricultural Soil Roughness for Radar Remote Sensing Studies. *IEEE Trans. Geosci. Remote Sens.* **2000**, *38*, 630–640. [CrossRef]
79. Picard, G.; Le Toan, T.; Mattia, F. Understanding C-Band Radar Backscatter from Wheat Canopy Using a Multiple-Scattering Coherent Model. *IEEE Trans. Geosci. Remote Sens.* **2003**, *41*, 1583–1591. [CrossRef]
80. Dobson, M.C.; Ulaby, F.T.; Hallikainen, M.T.; El-rayes, M.A. Microwave Dielectric Behavior of Wet Soil-Part II: Dielectric Mixing Models. *IEEE Trans. Geosci. Remote Sens.* **1985**, *GE-23*, 35–46. [CrossRef]
81. Khabbazan, S.; Vermunt, P.; Steele-Dunne, S.; Ratering Arntz, L.; Marinetti, C.; van der Valk, D.; Iannini, L.; Molijn, R.; Westerdijk, K.; van der Sande, C. Crop Monitoring Using Sentinel-1 Data: A Case Study from The Netherlands. *Remote Sens.* **2019**, *11*, 1887. [CrossRef]
82. Allen, S.T.; Whitsell, M.L.; Keim, R.F. Leaf Area Allometrics and Morphometrics in Baldcypress. *Can. J. For. Res.* **2015**, *45*, 963–969. [CrossRef]
83. He, L.; Tong, L.; Li, Y.; Chen, Y.; Tan, L.; Guo, C. Polarimetric Analysis of Radar Backscatter from Ground-Based Scatterometers and Wheat Biomass Monitoring with Advanced Synthetic Aperture Radar Images. *J. Appl. Remote Sens.* **2016**, *10*, 026008. [CrossRef]
84. Meier, U.; Bleiholder, H.; Buhr, L.; Feller, C.; Hack, H.; Heß, M.; Lancashire, P.D.; Schnock, U.; Stauß, R.; van den Boom, T.; et al. The BBCH System to Coding the Phenological Growth Stages of Plants–History and Publications. *J. FÜr Kult.* **2009**, *61*, 41–52. [CrossRef]
85. Riedel, T.; Liebeskind, P.; Schmullius, C. Seasonal and Diurnal Changes of Polarimetric Parameters from Crops Derived by the Cloude Decomposition Theorem at L-Band. In Proceedings of the IEEE International Geoscience and Remote Sensing Symposium, Toronto, ON, Canada, 24–28 June 2002; Volume 5, pp. 2714–2716. [CrossRef]
86. Riedel, T.; Pathe, C.; Thiel, C.; Herold, M.; Schmullius, C. Systematic Investigation on the Effect of Dew and Interception on Multifrequency and Multipolarimetric Radar Backscatter Signals. *Sao/NASA Astrophys. Data Syst.* **2002**, *475*, 99–104.
87. Way, J.; Paris, J.; Dobson, M.; McDonals, K.; Ulaby, F.; Weber, J.; Ustin, L.; Vanderbilt, V.; Kasischke, E. Diurnal Change in Trees as Observed by Optical and Microwave Sensors: The EOS Synergism Study. *IEEE Trans. Geosci. Remote Sens.* **1991**, *29*, 807. [CrossRef]

88. Hornbuckle, B.K.; Rowlandson, T.L.; Russell, E.; Kaleita, A.; Logsdon, S.; Kruger, A.; Yueh, S.; De Roo, R.D. Howdoes Dew Affect L-Band Backscatter? Analysis of Pals Data at the Iowa Validation Site and Implications for Smap. In Proceedings of the 2010 IEEE International Geoscience and Remote Sensing Symposium, Honolulu, HI, USA, 25–30 July 2010; pp. 4835–4838. [CrossRef]
89. Konings, A.G.; Piles, M.; Rötzer, K.; McColl, K.A.; Chan, S.K.; Entekhabi, D. Vegetation Optical Depth and Scattering Albedo Retrieval Using Time Series of Dual-Polarized L-Band Radiometer Observations. *Remote Sens. Environ.* **2016**, *172*, 178–189. [CrossRef]

remote sensing

MDPI

Article

Comparative Analysis of Landsat-8, Sentinel-2, and GF-1 Data for Retrieving Soil Moisture over Wheat Farmlands

Qi Wang [1], Jiancheng Li [1,2], Taoyong Jin [1,2,*], Xin Chang [1], Yongchao Zhu [3], Yunwei Li [1], Jiaojiao Sun [4] and Dawei Li [1]

1 School of Geodesy and Geomatics, Wuhan University, Wuhan 430079, China; qwangchoice@whu.edu.cn (Q.W.); jcli@whu.edu.cn (J.L.); xchang@whu.edu.cn (X.C.); yunweili@whu.edu.cn (Y.L.); dwli@sgg.whu.edu.cn (D.L.)

2 Key Laboratory of Geospace Environment and Geodesy, Ministry of Education, Wuhan 430079, China

3 Meteorological Observation Center, China Meteorological Administration, Beijing 100081, China; zhuyc@cma.gov.cn

4 City College, Southwest University of Science and Technology, Mianyang 621000, China; sunjiaojiao@ccswust.edu.cn

* Correspondence: tyjin@sgg.whu.edu.cn; Tel.: +86-1360-712-9613

Received: 14 July 2020; Accepted: 18 August 2020; Published: 21 August 2020

Abstract: Soil moisture is an important variable in ecological, hydrological, and meteorological studies. An effective method for improving the accuracy of soil moisture retrieval is the mutual supplementation of multi-source data. The sensor configuration and band settings of different optical sensors lead to differences in band reflectivity in the inter-data, further resulting in the differences between vegetation indices. The combination of synthetic aperture radar (SAR) data with multi-source optical data has been widely used for soil moisture retrieval. However, the influence of vegetation indices derived from different sources of optical data on retrieval accuracy has not been comparatively analyzed thus far. Therefore, the suitability of vegetation parameters derived from different sources of optical data for accurate soil moisture retrieval requires further investigation. In this study, vegetation indices derived from GF-1, Landsat-8, and Sentinel-2 were compared. Based on Sentinel-1 SAR and three optical data, combined with the water cloud model (WCM) and the advanced integral equation model (AIEM), the accuracy of soil moisture retrieval was investigated. The results indicate that, Sentinel-2 data were more sensitive to vegetation characteristics and had a stronger capability for vegetation signal detection. The ranking of normalized difference vegetation index (NDVI) values from the three sensors was as follows: the largest was in Sentinel-2, followed by Landsat-8, and the value of GF-1 was the smallest. The normalized difference water index (NDWI) value of Landsat-8 was larger than that of Sentinel-2. With reference to the relative components in the WCM model, the contribution of vegetation scattering exceeded that of soil scattering within a vegetation index range of approximately 0.55–0.6 in NDVI-based models and all ranges in NDWI1-based models. The threshold value of NDWI2 for calculating vegetation water content (VWC) was approximately an NDVI value of 0.4–0.55. In the soil moisture retrieval, Sentinel-2 data achieved higher accuracy than data from the other sources and thus was more suitable for the study for combination with SAR in soil moisture retrieval. Furthermore, compared with NDVI, higher accuracy of soil moisture could be retrieved by using NDWI1 (R^2 = 0.623, RMSE = 4.73%). This study provides a reference for the selection of optical data for combination with SAR in soil moisture retrieval.

Keywords: soil moisture; Sentinel-1/2; Landsat-8; GF-1; vegetation water content

1. Introduction

Soil moisture accounts for more than 0.05% of fresh water resources on the Earth's surface [1]. Soil moisture is an important foundation for water-heat transfer and energy exchange between terrestrial and atmospheric systems, as well as the key bond between surface and groundwater circulation and the carbon cycle between lands [2–4]. Therefore, soil moisture retrieval and monitoring over extensive areas is of great significance.

Methods of soil moisture acquisition include field monitoring [5], remote sensing data retrieval [6], land surface and hydrological model simulation, as well as data assimilation [7,8]. In particular, remote sensing observation has become an important means of obtaining soil moisture information on a global scale, owing to its wide coverage, long duration, low cost, better characterization of spatial distribution, and monitoring of surface changes [9].

Among several remote sensing monitoring methods, optical methods based on spectral reflectance indices and thermal infrared methods based on thermal inertia are not applicable, as vegetation canopy can cover soil radiation information and thus affect the accuracy of soil moisture retrieval [5,6]. Consequently, they are more suitable for soil moisture monitoring in areas with bare soil and sparse vegetation. In areas with moderate to lush vegetation coverage, the microwave retrieval method has become one of the effective methods for soil moisture retrieval, due to its advantages such as long wavelength, strong penetration, and no influence from clouds [9]. However, the influence of vegetation on microwave soil moisture retrieval is inevitable. In order to correct the uncertainty of microwave scattering caused by vegetation, optical data are incorporated to obtain the vegetation parameters, through which the scattering and attenuation characteristics of vegetation are estimated and the effect on the total backscatter is eliminated [10–12].

The Sentinel 1A (launched in 2014) and Sentinel 1B (launched in 2016) satellites provide c-band synthetic aperture radar (SAR) data free of charge. Meanwhile, optical data have been increasingly used in soil moisture estimation, such as data from Landsat-8 [13], GF-1 [14], Sentinel-2 [15], and HJ-1 [16]. The combination of SAR data with multi-source optical data has been widely used for soil moisture retrieval. With Sentinel-1, MODIS, and Landsat-8 data, Qiu et al. [11] explored the impact of different vegetation indices on the parameterization of vegetation water content (VWC) and the accuracy of soil moisture retrieval by combining the WCM and advanced integral equation model (AIEM). Tao et al. [14] used multi-temporal SAR data and GF-1 data to propose improved vegetation backscattering model, and the results showed that the model was fairly accurate for soil moisture estimation. The multi-temporal VV polarized SAR data of Sentinel-1 and normalized difference vegetation index (NDVI) data in Sentinel-2 optical image were used. Based on the water cloud model (WCM), the accuracy of soil moisture retrieval was verified with ground measurements and neural network moisture products [17]. Using the support vector regression (SVR) technique, Attarzadeh et al. [18] combined features extracted from Sentinel-1 and Sentinel-2 data to develop a soil moisture retrieval technology, based on object-based image analysis in vegetation areas. The WCM model was parameterized based on SAR observations of different frequencies, ground measurements, and NDVI data derived from Sentinel-2, and then the neural network was used to retrieve SAR signals of different frequencies and estimate the soil moisture. The results showed that the estimated soil moisture provided by the L-band was slightly lower than that by the C-band [19]. In these studies, optical data were used to estimate the vegetation scattering part of the model by calculating vegetation parameters, for the purpose of estimating the impact of vegetation and improving the accuracy of soil moisture retrieval. However, vegetation indices derived from optical data vary among different sensors, which could lead to differences in the estimation of the scattering part of vegetation in the model of vegetation microwave scattering, and thus affect the accuracy of soil moisture retrieval. Therefore, the selection of optical data for the estimation of vegetation scattering in the model is extremely important, indicating the necessity of comparative analysis and evaluation of vegetation indices derived from optical data for soil moisture retrieval to be conducted in this direction.

In the present study, we selected a vegetation microwave scattering model (semi-empirical WCM) coupled with a physically-based AIEM, and combined them with a look-up table algorithm to carry out adaptive evaluation of soil moisture retrieval, based on vegetation indices from different sources of optical data. To quantitatively evaluate the accuracy of different optical data in soil moisture retrieval, seven VWC values calculated according to seven vegetation indices based on Landsat-8, Sentinel-2, and GF-1 data were added to the coupled vegetation microwave scattering model. This study focused on only the wheat growing areas with limited in situ soil moisture measurements. The following section introduces an overview of the study area and describes the adopted dataset. Section 3 presents the radar scattering model, algorithm of retrieval, and vegetation indices. Comparative analysis for the influence of vegetation indices derived from different sources of optical data on retrieval accuracy is given in Section 4. Finally, the main conclusions for this study are given in Section 5.

2. Data and Pre-Processing

2.1. Study Area and Ground Measurements

The study area is located in the Dingxing County, north of Baoding City, Hebei Province (Figure 1). It is located in the middle latitudes and belongs to the Haihe plain area, with an open terrain. This area belongs to the semi-arid climate in the warm and temperate zone of the eastern monsoon, which is significantly characterized by high temperatures, humid and rainy summers, and cold winters. With a long history of irrigation, it has been a well-known water conservancy area since ancient times. The main crops are winter wheat and corn, where the growth of winter wheat is from early October to early June next year. The annual mean temperature is 11.7 °C, the hottest month is July, and the monthly mean temperature is 26.2 °C.

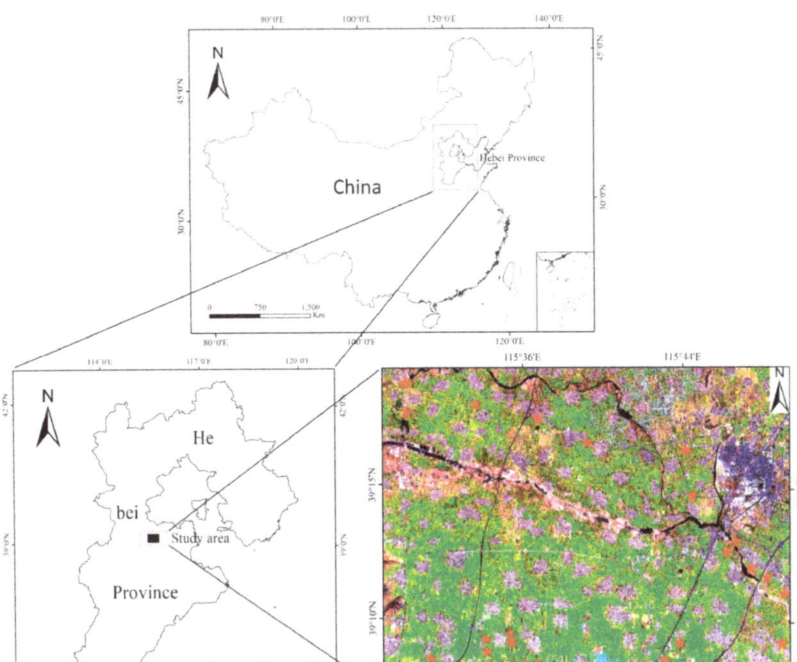

Figure 1. Location of the study area and the ground measured sites.

The soil moisture values of the 23 measured sites used in the study were all obtained by drying. Each measuring point of soil moisture was collected at a depth of 0–5 cm. First, a certain volume of soil was processed in a dryer, and then the change in soil weight before and after drying, was used to calculate the water content of the soil volume. The value for each measured station was the average of the 5 measurement points within 3 m of the station radius. The soil surface roughness at each site was measured from two directions, i.e., parallel and perpendicular to the wheat planting, using an 87 cm plate profiler and a camera. After digitizing the shooting results, the root mean square height (s) and correlation length (l) were calculated by the designed MATLAB program.

2.2. Sentinel-1 SAR Data Collection and Processing

The sentinel-1 constellation provides free C-band SAR data with high-imaging resolution (Table 1) and short revisit cycle (6 days) [18]. Compared with ESA's previous radar satellites (ERS-1, ERS-2, and ENVISAT), it combines features such as multi-polarization, multi-incident angle, and large width, with high-resolution ultra-wide mode and high-resolution wide interference mode. Compared to vertical transmission and horizontal reception (VH) polarization, vertical transmission and vertical reception (VV) polarization is less sensitive to vegetation cover volume scattering [20–22]. Zeng et al. [23] and Bao et al. [10] found that the backscattering coefficient of VV polarization is more suitable for soil moisture retrieval than that of VH polarization. As phase information was not considered, the ground distance imaging product, VV polarization, and imaging mode IW were selected.

Table 1. Related parameters of different observation modes.

Mode	Resolution/m	Polarization	Incidence Angle	Width(km)
Interferometric Wide Swath	5 × 20	VV/VH, HH/HV, VV, HH	29.1°–46°	250
Stripmap	5 × 5	VV/VH, HH/HV, VV, HH	18.3°–46.8°	80
Extra Wide Swath	20 × 40	VV/VH, HH/HV, VV, HH	18.9°–47°	400
Wave	5 × 5	VV, HH	21.6°–25.1° 34.8°–38°	20 × 20

The Sentinel-1 data in the study is a LEVEL-1 product. The software SNAP provided by the European Space Agency (http://www.esa.int/ESA) was used to process the original data. Preprocessing steps include Apply orbit file, Radiation calibration, Coherent speckle filtering (Refined Lee filter), and the Doppler terrain correction. The results to be preprocessed were converted to a logarithmic scale to obtain a true backscatter coefficient image.

2.3. Optical Data Collection and Processing

Sentinel-2 is a high spatial resolution (10, 20, 60 m), high temporal resolution (5 days), multispectral (13 bands) imaging satellite carrying a multispectral imager (MSI), comprising two satellites (2A and 2B) [19]. Sentinel-2A provides five levels of data products—L0, L1A, L1B, L1C, and L2A. ESA releases only L1C products to users. Sentinel-2 data were provided by the United States Geological Survey (https://glovis.usgs.gov/). Based on the plug-in-Sen2cor of the Snap software, pre-processing steps such as atmospheric correction and radiation calibration was performed to obtain the surface reflectance L2A product.

Landsat-8 is the eighth satellite in the Landsat satellite series, carrying two main payloads—Operational Land Image (OLI) and Thermal Infrared Sensor (TIRS) [13]. The satellite has a total of 11 bands. The spatial resolution is 30 m, except for the thermal infrared band (100 m) and the panchromatic band (15 m). Compared to other Landsat satellites, Landsat-8 is a significant improvement in terms of the number of bands, number of images, and level of data quantization. Two Landsat-8 images were downloaded from the geospatial data cloud (http://www.gscloud.cn/), and the original images were pre-processed with the ENVI5.3 software. The main operation steps include radiation calibration, atmospheric correction, and geometric correction.

Gaofen-1 is the first satellite of China's high-resolution Earth observation system, and was launched on April 26, 2013 [14]. The satellite is equipped with a panchromatic/multispectral camera (PMS sensor, 2m/8m) and a multispectral camera (WFV sensor, 16 m), with a short revisit cycle, high platform stability and a design life of no less than 5 years. The GF-1 WFV data in this study was downloaded from the China Centre for Resources Satellite Data and Application (http://www.cresda.com/CN/). Preprocessing steps, including radiation calibration, atmospheric correction, ortho-rectification, and geometric correction, were consecutively completed in the ENVI5.3 software.

The nearest neighbor sampling method has been widely used in many data comparison studies [24–26]. According to the requirement of the research of comparative analysis, preprocessed Landsat-8, Sentinel-2, and GF-1 data were sampled with a spatial resolution of 30 m, using the method of nearest neighbor sampling. As the vegetation in the agricultural area did not change greatly within one week, the difference in the number of days of the three types of optical image data corresponding to each SAR data was set within one week (Table 2).

Table 2. Sentinel-1 and optical data used in this study.

Sentinel-1 SAR Data			
Date	polarization	Observation Mode	Incidence Angle
16 November 2017	VV/VH	IW	39°–40.5°
9 April 2018	VV/VH	IW	39°–40.5°
Sentinel-2 MSI Data			
Date	Spatial Resolution(m)	Spectrum Range(m)	Width(km)
14 November 2017	10–60	0.4–2.4	290
8 April 2018	10–60	0.4–2.4	290
GF-1 WFV data			
Date	Spatial Resolution(m)	Spectrum Range(m)	Width(km)
16 November 2017	16	0.45–0.89	800
9 April 2018	16	0.45–0.89	800
Landsat-8 OLI Data			
Date	Spatial Resolution(m)	Spectrum Range(m)	Width(km)
15 November 2017	15–30	0.43–1.38	185
8 April 2018	15–30	0.43–1.38	185

3. Methodology

In this study, the structure of comparative analysis on the soil moisture retrieval based on GF-1, Landsat-8, and Sentinel-2 data is described by the flow chart in Figure 2.

The first step involved the processing of data used in this paper such as GF-1, Landsat-8, and Sentinel-1/2. Vegetation indices extracted from these processed optical data were used to calculate VWCs using the empirical VWC model. Based on WCM, the total backscatter coefficient achieved from processed Sentinel-1 SAR was combined with VWCs to obtain the bare soil backscatter coefficient. Referring to the range of in situ soil moisture and in situ s and l, results of the soil moisture retrieval based on the LUT algorithm and the database established by AIEM were obtained. Finally, results of soil moisture retrieval based on different sources of optical data were comparatively analyzed.

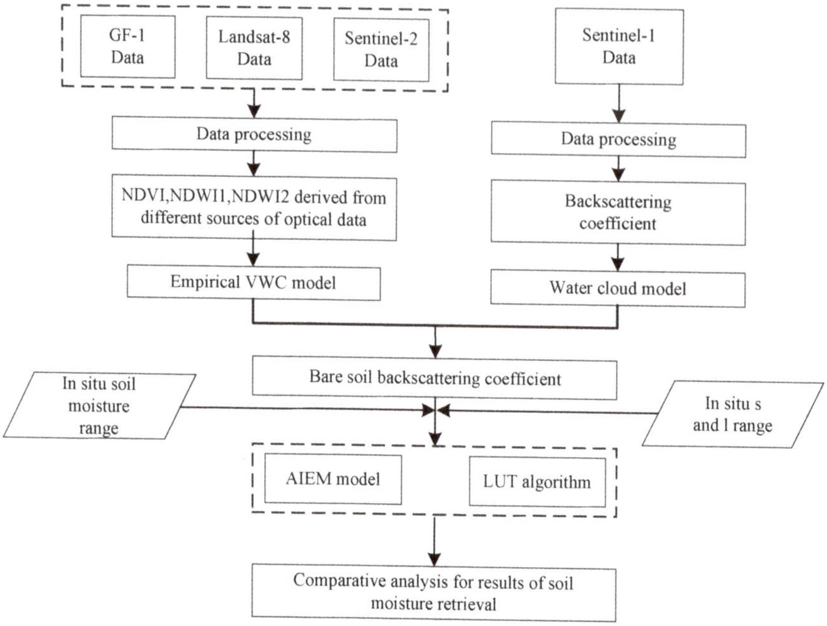

Figure 2. Flowchart of the soil moisture retrieval based on GF-1, Landsat-8, and Sentinel-1/2.

3.1. Microwave Scattering Model Based on Bare and Vegetation Cover

As this study focused on the retrieval of surface soil moisture in a wheat farmland, a WCM suitable for crop cover was selected [27]. This semi-empirical WCM model, which briefly described the scattering mechanism of crop cover, with few input parameter, had been widely used to simulate SAR data [28–30]. Under a given polarization mode, the total backscatter term (σ^0_{total}) of this model was composed of the body scattering term directly scattered by the vegetation layer (σ^0_{veg}) and the soil scattering term (σ^0_{soil}):

$$\sigma^0_{total} = \sigma^0_{veg} + \tau^2\sigma^0_{soil} \tag{1}$$

$$\sigma^0_{veg} = AE_1 cos(1 - \tau^2) \tag{2}$$

$$\tau^2 = \exp(-2BE_2/\cos\theta) \tag{3}$$

where E_1 and E_2 are descriptive parameters of vegetation canopy; θ is the radar incident angle (degrees); A and B are empirical coefficient which need to be adjusted according to radar configurations and vegetation types; τ^2 is the two-way attenuation.

In this study, the values of A and B refer to the results of Bindlish et al. [31], in which empirical parameters of WCMs with different land cover modes were investigated (Table 3). As the crop in the study area is wheat, winter wheat was selected, that is, A = 0.0018 and B = 0.1380. In most current studies, E1 and E2 are generally treated as equal, and they are usually characterized by NDVI, LAI, or VWC [21,28,32,33]. El Hajj et al. [9] reported similar accuracies for soil moisture retrieval when WCM was combined with vegetation description values such as LAI, VWC, biomass, and FAPAR. In this study, E1 = E2 = VWC (VWC values were obtained from the relationship between vegetation index and VWC; VWC is also wheat water content). According to the Sentinel-1 original data, the incident angle information was derived to minimize the influence of the incident angle on the retrieval accuracy.

Table 3. Empirical parameters of the water cloud model.

Parameter	All Vegetation	Rangeland	Winter Wheat	Pasture
A	0.0012	0.0009	0.0018	0.0014
B	0.091	0.032	0.138	0.084

The AIEM model developed on the basis of the IEM model [34,35], which can better simulate the backscattering of various bare surfaces [36,37], was used to replace the soil scattering term in Equation (1). In AIEM, the backscattering coefficient is a function of the dielectric constant, sensor parameters, radar frequency, incident angle, roughness parameter (l, s), and autocorrelation function (p). It can be expressed as follows:

$$\sigma_{pq}^{0} = \frac{k^2}{2} e^{-2k_z^2 s^2} \sum_{i=1}^{\infty} s^{2i} \left| I_{pq}^{i} \right|^2 \frac{W^i(-2k_x, 0)}{i!} \tag{4}$$

$$I_{pq}^{i} = (2k_z)^i f_{pq} e^{-k_z^2 s^2} + \frac{k_z^i}{2} \left[F_{pq}(-k_x, 0) + F_{pq}(k_x, 0) \right] \tag{5}$$

$$k_z = k\cos\theta \tag{6}$$

$$k_x = k\sin\theta \tag{7}$$

where pq stands for co-polarization or cross-polarization, k is the spatial wavenumber (=$2\pi f$), θ is the radar incident angle, $W^i(m, n)$ is the power Fourier transform of autocorrelation function, s is the root-mean-square height of the soil surface, and F_{pq} and f_{pq} are the complementary field and Kirchhoff coefficients, respectively. Here, the soil dielectric constant calculated by the Dobson model was selected as the input variable of the AIEM model. This model is a function of the incident wave frequency, soil moisture, soil texture, and soil temperature. It is used in both the AMSR-E/2 and SMOS soil moisture products algorithms [38–40].

3.2. Vegetation Indices and Vegetation Water Content

The VWC is an important parameter of the WCM model. We characterized it by the equation relationship between vegetation indices and VWC. In our study, Landsat-8, GF-1, and Sentinel-2 data were used for calculating NDVI and normalized difference water index (NDWI). As the Landsat-8 and Sentinel-2 have two short-wave infrared bands, they can provide NDWI1 and NDWI2 (NDWI1 corresponds to band 11 of Sentinel-2 and band 6 of Landsat-8, NDWI2 corresponds to band 12 of Sentinel-2 and band 7 of Landsat-8). Numerous studies have established linear, exponential, polynomial, and other relationships through VWC and vegetation indices [41–43]. Gao et al. [44] integrated a large number of existing vegetation moisture content models and proposed a more superior VWC model. To quantitatively describe the difference between different data sources, this study used an expression of VWC based on wheat [44]. The VWC expression was as follows:

$$\text{VWC} = 0.078 \times 10^{3.510*NDVI} \tag{8}$$

$$\text{VWC} = 2.45*NDWI1 + 0.57 \tag{9}$$

$$\text{VWC} = 12.38*NDWI2 - 3.26 \tag{10}$$

3.3. Look Up Table (LUT) Algorithm Creation

The AIEM model was used to simulate the backscattering coefficient values of the bare surface and generate a database. An optimal cost function was established to ensure that the backscatter coefficient simulated by the AIEM model best matched the backscatter coefficient of the radar data after removing the effects of vegetation. The final soil moisture retrieval result was achieved by finding the soil moisture information corresponding to the simulated backscatter coefficient that minimized the cost function. Soil roughness is an important factor affecting the accuracy of soil moisture retrieval, and many parameters had been developed to describe soil roughness. In this study, the threshold range of the roughness parameter was set in the LUT algorithm, which covers various possible conditions of the surface. Referring to the measured data, the root mean square height, correlation length, and soil moisture value of the study area were limited to [0, 4.0], [0, 13], and [0, 35], respectively. The increments for these three parameters were set as 0.1 cm, 1.0 cm, and 1%, respectively. The cost function was as follows:

$$S = \sqrt{\left(\sigma_{VV} - \sigma_{vv}^{0}\right)^{2}} \tag{11}$$

where S is the cost function, σ_{VV} is the backscattering coefficient of the VV polarization extracted from the radar image, and σ_{vv}^{0} is the backscattering coefficient simulated by the AIEM model.

4. Results and Discussion

4.1. Differences in Different Optical Data Indices

We systematically compared the differences between different optical data indices for the overall study area to the local measured sites (NDVI, NDWI1, NDWI2). Figure 3 and Table 4 provide the comparison results of 3 pairs of optical data (Since GF-1 did not have a shortwave infrared band, i.e., 7 pairs of indices).

Table 4. Statistics of three optical indices in the study area.

		GF-1	Landsat-8			Sentinel-2		
		NDVI	NDVI	NDWI1	NDWI2	NDVI	NDWI1	NDWI2
2017	Min	−0.111	−0.622	−0.363	−0.658	−0.445	−0.651	−0.821
	Max	0.591	0.778	0.851	0.860	0.817	0.584	0.728
	Mean	0.286	0.308	−0.011	0.101	0.316	−0.031	0.064
	Std dev	0.082	0.128	0.096	0.123	0.142	0.112	0.136
2018	Min	−0.032	−0.440	−0.269	−0.558	−0.385	−0.588	−0.577
	Max	0.683	0.714	0.410	0.631	0.806	0.471	0.625
	Mean	0.265	0.271	0.008	0.106	0.296	−0.029	0.040
	Std dev	0.099	0.124	0.088	0.121	0.145	0.096	0.134

From the statistical characteristics and pixel number distribution map (a–c) of the 3 pairs of NDVI, the mean values and standard deviation of sentinel-2 NDVI data appeared to always be higher than the mean values and standard deviation of GF-1 NDVI and Landsat-8 NDVI, regardless of differences in the remote sensor, season, and vegetation growth stage. GF-1 NDVI showed the lowest standard deviation and mean value. Sentinel-2 NDVI showed the most pixel counts in the high value part. The pixel number ratio (number of pixels for statistical threshold greater than 0.4/total number of pixels), was found to be the highest for Sentinel-2 NDVI (Table 5). The values of NDVI from the three sensors, ranked from low to high, were in the following order: GF-1 < Landsat-8 < Sentinel-2.

Table 5. Percentage statistics of NDVI threshold greater than 0.4.

	GF-1 NDVI	Sentinel-2 NDVI	Landsat-8 NDVI
2017	15%	28%	24%
2018	16%	23%	18%

As GF-1 did not have a short-wave infrared band, only NDWI1 and NDWI2 of Landsat-8 and Sentinel-2 were compared. According to the statistical characteristics of NDWI1 and NDWI2 in Table 4 and Figure 3d–g, the mean values and standard deviation of NDWI2 were found to always be greater than the mean values and the standard deviation of NDWI1, and the overall value of NDWI2 was higher than that of NDWI1. Compared to the NDWI1 and NDWI2 data of Landsat-8, the mean values of Sentinel-2 NDWI1 and NDWI2 data were smaller. The ratio of pixel number (number of pixels for statistical threshold greater than 0/total number of pixels), showed that Landsat-8 NDWI1 and NDWI2 accounted for a higher proportion (Table 6). The NDWI value (including NDWI1, NDWI2) of Landsat-8 was larger than that of Sentinel-2.

Table 6. Percentage statistics of NDWI threshold greater than 0.

	Landsat-8		Sentinel-2	
	NDWI1	NDWI2	NDWI1	NDWI2
2017	42%	80%	37%	64%
2018	45%	81%	35%	53%

These optical indices show a weak correlation over the measured sites (Figure 3h–n). If the time series of the measured sites are considered or a large number of points are selected for relationship fitting, a better fitting effect will be achieved [45,46]. At the same time, the comparison of sparse sites also reveals that there are some differences in the indices from different optical sensors.

The spectral response functions of the three sensors show different performances (Figure 4). In the infrared band, Landsat-8 and GF-1 were more stable, and Sentinel-2 exhibited a trough at 662 nm, indicating decreased ability to receive radiation signals. In the near-infrared band, due to the effect of water vapor absorption at 825 nm [47], the near-infrared band detection capability of Sentinel-2 and GF-1 are weakened. Sentinel-2 and Landsat-8 exhibit similar performances in the short-wave infrared band. The detection capabilities of the different band signals of each sensor vary, which is one of the reason for the difference between the indices [48,49]. Compared with the other two satellites, Sentinel-2 has a high spatial resolution and can detect ground objects in greater detail, which may be another reason for the difference in the indices [45]. Relevant research show that the difference in optical data is also related to transit time, preprocessing steps, etc. [47,50]. In this study, the values of Sentinel-2 NDVI were higher and the values of GF-1 NDVI were smaller, which is consistent with the relevant results of data comparison of Lessio et al [51] and He et al [52], respectively. According to the conclusion of Xu and Zhang [45] that differences in the mean values of NDVI reflect the strength of vegetation signal detection, the results show that Sentinel-2 data are more sensitive to vegetation characteristics and have a stronger ability to detect vegetation signals, which is consistent with the conclusion of Pan et al. [53] in the comparative study of optical data. By analyzing the spectral characteristic curve of vegetation (right in Figure 4), surface reflectance was found to gradually decrease with increasing wavelength of short-wave infrared, resulting in a larger NDWI2 value in the longer band (2100–2300 nm). It is interesting that Yantao et al. [46] compared the average reflectance of the two optical data bands. The average reflectance of Landsat-8 in the near-infrared band is higher than that of Sentinel-2. However, the short-wave infrared band tends to be equal to the increase in wavelength, which shows that the values of NDWI1 and NDWI2 of Sentinel-2 are less than the corresponding indices of Landsat-8, which is consistent with the results of this study.

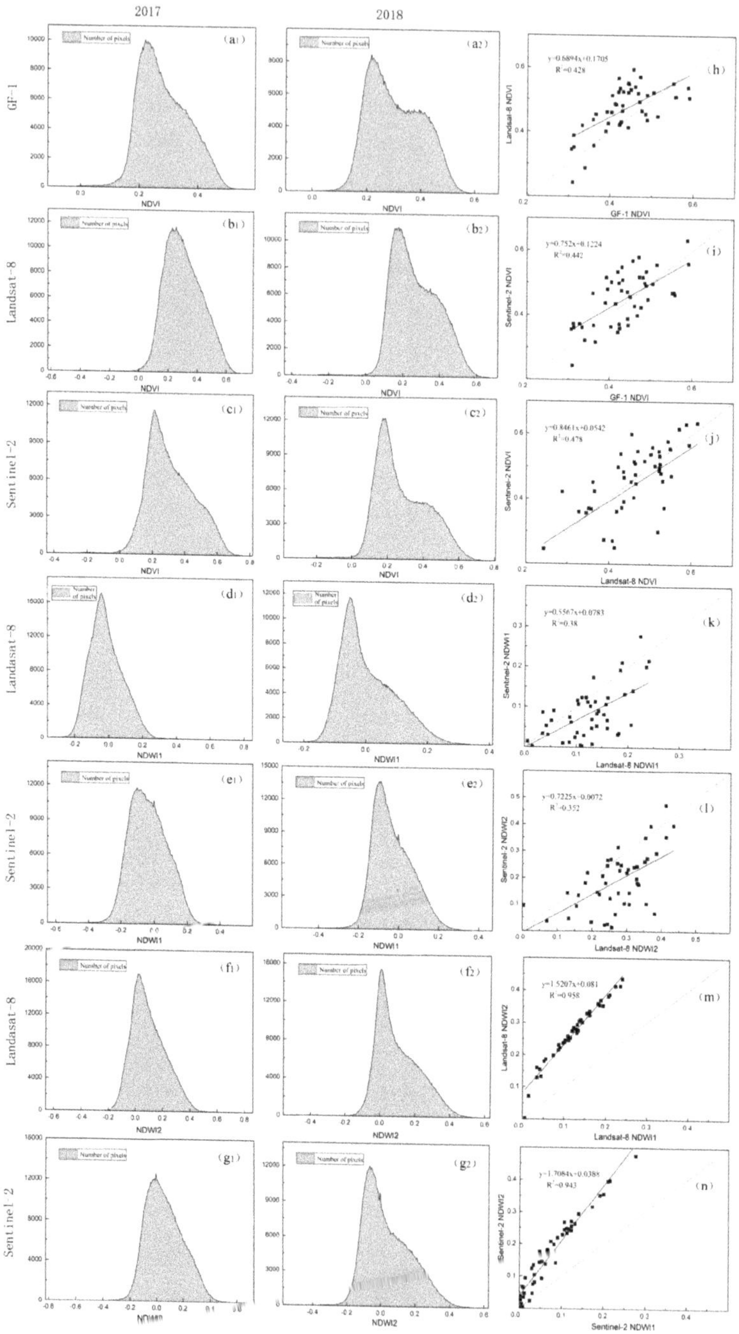

Figure 3. Pixel number distribution of each index data (left and middle) and relationship between different optical indices collected from 46 measured sites in two time periods (right) (2017 and 2018, represent the optical data used in this study in 2017 and the optical data used in 2018, respectively).

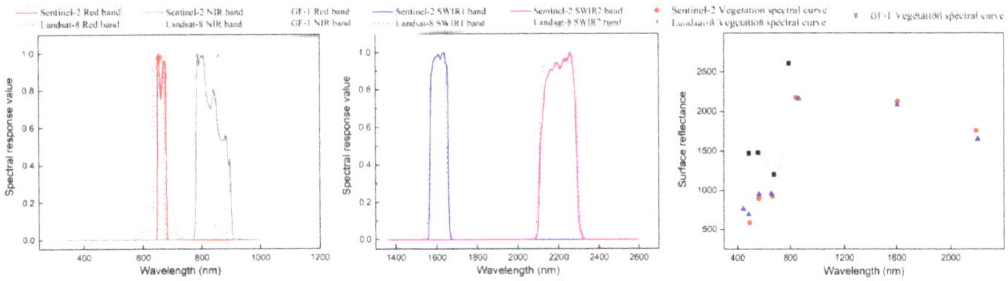

Figure 4. Spectral responses of the red, near-infrared (NIR), and short-wave infrared (SWIR1, SWIR2) bands for GF-1, Landsat-8 and Sentinel-2 (left and middle), and vegetation spectral curves of three optical sensors (right).

4.2. Contribution of Bare Soil Scattering to Total Scattering with Different Indices

Differences between indices cause deviations in vegetation descriptors, which in turn leads to a certain difference in the contribution of vegetation scattering described. In the model, the ratio of the contribution of soil scattering to total scattering can reflect contribution of vegetation scattering to a certain extent. Combining the data of the same indices from the same remote sensor in two time periods, a graph of the ratio of the contribution of soil scattering to the total scattering as a function of indices was prepared based on the measured sites (Figure 5). Among them, according to formula 10, when the calculation of VWC for NDWI2 satisfies NDWI2 > 0.27, VWC is a positive value. Accordingly, only data with threshold greater than 0.27 was selected in NDWI2. In Figure 5, the scattering ratio of all the measured sites gradually decrease with increasing values of the indices, indicating that the contribution of vegetation scattering increases with the increasing values of the indices, while the contribution of soil scattering shows the opposite trend.

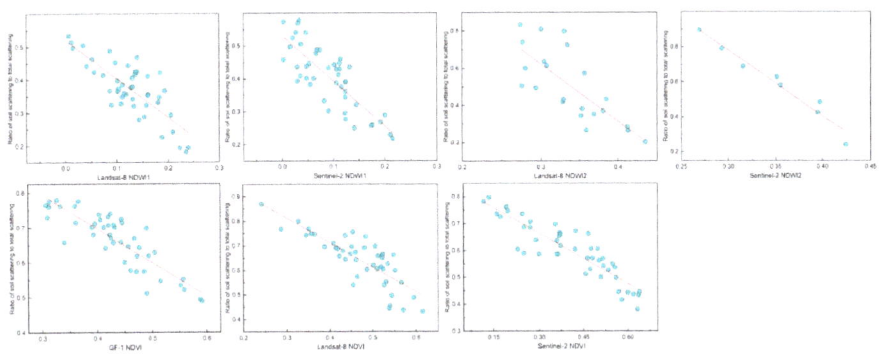

Figure 5. Ratio of bare soil contribution to total scattering ($\sigma^0_{soil}/\sigma^0_{total}$) for all sites using seven indices. Each graph fits a curve through the regression equation to show the trend.

Regardless of the sensor type, the threshold value for the contribution of vegetation scattering exceeded that of soil scattering within approximately 0.55–0.6 for NDVI and any range for NDWI1. Compared with NDVI, NDWI1 has a stronger response to vegetation scattering. In addition, Landsat-8 has the strongest response to vegetation scattering in NDWI1-based models. Sentinel-2 has the strongest response to vegetation scattering in NDVI-based models. In NDWI2 shown in Figure 3, regardless of the type of sensor, eligible points were extremely rare. This is because most pixels of NDWI2 were below 0.27. The investigation results show that when Landsat-8 NDVI > 0.48 and Sentinel-2 NDVI > 0.53, NDWI2 is above 0.27. A further review of the works of Maggioni et al. [54] and Yi et al. [55] revealed that the calculation of VWC for NDWI2 has certain limitations, that is, NDWI2 requires a certain threshold. Combined with the survey, the threshold at which a positive value can be achieved for VWC calculated with NDWI2 is approximately 0.4–0.55 for NDVI. VWC calculated with NDWI2 is meaningful in areas with medium and high vegetation coverage.

4.3. Effect of Indices from Different Optical Data on Accuracy of Soil Moisture Retrieval

The degree of elimination for the contribution of vegetation scattering can affect the accuracy of soil moisture retrieval. As NDWI2 for the study area cannot be used to effectively retrieve VWC, NDWI2 was excluded from further the comparative experiments of soil moisture retrieval. The retrieval results based on the indices from different optical data show that Sentinel-2 NDVI and Sentinel-2 NDWI1 achieve higher retrieval accuracy (Figure 6, Table 7). Nevertheless, the accuracy of soil moisture retrieval based on NDWI1 was higher than that based on NDVI in the Sentinel-2 data (R^2 = 0.623, RMSE = 4.73%).

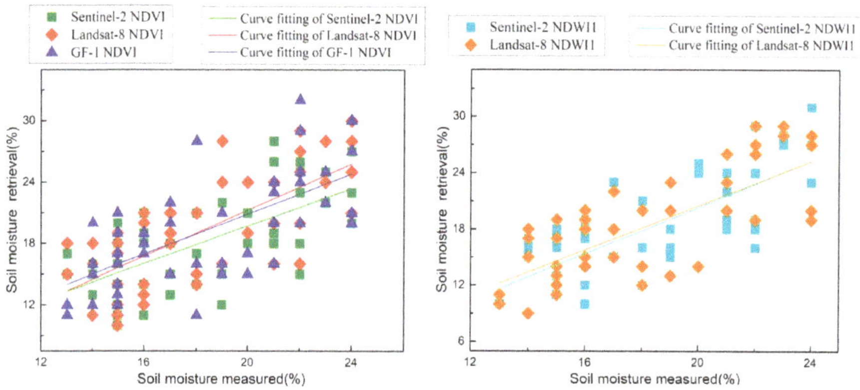

Figure 6. Scatter plot between the value of soil moisture retrieval and the value of measured soil moisture.

Table 7. Statistical parameters of soil moisture retrieval with different optical data indices.

	Sentinel-2		Landsat-8		GF-1
	NDVI	NDWI1	NDVI	NDWI1	NDVI
R^2	0.54	0.623	0.51	0.56	0.52
RMSE (%)	5.78	4.73	6.18	5.21	6.44

NDVI < 0.7 is more sensitive to soil moisture [56], making the accuracy of soil moisture retrieval in this interval more reliable. The NDVI values of each sensor were different, but they were all less than 0.7 in this study. Sentinel-2 NDVI has a stronger ability to detect vegetation signals, and more accurately describes the true situation of NDVI, thereby effectively eliminating the contribution of vegetation scattering and making soil moisture retrieval with Sentinel-2 data more accurate. Short-wave infrared is more sensitive to VWC [57,58], and NDWI1 had a higher correlation with VWC [59], providing more accurate elimination of the contribution of vegetation scattering and higher accuracy of soil moisture retrieval. NDWI derived from Sentinel-2 is more advantageous than that from Landsat-8 [46]. Although Landsat-8 is slightly higher in band reflectivity than Sentinel-2, it may cause excessive elimination of vegetation scattering contribution, resulting in lower accuracy of soil moisture retrieval.

5. Conclusions

The removal of contribution of vegetation scattering in vegetation coverage areas has a significant impact on the retrieval accuracy. As optical remote sensing satellites are continuously being launched into the space, the diversity and difference of optical data will affect the degree of contribution of vegetation scattering, which will affect the retrieval accuracy. Selecting appropriate optical data to achieve the goal of better eliminating the contribution of vegetation scattering is one of the key factors for improving the reliability of retrieval results. For wheat vegetation, a comparative study on the accuracy of seven indices derived from three sources of optical data for soil moisture retrieval was carried out. Based on the WCM vegetation scattering model and the AIEM soil scattering model, the effectiveness of different types of optical data for soil moisture retrieval was investigated.

Compared with GF-1 and Landsat-8, the mean value and standard deviation of Sentinel-2 NDVI were the highest, indicating that sentinel-2 data was more sensitive to vegetation characteristics and had stronger capability for vegetation signal detection. The ranking of NDVI values from the three sensors followed the order: GF-1 < Landsat-8 < Sentinel-2. The NDWI1 and NDWI2 values of Landsat-8 were larger than those of Sentinel-2.Differences between the indices data derived from different sources of optical data lead to differences in the contribution of vegetation scattering. Regardless of the sensor type, the threshold value for the contribution of vegetation scattering exceeded that of soil scattering within approximately 0.55–0.6 for NDVI and any range for NDWI1. The threshold value at which NDWI2 can be used to calculate VWC was approximately 0.4–0.55 for NDVI. VWC calculated with NDWI2 was meaningful over areas with medium and high vegetation coverage. In the soil moisture retrieval, Sentinel-2 data achieved better retrieval accuracy for both NDVI and NDWI1 in the estimation of the contribution of vegetation scattering. And Sentinel-2 data was suitable for retrieval of soil moisture than data from the other sources. Furthermore, compared with NDVI, NDWI1 showed higher accuracy of soil moisture retrieval ($R^2 = 0.623$, RMSE = 4.73%).

Owing to the limited data available for this study, only factors such as vegetation growth stage and seasonal difference were considered in the comparative analysis of optical data. In the future, topography, climate and other factors will be considered for the comprehensive assessment of optical data, so as to provide more scientific reference for the selection of optical data for soil moisture retrieval. In addition, SAR data in the L-band shows great potential in soil moisture retrieval, and the use of SAR data in the L-band for studies of soil moisture retrieval will also be considered in the future.

Author Contributions: Q.W., J.S. and Y.Z. collected the original data; Q.W. and T.J. jointly designed the study and wrote the manuscript; X.C., D.L. and Y.L. helped on the revision and discussion; T.J. and J.L. provided supervision. All authors have read and agreed to the published version of the manuscript.

Funding: This research was funded by National Natural Science Foundation of China under grant 41721003, Natural Science Foundation of Hubei Province of China under grant 2019CFB427, National Basic Research Program of China under grant 2013CB733302, Scientific Research Project of City College, Southwest University of Science and Technology(Grant No. 2020XJXM03); the Key Laboratory of Geospace Environment and Geodesy, Ministry of Education, Wuhan University, under grant 17-02-07 and the Key Laboratory of Surveying and Mapping Science and Geospatial Information Technology of Ministry of Natural Resources (2020-1-3).

Acknowledgments: All the respectable reviewers and editors are acknowledged for their fruitful comments and suggestions about the paper.

Conflicts of Interest: The authors declare no conflict of interest.

References

1. Gleick, P.H. Water resources. In *Encyclopedia of Climate and Weather*; Schneider, S.H., Ed.; Oxford University Press: New York, NY, USA, 1996; pp. 817–823.
2. Scholze, M.; Kaminski, T.; Knorr, W.; Blessing, S.; Vossbeck, M.; Grant, J.P.; Scipal, K. Simultaneous assimilation of SMOS soil moisture and atmospheric CO2 in-situ observations to constrain the global terrestrial carbon cycle. *Remote Sens. Environ.* **2016**, *180*, 334–345. [CrossRef]
3. Larson, K.M.; Small, E.E.; Gutmann, E.D.; Bilich, A.L.; Braun, J.J.; Zavorotny, V.U. Use of GPS receivers as a soil moisture network for water cycle studies. *Geophys. Res. Lett.* **2008**, *35*, 24. [CrossRef]
4. Petrone, R.M.; Price, J.S.; Waddington, J.M.; Von Waldow, H. Surface moisture and energy exchange from a restored peatland, Quebec, Canada. *J. Hydrol.* **2004**, *295*, 198–210. [CrossRef]
5. Zhang, D.; Zhou, G. Estimation of soil moisture from optical and thermal remote sensing: A. review. *Sensors* **2016**, *16*, 1308. [CrossRef] [PubMed]

6. Baghdadi, N.N.; El Hajj, M.; Zribi, M.; Fayad, I. Coupling SAR C-band and optical data for soil moisture and leaf area index retrieval over irrigated grasslands. *IEEE J. Sel. Top. Appl. Earth Obs. Remote Sens.* **2015**, *9*, 1229–1243. [CrossRef]
7. Houser, P.R.; Shuttleworth, W.J.; Famiglietti, J.S.; Gupta, H.V.; Syed, K.H.; Goodrich, D.C. Integration of soil moisture remote sensing and hydrologic modeling using data assimilation. *Water Resour. Res.* **1998**, *34*, 3405–3420. [CrossRef]
8. Draper, C.; Mahfouf, J.; Calvet, J.C.; Martin, E.L.; Wagner, W. Assimilation of ASCAT near-surface soil moisture into the SIM hydrological model over France. *Hydrol. Earth Syst. Sci.* **2011**, *15*, 3829. [CrossRef]
9. El Hajj, M.; Baghdadi, N.; Zribi, M.; Belaud, G.; Cheviron, B.; Courault, D.; Charron, F. Soil moisture retrieval over irrigated grassland using X-band SAR data. *Remote Sens. Environ.* **2016**, *176*, 202–218. [CrossRef]
10. Bao, Y.; Lin, L.; Wu, S.; Deng, K.A.; Petropoulos, G.P. Surface soil moisture retrievals over partially vegetated areas from the synergy of Sentinel-1 and Landsat 8 data using a modified water-cloud model. *Int. J. Appl. Earth Obs. Geoinf.* **2018**, *72*, 76–85. [CrossRef]
11. Qiu, J.; Crow, W.T.; Wagner, W.; Zhao, T. Effect of vegetation index choice on soil moisture retrievals via the synergistic use of synthetic aperture radar and optical remote sensing. *Int. J. Appl. Earth Obs. Geoinf.* **2019**, *80*, 47–57. [CrossRef]
12. Zribi, M.; Muddu, S.; Bousbih, S.; Bitar, A.A.; Tomer, S.K.; Baghdadi, N.; Bandyopadhyay, S. Analysis of L-band SAR data for soil moisture estimations over agricultural areas in the tropics. *Remote Sens.* **2019**, *11*, 1122. [CrossRef]
13. Zhao, X.; Huang, N.; Niu, Z.; Raghavan, V.; Song, X. Soil moisture retrieval in farmland using C-band SAR and optical data. *Spat. Inf. Res.* **2017**, *25*, 431–438. [CrossRef]
14. Tao, L.; Wang, G.; Chen, W.; Chen, X.; Li, J.; Cai, Q. Soil moisture retrieval from SAR and optical data using a combined model. *IEEE J. Sel. Top. Appl. Earth Obs. Remote Sens.* **2019**, *12*, 637–647. [CrossRef]
15. Attarzadeh, R.; Amini, J. Towards an object-based multi-scale soil moisture product using coupled Sentinel-1 and Sentinel-2 data. *Remote Sens. Lett.* **2019**, *10*, 619–628. [CrossRef]
16. He, B.; Xing, M.; Bai, X. A synergistic methodology for soil moisture estimation in an alpine prairie using radar and optical satellite data. *Remote Sens.* **2014**, *6*, 10966–10985. [CrossRef]
17. Bousbih, S.; Zribi, M.; El Hajj, M.; Baghdadi, N.; Lilichabaane, Z.; Gao, Q.; Fanise, P. Soil moisture and irrigation mapping in A semi-arid region, based on the synergetic use of Sentinel-1 and Sentinel-2 data. *Remote Sens.* **2018**, *10*, 1953. [CrossRef]
18. Attarzadeh, R.; Amini, J.; Notarnicola, C.; Greifeneder, F. Synergetic use of Sentinel-1 and Sentinel-2 data for soil moisture mapping at plot scale. *Remote Sens.* **2018**, *10*, 1285. [CrossRef]
19. El Hajj, M.; Baghdadi, N.; Zribi, M. Comparative analysis of the accuracy of surface soil moisture estimation from the C-and L-bands. *Int. J. Appl. Earth Obs. Geoinf.* **2019**, *82*, 101888. [CrossRef]
20. Patel, P.; Srivastava, H.S.; Panigrahy, S.; Parihar, J.S. Comparative evaluation of the sensitivity of multi-polarized multi-frequency SAR backscatter to plant density. *Int. J. Remote Sens.* **2006**, *27*, 293–305. [CrossRef]
21. Baghdadi, N.; El Hajj, M.; Zribi, M.; Bousbih, S. Calibration of the water cloud model at C-band for winter crop fields and grasslands. *Remote Sens.* **2017**, *9*, 969. [CrossRef]
22. Chauhan, S.; Srivastava, H.S. Comparative evaluation of the sensitivity of multi-polarised sar and optical data for various land cover classes. *Int. J. Adv. Remote Sens. GIS Geogr.* **2016**, *4*, 1–14.
23. Zeng, X.; Xing, Y.; Shan, W.; Zhang, Y.; Wang, C. Soil water content retrieval based on Sentinel-1A and Landsat 8 image for Bei'an-Heihe Expressway. *Chin. J. Eco. Agric.* **2017**, *25*, 118–126.
24. Cui, C.; Xu, J.; Zeng, J.; Chen, K.; Bai, X.; Lu, H.; Chen, Q.; Zhao, T. Soil moisture mapping from satellites: An intercomparison of SMAP, SMOS, FY3B, AMSR2, and ESA CCI over two dense network regions at different spatial scales. *Remote Sens.* **2018**, *10*, 33. [CrossRef]
25. Kim, H.; Parinussa, R.; Konings, A.G.; Wagner, W.; Cosh, M.H.; Lakshmi, V.; Zohaib, M.; Choi, M. Global-scale assessment and combination of SMAP with ASCAT (active) and AMSR2 (passive) soil moisture products. *Remote Sens. Environ.* **2018**, *204*, 260–275. [CrossRef]
26. Ma, H.; Zeng, J.; Chen, N.; Zhang, X.; Cosh, M.H.; Wang, W. Satellite surface soil moisture from SMAP, SMOS, AMSR2 and ESA CCI: A comprehensive assessment using global ground-based observations. *Remote Sens. Environ.* **2019**, *231*, 111215. [CrossRef]
27. Attema, E.P.W.; Ulaby, F.T. Vegetation modeled as a water cloud. *Radio Sci.* **1978**, *13*, 357–364. [CrossRef]

28. Paloscia, S.; Pettinato, S.; Santi, E.; Notarnicola, C.; Pasolli, L.; Reppucci, A. Soil moisture mapping using Sentinel-1 images: Algorithm and preliminary validation. *Remote Sens. Environ.* **2013**, *134*, 234–248. [CrossRef]
29. Yang, G.; Shi, Y.; Zhao, C.; Wang, J. Estimation of soil moisture from multi-polarized SAR data over wheat coverage areas. In Proceedings of the 2012 First International Conference on Agro-Geoinformatics (Agro-Geoinformatics), Shanghai, China, 2–4 August 2012.
30. Zribi, M.; Chahbi, A.; Shabou, M.; Lilichabaane, Z.; Duchemin, B.; Baghdadi, N.; Amri, R.; Chehbouni, A. Soil surface moisture estimation over a semi-arid region using ENVISAT ASAR radar data for soil evaporation evaluation. *Hydrol. Earth Syst. Sci. Discuss.* **2011**, *15*, 345–358. [CrossRef]
31. Bindlish, R.; Barros, A.P. Parameterization of vegetation backscatter in radar-based, soil moisture estimation. *Remote Sens. Environ.* **2001**, *76*, 130–137. [CrossRef]
32. Sikdar, M.; Cumming, I. A Modified Empirical Model for Soil Moisture Estimation in Vegetated Areas Using SAR Data. In Proceedings of the IGARSS 2004. 2004 IEEE International Geoscience and Remote Sensing Symposium, Anchorage, AK, USA, 20–24 September 2004; IEEE: Piscataway, NJ, USA, 2004; Volume 2, pp. 803–806.
33. Prevot, L.; Champion, I.; Guyot, G. Estimating surface soil moisture and leaf area index of a wheat canopy using a dual-frequency (C and X bands) scatterometer. *Remote Sens. Environ.* **1993**, *46*, 331–339. [CrossRef]
34. Chen, K.S.; Wu, T.D.; Tsay, M.K.; Fung, A.K. Note on the multiple scattering in an IEM model. *IEEE Trans. Geosci. Remote Sens.* **2000**, *38*, 249–256. [CrossRef]
35. Fung, A.K.; Liu, W.Y.; Chen, K.S.; Tsay, M. An improved IEM model for bistatic scattering from rough surfaces. *J. Electromagn. Waves Appl.* **2002**, *16*, 689–702. [CrossRef]
36. Chen, L.; Shi, J.; Wigneron, J.P.; Chen, K. A parameterized surface emission model at L-band for soil moisture retrieval. *IEEE Geosci. Remote Sens. Lett.* **2009**, *7*, 127–130. [CrossRef]
37. Nearing, G.S.; Moran, M.S.; Thorp, K.R.; Collins, C.D.; Slack, D.C. Likelihood parameter estimation for calibrating a soil moisture model using radar bakscatter. *Remote Sens. Environ.* **2010**, *114*, 2564–2574. [CrossRef]
38. Njoku, E.G.; Jackson, T.J.; Lakshmi, V.; Chan, T.K.; Nghiem, S.V. Soil moisture retrieval from AMSR-E. *IEEE Trans. Geosci. Remote Sens.* **2003**, *41*, 215–229. [CrossRef]
39. Njoku, E.G.; Chan, S.K. Vegetation and surface roughness effects on AMSR-E land observations. *Remote Sens. Environ.* **2006**, *100*, 190–199. [CrossRef]
40. Jacquette, E.; Al Bitar, A.; Mialon, A.; Kerr, Y.; Quesney, A.; Cabot, F.; Richaume, P. SMOS CATDS level 3 global products over land[C]//Remote Sensing for Agriculture, Ecosystems, and Hydrology XII. *Int. Soc. Opt. Photonics* **2010**, *7824*, 78240K.
41. Huang, J.; Chen, D.; Cosh, M.H. Sub-pixel reflectance unmixing in estimating vegetation water content and dry biomass of corn and soybeans cropland using normalized difference water index (NDWI) from satellites. *Int. J. Remote Sens.* **2009**, *30*, 2075–2104. [CrossRef]
42. Jiang, J.; Hu, D.; Li, Y.; Tang, X.; Li, J. Research of soil moisture retrieval model of wheat covered surface based on MIMICS model. *J. Triticeae Crop.* **2015**, *35*, 707–713.
43. Cheng, X.; Yang, G.; Xu, X.; Chen, T.; Wang, D. Inferred water content of winter wheat based on ground hyperspectral and remote sensing data of TM5. *J. Triticeae Crop.* **2014**, *34*, 227–233.
44. Gao, Y.; Walker, J.P.; Allahmoradi, M.; Monerris, A.; Ryu, D.; Jackson, T.J. Optical sensing of vegetation water content: A synthesis study. *IEEE J. Sel. Top. Appl. Earth Obs. Remote Sens.* **2015**, *8*, 1456–1464.
45. Xu, H.; Zhang, T. Assessment of consistency in forest-dominated vegetation observations between ASTER and Landsat ETM+ images in subtropical coastal areas of southeastern China. *Agric. For. Meteorol.* **2013**, *168*, 1–9. [CrossRef]
46. Yantao, X.I.; Thinh, N.X.; Cheng, L.I. Preliminary comparative assessment of various spectral indices for built-up land derived from Landsat-8 OLI and Sentinel-2A MSI imageries. *Eur. J. Remote Sens.* **2019**, *52*, 240–252.
47. Wu, X.P.; Xu, H.Q.; Jiang, Q.L. Cross-Comparison of GF-1, GF-2 and landsat-8 OLI sensor data. *Geomat. Inf. Sci. Wuhan Univ.* **2020**, *45*, 150–158. [CrossRef]
48. Mandanici, E.; Bitelli, G. Preliminary comparison of Sentinel-2 and landsat 8 imagery for a combined use. *Remote Sens.* **2016**, *8*, 1014. [CrossRef]

49. Arekhi, M.; Goksel, C.; Sanli, F.B.; Senel, G. Comparative evaluation of the spectral and spatial consistency of Sentinel-2 and Landsat-8 OLI data for igneada longos forest. *ISPRS Int. J. Geo Inf.* **2019**, *8*, 56. [CrossRef]
50. Zhao, J.; Li, J.; Liu, Q.; Wang, H.; Chen, C.; Xu, B.; Wu, S. Comparative analysis of Chinese HJ-1 CCD, GF-1 WFV and ZY-3 MUX sensor data for leaf area index estimations for maize. *Remote Sens.* **2018**, *10*, 68. [CrossRef]
51. Lessio, A.; Fissore, V.; Borgognomondino, E. Preliminary tests and results concerning integration of Sentinel-2 and Landsat-8 OLI for crop monitoring. *J. Imaging* **2017**, *3*, 49. [CrossRef]
52. He, L.I.; Chen, Z.; Jiang, Z.; Wenbin, W.U.; Ren, J.; Liu, B.; Tuya, H. Comparative analysis of GF-1, HJ-1, and Landsat-8 data for estimating the leaf area index of winter wheat. *J. Integr. Agric.* **2017**, *16*, 266–285.
53. Pan, Y.Y.; Li, C.C.; Ma, X.X.; Wang, B.S.; Fang, X. Atmospheric correction method of sentinel-2A satellite and result analysis. *Remote Sens. Inf.* **2018**, *33*, 41–48.
54. Maggioni, R.; Panciera, J.P.; Walker, M.; Rinaldi, V. A multi-sensor approach for high resolution airborne soil moisture mapping. In In Proceedings of the 30th Hydrology and Water Resource Symp[CD-ROM], Hobart, Australia, 4–8 December 2006.
55. Yi, Y.; Yang, D.; Chen, D.; Huang, J. Retrieving crop physiological parameters and assessing water deficiency using MODIS data during the winter wheat growing period. *Can. J. Remote Sens.* **2007**, *33*, 189–202. [CrossRef]
56. Elhajj, M.; Baghdadi, N.; Bazzi, H.; Zribi, M. Penetration analysis of sar signals in the c and l bands for wheat, maize, and grasslands. *Remote Sens.* **2019**, *11*, 31. [CrossRef]
57. Ceccato, P.; Gobron, N.; Flasse, S.P.; Pinty, B.; Tarantola, S. Designing a spectral index to estimate vegetation water content from remote sensing data: Part 1: Theoretical approach. *Remote Sens. Environ.* **2002**, *82*, 188–197. [CrossRef]
58. Ceccato, P.; Flasse, S.P.; Gregoire, J.M. Designing a spectral index to estimate vegetation water content from remote sensing data: Part 2. Validation and applications. *Remote Sens. Environ.* **2002**, *82*, 198–207. [CrossRef]
59. Chuvieco, E.; Riano, D.; Aguado, I.; Cocero, D. Estimation of fuel moisture content from multitemporal analysis of Landsat Thematic Mapper reflectance data: Applications in fire danger assessment. *Int. J. Remote Sens.* **2002**, *23*, 2145–2162. [CrossRef]

remote sensing

MDPI

Article

Spatial Gap-Filling of ESA CCI Satellite-Derived Soil Moisture Based on Geostatistical Techniques and Multiple Regression

Ricardo M. Llamas [1], Mario Guevara [1], Danny Rorabaugh [2], Michela Taufer [2] and Rodrigo Vargas [1,*]

1 Department of Plant and Soil Sciences, University of Delaware, Newark, DE 19716, USA; rllamas@udel.edu (R.M.L.); mguevara@udel.edu (M.G.)

2 Department of Electrical Engineering and Computer Science, University of Tennessee, Knoxville, TN 37996, USA; dror@utk.edu (D.R.); taufer@utk.edu (M.T.)

* Correspondence: rvargas@udel.edu; Tel.: +1-302-831-1386

Received: 28 January 2020; Accepted: 11 February 2020; Published: 18 February 2020

Abstract: Soil moisture plays a key role in the Earth's water and carbon cycles, but acquisition of continuous (i.e., gap-free) soil moisture measurements across large regions is a challenging task due to limitations of currently available point measurements. Satellites offer critical information for soil moisture over large areas on a regular basis (e.g., European Space Agency Climate Change Initiative (ESA CCI), National Aeronautics and Space Administration Soil Moisture Active Passive (NASA SMAP)); however, there are regions where satellite-derived soil moisture cannot be estimated because of certain conditions such as high canopy density, frozen soil, or extremely dry soil. We compared and tested three approaches, ordinary kriging (OK), regression kriging (RK), and generalized linear models (GLMs), to model soil moisture and fill spatial data gaps from the ESA CCI product version 4.5 from January 2000 to September 2012, over a region of 465,777 km^2 across the Midwest of the USA. We tested our proposed methods to fill gaps in the original ESA CCI product and two data subsets, removing 25% and 50% of the initially available valid pixels. We found a significant correlation (r = 0.558, RMSE = 0.069 m^3m^{-3}) between the original satellite-derived soil moisture product with ground-truth data from the North American Soil Moisture Database (NASMD). Predicted soil moisture using OK also had significant correlation with NASMD data when using 100% (r = 0.579, RMSE = 0.067 m^3m^{-3}), 75% (r = 0.575, RMSE = 0.067 m^3m^{-3}), and 50% (r = 0.569, RMSE = 0.067 m^3m^{-3}) of available valid pixels for each month of the study period. RK showed comparable values to OK when using different percentages of available valid pixels, 100% (r = 0.582, RMSE = 0.067 m^3m^{-3}), 75% (r = 0.582, RMSE = 0.067 m^3m^{-3}), and 50% (r = 0.571, RMSE = 0.067 m^3m^{-3}). GLM had slightly lower correlation with NASMD data (average r = 0.475, RMSE = 0.070 m^3m^{-3}) when using the same subsets of available data (i.e., 100%, 75%, 50%). Our results provide support for using geostatistical approaches (OK and RK) as alternative techniques to gap-fill missing spatial values of satellite-derived soil moisture.

Keywords: soil moisture; remote sensing; geostatistics; gap-filling; mesonet

1. Introduction

Addressing global environmental challenges requires knowledge and information derived from the most accurate and complete available datasets. Soil moisture has an important role in the water and energy cycles and is regarded as one of the essential terrestrial climate variables [1] due to its influence on soil and atmosphere feedbacks. Furthermore, soil moisture is a critical input variable for applications such as climate modeling [2–4], agricultural planning [5,6], and carbon budget analyses [7,8]. Because of the importance of soil moisture, there are many in situ monitoring networks, organized at the global [9], regional [10,11], or national-scale [12–15]. Despite these national to global efforts, there is still a challenge to represent spatially explicit soil moisture information across large regions related to spatial limitations of in situ ground measurements.

Soil moisture can be estimated using remote sensors (e.g., spaceborne radiometers and radar sensors) to provide coarse-scale estimates on a regular basis [9,16]. Examples of remote sensing soil moisture monitoring systems include NASA's Soil Moisture Active Passive (SMAP) [16], ESA's Soil Moisture and Ocean Salinity (SMOS) [17], and the European Space Agency Climate Change Initiative (ESA CCI) [11,18] that deliver publicly available data for a wide range of applications. Despite advances in remote sensing technology, there are still large areas where soil moisture information is not regularly acquired, yielding information gaps in time and space across the world. Missing information arises from certain circumstances such as high canopy density, snow and ice cover, extremely dry surface conditions, or frozen soil [11]. These factors hinder radiometers or radar sensors in measuring the dielectric constant in the top layer of soil in order to estimate the water content [19].

Consequently, there is a need to develop gap-filling strategies to provide spatially complete satellite-derived soil moisture data across the world. In the most recent version of the ESA CCI product (version 4.5), soil moisture values are derived from the combination of active and passive sensors based on a weighted mean, proportional to the signal-to-noise ratios (SNRs) [20]. In areas where soil moisture information cannot be derived using SNRs, values are estimated using a polynomial regression between the signal-to-noise ratios [20]. Version 4.5 masks areas of dense vegetation using vegetation optical depth layers and flags measurements under frozen conditions [21]; consequently the product has multiple gaps across the world [22].

Other statistical methods (e.g., discrete cosine transformations and singular spectrum analysis) have been applied to fill spatial gaps for satellite-derived geophysical datasets, as well as soil moisture from field measurements [23–25]. These approaches are focused either on the statistical distribution of the data or three-dimension information, which includes both space and time. We postulate that alternative gap-filling methods could take advantage of the information contained in the spatial distribution of soil moisture or its spatial and linear relationships with key geophysical variables, such as temperature and precipitation [3,9,26].

In this research, we test the performance of three methods to gap-fill satellite-derived soil moisture in ESA CCI product version 4.5. Although version 4.5 includes a gap-filling strategy (as described above), this version still contains gaps across many regions of the world [22]. Our research aims to offer alternative strategies to provide spatially complete soil moisture estimates to complement the methods applied in the ESA CCI product, version 4.5 [21].

We tested three approaches. The first one is based on ordinary kriging (OK) spatial interpolation [27–29] to take advantage of the spatial autocorrelation of satellite-derived soil moisture on gridded surfaces. The second one performs regression kriging, which combines the principles of kriging interpolation and linear regression with covariates [27,30] that are used to solve kriging weights [31]. In this work, RK relies on the relation between soil moisture (response variable) with precipitation and minimum air temperature (explanatory variables). Our last approach is based on the application of generalized linear models (GLMs) to explore the relationship between soil moisture and the same explanatory variables integrated in our RK analyses. We tested these three methods because: (a) OK has the advantage of requiring solely spatial soil moisture information; (b) GLM has the advantage of benefiting from the inclusion of geophysical covariates (i.e., independent explanatory variables); and (c) RK incorporates both linear relationships and geospatial distribution of explanatory variables.

We focused our study over a region in the Midwestern United States (with abundant satellite-data estimates and in situ measurements) between 2000 and 2012. We evaluated the outcome of our gap-filling approaches with ground-truth information using in situ measurements from the North American Soil Moisture Database (NASMD) [15]. Our results show that the overall correlations between OK or RK with field data (i.e., NASMD) were slightly higher than those using GLM. These results provide support for alternative techniques to complement other approaches aimed to gap-fill satellite-derived geophysical datasets [23,24] and highlight the potential of using geostatistical techniques. Furthermore, methods based on the spatial distribution of soil moisture, such as OK, which does not require information from geophysical covariates, are useful when covariate information (e.g., precipitation and air temperature) is missing in different regions across the world.

Section 2 provides a description of the region of interest as well as the parameters to select our time frame. Data acquisition, preprocessing, selection of the geophysical covariates, application of proposed gap-filling approaches, and the validation strategy are also described in Section 2. Section 3 describes the performance of OK, RK, and GLM techniques, as well as the results of cross-validation for the three models. Validation using reference correlation between original satellite data and ground-truth soil moisture information is also described in Section 3 and is compared with model outputs. Section 3 additionally shows the capability of our methods to reproduce the spatial soil moisture patterns shown by the original ESA CCI product. Section 4 proceeds with the discussion of our findings and their implications in providing spatially complete soil moisture information derived from ESA CCI satellite estimates from version 4.5. Section 5 summarizes the remarks of our work and their implications in providing soil moisture information for specific applications.

2. Materials and Methods

2.1. Region of Interest

The selected region of interest was an area of 465,777 km^2 (Figure 1a) centered in the state of Oklahoma (180,986 km^2) and covering some areas of surrounding states within Midwestern USA: Texas (159,489 km^2), Colorado (11,210 km^2), Kansas (61,343 km^2), Missouri (10,844 km^2), New Mexico (18,550 km^2), and Arkansas (23,356 km^2). The region of interest shows a variety of environmental conditions, both natural and human-driven, that allowed us to test the spatial performance of our gap-filling frameworks. This diversity mitigates bias due to specific environmental conditions (e.g., homogenous land cover, uniform topographic features), which are not the attention of this present study. The region of interest for this study was selected in response to the availability of ground-truth data in that area, mainly over Oklahoma, where mesonet [15] provides a robust set of historical soil moisture records [32]. Additionally, soil moisture data availability in northern Texas and the remaining areas in the region of interest are consistently represented by the NASMD. We highlight that the NASMD integrates data from several monitoring networks including mesonet [15].

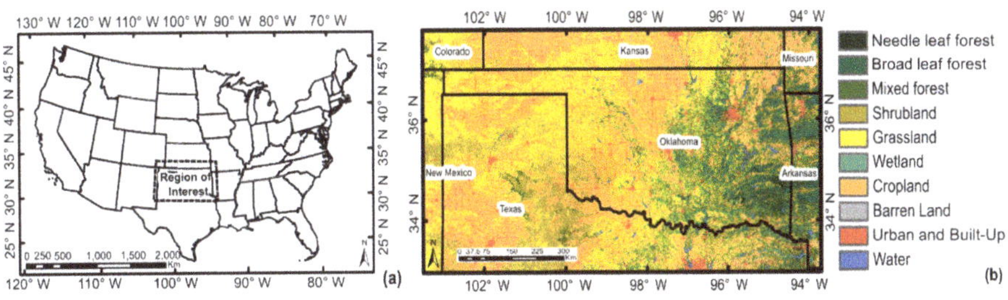

Figure 1. (a) Region of interest in the Midwestern USA, where soil moisture gap-filling methods were performed; (b) Land cover types over the region of interest (30 m), level 1 NALCMS classification [33].

The region of interest (Figure 1a) includes a wide variety of land cover types (Figure 1b) dominated by grassland (35.5%), cropland (31.9%), and shrubland (11.0%) in the central and western areas, whereas forested areas are mostly located in the eastern portion, distributed across needleleaf (2.2%), broadleaf (10.9%), and mixed forests (0.6%) [33].

2.2. Data

2.2.1. Satellite-Derived Soil Moisture

For this study, we used the ESA CCI soil moisture product version 4.5 (Table 1) that has gathered historical records from active and passive remote sensors [11,18,20]. This product provides soil moisture estimates at 0.25 degrees of spatial resolution on a daily basis, from November 1978 to December 2018 [20]. Active and passive sensors are combined by means of a weighted mean, being proportional to the signal-to-noise ratio (SNR) [20]. These ratios are estimated using triple collocation analysis, which is a method that estimates random error variances of three collocated datasets of soil moisture estimates [21]. In areas where no triple collocation analysis estimates are available, soil moisture values are estimated using a polynomial regression between the signal-to-noise ratios [20].

Table 1. Main characteristics of ESA CCI soil moisture version 4.5 [20].

Title ESA CCI Soil Moisture Version 4.5	
Release date	October 2019
Available products	Active Passive Combined
Scatterometer sensors used	SMMR SSM/I TMI AMI WS ASCAT
Radiometer sensors used	Windsat AMSR-E AMSR2 SMOS
Available time span	November 1978 to December 2018 (Combined product) August 1991 to December 2018 (Active product)

The ESA CCI product was developed in collaboration with Vienna University of Technology (TU Wien) and focuses on the use of data derived from C-band scatterometers, such as European Remote Sensing Satellites (ERS-1/2) and METOP, as well as the use of data from multi-frequency radiometers such as the Scanning Multichannel Microwave Radiometer (SMMR), Special Sensor Microwave Imager (SSM/I), Microwave Imager (TMI), Advanced Microwave Scanning Radiometer (AMSR-E), and Windsat [3]. These sensors are characterized for the suitability for soil moisture retrieval [3].

Daily soil moisture global records from the ESA CCI product were acquired and then cropped to the region of interest. Daily estimates were merged into monthly soil moisture spatial layers using mean and median values; in this way, we tackled the lack of daily coverage in areas out of the satellites' swath. Monthly mean values initially reduced the number of gaps in daily products but still provided reliable information to identify spatial patterns and trends in our study period. These values then were used to explore their relationship with different geophysical covariates (Supplementary Material S1). Monthly values can describe soil moisture variability over a few weeks due to soil moisture memory effects, as water content derived from sudden excessive rainfall or lack of water onset can generate wetness or dryness conditions that might last for a couple of weeks [2].

An important step in preparing the soil moisture data for analysis is identifying the most relevant summary statistics, such as the mean or median. The median value is more useful when data are concentrated on a brief period of the month (because of long data gaps) with an uneven distribution of data [34]. However, mean monthly soil moisture values showed higher correlation with the tested set of geophysical covariates (Supplementary Material S1). For our region of interest, Figure 2 shows the spatial distribution and number of soil moisture gaps (ESA CCI soil moisture version 4.5) during the study period (January 2000 to September 2012) where no mean values were calculated due to a lack of valid pixels. A pixel is considered valid when soil moisture estimates are available from the ESA CCI product over the region of interest. Figure 3 shows the number of gaps per monthly layer, regarding 741 pixels of 0.25 × 0.25 degrees in our region of interest.

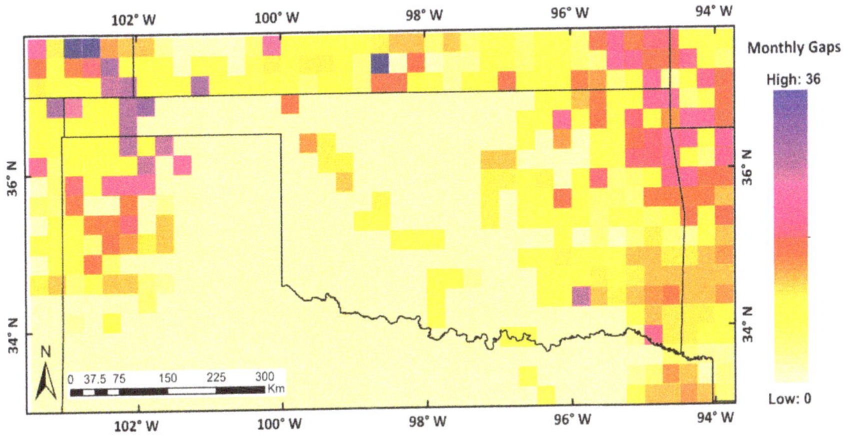

Figure 2. Number of gaps in monthly soil moisture estimates over the region of interest, derived from the ESA CCI product (version 4.5), January 2000–September 2012.

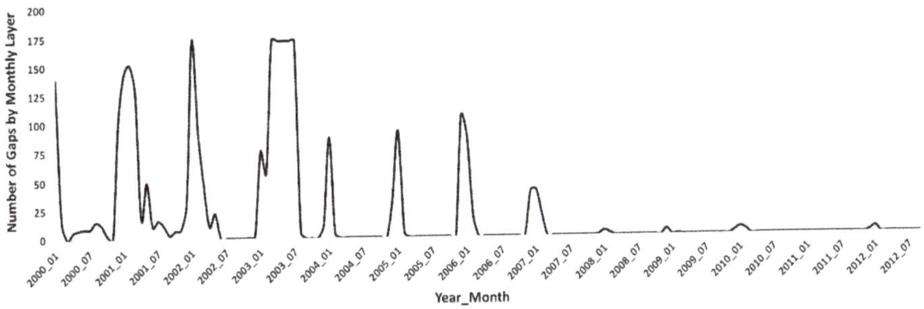

Figure 3. Distribution of gaps during the study period in monthly steps; each number represents the quantity of pixels without data, out of 741 pixels in the region of interest.

2.2.2. Soil Moisture Covariates

For RK and GLM gap-filling approaches, we explored the relationships between soil moisture and some geophysical variables. Monthly layers were generated for precipitation, atmospheric temperature, and static values of soil texture and the topographic wetness index (TWI). These selected variables are known to work as drivers for water input in soil [2,3].

Meteorological data were acquired at 1-km spatial resolution monthly layers produced by the Daily Surface Weather and Climatological Summaries (DAYMET) [35]. Total monthly precipitation and monthly averages of minimum and maximum air temperature raster layers from January 2000 to September 2012 were cropped to the region of interest, projected to the WGS84 Lat.–Long. coordinate system, and resampled to 0.25 degrees by means of the nearest neighbor method (ngb) [36].

Soil texture was obtained from the US soil survey geographic database [37], and we classified all classes into four general categories based on the texture triangle from the US Department of Agriculture (USDA) [38]: coarse, medium, medium fine, and fine. Soil texture then was resampled to 0.25 degrees resolution using ngb [36]. We calculated TWI using SAGA GIS [39] with a digital elevation model at 250 meters resolution [27] and then resampled the output to 0.25 degrees using ngb [36]. Detailed information on the definition of geophysical variables for this work and their further processing are given in the Supplementary Material S1.

2.2.3. Validation Data

In order to establish a reference value that describes the spatial distribution pattern of soil moisture over our region of interest, we acquired records from the North American Soil Moisture Database (NASMD). NASMD provides the densest possible soil moisture network that integrates field measurements across North America [15]. By 2015, the NASMD had integrated 33 observation networks and two short-term soil moisture campaigns, providing ground-truth data for over 1800 observation sites in the USA, Canada, and Mexico [15]. Some of the densest regional networks integrated by NASMD offer soil moisture data in our region of interest (e.g., MESONET), and records at 5-cm depth, where the soil layer closely interacts with the atmosphere and it is sensed by satellites [40]. We extracted all information available from the NASMD over our region of interest that comprised records at 5-cm depth, from January 2000 to September 2012. Finally, we transformed these data to georeferenced point layers to be integrated in our ground-truth validation approach.

2.3. *Gap-Filling Methods*

Our first two gap-filling approaches were based on kriging interpolation (OK and RK). These techniques lead to high uncertainty over areas with very large continuous spatial gaps because they rely on the spatial autocorrelation of available data. Consequently, we also tested a third approach based on GLM to test the relationship between soil moisture and geophysical covariates. We clarify that the GLM approach does not depend on the spatial autocorrelation of available data.

The OK interpolation strategy depends solely on the separation distance between sampled locations and not on an absolute position [29]. This offers a feasible strategy to fill spatial gaps in areas where no other information is available to be included in similar interpolation methods such as cokriging or regression kriging. This is the most popular among all kriging methods, as it works in almost any situation and its assumptions are easily filled [29].

Regression kriging also depends on the spatial location of soil moisture values but incorporates the location of information from covariates as well [27]. Regression kriging yields to a better representation of the spatial patterns depicted by the covariates known as be correlated with the response variable [30].

Generalized linear models (GLMs), as an alternative approach, represent multivariate regression models [41]. In this approach, we assume linear relationships between the dependent variable (soil moisture) and the predefined covariates (precipitation, minimum air temperature) before considering relationships that are more complex. These relationships have also been explored in previous studies of soil moisture derived from field measurements, integrating predictors such as vegetation indices, precipitation, and temperature [42,43]. However, GLM represents an approach that can be applied to satellite-derived soil moisture estimates to fill spatial gaps over large areas.

Soil moisture spatial-gaps in the region of interest are not always sufficient to test interpolation methods, as in some months there are no gaps over the region of interest. Thus, we decided to randomly remove valid data from each soil moisture monthly layer as well as their correspondent locations on the geophysical covariates layers. Therefore, OK, RK, and GLM were performed on 100%, 75%, and 50% of available valid pixels in each month, similar to gap filling analyses in previous studies [23].

The overall process for soil moisture prediction (Figure 4), derived from the proposed modeling techniques, was evaluated using cross-validation and ground-truth data from the NASMD available from January 2000 to September 2012. An extensive description of the workflow and a sample process for one month are provided in the Supplementary Material S2.

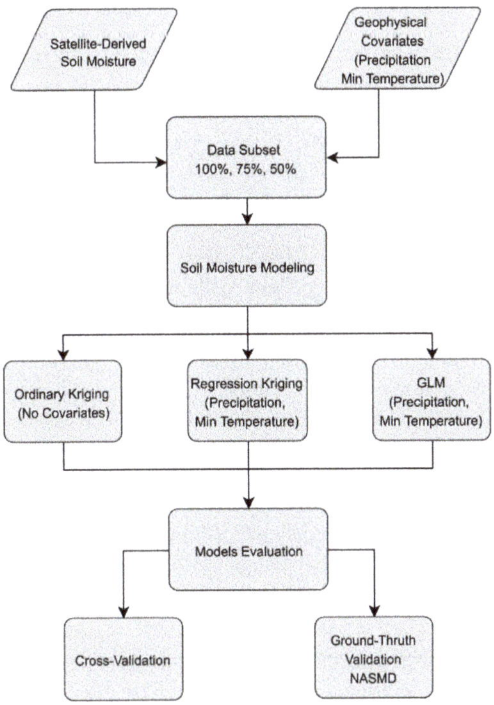

Figure 4. Workflow for soil moisture modeling and the gap-filling over the region of interest, regarding 100%, 75%, and 50% of available valid pixels in each monthly layer. Cross-validation as well as ground-truth validation is also described.

2.3.1. Ordinary Kriging

OK was performed using the AutoMap package developed for the R statistical platform [44]. By means of the autofit-variogram tool, the best-fitted variogram model was automatically selected to generate independent predictions over each month. Five different variogram models (i.e., spherical, exponential, Gaussian, Matern and Stein's parameterization) were evaluated, and the one with the smallest residual sum of the squares was selected [44]. The prediction of values at unsampled locations is the linear combination of N variables, as expressed in Equation (1):

$$Z(u) = \sum_{i=1}^{N} \lambda_i \, Z(u_i) \tag{1}$$

where λ_i represents the original weighted values. Weights are calculated as a function of the distance between sampled and unsampled locations to be predicted. The weight sum must be equal to 1, thus estimations fulfill the unbiasedness requirement [45].

Derived from OK spatial interpolation, predicted values as well as their standard errors were obtained for each month, derived in three different cases from 100%, 75%, and 50% of available valid pixels. We applied 10-fold cross-validation [44] to OK outputs for the above-mentioned percentages of valid pixels using autoKrige.cv [44]. Finally, we assessed the spatial dependence found in each monthly layer using the nugget–sill ratio. Ratios of at most 0.25 represented strong spatial dependence; between 0.25 and 0.75, moderate spatial dependence; and at least 0.75, weak spatial dependence, as previously reported [46].

2.3.2. Regression Kriging

RK was performed with the R package GSIF [47], using the function fit.regModel. Individual regression models were fitted to each monthly layer, incorporating monthly precipitation and minimum temperature data from DAYMET [35]. We combined regression on soil moisture data and the preselected geophysical covariates with simple kriging of the regression residuals [31]. GSIF tools allowed us to select different regression techniques (e.g., random forest, GLM, quantile regression forest). We selected GLM to make RK a hybrid approach between our two other proposed methods (i.e., OK and GLM). In RK, a spatial trend is assumed instead of stationarity across the region of interest. Based on the residuals of the identified trend in regression analysis, spatial interpolation is applied through OK. Prediction over unsampled locations is equal to the estimated trend plus the error prediction as expressed in Equation (2):

$$Z(x) = m(x) + \varepsilon(x) \quad (2)$$

where $Z(x)$ is the target variable to be predicted, $m(x)$ is the trend (explanatory power) identified from the relationship with geophysical covariates, and $\varepsilon(x)$ represents the stochastic residuals. Unlike OK, in RK, the trend is no longer constant, but is a function of the explanatory variables [48].

As we did for OK, we derived predicted values and associated error based on 153 months, using 100% of available data, as well as 75% and 50%; this yielded 459 predicted soil moisture layers. Then, 10-fold cross-validation was performed, and nugget–sill ratios were calculated as in the OK approach to identify the level of spatial dependence [46] depicted in each monthly layer.

2.3.3. Generalized Linear Models

For GLM, we first tested the overall correlation between soil moisture (monthly mean and median values) and each one of the geophysical covariates (monthly precipitation, monthly maximum and minimum air temperature, soil texture, and TWI). Secondly, we extracted a time series for each valid pixel along the 153 monthly soil moisture layers and tested the pixel-individual correlation with each one of the covariates. Finally, we calculated the correlation coefficients of all valid pixels available for each monthly layer with the corresponding temporal layer for each one of the covariates. Based on these analyses, we established that the spatial values of mean monthly precipitation and minimum air temperature were the variables with the highest absolute correlation coefficient with mean monthly soil moisture (Supplementary Material S1). These geophysical covariates were used to predict soil moisture based on GLM, as shown in Equation (3):

$$Y_i = \beta_0 + \beta_1 X_{i1} + \beta_2 X_{i2} + \varepsilon_i \quad (3)$$

where Y_i represents the response variable, X_{i1} and X_{i2} represent the predictor variables, β_0, β_1 and β_2 are the parameters of the model, and ε_i is the error term [41].

Predictions were also performed for the three predefined subsets (100%, 75%, and 50%) of available valid data over the region of interest in each month of the study period. We used the GLM tool from the caret statistical package in R [49] to generate independent models for each month, as well as a 10-fold cross-validation process. For this purpose, we used 75% of the data in each independent monthly dataset as training data and 25% as test data.

2.4. Ground-Truth Validation

2.4.1. Reference Correlation between NASMD and Satellite-Derived Soil Moisture

First, we established a reference correlation value between original satellite-derived soil moisture and data from the NASMD. We extracted all available data from NASMD over the region of interest for each month during the study period and calculated the mean monthly value of soil moisture at 5-cm depth for each field station, thus capturing as much variation as possible from the upper soil layers sensed by the satellites. We tested the correlation between satellite-derived values over each spatially correspondent pixel with soil moisture information derived from the NASMD. This process was performed over the layers using 100%, 75%, and 50% of available valid pixels. When there was more than one NASMD station within one corresponding pixel of satellite-derived soil moisture, every station value from within the pixel area was accounted for in the correlation analysis with the satellite data. Overall, we used data from 157 stations in the months with the highest availability of field soil moisture records. The use of all NASMD available stations allowed us to retain the overall observation-estimation pairs. Figure 5 shows the distribution of available NASMD stations over the region of interest for the entire study period. Figure 6 shows the number of NASMD stations used in each month to validate the outputs of our models. Across the entire study period, all available stations provided 19,007 points to compare satellite-derived soil moisture estimates and ground-truth data.

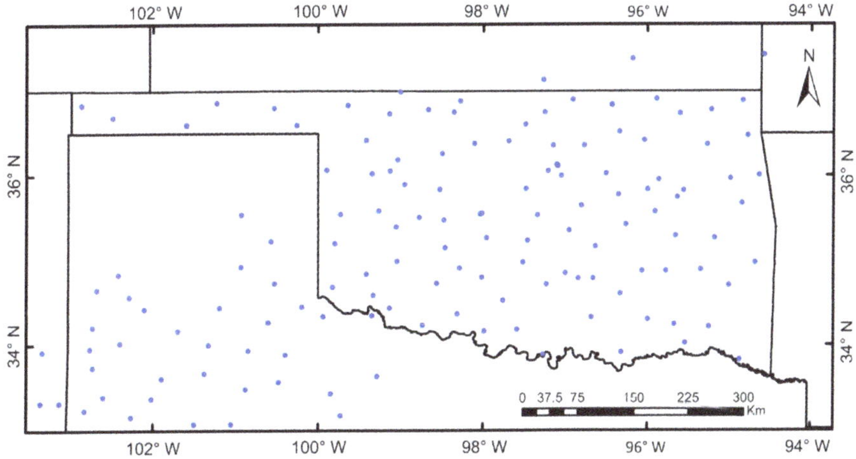

Figure 5. All NASMD stations available for the study period (157). Stations are broadly distributed over Oklahoma and northern Texas; however, ground-truth data are scarce over the surrounding states within the region of interest.

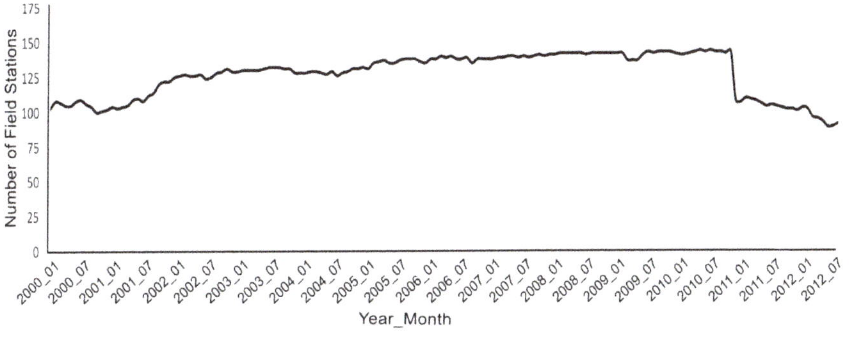

Figure 6. Number of ground-truth stations available per month for validation.

2.4.2. Correlation between Predicted Soil Moisture and NASMD

In order to validate our soil moisture predicted values, we looked for the closest similar correlation coefficient from our outputs and the NASMD to the correlation coefficient between the original ESA CCI estimates with NASMD, thus repeating the same value of a satellite estimate or predicted value for each field station that is located within the same cell. In this way, we take advantage of as much validation information as possible over our region of interest. We followed the same approach as in Section 2.4.1 to evaluate the soil moisture values derived from the modeling approaches with the NASMD. This allowed us to evaluate 19,411 pixels where we calculated the overall correlation coefficient (all months) and monthly correlation coefficients.

3. Results

3.1. OK and RK Models Selected for Soil Moisture Predictions

Variograms using Stein's parameterization [50] were the most common in OK across the 459 monthly layers (n = 402). Exponential (n = 53), spherical (n = 3), and Gaussian (n = 1) were used in a substantially lower number of predicted soil moisture layers. RK was based on exponential variogram models in all cases (459 monthly layers), regardless the percentage of valid data used (100%, 75%, or 50%). We found strong spatial dependence in 416 of the monthly layers (nugget–sill < 0.25) and moderate spatial dependence in the remaining 43 layers (0.25 < nugget–sill < 0.75) when using OK (Figure 7a). On the other hand, we found strong spatial dependence in 253 monthly layers out of 459 and moderate spatial dependence in 206 when using RK. The RMSE for predicted soil moisture layers with OK showed that Stein's parameterization [50] and spherical models had smaller minimum values. However, we found that the RMSE values were more distributed in Stein's parameterization than in spherical models. RK with exponential models had a higher RMSE value than OK, but the error distribution was less spread, with just a few extreme values (Figure 7b).

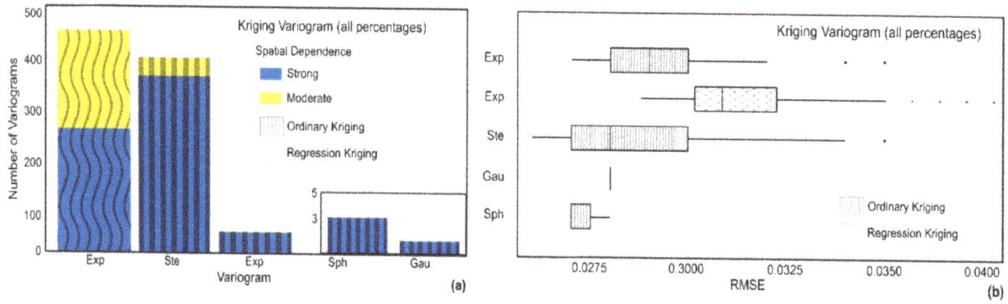

Figure 7. (**a**) Most fitted variogram models used to predict soil moisture; 459 models generated for both OK and RK in 153 monthly layers derived from all percentages (100%, 75%, and 50%) of valid pixels; (**b**) Boxplots of the RMSE for each predicted layer using the selected variograms.

3.2. Cross-Validation of Predicted Values

Overall, the three models had good cross-validation results, but OK and RK had consistently higher correlation coefficients and lower RMSE (Table 2). However, OK had slightly better performance than RK when a different percentage of available data was used.

Table 2. Cross-validation outputs for OK, RK, and GLM, and all predicted and observed values along the 153 monthly layers.

Method	Percentage of Data	Correlation	RMSE
OK	100%	0.886	0.029 m^3m^{-3}
	75%	0.886	0.029 m^3m^{-3}
	50%	0.886	0.029 m^3m^{-3}
RK	100%	0.886	0.029 m^3m^{-3}
	75%	0.878	0.030 m^3m^{-3}
	50%	0.869	0.031 m^3m^{-3}
GLM	100%	0.711	0.044 m^3m^{-3}
	75%	0.709	0.044 m^3m^{-3}
	50%	0.709	0.044 m^3m^{-3}

Additional cross-validation between predicted and observed values by month (January to December) was reported using Taylor diagrams (Figure 8), which simultaneously report the correlation coefficient, normalized standard deviation, and centered root mean squared error [51]. The Taylor diagrams [52] consistently showed that OK and RK had a higher correlation coefficient and lower centered RMSE and standard deviations, and consequently, were closer to the observations. These results were consistent regardless of the percentage of available data used. Overall, OK had a consistent correlation coefficient of 0.886, whereas RK ranged from 0.869 to 0.886 as the percentage of data to model values was lower. Finally, GLM values ranged between 0.711 to 0.709 with a lower percentage of valid data. Centered RMSE values between observed and predicted values with OK were consistently 0.029, RK ranged between 0.029 and 0.031, and GLM values were 0.044 m^3m^{-3} in all cases.

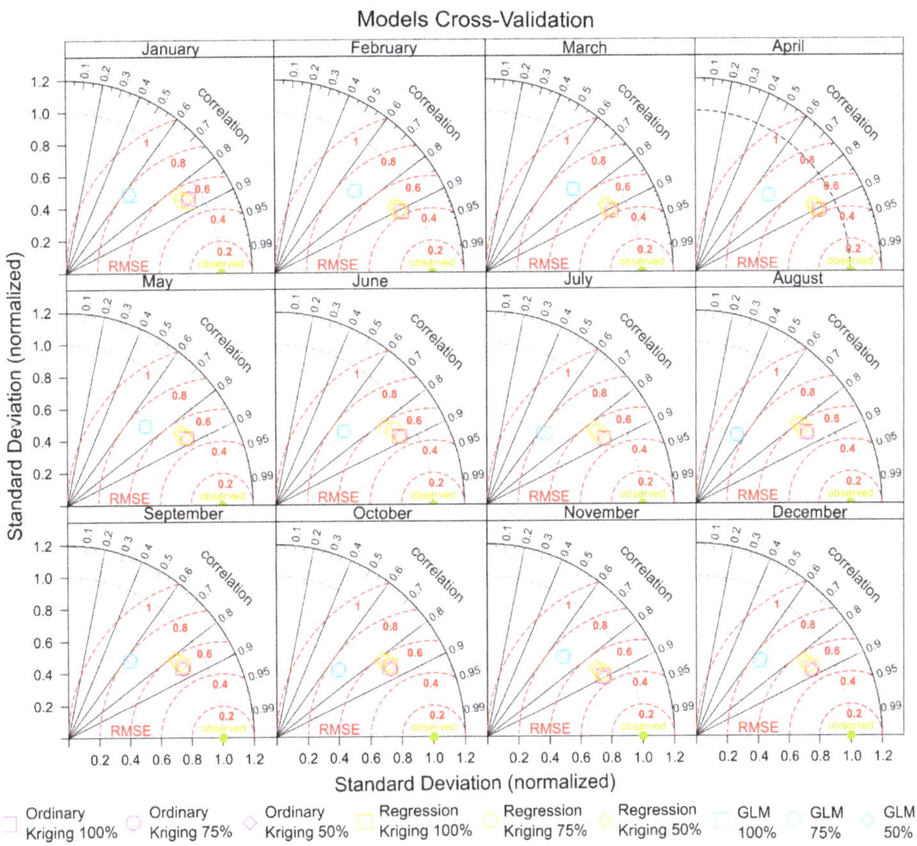

Figure 8. Taylor diagrams based on cross-validation by month across years; OK, RK, and GLM using 100%, 75%, and 50% of available valid data. Standard deviation and RMSE values for each output were normalized using the standard deviation of the observed values.

3.3. Ground-Truth Validation with NASMD

We found an overall correlation coefficient of r = 0.523 and an RMSE of 0.093 m^3m^{-3} between the original ESA CCI data and the available NASMD stations across the study period (153 months). These values served as a baseline and showed that values generated using OK with 100% and 75% of valid data were closer to the reference than those using RK and GLM (Table 3).

Table 3. Overall correlation coefficients between all ground-truth validation points and the CCI soil moisture product, as well as gap-filled outputs. Percentages show the data subset used to predict soil moisture values over the region of interest.

Method	Percentage of Data	Correlation	RMSE
CCI	100%	0.558	0.069 m^3m^{-3}
	100%	0.579	0.067 m^3m^{-3}
OK	75%	0.575	0.067 m^3m^{-3}
	50%	0.569	0.067 m^3m^{-3}
	100%	0.582	0.067 m^3m^{-3}
RK	75%	0.582	0.067 m^3m^{-3}
	50%	0.571	0.067 m^3m^{-3}
	100%	0.475	0.070 m^3m^{-3}
GLM	75%	0.475	0.070 m^3m^{-3}
	50%	0.475	0.070 m^3m^{-3}

We explored the temporal dynamics of the correlation coefficients and RMSEs by month throughout the study period. Figure 9 shows the R-squared values between the monthly correlation coefficients from ground-truth data and CCI products and the coefficients from ground-truth data and predicted values by our proposed methods (OK, RK, GLM). RMSE is reported in the same manner (Figure 9). OK correlation coefficients using 100% of available valid data with ground-truth data are the closest to the correlation coefficients used as a reference between validation data and the CCI product (Figure 9a). However, RK correlation coefficients show higher consistency when compared with the reference correlation coefficients across different percentages of available valid data (Figure 9b). In contrast, GLM outputs show lower general R-squared values between the outputs and the reference and are loosely fitted to the regression line (Figure 9c). In a similar way, R-squared values between RMSE from the CCI product and ground-truth data, as well RMSE from model outputs and ground-truth data, show closer relation for OK (Figure 9a) and RK (Figure 9b) outputs rather than for GLM (Figure 9c). Nevertheless, OK shows slightly better results than RK.

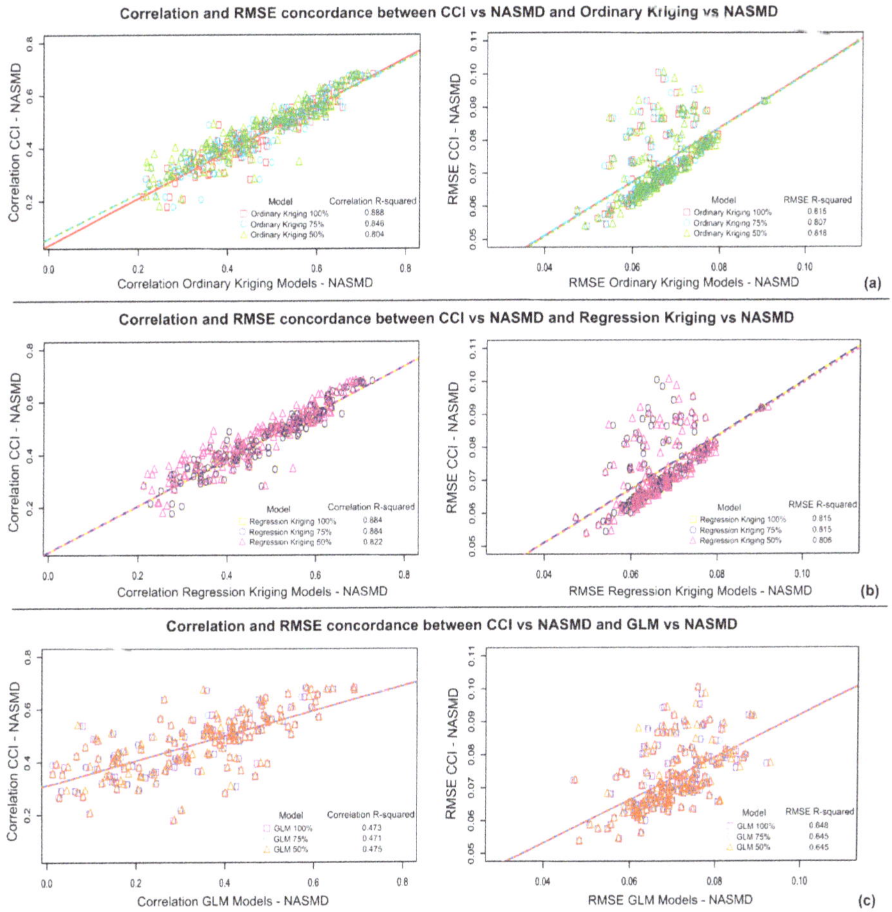

Figure 9. R-squared values showing the concordance between the reference validation (CCI-NASMD) and the validation of proposed gap-filling methods with NASMD. Each point represents one month, using 100%, 70%, or 50% of the available data, with the percentage indicated by the shape used. (**a**) Validation of ordinary kriging with NASMD, with correlation on the left and RMSE on the right; (**b**) Validation of regression kriging with NASMD, with correlation on the left and RMSE on the right; (**c**) Validation of GLM with NASMD, with correlation on the left and RMSE on the right. Regression lines between correlated datasets are shown in each plot.

3.4. Spatial Gap-Filling Performance of Modeling Methods

The comparison between the outputs of our modeling methods in contrast with the original ESA CCI soil moisture product shows that OK and RK approaches better reproduce the spatial pattern captured by satellite estimates. Figure 10a shows the mean soil moisture estimates from the ESA CCI product version 4.5 derived from 153 monthly layers in our region of interest, without any gap-filling technique. In comparison to the original spatial distribution of soil moisture, OK visually shows more similar patterns, independent of the percentage of valid pixels used for modeling (Figure 10b–d). RK visually shows very similar spatial patterns (Figure 10e–g) as OK. However, both methods, OK and RK, are challenged by extreme low and high values included in the original satellite product.

Conversely, GLM shows a lower performance in reproducing soil moisture spatial patterns, regardless of the percentage of valid pixels included in the modeling process (Figure 10h–j).

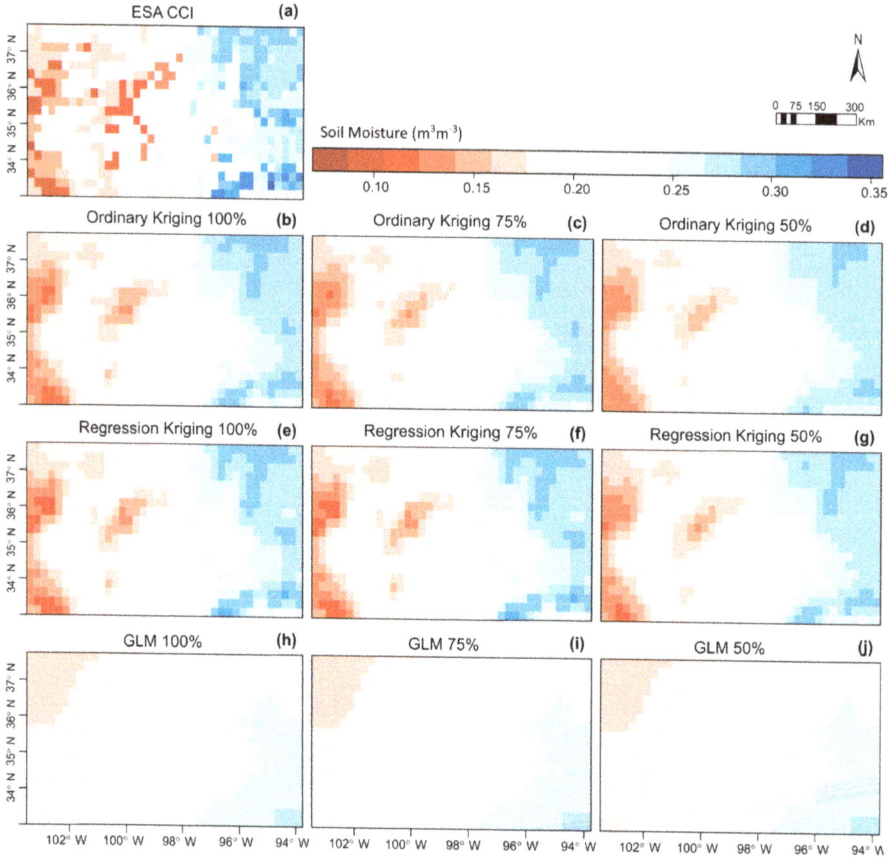

Figure 10. Mean soil moisture values during the study period (January 2000–September 2012) over the region of interest. (**a**) Mean values of original ESA CCI soil moisture estimates; no gap-filling methods applied; (**b**) Soil moisture mean values modeled using OK and 100% of available valid data; (**c**) Soil moisture mean values modeled using OK and 75% of available valid data; (**d**) Soil moisture mean values modeled using OK and 50% of available valid data; (**e**) Soil moisture mean values modeled using RK and 100% of available valid data; (**f**) Soil moisture mean values modeled using RK and 75% of available valid data; (**g**) Soil moisture mean values modeled using RK and 50% of available valid data; (**h**) Soil moisture mean values modeled using GLM and 100% of available valid data; (**i**) Soil moisture mean values modeled using GLM and 75% of available valid data; (**j**) Soil moisture mean values modeled using GLM and 50% of available valid data.

Finally, we found that the density distribution describing the mean soil moisture values during the study period in the original ESA CCI was better reproduced by the OK and RK approaches. The performances of OK and RK were similar, either using 100%, 75%, or 50% of available valid data (Figure 11a,b). In contrast, the GLM density distribution substantially deviated from the values of the original ESA CCI product (Figure 11c).

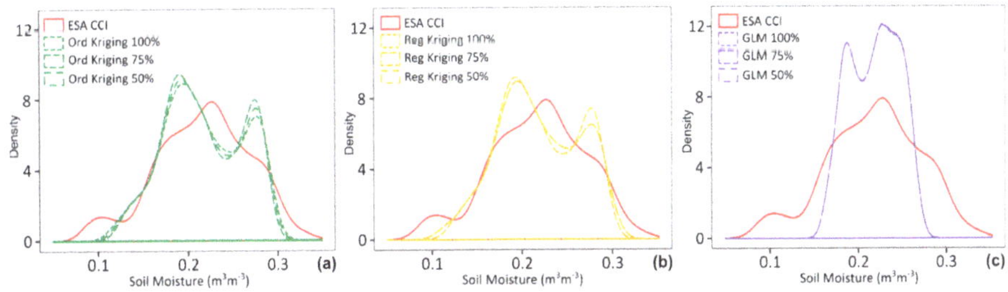

Figure 11. Density distribution of mean soil moisture values during the study period for 741 pixels over the region interest. (**a**) ESA CCI and modeled data using OK with 100%, 75%, and 50% of available valid data; (**b**) ESA CCI and modeled data using RK with 100%, 75%, and 50% of available valid data; (**c**) ESA CCI and modeled data using GLM with 100%, 75%, and 50% of available valid data.

4. Discussion

Our results showed that the OK, RK, and GLM techniques could be used as alternative approaches to gap-filling in soil moisture data derived from the ESA CCI product version 4.5. Our proposed methods can be used either in conjunction with geophysical covariates such as precipitation and temperature or using solely the spatial distribution of soil moisture estimates derived from the ESA CCI product. Furthermore, our results show that spatial patterns and temporal relations between satellite and ground-truth data are better preserved by using OK and RK, but we show the applicability of the GLM approach. The benefit of using different approaches would depend on the spatial structure of the missing data and the availability of covariates for applying OK, RK, or GLM approaches.

Precipitation and minimum air temperature were the strongest correlated environmental covariates with soil moisture (Supplementary Material S1). These relationships are likely influenced by the grid size (0.25 degrees), as the spatial influence of precipitation and air temperature represents regional and mesoscale climatic patterns [53]. Previous research showed that increasing spatial resolution yields more detail in the meteorological information but limited impacts on its forecasting skill [54]. It is known that from the plot to watershed scale, soil texture and topography are highly correlated with soil moisture [2,3], but these relationships may change at the coarse scale of the ESA CCI soil moisture product. Thus, these features were not included as geophysical covariates in our GLM or RK approaches.

Overall, our results provide support for OK, RK, and GLM as techniques to gap-fill spatial missing values of satellite-derived soil moisture products. However, overall performance indicates that OK and RK represent more reliable methods for soil moisture gap-filling in comparison with GLM. Previous studies have compared the advantages of OK and RK for interpolation of spatial soil moisture and other soil properties [27,55–58] but most analyses have been performed for spatial interpolation of soil properties based on field data [26,58–60]. OK has been regarded as an unbiased linear estimator [45], and our results support it as a feasible approach due to the spatial scale of the original ESA CCI estimates (0.25 degrees) under the gap scenarios tested in this work. At this coarse scale, soil moisture values represent a quasi-continuous matrix that meets basic assumptions of kriging analysis such as stationarity [45] and spatial dependence [58]. OK also incorporates spatial autocorrelation by using the variogram and providing the error variance estimation from predicted values, offering some advantages over deterministic methods such as inverse distance weighting (IDW), which may create noisy fields in interpolation processes. Similar to other kriging methods, OK is an exact interpolator, which ensures that values at sampled locations are exactly preserved. Thus, we aim to fill the spatial gaps by modeling the entire region of interest, while preserving original values where data existed previously. Additionally, OK performs value predictions based solely on spatial data distribution, offering a suitable approach in cases where no well represented covariates datasets are available over the region of interest, and it compensates for data clustering [61]. Additional evidence in support for OK is the fact that the nugget–sill ratio was less than 0.25 in 99% of the fitted variograms, which implies strong spatial dependence as discussed elsewhere [46].

RK on the other hand has been widely used to incorporate covariates to build a regression model with soil properties [27,62–64]. Whereas some authors do not find a better performance of OK in comparison with RK for the prediction of soil properties [27,62,63], our results support the use of RK, as it performed similarly to OK in our region of interest. As a hybrid method, RK has the advantage of incorporating spatially explicit information known to be correlated to the response variable [27,65]. The explicit correlation between soil properties and geophysical covariates provided good results when using terrain parameters [62,66,67] or other variables such as bare soil from remotely sensed sources, crop yield, temperature, and precipitation data [64,68] as predictors. Other authors highlight that RK performance depends on the relationships between soil and environmental factors [63,65]. This could explain the similar performance between OK and RK in our region of interest, as our selected covariates seem to account for similar influence at the coarse spatial scale of the ESA CCI product. Based on the spatial dependence depicted by the nugget–sill ratio in the fitted variograms using RK, we postulate that regardless of the similar performance using OK, our selected covariates did not have a consistent strong spatial dependence. Based on nugget–sill ratios, RK showed strong spatial dependence in 55% of the fitted variograms, while 45% showed moderate dependence when using the thresholds previously discussed [46]. Finally, it is possible that RK may not accurately describe the spatial patterns of soil properties when using coarse resolution geophysical covariates, but these covariates might help to improve prediction accuracy [30]. Thus, the incorporation of covariates may depend on the actual spatial dependence observed when modeling variograms using both OK and RK techniques.

The GLM approach allowed us to explore the most evident relationships between soil moisture in the upper layer of soil and the geophysical covariates that we found to be better correlated (Supplementary Material S1). We followed a parsimonious principle by means of the GLM technique, applying the simplest model with the fewest assumptions before assuming relationships that are more complex. This parsimonious reasoning and its applications to multivariate models have been explored in other studies [69].

The evaluation of our three approaches (OK, RK, and GLM) by means of cross-validation regarding their prediction capacity for actual satellite data shows similar correlation coefficients as those reported by [59] in the spatial interpolation of soil moisture and similar RMSE as reported by [58] for other soil properties. The cross-validation technique has been commonly used in other similar studies [58,59] and offers initial insights into modeling techniques without considering ground-truth data for validation. Our cross-validation strategy showed that OK and RK better predicted soil moisture values compared with GLM, in spite of pixel removal at different percentages. Regarding cross-validation for monthly grouped values, OK, RK, and GLM did not show an evident bias due to seasonality, as monthly correlation coefficients and RMSE values systematically describe the same patterns found when using data from the entire study period in a single dataset.

In spite of cross-validation results, ground-truth validation was performed to evaluate the suitability of each method (OK, RK, and GLM) to predict missing values in the ESA CCI product. We acknowledge the conceptual challenge of this data matching and the need of balancing ground-truth information in order to be representative of satellite-derived estimates. Representativeness challenges in validation of the ESA CCI product have been also acknowledged previously [40]. Two main problems are identified [40]: (1) Satellite sensors retrieve ground information from the upper soil layer (0.5–5-cm depth); this layer is directly exposed to the atmosphere; therefore, its physical characteristics may differ from the information provided by soil moisture sensors placed at 5-cm depth or deeper. Thus, satellite estimates represent a more variable soil layer, different from soil at deeper layers. (2) Even a spatially extensive soil moisture network cannot cover any area widely enough to provide scaling representativeness between point-scale measurements and satellite estimates. Field measurements depict soil characteristics in the range of a few square decimeters, while satellite products commonly cover a few kilometers per pixel (~27 km pixel sizes in the ESA CCI product). Additionally, other authors suggest the soil moisture representativeness, on a grid-scale domain, may be described based on three different methods [30]: (1) empiric methods, averaging all points within each single grid-cell; (2) upscaling methods based on time information; and (3) spatial interpolation by means of kriging methods to assign individual values to each center point in the grid-cell domain. In this regard, our work does not aim to provide strategies of accuracy assessment between field measurements and satellite estimates as explored by [70]. We seek to reproduce the spatial soil moisture patterns expressed by the satellite-derived soil moisture and its actual correlation with ground-truth data with the ultimate goal to gap-fill missing information.

As proposed by [57], the selection of reliable ground-truth stations and the definition of core validation sites (CSV) represent a step forward in the evaluation of remotely sensed soil moisture. However, regarding the limited availability of ground stations providing soil moisture information, we integrated all available ground-truth data for our region of interest instead of defining CSV. In this way, we took advantage of all available field soil moisture records over the region of interest. This approach might introduce uncertainty, as neighboring stations within the same 0.25 degrees pixels in some cases could be affected by different moisture conditions in large areas. However, as our approach aims to reproduce the spatial distribution of soil moisture showed by the satellite estimates based on the correlation with ground-truth data, we aim to retain all the variation offered by NASMD stations.

In order to define the best-tested soil moisture prediction model to fill the gaps in the ESA CCI product, version 4.5, correlation found with ground-truth data was set as a reference for our proposed models in every month of the study period. This yielded a more specific way to validate our proposed methods regarding different soil moisture estimate conditions in every month of the ESA CCI product. Given that our research aims to complete spatial information of ESA CCI, reference correlation coefficients helped us to define which model best reproduces the spatial pattern of the original product. OK and RK showed better results than GLM, as we found the higher the number of valid pixels to shape the variogram parameters, the closer the correlation coefficient to the reference. Furthermore, OK and RK performance does not significantly decrease even though valid pixels are artificially removed. On the other hand, GLM correlation with ground-truth data showed less similar values to the reference, independent of the percentage of valid pixels removed.

Given that the OK, RK, and GLM performance for our region of interest is not that different, GLM can be an alternative approach in similar regions where satellite-derived soil moisture estimates are spatially scarce or highly clustered, as GLM relies more on predictor availability than on spatial distribution. Besides, when OK and RK do not meet the best requirements, GLM can use input data from robust meteorological datasets [71,72] to obtain the geophysical covariates that we used in our analysis. Based on the correlation coefficient between the ESA CCI soil moisture product and NASMD ground-truth data, we found that OK and RK consistently better reproduce reference correlation coefficients and RMSE values. Nevertheless, GLM correlation coefficients and RMSE values with NASMD do not significantly decrease from the reference, which still makes this method an alternative approach to gap-filling. Finally, the analysis of the mean soil moisture spatial patterns during the study period showed that OK and RK outputs consistently better reproduced the spatial patterns in the original ESA CCI product. This can be visually distinguished on the mean soil moisture maps, as well as in the density distribution of the original product in comparison with OK, RK, and GLM outputs.

We acknowledge that OK and RK represent the best-tested methods for soil moisture prediction and gap-filling of the ESA CCI product over our region of interest, based on the analysis of the monthly mean values from January 2000 to September 2012. The application of these methods in other regions and under different conditions should consider availability and distribution of soil moisture estimates since in large discontinuous areas, stationary can be wrongly assumed, yielding high uncertainty in predicted values. We recognize the need to explore RK models at finer spatial scales, where linear relationships with geophysical covariates such as those explored in the Supplementary Material S1 might be stronger. In future research, it is necessary to explore ESA CCI gap-filling over larger areas such as the conterminous United States, where well spatially represented meteorological datasets are available and different scenarios of gaps distribution can be tested. Daily data must be also incorporated, as this is the temporal resolution in which original soil moisture estimates are delivered, thus opening the possibility to operationally fill the gaps in the original soil moisture estimates provided by the SA CCI soil moisture product (version 4.5). These implementations represent an upscaling need in computational capacities; therefore, high-performance computing (HPC) techniques must be considered.

5. Conclusions

For the region of interest, linear geostatistics techniques offer a suitable approach to fill the soil moisture spatial gaps of the ESA CCI product (version 4.5). Although the current version of the product follows different strategies to fill data gaps, our research highlights the incorporation of the spatial distribution of soil moisture, as well as the use of geophysical covariates to model missing values. Selected geophysical covariates to model soil moisture in this study, i.e., precipitation and minimum air temperature, can be easily integrated due to their historical availability across larger regions, e.g., the conterminous United States (CONUS). The selected region of interest provided a spatially extent set of valid pixels from January 2000 to September 2012, which allowed us to test our proposed methods under different scenarios of gap presence, due to natural conditions as well as artificial pixel removal.

The ordinary kriging method does not need to use any additional covariates, as it is built upon the spatial distribution of soil moisture data; on the other hand, RK benefits from relationships with geophysical covariates such as the ones explored in this work. However, these methods can be inconclusive over areas where reference data are highly sparse or clustered (i.e., data scenarios where we found weak spatial structure for satellite soil moisture). Generalized linear models, on the contrary, might offer an alternative to spatially model soil moisture and fill the gaps in the ESA CCI product, though their performance was lower than that of OK and RK in our region of interest. Soil moisture at a coarse scale can be significantly correlated with covariates such as precipitation and minimum air temperature, which can be easily inputted by predicting models over most of CONUS and other regions around the world.

Derived from cross-validation for each method and specific percentage of available data, the three proposed methods—ordinary kriging, regression kriging, and generalized linear models—showed a significant prediction performance with respect to soil moisture data. However, as we intended to reproduce the soil moisture spatial patterns of the ESA CCI product and its relationship with ground-truth soil moisture data, we considered field validation as the best approach to find the most suitable gap-filling method.

Besides offering information for a wide variety of applications by itself, spatially complete soil moisture information covering large areas can also be related to point-based soil moisture networks to jointly monitor ecological processes. Thus, gap-filled data can yield a better understanding of the role of soil moisture in water and carbon cycles, with important implications in plant and soil respiration, or plant growth, therefore influencing our capacity to predict climate change signals in soil moisture estimates from the regional to the global scale.

Supplementary Materials: The following are available online at http://www.mdpi.com/2072-4292/12/4/665/s1, Supplementary Materials S1 and S2 are submitted with this manuscript. Monthly gap-filled soil moisture layers derived from the approaches proposed in this work can be acquired at Hydroshare, https://bit.ly/31yxfQm, https://www.hydroshare.org/resource/f0091cf90bcc4487bf401ca19783d1eb/.

Author Contributions: R.M.L., M.G., and R.V. conceived and designed the research. R.M.L. and M.G. developed the code for processing and analyzing the data. R.M.L. wrote the first draft of the manuscript with input from R.V, M.G. and D.R. All authors contributed to interpretation of the results, reviewed, and approved the manuscript. R.V. and M.T. supervised and coordinated the research team. All authors have read and agreed to the published version of the manuscript.

Funding: This study was funded by a University of Delaware Strategic Initiative research grant and the National Science Foundation (OAC grant#1724843).

Acknowledgments: We thank Inder Tecuapetla for his input in the conceptual analysis of spatio-temporal correlation and Paula Olaya and Joe Teague for their comments to improve this work. MG acknowledges the Mexican National Council for Science and Technology (CONACyT) for a PhD Fellowship (#382790).

Conflicts of Interest: The authors declare no conflict of interest.

References

1. Ward, S. *The Earth Observation Handbook*; ESA Communication Production Office: Noordwijk, The Netherlands, 2008; ISBN 978-92-9221-408-1.
2. Koster, R.D.; Suarez, M.J. Soil Moisture Memory in Climate Models. *J. Hydrometeorol.* **2001**, *2*, 558–570. [CrossRef]
3. Seneviratne, S.I.; Corti, T.; Davin, E.L.; Hirschi, M.; Jaeger, E.B.; Lehner, I.; Orlowsky, B.; Teuling, A.J. Investigating soil moisture-climate interactions in a changing climate: A review. *Earth Sci. Rev.* **2010**, *99*, 125–161. [CrossRef]
4. Meehl, G.A.; Washington, W.M. A Comparison of Soil-Moisture Sensitivity in Two Global Climate Models. *J. Atmos. Sci.* **1988**, *45*, 1476–1492. [CrossRef]
5. Engman, E.T. Applications of microwave remote sensing of soil moisture for water resources and agriculture. *Remote Sens. Environ.* **1991**, *35*, 213–226. [CrossRef]
6. Narasimhan, B.; Srinivasan, R. Development and evaluation of Soil Moisture Deficit Index (SMDI) and Evapotranspiration Deficit Index (ETDI) for agricultural drought monitoring. *Agric. For. Meteorol.* **2005**, *133*, 69–88. [CrossRef]
7. Davidson, E.A.; Belk, E.; Boone, R.D. Soil water content and temperature as independent or confounded factors controlling soil respiration in a temperate mixed hardwood forest. *Glob. Chang. Biol.* **1998**, *4*, 217–227. [CrossRef]
8. Williams, C.A.; Albertson, J.D. Soil moisture controls on canopy-scale water and carbon fluxes in an African savanna. *Water Resour. Res.* **2004**, *40*, 1–14. [CrossRef]
9. Dorigo, W.A.; Wagner, W.; Hohensinn, R.; Hahn, S.; Paulik, C.; Xaver, A.; Gruber, A.; Drusch, M.; Mecklenburg, S.; Van Oevelen, P.; et al. The International Soil Moisture Network: A data hosting facility for global in situ soil moisture measurements. *Hydrol. Earth Syst. Sci.* **2011**, *15*, 1675–1698. [CrossRef]
10. Dorigo, W.A.; Xaver, A.; Vreugdenhil, M.; Gruber, A.; Hegyiová, A.; Sanchis-Dufau, A.D.; Zamojski, D.; Cordes, C.; Wagner, W.; Drusch, M. Global Automated Quality Control of In Situ Soil Moisture Data from the International Soil Moisture Network. *Vadose Zone J.* **2013**, *12*. [CrossRef]
11. Dorigo, W.A.; Gruber, A.; De Jeu, R.A.M.; Wagner, W.; Stacke, T.; Loew, A.; Albergel, C.; Brocca, L.; Chung, D.; Parinussa, R.M.; et al. Evaluation of the ESA CCI soil moisture product using ground-based observations. *Remote Sens. Environ.* **2015**, *162*, 380–395. [CrossRef]
12. Robock, A.; Mu, M.; Vinnikov, K.; Trofimova, I.V.; Adamenko, T.I. Forty-five years of observed soil moisture in the Ukraine: No summer desiccation (yet). *Geophys. Res. Lett.* **2005**, *32*, 1–5. [CrossRef]
13. Martínez-Fernández, J.; Ceballos, A. Temporal Stability of Soil Moisture in a Large-Field Experiment in Spain. *Soil Sci. Soc. Am. J.* **2003**, *67*, 1647. [CrossRef]
14. Schaefer, G.L.; Cosh, M.H.; Jackson, T.J. The USDA Natural Resources Conservation Service Soil Climate Analysis Network (SCAN). *J. Atmos. Ocean. Technol.* **2007**, *24*, 2073–2077. [CrossRef]
15. Quiring, S.M.; Ford, T.W.; Wang, J.K.; Khong, A.; Harris, E.; Lindgren, T.; Goldberg, D.W.; Li, Z. The North American Soil Moisture Database: Development and Applications. *Bull. Am. Meteorol. Soc.* **2016**, *97*, 1441–1459. [CrossRef]
16. Entekhabi, D.; Njoku, E.G.; O'Neill, P.E.; Kellogg, K.H.; Crow, W.T.; Edelstein, W.N.; Entin, J.K.; Goodman, S.D.; Jackson, T.J.; Johnson, J.; et al. The soil moisture active passive (SMAP) mission. *Proc. IEEE* **2010**, *98*, 704–716. [CrossRef]
17. ESA. SMOS Mission Summary. Available online: https://earth.esa.int/web/guest/missions/esa-eo-missions/smos/mission-summary (accessed on 13 July 2018).
18. Dorigo, W.; Wagner, W.; Albergel, C.; Albrecht, F.; Balsamo, G.; Brocca, L.; Chung, D.; Ertl, M.; Forkel, M.; Gruber, A.; et al. ESA CCI Soil Moisture for improved Earth system understanding: State-of-the art and future directions. *Remote Sens. Environ.* **2017**, *203*, 185–215. [CrossRef]

19. Topp, G.C.; Davis, J.L.; Annan, A.P. Electromagnetic Determination of Soil Water Content: Measruements in Coaxial Transmission Lines. *Water Resour. Res.* **1980**, *16*, 574–582. [CrossRef]
20. Chung, D.; Dorigo, W.; De Jeu, R.; Kidd, R.; Wagner, W. *ESA Climate Change Initiative Phase II – Soil Moisture, Product Specification Document (PSD)*; D.1.2.1 Version 4.4; Earth Observation Data Centre for Water Resources Monitoring (EODC) GmbH: Vienna, Austria, 2018; p. 49.
21. Gruber, A.; Dorigo, W.A.; Crow, W.; Wagner, W. Triple Collocation-Based Merging of Satellite Soil Moisture Retrievals. *IEEE Trans. Geosci. Remote Sens.* **2017**, *55*, 6780–6792. [CrossRef]
22. Gruber, A.; Scanlon, T.; van der Schalie, R.; Wagner, W.; Dorigo, W. Evolution of the ESA CCI Soil Moisture Climate Data Records and their underlying merging methodology. *Earth Syst. Sci. Data Discuss.* **2019**, 1–37. [CrossRef]
23. Wang, G.; Garcia, D.; Liu, Y.; de Jeu, R.; Johannes Dolman, A. A three-dimensional gap filling method for large geophysical datasets: Application to global satellite soil moisture observations. *Environ. Model. Softw.* **2012**, *30*, 139–142. [CrossRef]
24. Kondrashov, D.; Ghil, M.; Kondrashov, D.; Spatio-temporal, M.G. Spatio-temporal filling of missing points in geophysical data sets To cite this version: HAL Id: Hal-00331089 Nonlinear Processes in Geophysics Spatio-temporal filling of missing points in geophysical data sets. *Nonlinear Process. Geophys.* **2006**, *13*, 151–159. [CrossRef]
25. Mao, H.; Kathuria, D.; Duffield, N.; Mohanty, B.P. Gap Filling of High-Resolution Soil Moisture for SMAP/Sentinel-1: A Two-Layer Machine Learning-Based Framework. *Water Resour. Res.* **2019**, *55*, 6986–7009. [CrossRef]
26. DeLiberty, T.L.; Legates, D.R. Spatial Variability and Persistence of Soil Moisture in Oklahoma. *Phys. Geogr.* **2008**, *29*, 121–139. [CrossRef]
27. Hengl, T.; Heuvelink, G.B.M.; Stein, A. A generic framework for spatial prediction of soil variables based on regression-kriging. *Geoderma* **2004**, *120*, 75–93. [CrossRef]
28. Cressie, N. The origins of kriging. *Math. Geol.* **1990**, *22*, 239–252. [CrossRef]
29. Oliver, M.A.; Webster, R. *(Richard) Basics Steps in Geostatistics: The Variogram and Kriging*; SpringerBriefs in Agriculture; Springer: New York, NY, USA, 2015; ISBN 9783319158648.
30. Kang, J.; Jin, R.; Li, X. Regression kriging-based upscaling of soil moisture measurements from a wireless sensor network and multiresource remote sensing information over heterogeneous cropland. *IEEE Geosci. Remote Sens. Lett.* **2015**, *12*, 92–96. [CrossRef]
31. Hengl, T.; Heuvelink, G.B.M.; Rossiter, D.G. About regression-kriging: From equations to case studies. *Comput. Geosci.* **2007**, *33*, 1301–1315. [CrossRef]
32. Brock, F.V.; Crawford, K.C.; Elliott, R.L.; Cuperus, G.W.; Stadler, S.J.; Johnson, H.L.; Eilts, M.D. The Oklahoma Mesonet: A Technical Overview. *J. Atmos. Ocean. Technol.* **1995**, *12*, 5–19. [CrossRef]
33. Commission for Environmental Cooperation. 2010 Land Cover of North America at 30 Meters, Ed. 1. 2017. Available online: http://www.cec.org/tools-and-resources/north-american-environmental-atlas/land-cover-and-land-cover-change (accessed on 17 July 2019).
34. Llamas, R.M.; Bonifaz, R.; Valdés, M.; Riveros-Rosas, D.; Leyva-Contreras, A. Spatial and temporal variations of atmospheric aerosol optical thickness in northwestern Mexico. *Geofis. Int.* **2013**, *52*, 321–341. [CrossRef]
35. Thornton, M.M.; Thornton, P.E.; Wei, Y.; Mayer, B.W.; Cook, R.B.; Vose, R.S. *Daymet: Monthly Climate Summaries on A 1-Km Grid for North America*; Version 3; ORNL DAAC: Oak Ridge, TN, USA, 2018. [CrossRef]
36. Parker, J.A.; Kenyon, R.V.; Troxel, D.E. Comparison of Interpolating Methods for Image Resampling. *IEEE Trans. Med. Imaging* **1983**, *MI-2*. [CrossRef]
37. Soil Survey Staff. Gridded Soil Survey Geographic (gSSURGO) Database for the Conterminous United States. United States Department of Agriculture, Natural Resources Conservation Service, 2016. Available online: https://gdg.sc.egov.usda.gov/ (accessed on 10 August 2019).
38. Bertermann, D.; Bialas, C.; Morper-Busch, L.; Klug, H.; Rohn, J.; Stollhofen, H.; Psyk, M.; Jaudin, F.; Maragna, C.; Einarsson, G.M.; et al. ThermoMap—An Open-Source Web Mapping Application for Illustrating the very Shallow Geothermal Potential in Europe and selected Case Study Areas. *Eur. Geotherm. Congr.* **2013**, *2013*, 1–8.
39. Conrad, O.; Bechtel, B.; Bock, M.; Dietrich, H.; Fischer, E.; Gerlitz, L.; Wehberg, J.; Wichmann, V.; Böhner, J. System for Automated Geoscientific Analyses (SAGA); v.2.1.4. *Geosci. Model. Dev.* **2015**, *8*, 1991–2007. [CrossRef]

40. ESA. CCI ESA Soil Moisture CCI, Validation. Available online: https://www.esa-soilmoisture-cci.org/validation (accessed on 1 July 2019).
41. Kutner, M.H.; Nachtsheim, C.J.; Neter, J.; Li, W. *Applied Linear Statistical Models*, 5th ed.; McGraw-Hill: New York, NY, USA, 2005.
42. Jung, C.; Lee, Y.; Cho, Y.; Kim, S. A study of spatial soil moisture estimation using a multiple linear regression model and MODIS land surface temperature data corrected by conditional merging. *Remote Sens.* **2017**, *9*, 870. [CrossRef]
43. Cai, Y.; Zheng, W.; Zhang, X.; Zhangzhong, L.; Xue, X. Research on soil moisture prediction model based on deep learning. *PLoS ONE* **2019**, *14*, e0214508. [CrossRef] [PubMed]
44. Hiemstra, P. Automatic Interpolation Package. Available online: https://cran.r-project.org/web/packages/automap/index.html (accessed on 28 January 2020).
45. Isaaks, E.H.; Srivastava, R.M. Ordinary Kriging. In *Applied Geostatistics*; Oxford University Press: New York, NY, USA, 1989; pp. 278–322.
46. Cambardella, C.A.; Moorman, T.B.; Parkin, T.B.; Karlen, D.L.; Novak, J.M.; Turco, R.F.; Konopka, A.E. Field-Scale Variability of Soil Properties in Central Iowa Soils. *Soil Sci. Soc. Am. J.* **1994**, *58*, 1501. [CrossRef]
47. Hengl, T.; Kempen, B.; Heuvelink, G.; Malone, B. Available online: https://rdrr.io/rforge/GSIF/ (accessed on 28 January 2020).
48. Giraldo-Henao, R. Introducción a la Geoestadística, Teoría y Aplicación. Available online: https://geoinnova.org/blog-territorio/wp-content/uploads/2015/05/LIBRO_-DE-_GEOESTADISTICA-R-Giraldo.pdf (accessed on 28 January 2020).
49. Kuhn, M. Caret: Classification and Regression Training. Available online: http://topepo.github.io/caret/index.html (accessed on 28 January 2020).
50. Stein, M.L. *Interpolation of Spatial Data, Some Theory for Kriging*; Springer: Chicago, IL, USA, 1999.
51. Taylor, K.E. Summarizing multiple aspects of model performance in a single diagram. *J. Geophys. Res.* **2001**, *106*, 7183–7192. [CrossRef]
52. Lemon, J. Plotrix: A package in the red light district of R. *R News* **2006**, *6*, 8–12.
53. Tang, C.; Chen, D. Interaction between Soil Moisture and Air Temperature in the Mississippi River Basin. *J. Water Resour. Prot.* **2018**, *9*, 9–12. [CrossRef]
54. Salathé, E.P., Jr.; Steed, R.; Mass, C.F.; Zahn, P.H. A High-Resolution Climate Model for the U.S. Pacific Northwest: Mesoscale Feedbacks and Local Responses to Climate Change. *J. Clim.* **2008**, *21*, 5708–5726. [CrossRef]
55. Ding, Y.; Wang, Y.; Miao, Q. Research on the spatial interpolation methods of soil moisture based on GIS. Proceedings of 2011 International Conference on Information Science and Technology, Nanjing, China, 26–28 March 2011; pp. 709–711.
56. Chen, H.; Fan, L.; Wu, W.; Liu, H. Bin Comparison of spatial interpolation methods for soil moisture and its application for monitoring drought. *Environ. Monit. Assess.* **2017**, *189*. [CrossRef]
57. Zhang, M.; Li, M.; Wang, W.; Liu, C.; Gao, H. Temporal and spatial variability of soil moisture based on WSN. *Math. Comput. Model.* **2013**, *58*, 820–827. [CrossRef]
58. Cruz-Cárdenas, G.; López-Mata, L.; Ortiz-Solorio, C.A.; Villaseñor, J.L.; Ortiz, E.; Silva, J.T.; Estrada-Godoy, F. Interpolation of mexican soil properties at a scale of 1:1,000,000. *Geoderma* **2014**, *213*, 29–35. [CrossRef]
59. De Benedetto, D.; Castrignanò, A.; Quarto, R. A Geostatistical Approach to Estimate Soil Moisture as a Function of Geophysical Data and Soil Attributes. *Procedia Environ. Sci.* **2013**, *19*, 436–445. [CrossRef]
60. Abuduwaili, J.; Tang, Y.; Abulimiti, M.; Liu, D.; Ma, L. Spatial distribution of soil moisture, salinity and organic matter in Manas River watershed, Xinjiang, China. *J. Arid Land* **2012**, *4*, 441–449. [CrossRef]
61. Brusilovskiy, E. Spatial Interpolation: A Brief Introduction. Available online: http://www.bisolutions.us/A-Brief-Introduction-to-Spatial-Interpolation.php (accessed on 22 October 2019).
62. Odeh, I.O.A.; McBratney, A.B.; Chittleborough, D.J. Further results on prediction of soil properties from terrain attributes: Heterotopic cokriging and regression-kriging. *Geoderma* **1995**, *67*, 215–226. [CrossRef]
63. Zhu, Q.; Lin, H.S. Comparing Ordinary Kriging and Regression Kriging for Soil Properties in Contrasting Landscapes. *Pedosphere* **2010**, *20*, 594–606. [CrossRef]
64. Guevara, M.; Olmedo, G.F.; Stell, E.; Yigini, Y.; Duarte, Y.A.; Hernández, C.A.; Arévalo, G.E.; Arroyo-cruz, C.E.; Bolivar, A.; Bunning, S.; et al. No silver bullet for digital soil mapping: Country-specific soil organic carbon estimates across Latin America. *SOIL* **2018**, *4*, 173–193. [CrossRef]

65. Sun, W.; Minasny, B.; McBratney, A. Analysis and prediction of soil properties using local regression-kriging. *Geoderma* **2012**, *171–172*, 16–23. [CrossRef]
66. Zhang, S.; Huang, Y.; Shen, C.; Ye, H.; Du, Y. Spatial prediction of soil organic matter using terrain indices and categorical variables as auxiliary information. *Geoderma* **2012**, *171–172*, 35–43. [CrossRef]
67. Sumfleth, K.; Duttmann, R. Prediction of soil property distribution in paddy soil landscapes using terrain data and satellite information as indicators. *Ecol. Indic.* **2008**, *8*, 485–501. [CrossRef]
68. Bishop, T.F.A.; Mcbratney, A.B. A comparison of prediction methods for the creation of field-extent soil property maps. *Geoderma* **2001**, *103*, 149–160. [CrossRef]
69. Seasholtz, M.B.; Kowalski, B. The parsimony principle applied to multivariate calibration. *Anal. Chim. Acta* **1993**, *277*, 165–177. [CrossRef]
70. Colliander, A.; Jackson, T.J.; Bindlish, R.; Chan, S.; Das, N.; Kim, S.B.; Cosh, M.H.; Dunbar, R.S.; Dang, L.; Pashaian, L.; et al. Validation of SMAP surface soil moisture products with core validation sites. *Remote Sens. Environ.* **2017**, *191*, 215–231. [CrossRef]
71. Menne, M.J.; Durre, I.; Korzeniewski, B.; McNeal, S.; Thomas, K.; Yin, X.; Anthony, S.; Ray, R.; Vose, R.S.; Gleason, B.E.; et al. Global Historical Climatology Network—Daily (GHCN-Daily); Version 3. Available online: https://catalog.data.gov/dataset/global-historical-climatology-network-daily-ghcn-daily-version-3 (accessed on 28 January 2020).
72. Fick, S.E.; Hijmans, R.J. WorldClim 2: New 1-km spatial resolution climate surfaces for global land areas. *Int. J. Climatol.* **2017**, *37*, 4302–4315. [CrossRef]

remote sensing

MDPI

Technical Note

Development of a Multimode Field Deployable Lidar Instrument for Topographic Measurements of Unsaturated Soil Properties: Instrument Description

Sean E. Salazar [1,*], Cyrus D. Garner [2] and Richard A. Coffman [1]

[1] Department of Civil Engineering, University of Arkansas, 4190 Bell Engineering Center, Fayetteville, AR 72701, USA; rick@uark.edu

[2] Stantec Consulting Services, Inc., 3052 Beaumont Centre Circle, Lexington, KY 40513, USA; cyrus.garner@stantec.com

* Correspondence: ssalazar@uark.edu

Received: 7 January 2019; Accepted: 29 January 2019; Published: 1 February 2019

Abstract: The hydrological and mechanical behavior of soil is determined by the moisture content, soil water (matric) potential, fines content, and plasticity. However, these parameters are often difficult or impractical to determine in the field. Remote characterization of soil parameters is a non-destructive data collection process well suited to large or otherwise inaccessible areas. A ground-based, field-deployable remote sensor, called the soil observation laser absorption spectrometer (SOLAS), was developed to collect measurements from the surface of bare soils and to assess the in-situ condition and essential parameters of the soil. The SOLAS instrument transmits coherent light at two wavelengths using two, continuous-wave, near-infrared diode lasers and the instrument receives backscattered light through a co-axial 203-mm diameter telescope aperture. The received light is split into a hyperspectral sensing channel and a laser absorption spectrometry (LAS) channel via a multi-channel optical receiver. The hyperspectral channel detects light in the visible to shortwave infrared wavelengths, while the LAS channel filters and directs near-infrared light into a pair of photodetectors. Atmospheric water vapor is inferred using the differential absorption of the on- and off-line laser wavelengths (823.20 nm and 847.00 nm, respectively). Range measurement is determined using a frequency-modulated, self-chirped, coherent, homodyne detection scheme. The development of the instrument (transmitter, receiver, data acquisition components) is described herein. The potential for rapid characterization of physical and hydro-mechanical soil properties, including volumetric water content, matric potential, fines content, and plasticity, using the SOLAS remote sensor is discussed. The envisioned applications for the instrument include assessing soils on unstable slopes, such as wildfire burn sites, or stacked mine tailings. Through the combination of spectroradiometry, differential absorption, and range altimetry methodologies, the SOLAS instrument is a novel approach to ground-based remote sensing of the natural environment.

Keywords: instrument development; hyperspectral; spectroradiometry; LiDAR; soil

1. Introduction

Remote sensing is well suited for non-intrusive observation of bare soils, especially over large, hazardous, or inaccessible areas, such as a wildfire site. For example, spaceborne remote sensing techniques are commonly used to rapidly (1) establish wildfire perimeters, (2) assess the remaining vegetative cover, and (3) determine the burn severity after containment of the fire. Collected remotely sensed data (burn severity, extent) are often calibrated with ground-truthing methods, yet these proximal ground-truthing methods are often point-wise, spatially limited, and cannot easily cover vast areas. Moreover, information about the soil is not commonly collected in these areas following a wildfire. Characterization of soil in a wildfire-affected area commonly relies on regional, typified soils data from databases like the Soil Survey Geographic Database (SSURGO) and the State Soil Geographic Survey (STATSGO). These data, however, have insufficient resolution for reliable, site-specific, predictive modeling of post-wildfire hazards (e.g., debris flows) and do not capture the time-variability associated with meteorological and hydrological action. Because burned areas are ideally suited for study with remote sensing techniques, due to the absence of vegetation (fire-induced denudation), there is a need for methods to collect high-resolution, timely, and site-specific soils information.

To address this need, a ground-based, remote sensor, called the soil observation laser absorption spectrometer (SOLAS), was developed to rapidly infer soil properties at the field scale. The development of the SOLAS followed laboratory-based, proof-of-concept testing that successfully derived soil water characteristic curves (SWCC) as well as index properties (liquid limit (LL), plastic limit (PL), and clay fraction (CF)) for several soil types by using only non-contact, optical techniques. By combining spectroradiometric, differential laser absorption, and range altimetry techniques, the SOLAS instrument was designed to collect range-resolved information from bare soils, including soil surface moisture (an estimation of volumetric water content, θ_v), soil matric potential (ψ_m), burn severity, LL, PL, and CF. An initial description of the SOLAS instrument is provided herein; as such, the materials and methods used in the development of the instrument are detailed and described. Additionally, supporting background information about reflectance spectroradiometry, lidar altimetry, and differential laser absorption is provided. Measurement results from field-testing will be described by the authors in later articles.

2. Background

A variety of remote and proximal sensing techniques for obtaining soils information have been demonstrated. These techniques include passive imaging spectroradiometry (multispectral, hyperspectral, visible near-infrared (VNIR), shortwave infrared (SWIR), and mid-wave-infrared (MWIR)), active and passive microwave systems (synthetic- and real-aperture radar, ground-penetrating radar), and gamma-ray spectrometry [1]. Although the correlation between reflectance and soil moisture was studied as early as 1925 [2], advances in ground-based multispectral and hyperspectral measurement techniques of reflectance spectra, primarily in the VNIR (380–1000 nm) and SWIR (1000–2500 nm) ranges, have been utilized to estimate soil moisture content (SMC) in the laboratory setting [3–24]. In the aforementioned studies, the laboratory measurements were collected using carefully prepared or dilute soil specimens under controlled conditions. Fewer studies were conducted under field conditions [23,25,26]. Among the numerous developed soil reflectance correlations in the literature, other soil parameters of interest have included clay content [9,14,24,27–30], grain size [9,28,29,31], soil plasticity [24,32,33] and matric potential [24].

The SOLAS instrument that is described herein was designed based on other work previously performed at the University of Arkansas. For example, Garner [24] utilized a laboratory-based diffuse reflectance infrared Fourier transform (DRIFT) technique to develop an empirical relationship between reflectance spectra and soil plasticity for illite and kaolinite soil types, as well as for a commercial synthetic nepheline synetite material (Donna Fill Co., Little Rock, AR, USA). Garner [24] also developed a laser analysis of soil tension (LAST) technique to infer the SWCC for dilute pressure plate extractor (PPE) prepared soil specimens. The measurement technique utilized coherent illumination from two low-power, near-infrared laser diodes and data collection using a high radiometric-resolution spectrometer (ASD FieldSpec 4 Hi-Res; Malvern Panalytical, Longmont, CO, USA) to relate θ_v and ψ_m through the SWCC. The empirical relationships relied upon partial least squares and principle components regression techniques [9,24,34].

2.1. FMCW Lidar Altimetry

Among laser altimetry methods, coherent, frequency modulated continuous waveform (FMCW) lidar has been widely pursued [35–42]. A pulse compression technique has been applied to FMCW lidar systems, whereby a linear frequency sweep or "chirp" with a large bandwidth is used to modulate the optical carrier signal. As documented in the aforementioned FMCW lidar literature, range accuracy was maintained, while peak output power and receiver bandwidth requirements were reduced (over direct detection or conventional, pulsed, time-of-flight systems).

Adany et al. [39] demonstrated the advantages of a self-chirped, homodyne detection scheme for FMCW lidar. The simplified homodyne system offered significant advantages over direct detection and heterodyne detection methods through less complex receiver configuration. Furthermore, improved receiver sensitivity permitted better long-range lidar measurements. In the Adany et al. [39] configuration, the optical signal was intensity-modulated with a linear frequency modulated (FM) sweep (from frequency f_1 to f_2) with chirp bandwidth, B, equal to $f_2 - f_1$. For the Adany et al. [39] design, a portion of the carrier signal was used as the local oscillator (LO) in conjunction with a balanced photodetector (BPD). The range to the target was proportional to the frequency difference between the LO and the received signal (beat frequency, f_R). For FMCW lidar with self-chirped homodyne detection, like that proposed by Adany et al. [39], the range to target (R) should be calculated using Equations (1) and (2) [39], while the approximate range accuracy (σ_R) should be determined by using Equations (3) and (4) [36,43,44].

$$R = \frac{c \cdot f_R \cdot \tau}{2 \cdot (f_2 - f_1)} \quad (1)$$

$$f_R = \left(\frac{f_2 - f_1}{\tau}\right) \cdot \Delta t \quad (2)$$

$$\sigma_R = \frac{K \cdot c}{B\sqrt{SNR}} \quad (3)$$

$$SNR_{coh} = \frac{\Re \cdot P_r}{2 \cdot q \cdot B_{RX}} \quad (4)$$

In Equation (1), R is the range to target, c is the speed of light, f_R is the beat frequency, τ is the chirp duration, and $f_2 - f_1$ is the chirp bandwidth. In Equation (2), Δt is the time delay for roundtrip propagation through the atmosphere. In Equation (3), σ_R is the range accuracy, K is a chirp waveform constant, B is the signal bandwidth, and SNR is the signal to noise ratio of the receiver data. In Equation (4), SNR_{coh} is the signal to noise ratio for a shot-noise-dominant coherent detection process, $\Re$ is the photodetector responsivity, P_r is the received signal power, q is the electron charge (1.6×10^{-10} C), and B_{RX} is the bandwidth of the receiver.

2.2. Differential Absorption Measurements

The differential absorption lidar (DIAL) technique, sometimes also called (differential) laser absorption spectrometry (LAS), has been employed to determine the concentration of molecular species in the atmosphere by measuring the difference in light absorption between two transmitted laser wavelengths. DIAL theory was developed by Schotland [45] but has been advanced over the last six decades [46–54]. Moreover, during this time period, DIAL has become the most accurate measurement technique for tropospheric water vapor concentration [51,53,55,56]. A variety of DIAL instruments and measurement techniques have been developed to measure water vapor profiles and concentrations of other atmospheric greenhouse gases (e.g., carbon dioxide, methane). These measurements have been performed from ground-based platforms [50,57–66], airborne platforms [67–72], and proposed spaceborne platforms [55,73–76].

DIAL measurements are typically achieved by alternating the transmission of two laser wavelengths through the atmosphere along the same path to determine the water vapor concentration. The so-called on-line wavelength is tuned to correspond with a water vapor absorption feature, while the off-line wavelength is tuned to a nearby spectral region in which water vapor is not well absorbed. For accurate measurement, a spectral region of interest must be identified for which the on- and off-line wavelengths are adjacent and the temperature dependence of the DIAL measurement is minimal. Various wavelength ranges have been recommended in the literature for measurement of water vapor. For example, Grant [49] utilized the 720–730 nm wavelength range, while Machol et al. [60] used wavelengths near 823 nm. The water vapor density (ρ_v), averaged over distance (R), is commonly calculated using the DIAL equation proposed by Schotland [46] and presented in the form of Equations (5)–(7) [60]. For vertical measurements of the atmospheric water vapor concentration, the Voigt function (Λ) changes due to thermal- and pressure-broadening effects, which are typically extrapolated from ground measurements. The water vapor concentration is commonly calculated using Equations (8) and (9) [60].

$$\rho_v(R) = \frac{M_{H_2O}}{N_A} \cdot \frac{1}{2 \cdot \left(\sigma_{on} - \sigma_{off}\right) \cdot \Delta R} \cdot \left[ln \frac{P_{on} \cdot R \cdot P_{off} \cdot (R + \Delta R)}{P_{on} \cdot (R + \Delta R) \cdot P_{off} \cdot R} \right] \tag{5}$$

$$\sigma = S \cdot \Lambda \tag{6}$$

$$S(T) = S_0 \cdot \left(\frac{T_0}{T}\right)^{1.5} \cdot exp\left[-\frac{h \cdot c \cdot E''}{k_B} \left(\frac{1}{T} - \frac{1}{T_0}\right) \right] \tag{7}$$

$$\rho_v = \frac{e_s \cdot RH}{100 \cdot R_v \cdot T} \tag{8}$$

$$e_s = e_{s0} \cdot exp\left[\frac{L}{R_v} \left(\frac{1}{T_0} - \frac{1}{T} \right) \right] \tag{9}$$

In Equation (5), ρ_v is the water vapor density averaged over a distance ΔR at a range R, M_{H2O} is the molecular weight of water, N_A is Avogadro's constant, σ_{on} and σ_{off} are the on-line and off-line water vapor absorption cross-sections obtained from Equation (6), and P_{on} and P_{off} are the received on-line and off-line backscatter signals. In Equation (6), S is the temperature-dependent absorption line strength and Λ is the Voigt function. In Equation (7), S_0 and T_0 are the absorption line strength and temperature under standard conditions, T is the temperature, h is the Planck constant, c is the speed of light, E'' is the lower-state energy (in cm^{-1}), and k_B is the Boltzmann constant. In Equation (8), e_s is the saturation vapor pressure obtained from Equation (9), RH is the relative humidity ($RH \approx 100 \times e/e_s$), where $e = \rho_v \cdot R_v \cdot T$, and R_v is the water vapor gas constant equal to 461 $J \cdot kg^{-1} \cdot K^{-1}$. In Equation (9), e_{s0} is the saturation vapor pressure at T_0 = 273 K and is equal to 611 Pa, and L is the latent heat of vaporization and is equal to 2.5×10^6 $J \cdot kg^{-1}$.

3. Development of the SOLAS Concept

The SOLAS instrument was devised to collect range-resolved hyperspectral measurements of soils while also measuring water absorption, due to water vapor, over the measurement range. Moreover, the bench-scale studies conducted by Garner [24] indicated that under coherent illumination, empirical inference of soil matric potential (ψ_m), and volumetric water content (θ_v) was possible. The instrument therefore utilized laser transmission to achieve these metrics while collecting passive radiometric measurements across the VNIR to SWIR range (350–2500 nm). Based on water vapor absorption spectra published by the high-resolution transmission (HITRAN) molecular absorption database [77] and the availability of commercial off-the-shelf laser diodes, laser wavelengths of 823.20 nm (on-line) and 847.00 nm (off-line) were selected. For completeness, the on- and off-line wavelengths transmitted by the SOLAS instrument are transposed over a plot of the atmospheric absorption coefficient as a function of wavelength in Figure 1.

Because DIAL instruments have primarily been developed to measure vertical gas and aerosol profiles, there are limited examples of instruments operating in horizontal orientations or for topographic target returns [48,50,62,78,79]. Furthermore, DIAL instruments have typically utilized pulsed, rapid spectral-switching lasers to increase the accuracy of atmospheric volume sampling, especially over long vertical ranges (vertical measurements of atmospheric water vapor are extremely sensitive to pressure- and temperature-induced gradients). To provide coherent illumination to the target, while enabling simplified topographic ranging and differential absorption measurements, a diode-laser-based FMCW laser scheme was designed to switch between the on-line and off-line laser sources over short intervals (seconds). The use of a self-chirped, homodyne detection configuration (similar to [39]), has enabled range-resolved measurements.

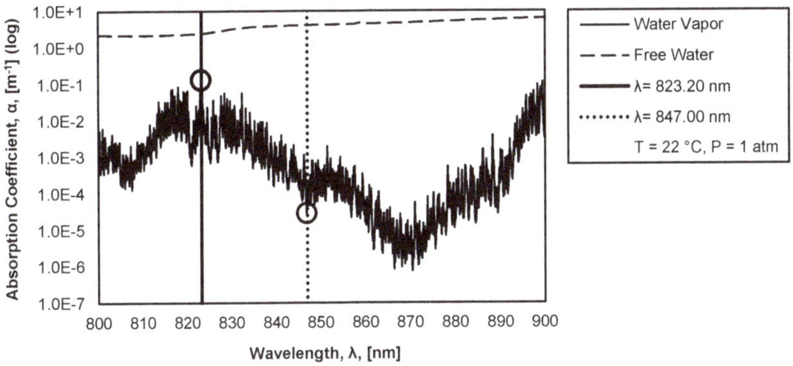

Figure 1. Absorption coefficient, as a function of wavelength, for free water and water vapor with transposed on-line (823.20 nm) and off-line (847.00 nm) laser wavelengths; raw data from [80–82].

4. Instrument Description

The SOLAS instrument combines range altimetry, differential absorption, and reflectance spectroradiometry technologies. The instrument is comprised of (1) a laser source and transmitting system, (2) a multi-channel receiving system (active LAS and passive hyperspectral sensing), and (3) a data acquisition and control system (signal processing and component control). A schematic of the major architecture of the SOLAS instrument is presented in Figure 2 and a table describing the technical specifications is presented as Table 1. Each of the instrument subsystems are further described in the following sections.

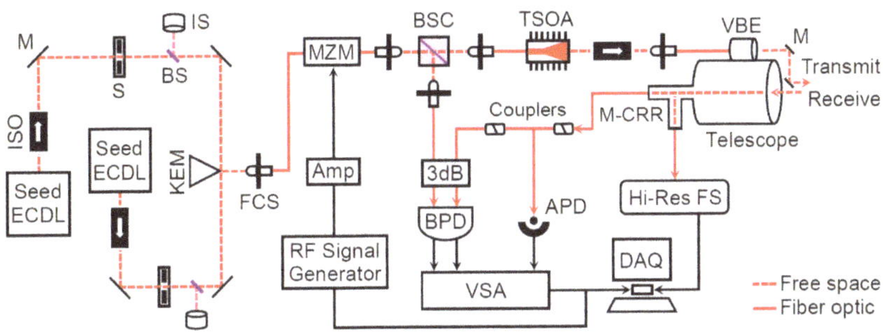

Figure 2. Schematic of the soil observation laser absorption spectrometer (SOLAS). Key: ECDL = External Cavity Diode Laser; ISO = Optical Isolator; M = Mirror; S = Shutter; BS = Beam Sampler; IS = Integrating Sphere; KEM = Knife-Edge Mirror; FCS = Fiber-Coupling Stage; MZM = Mach-Zehnder Modulator; Amp = Amplifier; BSC = Beamsplitter Cube; TSOA = Tapered Semiconductor Optical Amplifier; VBE = Variable Beam Expander; M-CRR = Multi-Channel Receiver Relay; Hi-Res FS = High-Resolution Field Spectroradiometer; DAQ = Data Acquisition; APD = Avalanche Photodetector; 3 dB = 3 dB 2 × 2 Optical Coupler; BPD = Balanced Photodetector; VSA = Vector Signal Analyzer.

Table 1. Specifications of the soil observation laser absorption spectrometer (SOLAS).

Transmitter	Unit	Specification	Receiver	Unit	Specification
Type	-	CW	Telescope	-	Schmidt-Cassegrain
On-line	[nm]	823.20	Primary diameter (D)	[mm]	203
Off-line	[nm]	847.00	Focal length (f)	[mm]	2032
Linewidth	[kHz]	≤200	N (f/D)	-	10
Modulation	-	Electrooptic AM	Field of view (FOV)		
Type	-	Linear RF chirp	Hyperspectral channel	[mrad]	0.32–0.61
Frequency	[MHz]	100–500	LAS channel	[mrad]	0.27
Transmitter power	[W]	≤0.5 (fiber-end)	Detectors		
Beam diameter	[mm]	2.0–8.0	Hyperspectral receiver	-	ASD FieldSpec 4 Hi-Res
Beam divergence	[mrad]	0.29	Type	-	Silicon, InGaAs
			Spectral range	[nm]	350–2500
Signal Processing	**Unit**	**Specification**	Number of bands	-	2151
			Balanced receiver (LO sig.)	-	New Focus 1607-AC-FC
Bandwidth (instantan.)	[MHz]	50	Type	-	Silicon
Frequency range	[GHz]	0.01–6.6	Bandwidth	[MHz]	650
Water vapor resolution	-	Column averaged	Avalanche photodetector	-	Thorlabs APD430A
Range to target resolution	[cm]	10 (theoretical)	Type	-	Silicon
			Bandwidth	[MHz]	400

Key: CW = Continuous-Wave; AM = Amplitude Modulation; RF = Radio Frequency; N = F-number; LAS = Laser Absorption Spectrometry; ASD = Analytical Spectral Devices; InGaAs = Indium Gallium Arsenide; LO = Local Oscillator.

4.1. Transmitter Design

The optical carrier signal is seeded by two New Focus TLB-6817 Vortex Littman–Metcalf external cavity diode lasers (ECDL) precision-tuned to center wavelengths of 823.20 nm and 847.00 nm, with fine tuning from 823.03 nm to 823.35 nm, and 846.84 nm to 847.14 nm, respectively (Newport Corporation; Irvine, CA, USA). Each laser is powered with a low noise controller (New Focus TLB-6800-LN), producing 17 mW to 26 mW outputs with narrow linewidths ($\leq$200 kHz). As previously presented in Figure 2, the laser transmission path is partially free space and partially fiber optic based. To protect each ECDL from back reflections, the laser beams pass through narrowband polarization-dependent Faraday isolators (Thorlabs IO-5-850-HP) that are tuned to match each respective wavelength (Thorlabs Inc.; Newton, NJ, USA). Optomechanical shutters in the free space laser paths provide a fail-safe (Thorlabs SH05). A sequence of dielectric mirrors direct each laser beam into a polarization-maintaining fiber optic cable via a Thorlabs PAF-X-5-B fiber-coupling stage. The light energy within the fiber optic cable is then coupled into a Jenoptik AM830 Mach–Zehnder modulator (MZM) where the optical signal is intensity modulated (Jenoptik Optical Systems GmbH; Jena, Germany). The modulation is achieved by utilizing a radio frequency (RF) signal generator to encode the transmitted light with a chirp. Seventy percent of the intensity-modulated optical signal continues along the transmitter path (into the tapered semiconductor optical amplifier (TSOA)) while the remaining 30% is reflected through a free-space beamsplitting cube and fiber-coupled into a 650 MHz bandwidth New Focus 1607-AC-FC balanced photodetector (BPD) to provide the local oscillator (LO) input signal. The carrier signal is fiber-coupled and amplified through a Thorlabs TPA830P10-SP butterfly package TSOA mounted to a thermoelectric-cooled (TEC) 205 TEC Butterfly LaserMount (Arroyo Instruments LLC, San Luis Obispo, CA, USA). The TSOA chip is tuned to a center wavelength (CWL) of 835 nm (centered between the 823.20 nm and 847.00 nm transmitting wavelengths). The amplified beam is subsequently shaped with a collimation package before exiting the TSOA output window in free space. The beam is then isolated (Faraday isolator tuned to a CWL of 835 nm) and coupled into a high-power, armored fiber optic cable. The laser output is transmitted into the atmosphere co-axial with the optical receiver (telescope) by means of a collimator (Thorlabs F280SMA-835), a variable beam expander (Thorlabs BE052-B), and a pair of mirrors, as depicted in Figures 2 and 3. The transmitter beam has an adjustable output diameter between 2.0 mm and 8.0 mm with an average beam divergence of 0.29 mrad (resulting in the diameter increasing to approximately 29 cm at a range of 1.0 km). The average beam diameter-dependent power density ranges from 10–160 mW·mm^{-2} at the source, with the density decreasing as a function of range.

4.2. Receiver Design

The receiving aperture for the instrument is a 203-mm diameter, 2032-mm equivalent focal length, Schmidt-Cassegrain catadioptric telescope (model LX200-ACF 203 mm f/10) from Meade Instruments (Irvine, CA, USA). As depicted in Figure 3, a custom-built, multi-channel, optical receiver relay is mounted to the rear of the telescope. The receiver was designed to gather, collimate, split, and focus the light from the telescope into two separate channels. On the primary channel (LAS channel), backscattered light is filtered (to isolate the on-line and off-line wavelengths and to reduce diffuse sunlight saturation), focused, and fiber-coupled into the SOLAS instrument. The optical signal is further divided through a multimode fiber optic coupler. Ten percent of the split light is directed into a 400 MHz bandwidth, variable gain Thorlabs APD430A silicon avalanche photodetector (APD) via a beam collimator and focuser. The remaining 90% of the light is coupled into the BPD via a 3 dB 2 $\times$ 2 fiber optic coupler. The signal is de-chirped (i.e., mixed with the LO signal) and the beat frequency is measured directly. On the secondary channel (hyperspectral channel), the light remains unfiltered and is focused and fiber-coupled into a high-resolution spectroradiometer instrument (ASD FieldSpec 4 Hi-Res). The spectral resolution of the secondary channel is 3nm in the VNIR range (350–1000 nm) and 8 nm in the SWIR range (1000–2500 nm). The sampling interval is 1.4 nm and 1.1 nm in the VNIR and SWIR ranges, respectively. The angular field of view (FOV) for the LAS channel is 0.27 mrad and the FOV for the hyperspectral channel is 0.32 mrad (VNIR range) and 0.61 mrad (SWIR range). Due to space limitations in this manuscript, the optical receiver is described in more detail in a separate publication.

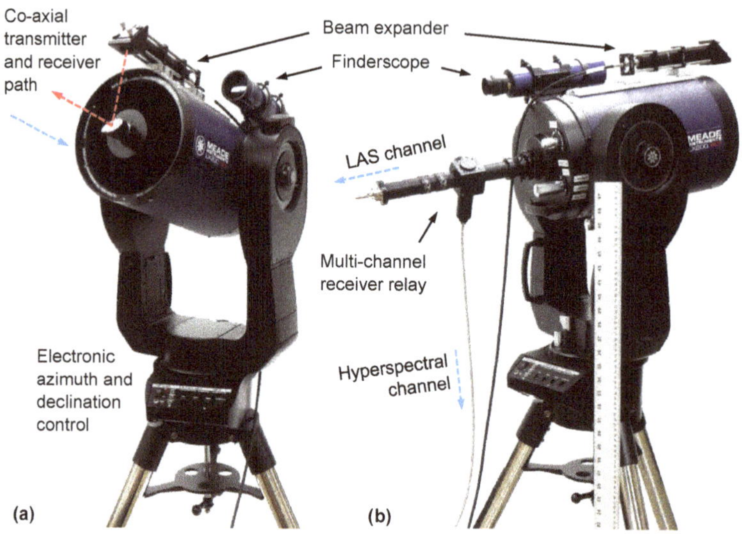

Figure 3. Annotated photograph of the (**a**) front, and (**b**) rear, of the receiver (scale for reference).

4.3. Data Acquisition and Control Design

Data acquisition and component control for the SOLAS are achieved via a computer that is mounted in a compact, module-based National Instruments (Austin, TX, USA) PXIe chassis (PXIe-8135 computer, PXIe-1082 chassis) via LabVIEW software in a Windows environment. Within the chassis are (1) a high frequency RF signal generator module (PXIe-5652), (2) a wide instantaneous bandwidth vector signal analyzer (PXIe-5663E) comprised of three parallel modules (PXIe-5601, PXIe-5622, PXIe-5652), and (3) a multifunction input/output module (PXI-6238). The LabVIEW software is used to generate the chirp signal (100 MHz to 500 MHz linear ramping signal with a chirp rate of 6 MHz/μs) that is amplified and directed into the MZM. The software is also used to (1) collect and interpret the de-chirped frequency from the BPD (to determine the range to the target), and to (2) collect and interpret data from the APD (to detect atmospheric water vapor en route to the target). The ASD RS^3 software is used to collect the reflectance spectra from the spectroradiometer and the ASD ViewSpec^{TM} Pro software is used to export the raw data for further processing. A flow diagram outlining the data acquisition and processing chain is presented in Figure 4.

4.4. Field Ruggedization

The majority of the components that were previously presented in Figure 2 are mounted within a hermetically sealed, nitrogen-purged box. The ECDL heads and MZM are mounted directly to the 12 mm thick aluminum floor of the box with thermal paste to enable the floor to act as a heat sink. The remaining power-emitting components (e.g., Thorlabs TPA830P10-SP amplifier) are actively regulated via thermoelectric cooling or are self-regulating (e.g., New Focus 1607-AC-FC and Thorlabs APD430A photodetectors). The floor of the box also acts as an optical bench for the bulk-optical components associated with the free space lasers. The transmitting and receiving fiber optic cables, RF signal cables, and component power cables are fed through one wall of the box via sealed cable glands. A plan view of the box interior is presented in Figure 5 and a photograph of the SOLAS instrument annotated with major assemblies is presented as Figure 6.

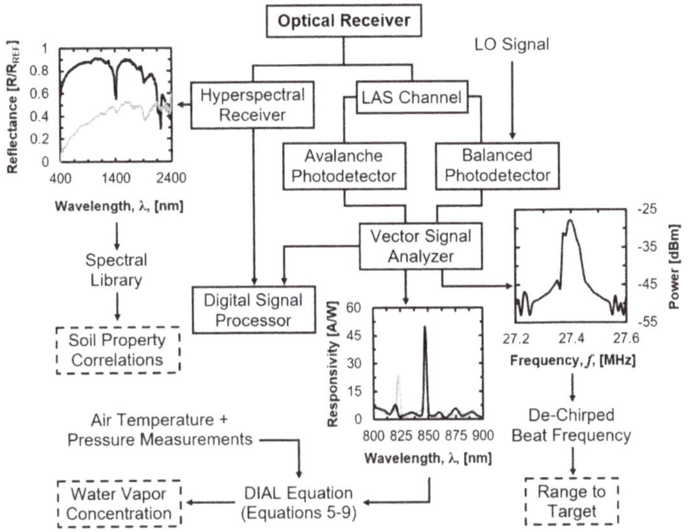

Figure 4. Data acquisition and processing chain for the soil observation laser absorption spectrometer (SOLAS) instrument (note: simulated data). Key: DIAL = Differential Absorption Lidar; LAS = Laser Absorption Spectrometry; LO = Local Oscillator.

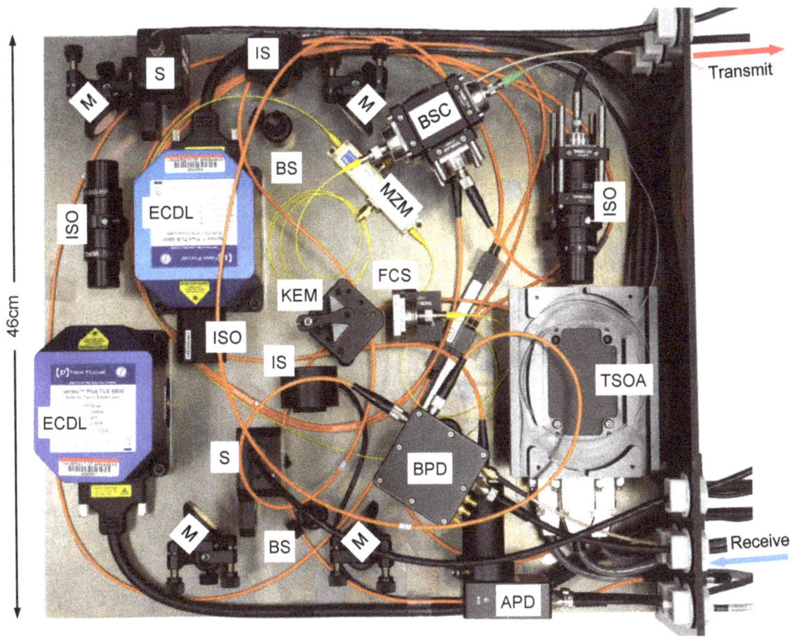

Figure 5. Annotated plan view of the hermetically sealed box depicting the major components of the transmitter and the primary laser absorption spectrometer (LAS) receiver channel. Key: ECDL = External Cavity Diode Laser; ISO = Optical Isolator; M = Dielectric Mirror; S = Shutter; BS = Beam Sampler; IS = Integrating Sphere; KEM = Knife-Edge Mirror; FCS = Fiber-Coupling Stage; BSC = Beamsplitter Cube; MZM = Mach-Zehnder Modulator; TSOA = Tapered Semiconductor Optical Amplifier; APD = Avalanche Photodetector; BPD = Balanced Photodetector.

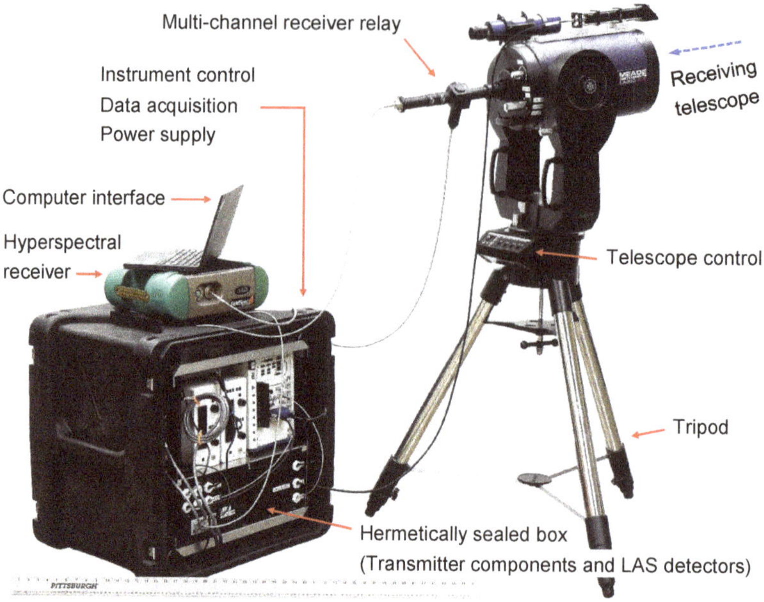

Figure 6. Annotated photograph of the soil observation laser absorption spectrometer (SOLAS) instrument with major assemblies (transmitter, receiver, data acquisition and control).

5. Discussion

The SOLAS instrument was designed to transmit on-line and off-line wavelengths of 823.20 nm and 847.00 nm, respectively. The difference between these wavelengths combined with continuous-wave transmission, necessitated the use of two separate seed lasers (whereas some dedicated DIAL instruments have achieved on- and off-line wavelength transmission with a single, widely tunable, pulsed laser source). The two lasers were aligned into a common transmitter system using readily-available, free-space bulk optics to ease customization, calibration, and implementation. Therefore, the efficiency of the laser delivery system may be improved using an all-fiber-based design in future iterations.

The collection of measurements in the field introduces additional complexity, primarily due to (1) viewing geometry (i.e., incidence and viewing angles), (2) the sensitivity of the hyperspectral measurements to changes in light conditions (solar irradiation intensity), and (3) environmental interferences (dust, water droplets, vegetative cover). To address these issues, the instrument observation location must be carefully selected and the spectroradiometer should be calibrated using a diffuse white reference panel (e.g., Spectralon®; Labsphere Inc., North Sutton, NH, USA) positioned at approximately the same incidence angle as the intended measurements. The manufacturer of the spectroradiometer recommends frequent recalibration (referencing of the diffuse reflector panel) when collecting typical proximal (<1 m distance) measurements in the laboratory or in the field. However, it would be possible to collect remote (up to 1 km distance, or greater) measurements for an extended period of time, without frequent recalibration, if careful considerations are made. The spectroradiometer, as well as other components (e.g., laser sources, data acquisition system, and telescope), should be allowed a warm-up period (to minimize instrument noise and temperature-induced drift). Furthermore, after initial calibration of the spectroradiometer, any changes in light conditions (e.g., temporary cloud cover over target) should be observed and, if necessary, the measurements should be repeated.

The data collected by the three receivers (spectroradiometer and two LAS channel detectors) must be synthesized for meaningful interpretation of a measurement. Reflectance spectra are compiled, averaged, and compared with spectral libraries for different soil types. The measurements require post-processing (empirical calibration and statistical analysis) to extract the soil properties of interest. While reflectance data is collected using the ASD software (native to the spectroradiometer), future development of the SOLAS instrument software will enable custom data collection and near real-time data interpretation. The reflectance measurements are susceptible to attenuation, due to atmospheric water vapor, especially at longer ranges or in conditions with higher relative humidity. To correct for the additional atmospheric absorption en route to the soil surface, the differential laser absorption measurements are used. The coherent signals also provide sub-meter range to target identification. Preliminary hyperspectral measurements have been collected for ranges greater than 100 m (laboratory setting) and 500 m field setting). Based on design calculations, measurements are possible for ranges of up to a kilometer or more (depending on atmospheric conditions), with spatial resolutions of 6 cm, 30 cm, and 60 cm (nadir) for ranges of 100 m, 500 m, and 1.0 km, respectively.

6. Conclusions

The development of a field-deployable, ground-based, remote sensing instrument for obtaining physical and hydro-mechanical soil properties was described herein. The soil observation laser absorption spectrometer (SOLAS) was designed to collect range-resolved hyperspectral backscatter data from bare soil surfaces across the visible to shortwave infrared spectral ranges (350–2500 nm). The SOLAS instrument transmits two near-infrared wavelength lasers (823.20 nm and 847.00 nm) to measure atmospheric water vapor by differential absorption along the transmitter path. Self-chirped, coherent detection of the same lasers provides target range measurements. The backscattered light is received through a 203-mm diameter telescope. The combination of high-resolution reflectance spectroradiometry and lidar (ranging and differential absorption) techniques has introduced a new ground-based approach to remote sensing of the natural environment. Envisioned applications for the instrument include rapid classification of soils on unstable slopes, mine tailings, or in wildfire-affected areas. Future improvements will enable long-range measurements, increased portability (lighter instrument components), or semi-autonomous measurements as part of a long-term monitoring installation (e.g., wildfire basin or mining operation).

Author Contributions: Conceptualization, R.A.C., C.D.G. and S.E.S.; Methodology, S.E.S., C.D.G. and R.A.C.; Software, S.E.S.; Validation, S.E.S. and R.A.C.; Formal Analysis, S.E.S.; Investigation, S.E.S. and R.A.C.; Resources, R.A.C.; Data Curation, S.E.S.; Writing—Original Draft Preparation, S.E.S.; Writing—Review and Editing, S.E.S., C.D.G. and R.A.C.; Visualization, S.E.S.; Supervision, R.A.C.; Project Administration, R.A.C.; Funding Acquisition, R.A.C. and S.E.S.

Funding: This project was funded by the U.S. Department of Transportation (USDOT) through the Office of the Assistant Secretary for Research and Technology (OST-R) under USDOT Cooperative Agreement No. OASRTRS-14-H-UARK. The views, opinions, findings and conclusions reflected in this publication are solely those of the authors and do not represent the official policy or position of the USDOT/OST-R, or any State or other entity. USDOT/OST-R does not endorse any third party products or services that may be included in this publication. This material is also based upon work supported by the National Science Foundation Graduate Research Fellowship Program under Grant No. DGE-1450079. Any opinions, findings, and conclusions or recommendations expressed in this material are those of the authors and do not necessarily reflect the views of the National Science Foundation.

Conflicts of Interest: The authors declare no conflict of interest. The funders had no role in the design of the study; in the collection, analyses, or interpretation of data; in the writing of the manuscript, or in the decision to publish the results.

Abbreviations

The following abbreviations are used in this manuscript:

dB	Decibel
AM	Amplitude Modulation
APD	Avalanche Photodetector

ASD	Analytical Spectral Devices Inc. (a Malvern Panalytical Company, Longmont, CO, USA)
B	Signal Bandwidth
BPD	Balanced Photodetector
B_{RX}	Bandwidth of the Receiver
BSC	Beamsplitter Cube
c	Speed of Light
CW	Continuous-Wave
CWL	Center Wavelength
CF	Clay Fraction
Δt	Roundtrip Time Delay for Atmospheric Propagation
D	Diameter of Primary Mirror (Telescope)
DIAL	Differential Absorption Lidar
DRIFT	Diffuse Reflectance Infrared Fourier Transform
e_s	Saturation Vapor Pressure
e_{s0}	Saturation Vapor Pressure at T0 = 273 K
E''	Lower State Energy
ECDL	External Cavity Diode Laser
FM	Frequency Modulated
FMCW	Frequency-Modulated Continuous-Wave
FOV	Field of View (Angular)
f	Focal Length
$f_2 - f_1$	Chirp Bandwidth
f_R	Beat Frequency
h	Planck Constant
HITRAN	High-Resolution Transmission Molecular Absorption Database
θ_v	Volumetric Water Content
InGaAs	Indium Gallium Arsenide
k_B	Boltzmann Constant
K	Chirp Waveform-Dependent Constant
λ	Wavelength
Λ	Voigt Function
L	Latent Heat of Vaporization
LAS	Laser Absorption Spectrometry
LAST	Laser Analysis of Soil Tension
LIDAR	Light Detection and Ranging (commonly Lidar)
LL	Liquid Limit
LO	Local Oscillator
M_{H2O}	Molecular Weight of Water
MWIR	Mid-Wave-Infrared
MZM	Mach-Zehnder Modulator
N	F-Number
N_A	Avogadro's Constant
NIR	Near-Infrared
ρ_v	Water Vapor Density
PL	Plastic Limit
PPE	Pressure Plate Extractor
$P_{n\text{-}RX}$	Receiver Power Noise
P_{off}	Received Off-Line Backscatter Signal
P_{on}	Received On-Line Backscatter Signal
P_r	Received Signal Power
P_{sig}	Detected Signal Power
PXIe	PCI (Peripheral Component Interconnect) Extensions for Instrumentation Express
q	Electron Charge

$\Re$	Photodetector Responsivity
R	Range to Target
RF	Radio Frequencies
RH	Relative Humidity
R_v	Water Vapor Gas Constant
σ_{off}	Off-Line Water Vapor Absorption Cross-section
σ_{on}	On-Line Water Vapor Absorption Cross-section
σ_R	Range Accuracy
S	Temperature-Dependent Absorption Line Strength
S_0	Absorption Line Strength under Standard Conditions
SMC	Soil Moisture Content
SNR	Signal-to-Noise Ratio
SNR_{coh}	Signal to Noise Ratio for Shot-Noise-Dominant Coherent Detection
$SOLAS$	Soil Observation Laser Absorption Spectrometer
$SSURGO$	Soil Survey Geographic Database
$STATSGO$	State Soil Geographic Survey
$SWCC$	Soil Water Characteristic Curve
$SWIR$	Shortwave Infrared
τ	Chirp Duration
T	Temperature
T_0	Absorption Temperature under Standard Conditions
TEC	Thermoelectric-Cooled
$TSOA$	Tapered Semiconductor Optical Amplifier
$UHTC$	Ultra-High Transmission Coating (Meade Instruments Corporation, Irvine, CA, USA)
$VNIR$	Visible Near-Infrared
ψ_m	Soil Water Matric Potential (Soil Suction)

References

1. Wulf, H.; Mulder, T.; Schaepman, M.E.; Keller, A.; Jörg, P.C. *Remote Sensing of Soils*; University of Zurich: Zurich, Switzerland, 2015; pp. 1–71.
2. Ångström, A. The Albedo of Various Surfaces of Ground. *Geogr. Ann.* **1925**, *7*, 323–342. [CrossRef]
3. Bowers, S.A.; Hanks, R.J. Reflection of Radiant Energy from Soils. *Soil Sci.* **1965**, *100*, 130–138. [CrossRef]
4. Park, J.K. A Soil Moisture Reflectance Model in Visible and Near IR Bands. In Proceedings of the International Symposium on Machine Processing of Remotely Sensed Data and Soil Information Systems and Remote Sensing and Soil Survey, Purdue University, West Lafayette, IN, USA, 3–6 June 1980.
5. Dalal, R.C.; Henry, R.J. Simultaneous Determination of Moisture, Organic Carbon, and Total Nitrogen by Near Infrared Reflectance Spectrophotometry. *Soil Sci. Soc. Am. J.* **1986**, *50*, 120–123. [CrossRef]
6. Whalley, W.R.; Leeds-Happison, P.B.; Bowman, G.E. Estimation of Soil Moisture Using Near Infrared Reflectance. *Hydrol. Process.* **1991**, *5*, 321–327. [CrossRef]
7. Bach, H.; Mauser, W. Modelling and Model Verification of the Spectral Reflectance of Soils Under Varying Moisture Conditions. In Proceedings of the International Geoscience and Remote Sensing Symposium (IGARSS'94), Pasadena, CA, USA, 8–12 August 1994; pp. 2354–2356. [CrossRef]
8. Ben-Dor, E.; Banin, A. Near-Infrared Analysis as a Rapid Method to Simultaneously Evaluate Several Soil Properties. *Soil Sci. Soc. Am. J.* **1995**, *59*, 364–372. [CrossRef]
9. Chang, C.-W.; Laird, D.; Mausbach, M.J.; Hurburgh, C.R. Near-Infrared Reflectance Spectroscopy-Principle Components Regression Analysis of Soil Properties. *Soil Sci. Soc. Am. J.* **2001**, *65*, 480–490. [CrossRef]
10. Hummel, J.W.; Sudduth, K.A.; Hollinger, S.E. Soil Moisture and Organic Matter Prediction of Surface and Subsurface Soils Using an NIR Soil Sensor. *Comput. Electron. Agric.* **2001**, *32*, 149–165. [CrossRef]
11. Liu, W.; Baret, F.; Gu, X.; Tong, Q.; Zheng, L.; Zhang, B. Relating Soil Surface Moisture to Reflectance. *Remote Sens. Environ.* **2002**, *81*, 238–246.
12. Liu, W.; Baret, F.; Gu, X.; Zhang, B.; Tong, Q.; Zheng, L. Evaluation of Methods for Soil Surface Moisture Estimation from Reflectance Data. *Int. J. Remote Sens.* **2003**, *24*, 2069–2083. [CrossRef]
13. Lobell, D.B.; Asner, G.P. Moisture Effects on Soil Reflectance. *Soil Sci. Soc. Am. J.* **2002**, *66*, 722–727. [CrossRef]

14. Islam, K.; Singh, B.; McBratney, A. Simultaneous Estimation of Various Soil Properties by Ultra-Violet, Visible, and Near-Infrared Reflectance Spectroscopy. *Aust. J. Soil Res.* **2003**, *41*, 1101–1114. [CrossRef]
15. Whiting, M.L.; Li, L.; Ustin, S.L. Predicting Water Content Using Gaussian Model on Soil Spectra. *Remote Sens. Environ.* **2004**, *89*, 535–552. [CrossRef]
16. Mouazen, A.M.; Karoui, R.; de Baerdemaeker, J.; Ramon, H. Characterization of Soil Water Content Using Measured Visible and Near Infrared Spectra. *Soil Sci. Soc. Am. J.* **2006**, *70*, 1295–1302. [CrossRef]
17. Lesaignoux, A.; Fabre, S.; Briottet, X. Influence of Soil Moisture Content on Spectral Reflectance of Bare Soils in the 0.4-14 μm Domain. *Int. J. Remote Sens.* **2013**, *34*, 2268–2285. [CrossRef]
18. Knadel, M.; Deng, F.; Alinejadian, A.; de Jonge, L.W.; Moldrup, P.; Greve, M.H. The Effects of Moisture Conditions—From Wet to Hyper Dry—On Visible Near-Infrared Spectra of Danish Reference Soils. *Soil Sci. Soc. Am. J.* **2014**, *78*, 422–433. [CrossRef]
19. Nolet, C.; Poortinga, A.; Roosjen, P.; Bartholomeus, H.; Ruessink, G. Measuring and Modeling the Effect of Surface Moisture on the Spectral Reflectance of Coastal Beach Sand. *PLoS ONE* **2014**, *9*, e112151:1–e112151:9. [CrossRef] [PubMed]
20. Fabre, S.; Briottet, X.; Lesaignoux, A. Estimation of Soil Moisture Content from the Spectral Reflectance of Bare Soils in the 0.4–2.5 μm Domain. *Sensors* **2015**, *15*, 3262–3281. [CrossRef]
21. Sadeghi, M.; Jones, S.B.; Philpot, W.D. A Linear Physically-Based Model for Remote Sensing of Soil Moisture Using Short Wave Infrared Bands. *Remote Sens. Environ.* **2015**, *164*, 66–76. [CrossRef]
22. Tian, J.; Philpot, W.D. Relationship Between Surface Soil Water Content, Evaporation Rate, and Water Absorption Band Depths in SWIR Reflectance Spectra. *Remote Sens. Environ.* **2015**, *169*, 280–289. [CrossRef]
23. Xu, C.; Zeng, W.; Huang, J.; Wu, J.; van Leeuwen, W.J.D. Prediction of Soil Moisture Content and Soil Salt Concentration from Hyperspectral Laboratory and Field Data. *Remote Sens.* **2016**, 8. [CrossRef]
24. Garner, C.D. Development of a Multiband Remote Sensing System for Determination of Unsaturated Soil Properties. Ph.D. Dissertation, University of Arkansas, Fayetteville, AR, USA, 2017.
25. Mouazen, A.M.; Maleki, M.R.; de Baerdemaeker, J.; Ramon, H. On-line Measurement of Some Selected Soil Properties Using a VIS-NIR Sensor. *Soil Till. Res.* **2007**, *93*, 13–27. [CrossRef]
26. Haubrock, S.N.; Chabrillat, S.; Lemmnitz, C.; Kaufmann, H. Surface Soil Moisture Quantification Models from Reflectance Data Under Field Conditions. *Int. J. Remote Sens.* **2008**, *29*, 3–29. [CrossRef]
27. Walvoort, A.; McBratney, A. Diffuse Reflectance Spectrometry as a Proximal Sensing Tool for Precision Agriculture. In Proceedings of the 3rd European Conference on Precision Agriculture, Montpellier, France, 18–20 June 2001; pp. 503–508.
28. Shepherd, K.D.; Walsh, M.G. Development of Reflectance Spectral Libraries for Characterization of Soil Properties. *Soil Sci. Soc. Am. J.* **2002**, *66*, 988–998. [CrossRef]
29. Cozzolino, D.; Moron, A. The Potential of Near-Infrared Reflectance Spectroscopy to Analyze Soil Chemical and Physical Characteristics. *J. Agric. Sci.* **2003**, *140*, 65–71. [CrossRef]
30. Knadel, M.; Stenberg, B.; Deng, F.; Thomsen, A.; Greve, M.H. Comparing Predictive Abilities of Three Visible-Near Infrared Spectrophotometers for Soil Organic Carbon and Clay Determination. *J. Near Infrared Spectrosc.* **2013**, *21*, 67–80. [CrossRef]
31. Leu, D.J. Visible and Near-Infrared Reflectance of Beach Sands: A Study on the Spectral Reflectance/Grain Size Relationship. *Remote Sens. Environ.* **1977**, *6*, 169–182. [CrossRef]
32. Yitagesu, F.A.; van der Meer, F.; van der Werff, H.; Zigterman, W. Quantifying Engineering Parameters of Expansive Soils from their Reflectance Spectra. *Eng. Geol.* **2009**, *105*, 151–160. [CrossRef]
33. Waruru, B.K.; Shepherd, K.D.; Ndegwa, G.M.; Kamoni, P.T.; Sila, A.M. Rapid Estimation of Soil Engineering Properties Using Diffuse Reflectance Near Infrared Spectroscopy. *Biosyst. Eng.* **2014**, *121*, 177–185. [CrossRef]
34. Janik, L.; Skjemstad, J. Characterization and Analysis of Soils Using Mid-Infrared Partial Least Squares: II. Correlations with Some Laboratory Data. *Aust. J. Soil Res.* **1995**, *33*, 637–650. [CrossRef]
35. Karlsson, C.J.; Olsson, F.Å.A. Linearization of the Frequency Sweep of a Frequency-Modulated Continuous-Wave Semiconductor Laser and the Resulting Ranging Performance. *Appl. Opt.* **1999**, *38*, 3376–3386. [CrossRef]
36. Allen, C.; Cobanoglu, Y.; Chong, S.K.; Gogineni, S. Performance of a 1319 nm Laser Radar Using RF Pulse Compression. In Proceedings of the International Geoscience and Remote Sensing Symposium (IGARSS 2001), Sydney, Australia, 9–13 July 2001; pp. 997–999. [CrossRef]

37. Pierrottet, D.; Amzajerdian, F.; Peri, F. Development of an All-Fiber Coherent Laser Radar for Precision Range and Velocity Measurements. In Proceedings of the Materials Research Society Symposium, Boston, MA, USA, 28 November–2 December 2005. [CrossRef]
38. Pierrottet, D.; Amzajerdian, F.; Petway, L.; Barnes, B.; Lockard, G.; Rubio, M. Linear FMCW Laser Radar for Precision Range and Vector Velocity Measurements. In Proceedings of the Materials Research Society Symposium, San Francisco, CA, USA, 24–28 March 2008. [CrossRef]
39. Adany, P.; Allen, C.; Hui, R. Chirped Lidar Using Simplified Homodyne Detection. *J. Lightw. Technol.* **2009**, *27*, 3351–3357. [CrossRef]
40. Barber, Z.W.; Babbitt, W.R.; Kaylor, B.; Reibel, R.R.; Roos, P.A. Accuracy of Active Chirp Linearization for Broadband Frequency Modulated Continuous Wave Ladar. *Appl. Opt.* **2010**, *49*, 213–219. [CrossRef] [PubMed]
41. Gao, S.; Hui, R. Frequency-Modulated Continuous-Wave Lidar Using I/Q Modulator for Simplified Heterodyne Detection. *Opt. Lett.* **2012**, *37*, 2022–2024. [CrossRef] [PubMed]
42. Mateo, A.B. Applications of High Resolution and Accuracy Frequency Modulated Continuous Wave Ladar. Master's Thesis, Montana State University, Bozeman, MT, USA, 2014.
43. Skolnik, M.I. Theoretical Accuracy of Radar Measurements. IRE T. *Aeron. Navig. Electr.* **1960**, 123–129. [CrossRef]
44. Jelalian, A. *Laser Radar Systems*; Artech House: Norwood, MA, USA, 1992; ISBN 9780890065549.
45. Schotland, R.M. Some Observations of the Vertical Profile of Water Vapor by a Laser Optical Radar. In Proceedings of the Fourth Symposium on Remote Sensing of the Environment, Ann Arbor, MI, USA, 12–14 April 1966; pp. 273–283.
46. Schotland, R.M. Errors in the Lidar Measurement of Atmospheric Gases by Differential Absorption. *J. Appl. Meteorol.* **1974**, *13*, 71–77.
47. Menzies, R.T.; Shumate, M.S. Remote Measurements of Ambient Air Pollutants with a Bistatic Laser System. *Appl. Opt.* **1976**, *15*, 2080–2084. [CrossRef] [PubMed]
48. Grant, W.B. Effect of Differential Spectral Reflectance on DIAL Measurements Using Topographic Targets. *Appl. Opt.* **1982**, *21*, 2390–2394. [CrossRef] [PubMed]
49. Grant, W.B. Differential Absorption and Raman Lidar for Water Vapor Profile Measurements: A Review. *Opt. Eng.* **1991**, *30*, 40–48. [CrossRef]
50. Hardesty, R.M. Coherent DIAL Measurement of Range-Resolved Water Vapor Concentration. *Appl. Opt.* **1984**, *23*, 2545–2553. [CrossRef]
51. Bösenberg, J. Ground-Based Differential Absorption Lidar for Water-Vapor and Temperature Profiling: Methodology. *Appl. Opt.* **1998**, *37*, 3845–3860. [CrossRef]
52. Wulfmeyer, V.; Bösenberg, J. Ground-Based Differential Absorption Lidar for Water-Vapor Profiling: Assessment of Accuracy, Resolution, and Meteorological Applications. *Appl. Opt.* **1998**, *37*, 3825–3844. [CrossRef] [PubMed]
53. Wulfmeyer, V.; Walther, C. Future Performance of Ground-Based and Airborne Water-Vapor Differential Absorption Lidar. I. Overview and Theory. *Appl. Opt.* **2001**, *40*, 5304–5320. [CrossRef] [PubMed]
54. Spuler, S.; Repasky, K.; Morley, B.; Moen, D.; Weckwerth, T.; Hayman, M.; Nehrir, A. Advances in Diode-Laser-Based Water Vapor Differential Absorption Lidar. In Proceedings of the 27th International Laser Radar Conference (ILRC 27), New York, NY, USA, 5–10 July 2015. [CrossRef]
55. Ismail, S.; Browell, E.V. Airborne and Spaceborne Lidar Measurements of Water Vapor Profiles: A Sensitivity Analysis. *Appl. Opt.* **1989**, *28*, 3603–3615. [CrossRef] [PubMed]
56. Weckwerth, T.M.; Weber, K.J.; Turner, D.D.; Spuler, S.M. Validation of a Water Vapor Micropulse Differential Absorption Lidar (DIAL). *J. Atmos. Ocean. Technol.* **2016**, *33*, 2353–2372. [CrossRef]
57. Prasad, N.S.; Geiger, A.R. Remote Sensing of Propane and Methane by Means of a Differential Absorption Lidar by Topographic Reflection. *Opt. Eng.* **1996**, *35*, 1105–1111. [CrossRef]
58. Wulfmeyer, V. Ground-Based Differential Absorption Lidar for Water-Vapor and Temperature Profiling: Development and Specifications of a High-Performance Laser Transmitter. *Appl. Opt.* **1998**, *37*, 3804–3824. [CrossRef]
59. Little, L.M.; Papen, G.C. Fiber-Based Lidar for Atmospheric Water-Vapor Measurements. *Appl. Opt.* **2001**, *40*, 3417–3427. [CrossRef]

60. Machol, J.; Ayers, T.; Schwenz, K.; Koenig, K.; Hardesty, R.; Senff, C.; Krainak, M.; Abshire, J.; Bravo, H.; Sandberg, S. Preliminary Measurements with an Automated Compact Differential Absorption LIDAR for Profiling Water Vapor. *Appl. Opt.* **2004**, *43*, 3110–3121. [CrossRef]
61. Nehrir, A.R. Water Vapor Profiling Using a Compact Widely Tunable Diode Laser Differential Absorption Lidar (DIAL). Master's Thesis, Montana State University, Bozeman, MT, USA, 2008.
62. Nehrir, A.R. Development of an Eye-Safe Diode-Laser-Based Micro-Pulse Differential Absorption Lidar (MP-DIAL) for Atmospheric Water-Vapor and Aerosol Studies. Ph.D. Dissertation, Montana State University, Bozeman, MT, USA, 2011.
63. Nehrir, A.R.; Repasky, K.S.; Carlsten, J.L.; Obland, M.D.; Shaw, J.A. Water Vapor Profiling Using a Widely Tunable, Amplified Diode-Laser-Based Differential Absorption Lidar (DIAL). *J. Atmos. Ocean. Technol.* **2009**, *26*, 733–745. [CrossRef]
64. Nehrir, A.R.; Repasky, K.S.; Carlsten, J.L. Eye-Safe Diode-Laser-Based Micropulse Differential Absorption Lidar (DIAL) for Water Vapor Profiling in the Lower Troposphere. *J. Atmos. Ocean. Technol.* **2011**, *28*, 131–147. [CrossRef]
65. Nehrir, A.R.; Repasky, K.S.; Carlsten, J.L. Micropulse Water Vapor Differential Absorption Lidar: Transmitter Design and Performance. *Opt. Express* **2012**, *20*, 25137–25151. [CrossRef] [PubMed]
66. Spuler, S.M.; Repasky, K.S.; Morley, B.; Moen, D.; Hayman, M.; Nehrir, A.R. Field-Deployable Diode-Laser-Based Differential Absorption Lidar (DIAL) for Profiling Water Vapor. *Atmos. Meas. Tech.* **2015**, *8*, 1073–1087. [CrossRef]
67. Remsberg, E.E.; Gordley, L.L. Analysis of Differential Absorption Lidar from the Space Shuttle. *Appl. Opt.* **1978**, *17*, 624–630. [CrossRef] [PubMed]
68. Bufton, J.L.; Itabe, T.; Grolemund, D.A. Airborne Remote Sensing Measurements with a Pulsed CO2 Dial System. In *Optical and Laser Remote Sensing*; Killinger, D.K., Mooradian, A., Eds.; Springer: Berlin/Heidelberg, Germany, 1983; pp. 2–9, ISBN 9783662157367.
69. Ehret, G.; Kiemle, C.; Renger, W.; Simmet, G. Airborne Remote Sensing of Tropospheric Water Vapor with a Near-Infrared Differential Absorption Lidar System. *Appl. Opt.* **1993**, *32*, 4534–4551. [CrossRef] [PubMed]
70. Bruneau, D.; Quaglia, P.; Flamant, C.; Meissonnier, M.; Pelon, J. Airborne Lidar LEANDRE II for Water-Vapor Profiling in the Troposphere. I. System Description. *Appl. Opt.* **2001**, *40*, 3450–3461. [CrossRef] [PubMed]
71. Bruneau, D.; Quaglia, P.; Flamant, C.; Pelon, J. Airborne Lidar LEANDRE II for Water-Vapor Profiling in the Troposphere. II. First Results. *Appl. Opt.* **2001**, *40*, 3462–3475. [CrossRef] [PubMed]
72. Browell, E.V.; Dobler, J.; Kooi, S.A.; Choi, Y.; Harrison, F.W.; Moore, B.; Zaccheo, T.S. Airborne Validation of Laser Remote Measurements of Atmospheric Carbon Dioxide. In Proceedings of the 25th International Laser Radar Conference (ILRC 25), St. Petersburg, Russia, 5–9 July 2010; pp. 779–782.
73. Megie, G.; Menzies, R.T. Complementarity of UV and IR Differential Absorption Lidar for Global Measurements of Atmospheric Species. *Appl. Opt.* **1980**, *19*, 1173–1183. [CrossRef] [PubMed]
74. Ehret, G.; Kiemle, C.; Wirth, M.; Amediek, A.; Fix, A.; Houweling, S. Space-Borne Remote Sensing of CO2, CH4, and N2O by Integrated Path Differential Absorption Lidar: A Sensitivity Analysis. *Appl. Phys. B* **2008**, *90*, 593–608. [CrossRef]
75. Wirth, M.; Fix, A.; Mahnke, P.; Schwarzer, H.; Schrandt, F.; Ehret, G. The Airborne Multi-Wavelength Water Vapor Differential Absorption Lidar WALES: System Design and Performance. *Appl. Phys. B* **2009**, *96*, 201–213. [CrossRef]
76. Barrientos-Barria, J.; Mammez, D.; Dherbecourt, J.-B.; Raybaut, M.; Melkonian, J.-M.; Pelon, J.; Godard, A.; Lefebvre, M. Progress on High Energy Optical Parametric Transmitter for Multiple Greenhouse Gases DIAL. In Proceedings of the International Conference on Space Optics (ICSO 2014), Tenerife, Canary Islands, Spain, 6–10 October 2014; Volume 10563. [CrossRef]
77. Gordon, I.E.; Rothman, L.S.; Hill, C.; Kochanov, R.V.; Tan, Y.; Bernath, P.F.; Birk, M.; Boudon, V.; Campargue, A.; Chance, K.V.; et al. The HITRAN2016 Molecular Spectroscopic Database. *J. Quant. Spectrosc. Radiat.* **2017**, *203*, 3–69. [CrossRef]
78. Sakaizawa, D.; Kawakami, S.; Nakajima, M.; Sawa, Y.; Matsueda, H. Ground-Based Demonstration of CO2 Remote Sensor using 1.57 μm Differential Laser Absorption Spectrometer with Direct Detection. *J. Appl. Remote Sens.* **2010**, *4*, 043548:1–043548:17. [CrossRef]

79. Ishii, S.; Koyama, M.; Baron, P.; Iwai, H.; Mizutani, K.; Itabe, T.; Sato, A.; Asai, K. Ground-Based Integrated Path Coherent Differential Absorption Lidar Measurement of CO_2: Foothill Target Return. *Atmos. Meas. Tech.* **2013**, *6*, 1359–1369. [CrossRef]
80. Kou, L.; Labrie, D.; Chylek, P. Refractive Indices of Water and Ice in the 0.65- to 2.5-μm Spectral Range. *Appl. Opt.* **1993**, *32*, 3531–3540. [CrossRef] [PubMed]
81. Pope, R.M.; Fry, E.S. Absorption Spectrum (380–700 nm) of Pure Water. II. Integrating Cavity Measurements. *Appl. Opt.* **1997**, *36*, 8710–8723. [CrossRef] [PubMed]
82. Rothman, L.S.; Gordon, I.E.; Babikov, Y.; Barbe, A.; Benner, D.C.; Bernath, P.F.; Birk, M.; Bizzocchi, L.; Boudon, V.; Brown, L.R.; et al. The HITRAN2012 Molecular Spectroscopic Database. *J. Quant. Spectrosc. Radiat.* **2013**, *130*, 4–50. [CrossRef]

remote sensing

MDPI

Article

Spatial Distribution of Soil Moisture in Mongolia Using SMAP and MODIS Satellite Data: A Time Series Model (2010–2025)

Enkhjargal Natsagdorj [1,2,*], Tsolmon Renchin [2], Philippe De Maeyer [1] and Bayanjargal Darkhijav [3]

1 Department of Geography, Faculty of Science, Ghent University, 9000 Ghent, Belgium; Philippe.Demaeyer@ugent.be
2 NUM-ITC-UNESCO Laboratory for Space Science and Remote Sensing, School of Arts and Sciences, National University of Mongolia, Ulaanbaatar 14200, Mongolia; tsolmon@num.edu.mn
3 Department of Applied Mathematics, School of Applied Science and Engineering, National University of Mongolia, Ulaanbaatar 14200, Mongolia; bayanjargal@seas.num.edu.mn
* Correspondence: enkhjargal.natsagdorj@ugent.be

Abstract: Soil moisture is one of the essential variables of the water cycle, and plays a vital role in agriculture, water management, and land (drought) and vegetation cover change as well as climate change studies. The spatial distribution of soil moisture with high-resolution images in Mongolia has long been one of the essential issues in the remote sensing and agricultural community. In this research, we focused on the distribution of soil moisture and compared the monthly precipitation/temperature and crop yield from 2010 to 2020. In the present study, Soil Moisture Active Passive (SMAP) and Moderate Resolution Imaging Spectroradiometer (MODIS) data were used, including the MOD13A2 Normalized Difference Vegetation Index (NDVI), MOD11A2 Land Surface Temperature (LST), and precipitation/temperature monthly data from the Climate Research Unit (CRU) from 2010 to 2020 over Mongolia. Multiple linear regression methods have previously been used for soil moisture estimation, and in this study, the Autoregressive Integrated Moving Arima (ARIMA) model was used for soil moisture forecasting. The results show that the correlation was statistically significant between SM-MOD and soil moisture content (SMC) from the meteorological stations at different depths ($p < 0.0001$ at 0–20 cm and $p < 0.005$ at 0–50 cm). The correlation between SM-MOD and temperature, as represented by the correlation coefficient (r), was 0.80 and considered statistically significant ($p < 0.0001$). However, when SM-MOD was compared with the crop yield for each year (2010–2019), the correlation coefficient (r) was 0.84. The ARIMA (12, 1, 12) model was selected for the soil moisture time series analysis when predicting soil moisture from 2020 to 2025. The forecasting results are shown for the 95 percent confidence interval. The soil moisture estimation approach and model in our study can serve as a valuable tool for confident and convenient observations of agricultural drought for decision-makers and farmers in Mongolia.

Keywords: soil moisture; time series; remote sensing; Mongolia

Citation: Natsagdorj, E.; Renchin, T.; Maeyer, P.D.; Darkhijav, B. Spatial Distribution of Soil Moisture in Mongolia Using SMAP and MODIS Satellite Data: A Time Series Model (2010–2025). *Remote Sens.* **2021**, *13*, 347. https://doi.org/10.3390/rs13030347

Received: 21 December 2020
Accepted: 19 January 2021
Published: 20 January 2021

Publisher's Note: MDPI stays neutral with regard to jurisdictional claims in published maps and institutional affiliations.

1. Introduction

Soil moisture (SM) plays an important role in the terrestrial water cycle and has been assessed in many field studies, e.g., in water management, agricultural irrigation management, crop production, vegetation cover, drought, and global climate change [1–4]. In addition, soil moisture indicates groundwater conditions and links the exchange of water and energy between the atmosphere and land surface. There are many ways to estimate soil moisture, including direct and indirect methods. The most accurate method is direct measurement in the field (gravimetric method) to estimate soil moisture by point measurement [5], but this is costly [6]. Therefore, remote sensing techniques have become popular for estimating soil moisture at a regional scale due to the sensing ability of the regional SM with low-resolution images. Microwave remote sensing methods have been used at the global and regional scale to establish models [7,8]. To date, some highly advanced SM products have been developed, e.g., soil moisture active passive (SMAP) from the National Aeronautics and Space Administration (NASA), soil moisture ocean salinity (SMOS), and climate change initiative (CCI) from European Space Agency (ESA). In optical and thermal remote sensing, many researchers have established methods based on the relationships between SM and soil reflection/soil temperature and vegetation cover [9–11]. Using a combined microwave and optical remote sensing data can give more precise information on soil moisture rather than estimation based only on data from one type of remote sensing.

Remote sensing technology is a powerful method for soil moisture monitoring at the regional level. Many studies have established that SMAP products generate accurate in situ measurements and can be used in various fields of study, such as agriculture, environmental monitoring, and hydrology [12]. They have also been intensively validated by several studies over the past few years [13–15]. For instance, Zeng et al. [16] approved a SMAP product for the preliminary evaluation of soil moisture compared to in situ measurements from the three networks that cover different climatic and land surface conditions; moreover, the results show that the SMAP product is in good agreement with in situ measurements. However, it has limited application to agricultural studies because SMAP products only provide 3, 9, or 36 km spatial resolution data at the global or regional scale. In this paper, we used SMAP products at a spatial resolution of 9 km for the development of a soil moisture model by combining SMAP and optical/thermal satellite images. The combination of optical/thermal and microwave remote sensing potentially expands the application possibilities [7].

In previous studies, various researchers have investigated and developed methods based on LST/NDVI, in view of vegetation types and topography and climate parameters, among other factors [9,17–20]. These approaches have mainly used reflectance to estimate SM in visible/thermal infrared sensors. Chelsea et al. [21] used LST/NDVI data to obtain an enhanced spatial resolution of soil moisture from SMAP at 9 km down to 1 km. Natsagdorj et al. [5] developed a model for soil moisture using multiple regression analysis, and the model showed that the type of soil moisture index from the satellite measurements depends on the LST, NDVI, elevation, slope, and aspect. The results indicate a good correlation between the developed model and ground truth measurements in the subprovinces of Mongolia. The lack of field measurements for SM makes it challenging to validate remote sensing SM estimates in Mongolia because the territory is so widespread (1565 million km^2).

Due to the characteristics of the Mongolian climate, agricultural production is strongly limited by a short growing season (generally 80 to 100 days, but varies from 70 to 130 days depending on the altitude and location), low precipitation, and high evaporation [22]. Mongolian steppe ecosystems are crucial for relieving regional and even global climate variation through their interaction with the atmosphere [23]. Many studies have shown that in Mongolia, due to the harsh continental climate and the distance from the sea, the processes of soil drying, desertification, and degradation are intensifying due to the loss of vegetation and changes in soil moisture due to global warming. Therefore, to study the impacts of climate change, there is an urgent need to consider soil moisture as one of its indicators. Few studies of soil moisture have been conducted with point-scale measurements [24,25]. In Mongolia, SM distribution data with a higher resolution are needed for practical applications such as agricultural management, water management, and flood and drought monitoring. Therefore, time series analysis of long-term soil moisture was conducted using the autoregressive integrated moving average (ARIMA) model [26].

Few previous studies have examined soil moisture and river flow forecasting [27]; on the other hand, various reviews have addressed drought monitoring [28–31]. The SM forecast data support farmers in organizing their resources for crop production. The ARIMA model is commonly used in time series models. There are many methods and criteria for ranking and selecting the autoregressive (AR), moving average (MA), or ARIMA models for a given purpose. These models are suitable for limited data values and short-term forecasting [27]. However, the main advantages of ARIMA model forecasting is that it only requires time series data [28]. In this study, we use the ARIMA model to investigate the time series analysis of soil moisture dynamics between 2010 and 2020 based on SMAP and MODIS satellite images. Remarkably, this research focused on that to have a higher spatial resolution (1 km) soil moisture map than SMAP (9 km) provide us then the SMAP data periods 2015–2020 was used in order to build a model. From the model, spatially distributed monthly soil moisture data will contribute return back into time (2010–2020), and it is towards the future by ARIMA model.

The main objectives of this research are to estimate a monthly soil moisture distribution map and to build appropriate models to forecast future trends. Because of the stochastic nature of monthly soil moisture, we used time series analysis for monthly soil moisture forecasting. A process is considered stationary if its statistical properties, such as the average and variance, do not change over time. The monthly soil moisture map was estimated from remote sensing data in Mongolia. The modeling and prediction of soil moisture were done through statistical methods based on ARIMA. In this paper, soil moisture modeling and forecasting was performed by means of the conventional method, the Box-Jenkins time series model. The monthly soil moisture distribution map has not yet been considered in previous studies in Mongolia and is expected to be useful for agriculture, hydrology, and climate science.

2. Study Area and Data Preprocessing

2.1. Study Area

Mongolia is a landlocked country situated in Central Asia, bordered by Russia and China and located between the latitudes of 41°35′ N and 52°09′ N and the longitudes of 87°44′ E and 119°56′ E, with a total area of 1565 million km^2 (Figure 1). Mongolia is covered by 73% agricultural land, 0.5% villages, and other settlements, with 0.35% land representing roads and networks, 9.2% forest and forest resources, 0.4% water and water resources, and 16.1% land for special needs (protected, historical and monumental natural beauty) [32].

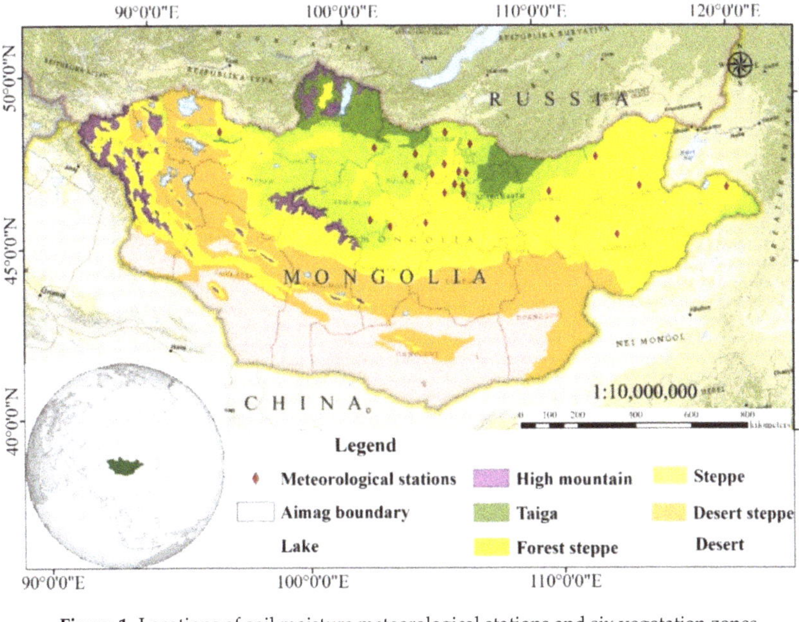

Figure 1. Locations of soil moisture meteorological stations and six vegetation zones.

The Mongolian climate is highly continental, with arid and semi-arid conditions [33], and has four distinct seasons, high temperature fluctuations, and little precipitation. About 85% of the total precipitation falls from April to September, of which about 50–60% falls during July and August [34]. The mean annual temperature is −8 °C in the northern areas and 6 °C in southern regions [34]. The total soil moisture decreases, going from the north to the south of Mongolia [35] due to the different regions of Mongolian vegetation [36]. The annual precipitation is 300–400 mm in the taiga, high mountain, and forest steppe zones: 150–250 mm in the steppe, 100–150 mm in the desert steppe, and 50–100 mm in the desert (Gobi) region. Most of the cropland is located in the north of Mongolia, which encompasses the forest steppe and steppe zones.

2.2. Remote Sensing Data

2.2.1. SMAP

On 31 January 2015, NASA launched SMAP, which has an initial L-band with both radar and radiometer to assess soil moisture [37]. The daily coverage started on 31 March 2015 at a spatial resolution of 3–36 km. The average monthly SMAP data were obtained using daily SMAP with 9 km resolution. We downloaded daily SMAP L3 Radiometer Global Daily 9 km EASE-Grid Soil moisture data (SPL3SMP_E.003) through the Application for Extracting and Exploring Analysis Ready Samples (AppEEARS) between 2015 and 2020 [38]. AppEEARS is a useful tool for time series analysis in specific regions and at certain scales. It provides data by enabling users to download only the information (geospatial datasets using spatial, temporal, and band/layer parameters) needed from several federal archives (https://lpdaacsvc.cr.usgs.gov/appeears/). The downloaded images were preprocessed with the ENVI 5.3 and ArcGIS 10.3 software to obtain the monthly soil moisture data.

2.2.2. MODIS

We used MODIS products over a 10-year period (2010–2020) to observe the dynamic range of the NDVI and LST. Zhang et al. [39] used a similar approach to detect anomalies using MODIS land products via time series analysis. Accordingly, monthly composites of 1 km spatial resolution MOD13A3 [40] and MOD11A2 [41] data from MODIS and the National Aeronautics and Space Administration (NASA) Earth Observing system (https://lpdaac.usgs.gov/product_search/) data were used. The MODIS vegetation indices (MOD13A3) version 6 data are provided monthly at 1 km spatial resolution in the sinusoidal projection [40]. The MOD11A2 version 6 product provides an average eight-day-per-pixel Land Surface Temperature and Emissivity (LST&E) with a 1 km spatial resolution [41]. We calculated eight-day LST as monthly averages using product version 6 (MOD11A1). The Application for Extracting and Exploring Analysis Ready Samples (AppEEARS) tool offers vegetation and LST products of MODIS for long-term data [42].

2.3. CRU Data and Meteorological Data

CRU TS (Climatic Research Unit gridded Time Series) data are broadly used in climate studies and are available at 0.5 × 0.5 degrees over the whole surface of the Earth. It provides a monthly land-based gridded high-resolution dataset from 1901. The CRU dataset is derived by the interpolation of monthly weather station observations of extensive networks. The database is updated annually [43]. The CRU TS global data freely downloadable and accessible online (https://crudata.uea.ac.uk/cru/data/). We applied CRU TS v4 to this research for available temperature and precipitation data between 2010 and 2019.

The meteorological station data were provided by the Information and Research Institute of Meteorology, Hydrology and Environment (IRIMHE) of Mongolia website (http://tsag-agaar.gov.mn/). There were limited stations for soil moisture measurement in croplands over Mongolia. Soil moisture content was acquired at depths of 0–20 cm and 0–50 cm at monthly intervals (the 7th, 17th, and 27th days of each month) from April to September due to seasonal conditions [44]. Soil moisture content was averaged over the monthly intervals from May to August (2015–2020). The selected meteorological stations are shown in Figure 1, and the locations of the stations have been categorized into two vegetation zones (Table 1).

Table 1. Location of the agricultural meteorological stations of soil moisture.

Aimag Name	Station Name	Vegetation Zones	Latitude (° N)	Longitude (° E)	Elevation (m)
Arkhangai	Tuvshruulekh	Forest steppe	47°23′12.9″ N	101°54′30.9″ E	1619
Khuvsgul	Tarialan	Forest steppe	49°36′32.98″ N	101°59′4.52″ E	1218
Selenge	Tsagaannuur	Forest steppe	50°6′37.83″ N	105°27′7.28″ E	786
Selenge	Eruu	Forest steppe	49°44′56.52″ N	106°39′40.48″ E	673
Selenge	Baruunkharaa	Forest steppe	48°54′47.21″ N	106°5′23.11″ E	811
Selenge	Orkhon	Steppe	49°8′37.57″ N	105°24′8.44″ E	756
Selenge	Orkhontuul	Steppe	48°50′7.6″ N	104°48′23.09″ E	831
Selenge	Zuunkharaa	Forest steppe	48°51′37.86″ N	106°26′35.31″ E	883
Bulgan	Ingettolgoi	Forest steppe	49°27′33.8″ N	103°59′5.20″ E	763
Bulgan	Bulgan	Forest steppe	48°49′5.39″ N	103°31′8.18″ E	1221
Dornod	Onon	Steppe	49°6′51.42″ N	112°39′29.35″ E	894
Dornod	Choibalsan	Steppe	48°4′53.16″ N	114°32′16.21″ E	747
Dornod	Khalkhgol	Steppe	47°37′48.86″ N	118°37′20.21″ E	987
Uvs	Baruunturuun	Steppe	49°39′31.10″ N	94°24′14.62″ E	1232
Uvurkhangai	Kharkhorin	Forest steppe	47°11′40.99″ N	102°49′47.78″ E	1480
Tuv	Erdenesant	Steppe	47°20′0.35″ N	104°29′34.13″ E	1364
Tuv	Ugtaal	Steppe	48°15′29.19″ N	105°24′19.01″ E	1161
Tuv	Jargalant	Forest steppe	48°31′35.61″ N	105°52′50.67″ E	1015
Tuv	Bayanchandmani	Forest steppe	48°13′37.57″ N	106°17′2.89″ E	1255
Tuv	Bornuur	Forest steppe	48°28′7.56″ N	106°15′37.16″ E	1023
Khentii	Gurvanbayan	Steppe	48°11′4.95″ N	110°19′22.03″ E	1207
Khentii	Undurkhaan	Steppe	47°18′29.62″ N	110°37′28.29″ E	1035
Sukhbaatar	Baruun-Urt	Steppe	46°40′21.92″ N	113°16′57.07″ E	981

2.4. Crop Yield Statistical Data

Soil moisture is one of the most important factors in the agricultural sectors of Mongolia. The National Statistical Organization (NSO) website (http://nso.mn/) of Mongolia provides information on crop yields every year in the subprovinces. The NSO has been accumulating data on croplands and harvests collected from agricultural enterprises and local farmers through the statistical departments and offices of each province. Crop yield information was applied for the validation of soil moisture distribution in Mongolia from 2010 to 2019. The total harvest includes the amount of potatoes, fodder crops, cereals, fruits, vegetables, etc. (in thousands of tons), and the crop yield information is averaged every year. Crop yield is shown as the amount of agricultural production per unit area (from 1 hectare). The crop yield per hectare is estimated as the ratio of total harvest to total sown area [45].

3. Methodology

In this study, the structure of the spatial distribution of the soil moisture and time series analysis based on SMAP and MODIS products is given in the flowchart in Figure 2.

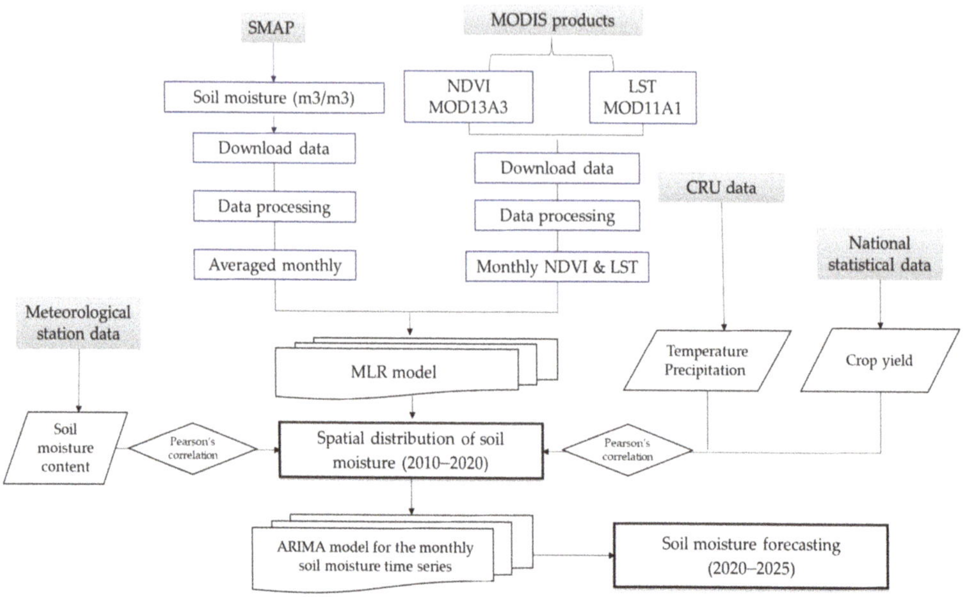

Figure 2. Flowchart of the soil moisture distribution of Mongolia based on satellite images.

The first involved downloading the data and processing it, e.g., by the SMAP satellite, NDVI, and LST from the MODIS satellite. After data processing, a multiple linear regression model was developed. Prepared monthly soil moisture content from the station, temperature/precipitation of CRU data, and yearly crop yield information from the NSO were applied for validation of the estimated soil moisture over Mongolia. Finally, time series analysis and forecasting model were used for the prediction of estimated monthly soil moisture.

3.1. Multiple Linear Regression—SM-MOD

Multiple linear regression models have often been used in natural resource studies, and these involve calculating the dependent variable Y by applying a linear combination of independent variables X_i. The linear regression form is shown in the following equation [46]:

$$Y_i = \beta_0 + \sum_{m=0}^{n} \beta_m * X_{m,i} \tag{1}$$

where β_0, β_m ($m > 0$) are constant terms corresponding to the connection, the regression coefficients, and the model error, $X_{m,i}$, respectively. The input variables (X_i) describe the output variable (Y_i) according to the results of the multiple regression model. Therefore, each input variable has different information that means all input variables should not be collinear. We used p-value statistics to estimate the significance of each variable for input variable selection [46]. The variance inflation factor (VIF) is applied to detect collinearity (also called multicollinearity) among predictors in regression models [47,48]. The VIF values are between 1 and 10, which means that there is no multicollinearity for the regression model. After these analyses, a combination of p-value and VIF measures was used. Additionally, the independent variables are normally distributed for the assumptions of the linear regression model. Table 2 shows the statistical variables computed from satellite images.

Table 2. Statistical variables of inputs, minimum (Min), maximum (Max), mean, and standard deviation (SD).

Variable	Unit	Min	Max	Mean	SD
SMAP (dependent)	m^3/m^3	0.037	0.157	0.090	0.024
LST (independent)	Celsius	−19.065	39.902	15.680	18.469
NDVI (independent)		0.024	0.381	0.186	0.097

3.2. ARIMA Model

The Box-Jenkins time series models are named after the statisticians George Box and Gwilym Jenkins [49]. These models generate forecast values based on the statistical parameters of observed time series data and are used in many fields. The Box-Jenkins Autoregressive Integrated Moving Average (ARIMA) model is a combination of the Autoregressive (AR), Integrated (I), and Moving Average (MA) terms. The Box-Jenkins model describes a wide class of models forecasting univariate time series that can be made stationary by applying transformations such as differencing of nonstationary series one or more times to achieve stationarity [50,51].

A seasonal ARIMA model is denoted by ARIMA (p, d, q), where p is the number of autoregressive terms, q is the number of moving average terms, and d represents the number of differences applied to the series [52].

The AR (p) model is defined as

$$X_t = c + \varphi_1 X_{t-1} + \varphi_2 X_{t-2} + \ldots + \varphi_p X_{t-p} + u_t = c + \sum_{i=1}^{p} \varphi_i X_{t-i} + u_t \quad (2)$$

where $\varphi_1, \varphi_2, \ldots, \varphi_n$ are the autoregressive coefficients, c is a constant, and u_t is white noise. In the autoregressive model of order p, the value of the time series at t, X_t depends upon its previous p-values and a random disturbance (the stochastic part).

The MA (p) model is defined as

$$u_t = \theta_1 z_{t-1} + \theta_2 z_{t-2} + \ldots + \theta_q z_{t-q} = z_t + \sum_{i=1}^{q} \theta_i z_{t-i} \quad (3)$$

where $\theta_1, \theta_2, \ldots, \theta_q$ are the moving average coefficients and $\{z_t\}$ is a white noise process with mean 0 and variance σ^2. Combing autoregressive and moving average models, we get Equation (4):

$$x_t = c + \varphi_1 X_{t-1} + \varphi_2 X_{t-2} + \ldots + \varphi_p X_{t-p} - \theta_1 z_{t-1} - \theta_2 z_{t-2} - \ldots - z_{t-q} \quad (4)$$

where x_t denotes the dth difference of X_t.

This defines the autoregressive moving average (ARIMA) process of p and q order and difference d, or ARIMA (p, q) [26].

There are many methods and criteria for selecting the order of an AR, MA, or ARIMA model. One of them is based on the so-called information criteria and computes the values of Akaike's information criterion (AIC) and the Bayesian information criterion (BIC) or Schwarz criterion, with smaller values of AIC and BIC preferred [53,54]. The most commonly used approach for checking the model's adequacy is to examine the residuals by using the autocorrelation function (ACF) and partial autocorrelation function (PACF) graphs. If the selected model is appropriate, the residual graphs of both correlation functions should be white noise, indicating no remaining correlation.

3.3. Model Validation

The model was implemented to evaluate the next step. Pearson's correlation (r) [55] values were applied for the comparison for estimated soil moisture and observed soil moisture and crop yield values. The coefficients of the Pearson's correlation (r) are given in Equation (5):

$$r = \frac{\sum_{i=1}^{n}(X_i - \overline{X})(Y_i - \overline{Y})}{\sqrt{\sum_i^n (X_i - \overline{X})^2}\sqrt{\sum_i^n (Y_i - \overline{Y})^2}} \quad (5)$$

where X_i and Y_i are the individual derivations and measurements of variables X and Y, respectively. $\overline{X}$ and $\overline{Y}$ are the means of X and Y, respectively. The correlation coefficient (r) ranges between -1 and 1. If r is equal to zero, this means that there is no linear association between the variables. If r is equal to 1, there is a perfect positive linear relationship between the variables, and all individuals sampled would be exactly on the same straight line with a positive slope. If $0 < r < 1$, this means a positive linear trend, but sampled individuals would be scattered around this common trend line; the smaller the absolute r value, the less well the data can be characterized by a single linear relationship. If r is positive and r values are close to 1, this describes a valuable relationship between variables [56]. Linear Pearson's correlation (r) was determined at a monthly timescale [44,55] for the satellite-derived and meteorological station/NSO data.

4. Results

4.1. SM-MOD—Multiple Linear Regression Model

The linear regression model in Equation (1) was applied to estimate soil moisture in Mongolia, with 1 km resolution. The multicollinearity tests for all variables (NDVI and LST) were determined as in Table 3. The VIF values were lower than five, which shows that there was no multicollinearity for the regression model. Also, histogram normality test has been checked by the Jarque-Bera test for the linear regression model. In this test, if the probability of Jarque-Bera greater than 0.05% or 5%, then we accepted the null hypothesis, which means that the residuals were normally distributed. Table 3 summarizes the multiple linear regression model coefficients, p-values, standard error, t-statistics, VIF statistics. The monthly NDVI and LST explained 78% of the variation in soil moisture. The F-statistic was less than 0.05, which means that this model can be used for soil moisture analysis.

Table 3. Result of the linear regression model.

Variable	Coefficient	Std. Error	*t*-Statistic	Prob.	Collinearity Statistics	
					Tolerance	VIF
Intercept	0.044	0.003721	11.79743	0.0000		
NDVI	0.289	0.027357	10.56239	0.0000	0.300	3.336
LST (Celsius)	−0.0005	0.000144	−3.499165	0.0009	0.247	4.041
R-squared	0.780878	Adjusted R-squared	0.773322	Mean dependent variable	0.089744	
SD dependent variable	0.023754	SE of regression	0.011310	Akaike info. criterion	−6.078392	
Prob (F-statistic)	0.000000					

In this research, we assume that SM is derived from the satellite and depends on the independent variables NDVI and LST, while SMAP is the dependent variable. From the assumption, a multiple regression model has been developed. Finally, the MLR model resulted in Equation (6):

$$SM_{MOD} = 0.044 + 0.289 * NDVI - 0.0005 * LST \quad (6)$$

where SM_{MOD} is the modeled soil moisture; constant coefficients were estimated from Table 3. Figure 3 shows graphs of the actual, fitted, and residual values of the linear regression. The figure suggests that in most of the studied months, the correlation between the real-life situation and the model was high.

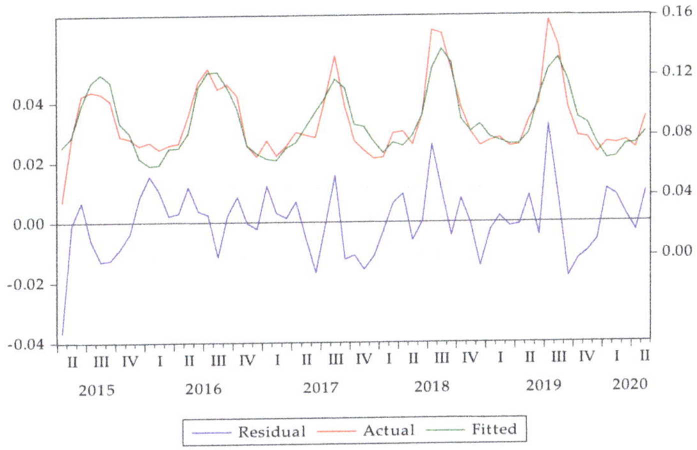

Figure 3. Actual, fitted, and residual values of the multiple regression model.

Soil moisture (SM-MOD) was calculated using Equation (6), with values in m^3/m^3. The lowest value of 0 indicates dry areas and the highest value of 0.35 m^3/m^3 indicates wet areas. Figure 4 represents the spatial distribution of monthly soil moisture over Mongolia. Monthly SM maps were averaged by month between 2010 and 2020. During the years 2010–2020, the winter had the lowest soil moisture (November, December, and January), and spring also had low soil moisture (February, March, and April). The summer showed high soil moisture in May, June, and July. However, the autumn showed the highest soil moisture in August, September, and October (Figure 4). Apparently, increased soil moisture is observed in the northern part, which is taiga, forest steppe, and steppe zones, while low soil moisture is restricted to the southern part of Mongolia, which is mostly desert steppe and desert vegetation.

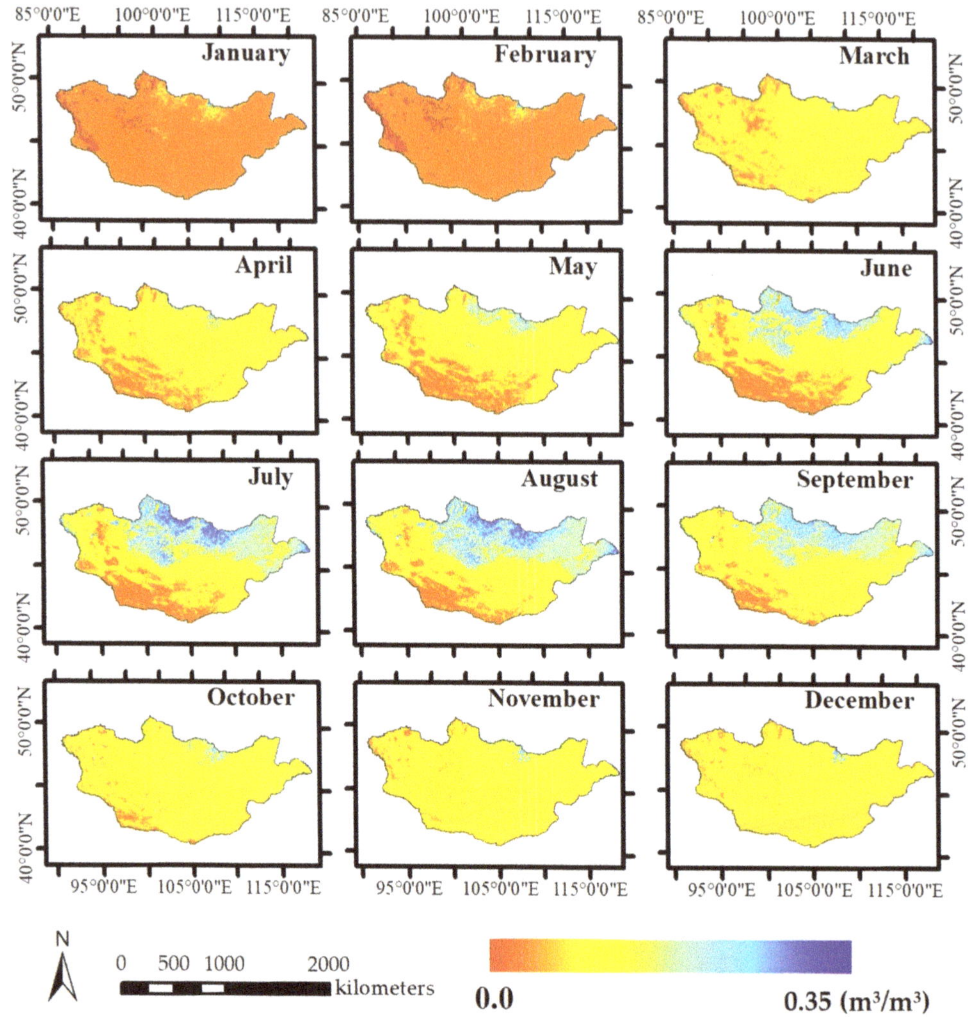

Figure 4. Spatial distribution of soil moisture content from the model (SM_{MOD}) (averaged monthly from 2010 to 2020).

4.2. Comparison between MLR Model and SMC from the Meteorological Station

In general, the estimation of soil moisture from the model was reasonably accurate, as confirmed by applying satellite images. Figure 5a–d describe the correlation between SMAP and SM-MOD with the SMC from the meteorological station at 0–20 and 0–50 cm depths. Table 4 shows the correlations of the SMAP and SM-MOD with the SMC from the meteorological stations at different depths. The correlation coefficients (r) between SMAP and SMC from meteorological stations were 0.279 (Figure 5a) and 0.181 (Figure 5b) at 0–20 and 0–50 cm depths, respectively. This was statistically significant, with root mean square error (RMSE) values of 0.094 m^3/m^3 and 0.098 m^3/m^3, as shown in Table 4a. Table 4b indicates that the values of correlation coefficients (r) were 0.191 between SM-MOD and SMC at 0–20 cm depth from the meteorological station, which was statistically significant, with an RMSE of 0.090 m^3/m^3 ($p < 0.0001$; Figure 5c), and 0.126 between SM-MOD and SMC at 0–50 cm depth from the meteorological station, which was also statistically significant, with an RMSE of 0.091 m^3/m^3 ($p < 0.005$; Figure 5d). The confidence intervals of 95% were between SM-MOD and SMC at 0–20 cm depth from the meteorological station from 0.104 to 0.276, and with SMC at 0–50 cm from the meteorological station from 0.038 to 0.213.

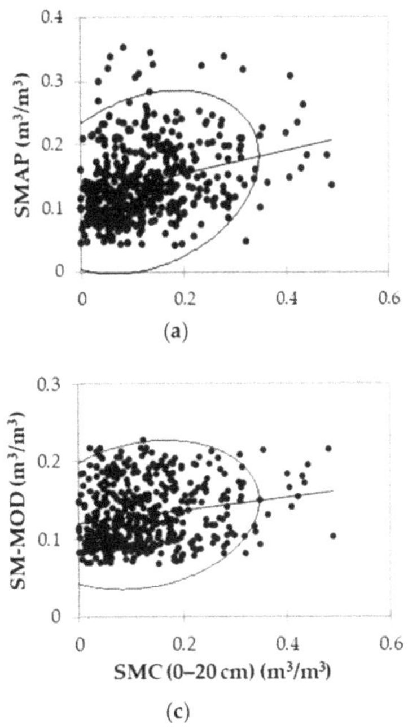

Figure 5. Scatter diagram of SMAP and SM-MOD with SM measurements from the meteorological stations for different depths from May to August of 2015–2020 over the study area: (**a**) SMAP and SMC from the meteorological stations at 0–20 cm depth; (**b**) SMAP and SMC from the meteorological stations at 0–50 cm depth; (**c**) SM-MOD and SMC from the meteorological stations at 0–20 cm depth; (**d**) SM-MOD and SMC from the meteorological stations at 0–50 cm depth.

Table 4. Correlation between (**a**) SMAP and (**b**) SM-MOD with the SMC from the meteorological stations at the different depths from May to August of 2015–2020.

	(a)	
SMAP (m^3/m^3)	**SMC/0–20 cm/(m^3/m^3)**	**SMC/0–50 cm/(m^3/m^3)**
p-values (Pearson)	<0.0001	0.005
Coefficients of determination (Pearson)	0.078	0.087
Correlation (Pearson)	0.279 **	0.181 **
RMSE	0.094	0.098
Bias	0.0016	0.0021
Confidence intervals (95%)	(0.194, 0.359)	(0.094, 0.266)
	(b)	
SM-MOD (m^3/m^3)	**SMC/0–20 cm/(m^3/m^3)**	**SMC/0–50 cm/(m^3/m^3)**
p-values (Pearson)	<0.0001	0.005
Coefficients of determination (Pearson)	0.037	0.016
Correlation (Pearson)	0.191 **	0.126 **
RMSE	0.090	0.091
Bias	0.0016	0.0020
Confidence intervals (95%)	(0.104, 0.276)	(0.038, 0.213)

** Correlation is significant at the 0.01 level (2-tailed).

4.3. Comparison between SM-MOD and CRU Data

We also examined the trends of monthly precipitation and temperature from the CRU data. Figure 6 displays the time series of monthly precipitation, temperature, and SM-MOD from 2010 to 2020 in Mongolia. The highest precipitation was observed in July 2018, and the highest SM-MOD value was in August 2018. From the comparison, when the precipitation was high, then the following month's soil moisture was high, which means that in Mongolia, the soil moisture directly depends on precipitation.

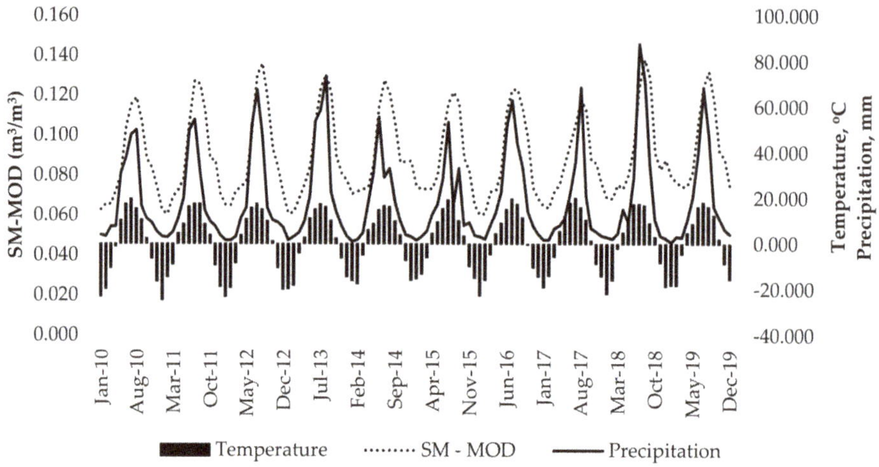

Figure 6. Comparison between monthly precipitation (mm), temperature (°C), and SM-MOD (m^3/m^3) from January 2010 to December 2019 in Mongolia.

Figure 7a–c describes the correlation between SM-MOD and the temperature and precipitation. Table 5 shows the correlations of the SM-MOD with the temperature and precipitation from the CRU data over Mongolia. It indicates that the values of correlation coefficients (r) were 0.802 between SM-MOD and temperature, which was statistically significant ($p < 0.0001$; Figure 7b), and 0.826 between SM-MOD and precipitation, which was also statistically significant ($p < 0.0001$; Figure 7c). The confidence intervals of 95% were between SM-MOD from the model and monthly temperature from 0.728 to 0.858, and with monthly precipitation from 0.759 to 0.876 over Mongolia.

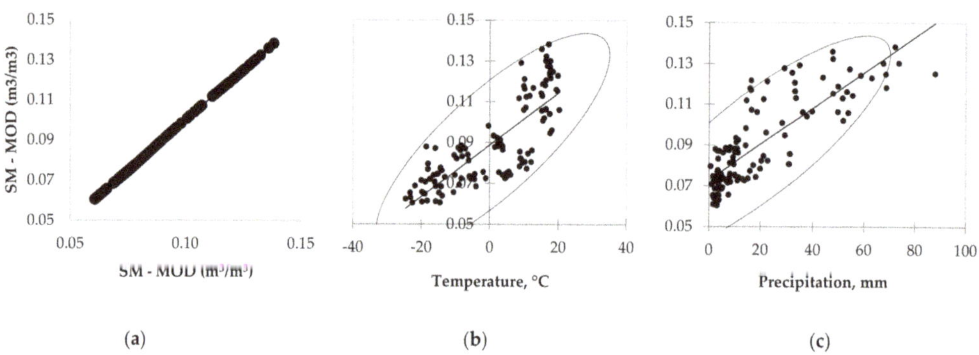

Figure 7. Scatter diagram of the monthly SM-MOD (m^3/m^3), monthly temperature (°C), and monthly precipitation (mm) from 2010 to 2020 over the study area: (**a**) SM-MOD and SM-MOD; (**b**) SM-MOD and temperature (°C); (**c**) SM-MOD and precipitation (mm).

Table 5. Correlation among the monthly SM-MOD with the monthly temperature (°C) and monthly precipitation (mm) from the CRU data between 2010 and 2020.

Variables	SM-MOD (m^3/m^3)	Temperature, °C	Precipitation, mm
Confidence intervals (95%)/lower bound	1	0.728	0.759
Confidence intervals (95%)/upper bound	1	0.858	0.876
Correlation matrix (Pearson)	**1**	**0.802**	**0.826**
p-values (Pearson)	**0**	**<0.0001**	**<0.0001**
Bias	0	0.025	0.026

Values in bold are different from 0, with a significance level alpha = 0.05.

4.4. Comparison between SM-MOD and Crop Yield

We considered the crop yield information for every year to correlate with SM-MOD from the model. The National Statistical Organization (NSO) provides every province's crop yield information since 2010. Figure 8 shows the trends of averaged SM-MOD from May to September of 2010–2019 and the observed total crop yield data for each year (2010–2019).

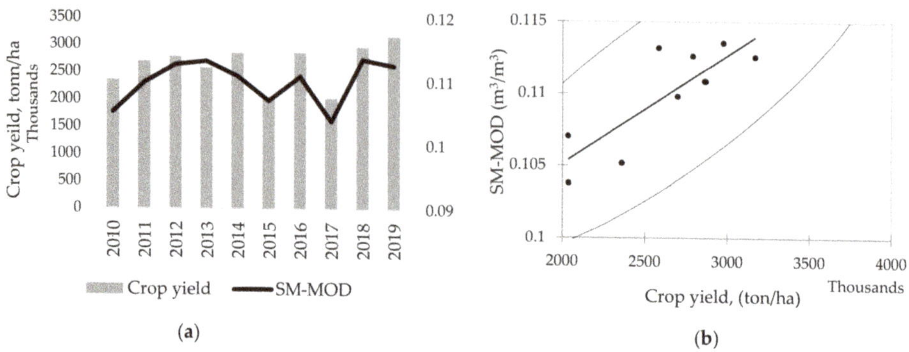

Figure 8. Comparison of the SM-MOD and crop yield information: (**a**) yearly crop yield (ton/ha) and averaged SM-MOD from May to September (2010–2019); (**b**) scatter diagram of SM-MOD and crop yield information.

To apply the model to the use, we examined the relationship between SM-MOD and crop yield over Mongolia. The results show that there is a significant trend in the SM-MOD. It was statistically significantly ($p < 0.003$) correlated with the crop yield, with $r = 0.835$ (Table 6).

Table 6. Correlation between the averaged SM-MOD from May to September for 2010–2019 and the yearly crop yield from NSO for 2010–2019.

Variables	SM-MOD (m^3/m^3)	Crop Yield (ton/ha)
Coefficients of determination (Pearson)	1	0.697
Correlation matrix (Pearson)	**1**	**0.835**
p-values (Pearson)	0	0.003
Bias	0	−0.005

Values in bold are different from 0, with a significance level alpha = 0.05.

4.5. ARIMA Model of Soil Moisture

We selected the most appropriate model for the time series from the possible models. We selected our model using these criteria: first, the most significant coefficients; second, the lowest volatility; third, the highest adjusted R-squared; and last, the lowest Akaike's information criterion (AIC)/Schwarz information criterion (SIC). Since the theory behind ARMA estimation is based on a stationary time series, we first used the transformation based on a logarithm and considered the first difference of the soil moisture time series. According to the unit root test, the differenced series is a stationary series, so we applied the plots of ACF and PACF to identify the structure of the model. The plots and statistical tests showed that the ARIMA (12, 1, 12) model was suitable for the log time series of soil moisture. The estimation results and the actual, fitted residual graphs of the model are shown in Table 7 and Figure 9, respectively. The residual diagnostics tests suggest that the estimation residuals are white noise. The final results indicate the good performance of the model, and we can say that about 82% of the variability in the soil moisture was predicted by the selected model (Table 7).

Table 7. Results of the time series analysis for soil moisture.

Dependent Variable: DLOG (SM-MOD)			Method: Least Squares	
Sample: 2010M02–2020M05			Included Observations: 124	
Failure to Improve Objective (Nonzero Gradients) after 90 Iterations				
Coefficient Covariance Computed Using Outer Product of Gradients				
Variable	**Coefficient**	**Std. Error**	***t*-Statistic**	**Prob.**
Constant (c)	0.000560	0.116782	0.004795	0.9962
AR (12)	0.999321	0.000118	8501.412	0.0000
AR (1)	0.000625	0.000160	3.899161	0.0002
MA (12)	−0.950003	0.020568	−46.18744	0.0000
SIGMASQ	0.003390	0.000392	8.655254	0.0000
R-squared	0.822436	Mean dependent variable		0.002194
Adjusted R-squared	0.816468	SD dependent variable		0.138724
SE of regression	0.059430	Akaike info. criterion		−2.485006
Sum squared residuals	0.420300	Schwarz criterion		−2.371285
Log likelihood	159.0703	Hannan-Quinn criterion		−2.438809
F-statistic	137.7955	Durbin-Watson statistic		2.030050
Prob (F-statistic)	0.000000			

Sample: 2010M01–2020M05 Included observations: 124

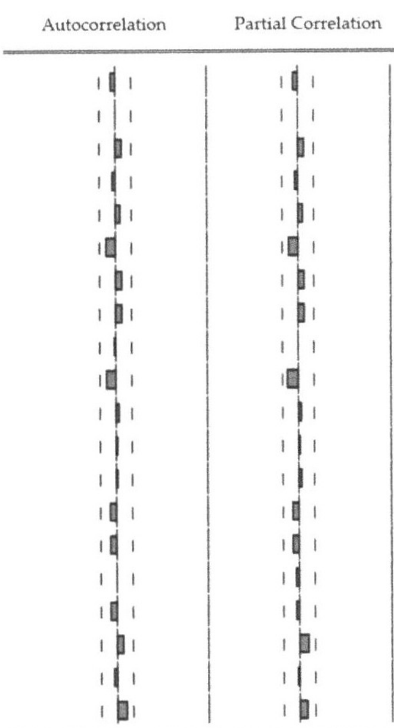

	AC	PAC	Q-Stat	Prob
1	-0.051	-0.051	0.3333	0.564
2	0.006	0.004	0.3384	0.844
3	0.066	0.066	0.8956	0.827
4	-0.040	-0.034	1.1075	0.893
5	0.053	0.049	1.4795	0.915
6	-0.107	-0.107	2.9887	0.810
7	0.077	0.073	3.7843	0.804
8	0.079	0.080	4.6240	0.797
9	-0.019	0.005	4.6748	0.862
10	-0.102	-0.128	6.1135	0.806
11	0.035	0.032	6.2821	0.854
12	0.021	0.016	6.3463	0.898
13	0.009	0.036	6.3567	0.932
14	-0.074	-0.082	7.1445	0.929
15	-0.066	-0.080	7.7684	0.933
16	0.000	-0.039	7.7684	0.955
17	-0.070	-0.031	8.4925	0.955
18	0.081	0.102	9.4620	0.948
19	-0.026	-0.024	9.5652	0.963
20	0.110	0.087	11.374	0.936

Figure 9. Correlogram of residuals squared of the autocorrelation function (ACF) and partial autocorrelation function (PACF).

The selected model is written as follows:

$$d(X_t) = 0.0006 + 0.9993X_{t-12} + 0.0006X_{t-1} - 0.95z_{t-12} \quad (7)$$

or, equivalently, as

$$X_t = 0.0006 + 0.9993X_{t-12} + 0.0006X_{t-1} - 0.95z_{t-12}. \quad (8)$$

where X_t is the log values of soil moisture at time t and z_t is the error term at time t.

The above model shows that the soil moisture at time t depends on the value of soil moisture of previous months and also on the error terms of 12 months ago.

Partial autocorrelation (PAC) measures the correlation between observations that are p periods apart after controlling for correlations at intermediate lags (i.e., lags less than p). The correlogram of the residuals is flat, which indicates that all information has been captured (Figure 9). A flat correlogram of the residuals is ideal. Therefore, the forecast will be based on this model.

One of the main purposes of the ARMA and ARIMA models is to provide short-term forecasts. Hence, we have predicted the values of soil moisture from 2020 to 2025 using the selected model. Figure 10 shows the forecasting results for soil moisture from 2020 to 2025 with a 95% confidence interval, which is the range within which the actual dependent value should fall a given percentage of the time (the level of confidence). For forecasting, the root mean squared error was 0.002, and the bias proportion was 0.044. Figure 11 compares the actual soil moisture and soil moisture forecasting (in m^3/m^3) from January to May 2020. The prediction from February and April is almost the same as the actual soil moisture and there were slight deviations when predicting March and May. Overall, the model demonstrated accurate forecasting of soil moisture.

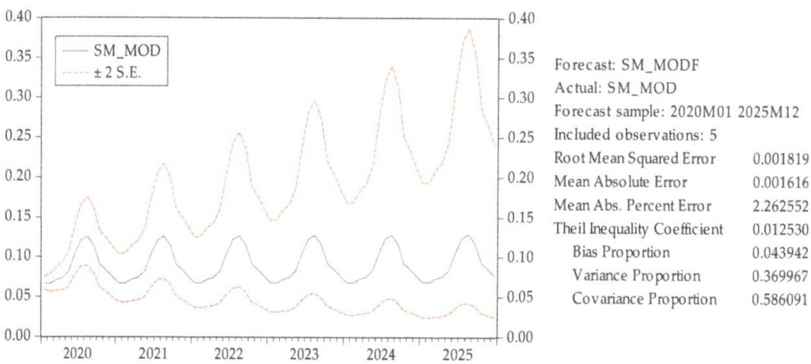

Figure 10. Soil moisture forecasting from the ARIMA model (m^3/m^3).

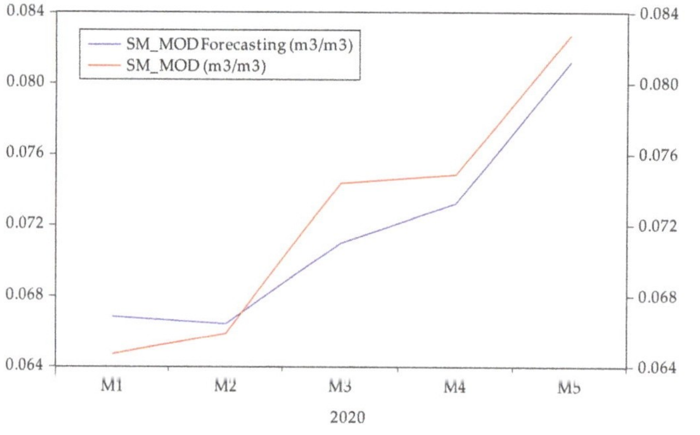

Figure 11. Comparison graph of the real soil moisture and soil moisture forecasting.

The essence of an appropriate ARIMA model is to forecast the future trends of series. Hence, we used the past information of the soil moisture series itself. The forecast was based on the final selected model, ARIMA (12, 1, 12).

The forecasting values of soil moisture are given in Figure 12, with the kernel density of the values on the y axis.

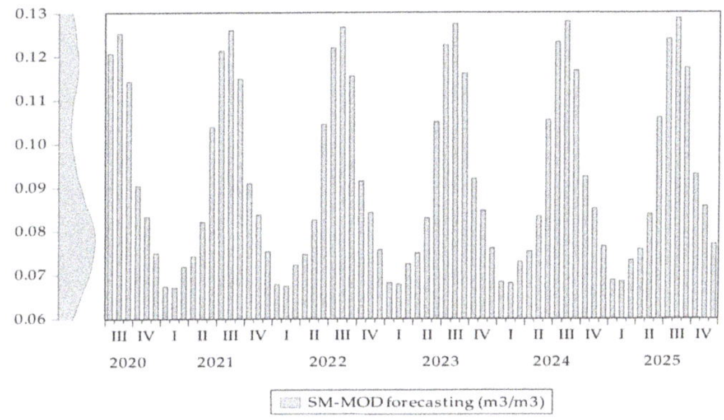

Figure 12. Predicting soil moisture trend until December 2025.

5. Discussion

In the present study, we used a linear regression model to estimte the spatial distribution of soil moisture over Mongolia by considering satellite images (SMAP and MODIS). We estimated the monthly (January–December) soil moisture during the period 2010–2020 in Mongolia. The SM model performance was validated by comparison with SMC from the agricultural meteorological stations, with data on precipitation, temperature, crop yield, etc. The correlation has shown that the model (SM-MOD) gives accurate information on the soil moisture for each month. Moreover, the present model has the advantage of recognizing soil moisture spatial distribution with a high spatial resolution (1 km); this is the first time such information has been gathered for Mongolia. Therefore, we established the ARIMA model for soil moisture forecasting based on estimated soil moisture between 2010 and 2020. The results provide the monthly spatial distribution of soil moisture, which is valuable data for use in numerous contexts, including agricultural management, drought monitoring, assessment of climate change, flooding, determining pasture and land degradation. Land degradation in central Mongolia is mostly caused by overgrazing; however, changes in summertime precipitation have also occurred [57]. Mongolian grassland is decreasing, and drought is increasing [58]. Our research on time series analysis for monthly SM-MOD forecasting is vital for the monitoring of land degradation and drought.

The results of correlation coefficients are low because of limited data availability at the agricultural meteorological station. However, the correlation was statistically significant at $p < 0.0001$ (0–20 cm) and $p < 0.005$ (0–50 cm), respectively, between SM-MOD and SMC from the meteorological stations at different depths. From Figure 6, we see that the previous month's precipitation directly impacted the soil moisture during the growing season (June–September). The correlation between SM-MOD and temperature had correlation coefficients (r) of 0.80 (statistically significant at $p < 0.0001$) and 0.83 (statistically significant at $p < 0.0001$). However, SM-MOD compared with the crop yield for each year (2010–2019) had a correlation coefficient (r) of 0.84.

Therefore, the time series analysis for monthly soil moisture forecasting was developed over Mongolia based on the established ARIMA model. From the study, we selected the ARIMA (12, 1, 12) model, which was most suitable for the SM-MOD time series, for prediction, in which the values of soil moisture (SM-MOD) were predicted from 2020 to 2025 using the selected model. Forecasting results are shown with a 95% confidence interval. Time series SM-MOD data will provide valuable information for decision-makers and researchers. SM-MOD time series and forecasting data are good sources of data for long-term agricultural management, planning, and climate change and drought monitoring.

In terms of applications, this multiple linear regression model is a practical tool for the reliable and timely monitoring of droughts; thus, the advantage of this research lies in providing valuable information for decision-makers and farmers. In further studies, we will investigate seasonal soil moisture in different vegetation zones using this method along with field measurements.

6. Conclusions

Soil moisture is an important factor for the agricultural land in Mongolia. The model used in this paper is suitable for use in agricultural areas and has useful applications for agricultural management (irrigation, pasture, and hayfield yield) and drought monitoring in Mongolia. Time series analysis is one of the main tools for analyzing and predicting future trends of soil moisture. Most previous studies have examined soil moisture by comparing it with climate factors that were analyzed based on correlation analysis and multilinear regression [7,59,60]. The LST/NDVI combination method may prove to be a robust method for estimating SM; this combination is easy to operate and has a strong physical basis [7].

In general, the model's performance in determining soil moisture was practically assessed using satellite images. This study took Mongolia as the study area, divided into six vegetation zones. The linear regression method was applied in soil moisture estimation using SMAP and MODIS satellite images. From the model, the spatial distribution of soil moisture was developed monthly from 2010 to 2020. The soil moisture was high in the north, while low soil moisture was observed in southern Mongolia, especially during the warm season. Then output maps were compared with the soil moisture content from the agricultural meteorological stations and precipitation/temperature from the CRU data. The results show that the estimated soil moisture was statistically significantly correlated with the actual soil moisture content reported by the station. Moreover, the estimated soil moisture (SM-MOD), when compared with the crop yield, showed a high correlation, though there is a need for more accurate, detailed ground-measured data. Finally, we performed a time series analysis of soil moisture from 2010 to 2020 and predicted soil moisture until 2025 in this study area. Overall, the developed SM model and time series method can both be used to investigate the changes in soil moisture in Mongolia, so it is reasonable to use them in agriculture, hydrology, and climate science. However, this linear regression model should be elaborated to suit each vegetation zone or eco-climate regions in the applied study area.

Author Contributions: E.N. analyzed all data with ArcGIS 10.3 & ENVI 4.7 and computed the statistical analysis with E-view 9.0 and XLSTAT2020 software. Also, E.N. wrote the first draft. T.R. and P.D.M. provided useful advice and revised the manuscript. B.D. helped the statistical analysis and time series methods. All authors contributed to the final manuscript. All authors have read and agreed to the published version of the manuscript.

Funding: This research received no external funding.

Institutional Review Board Statement: Not applicable.

Informed Consent Statement: Not applicable.

Acknowledgments: The first author is very grateful to the European Union ERASMUS-IMPAKT-2016 program scholarship, allowing her to pursue her study at the Ghent University, Belgium. Also, very thankful to the Department of Geography for research supports and very thankful to the co-authors, besides special thanks to Frieke Vancoillie of the Ghent University given the specific comments. We are grateful to the Information and Research Institute and Research Institute of Meteorology, Hydrology and Environment (IRIMHE) and National Statistical Office (NSO) of Mongolia for providing us soil moisture measurements and crop yield growth data used in this research. We acknowledge the anonymous reviewers for their valuable comments, which remarkably improved our paper.

Conflicts of Interest: The authors declare no conflict of interest.

References

1. Gao, X.; Wu, P.; Zhao, X.; Wang, J.; Shi, Y. Effects of land use on soil moisture variations in a semi-arid catchment: Implications for land and agricultural water management. *Land Degrad. Dev.* **2014**, *25*, 163–172. [CrossRef]
2. Deryng, D.; Sacks, W.J.; Barford, C.C.; Ramankutty, N. Simulating the effects of climate and agricultural management practices on global crop yield. *Glob. Biogeochem. Cycles* **2011**, *25*. [CrossRef]
3. Wang, X.; Wang, B.; Xu, X.; Liu, T.; Duan, Y.; Zhao, Y. Spatial and temporal variations in surface soil moisture and vegetation cover in the Loess Plateau from 2000 to 2015. *Ecol. Indic.* **2018**, *95*, 320–330. [CrossRef]
4. Park, S.; Im, J.; Park, S.; Rhee, J. Drought monitoring using high resolution soil moisture through multi-sensor satellite data fusion over the Korean peninsula. *Agric. Forest Meteorol.* **2017**, *237*, 257–269. [CrossRef]
5. Natsagdorj, E.; Tsolmon, R.; Martin, K.; Batchuluun, T.; Chimgee, D.; Oyunbileg., T.; Ulam-Orgikh, D. An integrated methodology for soil moisture analysis using multispectral data in Mongolia. *Geo-Spat. Inf. Sci.* **2017**, *20*, 46–55. [CrossRef]
6. Rahimzadeh-Bajgiran, P.; Berg, A. Soil moisture retrievals using optical/TIR methods. In *Satellite Soil Moisture Retrieval: Techniques and Applications*; Elsevier Inc.: Amsterdam, The Netherlands, 2016; pp. 47–72.
7. Zhang, D.; Zhou, G. Estimation of Soil Moisture from Optical and Thermal Remote Sensing: A Review. *Sensors* **2016**, *16*, 1308. [CrossRef] [PubMed]
8. Peng, J.; Loew, A. Recent Advances in Soil Moisture Estimation from Remote Sensing. *Water* **2017**, *9*, 530. [CrossRef]
9. Jung, C.; Lee, Y.; Cho, Y.; Kim, S. A Study of Spatial Soil Moisture Estimation Using a Multiple Linear Regression Model and MODIS Land Surface Temperature Data Corrected by Conditional Merging. *Remote Sens.* **2017**, *9*, 870. [CrossRef]
10. Vani, V.; Pavan Kumar, K.; Ravibabu, M.V. Temperature and vegetation indices based surface soil moisture estimation: A remote sensing data approach. In *Springer Series in Geomechanics and Geoengineering*; Springer: Berlin/Heidelberg, Germany, 2019; pp. 281–289.
11. Hong, Z.; Zhang, W.; Yu, C.; Zhang, D.; Li, L.; Meng, L. SWCTI: Surface Water Content Temperature Index for Assessment of Surface Soil Moisture Status. *Sensors* **2018**, *18*, 2875. [CrossRef]
12. Brocca, L.; Crow, W.T.; Ciabatta, L.; Massari, C.; Rosnay, P.D.; Enenkel, M.; Hahn, S.; Amarnath, G.; Camici, S.; Tarpanelli, A.; et al. A Review of the Applications of ASCAT Soil Moisture Products. *IEEE J. Sel. Top. Appl. Earth Obs. Remote Sens.* **2017**, *10*, 2285–2306. [CrossRef]
13. Chan, S.K.; Bindlish, R.; Njoku, E.; Jackson, T.; Colliander, A.; Chen, F.; Burgin, M.; Dunbar, S.; Piepmeier, J.; Yueh, S.; et al. Assessment of the SMAP Passive Soil Moisture Product. *IEEE Trans. Geosci. Remote Sens.* **2016**, *54*, 4994–5007. [CrossRef]
14. Zhang, X.; Zhang, T.; Zhou, P.; Shao, Y.; Gao, S. Validation Analysis of SMAP and AMSR2 Soil Moisture Products over the United States Using Ground-Based Measurements. *Remote Sens.* **2017**, *9*, 104. [CrossRef]
15. Chen, F.; Crow, W.T.; Colliander, A.; Cosh, M.H.; Jackson, T.J.; Bindlish, R.; Reichle, R.H.; Chan, S.K.; Bosch, D.D.; Starks, P.J.; et al. Application of Triple Collocation in Ground-Based Validation of Soil Moisture Active/Passive (SMAP) Level 2 Data Products. *IEEE J. Sel. Top. Appl. Earth Obs. Remote Sens.* **2017**, *10*, 489–502. [CrossRef]
16. Zeng, J.; Chen, K.S.; Bi, H.; Chen, Q. A Preliminary Evaluation of the SMAP Radiometer Soil Moisture Product over United States and Europe Using Ground-Based Measurements. *IEEE Trans. Geosci. Remote Sens.* **2016**, *54*, 4929–4940. [CrossRef]
17. Nemani, R.; Running, S. Land cover characterization using multitemporal Red, Near-Ir, and Thermal-Ir data from NOAA/AVHRR. *Ecol. Appl.* **1997**, *7*, 79–90. [CrossRef]
18. Lambin, E.F.; Ehrlich, D. The surface temperature-vegetation index space for land cover and land-cover change analysis. *Int. J. Remote Sens.* **1996**, *17*, 463–487. [CrossRef]
19. Chae, S.-H.; Park, S.-H.; Lee, M.-J. A Study on the Observation of Soil Moisture Conditions and its Applied Possibility in Agriculture Using Land Surface Temperature and NDVI from Landsat-8 OLI/TIRS Satellite Image. *Korean J. Remote Sens.* **2017**, *33*, 931–946. [CrossRef]
20. Saha, A.; Patil, M.; Goyal, V.C.; Rathore, D.S. Assessment and Impact of Soil Moisture Index in Agricultural Drought Estimation Using Remote Sensing and GIS Techniques. *Proceedings* **2018**, *7*, 2. [CrossRef]
21. Dandridge, C.; Fang, B.; Lakshmi, V. Downscaling of SMAP Soil Moisture in the Lower Mekong River Basin. *Water* **2019**, *12*, 56. [CrossRef]
22. Leary, N. *Climate Change and Vulnerability and Adaptation: Two Volume Set*; Routledge: Abingdon, UK, 2008.
23. Yatagai, A.; Yasunari, T. Interannual Variations of Summer Precipitation in the Arid/semi-arid Regions in China and Mongolia: Their Regionality and Relation to the Asian Summer Monsoon. *J. Meteorol. Soc. Jpn. Ser. II* **1995**, *73*, 909–923. [CrossRef]
24. Nandintsetseg, B.; Shinoda, M. Seasonal change of soil moisture in Mongolia: Its climatology and modelling. *Int. J. Clim.* **2011**, *31*, 1143–1152. [CrossRef]
25. Shinoda, M.; Nandintsetseg, B. Soil moisture and vegetation memories in a cold, arid climate. *Glob. Planet. Chang.* **2011**, *79*, 110–117. [CrossRef]
26. Box, G.E.P.; Jenkins, G.M.; Reinsel, G.C. *Time Series Analysis Forecasting and Gontrol*; John Willey & Sons Inc.: Hoboken, NJ, USA, 2016.

27. Singh, S.; Kaur, S.; Kumar, P. Forecasting Soil Moisture Based on Evaluation of Time Series Analysis. *Lect. Notes Electr. Eng.* **2020**, *609*, 145–156. [CrossRef]
28. Tian, M.; Wang, P.; Khan, J. Drought Forecasting with Vegetation Temperature Condition Index Using ARIMA Models in the Guanzhong Plain. *Remote Sens.* **2016**, *8*, 690. [CrossRef]
29. Karthika, M.; Thirunavukkarasu, V. Forecasting of meteorological drought using ARIMA model. *Indian J. Agric. Res.* **2017**, *51*, 103–111. [CrossRef]
30. Han, P.; Wang, P.X.; Zhang, S.Y.; Zhu, D.H. Drought forecasting based on the remote sensing data using ARIMA models. *Math. Comput. Model.* **2010**, *51*, 1398–1403. [CrossRef]
31. Mishra, A.K.; Desai, V.R. Drought forecasting using stochastic models. *Stoch. Environ. Res. Risk Assess.* **2005**, *19*, 326–339. [CrossRef]
32. Otgonbayar, M.; Atzberger, C.; Chambers, J.; Amarsaikhan, D.; Böck, S.; Tsogtbayar, J. Land Suitability Evaluation for Agricultural Cropland in Mongolia Using the Spatial MCDM Method and AHP Based GIS. *J. Geosci. Environ. Prot.* **2017**, *5*, 238–263. [CrossRef]
33. Nandintsetseg, B.; Greene, J.S.; Goulden, C.E. Trends in extreme daily precipitation and temperature near lake Hövsgöl, Mongolia. *Int. J. Clim.* **2007**, *27*, 341–347. [CrossRef]
34. Batima, P.; Natsagdorj, L.; Gombluudev, P.; Erdenetsetseg, B. Observed climate change in Mongolia. *Assess. Imp. Adapt. Clim. Chang. Work Pap.* **2005**, *12*, 1–26.
35. Natsagdorj, E.; Renchin, T. Determination of moisture in Mongolia using remotely sensned data. In Proceedings of the Asian Conference on Remote Sensing, Hanoi, Vietnam, 1–5 November 2010; p. 921.
36. Yunatov, A.A. Vegetation Map of the Mongolian Peoples' Republic 1:1,500,000. *Akad. Nauk SSSR Ai MNP* **1979**.
37. Entekhabi, D.; Njoku, E.G.; O'Neil, P.E.; Kellogg, K.H.; Crow, W.T.; Edelstein, W.N.; Entin, J.K.; Goodman, S.D.; Jackson, T.J.; Johnson, J.; et al. The soil moisture active passive (SMAP) mission. *Proc. IEEE* **2010**, *98*, 704–716. [CrossRef]
38. O'Neill, J.C.; Chan, S.; Njoku, E.G.; Jackson, T.; Bindlish, R.; NASA National Snow and Ice Data Center Distributed Active Archive Center. SMAP Enhanced L3 Radiometer Global Daily 9 km EASE-Grid Soil Moisture, Version 3. 2019. Available online: https://nsidc.org/data/SPL3SMP_E/versions/3 (accessed on 7 September 2020).
39. Zhang, J.; Roy, D.; Devadiga, S.; Zheng, M. Anomaly detection in MODIS land products via time series analysis. *Geo-Spat. Inf. Sci.* **2007**, *10*, 44–50. [CrossRef]
40. Didan, K.; NASA EOSDIS Land Processes DAAC. MOD13A3 MODIS/Terra Vegetation Indices Monthly L3 Global 1 km SIN Grid V006. 2015. Available online: https://lpdaac.usgs.gov/products/mod13a3v006/ (accessed on 8 June 2020).
41. Hulley, G.; Wan, Z.; Hook, S.; NASA EOSDIS Land Processes DAAC. MOD11A2 MODIS/Terra Land Surface Temperature/Emissivity 8-Day L3 Global 1km SIN Grid V006. 2015. Available online: https://lpdaac.usgs.gov/products/mod11a2v006/ (accessed on 30 May 2020).
42. AppEEARS Team; NASA EOSDIS Land Processes Distributed Active Archive Center (LP DAAC); USGS/Earth Resources Observation and Science (EROS) Center. Application for Extracting and Exploring Analysis Ready Samples (AppEEARS). 2020. Available online: https://lpdaacsvc.cr.usgs.gov/appeears/ (accessed on 1 June 2020).
43. Harris, I.; Osborn, T.J.; Jones, P.; Lister, D. Version 4 of the CRU TS monthly high-resolution gridded multivariate climate dataset. *Sci. Data* **2020**, *7*, 1–18. [CrossRef]
44. Natsagdorj, E.; Renchin, T.; de Maeyer, P.; Dari, C.; Tseveen, B. Long-term soil moisture content estimation using satellite and climate data in agricultural area of Mongolia. *Geocarto Int.* **2019**, *34*, 722–734. [CrossRef]
45. NSO. Agriculture. 2020. Available online: http://www.1212.mn/Stat.aspx?LIST_ID=976_L10_2&type=tables (accessed on 7 September 2020).
46. Weisberg, S. *Applied Linear Regression*; John Wiley & Sons, Inc.: Hoboken, NJ, USA, 2005.
47. Murray, L.; Nguyen, H.; Lee, Y.F.; Remmenga, M.D.; Smith, D.W. Models with variance inflation factors in regression models with dummy variables. In Proceedings of the Conference on Applied Statistics in Agriculture Kansas State University, Manhattan, KS, USA, 29–30 April 2012; pp. 161–177. [CrossRef]
48. Harrell, F.E. Multivariable modeling strategies. In *Regression Modeling Strategies*; Springer: Cham, Switzerland, 2001; pp. 53–85.
49. Box, G.E.P.; Jenkins, G.M. *Time Series Analysis: Forecasting and Control.*; Holden-Day: San Francisco, CA, USA, 1970.
50. Box, G.E.P.; Pierce, D.A. Distribution of residual autocorrelations in autoregressive-integrated moving average time series models. *J. Am. Stat. Assoc.* **1970**, *65*, 1509–1526. [CrossRef]
51. Box, G.E.P.; Jenkins, G.M.; MacGregor, J.F. Some Recent Advances in Forecasting and Control. *Appl. Stat.* **1974**, *23*, 158. [CrossRef]
52. Rahman, A.; Hasan, M.M. Modeling and Forecasting of Carbon Dioxide Emissions in Bangladesh Using Autoregressive Integrated Moving Average (ARIMA) Models. *Open J. Stat.* **2017**, *7*, 560–566. [CrossRef]
53. Adnan, R.M.; Yuan, X.; Kisi, O.; Curtef, V. Application of Time Series Models for Streamflow Forecasting. *Civ. Environ. Res.* **2017**, *9*, 56–63.
54. Hyndman, R.J.; Koehler, A.B. Another look at measures of forecast accuracy. *Int. J. Forecast.* **2006**, *22*, 679–688. [CrossRef]
55. Sedgwick, P. Pearson's correlation coefficient. *BMJ* **2012**, *345*, e4483. [CrossRef]
56. Puth, M.T.; Neuhäuser, M.; Ruxton, G.D. Effective use of Pearson's product-moment correlation coefficient. *Anim. Behav.* **2014**, *93*, 183–189. [CrossRef]
57. Hilker, T.; Natsagdorj, E.; Waring, R.H.; Lyapustin, A.; Wang, Y. Satellite observed widespread decline in Mongolian grasslands largely due to overgrazing. *Glob. Chang. Biol.* **2014**, *20*, 418–428. [CrossRef] [PubMed]

58. Sternberg, T. Investigating the presumed causal links between drought and dzud in Mongolia. *Nat. Hazards* **2018**, *92*, 27–43. [CrossRef]
59. Xia, L.; Song, X.; Leng, P.; Wang, Y.; Hao, Y.; Wang, Y. A comparison of two methods for estimating surface soil moisture based on the triangle model using optical/thermal infrared remote sensing over the source area of the Yellow River. *Int. J. Remote Sens.* **2019**, *40*, 2120–2137. [CrossRef]
60. Wang, Y.; Yang, J.; Chen, Y.; Wang, A.; de Maeyer, P. The Spatiotemporal Response of Soil Moisture to Precipitation and Temperature Changes in an Arid Region, China. *Remote Sens.* **2018**, *10*, 468. [CrossRef]

remote sensing

MDPI

Article

Improving Soil Moisture Estimation by Identification of NDVI Thresholds Optimization: An Application to the Chinese Loess Plateau

Lina Yuan [1], Long Li [1,2], Ting Zhang [1], Longqian Chen [1,*], Jianlin Zhao [3], Weiqiang Liu [4], Liang Cheng [1,5], Sai Hu [6], Longhua Yang [7] and Mingxin Wen [1]

1 School of Public Policy and Management, China University of Mining and Technology, Daxue Road 1, Xuzhou 221116, China; lnyuan@cumt.edu.cn (L.Y.); long.li@cumt.edu.cn or long.li@vub.be (L.L.); tingzhang@cumt.edu.cn (T.Z.); liang.cheng@cumt.edu.cn (L.C.); mingxin.wen@cumt.edu.cn (M.W.)
2 Department of Geography & Earth System Science, Vrije Universiteit Brussel, Pleinlaan 2, 1050 Brussels, Belgium
3 Department of Geology Engineering and Geomatics, Chang'an University, Yanta Road 120, Xi'an 710054, China; jianlin.zhao@chd.edu.cn
4 School of Environment and Spatial Informatics, China University of Mining and Technology, Daxue Road 1, Xuzhou 221116, China; weiqiang.liu@cumt.edu.cn
5 Henry Fok College of Biology and Agriculture, Shaoguan University, Daxue Road 26, Shaoguan 512005, China
6 School of Humanities and Law, Jiangsu Ocean University, Cangwu Road 59, Lianyungang 222005, China; saihu@jou.edu.cn
7 Department of Research and Development, Shanghai Gongjing Environmental Protection Co., Ltd., Yuanjiang Road 525, Shanghai 201100, China; yanglh3@163.com
* Correspondence: chenlq@cumt.edu.cn; Tel.: +86-516-8359-1327

Abstract: Accuracy soil moisture estimation at a relevant spatiotemporal scale is scarce but beneficial for understanding ecohydrological processes and improving weather forecasting and climate models, particularly in arid and semi arid regions like the Chinese Loess Plateau (CLP). This study proposed Criterion 2, a new method to improve relative soil moisture (RSM) estimation by identification of normalized difference vegetation index (NDVI) thresholds optimization based on our previously proposed iteration procedure of Criterion 1. Apparent thermal inertia (ATI) and temperature vegetation dryness index (TVDI) were applied to subregional RSM retrieval for the CLP throughout 2017. Three optimal NDVI thresholds ($NDVI_0$ was used for computing TVDI, and both $NDVI_{ATI}$ and $NDVI_{TVDI}$ for dividing the entire CLP) were firstly identified with the best validation results ($\overline{R}$) of subregions for 8-day periods. Then, we compared the selected optimal NDVI thresholds and estimated RSM with each criterion. Results show that NDVI thresholds were optimized to robust RSM estimation with Criterion 2, which characterized RSM variability better. The estimated RSM with Criterion 2 showed increased accuracy (maximum $\overline{R}$ of 0.82 ± 0.007 for Criterion 2 and of 0.75 ± 0.008 for Criterion 1) and spatiotemporal coverage (45 and 38 periods (8-day) of RSM maps and the total RSM area of 939.52×10^4 km^2 and 667.44×10^4 km^2 with Criterion 2 and Criterion 1, respectively) than with Criterion 1. Moreover, the additional NDVI thresholds we applied was another strategy to acquire wider coverage of RSM estimation. The improved RSM estimation with Criterion 2 could provide a basis for forecasting drought and precision irrigation management.

Keywords: MODIS; relative soil moisture; Chinese Loess Plateau; ATI; TVDI

Citation: Yuan, L.; Li, L.; Zhang, T.; Chen, L.; Zhao, J.; Liu, W.; Cheng, L.; Hu, S.; Yang, L.; Wen, M. Improving Soil Moisture Estimation by Identification of NDVI Thresholds Optimization: An Application to the Chinese Loess Plateau. *Remote Sens.* **2021**, *13*, 589. https://doi.org/10.3390/rs13040589

Academic Editor: Marion Pause
Received: 26 December 2020
Accepted: 4 February 2021
Published: 7 February 2021

1. Introduction

Accurate and timely soil moisture (SM) information has essential applications in different fields, such as flood/drought forecasting, climate and weather modeling, water resources, and agriculture management [1]. Proper water resource management is crucial in declining vulnerability to drought and other extreme events that may occur with increasing frequency because of climate change. This has been widely recognized in the arid and semi-arid regions of the Chinese Loess Plateau (CLP), which is particularly susceptible to soil erosion and water shortage due to intensive rainstorms, fractured and steep terrain, low vegetation cover, highly erodible loess soil, and a semiarid to arid climate [2–4].

Microwave sensors have been used for SM retrieval due to the direct relationship between microwave radiation and soil dielectric, though providing a coarse spatial resolution [5]. As for indirect approaches, SM estimation from visible and infrared data is based on land surface reflectance at much higher spatial resolutions [6]. To evaluate SM estimation, there are many studies on the comparison of SM products and modeled SM on different scales [7–12]. In order to explore the potential of optical and thermal remote sensing imagery for SM estimation, different indices (e.g., VCI—vegetation condition index, TVDI—temperature vegetation dryness index, TCI—temperature condition index, ATI—apparent thermal inertia) [13–17] and models (SVAT—soil vegetation atmosphere transfer, EF—evaporative fraction model) [18,19] have been applied to different climate conditions. The majority of previous studies derived SM from optical and infrared remote sensing imagery concentrating on vegetation-growing seasons [20–22]. In addition, existing SM estimation algorithms are not applicable to steep regions [23]. The continuous evaluation of SM throughout the year at a moderate 1-km resolution (compared with coarse resolution of a few tens of kilometers and fine resolution of the tens of meters [5,24,25]) over the CLP at a regional scale of 640,000 km^2 (local $\leq 10^4$ km^2, 10^4 km^2 < regional < 10^7 km^2, and global$\geq 10^7$ km^2 [26]) is, however, scarce.

Because geology and soil composition change only over a long period, short-term changes in thermal inertia (TI) can be associated with variations in SM [27]. Originally proposed by Price [28], ATI is a simplified calculation for TI and has been routinely used to estimate SM in bare soil and sparsely vegetated land [15,29,30]. For a densely vegetated surface, the triangular or trapezoidal NDVI (normalized difference vegetation index)—LST (land surface temperature) space, interpreted as TVDI, is widely used for SM estimation [31,32]. Moreover, a viable method for time series SM monitoring could overcome the limitations of a single method (PDI—perpendicular drought index or TVDI) [33], and the ATI and TVDI models were applied to retrieval the relative soil moisture (RSM) in Guangxi, south China [34].

RSM represents the percentage of SM that accounts for the moisture storage capacity and describes the SM levels in the present study. A detailed explanation regarding RSM is presented in Section 3.1.2. Yuan et al. [35] applied the MODIS-derived ATI and TVDI models to subregional RSM estimation for the CLP. They highlighted the identification of three optimal NDVI thresholds ($NDVI_0$ was for computing TVDI, both $NDVI_{ATI}$ and $NDVI_{TVDI}$ were used for dividing the whole CLP) and concluded that the ATI/TVDI joint models were more applicable (accounting for 36/38 8-day periods) and accurate ($\overline{R}$: 0.75 $\pm$ 0.008 on DOY—day of the year (hereafter referred to as DOY), 313) than the ATI-based and TVDI-based models. Here, the limiting condition ($NDVI_0 \leq NDVI_{ATI} \leq NDVI_{TVDI}$) when selecting optimal NDVI thresholds, as proposed by Yuan et al. [35], is regarded as Criterion 1. It failed to select the highest $\overline{R}$ in certain subregions, resulting in no corresponding subregional RSM maps being produced, thereby causing incomplete RSM maps for 8-day periods. To produce more complete RSM maps, additional criteria (Criterion 2) for determining optimal NDVI thresholds should be tested. Importantly, because the $NDVI_0$ and $NDVI_{ATI}$ thresholds with Criterion 2 do not influence each other, highest $\overline{R}$ for an individual subregion could have more opportunities to be selected. In this case, we could obtain more subregional RSM maps to produce more complete RSM maps.

The main objective of this study is to improve RSM estimation by applying Criterion 2 ($NDVI_{ATI} < NDVI_{TVDI}$) for optimizing the identification of NDVI thresholds. Three optimal NDVI thresholds were firstly identified with Criterion 2 for each 8-day period and 8-day RSM maps were generated using selected optimal NDVI thresholds. Then, in order to evaluate the accuracy and coverage of RSM estimation, we compared the selected optimal NDVI thresholds and estimated RSM with each criterion. Monthly, seasonal, and yearly RSM maps of the CLP in 2017 were produced via the 8-day RSM maps and examined lastly.

2. Study Area and Data

2.1. Study Area

The CLP (Chinese Loess Plateau) is located at 100°54′–114°33′E and 33°43′–41°16′N and has a total area of 64,000 km^2 covering seven northern Chinese provinces (Figure 1a). The study area has an arid to semi-arid temperate climate with a wet monsoon season. Mean annual precipitation is 420 mm and ranges from 200 mm in the northeast to 750 mm in the southeast, around 55–78% of which concentrates in the wet season from July to September [36,37]. It is considered one of the most seriously eroded landscapes in the world owing to its loose and erodible soil [4,38]. With various land cover types though, this region is mostly covered by grasslands and croplands (Figure 1b). There are 213 Chinese automatic soil moisture observation stations (CASMOSs) over the CLP except in Ningxia (Figure 1b).

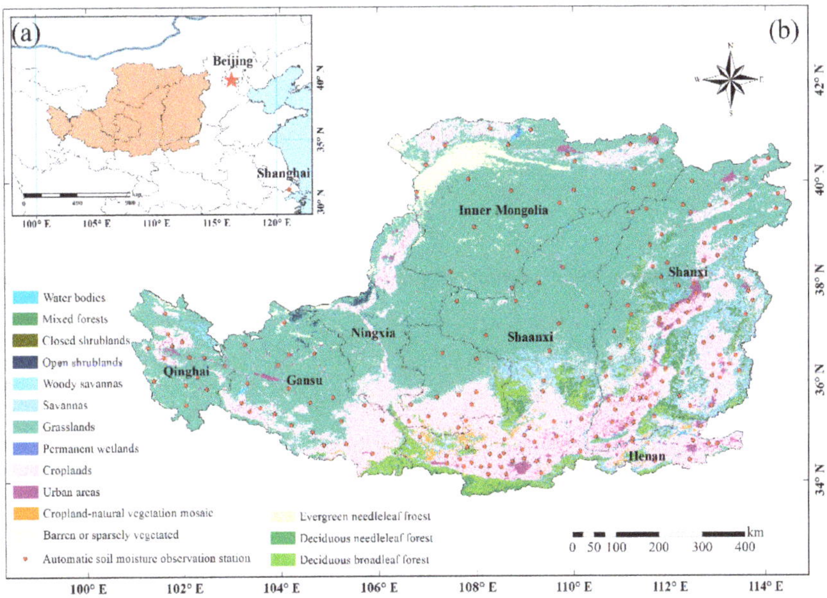

Figure 1. Study area: (**a**) Location of the Chinese Loess Plateau (CLP) in China; (**b**) spatial distribution of 213 automatic soil moisture observation stations used in the study. The background image shows the MODIS (MCD12Q1 Type2—the University of Maryland (UMD) land cover classification scheme) land cover product over the CLP. There are 16 different land cover types in the UMD land cover classification scheme and only 15 land cover types except the land cover of evergreen broadleaf forest over the CLP.

2.2. Satellite Data and Image Pre-Processing

The MODIS data used in this study are composed of MODIS/Terra 500-m resolution 8-day surface reflectance (MOD09A1), and MODIS/Terra 1-km resolution 8-day LST products (MOD11A2) in 2017 for calculating ATI, NDVI, and TVDI. Moreover, the MOD09GA 500-m daily products were used to extract the acquisition time, serving for collecting the corresponding time of in situ RSM observations. We also used the 1 km for 2017 MCD12Q1. Type2 land cover product from the University of Maryland (UMD) land cover classification scheme (15 different land cover types over the CLP in Figure 1b) masked water bodies. Here, the mask of water bodies we used covered the water bodies' extent derived from the normalized difference water index (NDWI) for each 8-day period. The NDWI was first proposed by McFeeters in 1996 to monitor changes related to water content in water bodies [39]. To cover the CLP, five granules of MODIS data were mosaicked, re-projected, and re-sampled at the resolution of 1/224° (~500 m) using the MODIS Reprojection Tool and were clipped in ArcGIS 10.2 (ESRI Inc., Redlands, CA, USA).

2.3. In Situ Observation Data

Hourly RSM—relative soil moisture (%) of the 20-cm soil layer provided by the 213 automatic soil moisture observation stations were used in this study. The number of in situ observations used for each 8-day period in this study is given in Section 4.1. To narrow the temporal gaps between the in situ observation data and the 8-day composite products, the 8-day average in situ RSM value at each station was calculated by averaging the corresponding daily RSM. In detail, the daily granule acquisition time of the MOD09GA products (from the beginning to ending date-time) were collected first to serve as the reference for selecting corresponding in situ RSM observations. Precipitation (in mm/h) and elevation data of these stations were provided by the China Meteorological Data Service Center (http://data.cma.cn/en).

3. Principles and Methods

3.1. Subregional RSM Estimation

Subregional RSM estimation was applied with the two criteria. Here, we took Criterion 2 as an example (Figure 2) and a detailed comparison of the two criteria is shown in Section 3.2.2. During the early stages of crop growth, the monitoring accuracy of ATI—apparent thermal inertia (defined in Section 3.1.1) is better than that of TVDI—temperature vegetation dryness index (defined in Section 3.1.2). However, as crop growth progresses, the advantages of TVDI become evident. The average of the ATI and TVDI value, ranging from 0 to 1, was calculated, where NDVI varied from $NDVI_{ATI}$ and $NDVI_{TVDI}$ in the ATI/TVDI subregion. The idea of averaging ATI and TVDI as an assigned value in the ATI/TVDI subregion was inspired by previous studies [33,34]. A model-level integrated approach was used to effectively retrieve regional-scale daily SM. The average value of SM from the ATI-based model and the TVDI-based model was regarded as the SM when NDVI ranged from 0.10 to 0.18 [40]. Similarly, SM was obtained by averaging the PDI-based model and TVDI-based model [33].

To estimate RSM, the whole CLP was divided into three subregions (the ATI subregion, the TVDI subregion, and the ATI/TVDI subregion) according to the NDVI of individual pixels. While the ATI-based models (ATI) were merely applied to the ATI subregion ($NDVI \leq NDVI_{ATI}$), the TVDI-based models (TVDI) were merely applied to the TVDI subregion ($NDVI > NDVI_{TVDI}$). Then, the ATI-based models and the TVDI-based models together (the ATI/TVDI joint models—the average of ATI and TVDI) were applied to the ATI/TVDI subregion ($NDVI_{ATI} < NDVI \leq NDVI_{TVDI}$). Here, as mentioned before, the ATI-based model is routinely used to estimate the RSM of bare soil and sparsely vegetated areas with low NDVI and the TVDI-based model is more suitable for dense vegetation coverage with high NDVI. Thus, the case of $NDVI_{ATI}$ greater than $NDVI_{TVDI}$ is not considered in an individual subregion.

The RSM estimation using the ATI-based, the TVDI-based, and the ATI/TVDI joint models should contain two key procedures. One is the iteration procedure applying three NDVI thresholds to calculate $\overline{R}$. The other is the identification of optimal NDVI thresholds for subregional RSM estimation. In this study, we proposed a new method to select optimal NDVI thresholds for improving RSM estimation involving increased accuracy and spatiotemporal coverage, which was defined as Criterion 2. Both Criterion 1 and Criterion 2 shared similar iteration procedures (with different maximum iterations) to obtain $\overline{R}$.

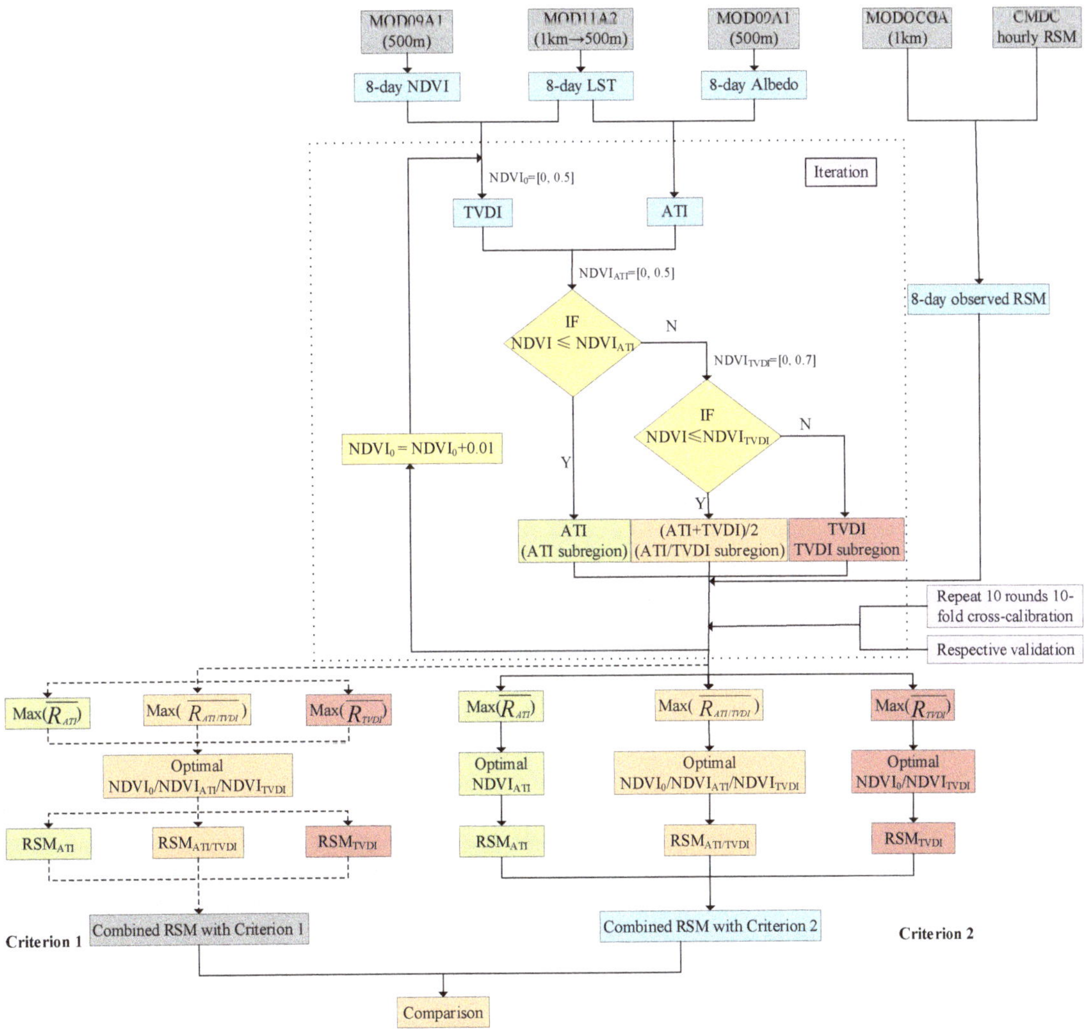

Figure 2. Flowchart of relative soil moisture (RSM) estimation by the apparent thermal inertia (ATI)-based models, the ATI/TVDI—temperature vegetation dryness index joint models, and the TVDI-based models with Criterion 2 (adapted from [35]). $NDVI_0$ was used for computing TVDI, and both $NDVI_{ATI}$ and $NDVI_{TVDI}$ were applied for dividing the entire Chinese Loess Plateau (CLP). The three subregions, namely the ATI subregion ($NDVI \leq NDVI_{ATI}$), the ATI/TVDI subregion ($NDVI_{ATI} < NDVI \leq NDVI_{TVDI}$), and the TVDI subregion ($NDVI > NDVI_{TVDI}$), were assigned by calculated ATI, the average of ATI and TVDI, and TVDI, respectively.

In addition, the NDVI threshold ($NDVI_0$), the lower limit of NDVI, below which the data are excluded when we derive dry/wet edges, would be used for generating TVDI [35]. To obtain optimal NDVI thresholds, $NDVI_0$ and $NDVI_{ATI}$, both ranging from 0 to 0.5, and $NDVI_{TVDI}$, ranging from 0 to 0.7 with an interval of 0.01, were successively tested in the iterative process (Figure 2). We carried out NDVI ranges of the iteration in the programing design used for loops statement and it was easy to run programming using the fixed large range for each 8-day period in 2017. For one iteration, linear relation analysis between an assigned value (e.g., ATI) and the corresponding in situ RSM observations was performed through the 10-fold cross-calibration in the calibration process. Detailed calibration and validation processes are presented in Section 3.2.2. After the completion of all iterations, three groups of optimal NDVI thresholds with a maximum $\overline{R}$ in validation were selected with Criterion 2 for one 8-day period rather than one group with Criterion 1. The overall 8-day RSM map was ultimately produced with the generated subregional RSM by applying the selected optimal thresholds. The comparison between the two criteria was conducted as follows.

3.1.1. Apparent Thermal Inertia (ATI)

Soil TI—thermal inertia is a thermal property of soil and describes the resistance of soil to temperature variations. TI proportionally increases as SM increases because moist soil has higher water thermal conductivity and heat capacity, thereby exhibiting a lower diurnal temperature fluctuation [27,29]. ATI—apparent thermal inertia is a simplified calculation for TI [28,41]. As one of the SM estimation methods, ATI can be easily calculated by measuring surface albedo and the diurnal temperature range as follows:

$$\mathrm{ATI} = \frac{1 - \mathrm{A}}{\Delta \mathrm{LST}} \tag{1}$$

where ATI is the apparent thermal inertia [K^{-1}], A is the broadband surface albedo, and ΔLST corresponds to the diurnal temperature range between day and night [K]. Theoretically, the MODIS-derived ATI was computed as the ratio of the daily surface albedo and the diurnal temperature range [42]. However, the availability of the MODIS-derived ATI on some days was mainly limited by the availability of satellite LST observations because LST was often absent due to the presence of cloud during the satellite overpass times. Thus, it was difficult to obtain a continuous time series of ATI. In detail, the merged Terra 8-day surface reflectance and temperature products (MOD09A1/MOD11A2) are distributed on 8-day synthesis periods of clear sky data accumulation and each 8-day composite pixel contains the best possible observation according to specified criteria [43]. In addition, previous studies applied MOD11A2 LST products and MOD09A1 surface reflectance data (time resolution of 8 days) to Equation (1) [15,34,44]. In this case, our study was also carried out for an 8-day temporal resolution to calculate ATI. ATI varies between 0 and 1. The broadband albedo can be computed in shortwave spectral ranges from Terra MODIS surface reflectance [45]:

$$\mathrm{A} = 0.16\mathrm{b}_1 + 0.291\mathrm{b}_2 + 0.243\mathrm{b}_3 + 0.11\mathrm{b}_4 + 0.112\mathrm{b}_5 + 0.081\mathrm{b}_7 - 0.0015 \tag{2}$$

where b_1–b_5 and b_7 are the reflectance of bands 1, 2, 3, 4, 5, and 7 of MODIS, respectively [15,42,46,47]. MODIS's six bands are excellent in making the broadband albedo conversions under the general atmospheric conditions [45].

3.1.2. Temperature Vegetation Dryness Index (TVDI)

As the scatter plot of NDVI against LST forms a triangle or trapezium (hereinafter called triangle), Sandholt et al. [48] proposed the concept of TVDI, which can represent RSM, which is formulated as follows:

$$\mathrm{TVDI} = \frac{\mathrm{LST} - \mathrm{LST_{min}}}{\mathrm{LST_{max}} - \mathrm{LST_{min}}} \tag{3}$$

$$\mathrm{LST_{max}} = \mathrm{a_{dry}} \times \mathrm{NDVI} + \mathrm{b_{dry}} \tag{4}$$

$$\mathrm{LST_{min}} = \mathrm{a_{wet}} \times \mathrm{NDVI} + \mathrm{b_{wet}} \tag{5}$$

where LST represents the MODIS-derived LST in each of the pixels, LST_{min} and LST_{max} refer to the minimum/maximum LST in the triangle space defining the wet/dry edge at a given NDVI, respectively. a_{wet}, a_{dry}, b_{wet}, and b_{dry} are the linear regression parameters (slope and intercept) of dry/wet edges, respectively. Based on TVDI, RSM can be related to LST_{min} and LST_{max} with the following equation [49,50]:

$$\frac{RSM_w - RSM}{RSM_w - RSM_d} = \frac{LST - LST_{min}}{LST_{max} - LST_{min}} \tag{6}$$

From Equation (6) RSM can be found as:

$$RSM = RSM_w - \frac{LST - LST_{min}}{LST_{max} - LST_{min}}(RSM_w - RSM_d) \tag{7}$$

$$RSM = RSM_w - TVDI(RSM_w - RSM_d) \tag{8}$$

where RSM is the relative soil moisture at any given pixel, RSM_w is the maximum RSM according to wet edge, and RSM_d is the minimum RSM corresponding to dry edge. The trend line of LST_{max} gives the dry edge and that of the LST_{min} represents the wet edge. The NDVI-LST triangle space defining the dry/wet edges on DOY—day of the year 113 over the CLP is shown in Figure 3.

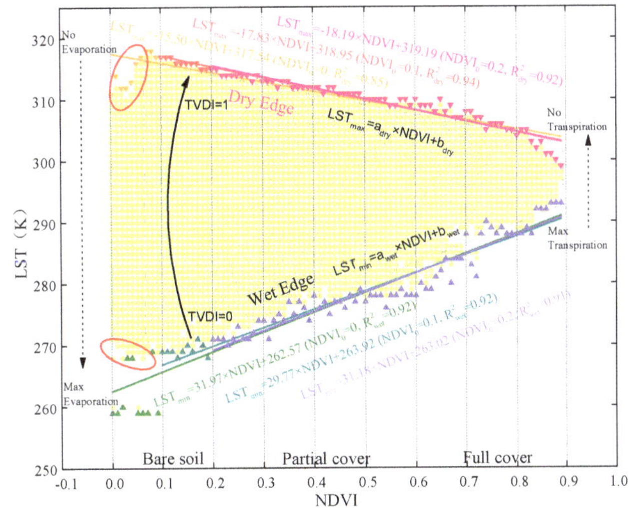

Figure 3. The NDVI-LST triangle space defining the dry/wet edges on DOY—day of the year 113 over the CLP (adapted from [35,48,51]). Theoretically, in the triangular figure (pink region area), the base edge of the triangle with maximum evapotranspiration pixels, and the top edge of the triangle with zero evapotranspiration pixels are displayed. As the NDVI increases, the maximum LST decreases and can be fitted to a negative slope using the least square method, which is defined as the dry edge in red color lines (LST_{max}). The base line of the triangle represents the wet edge in blue color lines (LST_{min}), which is calculated by averaging a group of points in the lower limits of the scatterplot. The TVDI increases from 0 to 1 (a black arrow going from TVDI = 0 to TVDI = 1), indicating a land surface change from extreme wetness to extreme drought. Linear equations were generated when $NDVI_0$ equals 0, 0.1, and 0.2, respectively. The linear regression coefficients (slopes, intercepts, and R^2) of dry/wet edges varied with different $NDVI_0$.

Generally, the TVDI value of the dry edge, referring to the driest region of the study area, equals 1, while that of the wet edge (being the most humid region) is close to 0 [52]. LST linearly changes with NDVI in the conditions of the same RSM. Between two edges (dry/wet edges), all intermediate conditions can occur, and all RSM conditions can consequently be represented within the NDVI-LST triangle space [31,53]. However, the maximum and minimum LST at lower NDVI in the scatterplot (two red circles in Figure 3) do not seem to contribute to the formation of the dry/wet edges, which has been noted by previous studies [54–56]. Thus, in our study, the $NDVI_0$ is the low limit of NDVI, below which the data are excluded when we derive dry/wet edges. The TVDI ($NDVI < NDVI_0$) with Criterion 2 was calculated using the linear regression parameters of derived dry/wet edges, and the TVDI ($NDVI < NDVI_0$) was not computed because of setting $NDVI_0$ smaller than $NDVI_{ATI}$ for Criterion 1. In other words, TVDI would only be used both in the ATI/TVDI subregion and the TVDI subregion, where NDVI was higher than $NDVI_{ATI}$, not to mention $NDVI_0$.

3.1.3. RSM Estimation with Criterion 2

The overall RSM was produced with three groups of selected optimal NDVI thresholds (Criterion 2) using MODIS-derived ATI, TVDI, and the mean of ATI and TVDI against in situ RSM observations for one 8-day period. The equations were used as follows:

$$RSM_{overall} = \begin{cases} RSM_{ATI} = a_{ATI} \times ATI + b_{ATI} & NDVI \in [0, NDVI_{ATI}] \\ RSM_{ATI/TVDI} = a_{ATI/TVDI} \times \frac{ATI+TVDI}{2} + b_{ATI/TVDI} & NDVI \in (NDVI_{ATI}, NDVI_{TVDI}] \\ RSM_{TVDI} = a_{TVDI} \times TVDI + b_{TVDI} & NDVI \in (NDVI_{TVDI}, 1] \end{cases} \quad (9)$$

where $RSM_{overall}$ represents the overall RSM and it is combined by three subregional RSM (RSM_{ATI}, $RSM_{ATI/TVDI}$, and RSM_{TVDI}). RSM_{ATI} and RSM_{TVDI} are the RSM estimated by the ATI-based and TVDI-based models, respectively, and $RSM_{ATI/TVDI}$ is the RSM estimated by the ATI/TVDI joint model. a_{ATI} and b_{ATI} are coefficients from fitting the ATI values and in situ RSM observations in the ATI subregion. a_{TVDI} and b_{TVDI} are coefficients from fitting the TVDI values and in situ RSM observations in the TVDI subregion. $a_{ATI/TVDI}$ and $b_{ATI/TVDI}$ are coefficients from fitting the average value of ATI and TVDI and in situ RSM observations in the ATI/TVDI subregion. $NDVI_{ATI}$ and $NDVI_{TVDI}$ are the selected optimal thresholds for generating subregions.

3.2. *Comparison of the Two Criteria*

3.2.1. Calibration and Validation Processes for the Two Criteria

The two criteria differ in how the optimal NDVI thresholds are identified, namely $NDVI_0 \leq NDVI_{ATI} \leq NDVI_{TVDI}$ for Criterion 1 and $NDVI_{ATI} < NDVI_{TVDI}$ for Criterion 2. In other words, the limit condition of Criterion 1 was stricter than Criterion 2 and the value of $NDVI_0$ might be greater/lower than $NDVI_{ATI}$ for Criterion 2. As we mentioned in Section 3.1.2., dry/wet edges were calculated when NDVI was greater than $NDVI_0$. In this case, the value of TVDI ($NDVI < NDVI_0$) was calculated based on the dry/wet edge derived from NDVI greater than $NDVI_0$. The size relationship between $NDVI_0$ and $NDVI_{ATI}$ is the main difference of the two criteria, which directly lead to a different assigned value in the ATI/TVDI subregion, thereby affecting the calibration process when using the ATI/TVDI joint models. The detailed differences between subregional RSM estimation with each criterion are represented in Figures 4 and 5. For Criterion 2, the average of ATI and TVDI was assigned to the pixels in the ATI/TVDI subregion, in which TVDI includes two intervals ($NDVI_{ATI} < NDVI < NDVI_0$ and $NDVI_0 \leq NDVI \leq NDVI_{TVDI}$) with Criterion 2 (Figure 5). TVDI with Criterion 1 is from only one interval, namely $NDVI_{ATI} < NDVI \leq NDVI_{TVDI}$ in the ATI/TVDI subregion from Figure 4.

Importantly, because optimal NDVI thresholds were identified based on the $\overline{R}$ values in the validation, both criteria performed the iteration procedure (calibration and validation processes) in three subregions (the ATI subregion, the TVDI subregion, and the ATI/TVDI subregion) to calculate $\overline{R}$ for 8-day periods (procedures with a purple background in Figure 2). In order to not limit the size of $NDVI_0$ and $NDVI_{ATI}$ in the iteration procedure, more iterations would be implemented with Criterion 2 (maximum of 97,546 iterations) than that of Criterion 1 (maximum of 48,620 iterations) [35]. The number of iterations was calculated by all combinations of the three thresholds.

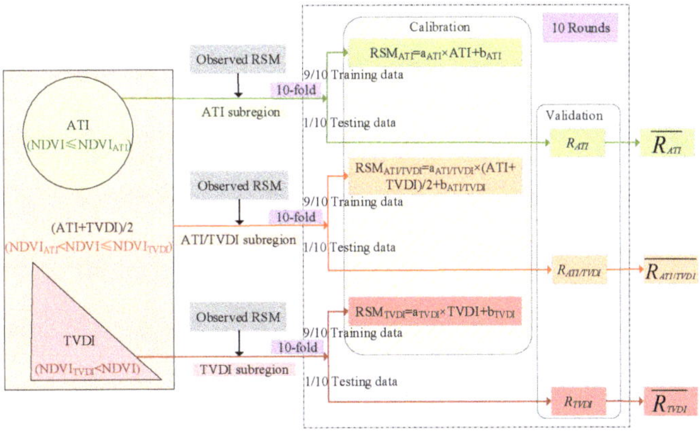

Figure 4. Schematic diagram of the calibration and validation processes with Criterion 1 ($NDVI_0 \leq NDVI_{ATI} \leq NDVI_{TVDI}$). $NDVI_{ATI}$ and $NDVI_{TVDI}$ were applied to divide the whole CLP. $NDVI_0$ was lower than $NDVI_{ATI}$ and merely used for calculating TVDI (not shown in the figure). ATI, the average of ATI and TVDI, and TVDI were assigned to the ATI subregion ($NDVI \leq NDVI_{ATI}$), the ATI/TVDI subregion ($NDVI_{ATI} < NDVI \leq NDVI_{TVDI}$), and the TVDI subregion ($NDVI > NDVI_{TVDI}$), respectively. RSM_{ATI} and RSM_{TVDI} were the RSM estimated by the ATI-based and TVDI-based models, respectively, and $RSM_{ATI/TVDI}$ was the RSM estimated by the ATI/TVDI joint model. a_{ATI}, b_{ATI}, a_{TVDI}, and b_{TVDI} in the equations were coefficients from fitting the ATI values and TVDI values with in situ RSM observations in the ATI subregion and the TVDI subregion, respectively. $a_{ATI/TVDI}$ and $b_{ATI/TVDI}$ were coefficients from fitting the average value of ATI and TVDI and in situ RSM observations in the ATI/TVDI subregion. After 10 rounds 10-fold cross-calibration, $\overline{R_{ATI}}$, $\overline{R_{ATI/TVDI}}$, and $\overline{R_{TVDI}}$ in validation were calculated by averaging R_{ATI}, $R_{ATI/TVDI}$, and R_{TVDI}, respectively.

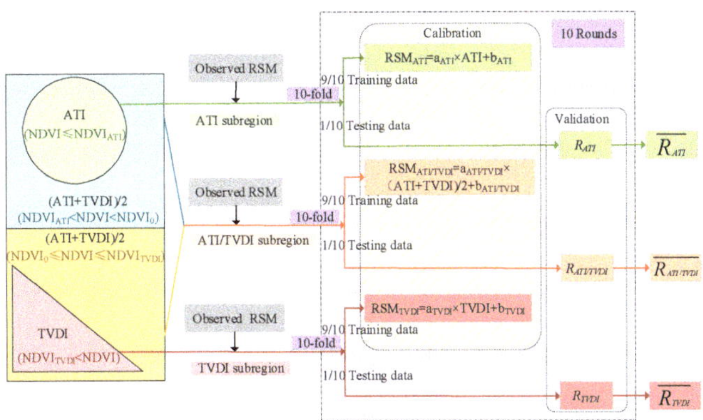

Figure 5. Schematic diagram of the calibration and validation processes with Criterion 2 ($NDVI_{ATI} < NDVI_{TVDI}$). $NDVI_{ATI}$ and $NDVI_{TVDI}$ were applied to divide the whole CLP. $NDVI_0$ might be lower/greater than $NDVI_{ATI}$ and appeared when $NDVI_0$ was greater than $NDVI_{ATI}$, indicating the calculated TVDI from two NDVI intervals (blue and light orange color regions) in the ATI/TVDI subregion. The value of TVDI with Criterion 2 in the blue color regions ($NDVI < NDVI_0$) was calculated based on the dry/wet edge derived from NDVI greater than $NDVI_0$. The computed TVDI with Criterion 2 in the light orange color regions ($NDVI \geq NDVI_0$) was the same as that of Criterion 1 (orange color regions in Figure 4). For the meaning of the other parameters (RSM_{ATI}, RSM_{TVDI}, $RSM_{ATI/TVDI}$, R_{ATI}, $R_{ATI/TVDI}$, R_{TVDI}, $\overline{R_{ATI}}$, $\overline{R_{ATI/TVDI}}$, and $\overline{R_{TVDI}}$), please refer to the caption of Figure 4.

For each subregion, to decrease the variability of the calibration, the linear fit between the assigned value (ATI, the average of ATI and TVDI, and TVDI, respectively) and observed RSM was performed using 10 rounds of 10-fold cross-calibration. To be more specific, after the random splitting of the paired assigned values and in situ RSM observations (when the number of available RSM observation stations was greater than 20) into 10 subsamples with nine as training data and one as testing data, acquired regression parameters (slopes and intercepts) in training data were applied to testing data. Then, one group of estimated RSM and in situ RSM observations was generated. After the 10th fold was accomplished, the validation data of 10 groups of estimated RSM and in situ RSM observations were computed, including R, as well as the *p*-value for a significance test (p-value < 0.05), which was just called "one-round". After 10 rounds of iterations, we averaged the R ($\overline{R}$) from the 10 rounds of validation as the reference to identify corresponding optimal NDVI thresholds.

3.2.2. Evaluation of Estimated RSM for the Two Criteria

The threshold qualification when identifying optimal NDVI thresholds for the two criteria was different, but both were based on the averaged validation results ($\overline{R}$), which directly represents the accuracy of RSM estimation. For an individual subregion, only the maximum $\overline{R}$ greater than 0.23 for Criterion 2 and slightly lower (0.17) for Criterion 1 should be selected; otherwise, no optimal NDVI thresholds were chosen for that subregion. For Criterion 1, the NDVI thresholds corresponding to maximum $\overline{R}$ among the three subregions were chosen as the optimal thresholds when the $NDVI_0$, $NDVI_{ATI}$, and $NDVI_{TVDI}$ thresholds existed simultaneously for one 8-day period [35]. In this way, they could not guarantee that maximum $\overline{R}$ would be reached for all the three subregions and the generated three subregions according to $NDVI_{ATI}$ and $NDVI_{TVDI}$ always excluding each other. Thus, only one group of optimal NDVI thresholds were chosen for subregional RSM estimation and then the overall RSM with Criterion 1 was combined by subregional RSM for each 8-day period [35].

However, for Criterion 2, more selected processes for optimal NDVI thresholds were carried out. The NDVI thresholds corresponding to maximum $\overline{R}$ in each subregion, namely the highest $\overline{R_{ATI}}$, $\overline{R_{ATI/TVDI}}$, and $\overline{R_{TVDI}}$, were chosen as the optimal thresholds. The optimal NDVI thresholds of the three subregions were not influenced by each other. Thus, three groups of optimal NDVI thresholds regarding three subregions were selected for subregional RSM estimation and then the overall RSM with Criterion 2 was combined by that subregional RSM for each 8-day period. In this case, the three subregions for one 8-day period might be overlapped due to the three groups of optimal NDVI thresholds individually selected instead of always excluding each other for Criterion 1. Based on the relationships of the selected optimal NDVI thresholds of the subregions, we list all cases in terms of different combinations of models used to produce an overall RSM map for one 8-day period (Table 1). Overlaps might exist when combining different models for RSM estimation for one 8-day period (see cases 4, 5, 6, and 7 in Table 1).

We combined the subregional RSM generated using the corresponding models with the selected optimal NDVI thresholds to obtain an overall RSM map for one 8-day period. Then, we evaluated and compared the selected optimal NDVI thresholds with each criterion in terms of the validation outcomes (Figure 6). Furthermore, the performance of the selected optimal NDVI thresholds for RSM retrieval, involving accuracy and spatiotemporal coverage using the two criteria, was compared. Monthly, seasonal, and yearly RSM maps were generated by combining 8-day RSM maps under Criterion 2 and were eventually examined.

Table 1. Cases of the model used with the relationships of selected optimal NDVI thresholds with Criterion 2.

Cases	Model Used [1]	Subregions	Optimal NDVI Thresholds [2]	Relationships [3]	If Overlaps
1	ATI	ATI	$NDVI_{ATI}$	-	No
2	ATI/TVDI	ATI/TVDI	$NDVI_0/NDVI_{ATI}/NDVI_{TVDI}$	$NDVI_{ATI} < NDVI_{TVDI}$	No
3	TVDI	TVDI	$NDVI_0/NDVI_{TVDI}$	-	No
4	ATI	ATI	$NDVI_{ATI}$	$NDVI_{ATI} < NDVI_{TVDI}$	No
	TVDI	TVDI	$NDVI_0/NDVI_{TVDI}$	$NDVI_{ATI} \geq NDVI_{TVDI}$	Yes
5	ATI	ATI	$NDVI_{ATI\_a}$	$NDVI_{ATI\_a} < NDVI_{ATI\_j}$	No
	ATI/TVDI	ATI/TVDI	$NDVI_0/NDVI_{ATI\_j}/NDVI_{TVDI}$	$NDVI_{ATI\_a} \geq NDVI_{ATI\_j}$	Yes
6	ATI/TVDI	ATI/TVDI	$NDVI_{0\_j}/NDVI_{ATI}/NDVI_{TVDI\_j}$	$NDVI_{TVDI\_j} < NDVI_{TVDI\_t}$	No
	TVDI	TVDI	$NDVI_{0\_t}/NDVI_{TVDI\_t}$	$NDVI_{TVDI\_j} \geq NDVI_{TVDI\_t}$	Yes
7	ATI ATI/TVDI TVDI	ATI ATI/TVDI TVDI	$NDVI_{ATI\_a}$ $NDVI_{0\_j}/NDVI_{ATI\_j}/NDVI_{TVDI\_j}$ $NDVI_{0\_t}/NDVI_{TVDI\_t}$	$NDVI_{ATI\_a} < NDVI_{ATI\_j}$ & $NDVI_{TVDI\_j} < NDVI_{TVDI\_t}$	No
				$NDVI_{TVDI\_t} \geq NDVI_{ATI\_a} \geq NDVI_{ATI\_j}$	Yes
				$NDVI_{TVDI\_j} \geq NDVI_{TVDI\_t} \geq NDVI_{ATI\_a}$	Yes
				$NDVI_{ATI\_a} \geq NDVI_{TVDI\_t}$	Yes

[1] One type model is considered to be used once in one case. [2] The subscripts a in the $NDVI_{ATI\_a}$ and j in the $NDVI_{ATI\_j}$ represent the selected optimal $NDVI_{ATI}$ thresholds in the ATI subregion and the ATI/TVDI subregion, respectively. The subscripts j in $NDVI_{0\_j}$ and t in $NDVI_{0\_t}$ mean the selected optimal $NDVI_0$ in the ATI/TVDI subregion and the TVDI subregion, respectively. The subscripts j in $NDVI_{TVDI\_j}$ and t in $NDVI_{TVDI\_t}$ refer to the selected optimal $NDVI_{TVDI}$ in the ATI/TVDI subregion and the TVDI subregion, respectively. [3] $NDVI_{ATI} < NDVI_{TVDI}$ for an individual subregion because the ATI-based model was suitable for regions with low NVDI and the TVDI-based model was applicable for regions with high NDVI [15,57].

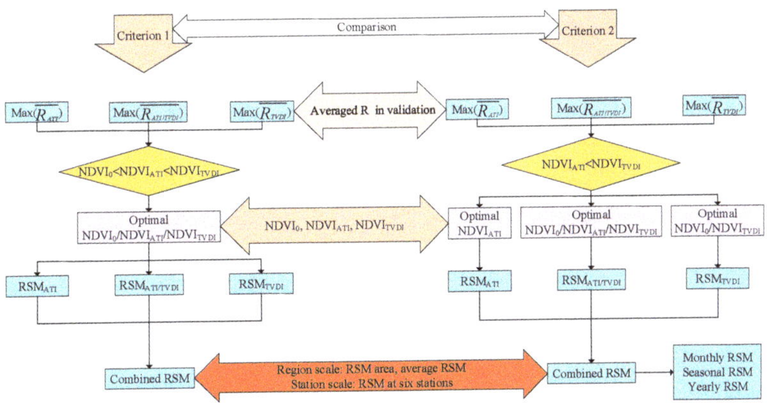

Figure 6. Flowchart for comparison of the two criteria.

4. Results and Discussion

4.1. Evaluation of the Optimal NDVI Thresholds

4.1.1. Comparison of Validation Results

The measure $\overline{R}$ in validation reflects the accuracy of the estimated RSM compared to the observed RSM. Lower $\overline{R}$ represents poorer estimated RSM. Optimal NDVI thresholds were identified corresponding to the highest $\overline{R}$ in subregions. Then, RSM maps could be generated using the selected optimal NDVI thresholds for each 8-day period. Eventually, we selected 45 8-day periods (out of 46 except for DOY 73 in 2017) with Criterion 2 and only 38 8-day periods with Criterion 1 (Figure 7) [35]. Thus, 45 8-day periods of estimated RSM maps (improved 7 8-day periods compared to with Criterion 1) would be generated with Criterion 2. The $\overline{R}$ values with Criterion 2 were slightly higher than those with Criterion 1 but roughly showed a similar trend throughout the year. These values fluctuated irregularly with peak values (0.82 ± 0.007 with Criterion 2 and 0.75 ± 0.008 with Criterion 1, respectively) in winter [35]. In this case, the accuracy of RSM estimation with Criterion 2 was improved.

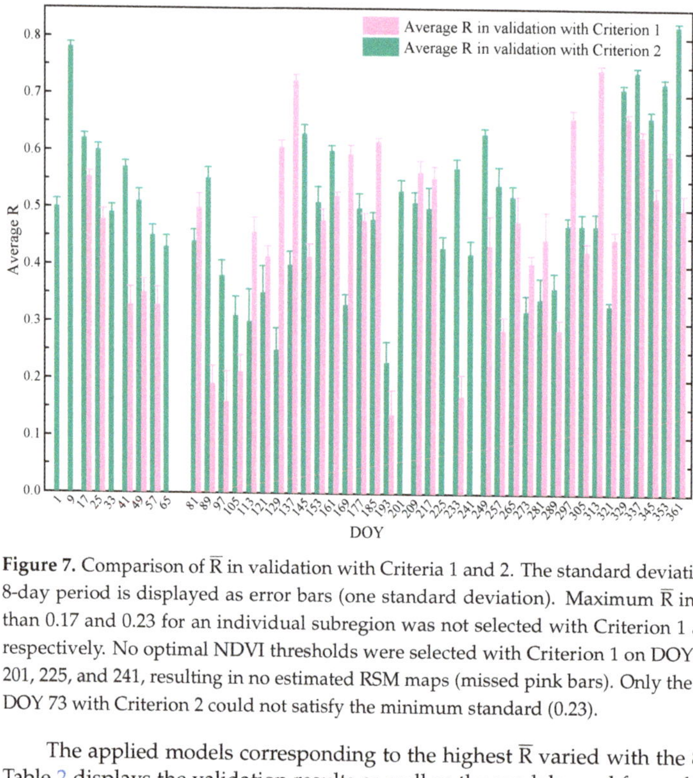

Figure 7. Comparison of $\overline{R}$ in validation with Criteria 1 and 2. The standard deviation of $\overline{R}$ for each 8-day period is displayed as error bars (one standard deviation). Maximum $\overline{R}$ in validation less than 0.17 and 0.23 for an individual subregion was not selected with Criterion 1 and Criterion 2, respectively. No optimal NDVI thresholds were selected with Criterion 1 on DOYs 1, 9, 33, 65, 73, 201, 225, and 241, resulting in no estimated RSM maps (missed pink bars). Only the 8-day period of DOY 73 with Criterion 2 could not satisfy the minimum standard (0.23).

The applied models corresponding to the highest $\overline{R}$ varied with the 8-day periods. Table 2 displays the validation results as well as the models used for each 8-day period with Criterion 2. The seven rows in blue (DOYs 1, 9, 33, 65, 201, 225, and 241) indicate the new validation results with Criterion 2 for RSM retrieval compared to Criterion 1. The ATI/TVDI joint models were used in almost all 8-day periods (34/45 8-day periods), which was well in line with Criterion 1 [35]. The highest $\overline{R}$ (0.82 ± 0.007 on DOY 361) with Criterion 2 was higher than that of Criterion 1 (0.75 ± 0.008 on DOY 313). The validation score of $\overline{R}$ was better than previous results in the study area [35].

For DOYs 145 and 305, the ATI/TVDI joint models were applied twice in the two complementary ATI/TVDI subregions with different NDVI ranges ($0.00 < \text{NDVI} \leq 0.20$ with $\overline{R}$ of 0.61 ± 0.044 and $0.25 < \text{NDVI} \leq 0.27$ with $\overline{R}$ of 0.63 ± 0.017 on DOY 145, and $0.13 < \text{NDVI} \leq 0.17$ with $\overline{R}$ of 0.47 ± 0.021 and $0.30 < \text{NDVI} \leq 0.34$ with $\overline{R}$ of 0.44 ± 0.020 on DOY 305, respectively). In the ATI/TVDI subregions, different NDVI thresholds ($NDVI_{ATI}$ and $NDVI_{TVDI}$) means the generated RSM maps with these NDVI intervals. Here, higher $\overline{R}$, not merely the highest $\overline{R}$, could be selected as well and their corresponding thresholds were regarded as additional thresholds for RSM retrieval. In this context, the two complementary ATI/TVDI subregional RSM could be merged to produce the overall RSM map.

Thus, the more additional thresholds we selected, the more completed the obtained RSM map would be. The additional NDVI thresholds we applied, here, was another improved strategy to acquire wider coverage of RSM estimation. The number of stations used with each criterion was the same for each 8-day period.

Table 2. Validation results for selecting optimal NDVI thresholds with Criterion 2 (blue color rows represent new 8-day periods of validation results compared to Criterion 1).

DOY	Month/Season	No. of Stations	Models	$\bar{R} \pm$ Std	Optimal $NDVI_0$	Optimal $NDVI_{ATI}$	Optimal $NDVI_{TVDI}$
1	Jan/Winter	72	ATI/TVDI	0.50 ± 0.015	0.01	0.19	0.45
9	Jan/Winter	69	ATI/TVDI	0.78 ± 0.009	0.05	0.19	0.54
17	Jan/Winter	75	ATI/TVDI	0.62 ± 0.017	0.34	0.22	0.55
			TVDI	0.56 ± 0.010	0.00		0.12
25	Jan/Winter	74	ATI/TVDI	0.60 ± 0.012	0.09	0.15	0.44
			TVDI	0.49 ± 0.015	0.08		0.14
			ATI	0.40 ± 0.021		0.16	
			ATI/TVDI	0.78 ± 0.009	0.05	0.19	0.54
41	Feb/Winter	86	ATI/TVDI	0.57 ± 0.012	0.10	0.20	0.35
49	Feb/Winter	101	ATI/TVDI	0.51 ± 0.022	0.03	0.18	0.24
			TVDI	0.43 ± 0.026	0.13		0.19
57	Mar/Spring	157	ATI/TVDI	0.45 ± 0.019	0.00	0.19	0.27
			TVDI	0.44 ± 0.012	0.07		0.17
65	Mar/Spring	196	ATI/TVDI	0.43 ± 0.021	0.14	0.13	0.15
81	Mar/Spring	209	ATI/TVDI	0.44 ± 0.022	0.44	0.29	0.68
89	Apr/Spring	209	ATI/TVDI	0.55 ± 0.021	0.03	0.29	0.56
97	Apr/Spring	210	ATI/TVDI	0.38 ± 0.028	0.04	0.11	0.14
105	Apr/Spring	210	ATI/TVDI	0.31 ± 0.034	0.50	0.47	0.63
113	Apr/Spring	210	ATI/TVDI	0.30 ± 0.058	0.00	0.08	0.14
121	May/Spring	210	ATI/TVDI	0.35 ± 0.049	0.29	0.30	0.39
129	May/Spring	211	ATI/TVDI	0.25 ± 0.040	0.15	0.22	0.25
137	May/Spring	211	ATI/TVDI	0.40 ± 0.025	0.12	0.02	0.19
145	May/Spring	211	ATI/TVDI	0.63 ± 0.017	0.15	0.25	0.27
			ATI/TVDI	0.61 ± 0.044	0.19	0.00	0.20
153	Jun/Summer	211	ATI/TVDI	0.51 ± 0.028	0.05	0.00	0.24
161	Jun/Summer	208	ATI	0.60 ± 0.010		0.28	
169	Jun/Summer	211	ATI/TVDI	0.33 ± 0.019	0.10	0.22	0.34
177	Jun/Summer	211	ATI/TVDI	0.48 ± 0.028	0.01	0.07	0.24
			TVDI	0.50 ± 0.026	0.01		0.57
185	Jul/Summer	213	ATI/TVDI	0.48 ± 0.014	0.12	0.21	0.32
193	Jul/Summer	212	ATI	0.23 ± 0.036		0.27	
201	Jul/Summer	212	ATI/TVDI	0.53 ± 0.020	0.17	0.50	0.59
209	Aug/Summer	212	ATI/TVDI	0.51 ± 0.019	0.03	0.26	0.32
217	Aug/Summer	213	ATI/TVDI	0.50 ± 0.037	0.25	0.29	0.33
225	Aug/Summer	213	ATI/TVDI	0.43 ± 0.021	0.29	0.32	0.38
233	Aug/Summer	213	TVDI	0.57 ± 0.017	0.43		0.55
241	Sep/Autumn	213	TVDI	0.42 ± 0.024	0.01		0.52
249	Sep/Autumn	213	ATI/TVDI	0.63 ± 0.010	0.29	0.22	0.32
			ATI/TVDI	0.52 ± 0.020	0.16	0.14	0.28
257	Sep/Autumn	213	ATI	0.54 ± 0.033		0.26	
			TVDI	0.51 ± 0.017	0.47		0.54
265	Sep/Autumn	212	ATI/TVDI	0.49 ± 0.017	0.16	0.35	0.41
			TVDI	0.52 ± 0.020	0.10		0.52
273	Oct/Autumn	212	ATI/TVDI	0.32 ± 0.028	0.38	0.38	0.48
281	Oct/Autumn	212	ATI/TVDI	0.34 ± 0.038	0.00	0.33	0.39
289	Oct/Autumn	212	ATI/TVDI	0.36 ± 0.027	0.01	0.20	0.23
297	Oct/Autumn	212	ATI/TVDI	0.47 ± 0.015	0.05	0.21	0.25
305	Nov/Autumn	212	ATI/TVDI	0.47 ± 0.021	0.06	0.13	0.17
			ATI/TVDI	0.44 ± 0.020	0.10	0.30	0.34
313	Nov/Autumn	213	ATI/TVDI	0.47 ± 0.022	0.05	0.15	0.18
321	Nov/Autumn	189	TVDI	0.33 ± 0.007	0.10		0.10
329	Nov/Autumn	185	ATI/TVDI	0.71 ± 0.008	0.02	0.18	0.20
			TVDI	0.59 ± 0.003	0.00		0.12
337	Dec/Winter	185	ATI/TVDI	0.74 ± 0.009	0.20	0.24	0.31
			TVDI	0.56 ± 0.005	0.01		0.11
345	Dec/Winter	185	ATI/TVDI	0.66 ± 0.012	0.03	0.24	0.34
			TVDI	0.58 ± 0.003	0.01		0.09
353	Dec/Winter	185	ATI/TVDI	0.71 ± 0.010	0.30	0.23	0.38
			TVDI	0.72 ± 0.009	0.23		0.23
361	Dec/Winter	182	ATI/TVDI	0.82 ± 0.007	0.12	0.19	0.24
			TVDI	0.76 ± 0.007	0.19		0.19

4.1.2. The Optimal NDVI Thresholds

The thresholds $NDVI_{ATI}$ and $NDVI_{TVDI}$ were used to divide the entire study area into three subregions for subregional RSM estimation. In terms of the fixed ranges of $NDVI_0$, $NDVI_{ATI}$, and $NDVI_{TVDI}$ we used during iteration, to cover more combinations of NDVI thresholds, the variation of NDVI for all seasons was taken into consideration and maximum possible ranges were selected throughout the year. Indeed, the NDVI ranges we used were a little bit wide compared to other studies, to some extent, resulting in many redundant iterations [33,40]. There was no significant trend in these optimal thresholds throughout the year. It would be explained by taking into account seasonal change (or phenological variation) effects on prediction. To better explain the variation of the NDVI threshold, the construction of seasonal models should be considered over long periods for further studies. Moreover, the NDVI threshold could improve the application of other phenology-based RSM estimation methods meant basically for the growing season [58–60].

The selected minimum and maximum $NDVI_{ATI}$ (0.00 and 0.50) and $NDVI_{TVDI}$ (0.09 and 0.68) are shown in Table 2. The obvious difference between the two criteria was the identification of $NDVI_0$, which not only determines the value of TVDI but also affects the value of pixels in the ATI/TVDI subregion. The pixels with $NDVI_{ATI} < NDVI \leq NDVI_{TVDI}$ were assigned the average of ATI and TVDI in the ATI/TVDI subregion for two criteria, but some of the TVDI ($NDVI_{ATI} < NDVI < NDVI_0$) were calculated by the wet/dry edges produced from those NDVI higher than $NDVI_0$ for Criterion 2. The identification of $NDVI_{ATI}$ was not affected by the other two thresholds with Criterion 2 and the biggest $NDVI_{ATI}$ was 0.50 solely using the ATI-based models on DOY 201. The selected $NDVI_{TVDI}$ with Criterion 2 was up to 0.68, approaching the upper limited value (0.70) on DOY 81 using the ATI/TVDI joint models.

In terms of $NDVI_0$ for the NDVI–LST scatter plots, the value of selected $NDVI_0$ with Criterion 2 fluctuated over 8-day periods but remained relatively low and stable in winter (DOYs 1–49 and 337–361). Most of R^2_{dry} were higher than R^2_{wet}, particularly in summer (DOYs 153–233), and R^2_{wet} peaked in spring and autumn (Figure 8b). The slopes (a_{dry} and a_{wet}) and intercepts (b_{dry} and b_{wet}) of the regression equations changed symmetrically (Figure 8a) and showed similar trends regardless of the criterion [35].

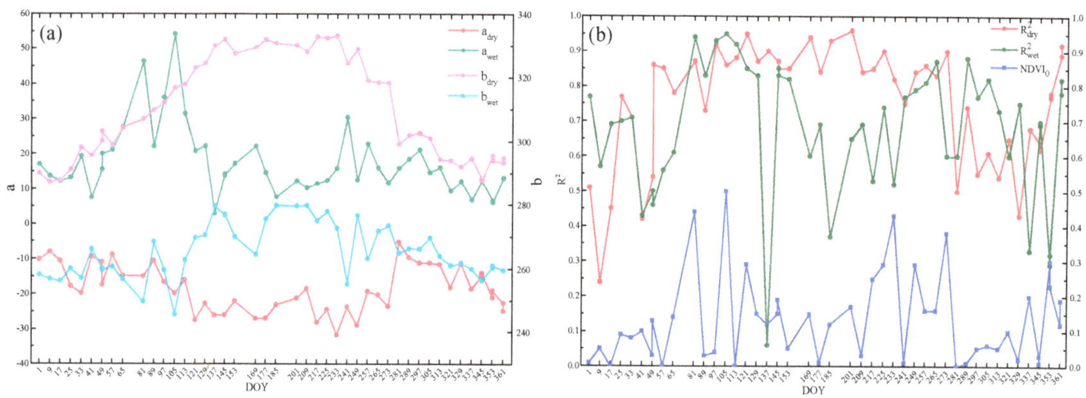

Figure 8. The parameters of the dry and wet edges for 8-day periods in 2017: (**a**) Slopes and intercepts corresponding to selected $NDVI_0$: a_{dry}, a_{wet}, b_{dry}, and b_{wet}; (**b**) correlation coefficients with selected $NDVI_0$: R^2_{dry} and R^2_{wet}.

4.2. Comparison of Estimated RSM

4.2.1. Evaluation of Estimated RSM at the Regional Scale

The estimated RSM area and average RSM obtained with each of the two criteria varied throughout the year (Figure 9). There were 45 8-day periods of the RSM maps produced in 2017 with Criterion 2, compared to 38 with Criterion 1 [35] This clearly shows that using Criterion 2 improved the temporal coverage of RSM estimation.

In terms of spatial coverage, the two criteria also led to different RSM estimation results. The total area of estimated RSM with Criterion 2 was 939.52 × 10^4 km^2 in 2017, which was 40.76% higher than with Criterion 1 (667.44 × 10^4 km^2). Among the 45 8-day periods, there were 24 8-day periods for which Criterion 2 produced a larger geographical area of estimated RSM than Criterion 1. The increased area of estimated RSM ranged from 0.30 × 10^4 km^2 on DOY 81 to 52.58 × 10^4 km^2 on DOY 321 and the average increased area was 19.77 × 10^4 km^2 [35]. Therefore, RSM estimation with Criterion 2 also improved the spatial coverage compared with Criterion 1.

Despite such differences in spatiotemporal coverage, estimated RSM with each criterion also shared some similarities. Both average RSM were highest (~22%) in autumn (DOYs 241–329), which agrees with the study by Jiao et al. (2016) that reported autumn had higher average SM than other seasons in 1998–2000 over the CLP [61]. In addition, we also found that spring (DOYs 57–145) had higher RSM than winter (DOYs 1–49 and 337–361) and summer (DOYs 153–233) (Figure 9).

It should be noted that the area produced by Criterion 1 could occasionally be greater than that of Criterion 2 (e.g., DOYs 49, 105, 153, 161, 249, and 289). The used optimal $NDVI_{ATI}$, $NDVI_{TVDI}$, $\overline{R}$ in validation, and estimated RSM area with each criterion for these 8-day periods are shown in Table 3. Models performed better with Criterion 2 (higher $\overline{R}$ in validation) despite a greater estimated RSM area with Criterion 1. Thus, we could not conclude that Criterion 1 was better than Criterion 2 by merely considering the spatial coverage. In addition, as we mentioned in Section 4.1.1, to acquire wider coverage of RSM estimation, we could choose the additional NDVI thresholds to estimate RSM, which corresponds to higher $\overline{R}$ rather than the highest $\overline{R}$. Therefore, we might test more additional NDVI thresholds in subregions to combine the overall RSM map with the desired accuracy for further study.

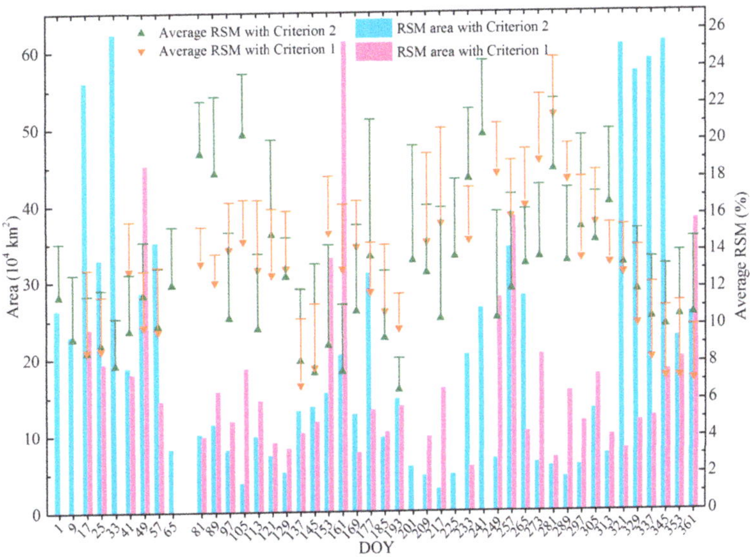

Figure 9. Comparison of the estimated RSM area and of the average RSM with each criterion. The standard deviation of RSM of all pixels for each 8-day period is shown as error bars (one standard deviation).

Table 3. Comparison selected optimal $NDVI_{ATI}$, $NDVI_{TVDI}$, $\overline{R}$, and estimated RSM area of two criteria (area produced by Criterion 1 was greater than that of Criterion 2). Models that performed better for 8-day periods are in bold.

Periods (DOY)	Criterion	Model	$NDVI_{ATI}$	$NDVI_{TVDI}$	$\overline{R} \pm$ Std	Area (10^4 km^2)
49	1	ATI/TVDI	0.14	0.64	0.38 ± 0.023	45.14
	2	ATI/TVDI	0.18	0.24	**0.51 ± 0.022**	28.58
		TVDI		0.19	0.43 ± 0.026	
105	1	ATI/TVDI	0.21	0.35	0.25 ± 0.028	18.59
	2	ATI/TVDI	0.47	0.63	**0.31 ± 0.034**	3.70
153	1	ATI	0.34	0.41	0.35 ± 0.016	32.93
		ATI/TVDI			0.50 ± 0.020	
	2	ATI/TVDI	0.00	0.24	**0.51 ± 0.028**	15.43
161	1	ATI	0.49	0.69	0.54 ± 0.008	61.17
		ATI/TVDI			0.23 ± 0.044	
		TVDI			0.34 ± 0.044	
	2	ATI	0.28		**0.60 ± 0.010**	20.38
249	1	ATI	0.36	0.50	0.46 ± 0.049	27.78
		ATI/TVDI			0.37 ± 0.040	
	2	ATI/TVDI	0.22	0.32	**0.63 ± 0.010**	6.77
289	1	ATI/TVDI	0.26	0.38	0.32 ± 0.018	15.51
	2	ATI/TVDI	0.20	0.23	**0.36 ± 0.027**	4.37

4.2.2. Evaluation of Estimated RSM at the Station Scale

At the station scale, in order to reflect the overall result, stations should be uniformly distributed as much as possible with diverse geographical characterization (e.g., precipitation, elevation, and land cover). In addition, stations with more than 23 8-day periods (half of the 46 8-day periods) of estimated RSM could be considered to better reveal the variation of the estimated RSM and the observed RSM. In this case, six stations (station 53,553, station 53,771, station 53,845, station 53,857, station 57,031, and station 57,048) were selected from those stations as samples. The estimated RSM with each criterion and observed RSM with the total 8-day precipitation of six stations (described in Table 4) are presented in Figure 10. A detailed comparison reveals that despite missing some of the estimated RSM, especially for Criterion 1, the values of estimated and observed RSM have the same tendency. Importantly, the estimated RSM (closer to the observation) with Criterion 2 kept a better trend with the observed RSM at the station scale. Moreover, more estimated RSM with Criterion 2 was observed than with Criterion 1 throughout the period among six stations (the blue lines look more consistent than the red lines in Figure 10). The maximum estimated RSM values (near DOY 281) were observed in autumn among the six stations, consistent with heavy rainfall at that time.

Table 4. Descriptions of the six automatic RSM observation stations.

Station	Longitude (°N)	Latitude (°E)	Elevation (m)	Land cover	Total Precipitation in 2017 (mm)
53,553	111.22	39.85	1221.40	Grassland	410
53,771	112.05	37.41	750.00	Cropland	616
53,845	109.50	36.60	1180.50	Urban area	750.8
53,857	110.18	36.06	110.18	Cropland	549
57,031	108.55	34.83	1012.70	Cropland	662
57,048	108.72	34.40	472.80	Cropland	708

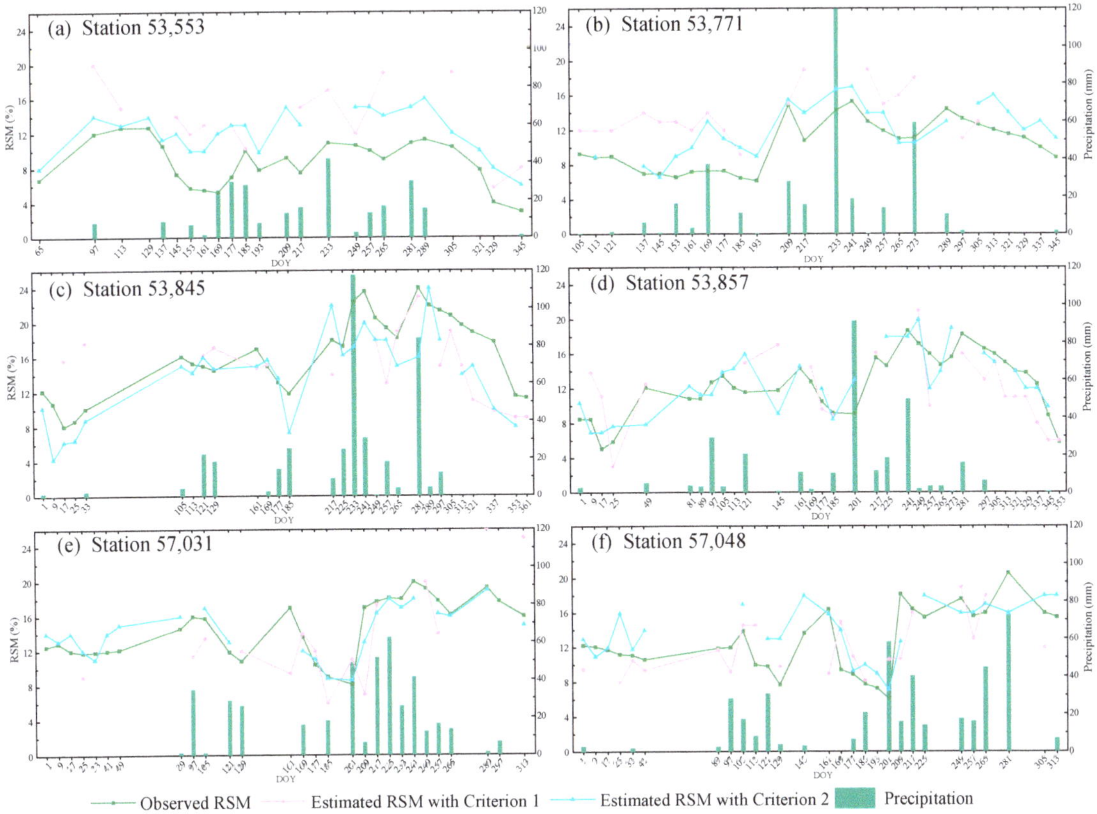

Figure 10. Comparison of estimated and observed RSM with precipitation at six observation stations in 2017.

With increased precipitation, RSM increased at station 53,553 on DOYs 185, 233, and 281; at station 53,771 on DOYs 169, 209, and 233; at station 53,845 on DOYs 233 and 281; at station 53,857 on DOYs 49 and 241; at station 57,031 on DOYs 97 and 225; and at station 57,048 on DOYs 105 and 265 (Figure 10). In addition, the lag-effect impact of precipitation on RSM reflected at station 53,771 from DOYs 153 to 161 and from DOYs 233 to 241 (less precipitation but higher RSM on DOYs 161 and 241 compared with DOYs 153 and 233, respectively), at station 53,845 from DOYs 233 to 241, at station 57,031 from DOYs 201 to 209, and at station 57,048 from DOYs 201 to 209. The impact of precipitation's lag-effect on the estimated RSM has been studied [62] and RSM might be increased dramatically after rainfall due to a delayed response of RSM changes to rainfall [13]. Moreover, high estimated RSM was observed in the periods when there were records of rainfall events. Meanwhile, one limitation of this study was that there were some temporal differences between in situ observations and the remote sensing imagery used.

4.3. Evaluation of Estimated RSM with Criterion 2

4.3.1. Estimated Monthly RSM

The maps of monthly estimated RSM via Criterion 2 are demonstrated in Figure 11. The change of color from orange, red, green, and blue represents the gradual increase of RSM. RSM was higher in the southern part of the CLP in winter (December, January, and February), and in the western area in spring (March, April, and May) based on the clustering of the blue color. The soil in autumn (September, October, and November) was generally wetter than other seasons because most of the area was covered by green.

The wetter soil might be mainly due to concentrated precipitation in autumn. Obviously, the estimated RSM maps in winter were more complete. Some researchers applied other dryness indices (e.g., PDI) or modified TVDI to estimate RSM [16,33,63,64]. To obtain more complete (monthly) RSM maps, ATI and TVDI could be replaced by other dryness indices to estimate RSM with Criterion 2 in our future research.

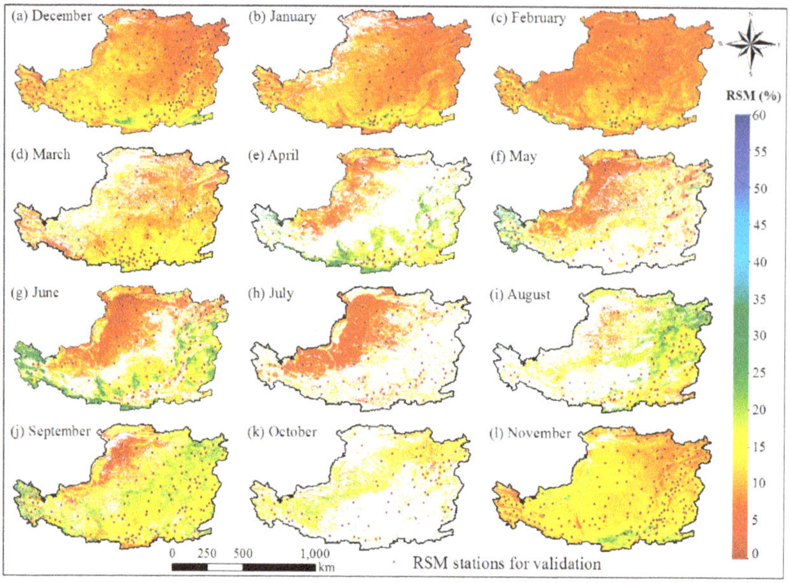

Figure 11. Spatiotemporal pattern of monthly RSM over the CLP in 2017. The location of available RSM observation stations for validation is displayed on each RSM map. White color over the CLP for each monthly RSM map means no value of RSM calculated with Criterion 2.

Importantly, the number of RSM stations for RSM verification varied among months. The available RSM stations are related not only to observation throughout the month but also to estimation in that month. In this case, there are fewer RSM stations for validation in January and February because of no RSM observations in the frozen soil and in April and October because of relatively incomplete RSM maps, respectively. The maximum station for validation was 180 in November and the minimum station was 59 in January and April (Figure 12). There is a rule of thumb for interpreting the size of the correlation coefficient using the absolute value of the Pearson's r: 0.00–0.30, very weak; 0.30–0.50, weak; 0.50–0.70, moderate; 0.70–0.90, strong; and 0.90–1.00, very strong [65,66]. Estimated RSM had a weak correlation with the observed RSM in most months and the highest Pearson's r was 0.68 in January. The root mean square error (RMSE) varied from 3.77% in April to 6.10% in October. The highest mean absolute error (MAE) also appeared in October (4.97%) and the lowest MAE was 3.02% in March.

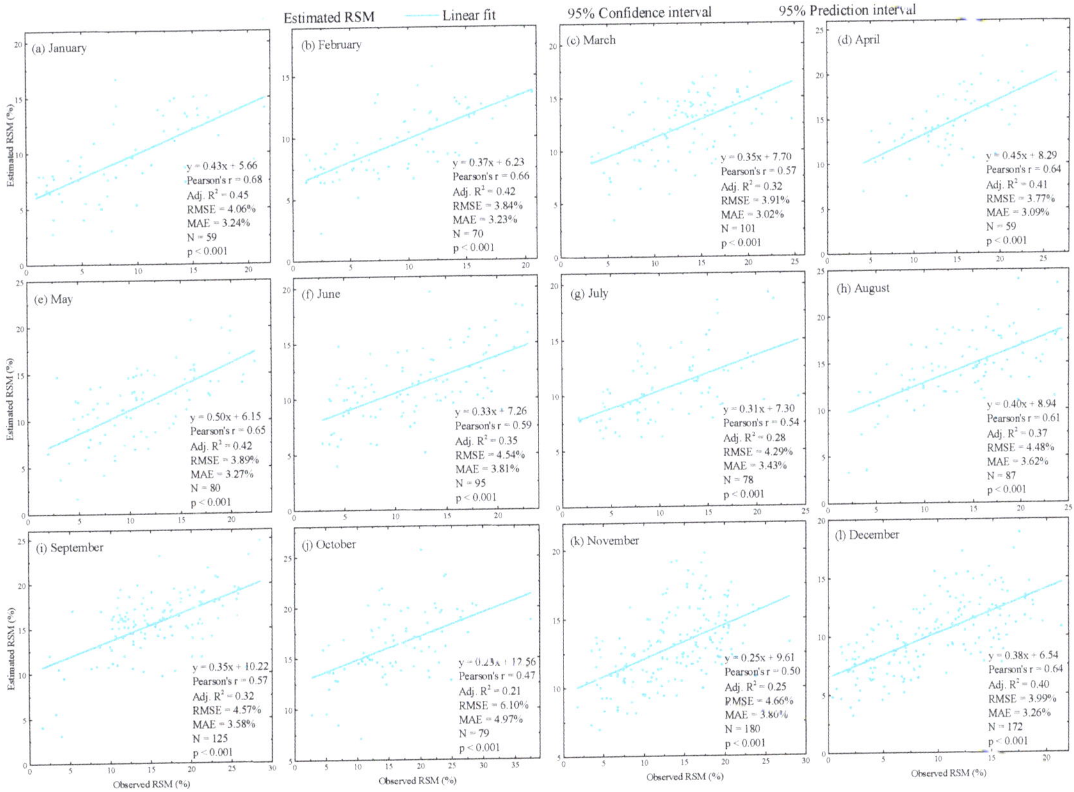

Figure 12. Scatter plots of the observed and estimated RSM at the monthly time scale (linear fitting shown by blue lines with 95% confidence interval and 95% prediction interval shaded in pink and orange, respectively). Scores (Pearson's correlation coefficient (Pearson's r), adjusted R^2, root mean square error (RMSE), and mean absolute error (MAE)) were computed using data included in the corresponding subplot boundary. N represents the number of available RSM observation station samples for each month. The associated *p*-values (*p* in the subplots) with the correlation coefficients are <0.001.

The estimated monthly RSM and their area is illustrated in Figure 13. The average RSM varied from 8.64% in February to 16.42% in August and the highest and lowest maximum RSM (60.96% and 20.93%) appeared in September and January, respectively. The minimum RSM value was 0.00% for most months. However, there were two peaks in spring and the average RSM was clustered in the peak valley (11.42% in March, 14.18% in April, and 10.50% in May, respectively). The total area of the generated RSM area in February was the greatest (62.39×10^4 km^2). The area of generated RSM in April, July, and October was quite small (less than half the area of the CLP). Through comparison of the area distribution of the estimated RSM to the frequency of the observed RSM at available stations, similar tendencies were demonstrated for most months except in January and February. This might be attributed to the scarce RSM observations at that time, thus failing to reflect the spatial distribution of the whole CLP.

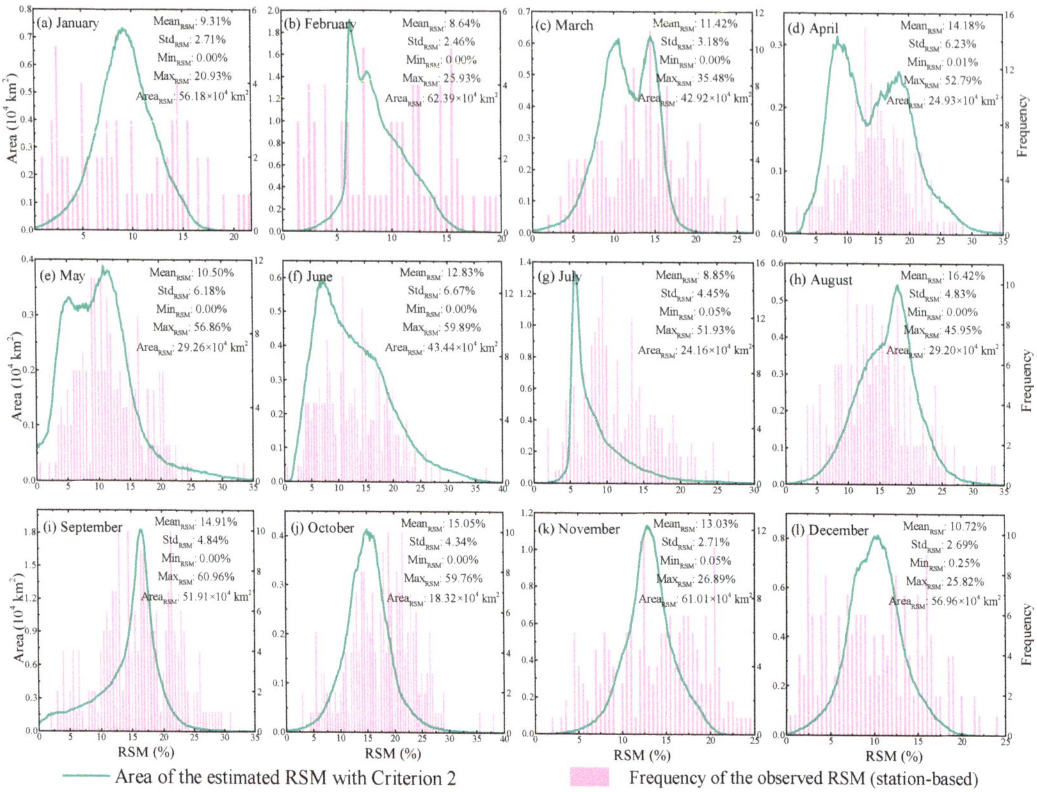

Figure 13. The plots of monthly RSM with its area and frequency of the observed RSM (station-based) in 2017. The average RSM ($Mean_{RSM}$), standard deviation (Std_{RSM}) of RSM, minimum RSM (Min_{RSM}), maximum RSM (Max_{RSM}) of all pixels, and the area of the generated RSM were computed monthly.

4.3.2. Estimated Seasonal and Yearly RSM

The seasonal and yearly RSM maps with the statistics of RSM are shown in Figure 14. Although estimated RSM had a weak correlation with the observed RSM among seasons (Pearson's r ranges from 0.53 in spring to 0.67 in winter and autumn), there was a moderate correlation (0.73) for annual RSM. RMSE varied from 3.74% in winter to 4.41% in autumn. The highest MAE also appeared in autumn (3.64%) and the lowest MAE was 3.00% for annual RSM. We found that autumn had the greatest error among the four seasons, which was highly related to the corresponding months (September, October, and November) with the higher error of RSM estimation (Figure 12). In addition, the generated RSM area of these months in autumn was small and the merged seasonal RSM map would also have a larger error because the seasonal RSM of a certain pixel might be computed by the estimated RSM of a specific month instead of all months in autumn.

The available RSM stations for validation were 49 in the year 2017, which means only 49 observation stations have continuous RSM observations throughout the year. Meanwhile, the scarce RSM observations in winter and 2017 could not reflect the spatial distribution of the estimated RSM for the whole CLP (Figure 14(a3,e3)). The seasonal difference of RSM over the CLP is distinct in 1998–2000 and 2008–2010 [61], which agrees with our results especially for summer (dryer in the southeastern area and wetter in the remaining regions). The southeastern regions were affected by drought and extreme drought after analyzing the drought variation trends in different subregions of the CLP over four decades [67]. However, mean RSM was highest in autumn and lowest in spring [61], which is slightly different from our results (highest (13.91%) in autumn but lowest (9.08%) in winter). The reason for such a difference might be that they just compared SM in three seasons (spring, summer, and autumn) while we evaluated all four seasons [61].

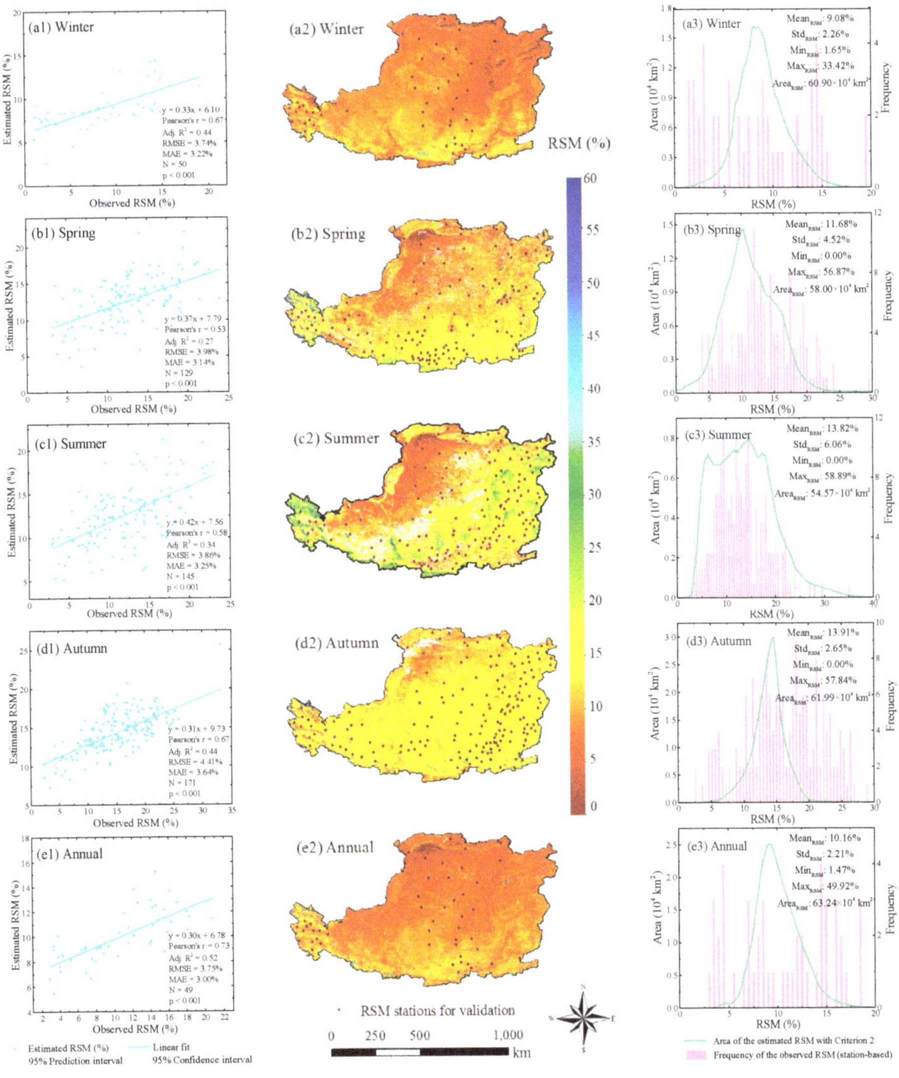

Figure 14. Seasonal and yearly RSM: (**a1–e1**) Scatter plots of the observed and estimated RSM at a seasonal and yearly time scale (linear fitting shown by blue lines with 95% confidence interval and 95% prediction interval shaded in pink and orange, respectively). Scores (Pearson's r, Adjusted R^2, RMSE, and MAE) were computed using observed and estimated RSM. N represents the number of available RSM observation station samples for each period. The associated *p*-values with the correlation coefficients are <0.001; (**a2–e2**) spatiotemporal pattern of seasonal and yearly RSM over the CLP in 2017. The location of available RSM observation stations for validation is displayed on each RSM map. The white color of seasonal and yearly maps over the CLP means no value of RSM calculated with Criterion 2; (**a3–e3**) the plots of seasonal and yearly RSM with its area and frequency of the observed RSM (station-based) in 2017. The average RSM ($Mean_{RSM}$), standard deviation (Std_{RSM}) of RSM, minimum RSM (Min_{RSM}), and maximum RSM (Max_{RSM}) of all pixels and the area of the generated RSM were computed.

5. Conclusions

This study aimed to improve the relative soil moisture (RSM) estimation that is based on the apparent thermal inertia (ATI) and temperature vegetation dryness index (TVDI). By optimizing the identification of NDVI thresholds, Criterion 2 ($NDVI_{ATI}<NDVI_{TVDI}$) improved both the accuracy and spatiotemporal coverage of estimation compared with Criterion 1 ($NDVI_0 \leq NDVI_{ATI} \leq NDVI_{TVDI}$). In addition, monthly, seasonal, and yearly RSM maps of the Chinese Loess Plateau (CLP) in 2017 were produced via the 8-day RSM maps and then examined. From our results, we conclude that:

- The ATI/TVDI joint models not only have higher applicability than the ATI-based and TVDI-based models for all 8-day periods but also for simultaneous use within different NDVI ranges in the ATI/TVDI subregions for one 8-day period. Thus, in addition to the optimal NDVI thresholds, the additional NDVI thresholds we applied was another improved strategy to acquire wider spatial coverage of RSM estimation.
- NDVI thresholds were optimized for robust RSM estimation with Criterion 2 for each 8-day period over the CLP and the selected optimal thresholds constantly changed throughout the study period. The applicability of Criterion 2, involving spatiotemporal coverage (45 and 38 8-day periods of RSM maps and the total RSM area of 939.52×10^4 km^2 and 667.44×10^4 km^2 with Criterion 2 and Criterion 1, respectively) and the accuracy (maximum $\overline{R}$ of 0.82 ± 0.007 for Criterion 2 and of 0.75 ± 0.008 for Criterion 1) of estimated RSM, was better than that of Criterion 1.
- The estimated RSM (closer to the observation) with Criterion 2 kept a better trend with the observed RSM at the station scale. Moreover, more estimated RSM with Criterion 2 was observed than with Criterion 1 throughout the period.
- High estimated RSM was observed in the periods when there were records of rainfall events, especially in autumn (mean RSM of $13.91 \pm 2.65\%$)—wetter than other seasons. With a mean annual RSM of $10.16 \pm 2.21\%$, the annual RSM map shows dryer areas in the southeastern part of the CLP.

This study focused on improving RSM estimation from MODIS imagery with Criterion 2, and such effort is still fundamental to the general research of SM remote sensing. We improved the mapping of RSM using Criterion 2. The next steps following this research would extend the retrieval to other sensors like Landsat and Sentinel satellites, to perform reliable estimates of SM at a high spatiotemporal resolution.

Author Contributions: Conceptualization, L.Y. (Lina Yuan) and L.L.; methodology, L.Y. (Lina Yuan); software, W.L. and S.H.; validation, L.Y. (Lina Yuan) and J.Z.; resources, T.Z., L.Y. (Longhua Yang) and L.C. (Liang Cheng); writing—original draft preparation, L.Y. (Lina Yuan) and T.Z.; writing—review and editing, L.L. and L.C. (Longqian Chen); visualization, M.W. and W.L.; supervision, L.C. (Longqian Chen) and L.L. All authors have read and agreed to the published version of the manuscript.

Funding: This research was supported by the Fundamental Research Funds for the Central Universities (Grant No.: 2018ZDPY07).

Institutional Review Board Statement: Not applicable for this study.

Informed Consent Statement: Not applicable for this study.

Data Availability Statement: The input data used in this research can be accessed freely from online sources.

Acknowledgments: The authors would like to thank the China Meteorological Data Service Center (CMDC, http://data.cma.cn/) for providing in situ RSM observations data over the CLP. Remote sensing data were freely provided by the NASA Land Processes Distributed Active Archive Center (LP DAAC, https://lpdaac.usgs.gov/). Furthermore, we appreciate the editors and reviewers for their constructive comments and suggestions.

Conflicts of Interest: The authors declare no conflict of interest.

References

1. Kumar, S.V.; Dirmeyer, P.A.; Peters-Lidard, C.D.; Bindlish, R.; Bolten, J. Information Theoretic Evaluation of Satellite Soil Moisture Retrievals. *Remote Sens. Environ.* **2018**, *204*, 392–400. [CrossRef]
2. Li, Z.; Lin, X.; Coles, A.E.; Chen, X. Catchment-Scale Surface Water-Groundwater Connectivity on China's Loess Plateau. *Catena* **2017**, *152*, 268–276. [CrossRef]
3. Hou, J.; Zhu, H.; Fu, B.; Lu, Y.; Zhou, J. Functional Traits Explain Seasonal Variation Effects of Plant Communities on Soil Erosion in Semiarid Grasslands in the Loess Plateau of China. *Catena* **2020**, *194*, 104743. [CrossRef]
4. Zhao, J.; Van Oost, K.; Chen, L.; Govers, G. Moderate Topsoil Erosion Rates Constrain the Magnitude of the Erosion-Induced Carbon Sink and Agricultural Productivity Losses on the Chinese Loess Plateau. *Biogeosciences* **2016**, *13*, 4735–4750. [CrossRef]
5. Djamai, N.; Magagi, R.; Goïta, K.; Merlin, O.; Kerr, Y.; Roy, A. A Combination of DISPATCH Downscaling Algorithm with CLASS Land Surface Scheme for Soil Moisture Estimation at Fine Scale during Cloudy Days. *Remote Sens. Environ.* **2016**, *184*, 1–14. [CrossRef]
6. Hardy, A.J.; Barr, S.L.; Mills, J.P.; Miller, P.E. Characterising Soil Moisture in Transport Corridor Environments Using Airborne LIDAR and CASI data. *Hydrol. Process.* **2011**, *26*, 1925–1936. [CrossRef]
7. Yee, M.S.; Walker, J.P.; Rüdiger, C.; Parinussa, R.M.; Koike, T.; Kerr, Y.H. A Comparison of SMOS and AMSR2 Soil Moisture Using Representative Sites of the OzNet Monitoring Network. *Remote Sens. Environ.* **2017**, *195*, 297–312. [CrossRef]
8. Zawadzki, J.; Kedzior, M. Soil Moisture Variability over Odra Watershed: Comparison between SMOS and GLDAS Data. *Int. J. Appl. Earth Obs. Geoinf.* **2016**, *45*, 110–124. [CrossRef]
9. Ebrahimi-Khusfi, M.; Alavipanah, S.K.; Hamzeh, S.; Amiraslani, F.; Samany, N.N.; Wigneron, J.-P. Comparison of Soil Moisture Retrieval Algorithms Based on the Synergy Between SMAP and SMOS-IC. *Int. J. Appl. Earth Obs. Geoinf.* **2018**, *67*, 148–160. [CrossRef]
10. Escorihuela, M.J.; Quintana-Seguí, P. Comparison of Remote Sensing and Simulated Soil Moisture Datasets in Mediterranean Landscapes. *Remote Sens. Environ.* **2016**, *180*, 99–114. [CrossRef]
11. Pardé, M.; Wigneron, J.-P.; Chanzy, A.; Waldteufel, P.; Kerr, Y.; Huet, S. Retrieving Surface Soil Moisture over a Wheat Field: Comparison of Different Methods. *Remote Sens. Environ.* **2003**, *87*, 334–344. [CrossRef]
12. Holgate, C.M.; De Jeu, R.; Van Dijk, A.; Liu, Y.; Renzullo, L.; Vinodkumar; Dharssi, I.; Parinussa, R.; Van Der Schalie, R.; Gevaert, A.; et al. Comparison of Remotely Sensed and Modelled Soil Moisture Data Sets across Australia. *Remote Sens. Environ.* **2016**, *186*, 479–500. [CrossRef]
13. Agutu, N.; Awange, J.; Zerihun, A.; Ndehedehe, C.; Kuhn, M.; Fukuda, Y. Assessing Multi-Satellite Remote Sensing, Reanalysis, and Land Surface Models' Products in Characterizing Agricultural Drought in East Africa. *Remote Sens. Environ.* **2017**, *194*, 287–302. [CrossRef]
14. Hu, T.; Van Dijk, A.I.; Renzullo, L.J.; Xu, Z.; He, J.; Tian, S.; Zhou, J.; Li, H. On Agricultural Drought Monitoring in Australia Using Himawari-8 Geostationary Thermal Infrared Observations. *Int. J. Appl. Earth Obs. Geoinf.* **2020**, *91*, 102153. [CrossRef]
15. Yu, Y.; Wang, J.; Cheng, F.; Chen, Y. Soil Moisture by Remote Sensing Retrieval in the Tropic of Cancer of Yunnan Province. *Pol. J. Environ. Stud.* **2020**, *29*, 1981–1993. [CrossRef]
16. Amani, M.; Salehi, B.; Mahdavi, S.; Masjedi, A.; Dehnavi, S. Temperature-Vegetation-soil Moisture Dryness Index (TVMDI). *Remote Sens. Environ.* **2017**, *197*, 1–14. [CrossRef]
17. Babaeian, E.; Sadeghi, M.; Franz, T.E.; Jones, S.B.; Tuller, M. Mapping Soil Moisture with the OPtical TRApezoid Model (OPTRAM) Based on Long-Term MODIS Observations. *Remote Sens. Environ.* **2018**, *211*, 425–440. [CrossRef]
18. Hashemian, M.; Ryu, D.; Crow, W.T.; Kustas, W.P. Improving Root-Zone Soil Moisture Estimations Using Dynamic Root Growth and Crop Phenology. *Adv. Water Resour.* **2015**, *86*, 170–183. [CrossRef]
19. Rahimzadeh-Bajgiran, P.; Berg, A.A.; Champagne, C.; Omasa, K. Estimation of Soil Moisture Using Optical/Thermal Infrared Remote Sensing in the Canadian Prairies. *ISPRS J. Photogramm. Remote Sens.* **2013**, *83*, 94–103. [CrossRef]
20. Zhang, F.; Zhang, L.-W.; Shi, J.-J.; Huang, J.-F. Soil Moisture Monitoring Based on Land Surface Temperature-Vegetation Index Space Derived from MODIS Data. *Pedosphere* **2014**, *24*, 450–460. [CrossRef]
21. Taktikou, E.; Papaioannou, G.; Bourazanis, G.; Kerkides, P. Soil Moisture Assessment from MODIS data. *Desalination Water Treat.* **2017**, *99*, 59–71. [CrossRef]
22. Phillips, A.J.; Newlands, N.K.; Liang, S.H.; Ellert, B.H. Integrated Sensing of Soil Moisture at the Field-Scale: Measuring, Modeling and Sharing for Improved Agricultural Decision Support. *Comput. Electron. Agric.* **2014**, *107*, 73–88. [CrossRef]
23. Zhang, T.; Wen, J.; Su, Z.; Van Der Velde, R.; Timmermans, J.; Liu, R.; Liu, Y.; Li, Z. Soil Moisture Mapping over the Chinese Loess Plateau Using ENVISAT/ASAR data. *Adv. Space Res.* **2009**, *43*, 1111–1117. [CrossRef]
24. Baghdadi, N.; Zribi, M.; Loumagne, C.; Ansart, P.; Anguela, T. Analysis of TerraSAR-X Data and Their Sensitivity to Soil Surface Parameters over Bare Agricultural Fields. *Remote Sens. Environ.* **2008**, *112*, 4370–4379. [CrossRef]
25. Jin, Y.; Ge, Y.; Wang, J.; Heuvelink, G.B. Deriving Temporally Continuous Soil Moisture Estimations at Fine Resolution by Downscaling Remotely Sensed Product. *Int. J. Appl. Earth Obs. Geoinf.* **2018**, *68*, 8–19. [CrossRef]
26. Mulder, V.; De Bruin, S.; Schaepman, M.; Mayr, T. The Use of Remote Sensing in Soil and Terrain Mapping—A Review. *Geoderma* **2011**, *162*, 1–19. [CrossRef]

27. Van Doninck, J.; Peters, J.; De Baets, B.; De Clercq, E.M.; Ducheyne, E.I.; Verhoest, N.E.C. The Potential of Multitemporal Aqua and Terra MODIS Apparent Thermal Inertia as a Soil Moisture Indicator. *Int. J. Appl. Earth Obs. Geoinf.* **2011**, *13*, 934–941. [CrossRef]
28. Price, J.C. On the Analysis of Thermal Infrared Imagery: The Limited Utility of Apparent Thermal Inertia. *Remote. Sens. Environ.* **1985**, *18*, 59–73. [CrossRef]
29. Taktikou, E.; Bourazanis, G.; Papaioannou, G.; Kerkides, P. Prediction of Soil Moisture from Remote Sensing Data. *Procedia Eng.* **2016**, *162*, 309–316. [CrossRef]
30. Lu, L.; Luo, G.-P.; Wang, J.-Y. Development of an ATI-NDVI Method for Estimation of Soil Moisture from MODIS data. *Int. J. Remote. Sens.* **2014**, *35*, 3797–3815. [CrossRef]
31. Rahimzadeh-Bajgiran, P.; Omasa, K.; Shimizu, Y. Comparative Evaluation of the Vegetation Dryness Index (VDI), the Temperature Vegetation Dryness Index (TVDI) and the improved TVDI (iTVDI) for Water Stress Detection in Semi-arid Regions of Iran. *ISPRS J. Photogramm. Remote Sens.* **2012**, *68*, 1–12. [CrossRef]
32. Liu, Y.; Yue, H. The Temperature Vegetation Dryness Index (TVDI) Based on Bi-Parabolic NDVI-Ts Space and Gradient-Based Structural Similarity (GSSIM) for Long-Term Drought Assessment Across Shaanxi Province, China (2000–2016). *Remote Sens.* **2018**, *10*, 959. [CrossRef]
33. Wang, H.; He, N.; Zhao, R.; Ma, X. Soil Water Content Monitoring Using Joint Application of PDI and TVDI Drought Indices. *Remote Sens. Lett.* **2020**, *11*, 455–464. [CrossRef]
34. Lu, X.J.; Zhou, B.; Yan, H.B.; Luo, L.; Huang, Y.H.; Wu, C.L. Remote Sensing Retrieval of Soil Moisture in Guangxi Based on Ati and Tvdi Models. *ISPRS—Int. Arch. Photogramm. Remote Sens. Spat. Inf. Sci.* **2020**, *42*, 895–902. [CrossRef]
35. Yuan, L.; Li, L.; Zhang, T.; Chen, L.; Zhao, J.; Hu, S.; Cheng, L.; Liu, W. Soil Moisture Estimation for the Chinese Loess Plateau using MODIS-derived ATI and TVDI. *Remote Sens.* **2020**, *12*, 3040. [CrossRef]
36. Su, C.; Fu, B. Evolution of Ecosystem Services in the Chinese Loess Plateau under Climatic and Land Use Changes. *Glob. Planet. Chang.* **2013**, *101*, 119–128. [CrossRef]
37. Tasumi, M.; Kimura, R. Estimation of Volumetric Soil Water Content over the Liudaogou River Basin of the Loess Plateau Using the SWEST Method with Spatial and Temporal Variability. *Agric. Water Manag.* **2013**, *118*, 22–28. [CrossRef]
38. Xin, Z.; Yu, X.; Li, Q.; Lu, X.X. Spatiotemporal Variation in Rainfall Erosivity on the Chinese Loess Plateau during the Period 1956–2008. *Reg. Environ. Chang.* **2010**, *11*, 149–159. [CrossRef]
39. McFeeters, S.K. The use of the Normalized Difference Water Index (NDWI) in the Delineation of Open Water Features. *Int. J. Remote Sens.* **1996**, *17*, 1425–1432. [CrossRef]
40. Zhao, J.-P.; Zhang, X.; Bao, H.-Y.; Tong, Q.-X.; Wang, X.-Y.; Liao, C.-H. Monitoring Land Surface Soil Moisture: Co-inversion of Visible, Infrared and Passive Microwave Sensing Data. *J. Infrared Millim. Waves* **2012**, *31*, 137–142. [CrossRef]
41. Claps, P.; LaGuardia, G. Assessing Spatial Variability of Soil Water Content through Thermal Inertia and NDVI. *Remote Sens. Agric. Ecosyst. Hydrol. V* **2004**, *5232*, 378–388. [CrossRef]
42. Pablos, M.; González-Zamora, Á.; Sanchez, N.; Martínez-Fernández, J. Assessment of Root Zone Soil Moisture Estimations from SMAP, SMOS and MODIS Observations. *Remote Sens.* **2018**, *10*, 981. [CrossRef]
43. Vermote, E.F.; Roger, J.C.; Ray, J.P. MODIS Surface Reflectance User's Guide Collection 6. Available online: http://modis-sr.ltdri.org (accessed on 15 November 2018).
44. Qin, J.; Yang, K.; Lu, N.; Chen, Y.; Zhao, L.; Han, M. Spatial Upscaling of in-Situ Soil Moisture Measurements Based on MODIS-Derived Apparent Thermal Inertia. *Remote Sens. Environ.* **2013**, *138*, 1–9. [CrossRef]
45. Liang, S.; Shuey, C.J.; Russ, A.L.; Fang, H.; Chen, M.; Walthall, C.L.; Daughtry, C.S.; Hunt, R. Narrowband to Broadband Conversions of Land Surface Albedo: II. Validation. *Remote Sens. Environ.* **2003**, *84*, 25–41. [CrossRef]
46. Verstraeten, W.W.; Veroustraete, F.; Van Der Sande, C.J.; Grootaers, I.; Feyen, J. Soil Moisture Retrieval Using Thermal Inertia, Determined with Visible and Thermal Spaceborne Data, Validated for European Forests. *Remote Sens. Environ.* **2006**, *101*, 299–314. [CrossRef]
47. Minacapilli, M.; Cammalleri, C.; Ciraolo, G.; D'Asaro, F.; Iovino, M.; Maltese, A. Thermal Inertia Modeling for Soil Surface Water Content Estimation: A Laboratory Experiment. *Soil Sci. Soc. Am. J.* **2012**, *76*, 92–100. [CrossRef]
48. Sandholt, I.; Rasmussen, K.; Andersen, J. A simple interpretation of the surface temperature/vegetation index space for assessment of surface moisture status. *Remote Sens. Environ.* **2002**, *79*, 213–224. [CrossRef]
49. Shen, R.; Yu, P.; Yan, J.; Wang, Y. Retrieving Soil Moisture by TVDI Based on Different Vegetation Index: A Case Study of Shanxi Province. In Proceedings of the International Conference on Computer Science and Electronics Engineering, Hangzhou, China, 23–25 March 2012; pp. 418–422. [CrossRef]
50. Runping, S.; Jing, Y. Soil Moisture Retrieval through Satellite Data for Gansu and Xinjiang Region of China. *Pak. J. Meterorol.* **2012**, *9*, 93–98.
51. Tagesson, T.; Horion, S.; Nieto, H.; Fornies, V.Z.; González, G.M.; Bulgin, C.; Ghent, D.; Fensholt, R. Disaggregation of SMOS Soil Moisture over West Africa Using the Temperature and Vegetation Dryness Index Based on SEVIRI land Surface Parameters. *Remote Sens. Environ.* **2018**, *206*, 424–441. [CrossRef]
52. Li, H.; Li, C.; Lin, Y.; Lei, Y. Surface Temperature Correction in TVDI to Evaluate Soil Moisture over a Large Area. *J. Food Agric. Environ.* **2010**, *8*, 1141–1145.

53. Holidi, H.; Armanto, M.E.; Damiri, N.; Putranto, D.D.A. Characteristics of Selected Peatland uses and Soil Moisture Based on TVDI. *J. Ecol. Eng.* **2019**, *20*, 194–200. [CrossRef]
54. Chen, J.; Wang, C.; Jiang, H.; Mao, L.; Yu, Z. Estimating Soil Moisture Using Temperature–Vegetation Dryness Index (TVDI) in the Huang-huai-hai (HHH) plain. *Int. J. Remote Sens.* **2011**, *32*, 1165–1177. [CrossRef]
55. Peng, W.; Wang, J.; Zhang, J.; Zhang, Y. Soil Moisture Estimation in the Transition Zone from the Chengdu Plain Region to the Longmen Mountains by Field Measurements and LANDSAT 8 OLI/TIRS-Derived Indices. *Arab. J. Geosci.* **2020**, *13*, 1–13. [CrossRef]
56. Yang, R.; Wang, H.; Hu, J.; Cao, J.; Yang, Y. An Improved Temperature Vegetation Dryness Index (iTVDI) and Its Applicability to Drought Monitoring. *J. Mt. Sci.* **2017**, *14*, 2284–2294. [CrossRef]
57. Song, Y.; Fang, S.; Yang, Z.; Shen, S. Drought Indices Based on MODIS Data Compared over a Maize-Growing Season in Songliao Plain, China. *J. Appl. Remote Sens.* **2018**, *12*, 046003. [CrossRef]
58. Pradhan, N.R.; Floyd, I.; Brown, S. Satellite Imagery-Based SERVES Soil Moisture for the Analysis of Soil Moisture Initialization Input Scale Effects on Physics-Based Distributed Watershed Hydrologic Modelling. *Remote Sens.* **2020**, *12*, 2108. [CrossRef]
59. Aktas, A.; Ustundag, B.B. Soil Moisture Monitoring of the Plant Root Zone by Using Phenology as Context in Remote Sensing. *IEEE J. Sel. Top. Appl. Earth Obs. Remote Sens.* **2020**, *13*, 6051–6063. [CrossRef]
60. Han, J.; Mao, K.; Xu, T.; Guo, J.; Zuo, Z.; Gao, C. A Soil Moisture Estimation Framework Based on the CART Algorithm and Its Application in China. *J. Hydrol.* **2018**, *563*, 65–75. [CrossRef]
61. Jiao, Q.; Li, R.; Wang, F.; Mu, X.; Li, P.; An, C. Impacts of Re-Vegetation on Surface Soil Moisture over the Chinese Loess Plateau Based on Remote Sensing Datasets. *Remote Sens.* **2016**, *8*, 156. [CrossRef]
62. Zhang, X.; Zhao, J.; Sun, Q.; Wang, X.; Guo, Y.; Li, J. Soil Moisture Retrieval From AMSR-E Data in Xinjiang (China): Models and Validation. *IEEE J. Sel. Top. Appl. Earth Obs. Remote Sens.* **2011**, *4*, 117–127. [CrossRef]
63. Qin, Q.; Ghulam, A.; Zhu, L.; Wang, L.; Li, J.; Nan, P. Evaluation of MODIS Derived Perpendicular Drought Index for Estimation of Surface Dryness over Northwestern China. *Int. J. Remote Sens.* **2008**, *29*, 1983–1995. [CrossRef]
64. Zhang, L.; Jiao, W.; Zhang, H.; Huang, C.; Tong, Q. Studying Drought Phenomena in the Continental United States in 2011 and 2012 using various drought indices. *Remote Sens. Environ.* **2017**, *190*, 96–106. [CrossRef]
65. Mukaka, M.M. Statistics Corner: A Guide to Appropriate Use of Correlation Coefficient in Medical Research. *Malawi Med. J.* **2012**, *24*, 69–71.
66. Li, L.; Bakelants, L.; Solana, C.; Canters, F.; Kervyn, M. Dating Lava Flows of Tropical Volcanoes by Means of Spatial Modeling of Vegetation Recovery. *Earth Surf. Process. Landforms* **2017**, *43*, 840–856. [CrossRef]
67. Zhang, B.; Wu, P.; Zhao, X.; Wang, Y.; Wang, J.; Shi, Y. Drought Variation Trends in Different Subregions of the Chinese Loess Plateau over the past Four Decades. *Agric. Water Manag.* **2012**, *115*, 167–177. [CrossRef]

remote sensing

MDPI

Article

New Downscaling Approach Using ESA CCI SM Products for Obtaining High Resolution Surface Soil Moisture

Jovan Kovačević *, Željko Cvijetinović, Nikola Stančić, Nenad Brodić and Dragan Mihajlović

Faculty of Civil Engineering, University of Belgrade, Bulevar kralja Aleksandra 73, 11000 Belgrade, Serbia; zeljkoc@grf.bg.ac.rs (Ž.C.); nstancic@grf.bg.ac.rs (N.S.); nbrodic@grf.bg.ac.rs (N.B.); draganm@grf.bg.ac.rs (D.M.)

* Correspondence: jkovacevic@grf.bg.ac.rs

Received: 18 February 2020; Accepted: 30 March 2020; Published: 1 April 2020

Abstract: ESA CCI SM products have provided remotely-sensed surface soil moisture (SSM) content with the best spatial and temporal coverage thus far, although its output spatial resolution of 25 km is too coarse for many regional and local applications. The downscaling methodology presented in this paper improves ESA CCI SM spatial resolution to 1 km using two-step approach. The first step is used as a data engineering tool and its output is used as an input for the Random forest model in the second step. In addition to improvements in terms of spatial resolution, the approach also considers the problem of data gaps. The filling of these gaps is the initial step of the procedure, which in the end produces a continuous product in both temporal and spatial domains. The methodology uses combined active and passive ESA CCI SM products in addition to in situ soil moisture observations and the set of auxiliary downscaling predictors. The research tested several variants of Random forest models to determine the best combination of ESA CCI SM products. The conclusion is that synergic use of all ESA CCI SM products together with the auxiliary datasets in the downscaling procedure provides better results than using just one type of ESA CCI SM product alone. The methodology was applied for obtaining SSM maps for the area of California, USA during 2016. The accuracy of tested models was validated using five-fold cross-validation against in situ data and the best variation of model achieved RMSE, R^2 and MAE of 0.0518 m^3/m^3, 0.7312 and 0.0374 m^3/m^3, respectively. The methodology proved to be useful for generating high-resolution SSM products, although additional improvements are necessary.

Keywords: soil moisture; downscaling; random forest; ESA CCI SM

1. Introduction

Soil moisture is a crucial component in Earths' system with great impact on interactions between the land surface and the atmosphere [1]. Consequently, using soil moisture information is critical to many applications such as hydrogeological monitoring [2,3], meteorology [4] and water resource management [5,6]. Soil moisture also plays an important role in evapotranspiration process [7], which subsequently influences precipitation occurrences [8]. Soil moisture also indirectly affects environment where its relationship with forest fires has been recognized [9]. Importance of soil moisture has also been recognized institutionally as it is listed as one of the 50 Essential Climate Variables within the Global Climate Observing System (GCOS) [10,11].

Soil moisture can be defined as a mass or volume of water stored between the earth particles in the upper unsaturated soil layer. It is usually distinguished as the surface soil moisture (SSM), which represents the topsoil water content (0–5 cm depth), and the root zone soil moisture (RSM), which accounts for water available to the plants' root system (<2 m depth) [12,13]. Soil moisture content is traditionally measured using ground instruments and techniques based on: (1) sampling and drying; (2) electrical resistance; (3) neutron scattering; (4) gamma-ray absorption; or (5) time-domain reflectometry [12]. This way, both SSM and RSM can be obtained in a form of point measurements and their spatiotemporal characteristics over the wider area have to be modeled, usually using geostatistical methods [14–16]. With the advancements of the satellite remote sensing, an alternative method for the retrieval of the soil moisture came to attention. Satellite observations provided a way of obtaining soil moisture content over the regional and global scales with the temporal resolution in a matter of days. Based on the part of the electromagnetic spectrum being used, the following satellite sensors proved to be useful for soil moisture mapping: (1) microwave (active and passive); (2) optical; and (3) thermal [1]. Unfortunately, due to the penetration depth of the electromagnetic waves through the soil, only SSM can be obtained from the satellite remote sensing [17], while RSM has to be obtained through vertical extrapolation [18]. Very comprehensive and recent reviews on possibilities of generating SSM from the satellite remote sensing data were done by Sabaghy et al. [19] and Peng et al. [20].

Numerous microwave remote sensing sensors have been developed and used for mapping soil moisture content. These include the Advanced Microwave Scanning Radiometer—Earth Observing System (AMSR-E) [21], Soil Moisture and Ocean Salinity (SMOS) satellite [22], Soil Moisture Active Passive (SMAP) mission [23], the Advanced Scatterometer (ASCAT) [24], ESA Sentinel-1 satellites [25] and many more. To achieve the optimal temporal and spatial coverage and to produce the long time series of soil moisture data, all these sources need to be synchronized and merged in the data assimilation process. During this procedure, the differences in operational, spatial, temporal and retrieval algorithm aspects of the used sources must be taken into account. The European Space Agency (ESA) produces such merged microwave soil moisture products as part of the Climate Change Initiative (CCI)—ESA CCI SM [26]. Although the ESA CCI SM product provides very good spatial coverage, there are still data gaps in some places. Another disadvantage is that the product has coarse spatial resolution of 0.25° (≈25 km), which is insufficient for many regional and local applications.

Several studies have aimed at improving the spatial resolution and filling the data gaps of coarse resolution SSM products [27–29]. Machine learning (ML) techniques proved to be a very useful tool for such purpose [30,31]. Studies have shown that Random Forest (RF) is one of the many available ML techniques that yields very good results in downscaling and filling data gaps thanks to its flexibility through randomization and ensemble approach [32]. This study successfully implemented a two-step approach to produce SSM product without missing data and with high spatial resolution (1 km). The first step is used as a data engineering tool and its output is used as the input for the second step. Bilinear interpolation and random forest model are considered as data engineering tools in the first step, and, in the second step, additional random forest regression is used. The methodology was tested over the study area of California, USA for the year 2016. ESA CCI SM products, together with the auxiliary products (Normalized Difference Vegetation Index (NDVI), Land Surface Temperature (LST), NWS Precipitation, and Köppen–Geiger climate classification map), were used within the prediction model. The approach described in this paper is novel in a method that considers the synergic use of multiple ESA CCI SM products instead of a single one in order to obtain high resolution SSM maps.

2. Materials

2.1. Study Area

The study area covers 423,967 km^2—the complete state of California, USA (Figure 1). The area's relief is dominated by the Central Valley, which runs 725 km through the state between the Coast Ranges to the west and the Sierra Nevada to the east and bounded by the Cascades in the north and Tehachapi Mountains to the south. California's land cover is diverse, where forests cover almost half of the state's area and with barren plains in the northern and desert area in the east-central parts. Climate conditions in California vary from polar to subtropical. The biggest part of the state has a Mediterranean climate and in the northeastern part the temperate climate is present. The climate also changes rapidly with elevation, where the alpine climate can be found in the higher mountains.

Different parts of the state receive various amount of precipitation, which ranges from more than 4300 mm in the northwest to small traces in the southeastern desert. Coastal areas are different too, where moderate temperatures and moderate rainfall prevail.

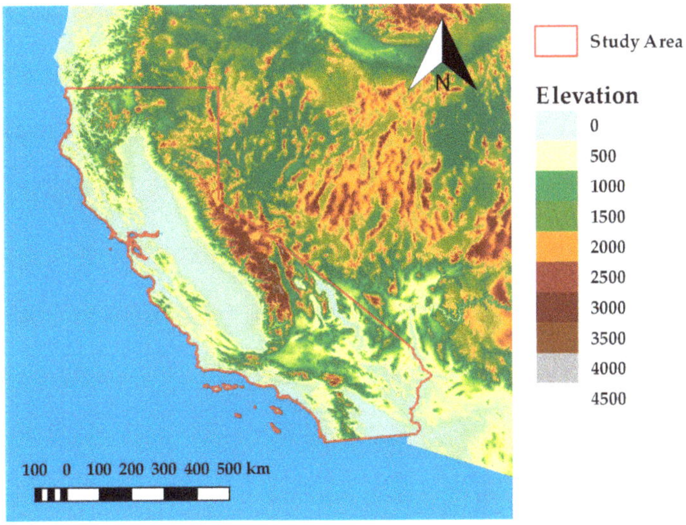

Figure 1. The study area—the state of California, USA.

The state is a major agriculture contributor accounting for over 13% of the USA's total agricultural value in 2018. It produces more than 400 commodities, with more than a third of the country's vegetables and two-thirds of the country's fruits and nuts being grown in California [33]. Such extensive agriculture production requires careful and smart water management, which can benefit significantly from high quality soil moisture maps.

2.2. European Space Agency Soil Moisture Products - ESA CCI SM

ESA CCI SM products are generated using soil moisture observations from active (ERS1-2 SCAT and MetOp ASCAT A-B) and passive (SMMR, SSM/I, TMI, AMSR-E, WindSat, AMSR2 and SMOS) microwave satellite sensors. Three groups of soil moisture products are generated in the assimilation process: active (ESA CCI SM A), passive (ESA CCI SM P) and combined (ESA CCI SM C). The active soil moisture products are generated from the C-band scatterometers using the change detection algorithm. The passive products are handled using the Land Parameter Retrieval Model (LPRM), which successfully translates the microwave observed land surface brightness temperature (Tb) to the soil moisture content. The combined product is obtained through the assimilation process of the previous two, with the appropriate weights assigned to each source [26]. All products provide daily global coverage with the spatial resolution of 0.25° (≈25 km). Active soil moisture products are expressed in the percentage of saturation (%), whereas passive and combined soil moisture products are expressed in volumetric units (m^3/m^3). In the latest version of ESA CCI SM products (04.5), the temporal range has been extended and covers 1978–2018. In this research, all three types of products (passive, active and combined) for 2016 were obtained from the ESA data archive (https://www.esa-soilmoisture-cci.org/).

2.3. PBO_H2O in Situ Soil Moisture Observations

PBO_H2O, a project that was operational from 2004 to 2017, implemented GPS interferometric reflectometry for the measurement of SSM. The observations represent volumetric soil moisture content in the topsoil layer (0–5 cm) with spatial scale of ~1 km^2 and accuracy of 0.04 m^3/m^3 [34].

PBO_H2O data can be obtained from the International Soil Moisture Network (ISMN) data archive (https://ismn.geo.tuwien.ac.at/en/), as it was done for the whole 2016. The complete dataset consists of 159 stations with hourly measurements. For each station, the observations were firstly aggregated to obtain the mean daily value. In the next step, the locations with the multiple sensors (same latitude and longitude) were averaged. fifty-six were stations left after cropping locations to the study area (Figure 2a), with a total of 18,307 daily surface soil moisture observations (Figure 2b).

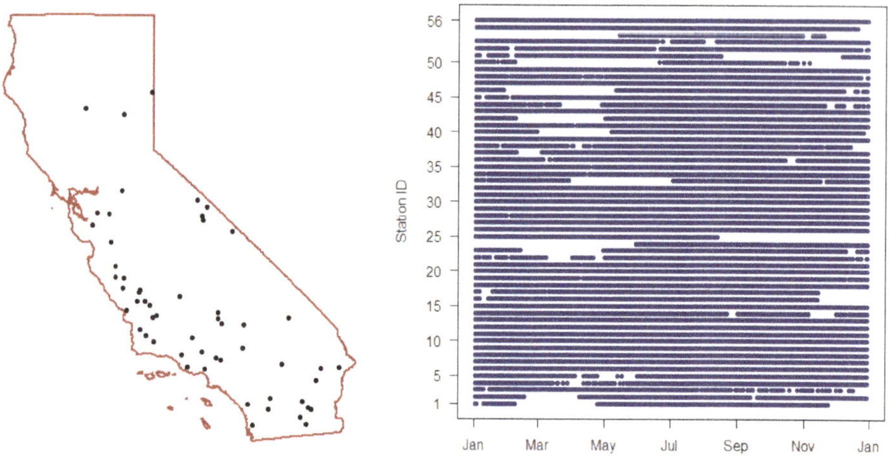

(a) (b)

Figure 2. (**a**) Spatial distribution of the PBO_H2O soil moisture stations in the study area; and (**b**) soil moisture observations per station during the 2016.

2.4. Auxiliary Data

2.4.1. Normalized Difference Vegetation Index (NDVI) and Land Surface Temperature (LST)

The connection between land water content and NDVI and LST has been widely used for downscaling coarse resolution remotely-sensed soil moisture [35–38]. The main advantage of using such data for downscaling is their fine spatial resolution, good temporal coverage and the many available satellite missions that collect them. However, the cloud contamination is a big problem for all optical sensors, making these products unavailable in certain places [1]. Moderate Resolution Imaging Spectroradiometer (MODIS) is one of the most commonly used sources for such products and therefore it was chosen as the provider of NDVI and LST.

NDVI was taken from MODIS Vegetation Indices 16-day Level 3 Global 1 km Version 6 products, from both Terra (MOD13A2) and Aqua (MYD13A2) satellites. The temporal coverage of NDVI included 2016 and 2017 with 46 Terra and 46 Aqua products. Each product was generated in WGS84 coordinate reference system. The data coverage was extended to include 2017 because it was necessary for generating and later improving daily NDVI products.

The LST data were induced from MODIS Land Surface Temperature/Emissivity Daily L3 Global 1km Version 6 products from both Terra (MOD11A1) and Aqua (MYD11A1) satellites. LST_{DAY} and LST_{NIGHT} land surface temperature maps in the form of rasters in WGS84 coordinate reference system were generated for each satellite and for each date of 2016 (ideally, four rasters for each date). In the next preprocessing step, for each date, Terra and Aqua products were merged by taking average of corresponding pixels, so that, in the end, single LST_{DAY} and LST_{NIGHT} rasters were produced for each date of 2016. Since the data for Terra products DOY 50-58 were missing, only the Aqua products were used for producing LST_{DAY} and LST_{NIGHT} rasters during these days.

2.4.2. NWS Precipitation Data

As a part of the natural water cycle, atmospheric water is transferred to the land through precipitation. The correlation between precipitation and soil moisture spatial and temporal patterns has been observed by many studies [2,8]. Since precipitation datasets are of higher spatial resolution, it has been used in the process of downscaling coarse resolution soil moisture [32,39].

National Weather Service (NWS) produces daily precipitation estimate maps for the whole USA from the combined sensor inputs: radar and rain gauge. The data represent 24-h accumulation and they are disseminated in the Hydrologic Rainfall Analysis Project (HRAP) grid coordinate system. Although the spatial resolution of the data is considered roughly ≈4 km over continental USA, the spatial resolution of the product over the study area is closer to ≈5 km due to the characteristics of the HRAP grid. After obtaining the data for 2016 (https://water.weather.gov/precip/), each file was preprocessed to ≈0.05° (≈5 km) in WGS84 coordinate reference system. Each HRAP grid point was assigned to the closest WGS84 pixel during the preprocessing.

2.4.3. Köppen–Geiger Climate Classification Map

Climate types are defined using average weather conditions over a long time. The certain climate type is directly or indirectly related to the precipitation amount, the dominant vegetation density/types and the land surface temperature [2]. Therefore, it can be expected that it can be useful for the downscaling procedure. To the authors' knowledge, no other studies used climate data for downscaling soil moisture.

Köppen–Geiger climate classification map is the most frequently used climate classification map created by Wladimir Köppen and it was presented in its latest version in 1961 by Rudolf Geiger. In this research, the updated and re-analyzed Köppen–Geiger map produced by Climate Change & Infectious Diseases was used [40]. The spatial resolution of the map is 5′ and it can be obtained from the group's website (http://koeppen-geiger.vu-wien.ac.at/). The climate classification map was additionally reclassified to first level of classification scheme with five different climate groups: A (Tropical), B (Arid), C (Temperate), D (Continental) and E (Polar).

3. Methods

3.1. Bilinear Interpolation

Bilinear interpolation is a widely popular two-dimensional interpolation method that uses the values of four closest points in order to estimate an output value [41]. The interpolation function that is used to fit a bilinear surface through these four points is given by the equation:

$$z = f(x, y) = a_0 + a_1x + a_2y + a_3xy. \quad (1)$$

When applied to a raster image, this interpolation method considers the known values of the four nearest pixels located in diagonal directions from the position of a new pixel. A new pixel value is calculated as a weighted average of these four pixel values from the original image. This resampling method can be used both as an aggregation or disaggregation raster tool. In this research, it was considered as a disaggregation tool used for downscaling remote sensing products from coarse to finer spatial resolution. Due to its vast popularity, the bilinear interpolation was taken for comparison purposes, that is, to compare its results with the results of the methods that are more sophisticated.

3.2. Random Forest Regression

Random forest is an ensemble approach machine learning technique which can be applied for both regression and classification problems. The technique proposed by Breiman [42] uses multiple decision trees built during the training phase from which mean prediction is taken as an output of the model. Each tree is built from the bootstrap sample created from some portion of the input training data, while the remaining data are used for the performance evaluation of each tree. This feature (also known as bootstrap aggregation) provides powerful tool for modeling nonlinear relationships while reducing the chance of overfitting and improving generalization [42].

In this study, random forest regression implemented in ranger R package was used [43]. The number of trees was set to 200 because a larger number did not produce significant error improvement, but increased the computation time. The split rule was set to "MaxStat" instead of the more usual default "Variance" split rule. All other parameters were left to their default values.

4. Methodology

The methodology used in this research consists of several steps (Figure 3). First, the input datasets were processed to fill gaps in the data in both temporal and spatial domains. Next, the created datasets were used to downscale coarse resolution ESA CCI SM products to high spatial resolution of 1 km (Data engineering). Since downscaled products still have large bias against the in situ soil moisture observations, additional processing was necessary. This was covered in the final step (Random forest), where all previously created downscaled datasets in congregation with in situ data were used to produce output SSM maps of high spatial resolution. The following sections describe all these steps in detail.

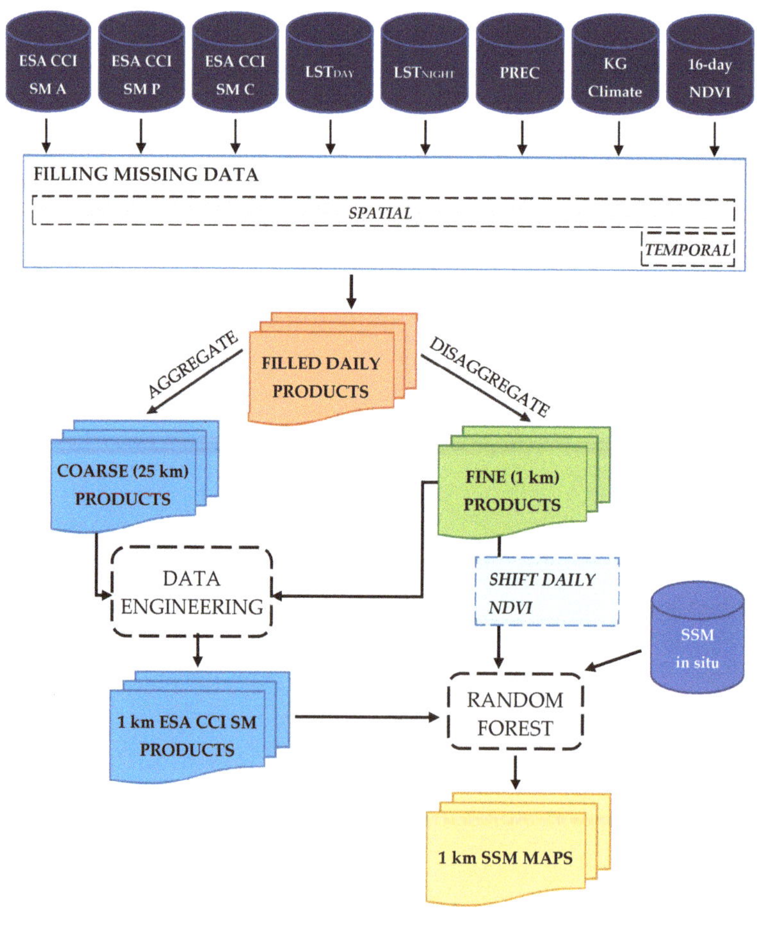

Figure 3. The flowchart diagram of the data processing steps.

4.1. Filling Spatial and Temporal Data Gaps

All input raster datasets (except climate classification map) have some spatial gaps. Gaps in ESA CCI SM products are caused by the lack of microwave soil moisture sources and their spatial coverage for some specific day; gaps in MODIS datasets are caused by clouds and/or other atmospheric conditions; and NSW precipitation has some missing data left after the transformation from HRAP grid to WGS84. To fill all missing data pixels in the study area, universal kriging interpolation technique is used. Universal kriging showed good performances compared to other commonly used interpolation techniques, almost as good as kriging with an external drift [44]. The advantage is that universal kriging does not require additional variables within the interpolation process. This enables that each of the input datasets can be filled independently of the other datasets.

The sample variogram is generated and used for fitting the spherical variogram model, where each raster pixel is considered as observation point. For the computational effectiveness, sample variogram was modeled using the 0.25° (≈25 km) spatial resolution, meaning that all datasets that have different spatial resolution (LST, NDVI and NWS Precipitation) have to be aggregated by the mean value before the variogram modeling. Using this technique, missing data for each input raster are independently filled. Additionally, for each NDVI 16-day composite raster, the raster that represents the day of the year that NDVI pixel corresponds to is generated and its spatial gaps are filled (NDVI_DOY). NDVI_DOY pixel values are rounded to avoid decimal values.

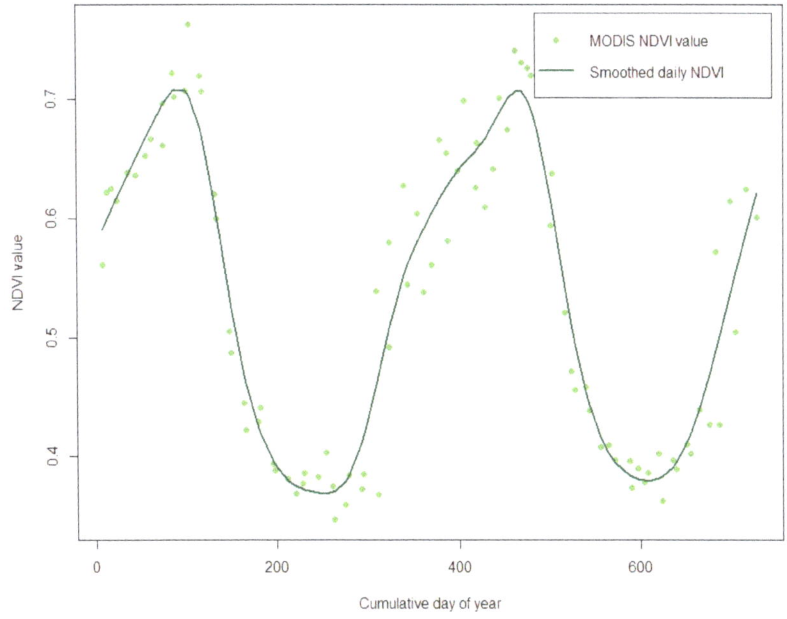

Figure 4. An example of the fitted cubic smoothing spline for NDVI temporal gap filling for the period 2016–2017.

Temporal gaps are a big problem for NDVI data, which are 16-day composites. Even with both satellites used synergically, the NDVI observations are ideally available every 9 days, which is too sparse for modeling daily SSM. The smoothing methods provide relatively simple, yet effective way for reconstructing NDVI time-series [45]. No smoothing method can be recommended more than others. However, spline smoothing provides rather good results and its parameters can be well tuned through cross-validation [45]. Therefore, temporal gap filling is done using NDVI and

NDVI_DOY information pixel-wise, by fitting a cubic smoothing spline. Since no ground NDVI dataset is available for determining the optimal spline parameters through cross-validation, these parameters were determined empirically by visual inspection of the smoothing curves. The curve is fitted in a way that the changes of NDVI are gradual, without unusual spikes or drops (Figure 4). This approach leads to smooth NDVI. Although this can lead to a smooth soil moisture time series, it is expected that such behavior will be avoided by the use of other daily available predictors. The values for each day are then generated after the spline fitting.

Correlation between Daily Filled Predictors and in Situ Soil Moisture Observations

Before proceeding to the next step, the relevance and quality of each daily filled predictor is assessed. This is done by calculating Pearson correlation coefficient between available in situ soil moisture observations and the daily filled values of the predictors to be used in the prediction model. As shown in Figure 5, strong positive correlation exists between the in situ soil moisture and all three types of ESA CCI SM products along with NDVI. Among others predictors, only LST_{DAY} shows strong negative correlation, while all others (LST_{NIGHT}, PREC and Climate) show medium correlation with the in situ data. Such correlation values indicate that filling data gaps was successful. Since at least the medium correlation exists, using the chosen set of predictors in the downscaling procedure is justified.

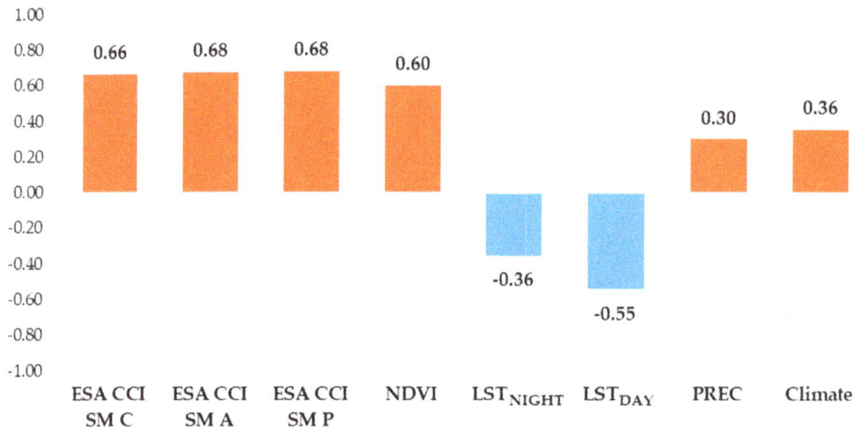

Figure 5. Pearson correlation coefficients between in situ soil moisture observations and each predictor.

4.2. Downscaling ESA CCI SM Products

Downscaling of the ESA CCI SM products is done using the previously generated datasets. No reprojection is necessary since all data are already in the common coordinate reference system—WGS84. The downscaling is performed independently using the bilinear interpolation technique (BIL) (Data engineering) and using the random forest (RF) method. The RF model (Data engineering) is defined as:

$$ESA_{SM-D} = DOY + ESA_{SM-1} + ESA_{SM-2} + NDVI + LST_{DAY} + LST_{NIGHT} + PREC + Climate \quad (2)$$

where ESA_{SM-D} is the downscaled ESA CCI SM product, ESA_{SM-1} and ESA_{SM-2} are the remaining two types of ESA CCI SM products and DOY, PREC and Climate represent day of the year, amount of precipitation and climate zone, respectively. The RF regression model is trained over the coarse spatial resolution of 25 km where all NDVI, LST_{DAY}, LST_{NIGHT} and PREC rasters are aggregated firstly. The trained model is then applied for the generation of 1 km ESA CCI SM rasters using two other ESA CCI SM products, NDVI, LST_{DAY}, LST_{NIGHT}, PREC and Climate 1 km predictors. In cases where there are no 1 km predictors available (ESA CCI SM products, PREC and Climate), they are disaggregated from coarser to the desired 1 km spatial resolution. For ESA CCI SM products and PREC raster, this is done using standard bilinear interpolation. Considering that the climate raster is the categorical raster map, it is disaggregated to 1 km spatial resolution using the nearest neighbor interpolation.

4.3. Generating Surface Soil Moisture Maps of High Spatial Resolution

4.3.1. Shifting NDVI Values

Figure 4 shows that there are some differences between observed and modeled NDVI values due to the spline fitting. The SSM has shifted, i.e., a delayed effect on vegetation, with the time lag of about half a month [35]. It can be expected that spline smoothing further emphasizes the delayed effect of SSM and NDVI. Therefore, even though it is useful for downscaling ESA CCI SM products, such NDVI product might not match well with the in situ soil moisture observations. Although the strong correlation (0.60) already exists between the in situ soil moisture observations and the daily filled NDVI, the removal of the time shift caused by smooth spline should further increase it.

Using this approach, the best shift value has been determined for each available in situ location. It is assumed that the best shift is represented by the shift value that corresponds to the highest correlation value between NDVI and in situ soil moisture observations (Figure 6a) obtained for each station independently. A range of shifts between −45 and +5 was tested against in situ soil moisture observations. The final shift value for the whole study area was then determined as a median value of all individual best shifts (Figure 6b). In some cases, the shift value did not converge to the local minimum (its value corresponded to the edge shift values). Therefore, these shift values were omitted from the median calculation. Because the shift is expected to be negative (soil moisture content affects the vegetation in the future), the NDVI data for both 2016 and 2017 have to be used. It should be noted that correlation of NDVI and in situ changes monthly, thus it always has to be calculated for the same time interval, in order to make correlation values comparable over the shifting range. That is why the calculation is always calculated for 2016, no matter the shifting range being examined.

The final shift value for the study area determined using the previously explained method is -24 days. This is larger than the reported time shift, probably due to the smoothing effect. This way, the correlation of the shifted NDVI was increased to 0.65, almost as high as of the ESA CCI SM products.

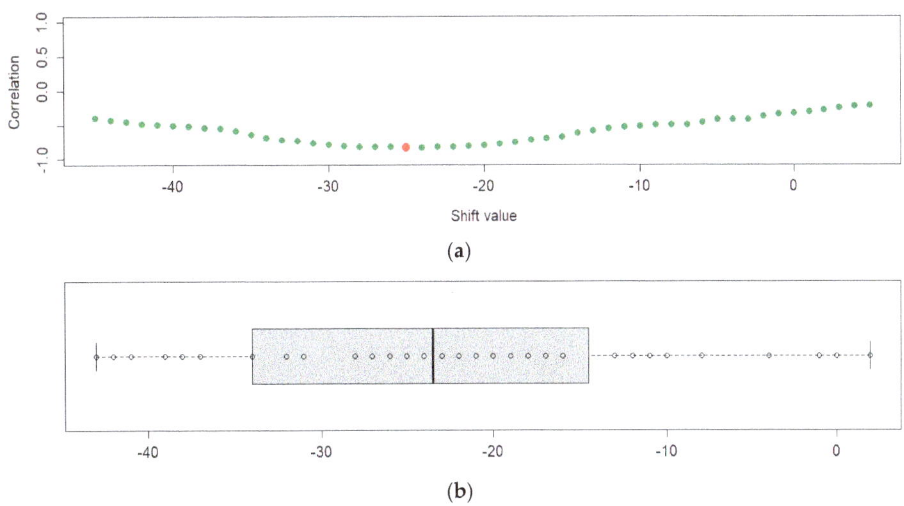

Figure 6. (**a**) An example of the determined best shift value compared to all correlation values from tested shift range; and (**b**) boxplot chart with individual best shift values from which the global shift value has been determined as median value.

4.3.2. Training Second RF Model

The second RF model uses all previously generated 1 km datasets, in addition to in situ soil moisture observations. The model (Random forest) is defined as:

$$SM_{in\ situ} = DOY + ESA_{SM-down} + NDVI_{SHIFT} + LST_{DAY} + LST_{NIGHT} + PREC + Climate \quad (3)$$

All combinations of ESA CCI SM products were examined to determine the optimal one. Every model was trained using all available surface soil moisture observations and corresponding predictors for each location and each date. Because in situ soil moisture observations and the data for all predictors have the same scale, it is possible to implement extracting by using the value of the pixel that in situ location falls within. After successfully training the RF model, it was used to produce 1 km soil moisture maps for each day of 2016.

4.4. Validation of the Results

The model validation was performed using five-fold cross-validation, where complete fold locations were left out of the model training set and were only used for the model validation. Five-fold cross-validation was repeated 10 times and the output predictions were calculated as the mean values. These were further used to determine the validation metrics. The metrics included root mean square error (RMSE), coefficient of determination (R^2) and mean absolute error (MAE) calculated between the observed soil moisture and the soil moisture generated from the cross-validation model output.

5. Results

The downscaling RF model (Data engineering) was trained using aggregated 25-km products for 2016. Since the data have no gaps, the 243,024 data entities (pixel stacks) are available for building the model. Two combinations were tested, one with and one without using other two ESA CCI SM products in the downscaling models. The version without using other two ESA CCI SM products was used as a benchmark, to determine the produced effect which these two products bring into the prediction model. All downscaled ESA CCI SM products were compared to the in situ data, where the passive product proved to be the best one by all metrics (displayed in Table 1).

Table 1. The validation metrics of downscaled ESA CCI SM 1-km products obtained by using in situ observations.

	Downscaled Using Standard Bilinear Interpolation			Downscaled without Using Other Two ESA CCI SM Products			Downscaled Using Other Two ESA CCI SM Products		
	RMSE [m^3/m^3]	R^2	MAE [m^3/m^3]	RMSE [m^3/m^3]	R^2	MAE [m^3/m^3]	RMSE [m^3/m^3]	R^2	MAE [m^3/m^3]
ESA CCI SM C	0.0745	0.4930	0.0591	0.0707	0.6095	0.0556	0.0694	0.6273	0.0557
ESA CCI SM A [1]	/	0.4860	/	/	0.5816	/	/	0.6199	/
ESA CCI SM P	0.0728	0.5092	0.0543	0.0613	0.6280	0.6280	**0.0591**	**0.6495**	**0.0434**

[1] RMSE and MAE cannot be determined because of the different unit system.

The second RF model was trained using all previously created daily 1 km predictors (ESA CCI SM products, NDVI, LST_{DAY}, LST_{NIGHT}, PREC and Climate) and 18,307 surface PBO_H2O soil moisture observations. All three versions of the downscaled ESA CCI SM products (without mixing them) were tested to determine the optimal combination. Since the downscaling step also introduces errors (see Table 1); testing all combinations helps understand the way these errors propagate in the following steps. The extracted validation metrics from the five-fold cross-validation are presented in Table 2.

Table 2. Validation metrics for each set of tested predictors (results after applying second RF).

	Predictor Combination [1]	Without NDVI Shift			Using NDVI Shift		
		RMSE [m³/m³]	R^2	MAE [m³/m³]	RMSE [m³/m³]	R^2	MAE [m³/m³]
Downscaled using bilinear interpolation	*C+A+P*	0.0528	0.7195	0.0382	0.0528	0.7191	0.0384
	C+A	0.0539	0.7101	0.0389	0.0538	0.7110	0.0390
	C+P	0.0543	0.7075	0.0391	0.0541	0.7089	0.0391
	A+P	0.0522	0.7294	0.0378	0.0521	0.7301	0.0378
	C	0.0560	0.6877	0.0403	0.0556	0.6916	0.0401
	A	0.0544	0.7058	0.0393	0.0543	0.7068	0.0394
	P	0.0546	0.7055	0.0392	0.0543	0.7087	0.0392
Downscaled using RF1 without other two ESA CCI SM products	*C+A+P*	0.0569	0.6729	0.0406	0.0561	0.6822	0.0402
	C+A	0.0568	0.6760	0.0404	0.0561	0.6843	0.0400
	C+P	0.0568	0.6763	0.0404	0.0561	0.6846	0.0400
	A+P	0.0566	0.6781	0.0405	0.0561	0.6837	0.0403
	C	0.0574	0.6705	0.0407	0.0566	0.6796	0.0403
	A	0.0574	0.6710	0.0410	0.0567	0.6779	0.0407
	P	0.0573	0.6717	0.0410	0.0567	0.6776	0.0408
Downscaled using RF1 with other two ESA CCI SM products	*C+A+P*	0.0525	0.7217	0.0379	0.0523	0.7238	0.0379
	C+A	0.0521	0.7279	0.0375	**0.0518**	**0.7312**	**0.0374**
	C+P	0.0521	0.7286	0.0375	0.0519	0.7305	0.0375
	A+P	0.0537	0.7104	0.0388	0.0536	0.7115	0.0387
	C	0.0524	0.7276	0.0376	0.0520	0.7310	0.0375
	A	0.0544	0.7051	0.0392	0.0541	0.7078	0.0390
	P	0.0543	0.7068	0.0391	0.0542	0.7068	0.0392

[1] C, A and P represent combined, active and passive ESA CCI SM products, respectively.

The improvements after each processing step can be clearly seen in Figure 7. The step with filling missing data produces continuous product, with both smaller and larger areas of missing data successfully reconstructed. As expected, the gap filling is less successful over the large missing data areas, where SSM content does not show reasonable changes. This is successfully covered in the next steps, where the gaps are hardly noticeable in the downscaled products, while they cannot be identified at all in the final products. Downscaling by BIL mostly fails to provide new spatial information. On the contrary, the main improvement of the downscaling process by RF is the spatial richness that has been obtained. Fine details which were previously hidden behind the coarse resolution can be differentiated in both the downscaled and the final output products. The local extremes present in the coarse resolution products are successfully adjusted in the following steps, albeit there are differences in soil moisture content between the downscaled and the final output products. Visually inspected, both types of downscaling, and especially bilinear interpolation, reduce extremes and produce smoother soil moisture products, while the in situ modeled products emphasize abrupt changes. This can be attributed to the bias that exists between the remotely-sensed soil moisture and in situ observations. As the ESA CCI SM products proved to be the most correlated predictor, the bias is successfully adjusted only when in situ observations are included in the model. This is done in the second RF regression model.

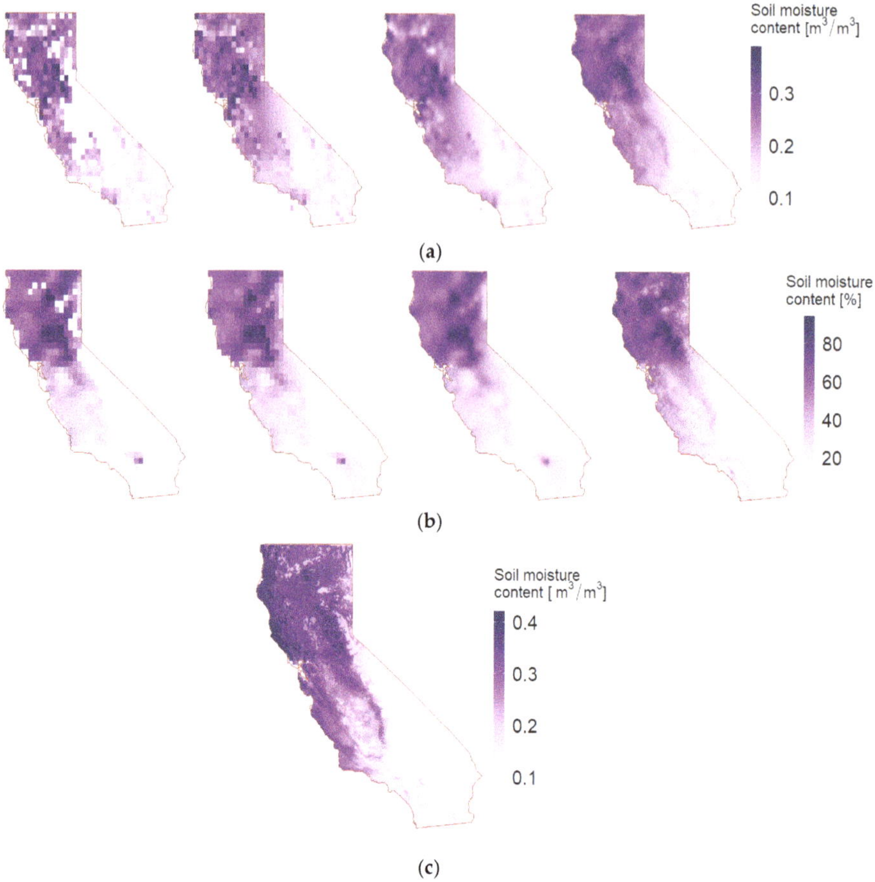

Figure 7. Output products after each processing step for DOY 81 2016: (**a**) combined ESA CCI SM input, daily filled, downscaled by bilinear interpolation (BIL) (Data engineering) and by RF products; (**b**) active ESA CCI SM input, daily filled, downscaled by BIL and by RF; and (**c**) final output SSM product.

6. Discussion

6.1. Determining the Best ESA CCI SM Predictor Combination

The supremacy of downscaling over standard bilinear interpolation as data engineering tool is as expected. Both types of RF models provided improvements across all validation metrics. However, there are still large errors present in all three types of spatial improvements. Incorporating ESA CCI SM products within the model leads to better performances across all metrics. Passive ESA CCI SM downscaled products outperform the others, but it still cannot be said with certainty which predictors should be chosen as an input for the second random forest model.

As displayed in Table 2, using RF downscaled products does not always contribute to a better prediction model. Although downscaling ESA CCI SM products using RF appeared to be a superior solution compared to the bilinear interpolation, it turns out that this is not always true. The use of first-step random forest only leads to marginally improved (or even deteriorated) accuracy compared to the bilinear interpolation method. The use of the RF downscaling procedure that excludes ESA CCI SM products provides the worst results across all metrics. In this case, there are no significant differences regarding the set of used products in the second RF model. Using bilinear interpolation method outperforms all variants of RF downscaling procedure that excludes ESA CCI SM products. Some of the best results from all the tested combinations are provided this way. Out of all model variants that use bilinearly interpolated products, the variant that uses the combination of active and passive ESA CCI SM products provides the best accuracy. Such combination provides the third best of all metrics from variants without NDVI shift and the fourth best from variants that use NDVI shift. The use of the downscaling through RF outperforms the bilinear interpolation only when other ESA CCI SM products are incorporated as model predictors. The best results across all metrics, for all tested combinations, are obtained in the case where combined and active ESA CCI SM RF downscaled products with incorporating ESA CCI SM products in the first RF model are used. The combination that uses combined and passive products and the one using just the combined product follow closely. Additionally, shifting NDVI values to obtain better matching with in situ data also introduces improvements across all ESA CCI SM combinations. The average improvement after NDVI shift in RMSE, R^2 and MAE is 0.0004 m^3/m^3, 0.0038 and 0.0001 m^3/m^3, respectively.

When comparing the downscaling using RF model with other two ESA CCI SM products and the downscaling using bilinear interpolation, the first one outperforms the other for most combinations. The differences in validation metrics vary with the combination of used ESA CCI SM predictors. Differences can be marginally small or as large as 0.0036 m^3/m^3, 0.0399 and 0.0027 m^3/m^3 for RMSE, R^2 and MAE, respectively. The metrics' differences are also not largely affected by the NDVI shift. These results suggest that the main work is done by the second RF model, while the method of downscaling of ESA CCI SM products in the first step has limited effect. All downscaling methods introduce additional errors which are not always successfully modeled in the second RF model. This is particularly seen in the case of the downscaling step without other ESA CCI SM predictors, where it actually deteriorates the quality of the final output.

In all tested combinations, the use of several ESA CCI SM products yielded better results than the use of only one product. Although the combined product is generated from active and passive data, it turns out that some variability between these three is left unaccounted for in the assimilation process. In congregation with other predictors, such variability can be successfully exploited by the two-step downscaling procedure. Since the RF downscaling procedure is complex and with significant requirements for memory and processing power, in some cases, the bilinear interpolation method might be preferred over it in the data engineering step. This can especially happen over larger areas, where RF model might become too heavy for standard uses and when simplicity of the bilinear interpolation is useful.

Considering that the usage of active and combined ESA CCI SM downscaled product (with ESA CCI SM products in RF model 1) has proved to be the best solution for generating high-resolution soil moisture maps, all additional validation was done only for that one, instead for all tested predictor combinations.

6.2. Predictor Importance

The predictor's relative importance was determined using the percentage of increase in RMSE that its omittance produces. The predictor being tested is omitted from all processing steps (from both RF models) and the RMSE is afterwards determined using the same validation technique as before. All ESA CCI SM products were grouped and treated as a single predictor to generalize interconnections and dependencies that exist between the two RF models. The NDVI shift was also considered and the relative importance was determined in both cases—with and without using NDVI shift.

Figure 8 shows that the group of ESA CCI SM products is by far the most important predictor. If omitted, at least two times bigger RMSE increase is to be expected, compared to the omittance of other predictors. Day of year (DOY) and NDVI can be classified as medium important predictors with increase of RMSE between 4% and 5% and the remaining ones are the least important predictors, having the increase of less than 1%. The NDVI shift slightly increased NDVI, DOY and LST_{DAY} importance, but it also decreased the relative importance of all the other predictors.

Such predictor importance corresponds to the observed correlation coefficient between the in situ soil moisture observations and the used predictors, except for the LST_{DAY} predictor. LST_{DAY} shows significant negative correlation with the in situ data (−0.55), but its importance is the smallest among all the predictors. The explanation for this can be that this is the result of the existence of two LST predictors. This way, the importance of each LST predictor is independently small, yet they have their share in the performance of the prediction model. On the other hand, the correlation coefficient of the DOY predictor is minor (−0.20), but it is the one of the top-three predictors by importance. Because some of the used predictors have delayed effect on the soil moisture content, such behavior can be better modeled by introducing the time information explicitly through DOY predictor. DOY information also helps the RF model to capture the yearly weather seasons, which have strong effect on the soil moisture content.

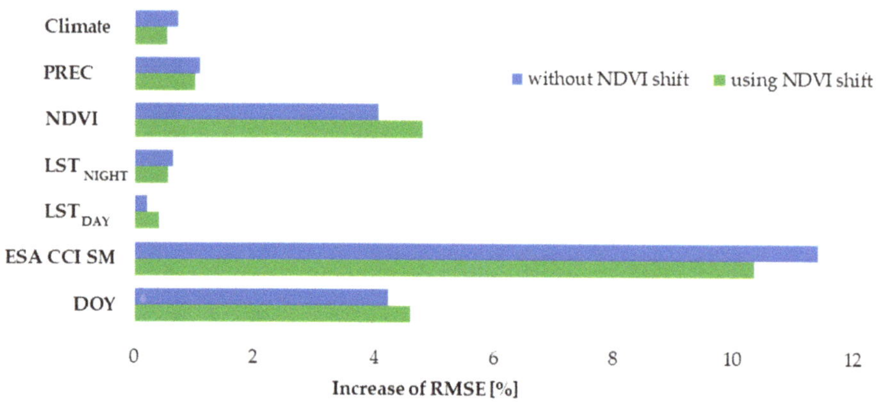

Figure 8. Relative variable importance based on increase of RMSE.

6.3. Spatial Patterns of the High-Resolution Soil Moisture Maps

The validation metrics have been calculated for every station independently (all metrics are available in Appendix A, Table A1). The calculated metrics have a wide range of values, with RMSE ranging 0.0182–0.1102 m^3/m^3, R^2 ranging 0.0000–0.9674 and MAE ranging 0.0141–0.0825 m^3/m^3 for results without the NDVI shift. When the NDVI shift is included, the metrics' upper boundaries are slightly improved, with RMSE of 0.0186–0.1065 m^3/m^3, R^2 of 0.0000–0.9694 and MAE of 0.0139–0.0795 m^3/m^3. Individually, if the RMSE threshold is defined as 0.04 m^3/m^3, only 19 stations (with and without the NDVI shift) reach this threshold. This is rather low performance, although it needs to be noted that low RMSE is in a way compensated by high R^2. From the stations that fail to reach the threshold, two thirds of them have R^2 higher than 0.7 and almost half of them have R^2 values higher than 0.8. No statistical relationship between RMSE and R^2 values has been detected. The number of observations per stations does not affect the metrics either, which suggests that the reason for such behavior needs to be examined regarding its spatial and climate characteristics.

Spatial patterns are examined by creating the bubble plots of the previously calculated metrics per station. As shown in Figure 9, the larger values across all metrics are more present in the coastal regions, while the values are generally smaller in the mainland. The NDVI shift has limited effect on spatial patterns, where individual values are changed, but the trend along the coast is still present. Such spatial patterns are attributed to the California's relief (Figure 1), which in a way creates a "wall" that limits the influence of the ocean and its effect on the precipitation and other climate conditions.

The climate zones by Köppen–Geiger are also differentiated by the "mountain wall". The ocean heavily impacts areas that are located between the coast and the mountain wall, while the mainland behind the mountains has its own climate conditions. The proximity of the ocean affects the soil moisture patterns because it influences precipitation amounts and the precipitation is taken as the main source of the soil moisture change. Near the coast, the tropical climate is present with larger precipitation amounts. Since the perception is used as a predictor in the model, the changes of SSM are successfully modeled near the coast (high R^2 values). Nevertheless, the RMSE has larger values because the precipitation has a spatial resolution of only 5 km and its additional improvements are necessary in order to reduce the RMSE values. On the contrary, the arid climate in the mainland with lower precipitation amounts has small variations of SSM content, which are not primarily caused by precipitation. Such variations are not modeled properly (lower R^2 values), but the remaining predictors still model total amount of SSM reasonably well, which provides lower RMSE values.

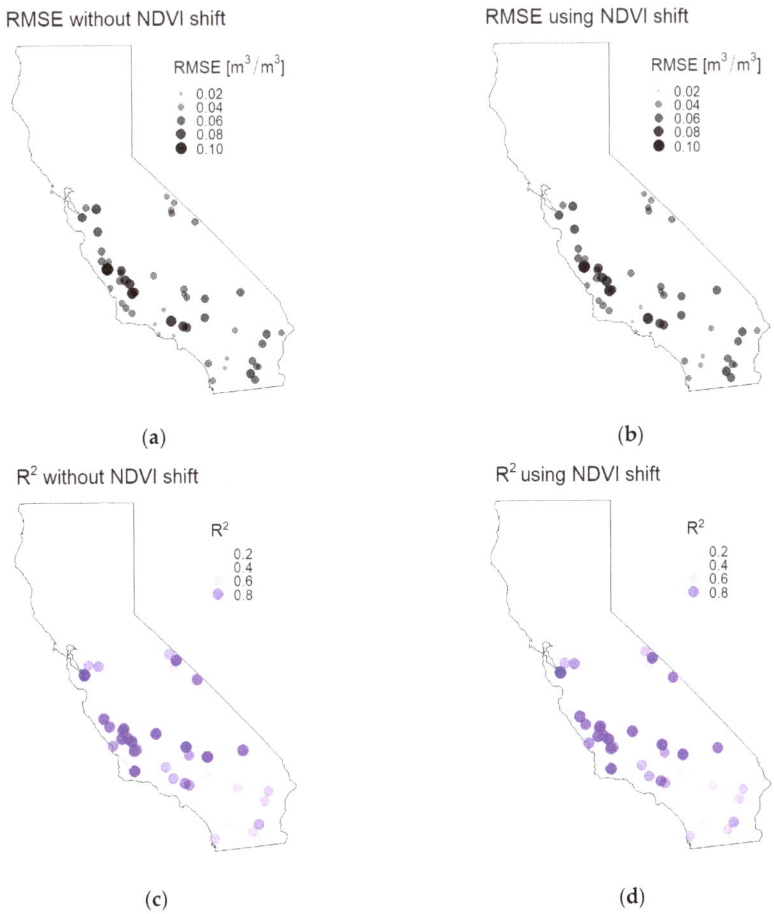

Figure 9. *Cont.*

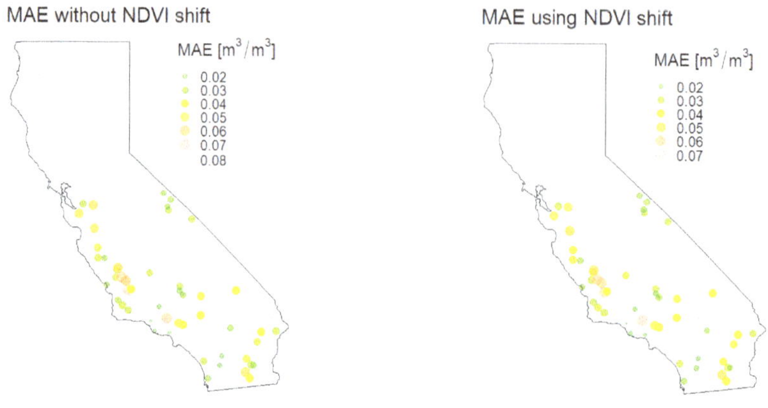

(e) (f)

Figure 9. Spatial distribution of the calculated metrics presented by bubble plots without and with using NDVI shift: (**a**,**b**) RMSE; (**c**,**d**) R^2; and (**e**,**f**) MAE.

These assumptions are confirmed by creating the boxplot charts of calculated metrics per climate type. Since spatial patterns are almost identical for results with and without NDVI shift, the boxplots of metrics are determined only for the results without NDVI shift. It can be clearly seen in Figure 10 that Climate Class B (desert and semi-arid climates) that covers most of the mainland has lower values of all metrics when compared to Climate Class C (tropical/megathermal climates). It can be said that the model is more accurate in the desert and semi-arid climate. This is more due to the lesser variations in the soil moisture than the efficiency of the model. For the tropical/megathermal climates, the situation is reversed, that is, the model is more efficient, but the overall accuracy of the downscaled data is lower. Although the climate is included in the predictor set as an attempt to differentiate such areas, it is clear that this approach was not successful. The potential solution might be to use this information as a way to split up the area of interest into segments and to build independent models for each segment separately. A similar approach has been done by soil types, and it proved to be useful [32]. In addition, the additional predictors should be checked and included in order to reduce RMSE in tropical/megathermal climates, but with the preservation of high R^2 value.

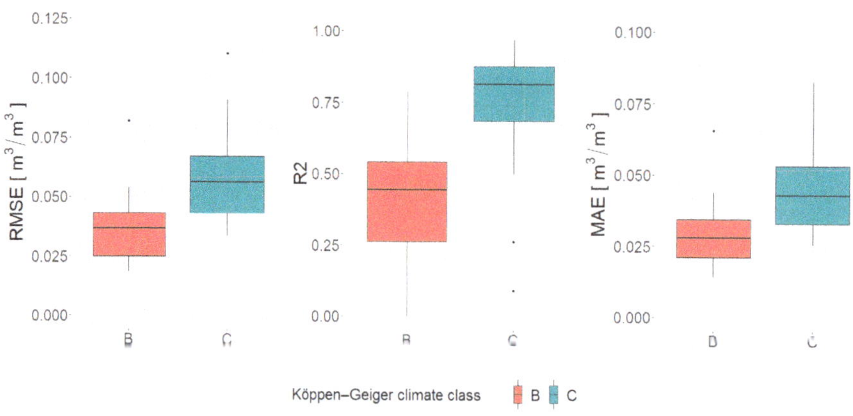

Figure 10. Boxplot of the calculated metrics (without NDVI shift) per stations differentiated by climate classes.

The spatial patterns are further examined regarding the land cover type by creating the boxplot charts per land cover type. For that purpose, the MODIS Land Cover Type Yearly L3 Global 500 m 2016 and its University of Maryland (UMD) classification scheme were used. Unfortunately, only a few land cover classes are well represented. Grassland, open shrubland and barren or sparsely vegetated land cover classes are represented with 25, 11 and 10 stations, respectively. All other land cover classes have three or fewer stations across the study area and therefore have to be left out from the boxplot, since there are insufficient data for analysis. As shown in Figure 11, the performance of the remaining three land cover classes varies per class. As expected, the barren or sparsely vegetated class and open shrubland class have lower RMSE and MAE than stations over grasslands. On the other hand, R^2 values over grasslands are very high (above 0.75) while R^2 of the other two land cover classes are below 0.5. Such values correspond to the findings of the other researches, where modeling soil moisture content becomes harder as the amount of vegetation increases [1].

Unfortunately, this highlights the main disadvantage of using PBO_H2O soil moisture network for modeling soil moisture content. Since it is a GPS based method for determining soil moisture content, only the land cover classes that provide open-sky conditions necessary for GPS signal can be used. Consequently, not all land cover classes (especially forests and other dense vegetation) can be covered. Therefore, since such land cover classes are not used during the training of the RF model, it is unlikely that the model will be able to provide good performance for these areas.

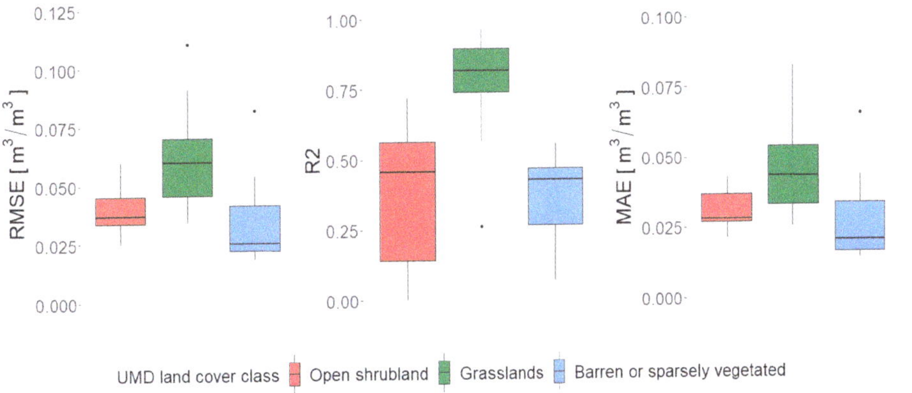

Figure 11. Boxplot of the calculated metrics (without NDVI shift) per stations differentiated by MODIS land cover classes (UMD classification scheme).

6.4. Temporal Patterns of the High-Resolution Soil Moisture Maps

The temporal line plot for each station is available in Appendix B Figure A1. Only the extreme ones are discussed in this section. The extreme values of RMSE and R^2 for the stations with more than 280 observations in 2016 are presented in Figure 12.

The modeled soil moisture is generally smoother when compared to the in situ values. Small variations are usually omitted, but most of the larger leaps are still successfully captured by the model. For most of the stations with larger RMSE values, the modeled soil moisture content is smaller than the in situ soil moisture content. The differences are larger than 0.2 m^3/m^3 and they mostly occur after the change in the soil moisture content (due to the precipitation). This suggests that, even though the precipitation is included in the predictors set, the model does not exploit that information well enough. The main advantage of the modeled soil moisture is that there are no missing data, which is a significant problem for some stations where several months of missing data occur. Although such results cannot be confirmed, by looking at the plotlines, the change of soil moisture contents in the time windows with missing in situ data seems to be reasonable, without unusual spikes and downs.

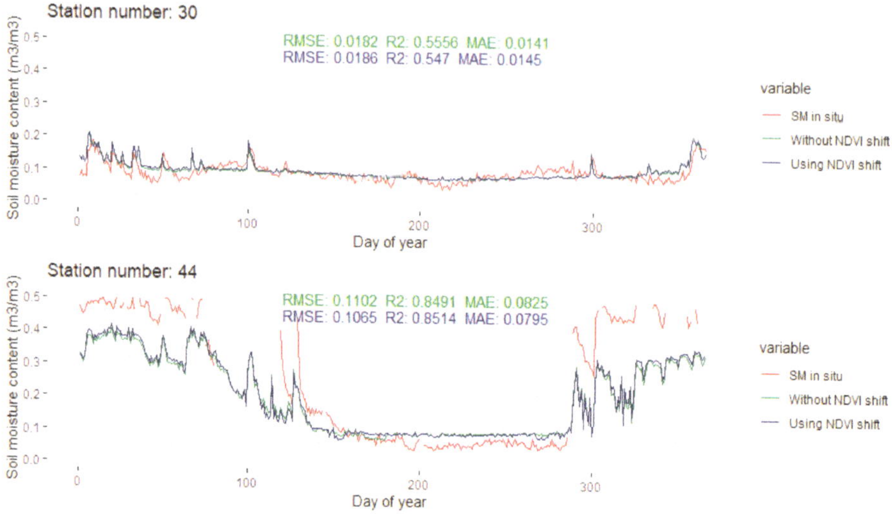

(a)

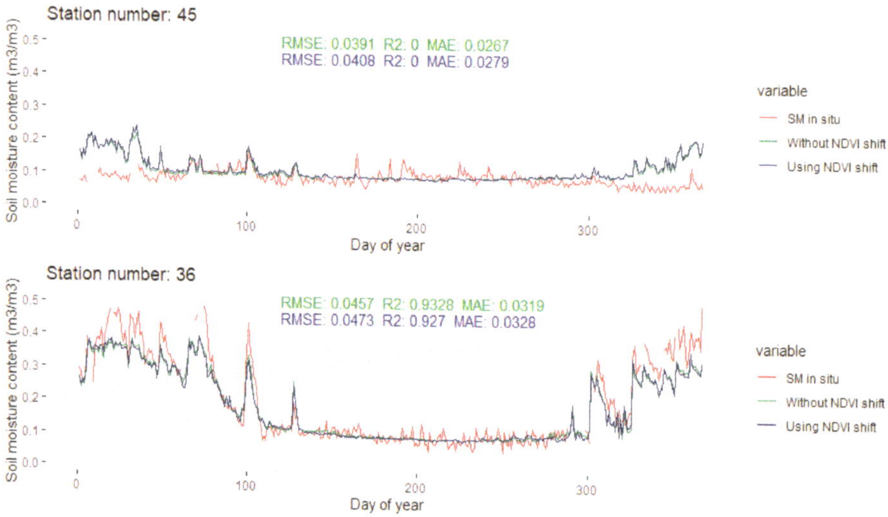

(b)

Figure 12. The temporal line plots of the stations with extreme values of RMSE and R^2: (**a**) the minimum value of RMSE for Station 30 and the maximum value of RMSE for Station 44; and (**b**) the minimum value of R^2 for Station 45 and the maximum value of R^2 for Station 36.

Temporal patterns of the soil moisture were also evaluated for every month of 2016. The in situ and the predicted soil moisture content have range of variations, which differ by month, as shown in Figure 13. The in situ variations are smallest in the June–September period, while they are larger throughout the rest of the year. Such patterns can also be observed in the predicted soil moisture content, although those variations are smaller. This corresponds to the smoothing introduced by the prediction model. Such behavior is especially present in the June–September period.

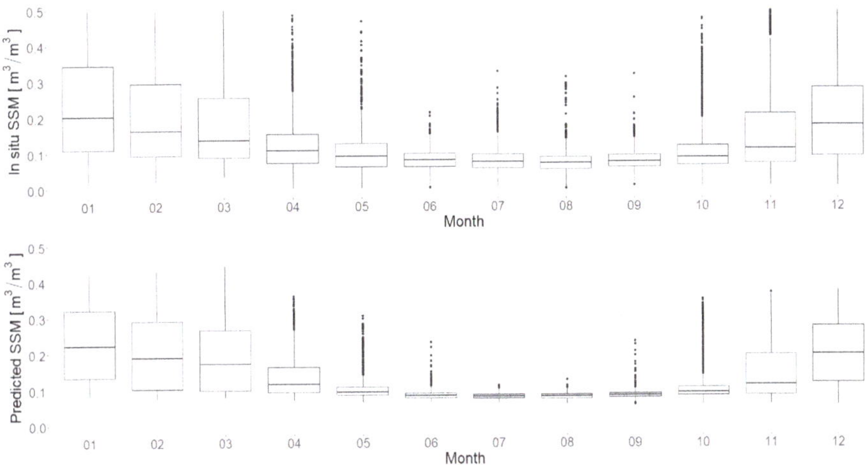

Figure 13. The boxplot charts of the in situ soil moisture observations and the five-fold cross-validations results (without NDVI shift), for every month of 2016.

The temporal monthly patterns were further inspected by calculating monthly validation metrics (Figure 14). As expected, the model efficiency differs during the year, with June–September RMSE and MAE being significantly lower than for the rest of the months. R^2 is very low for the same months and with improvements during the rest of the year.

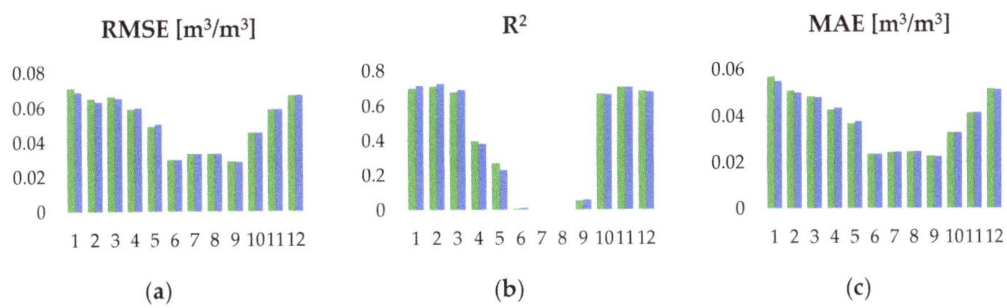

Figure 14. Validation metrics calculated over each month of 2016: (**a**) RMSE; (**b**) R^2; and (**c**) MAE;.

The monthly metrics correspond to the observed monthly variations. The months with smaller variations in SSM content in June–September period have relatively large measurement uncertainty. This is due to the reported accuracy of the in situ observations of 0.04 m^3/m^3. Although no observation weights are included in the RF model, this uncertainty is successfully characterized by RF model, producing smooth observations for this period. The predicted smooth lines mostly correspond to the average of in situ observations for these days. RMSE values under 0.04 m^3/m^3 for these months indicate that the smooth averaged predictions of the soil moisture are sufficiently accurate. On the other hand, the months with larger variations in SSM content show systematic behavior. The prediction model can capture this behavior successfully, which is proved by the strong correlation. Unfortunately, only the bigger changes are successfully modeled, while the smaller ones are omitted due to the same effect that exists in the June–September period. This produces larger RMSE values for these periods. Additionally, more snow and cloud cover is expected in these months, thus large areas of missing data might occur. The gap filling step is less effective in these conditions, which could also have an impact on higher values of RMSE for these months.

All these differences in RMSE and R^2 over the year imply that the model has limitations regarding the soil moisture variations that can be successfully captured. These limitations are primarily caused by the accuracy of the soil moisture observations, which needs to be accounted for in the prediction model.

6.5. Independent Validation of the High-Resolution Soil Moisture Maps

Independent validation of the generated soil moisture maps was performed using the in situ soil moisture observations available from the SCAN and USCRN soil networks. These datasets were pulled from the ISMN data archive. Only the top-soil soil moisture observations at 5 cm depth were taken into the account, which makes 15 SCAN stations and 6 USCRN stations available over the study area (Figure 15). Each station has daily soil moisture measurements for the whole 2016, resulting in 7686 soil moisture observations being available for the validation. For each station, the same validation metrics were calculated as in the five-fold cross-validation.

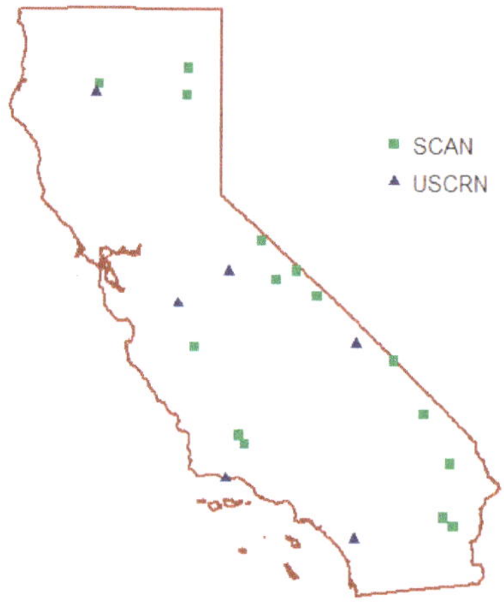

Figure 15. Spatial distribution of the SCAN and USCRN soil moisture stations over the study area used for the independent validation.

The calculated metrics (Table 3) show rather poor results of the produced high-resolution soil moisture maps against SCAN and USCRN stations data. Very large RMSE and MAE and low R^2 values over both networks suggest that these two sources are not at all comparable with the high-resolution soil moisture maps. The reason for this can be attributed to the differences in the soil moisture measurement depth and the mismatch of data sources regarding their spatial resolution. Both SCAN and USCRN in situ soil moisture observations provide soil moisture content at certain soil depth (5 cm depth used for the validation). Since second RF model is modeled using PBO_H2O in situ data that represent top-soil interval of 0–5 cm, the output high-resolution soil moisture maps also correspond to the 0–5 cm soil depth interval. Additionally, SCAN and USCRN in situ measurements are point measurements, while the output soil moisture maps have spatial resolution of 1 km. Point-scale in situ measurements need to be upscaled to the desired spatial resolution [46,47] before these two datasets can be compared. Such discrepancy in the spatial resolution is not present while building the second RF model since PBO_H2O spatial resolution of 1 km matches the spatial resolution of the used predictors. Conversion from the point-scale in situ measurement at certain depth to the interval measurements of 1 km spatial resolution is beyond the scope of this research. That is why the authors believe that the poor validation metrics against SCAN and USCRN stations should not be taken as a true quality assessment of the created high-resolution soil moisture maps and that the five-fold cross-validation metrics provide more realistic quality assessment.

Table 3. Validation metrics of the produced high-resolution soil moisture obtained by using SCAN and USCRN soil moisture stations data.

Netwok	Stations Count	Without NDVI Shift			Using NDVI Shift		
		RMSE [m^3/m^3]	R^2	MAE [m^3/m^3]	RMSE [m^3/m^3]	R^2	MAE [m^3/m^3]
SCAN	15	0.0885	0.2584	0.0689	0.0891	0.2631	0.0694
USCRN	6	0.0908	0.5666	0.0757	0.0938	0.5633	0.0776

7. Conclusions

The methodology used in this study proved to be a good solution for creating high resolution surface soil moisture maps over the area of California, USA. Within the research, downscaled SSM maps were produced for 2016. The output product spatial resolution was improved to 1 km. The proposed approach also considered filling data gaps as an initial step of the procedure, which in the end produced continuous product in both temporal and spatial domains. The filling of the missing data was performed using the universal kriging in the spatial domain and by applying spline fitting and interpolation in the temporal domain. The daily datasets without missing data were then used to produce 1 km soil moisture maps using two-step procedure. The first step was used as a data-engineering tool and its output was used as the input for the second step. The ESA CCI SM products and PBO_H2O in situ soil moisture observations were used as a main data input in the congregation with NDVI, LST, precipitation and climate zones as auxiliary datasets. The validation metrics were calculated for several tested models to determine the optimal one.

Comparison of the model results and soil moisture observations from SCAN and USCRN soil networks yielded rather poor results, which suggest that these two sources are not comparable due to the differences in the soil moisture measurement depth and the spatial resolution. That is why the model performance was evaluated through five-fold cross-validation. The best prediction model obtained soil moisture with RMSE of 0.0518 m^3/m^3 and R^2 of 0.7312, which is comparable to the results made by similar studies [19].

Our study found that both bilinear interpolation and RF downscaling procedure could be used as a data engineering tool for providing the additional predictors in soil moisture prediction models. As the calculated validation metrics indicate, the optimal soil moisture prediction model uses RF downscaled combined and active ESA CCI SM products in congregation with other auxiliary datasets and in situ soil moisture observations. The models that use bilinear interpolation as a data engineering tool provided results that are only marginally deteriorated. This is because the RF regression model in the second step does most of the work. The study showed that, when downscaling one type of ESA CCI SM product, the remaining two types of ESA CCI SM products in congregation with other predictors should be used. Although a combined product is generated from active and passive data, it turns out that, in the assimilation process, some variability between these three is left unaccounted for. The study also implemented the NDVI shift, due to its delayed effect on the SSM, in order to boost its correlation with in situ soil moisture observations. This proved to be useful, with the improvements of all metrics across all model variations.

The study also highlights the pros and cons of using PBO_H2O in situ soil moisture observations for soil moisture downscaling. Since it is a GPS based method, only the land cover classes that provide open-sky conditions necessary for GPS signal are available. Consequently, not all land cover classes (especially forests and other dense vegetation) can be covered. The accuracy of the soil moisture observations limits the amount of variations that could be successfully modeled in the downscaling procedure. On the other hand, the observations' spatial resolution of 1 km matches the commonly desired SSM output. This way, there is no need for in situ upscaling procedure which can induce additional prediction errors.

By analyzing the spatial patterns of the validation metrics, it is concluded that the model performances vary for different climate zones, even though the climate is included as a model predictor. The climate information should be further inspected and possibly used as a way to divide the area of interest into segments. The spatial patterns have also been examined regarding the land cover class. The model performed best over the barren or sparsely vegetated and open shrubland areas and it had lower performance over grasslands. Individually per station, higher RMSE value is followed by high R^2 value and vice versa. The model also has temporal variability, with lower RMSE values and low R^2 values over the June–September period, and higher RMSE and R^2 values for other months.

Unfortunately, PBO_H2O in situ soil moisture observations are no longer available because the project ended in 2017. Similar projects in the future are encouraged and welcomed, especially with improved accuracy of the SSM observations. In the meantime, the more common point-scale in situ observations might be usable, but this requires further testing. If so, the methodology can easily be transferable to other study sites, as long as some of the in situ soil moisture observations are available. All other used datasets are globally available, except the precipitation dataset which should be replaced by some of the alternative dataset.

Some additional sources of the used products (e.g., The Copernicus Global Land Service for NDVI and LST) should also be considered and incorporated within the gap filling procedure. The gap filling can be additionally improved by using the kriging with an external drift instead of the universal kriging. Beside these improvements, new predictors, such as soil characteristics, albedo, topography, etc., can be added to the model in the future, which could further improve soil moisture predictions. Additionally, downscaling of the precipitation dataset could also be considered, since its effect on the SSM variability is significant.

Author Contributions: All authors worked equally on the research conceptualization, investigation, discussion, and data modeling. J.K. was involved in data preparation and case study implementation. Ž.C., D.M., N.S. and N.B. participated in the writing of the manuscript, however J.K. took the lead. All authors have read and agreed to the published version of the manuscript.

Funding: This research received no external funding.

Acknowledgments: This study was supported by the Serbian Ministry of Education, Science and Technological Development, project TR 36020.

Conflicts of Interest: The authors declare no conflict of interest.

Appendix A

Table A1. Five-fold cross-validation metrics aggregated over 10 independent splits for each station.

Station ID	KG Climate Class	Land Cover Class (UMD)	Obs. Count	Without NDVI Shift			Using NDVI Shift		
				RMSE [m^3/m^3]	R^2	MAE [m^3/m^3]	RMSE [m^3/m^3]	R^2	MAE [m^3/m^3]
1	B	10	239	0.0527	0.6981	0.0390	0.0538	0.7097	0.0399
2	B	7	311	0.0365	0.6583	0.0281	0.0341	0.6468	0.0269
3	B	16	281	0.0538	0.4969	0.0437	0.0546	0.4744	0.0442
4	C	10	325	0.0674	0.6202	0.0539	0.0648	0.6385	0.0516
5	B	16	323	0.0253	0.4741	0.0202	0.0264	0.4875	0.0216
6	B	7	366	0.0341	0.1623	0.0275	0.0308	0.2232	0.0252
7	B	9	364	0.0445	0.7491	0.0341	0.0442	0.7477	0.0340
8	B	13	363	0.0419	0.3098	0.0327	0.0420	0.3086	0.0327
9	C	7	365	0.0490	0.0911	0.0386	0.0508	0.0453	0.0402
10	B	16	366	0.0247	0.4520	0.0209	0.0251	0.4276	0.0211

Table A1. *Cont.*

Station ID	KG Climate Class	Land Cover Class (UMD)	Obs. Count	Without NDVI Shift			Using NDVI Shift		
				RMSE [m^3/m^3]	R^2	MAE [m^3/m^3]	RMSE [m^3/m^3]	R^2	MAE [m^3/m^3]
11	C	10	364	0.0516	0.6515	0.0373	0.0532	0.6577	0.0386
12	B	16	365	0.0208	0.2470	0.0159	0.0216	0.2399	0.0163
13	C	9	366	0.0568	0.6773	0.0430	0.0563	0.6787	0.0436
14	B	16	330	0.0222	0.3164	0.0177	0.0231	0.2736	0.0187
15	B	16	366	0.0375	0.4366	0.0344	0.0356	0.4890	0.0325
16	C	7	309	0.0334	0.6184	0.0273	0.0314	0.5999	0.0256
17	C	9	305	0.0646	0.5597	0.0507	0.0652	0.5554	0.0511
18	C	10	366	0.0623	0.7811	0.0434	0.0629	0.7564	0.0429
19	C	10	351	0.0701	0.8165	0.0490	0.0660	0.8465	0.0465
20	B	16	365	0.0219	0.4168	0.0148	0.0212	0.4439	0.0139
21	C	10	362	0.0837	0.7437	0.0648	0.0848	0.7546	0.0664
22	C	7	317	0.0598	0.4977	0.0430	0.0595	0.5056	0.0426
23	C	10	277	0.0837	0.8414	0.0600	0.0859	0.8255	0.0615
24	C	10	215	0.0413	0.9163	0.0330	0.0417	0.9163	0.0333
25	B	16	226	0.0818	0.0706	0.0657	0.0809	0.0594	0.0643
26	B	7	366	0.0247	0.7200	0.0214	0.0243	0.7197	0.0211
27	B	7	366	0.0433	0.3089	0.0355	0.0437	0.2903	0.0358
28	B	7	363	0.0369	0.5036	0.0296	0.0380	0.4769	0.0305
29	C	10	358	0.0553	0.9235	0.0416	0.0540	0.9114	0.0405
30	B	16	366	0.0182	0.5556	0.0141	0.0186	0.5470	0.0145
31	C	10	364	0.0428	0.7395	0.0299	0.0456	0.7133	0.0314
32	B	7	366	0.0319	0.1234	0.0248	0.0317	0.1050	0.0249
33	C	10	260	0.0909	0.8956	0.0695	0.0850	0.9015	0.0649
34	C	10	361	0.0544	0.8733	0.0445	0.0547	0.8596	0.0438
35	C	10	363	0.0600	0.8106	0.0426	0.0588	0.8049	0.0419
36	C	10	340	0.0457	0.9328	0.0319	0.0473	0.9270	0.0328
37	B	10	340	0.0344	0.7878	0.0255	0.0335	0.7994	0.0250
38	C	10	319	0.0739	0.9224	0.0590	0.0760	0.9126	0.0597
39	C	10	359	0.0457	0.8704	0.0334	0.0454	0.8801	0.0333
40	C	10	290	0.0732	0.8691	0.0634	0.0720	0.8619	0.0621
41	C	12	361	0.0403	0.9081	0.0327	0.0431	0.8891	0.0353
42	C	12	284	0.0399	0.8071	0.0312	0.0462	0.7651	0.0352
43	C	6	366	0.0660	0.8704	0.0544	0.0646	0.8797	0.0535
44	C	10	285	0.1102	0.8491	0.0825	0.1065	0.8514	0.0795
45	B	7	352	0.0391	0.0000	0.0267	0.0408	0.0000	0.0279
46	C	13	239	0.0485	0.8772	0.0360	0.0502	0.8786	0.0371
47	B	7	366	0.0473	0.4567	0.0381	0.0486	0.4478	0.0398
48	C	10	348	0.0627	0.2619	0.0514	0.0621	0.2670	0.0508
49	C	12	364	0.0433	0.8493	0.0327	0.0425	0.8563	0.0324
50	C	10	152	0.0649	0.9674	0.0506	0.0625	0.9694	0.0496
51	B	16	239	0.0426	0.0886	0.0321	0.0430	0.0779	0.0321
52	C	10	344	0.0363	0.5600	0.0272	0.0375	0.5487	0.0280
53	C	10	343	0.0699	0.7566	0.0538	0.0630	0.8017	0.0489
54	C	8	178	0.0419	0.7131	0.0331	0.0430	0.7241	0.0338
55	C	10	353	0.0373	0.9127	0.0277	0.0366	0.9242	0.0270
56	C	10	365	0.0345	0.7206	0.0255	0.0349	0.7032	0.0260

Appendix B

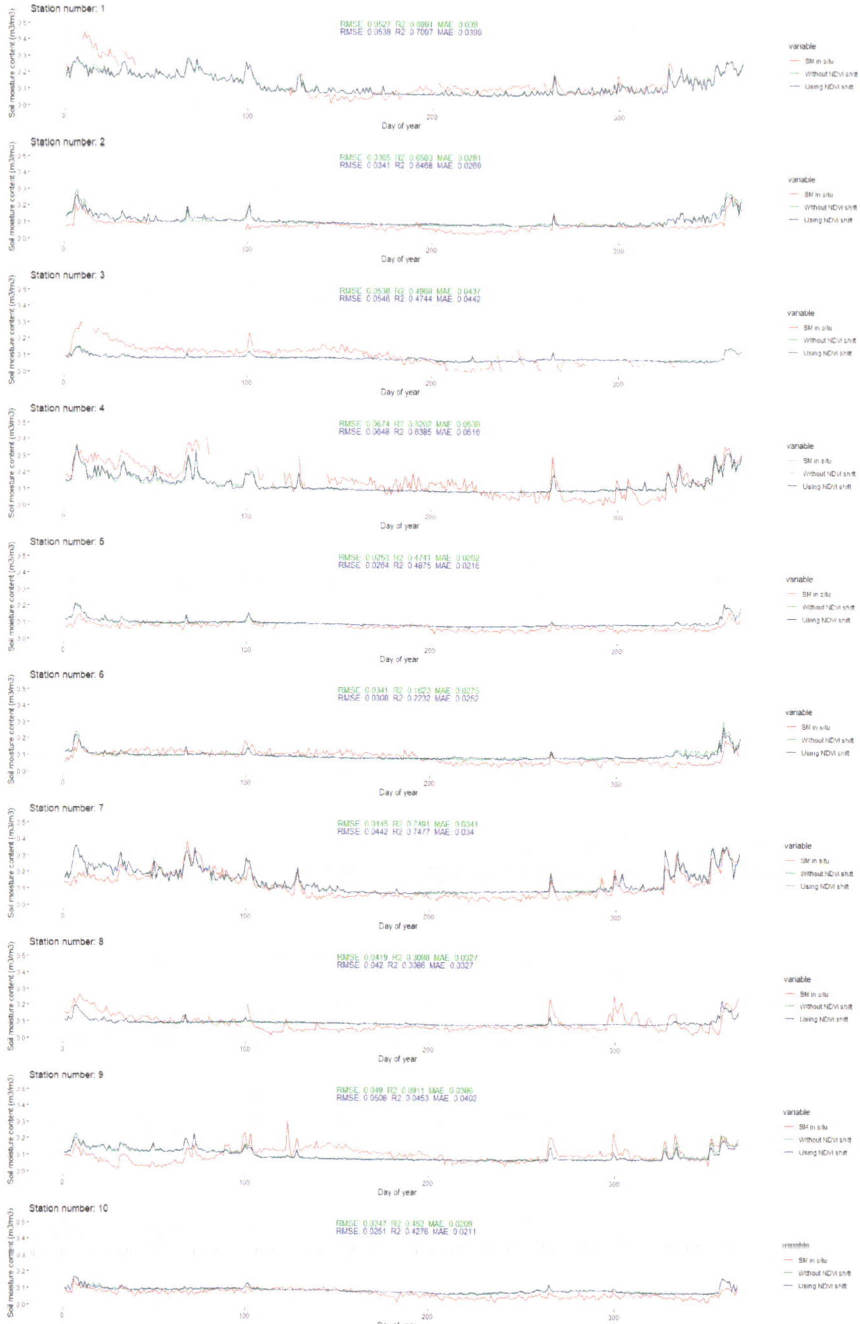

Figure A1. *Cont.*

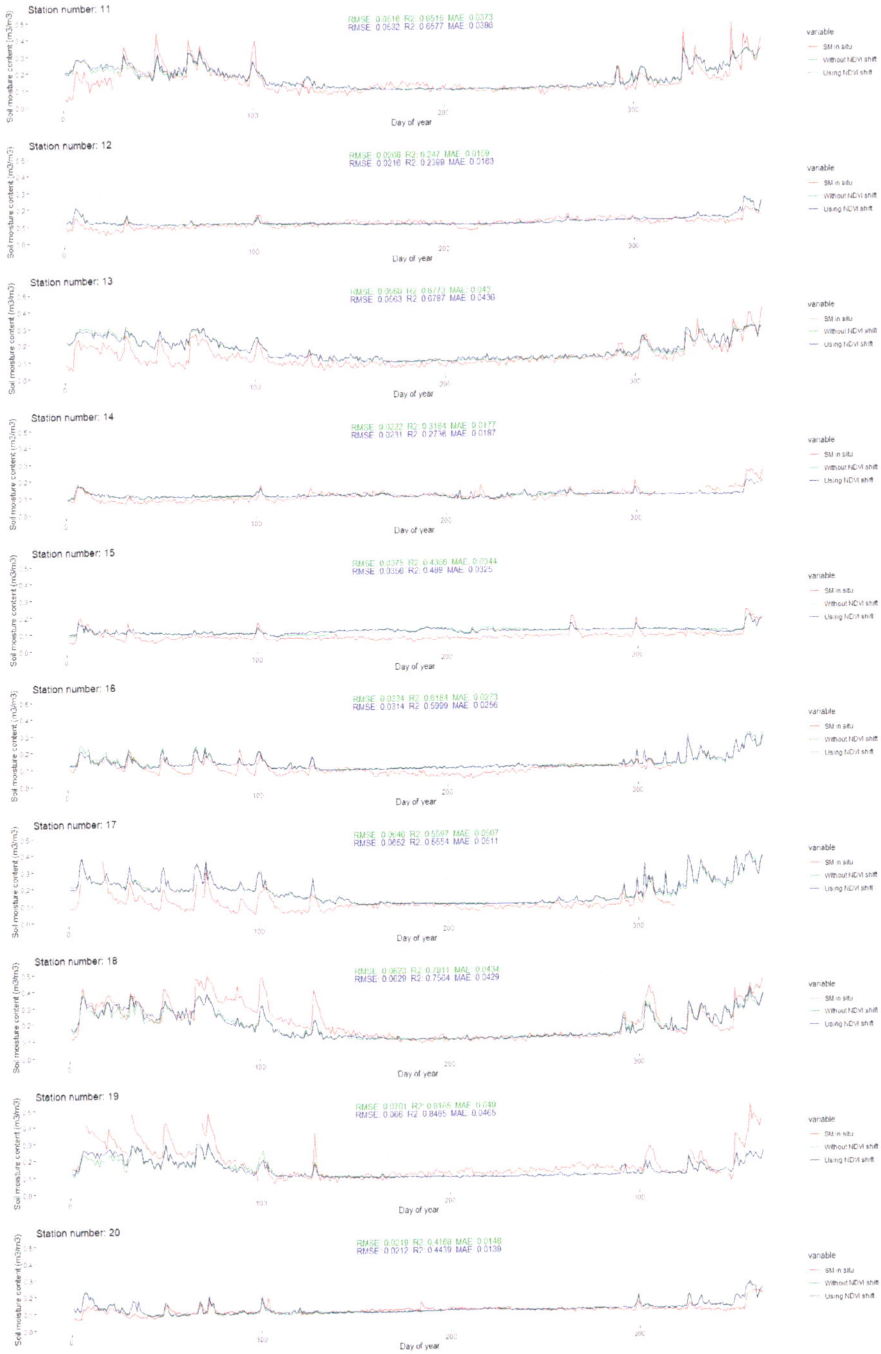

Figure A1. *Cont.*

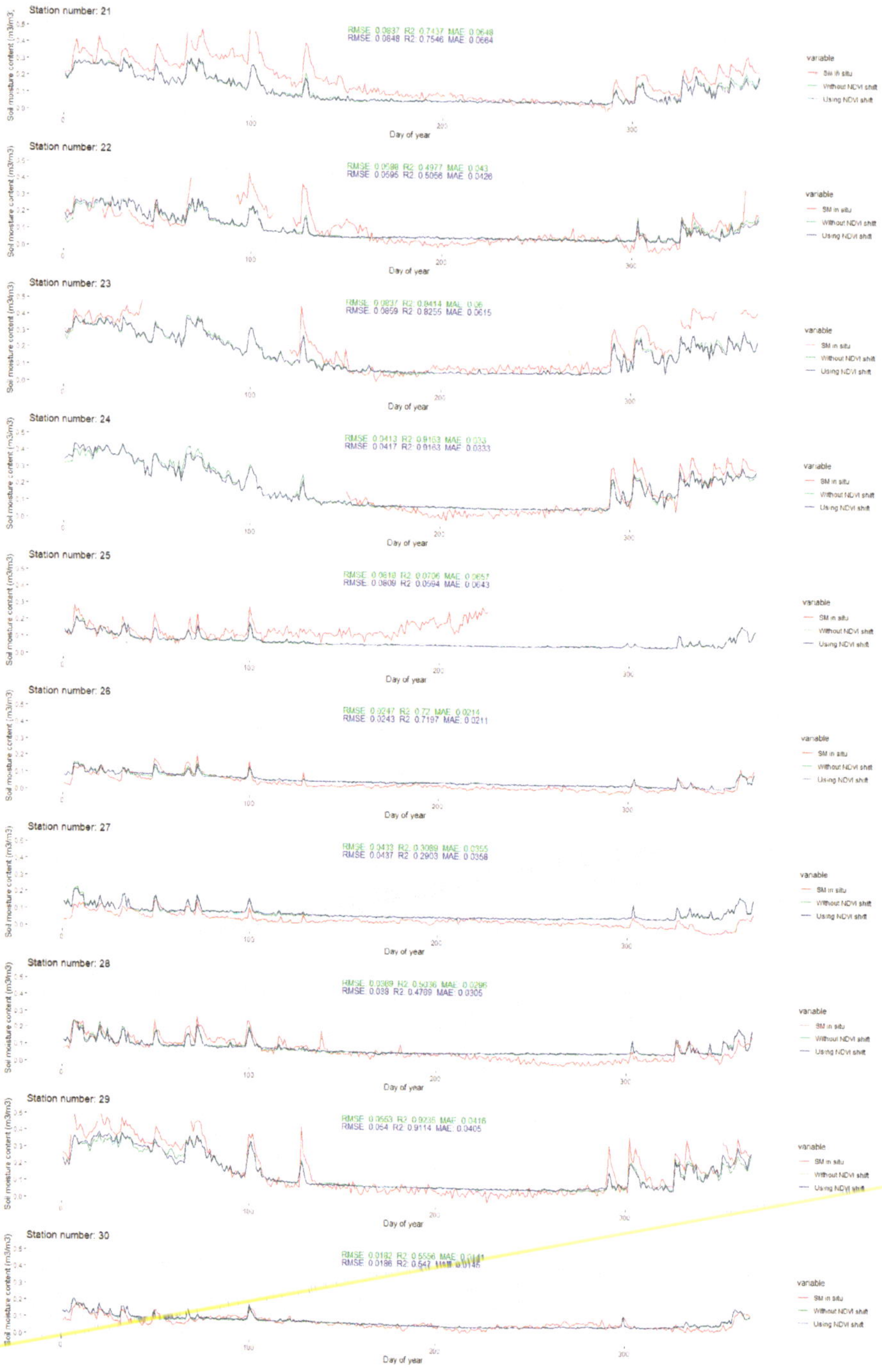

Figure A1. *Cont.*

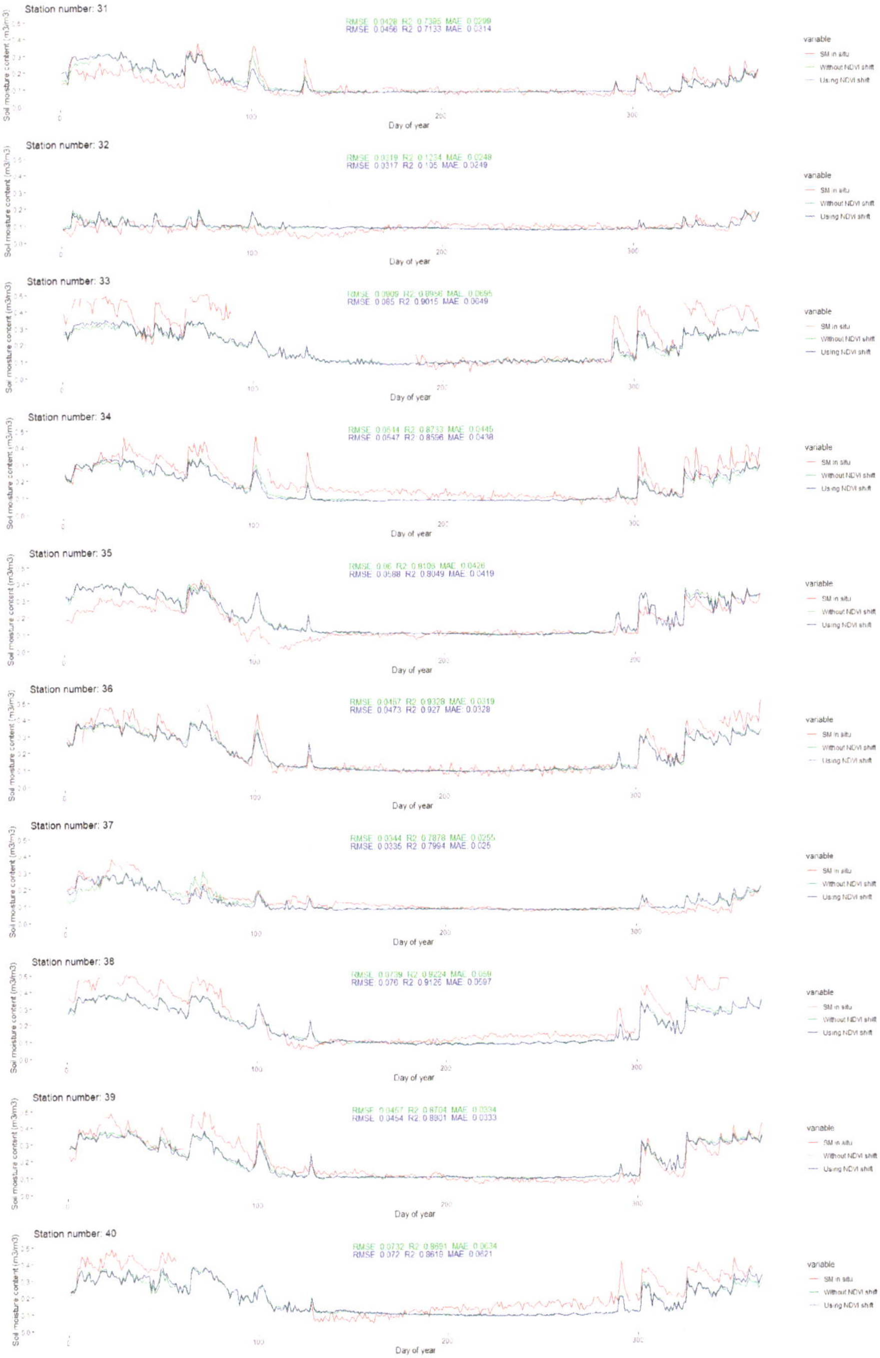

Figure A1. *Cont.*

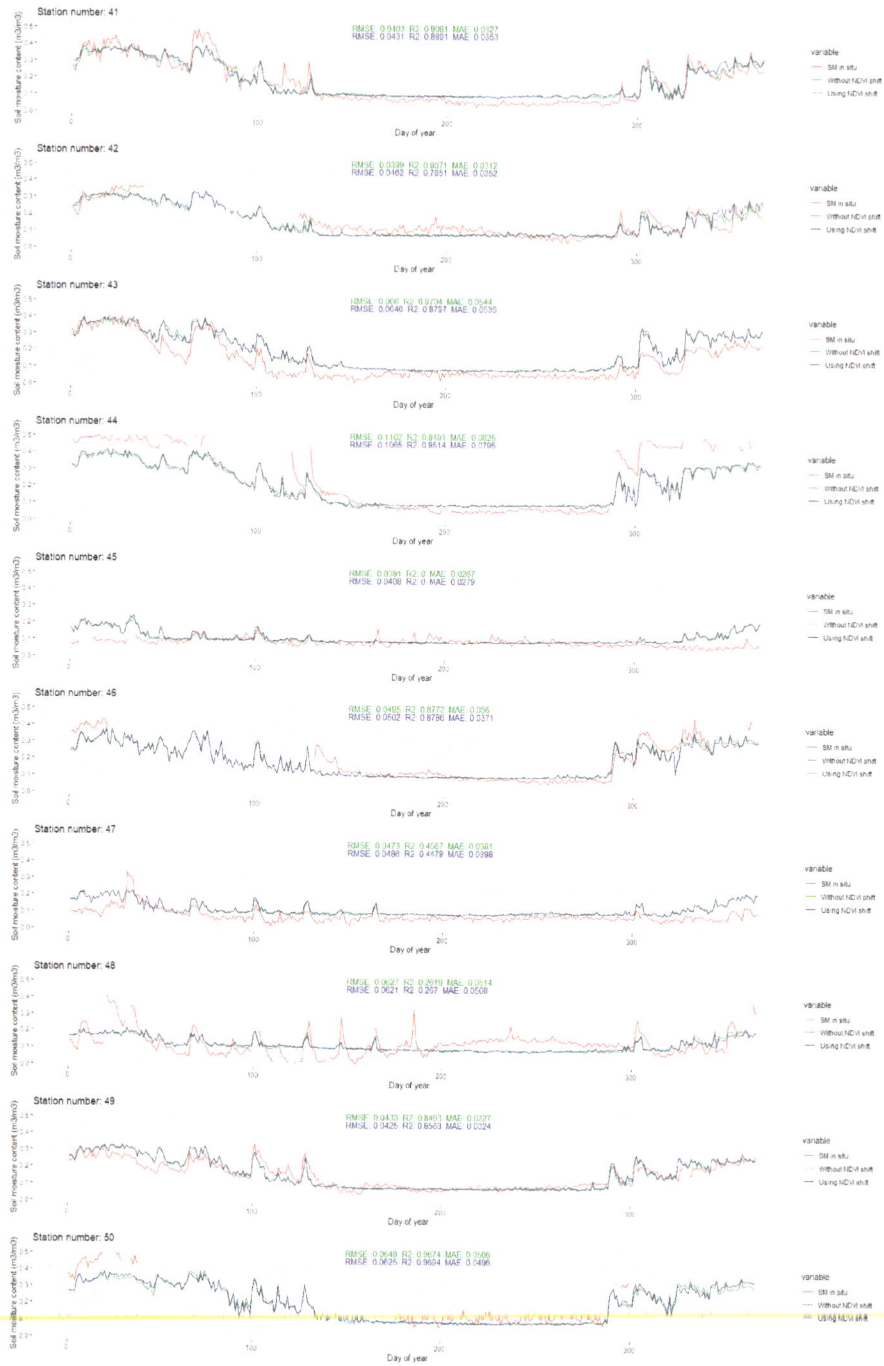

Figure A1. *Cont.*

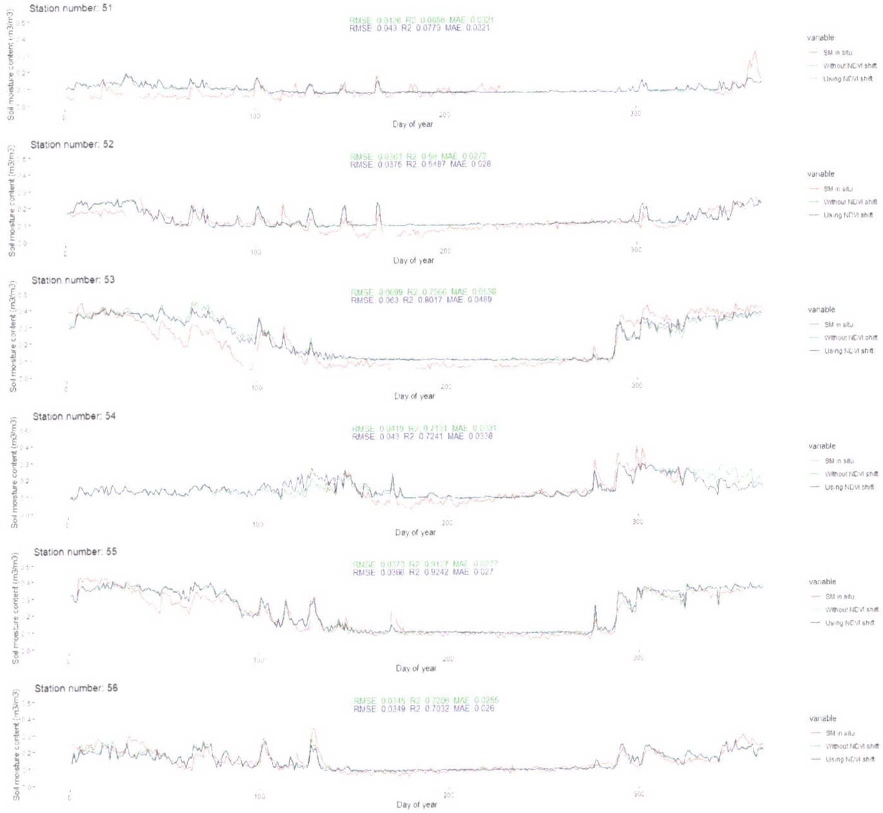

Figure A1. Temporal dynamics of modeled soil moisture against in situ values for each station in the study area.

References

1. Srivastava, P.K.; Petropoulos, G.P.; Kerr, Y.H. *Satellite soil Moisture Retrieval: Techniques and Applications*; Elsevier: Amsterdam, The Netherlands, 2016; ISBN 978-0-12-803388-3.
2. Seneviratne, S.I.; Corti, T.; Davin, E.L.; Hirschi, M.; Jaeger, E.B.; Lehner, I.; Orlowsky, B.; Teuling, A.J. Investigating soil moisture–climate interactions in a changing climate: A review. *Earth-Sci. Rev.* **2010**, *99*, 125–161. [CrossRef]
3. Corradini, C. Soil moisture in the development of hydrological processes and its determination at different spatial scales. *J. Hydrol.* **2014**, *516*, 1–5. [CrossRef]
4. Entekhabi, D.; Rodriguez-Iturbe, I.; Castelli, F. Mutual interaction of soil moisture state and atmospheric processes. *J. Hydrol.* **1996**, *184*, 3–17. [CrossRef]
5. Osenga, E.C.; Arnott, J.C.; Endsley, K.A.; Katzenberger, J.W. Bioclimatic and soil moisture monitoring across elevation in a mountain watershed: Opportunities for research and resource management. *Water Resour. Res.* **2019**, *55*, 2493–2503. [CrossRef]
6. Srivastava, P.K. Satellite soil moisture: review of theory and applications in water resources. *Water Resour. Manag.* **2017**, *31*, 3161–3176. [CrossRef]
7. Vivoni, E.R.; Moreno, H.A.; Mascaro, G.; Rodriguez, J.C.; Watts, C.J.; Garatuza-Payan, J.; Scott, R.L. Observed relation between evapotranspiration and soil moisture in the North American monsoon region. *Geophys. Res. Lett.* **2008**, *35*. [CrossRef]

8. Nagler, P.L.; Glenn, E.P.; Kim, H.; Emmerich, W.; Scott, R.L.; Huxman, T.E.; Huete, A.R. Relationship between evapotranspiration and precipitation pulses in a semiarid rangeland estimated by moisture flux towers and MODIS vegetation indices. *J. Arid Environ.* **2007**, *70*, 443–462. [CrossRef]
9. Bartsch, A.; Balzter, H.; George, C. The influence of regional surface soil moisture anomalies on forest fires in Siberia observed from satellites. *Environ. Res. Lett.* **2009**, *4*, 045021. [CrossRef]
10. World Meteorological Organization (WMO); United Nations Educational, Scientific and Cultural Organization; United Nations Environment Programme; International Council for Science. *GCOS, 138. Implementation Plan for the Global Observing System for Climate in Support of the UNFCCC*; WMO: Geneva, Switzerland, 2010.
11. World Meteorological Organization (WMO); United Nations Educational, Scientific and Cultural Organization; United Nations Environment Programme; International Council for Science. *GCOS, 154. Systematic Observation Requirements for Satellite-based Products for Climate Supplemental details to the satellite-based component of the Implementation Plan for the Global Observing System for Climate in Support of the UNFCCC*; WMO: Geneva, Switzerland, 2011.
12. Hillel, D. *Introduction to Environmental Soil Physics*; Elsevier: Amsterdam, The Netherlands, 2003; ISBN 978-0-12-348655-4.
13. Zhu, H.; Liu, T.; Xue, B.; Wang, G. Modified Richards' equation to improve estimates of soil moisture in two-layered soils after infiltration. *Water* **2018**, *10*, 1174. [CrossRef]
14. Moran, M.S.; Peters-Lidard, C.D.; Watts, J.M.; McElroy, S. Estimating soil moisture at the watershed scale with satellite-based radar and land surface models. *Can. J. Remote Sens.* **2004**, *30*, 22. [CrossRef]
15. Kumar, K.; Arora, M.K.; Hariprasad, K.S. Geostatistical analysis of soil moisture distribution in a part of solani river catchment. *Appl. Water Sci.* **2016**, *6*, 25–34. [CrossRef]
16. Kathuria, D.; Mohanty, B.P.; Katzfuss, M. A nonstationary geostatistical framework for soil moisture prediction in the presence of surface heterogeneity. *Water Resour. Res.* **2019**, *55*, 729–753. [CrossRef]
17. Wagner, W.; Blöschl, G.; Pampaloni, P.; Calvet, J.-C.; Bizzarri, B.; Wigneron, J.-P.; Kerr, Y. Operational readiness of microwave remote sensing of soil moisture for hydrologic applications. *Hydrol. Res.* **2007**, *38*, 1–20. [CrossRef]
18. Reichle, R.H.; Crow, W.T.; Koster, R.D.; Sharif, H.O.; Mahanama, S.P.P. Contribution of soil moisture retrievals to land data assimilation products. *Geophys. Res. Lett.* **2008**, *35*. [CrossRef]
19. Sabaghy, S.; Walker, J.P.; Renzullo, L.J.; Jackson, T.J. Spatially enhanced passive microwave derived soil moisture: Capabilities and opportunities. *Remote Sens. Environ.* **2018**, *209*, 551–580. [CrossRef]
20. Peng, J.; Loew, A.; Merlin, O.; Verhoest, N.E.C. A review of spatial downscaling of satellite remotely sensed soil moisture: Downscale satellite-based soil moisture. *Rev. Geophys.* **2017**, *55*, 341–366. [CrossRef]
21. Kawanishi, T.; Sezai, T.; Ito, Y.; Imaoka, K.; Takeshima, T.; Ishido, Y.; Shibata, A.; Miura, M.; Inahata, H.; Spencer, R.W.; et al. The advanced microwave scanning radiometer for the earth observing system (AMSR-E), NASDA's contribution to the EOS for global energy and water cycle studies. *IEEE Trans. Geosci. Remote Sens.* **2003**, *41*, 184–194. [CrossRef]
22. Kerr, Y.H.; Waldteufel, P.; Wigneron, J.-P.; Delwart, S.; Cabot, F.; Boutin, J.; Escorihuela, M.-J.; Font, J.; Reul, N.; Gruhier, C.; et al. The SMOS Mission: New Tool for Monitoring Key Elements of the Global Water Cycle. *Proc. IEEE* **2010**, *98*, 666–687. [CrossRef]
23. Entekhabi, D.; Njoku, E.G.; O'Neill, P.E.; Kellogg, K.H.; Crow, W.T.; Edelstein, W.N.; Entin, J.K.; Goodman, S.D.; Jackson, T.J.; Johnson, J.; et al. The Soil Moisture Active Passive (SMAP) Mission. *Proc. IEEE* **2010**, *98*, 704–716. [CrossRef]
24. Figa-Saldaña, J.; Wilson, J.J.W.; Attema, E.; Gelsthorpe, R.; Drinkwater, M.R.; Stoffelen, A. The advanced scatterometer (ASCAT) on the meteorological operational (MetOp) platform: A follow on for European wind scatterometers. *Can. J. Remote Sens.* **2002**, *28*, 404–412. [CrossRef]
25. Bauer-Marschallinger, B.; Cao, S.; Schaufler, S.; Paulik, C.; Naeimi, V.; Wagner, W. 1km Soil Moisture from Downsampled Sentinel-1 SAR Data: Harnessing Assets and Overcoming Obstacles. *Geophys. Res.* **2017**, *19*, 17330.
26. Dorigo, W.; Wagner, W.; Albergel, C.; Albrecht, F.; Balsamo, G.; Brocca, L.; Chung, D.; Ertl, M.; Forkel, M.; Gruber, A.; et al. ESA CCI soil moisture for improved earth system understanding: State-of-the art and future directions. *Remote Sens. Environ.* **2017**, *203*, 185–215. [CrossRef]

27. Jin, Y.; Ge, Y.; Wang, J.; Heuvelink, G.; Wang, L. Geographically weighted area-to-point regression kriging for spatial downscaling in remote sensing. *Remote Sens.* **2018**, *10*, 579. [CrossRef]
28. Montzka, C.; Rötzer, K.; Bogena, H.; Sanchez, N.; Vereecken, H. A new soil moisture downscaling approach for SMAP, SMOS, and ASCAT by predicting sub-grid variability. *Remote Sens.* **2018**, *10*, 427. [CrossRef]
29. Chen, S.; She, D.; Zhang, L.; Guo, M.; Liu, X. Spatial downscaling methods of soil moisture based on multisource remote sensing data and its application. *Water* **2019**, *11*, 1401. [CrossRef]
30. Torres-Rua, A.; Ticlavilca, A.; Bachour, R.; McKee, M. Estimation of surface soil moisture in irrigated lands by assimilation of landsat vegetation indices, surface energy balance products, and relevance vector machines. *Water* **2016**, *8*, 167. [CrossRef]
31. Liu, Y.; Yang, Y.; Jing, W.; Yue, X. Comparison of different machine learning approaches for monthly satellite-based soil moisture downscaling over northeast China. *Remote Sens.* **2017**, *10*, 31. [CrossRef]
32. Abbaszadeh, P.; Moradkhani, H.; Zhan, X. Downscaling SMAP radiometer soil moisture over the CONUS using an ensemble learning method. *Water Resour. Res.* **2019**, *55*, 324–344. [CrossRef]
33. CDFA–Statistics. Available online: https://www.cdfa.ca.gov/statistics/ (accessed on 31 January 2020).
34. Larson, K.M. GPS interferometric reflectometry: Applications to surface soil moisture, snow depth, and vegetation water content in the western United States. *WIREs Water* **2016**, *3*, 775–787. [CrossRef]
35. Nandintsetseg, B.; Shinoda, M.; Kimura, R.; Ibaraki, Y. Relationship between soil moisture and vegetation activity in the Mongolian steppe. *SOLA* **2010**, *6*, 29–32. [CrossRef]
36. Vicente -Serrano, S.M.; Pons-Fernández, X.; Cuadrat-Prats, J.M. Mapping soil moisture in the central Ebro river valley (northeast Spain) with landsat and NOAA satellite imagery: A comparison with meteorological data. *Int. J. Remote Sens.* **2004**, *25*, 4325–4350. [CrossRef]
37. Piles, M.; Camps, A.; Vall-llossera, M.; Corbella, I.; Panciera, R.; Rudiger, C.; Kerr, Y.H.; Walker, J. Downscaling SMOS-derived soil moisture using MODIS visible/infrared data. *IEEE Trans. Geosci. Remote Sens.* **2011**, *49*, 3156–3166. [CrossRef]
38. Liang, M.; Pause, M.; Prechtel, N.; Schramm, M. Regionalization of coarse scale soil moisture products using fine-scale vegetation indices—Prospects and case study. *Remote Sens.* **2020**, *12*, 551. [CrossRef]
39. Long, D.; Bai, L.; Yan, L.; Zhang, C.; Yang, W.; Lei, H.; Quan, J.; Meng, X.; Shi, C. Generation of spatially complete and daily continuous surface soil moisture of high spatial resolution. *Remote Sens. Environ.* **2019**, *233*, 111364. [CrossRef]
40. Kottek, M.; Grieser, J.; Beck, C.; Rudolf, B.; Rubel, F. World map of the Köppen-Geiger climate classification updated. *metz* **2006**, *15*, 259–263. [CrossRef]
41. Bovik, A.C. Chapter 3–Basic gray level image processing. In *The Essential Guide to Image Processing*; Bovik, A., Ed.; Academic Press: Boston, MA, USA, 2009; pp. 43–68. ISBN 978-0-12-374457-9.
42. Breiman, L. Random forests. *Mach. Learn.* **2001**, *45*, 5–32. [CrossRef]
43. Wright, M.N.; Ziegler, A. Ranger: A fast implementation of random forests for high dimensional data in C++ and R. *J. Stat. Softw.* **2017**, *77*, 1–17. [CrossRef]
44. Srivastava, P.K.; Pandey, P.C.; Petropoulos, G.P.; Kourgialas, N.N.; Pandey, V.; Singh, U. GIS and remote sensing aided information for soil moisture estimation: A comparative study of interpolation techniques. *Resources* **2019**, *8*, 70. [CrossRef]
45. Cai, Z.; Jönsson, P.; Jin, H.; Eklundh, L. Performance of smoothing methods for reconstructing NDVI time-series and estimating vegetation phenology from MODIS data. *Remote Sens.* **2017**, *9*, 1271. [CrossRef]
46. Crow, W.T.; Berg, A.A.; Cosh, M.H.; Loew, A.; Mohanty, B.P.; Panciera, R.; de Rosnay, P.; Ryu, D.; Walker, J.P. Upscaling sparse ground-based soil moisture observations for the validation of coarse-resolution satellite soil moisture products. *Rev. Geophys.* **2012**, *50*, RG2002. [CrossRef]
47. Wang, J.; Ge, Y.; Heuvelink, G.B.M.; Zhou, C. Upscaling in situ soil moisture observations to pixel averages with spatio-temporal geostatistics. *Remote Sens.* **2015**, *7*, 11372–11388. [CrossRef]

remote sensing

MDPI

Article

Improving the AMSR-E/NASA Soil Moisture Data Product Using In-Situ Measurements from the Tibetan Plateau

Qiuxia Xie [1,2], Massimo Menenti [1,3] and Li Jia [1,*]

1 State Key Laboratory of Remote Sensing Science, Institute of Remote Sensing and Digital Earth, Chinese Academy of Sciences, Beijing 100101, China; xieqx@radi.ac.cn (Q.X.); m.menenti@radi.ac.cn (M.M.)
2 University of Chinese Academy of Sciences, Beijing 100049, China
3 Department of Geoscience and Remote Sensing, Delft University of Technology, 2628 CN Delft, The Netherlands
* Correspondence: jiali@radi.ac.cn

Received: 16 October 2019; Accepted: 15 November 2019; Published: 22 November 2019

Abstract: The daily AMSR-E/NASA (the Advanced Microwave Scanning Radiometer-Earth Observing System/the National Aeronautics and Space Administration) and JAXA (the Japan Aerospace Exploration Agency) soil moisture (SM) products from 2002 to 2011 at 25 km resolution were developed and distributed by the NASA National Snow and Ice Data Center Distributed Active Archive Center (NSIDC DAAC) and JAXA archives, respectively. This study analyzed and evaluated the temporal changes and accuracy of the AMSR-E/NASA SM product and compared it with the AMSR-E/JAXA SM product. The accuracy of both AMSR-E/NASA and JAXA SM was low, with RMSE (root mean square error) > 0.1 cm^3 cm^{-3} against the in-situ SM measurements, especially the AMSR-E/NASA SM. Compared with the AMSR-E/JAXA SM, the dynamic range of AMSR-E/NASA SM is very narrow in many regions and does not reflect the intra- and inter-annual variability of soil moisture. We evaluated both data products by building a linear relationship between the SM and the Microwave Polarization Difference Index (MPDI) to simplify the AMSR-E/NASA SM retrieval algorithm on the basis of the observed relationship between samples extracted from the MPDI and SM data. We obtained the coefficients of this linear relationship (i.e., A_0 and A_1) using in-situ measurements of SM and brightness temperature (T_B) data simulated with the same radiative transfer model applied to develop the AMSR-E/NASA SM algorithm. Finally, the linear relationships between the SM and MPDI were used to retrieve the SM monthly from AMSR-E T_B data, and the estimated SM was validated using the in-situ SM measurements in the Naqu area on the Tibetan Plateau of China. We obtained a steeper slope, i.e., A_1 = 8, with the in-situ SM measurements against A_1 = 1, when using the NASA SM retrievals. The low A_1 value is a measure of the low sensitivity of the NASA SM retrievals to MPDI and its narrow dynamic range. These results were confirmed by analyzing a data set collected in Poland. In the case of the Tibetan Plateau, the higher value A_1 = 8 gave more accurate monthly AMSR-E SM retrievals with RMSE = 0.065 cm^3 cm^{-3}. The dynamic range of the improved retrievals was more consistent with the in-situ SM measurements than with both the AMSR-E/NASA and JAXA SM products in the Naqu area of the Tibetan Plateau in 2011.

Keywords: soil moisture; AMSR-E; the microwave polarization difference index

1. Introduction

Soil moisture is a key variable for energy balance research and climate change analysis. Particularly, long time series of soil moisture at a global level is very useful to understand the land-atmosphere exchange of energy and water [1]. We consider that a "long time series" must span a period of at least

To compare the NASA and the JAXA AMSR-E SM retrievals, it should be taken into account that the JAXA retrieval approach is rather different. An improved RTM, i.e., the fully physically based RTM, is applied to simulate brightness temperature data at multiple frequencies and polarizations, which are then used to generate an LUT [9]. The LUT establishes a relationship between the brightness temperature observations and the bio-geophysical variables to be retrieved, i.e., soil moisture and vegetation water content [9,10,18]. To compare the sensitivity of the JAXA and NASA SM products to actual variability in hydrological conditions, we have followed the approach summarized here and described in detail in Section 3, Methodology. This approach led us to a simple yet effective way to improve the NASA SM product. We first show that the relationship between the SM and MPDI applied in the NASA algorithm is quasi-linear. The slope of this relationship is a measure of the sensitivity. When the higher values of the slope are applied to MPDI in the SM retrieval, the agreement with in-situ SM measurements improves significantly. Finally, this conclusion suggests that the NASA SM algorithm may be improved by using reliable estimates of the slope to calibrate the parameters in the relationship between SM and MPDI.

The objectives of this study are then (1) to evaluate the spatiotemporal variability of the AMSR-E/NASA SM and compare it with the AMSR-E/JAXA SM for the Tibetan Plateau; (2) to evaluate the accuracy of both the AMSR-E/NASA and AMSR-E/JAXA SM against in-situ SM measurements on the Tibetan Plateau; (3) to explain the very narrow range of the AMSR-E/NASA SM; (4) to improve the AMSR-E/NASA SM product by using the slope of the linear regression in the Naqu area on the Tibetan Plateau of China.

2. Study Area and Data

2.1. *Study Area and In-Situ SM Data*

In-situ SM data were collected in 2011 at the Naqu site on the Tibetan Plateau, included in the ISMN (International Soil Moisture Network) (Download link: https://ismn.geo.tuwien.ac.at/en/) [16]. A total of 56 locations within the Naqu site are currently available, and data from 50 locations are used in this study. In this study, the 50 sub-sites were divided into 12 groups (named Pixel 1, Pixel 2 . . . , Pixel 12) according to the pixel boundaries of the AMSR-E/NASA SM product (Figure 2). Within Pixel 1, the variability of land cover types at the locations of our in-situ SM measurements (Figure 2A) was similar to the variability of land cover types in the 25 km × 25 km grid of the AMSR-E retrievals (Figure 2B). Most of our in-situ measurements were located in Pixel 1. We concluded that the in-situ SM measurements within Pixel 1 provided a reliable reference for our evaluation of the NASA and JAXA SM products. In Pixel 1 in 2011, there were 14 in-situ SM sites and fewer (i.e., <4) in-situ SM sites in the other pixels. This allowed us to divide the in-situ SM measurements in Pixel 1 into two subsets: one subset, including 7 sites, was used to validate the linear model and one sub-set, including the remaining 7 sites, was combined with sites in other pixels to estimate slope and offset of the linear relationship, i.e., 43 sites were used in the regression analysis.

The Naqu study area lies on the central Tibetan Plateau and it is hilly and mountainous, but the slopes are gentle [18–20]. The climate is cold semi-arid and it is affected by the Southeast Asian monsoon [12,16]. The annual mean temperature varies from −0.9 to −3.3 °C. The annual mean relative humidity ranges from 48% to 51%. The annual precipitation amount is about 400–500 mm [16,18]. The period from November to April is dry and windy [21] with low temperature. From May to September it is relatively warm, windy and sunny with precipitation accounting for 80% of the yearly total [16,22]. Therefore, there is an obvious seasonality in the evolution of soil moisture.

The GLOBCOVER 2009 map was released by ESA and the Université Catholique de Louvain (UCL) in 2010. In this study area, the GLOBCOVER 2009 map was used to show the land cover types on the Tibetan Plateau. The main land cover type in the Naqu area is low biomass alpine grasslands and shrubland, and accounts for 90% of the total study area, as seen in Figure 2 [23]. Thus, the attenuation of the microwave signal by vegetation is rather small [24], and atmospheric attenuation is also small due to low air mass and moisture. In addition, the study area is very sparsely populated; thus, radio frequency interference (RFI) is small [23], making SM retrieval easier. Analyses of soil texture at Naqu show that the sand and silt contents are relatively high, i.e., 50% and 46%, respectively, while the clay content is lower, i.e., about 10% on average. The organic carbon content is low, i.e., 3.6% [23].

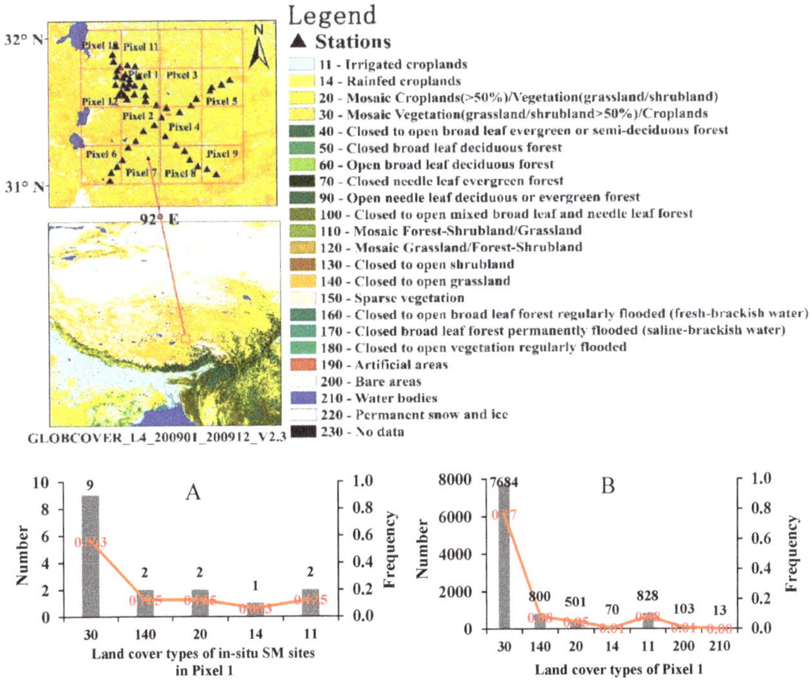

Figure 2. Location and land cover types of the study area and in-situ sites in Naqu on the Tibetan Plateau. **A**: histogram of land cover types at the locations of in-situ SM measurements in Pixel 1; **B**: histogram of land cover types of Pixel 1 with 25 km × 25 km; land cover based on the GLOBCOVER 2009 map at 300 m × 300 m spatial resolution.

2.2. AMSR-E/NASA and JAXA SM Products

Aqua is a NASA Earth Science satellite mission launched on 4 May 2002 that has six observing instruments on board including the AIRS (the Atmospheric Infrared Sounder), AMSU (the Advanced Microwave Sounding Unit), CERES (the Clouds and the Earth's Radiant Energy System), MODIS (the Moderate Resolution Imaging Spectroradiometer), AMSR-E (the Advanced Microwave Scanning Radiometer-Earth Observing System), and HSB (the Humidity Sounder for Brazil) [25]. The objective of the NASA Aqua satellite mission is primarily to collect information about Earth's water cycle including precipitation, evaporation, water vapor in the atmosphere, ice, snow cover, and soil moisture. In addition, observations on the vegetation cover, radiative energy fluxes, aerosols, and temperature can be retrieved from the AMSR-E data collected by Aqua [25]. The AMSR-E sensor onboard the NASA Aqua satellite is a microwave radiometer with 6 bands (6.92, 10.65, 18.7, 23.8, 36.5 and 89 GHz). The sun-synchronous orbit has equator overpasses at 1:30 AM and 1:30 PM local time. The AMSR-E sensor has dual polarization, H (horizontal) and V (vertical), and is widely used in the SM retrieval [8,9,26]. The level-1 (L1) brightness temperature data are generated by JAXA and then used to develop and generate the level-2 (L2) and level-3 (L3) data products [17]. In our study, the AMSR-E SM dataset is the L3 product based on brightness temperature observations along both the ascending and descending passes [27]. There are two data sources for the higher level data products, the NASA National Snow and Ice Data Center Distributed Active Archive Center (NSIDC DAAC, https://nsidc.org/data/amsre/data_summaries/index.html, AMSR-E/NASA) and the Japan Aerospace Exploration Agency (JAXA, https://gcom-w1.jaxa.jp/auth.html, AMSR-E/JAXA) [3,6,28,29].

The algorithms applied to retrieve the AMSR-E/NASA and AMSR-E/JAXA SM are different (Figure 1), although based initially on the same microwave RTM to describe the microwave emission by the land surface (Figure 1). This RTM includes many parameters describing soil, e.g., soil moisture and temperature, and vegetation properties, e.g., water content, optical depth, single scattering albedo, and temperature. The two SM retrieval algorithms differ in how these parameters are taken into account to retrieve land surface SM (Figure 1). The current AMSR-E/NASA SM is retrieved by applying a simplified RTM derived from a detailed microwave RTM by assuming: (a) negligible influence of atmospheric moisture and (b) canopy temperature equal to the soil temperature. This method uses the MPDI to effectively eliminate or reduce surface temperature effects. The algorithm first computes a vegetation/roughness parameter using the MPDI at both 10.7 GHz and 18.7 GHz [3]. Soil moisture is then computed using anomalies in the MPDI at 10.7 GHz from a baseline reference value. The baseline values for MPDI at 10.7 GHz are the observed minimum values at each grid during an annual cycle [30]. Conversely, the AMSR-E/JAXA SM is retrieved by applying an LUT generated using a fully physically based RTM, i.e., a forward model derived from the microwave RTM. The latter is derived by improving the original RTM to take into account volume scattering in the soil and the effect of surface roughness [10,31]. The LUT is generated for a large number of possible values of variables, e.g., soil moisture content, soil temperature, vegetation water content. Finally, the LUT is applied to determine these variables simultaneously, given the observations of brightness temperature at multiple frequencies and polarizations. Since the AMSR-E sensor malfunctioned in October 2011, the two SM products are available from 2002 to September 2011 only. Both SM products were resampled to a 25 km × 25 km grid. Additional information on the daily SM data set from 2002 to 2011 is provided in Table 1.

3. Methodology

In this study, the evaluation of the AMSR-E/NASA SM data product was done by applying two procedures. The first procedure was designed to compare the intra- and inter annual variability of AMSR-E/NASA and JAXA SM on the Tibetan Plateau. We also compared the AMSR-E/NASA and AMSR-E/JAXA SM data with in-situ SM measurements (Section 3.3). The second procedure was designed to analyze the possible reasons of the narrow intra- and inter-annual variation of the AMSR-E/NASA SM by estimating the relationship between the MPDI and SM in different cases, with and without using the in-situ SM measurements (Sections 3.1 and 3.2).

3.1. The Simplified RTM and the AMSR-E/NASA Algorithm to Retrieve SM

The brightness temperature observed by a space-borne radiometer, T_B, includes contributions from both the land surface (soil and vegetation) and the atmosphere [3], both the upwelling and down-welling atmospheric emittance [4,32]. The T_B can be written as:

$$T_B = T_u + t_{atm} \times \left(T_{sf} + r_{sf} \times T_d\right) \quad (1)$$

where T_{sf} is the at-surface brightness temperature (K), T_u and T_d are the respective upwelling and downwelling atmospheric emittance, t_{atm} is the atmospheric transmittance, r_{sf} is the surface reflectivity. Equation (1) is the initial RTM in Figure 1. At 6.9, 10.6 and 18 GHz, liquid cloud water affects the TOA (top-of-atmosphere) brightness temperature T_B by less than 2 K [3]. Thus, the bias in the retrieved SM due to atmospheric effects is rather small. Neglecting the contribution of atmosphere and assuming t_{atm} = 1, T_u and T_d = 0 [5], T_B in Equation (1) is equal to T_{sf}.

Assuming that the surface is homogeneous, the at-surface brightness temperature is related to the contributions from soil and vegetation through a linear mixture model (i.e., the zeroth order radiative transfer model) as:

$$T_{sf} = T_s \times t_v \times (1 - r_s) + T_v \times (1 - a) \times (1 - t_v) \times (1 + r_s \times t_v) \tag{2}$$

where T_s and T_v are the temperatures of soil and vegetation canopy, respectively (K), a is the single scattering albedo of vegetation, t_v is the vegetation transmittance, r_s is the soil reflectivity.

The vegetation transmittance (t_v) can be related to the water content of vegetation (w_v) as follows:

$$t_v = exp(-b_v \times w_v \times sec\theta) \tag{3}$$

where b_v is a coefficient, θ is the incident angle.

The soil reflectivity (r_s) can be deduced from the specular reflectivity (r_{sp}) and the surface roughness parameter (h) as follows:

$$r_s = r_{sp} \times exp\left(-h \times cos_\theta^2\right) \tag{4}$$

where h is a roughness parameter calculated as the surface root mean square height (m), r_{sp} is the specular reflectivity, equal to either $r_{sp,H}$ or $r_{sp,V}$:

$$r_{sp,H} = \left| \frac{cos\theta - \sqrt{\varepsilon_s - sin\theta^2}}{cos\theta + \sqrt{\varepsilon_s - sin\theta^2}} \right|^2 \tag{5}$$

$$r_{sp,V} = \left| \frac{\varepsilon_s \times cos\theta - \sqrt{\varepsilon_s - sin\theta^2}}{\varepsilon_s \times cos\theta + \sqrt{\varepsilon_s - sin\theta^2}} \right|^2 \tag{6}$$

where $r_{sp,H}$ and $r_{sp,V}$ are the specular reflectivity for horizontal (H) and vertical (V) polarization, respectively. Equations (5) and (6) are the Fresnel reflectivity equations. $r_{sp,H}$ and $r_{sp,V}$ are related to the dielectric constant of soil (i.e., ε_s). In this study, the Dobson dielectric model was used to simulate the relationship between the SM and ε_s [28].

The original AMSR-E/NASA SM algorithm is a multi-frequency-polarization method, where the coefficients are determined by minimizing the difference between the brightness temperature simulated with the simplified RTM (Equations (2)–(6)) and the AMSR-E brightness temperature observations [3]. Njoku, et al. 2004 developed a simplified method to retrieve SM from AMSR-E brightness temperature data [17,28,29]. The updated AMSR-E NASA SM algorithm applies the MPDI values according to Equations (7)–(9) [17]. In this algorithm, MPDI (Equation (7)) is used to retrieve SM. Using a baseline MPDI value under a dry condition, i.e., MPDI*, and three empirical coefficients (a_0, a_1 and a_2), the SM can be calculated using the MPDI value at 10.7 GHz from the AMSR-E brightness temperature observations (Equation (7)). The value of MPDI* is the minimum value in each grid and each month and it is calculated by the AMSR-E NASA SM algorithm [5]. The SM can be retrieved as a function of MPDI:

$$SM^t - SM^{dry} = a_0 \times g^* + a_1 \times \left(MPDI_{10.7}^t - MPDI_{10.7}^{dry}\right) \times exp(a_2 \times g^*) \tag{7}$$

$$MPDI_{10.7} = \left(T_{B(10.7V)} - T_{B(10.7H)}\right) / \left(T_{B(10.7V)} + T_{B(10.7H)}\right) \tag{8}$$

$$g^* = \beta_0 + \beta_1 \times \left(MPDI_{10.7}^*\right) \tag{9}$$

where t is time in days, SM^t is the time varying soil moisture; SM^{dry} is the minimum soil moisture value, i.e. 0.05 cm^3/cm^3; $MPDI_{10.7}^t$ is the MPDI value at 10.7 GHz on day t; $MPDI_{10.7}^{dry}$ is the annual minimum baseline MPDI for dry soil conditions. Here, g^* is the so-called baseline parameter to account for the effects of leaf water content and surface roughness; it is estimated using the $MPDI^*$ values and can be interpreted as an equivalent vegetation water content (kg/m^2). V and H indicate respective vertical and horizontal polarization; a_0, a_1, a_2, β_0 and β_1 are empirical coefficients. Equations (7)–(9) constitute the simplified algorithm to retrieve the AMSR-E/NASA daily SM.

To retrieve the daily SM using Equations (7)–(9), there are five unknown parameters (a_0, a_1, a_2, β_0, and β_1). Jackson et al. 2011 determined the values of a_0, a_1, a_2, β_0, and β_1 by calibration. The AMSR-E observations were collected over a region of naturally varying vegetation and roughness, with approximately uniform dry soil, that included portions of Chad, Sudan, and the Central African Republic [5]. More precisely, AMSR-E observations within these domains for a dry month (i.e., March 2004) with an assumed uniform value of soil moisture of 0.1 cm^3/cm^3 were used to estimate a_0, a_1, a_2, β_0, and β_1 [5,33].

3.2. Evaluation of the AMSR-E/NASA SM Retrieval Algorithm

We rearranged Equation (7) as:

$$SM^t = SM^{dry} + a_0 \times g^* - a_1 \times exp(a_2 \times g^*) \times MPDI_{10.7}^{dry} + a_1 \times exp(a_2 \times g^*) \times MPDI_{10.7}^{t} \quad (10)$$

and rewrote it as a linear function of MPDI at 10.7 GHz as:

$$SM^t = A_0 + A_1 \times MPDI_{10.7}^{t}, \ (t = 1, 2, \ldots, 365) \quad (11)$$

where $A_0 = SM^{dry} + a_0 \times g^* - a_1 \times exp(a_2 \times g^*) \times MPDI_{10.7}^{dry}$, $A_1 = a_1 \times exp(a_2 \times g^*)$, t is the day of year (DoY).

We explored ex-post the relationship between the observed MPDI and retrieved AMSR-E/NASA SM by plotting monthly values (Figure 3).

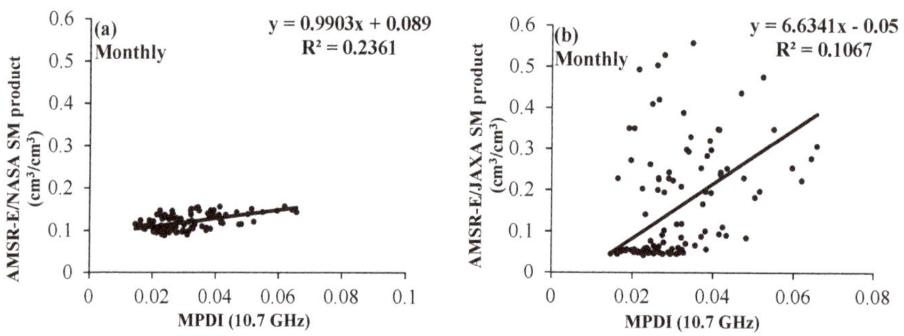

Figure 3. Relationships between the monthly AMSR-E/NASA and JAXA SM values and monthly MPDI in the Naqu area of the Tibetan Plateau in 2011: (**a**) NASA SM, all pixels; (**b**) JAXA SM, all pixels.

This shows that especially the monthly averaged SM is a nearly linear function of the monthly averaged MPDI, i.e., the number of unknown parameters in retrieving soil moisture using Equations (7)–(9) can be reduced from five to two (A_0 and A_1 in Equation (12)) as shown below. Further, the observations show (Figure 3) that a linear relationship applies, notwithstanding the intra-annual variability in SM. Clearly, the value of A_1 applied to the JAXA SM is much higher, i.e., more sensitive to MPDI, than the value applied to the NASA SM, at least in our study area (superscript t represents the month of the year, i.e., MoY):

$$SM^t = A_0 + A_1 \times MPDI_{10.7}^{t}, \ (= 1, 2, \ldots, 12) \quad (12)$$

Here, if the values of the A_0 and A_1 coefficients can be determined, SM^t can be obtained from Equation (12). According to Equation (12), SM^t is related only to the A_0 and $A_1 \times MPDI_{10.7}^{t}$. Therefore, the intra-annual variation of SM^t is described correctly as a linear function of MPDI, i.e., the value of A_1 determines the amplitude of the yearly variation in soil moisture. The value of A_0 is the initial value of SM in one year. When the inter- and intra-annual variability in SM is low, a special case may occur where A_0 is equal to 0 and A_1 is equal to 1, i.e., the value of SM^t is equal to the $MPDI^t_{10.7}$.

3.3. Comparison Strategy

In order to evaluate AMSR-E/NASA and AMSR-E/JAXA SM retrievals, we applied the mean absolute error (MAE) (Equation (13)), the RMSE (Equation (14)), and the correlation coefficient (R) (Equation (15)) to compare the satellite SM products with in-situ SM data [34],

$$MAE = \frac{\sum_{t=1}^{N}\left(SM_t^E - SM_t^O\right)}{N} \tag{13}$$

$$RMSE = \sqrt{\frac{\sum_{t=1}^{N}\left(SM_t^E - SM_t^O\right)^2}{N}} \tag{14}$$

$$R = \frac{Cov\left(SM_t^E, SM_t^O\right)}{\sigma_{SM_t^E} \times \sigma_{SM_t^O}}, \begin{cases} Cov : covariance \\ \sigma : standard\ deviation \end{cases} \tag{15}$$

where SM_t^E is the retrieved AMSR-E/NASA or AMSR-E/JAXA SM on day t; SM_t^O is the in-situ measured SM on day t; N is the number of total days of measurements. To ensure the reliability of validation, only sets of observations with more than 10 days (N >=10) were used in order to remove random errors. We also calculated the daily NASA, JAXA and in-situ SM relative anomalies (SM'_t) as follows:

$$SM'_t = \left(SM_t - \overline{SM}\right)/\overline{SM} \tag{16}$$

where t represents the day of year; $\overline{SM}$ is the average NASA SM or JAXA SM or in-situ SM in 2011.

In this study, we applied two different procedures to evaluate the AMSR-E NASA SM (Figure 4). In the first procedure, we compared the intra- and inter-annual variations and accuracies of AMSR-E/NASA and JAXA SM with in-situ SM measurements in the Naqu area (Procedure 1 in Figure 4). In the second procedure, we used two sets of brightness temperature data obtained in two different ways to calculate $MPDI_{10.7}$ (MPDI Datasets 1 and 2 in Figure 4). Dataset 1 is simulated brightness temperature by applying the simplified RTM (Equations (2)–(6)) to the in-situ SM measurements. Dataset 2 is the AMSR-E Level 2A brightness temperature data that was resampled to a grid (i.e., the Equal-Area Scalable Earth, EASE-Grid) of approximately 25 km × 25 km using the distance-weighting method applied to AMSR-E L1 brightness temperature data. EASE-Grid is a global, cylindrical, equal-area projection, with 1383 columns × 586 rows. Then, the linear regressions (Equation (12)) between the SM and MPDI based on in-situ SM measurements (Dataset 1) or on AMSR-E/NASA SM (Dataset 2) were determined.

In this way, (A_0, A_1) were obtained for four different pairs: (a) AMSR-E/NASA MPDI + NASA SM; (b) AMSR-E/NASA MPDI + in-situ SM; (c) RTM MPDI + NASA SM; (d) RTM MPDI + in-situ SM. A different sensitivity to the seasonality in precipitation between AMSR-E/NASA SM and the measured SM will appear as a very different A_1 value in the cases (a) and (c) compared with (b) and (d). Finally, monthly SM from AMSR-E Level 2A brightness temperature data were retrieved using all (A_0, A_1) pairs and evaluated against the in-situ SM measurements set aside for this evaluation (Procedure 2 in Figure 4).

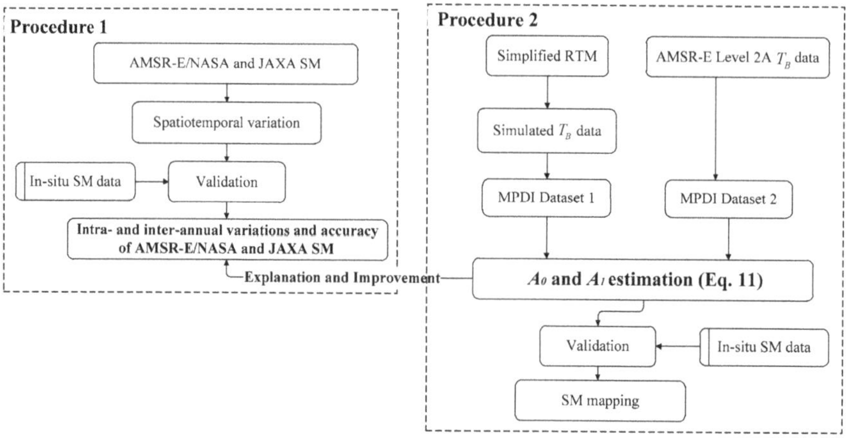

Figure 4. Flow chart of AMSR-E/NASA, JAXA SM product analysis and improving the AMSR-E SM product.

4. Results

4.1. Intra- and Inter Annual Variation of AMSR-E/NASA and AMSR-E JAXA SM

Both the AMSR-E/NASA and the AMSR-E/JAXA SM data are daily from 2002 to 2011 with 25 km × 25 km resolution. The mean value obtained with the AMSR-E/NASA SM is almost constant with a value of 0.12 in Figure 5a,b. The intra-annual amplitude is very small and also almost constant through the years. AMSR-E/JAXA SM changes with the season as a sine or cosine function, with a mean value of 0.08 cm^3/cm^3 and an amplitude of 0.04 cm^3/cm^3. At times, the standard deviation of the AMSR-E/JAXA SM data is higher than that of the AMSR-E/NASA data, which implies a larger spatial variability in the response to the seasonality of hydrologic conditions. The standard deviation of the AMSR-E/NASA data remains close to 0.06 cm^3/cm^3 over time, while it changes with time and season in the AMSR-E/JAXA data with an average of 0.05 cm^3/cm^3 and an amplitude of 0.035 cm^3/cm^3 (Figure 5c,d). Therefore, the AMSR-E/JAXA SM data capture both the intra- and inter-annual variability better than the AMSR-E/NASA data. The AMSR-E/NASA SM has an unrealistically narrow dynamic range, given the hydrological conditions in the study area.

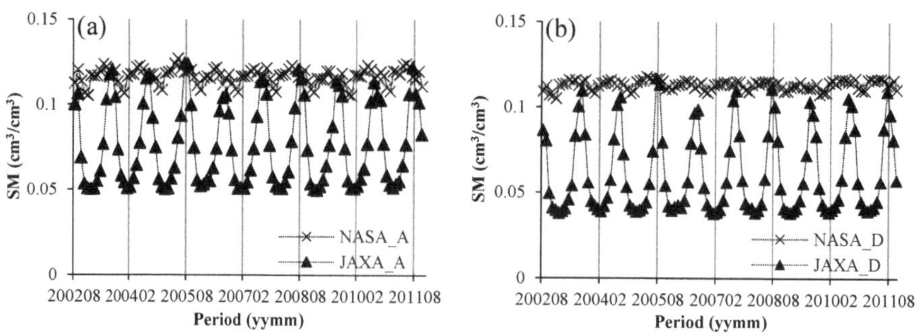

Figure 5. *Cont.*

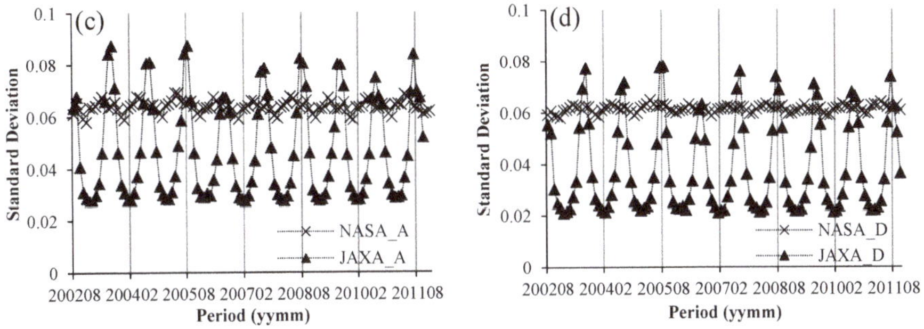

Figure 5. The monthly mean SM and standard deviation of AMSR-E/NASA and AMSR-E/JAXA retrievals for the same 2000 pixels for the Tibetan Plateau from 2002 to 2011: (**a**) SM retrieved with data collected during ascending orbits and (**b**) descending orbits; (**c**) SM standard deviation for ascending orbits, and (**d**) descending orbits.

4.2. Evaluation of AMSR-E/NASA and AMSR-E/JAXA SM Products

We first applied Procedure 1 (see Figure 4) in our evaluation. There are large differences between the two data products in reflecting the actual intra- and inter-annual variation of SM (Figure 5). In this study, the AMSR-E/NASA or AMSR-E/JAXA SM products were compared with in-situ SM measurements in 2011 in the Naqu study area in space and time (Figures 6 and 7) by evaluating the MAE, RMSE and R metrics (Table 2).

The AMSR-E/NASA SM hardly changes with the in-situ SM (Figure 6a,b), i.e., the AMSR-E/NASA SM is insensitive to actual soil moisture conditions. The AMSR-E/NASA SM is always smaller than 0.2 cm^3 cm^{-3} during both ascending and descending orbits. Contrariwise, the spatial SM dynamic range of in-situ SM measurements is from 0 to 0.6 cm^3 cm^{-3}. Therefore, the SM dynamic range of the AMSR-E/NASA SM is inconsistent with the in-situ SM measurements. Compared to the AMSR-E/NASA SM product, the spatial SM dynamic range of AMSR-E/JAXA SM is larger, from 0 to 0.6 cm^3 cm^{-3} (Figure 6c,d).

The temporal variability between AMSR-E/JAXA SM and in-situ measurements is similar, while the one of the AMSR-E/NASA SM is very different in all pixels. In this study, the in-situ measurements in Pixel 1 were used to illustrate the difference of temporal variability between AMSR-E/NASA SM and in-situ measurements (Figure 7). The in-situ measurements show that the SM content was very low in this area from January to April, i.e., less than 0.15 cm^3 cm^{-3}. Since April, SM increased gradually. The first small peak in SM appeared in May. This may be due to the melting of frozen soil as the temperature increased. The highest SM content was from June to August, and maximum SM was about 0.4 cm^3 cm^{-3}. The AMSR-E/JAXA SM largely overestimated the in-situ SM, especially from July to August, while underestimating the in-situ SM from January to April; AMSR-E/JAXA SM was more consistent with in-situ SM data. The AMSR-E/NASA SM slightly changes with time. The range of AMSR-E/NASA SM is from 0.05 to 0.20 cm^3 cm^{-3}.

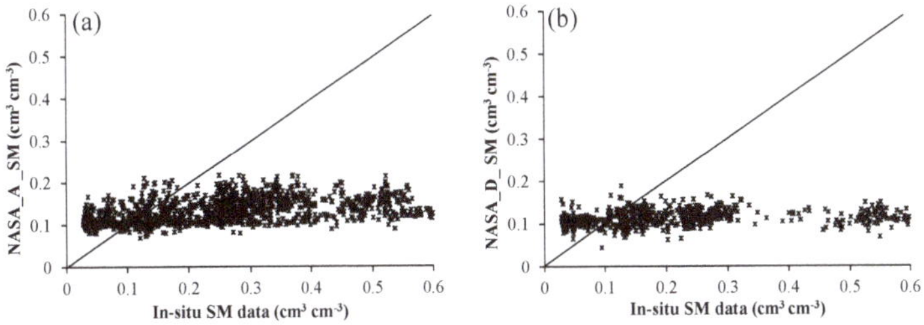

Figure 6. *Cont.*

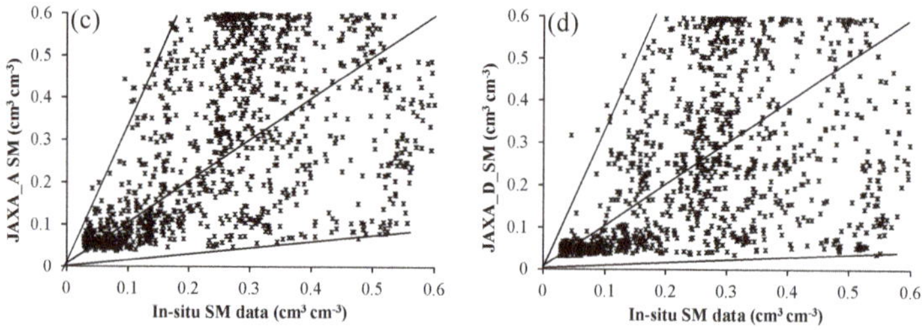

Figure 6. Comparison of AMSR-E/NASA or JAXA SM products with in-situ SM measurements data in 2011: (**a**) AMSR-E/NASA during ascending orbits and (**b**) descending orbits; (**c**) AMSR-E/JAXA during ascending orbits, and (**d**) descending orbits.

Differences in the relative anomalies calculated using Equation (16) of AMSR-E/NASA, JAXA and in-situ SM were significant. The relative anomaly range for the AMSR-E/NASA SM, i.e., −0.4 to 0.7 is smaller than the relative anomaly range of JAXA SM or in-situ SM, which was in the range −0.9 to 2.1 (Figure 7). Conversely, the relative anomalies in the JAXA SM product are generally comparable with the in-situ SM and slightly larger in summer (Figure 7C).

We used the MAE, RMSE, and R metrics to evaluate the AMSR-E/NASA and AMSR-E/JAXA SM data (Table 2). The mean SM value of in-situ measurements within each pixel (Figure 2) was taken as the true SM value.

The RMSEs of AMSR-E/NASA and AMSR-E/JAXA SM were both higher than 0.06 cm^3 cm^{-3} except in Pixel 6 where the RMSE of AMSR-E/JAXA SM was 0.04 cm^3 cm^{-3}. The minimum MAE for AMSR-E/JAXA SM was in Pixel 6, i.e., 0.03 cm^3 cm^{-3} and the maximum R was 0.91 in Pixel 1. As regards the AMSR-E/NASA SM, the minimum RMSE was 0.07 cm^3 cm^{-3} in Pixel 6, while the maximum R was 0.72 in pixels 1, 2 and 12. Overall, the accuracies of both SM data products were poor, especially of the AMSR-E/NASA SM. The RMSE averaged over all pixels of the AMSR-E/JAXA SM was less than that of the AMSR-E/NASA SM, i.e., 0.11 and 0.16 cm^3 cm^{-3}, respectively. Likewise, the mean R of the AMSR-E/JAXA SM product was higher than that of the AMSR-E/NASA SM, i.e., 0.85 and 0.62, respectively. The AMSR-E/JAXA SM, therefore, was more accurate than the AMSR-E/NASA SM, at least as regards the Naqu study area in 2011. All the standard deviation (std. dev) values of AMSR-E/NASA SM are very low, i.e., <0.019 cm^3 cm^{-3} with an average of 0.015 cm^3 cm^{-3}. The std. dev values of AMSR-E/JAXA SM were very high, i.e., up to 0.078 cm^3 cm^{-3} with an average of 0.129 cm^3 cm^{-3}. This also illustrates the narrow dynamic range of AMSR-E/NASA SM.

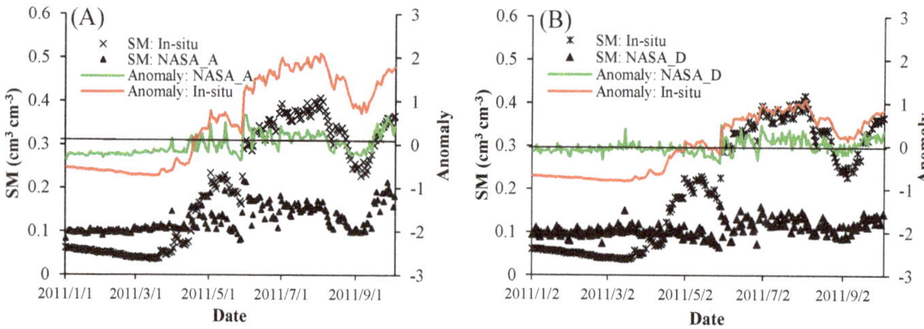

Figure 7. *Cont.*

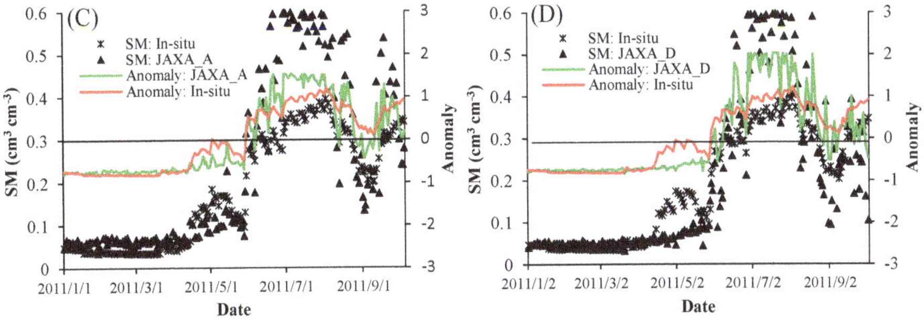

Figure 7. Intra-annual variation and relative anomaly of AMSR-E/NASA, JAXA SM data and in-situ SM measurements in Pixel 1, 2011: (**A**) AMSR-E/NASA during ascending orbits and (**B**) descending orbits; (**C**) AMSR-E/JAXA during ascending orbits, and (**D**) descending orbits.

Table 2. Calculated MAE, RMSE, and R of SM data in the Naqu area against in-situ SM measurements in 2011; st. dev are the standard deviation of each SM data product for each pixel, respectively.

	AMSR-E/NASA				AMSR-E/JAXA			
	MAE	**RMSE**	**R**	**Std. Dev**	**MAE**	**RMSE**	**R**	**Std. Dev**
Pixel 1	0.09	0.11	0.72	0.017	0.06	0.10	0.91	0.128
Pixel 2	0.08	0.10	0.72	0.016	0.07	0.09	0.88	0.143
Pixel 3	0.09	0.12	0.66	0.014	0.07	0.11	0.80	0.175
Pixel 4	0.12	0.16	0.64	0.019	0.09	0.10	0.88	0.123
Pixel 5	0.20	0.25	0.48	0.013	0.12	0.15	0.82	0.175
Pixel 6	0.06	0.07	0.67	0.017	0.03	0.04	0.87	0.101
Pixel 7	0.12	0.15	0.61	0.018	0.07	0.10	0.83	0.105
Pixel 8	0.15	0.18	0.54	0.014	0.09	0.11	0.86	0.118
Pixel 9	0.15	0.18	0.60	0.012	0.12	0.15	0.88	0.085
Pixel 10	0.13	0.15	0.41	0.015	0.08	0.10	0.86	0.078
Pixel 11	0.20	0.22	0.62	0.014	0.10	0.12	0.84	0.167
Pixel 12	0.19	0.20	0.72	0.018	0.07	0.09	0.79	0.145
Average	0.13	0.16	0.62	0.015	0.08	0.11	0.85	0.129

4.3. Improvement and Mapping of the AMSR-E/NASA SM Product

We applied Procedure 2 (Figure 4) to explore the possible causes of the poor accuracy of the AMSR-E/NASA SM. The analyses of the intra- and inter- annual variations of the AMSR-E/NASA and AMSR-E/JAXA SM products show that the shortcomings in the AMSR-E/NASA SM data are two-fold. On the one hand, the dynamic range of AMSR-E/NASA SM is very narrow, which does not reflect the actual intra- and inter-annual variation of precipitation (Figures 5–7). On the other hand, the accuracy is poor (Table 2), with RMSE higher than 0.1 cm^3 cm^{-3}. The high RMSE values of AMSR-E/NASA are not only caused by the small dynamic range of AMSR-E/NASA soil moisture. The RMSE might have been smaller if the NASA SM would have been close to either high or low in-situ SM. The high RMSE shows that the accuracy is poor, i.e., RMSE = 0.16 cm^3 cm^{-3}, because the NASA SM is different from the in-situ SM throughout the year. It is necessary, therefore, to revisit the retrieval algorithm of the AMSR-E/NASA SM to identify the likely cause of such poor performance.

As explained in Section 3.1, we applied a simplified RTM (Equations (2)–(6)) to simulate the brightness temperature for both H and V polarized emission at 10.7 GHz. To model brightness temperature, some target properties and the observation geometry must be known. The sand and clay fractions were determined for soil samples. The soil texture is rather uniform at Naqu, and we used the mean soil textural fractions over all pixels in our numerical experiments [16]. According to the parameters of the AMSR-E sensor in Table 1, the incident angle is set as 55°. Because the Naqu area is fairly smooth with rolling hills, we estimated the surface roughness to be small, i.e., 0.03 m [16]. In summary, the area is characterized by low biomass, low vegetation water content, and similar temperature of soil and vegetation canopy. Thus, we can assume attenuation by the vegetation canopy to be low. Further, we assumed the water content of vegetation to be 1 kg/m^2, the attenuation coefficient = 0.3, and single scattering albedo = 0. The in-situ measurements of SM were used in the simulation of MPDI by the simplified RTM (Equations (2)–(6)).

The monthly MPDI was calculated by applying Equation (12) to the brightness temperature data in Datasets 1 and 2 (Section 3.2 and Figure 4). As explained in Section 3.2, four different pairs of (A_0, A_1) were obtained (Table 3).

Table 3. Values of A_0 and A_1 parameters (see text for details on the estimation procedure).

Cases	A_1	A_0
(a) AMSR-E/NASA MPDI + NASA SM	1	0.09
(b) AMSR-E/NASA MPDI + in-situ SM	8	−0.15
(c) RTM MPDI + NASA SM	1	0.06
(d) RTM MPDI + in-situ SM	8	−0.36

We used the in-situ SM data in Pixel 1, Pixel 2, Pixel 3, and Pixel 4, where we had most of our measurements, to analyze the relationship with MPDI (Figure 8). By fitting Equation (12) to both the AMSR-E/NASA and the in-situ SM data, we estimated the parameters A_0 and A_1 and further retrieved four sets of SM data (Figure 8). As regards the AMSR-E/NASA SM data, the A_0 and A_1 values were approximately the same in all pixels and equal to 1 and 0.06, respectively. As regards the in-situ data, however, we obtained A_1 = 8. In other words, the value of A_1 obtained by fitting Equation (12) to the AMSR-E/NASA SM product is too small and it explains the very small dynamic range. Accordingly, the model SM = 8*MPDI-0.36, i.e., (A_0, A_1) values obtained with the RTM MPDI + in-situ SM data (Table 1), is expected to perform better when applied to retrieve SM in our study area.

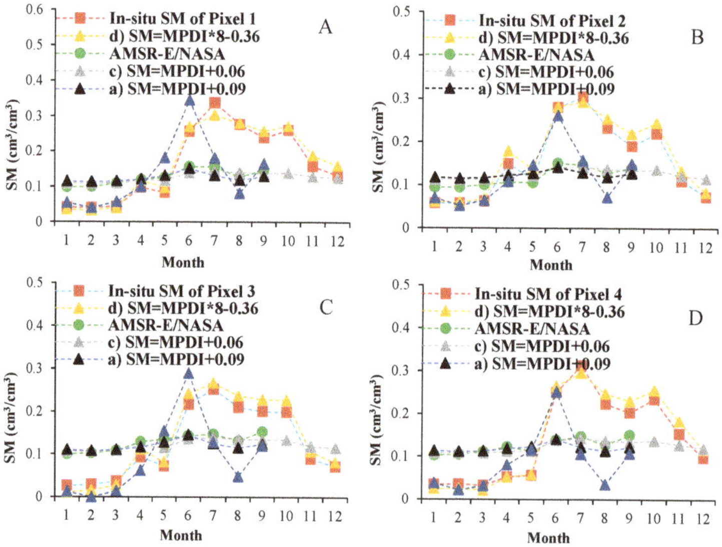

Figure 8. Monthly in-situ SM measurements and AMSR-E/NASA SM for the four cases in Table 1; Pixel 1 through Pixel 4 (**A**: Pixel 1; **B**: Pixel 2; **C**: Pixel 3; **D**: Pixel 4) in the Naqu area in 2011.

In principle, there are two different factors that can lead to large differences between the AMSR-E SM retrievals and in-situ SM measurements. First, the algorithm to retrieve SM may not be accurate, although the radiometry, i.e., the brightness temperature values, is correct. Second, the values of brightness temperature are not correct and, therefore, the MPDI values are also not correct. In this study, we assumed the brightness temperature observed by the AMSR-E sensor is correct and analyzed the AMSR-E/NASA SM retrieval algorithm.

The case (d) SM is closest to the in-situ SM data (Figure 8). In addition, by comparing case (b) with case (d), we found that the calculated SM values were close to in-situ SM measurements from January to June. From July to September, however, the SM values calculated by using case (b) SM = 8*MPDI-0.15 were much lower than the SM values calculated according to case (d) SM = 8*MPDI-0.36. The reason is that the MPDI values calculated with the AMSR-E Level 2A brightness temperature data are much lower than the MPDI values from simulated brightness temperature. The values of the AMSR-E Level 2A brightness temperature data, therefore, may not be accurate from July to September for the Tibetan Plateau. The reason may be that the TOA (top-of-atmosphere) radiance measured by AMSR-E is severely attenuated by vegetation, leading to a small difference between H and V polarization in the summer on the Tibetan Plateau.

To retrieve SM from AMSR-E Level 2A brightness temperature data from July to September, a correction for attenuation by vegetation or ex-post calibration must be applied. The good agreement of retrieved with measured SM in case (d) shows that the linear regression between the MPDI and in-situ SM data yields accurate SM retrievals and the MPDI values capture the seasonality in SM. In case (d), the brightness temperature data were simulated with the simplified RTM by applying a constant attenuation by vegetation. The same linear regression, i.e., Equation (17), applied to the AMSR-E brightness temperature observations yields accurate SM retrievals until June only. Our hypothesis is that in summer months, i.e., from July to September, vegetation biomass increases in response to snowmelt and monsoon precipitation. To estimate the correction, we opted for the simplest possible approach, i.e., by fitting a separate linear relationship to the AMSR-E MPDI values and SM measurements for all the calibration pixels and the summer months. This gave (Equation (17), July through September) a much higher and positive offset that provided the required correction. In summary, two different linear relationships must be applied to retrieve SM from the AMSR-E MPDI observations:

$$\begin{cases} SM^t = -0.15 + 8 \times MPDI^t_{10.7}, \ (t = 1,2,3,4,5,6) \\ SM^t = 0.05 + 8 \times MPDI^t_{10.7}, \ (t = 7,8,9) \end{cases} \quad (17)$$

where $MPDI^t_{10.7}$ is the MPDI calculated from the AMSR-E Level 2A brightness temperature data; t is the month.

The SM retrieved from AMSR-E/NASA MPDI data by applying Equation (17) is indicated as I_AMSR-E SM. The three SM datasets, i.e., AMSR-E/NASA, AMSR-E/JAXA, and I_AMSR-E, were mapped in our Naqu study area in January, March, June, July, and September (Figure 9). There was almost no rainfall in January and March (Figure 9), while rainfall was higher in June and July. The AMSR-E/NASA SM, however, shows very limited changes from winter to summer, and SM is low from January to September. The seasonal variation of precipitation is not captured by the AMSR-E/NASA SM, while it is very evident in both the AMSR-E/JAXA SM and I_AMSR-E SM. In term of capturing the seasonal variation of soil moisture in the Naqu area, the AMSR-E/JAXA SM and I_AMSR-E SM performed better than the AMSR-E/NASA SM. In July, the I_AMSR-E SM was highest, followed by June and September. In January and March, SM was low. The I_AMSR-E SM, which fits the in-situ measurements rather well (Figure 8), remained lower than the AMSR-E/JAXA SM in June and July; this implies that our I_AMSR-E SM avoids both over- and underestimation in summer.

There are large differences in the spatial SM pattern (Figure 9) between AMSR-E/NASA, AMSR-E/JAXA and I_AMSR-E SM. AMSR-E/NASA SM is very flat and uniform in spatial distribution in the Naqu area. Compared with the AMSR-E/NASA SM spatial distribution in the Naqu area in June, the spatial dynamic range of AMSR-E/JAXA and I_AMSR-E SM is larger from 0.05 to 0.5 cm^3 cm^{-3}. Compared with the spatial pattern of precipitation in June, the precipitation is higher in the NE (Northeast) than in the NW (Northwest) portion of the Naqu area. On the other hand, both the AMRS-E/JAXA and I_AMSR-E SM are lower in the NE than in the NW portion of the Naqu area.

The difference in the AMSR-E/NASA SM between the NE and NW portion is small. In comparing the spatial patterns of SM with precipitation, however, it should be taken into account that the NW portion is flatter with a large presence of water bodies, which explains the differences observed in Figure 9. We further note that our improved SM estimates (I_AMSR-E) correctly reflect the terrain, with a lower elevation catchment in the NW portion of the Naqu area.

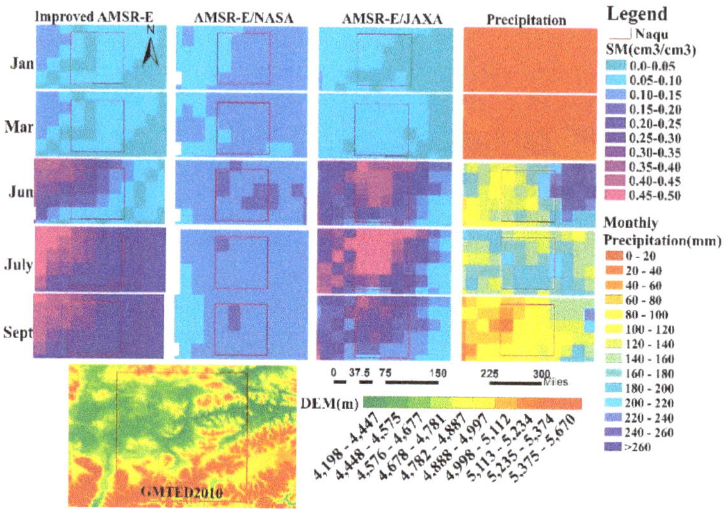

Figure 9. Spatial and temporal comparison of the monthly AMSR-E/NASA, AMSR-E/JAXA, I_AMSR-E SM and precipitation in January, March, June, July and September in 2011, the Naqu area (upper panel); Digital Elevation Model (DEM, lower panel).

In-situ SM measurements in Pixel 1 included 14 locations in 2011, which were divided into two sets: one subset was used to determine the linear regression (Figure 8 and Table 3), and the other subset was used for evaluation. Equation (12) was fitted to the 1st subset, i.e., 7 stations (Figure 8), while the 2nd subset was used to evaluate the I_AMSR-E SM and gave RMSE = 0.065 cm^3 cm^{-3} (Figure 10). In conclusion, the I_AMSR-E SM compares to in-situ SM measurements better than both the AMSR-E/NASA SM (RMSE = 0.11 cm^3 cm^{-3}, Table 2) and AMSR-E/JAXA SM (RMSE = 0.10 cm^3 cm^{-3}, Table 2).

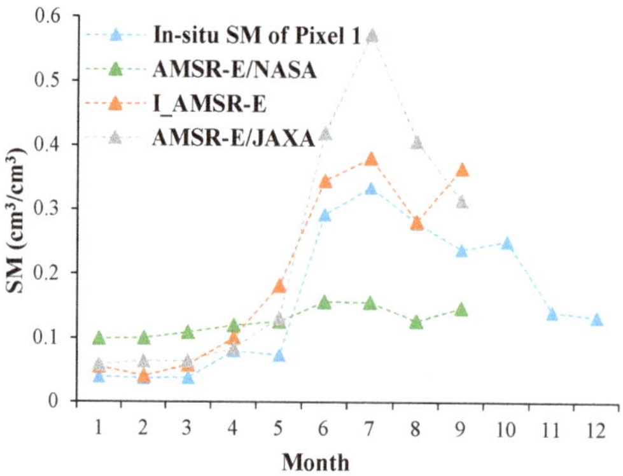

Figure 10. Monthly I_AMSR-E, AMSR-E/NASA and JAXA SM retrievals and in-situ SM measurements in Pixel 1 in the Naqu area of the Tibetan Plateau in 2011.

5. Discussion

The AMSR-E/NASA SM data product has an unrealistically narrow dynamic range given the hydrological conditions on the Tibetan Plateau [35,36]. By comparing the AMSR-E/NASA SM with the AMSR-E/JAXA SM, it appears that the AMSR-E/NASA SM is not sensitive to the large variations in hydrologic conditions characteristic of the Naqu site. The SM dynamic range of AMSR-E/NASA SM is very narrow and inconsistent with the in-situ measurements of SM. Our findings are confirmed by the evidence provided by [11,12]. Zeng et al., 2015 concluded that the AMSR-E/NASA SM product does not capture the soil moisture dynamics and that both AMSR-E/NASA and JAXA SM products gave relatively large RMSE values against in-situ SM measurements [11]. Chen et al., 2013 showed that the AMSR-E/NASA algorithm gave a dampened dynamic range of soil moisture while the AMSR-E/JAXA algorithm did reflect the seasonal variation of soil moisture but with too large amplitude. This implies that the AMSR-E/NASA and JAXA SM retrieval algorithms need to be improved to be applicable to the Tibetan Plateau [12]. The linear relationships between the monthly MPDI and NASA (JAXA) SM in Pixel 1 and Pixel 2 (Figure 11) show that the A_1 value of NASA SM is approximately equal to 1, and the A_1 value of JAXA SM is approximately equal to 8. The A_1 values of 12 pixels (Figure 11e) of the relationships between the JAXA SM and MPDI were higher, i.e., over 7.67. On the other hand, for our 12 pixels, the A_1 values of the relationships between the NASA SM and MPDI were very small, i.e., <1.924. This also illustrates that the AMSR-E/NASA SM does not reflect the dynamics of soil moisture but AMSR-E/JAXA can. Although the AMSR-E/NASA SM is considered a reference data set by the National Snow & Ice Data Center (NSIDC), the narrow dynamic range of SM seriously limits its application to drought and environmental change monitoring and other applications [13].

Two results given above point to the same possible cause. We have shown that a simple linear relationship (Equation (17)) with a slope equal to 8 gives accurate SM retrievals when the brightness temperature is calculated from in-situ SM measurements (Case d in Figure 8 and Table 3). When applying the same linear relationship to the AMSR-E brightness temperature data, the retrieved SM is severely underestimated in summer, i.e., under wet conditions, and the offset parameter in our linear relationship (Equation (17)) had to be recalculated to improve accuracy of the retrieved SM. Both slope and offset in Equation (17) are related to the g* parameter (Equation (9)), which should be evaluated for each month and location. We note that a time-dependent g* would correct for seasonal variations in attenuation by the vegetation canopy of soil emittance. Our results (Table 3) on the slope parameter A_1 of the linear relationship (Equation (17)) show that the sensitivity of the NASA SM retrieval algorithm to actual soil moisture changes may not be adequate. These results indicated that the value of A_1 in Equation (17) should be much higher, i.e., 8 instead of 1. This difference explains the unrealistically narrow dynamic range of the AMSR-E/NASA SM for the Tibetan Plateau. Zeng et al. (2015) noted that the parameters of the AMSR-/NASA SM retrieval algorithm were determined by calibration in specific regions, with the risk of the algorithm being not suitable to retrieve SM in other regions [11,12]. We are aware that the AMSR-E/NASA SM algorithm performed relatively well in Mongolia and the United States [5], although it did not provide accurate SM retrievals for the Tibetan Plateau [11,37]. It has been suggested by other authors that these parameters (Equations (7)–(9)) should be recalibrated in other areas besides the Tibetan Plateau to improve the accuracy of SM retrievals in these regions [11,13].

We have evaluated the AMSR-E/NASA SM just in one study area, i.e., Naqu on the Tibetan Plateau, but we concluded that the reason leading to the unrealistically narrow dynamic range of AMSR-E/NASA SM was the very low value, i.e., 1 instead of 8, of the parameter $A_1 = a_1 \cdot \exp(a_2 \cdot g^*)$ which is related as shown to the parameters in the AMSR-E/NASA SM algorithm. We believe there might be other areas, besides Naqu, potentially affected by the same problem; thus, we repeated our evaluation in a second experiment.

We chose an area in Poland, where the dynamic range of the AMSR-E/NASA SM is also small, notwithstanding a large seasonal variation in precipitation, and estimated our linear regression Equation (12) using in-situ SM measurements. More specifically, we evaluated the cases (a) and

idea might be useful to calibrate the A_1 value in other regions when no in-situ SM measurements are available. A broader perspective along the same line of thinking is to make use of the in-situ SM measurements collected at the ISMN sites and produce a global map of A_1 towards improved accuracy of the NASA SM data.

Author Contributions: This study was designed and completed through collaboration among all the authors, Q.X., M.M. and L.J., Q.X. prepared and processed the AMSR-E/NASA, AMSR-E/JAXA and in-situ measurements datasets, performed the main analysis and wrote this manuscript. M.M. contributed important ideas, revising and editing multiple versions of this manuscript. L.J. provided key comments and revised this manuscript including figures and formulas.

Funding: This work was jointly supported by the Strategic Priority Research Program of the Chinese Academy of Sciences (Grant No. XDA19030203), the National Natural Science Foundation of China project (Grant No. 41661144022), the NRSCC—ESA Dragon—4 project (Grant N0.32439), and the MOST High Level Foreign Expert program (Grant No. G20190161018). And the APC was funded by the Strategic Priority Research Program of the Chinese Academy of Sciences (Grant No. XDA19030203).

Acknowledgments: Thanks are extended to Kun Yang et al. for providing in-situ soil moisture data for the Naqu area. The authors would like to thank the NSIDC for the AMSR-E/NASA soil moisture product and the Globe Portal System (G-Portal) for the AMSR-E/JAXA soil moisture product. We were very thankful the valuable comments from the reviewers that helped us to improve our manuscript.

Conflicts of Interest: All authors declare no conflict of interest.

References

1. Karthikeyan, L.; Pan, M.; Wanders, N.; Kumar, D.N.; Wood, E.F. Four decades of microwave satellite soil moisture observations: Part 1. A review of retrieval algorithms. *Adv. Water Resour.* **2017**, *109*, 106–120. [CrossRef]
2. Karthikeyan, L.; Pan, M.; Wanders, N.; Kumar, D.N.; Wood, E.F. Four decades of microwave satellite soil moisture observations: Part 2. Product validation and inter-satellite comparisons. *Adv. Water Resour.* **2017**, *109*, 236–252. [CrossRef]
3. Njoku, E.G.; Jackson, T.J.; Lakshmi, V.; Chan, T.K.; Nghiem, S.V. Soil moisture retrieval from AMSR-E. *IEEE Trans. Geosci. Remote Sens.* **2003**, *41*, 215–229. [CrossRef]
4. Njoku, E.; Koike, T.; Jackson, T.; Paloscia, S. *Retrieval of Soil Moisture from AMSR Data*; Vsp Publishing: Utrecht, The Netherlands, 1999; pp. 525–533.
5. Jackson, T.J.; Cosh, M.H.; Bindlish, R.; Starks, P.J.; Bosch, D.D.; Seyfried, M.; Goodrich, D.C.; Moran, M.S.; Du, J.Y. Validation of advanced microwave scanning radiometer soil moisture products. *IEEE Trans. Geosci. Remote Sens.* **2010**, *48*, 4256–4272. [CrossRef]
6. Koike, T.; Njoku, E.; Jackson, T.J.; Paloscia, S. Soil moisture algorithm development and validation for the ADEOS-II/AMSR. In Proceedings of the IGARSS 2000, IEEE 2000 International Geoscience and Remote Sensing Symposium, Taking the Pulse of the Planet: The Role of Remote Sensing in Managing the Environment. Proceedings (Cat. No.00CH37120), Honolulu, HI, USA, 24–28 July 2000.
7. Paloscia, S.; Macelloni, G.; Santi, E.; Koike, T. A multifrequency algorithm for the retrieval of soil moisture on a large scale using microwave data from SMMR and SSM/I satellites. *IEEE Trans. Geosci. Remote Sens.* **2001**, *39*, 1655–1661. [CrossRef]
8. Cho, E.; Su, C.; Ryu, D.; Kim, H.; Choi, M. Does AMSR2 produce better soil moisture retrievals than AMSR-E over Australia? *Remote Sens. Environ.* **2017**, *188*, 95–105. [CrossRef]
9. Feng, X.; Li, J.; Cheng, W.; Fu, B.; Wang, Y.; Lü, Y.; Shao, M. Evaluation of AMSR-E retrieval by detecting soil moisture decrease following massive dryland re-vegetation in the Loess Plateau, China. *Remote Sens. Environ.* **2017**, *196*, 253–264. [CrossRef]
10. Lu, H.; Koike, T.; Fujii, H.; Ohta, T.; Tamagawa, K. Development of a Physically-based Soil Moisture Retrieval Algorithm for Spaceborne Passive Microwave Radiometers and its Application to AMSR-E. *J. Remote Sens. Soc. Jpn.* **2009**, *29*, 253–262.
11. Zeng, J.; Zhen, L.; Quan, C.; Bi, H.; Qiu, J.; Zou, P. Evaluation of remotely sensed and reanalysis soil moisture products over the Tibetan Plateau using in-situ observations. *Remote Sens. Environ.* **2015**, *163*, 91–110. [CrossRef]

12. Chen, Y.; Yang, K.; Qin, J.; Zhao, L.; Tang, W.; Han, M. Evaluation of AMSR-E retrievals and GLDAS simulations against observations of a soil moisture network on the central Tibetan Plateau. *J. Geophys. Res. Atmos.* **2013**, *118*, 4466–4475. [CrossRef]
13. Zeng, J.; Zhen, L.I.; Chen, Q.; Haiyun, B.I. A simplified physically-based algorithm for surface soil moisture retrieval using AMSR-E data. *Front. Earth Sci.* **2014**, *8*, 427–438. [CrossRef]
14. Lacava, T.; Brocca, L.; Faruolo, M.; Matgen, P.; Moramarco, T.; Pergola, N.; Tramutoli, V. A multi-sensor (SMOS, AMSR-E and ASCAT) satellite-based soil moisture products inter-comparison. In Proceedings of the 2012 IEEE International Geoscience and Remote Sensing Symposium, Munich, Germany, 22–27 July 2012.
15. Kang, C.S.; Kanniah, K.D. Validation of AMSR-E soil moisture product and the future perspective of soil moisture estimation using SMOS data over tropical region. In Proceedings of the 2013 IEEE International Geoscience and Remote Sensing Symposium—IGARSS, Melbourne, Australia, 21–26 July 2013.
16. Yang, K.; Qin, J.; Zhao, L.; Chen, Y.; Han, M. A Multi-Scale Soil Moisture and Freeze-Thaw Monitoring Network on the Third Pole. *Bull. Am. Meteorol. Soc.* **2013**, *94*, 1907–1916. [CrossRef]
17. Njoku, E.G.; Chan, S.K. Vegetation and surface roughness effects on AMSR-E land observations. *Remote Sens. Environ.* **2006**, *100*, 190–199. [CrossRef]
18. Chen, Y.; Yang, K.; Qin, J.; Cui, Q.; Lu, H.; La, Z.; Han, M.; Tang, W. Evaluation of SMAP, SMOS, and AMSR2 soil moisture retrievals against observations from two networks on the Tibetan Plateau. *J. Geophys. Res. Atmos.* **2017**, *122*, 5780–5792. [CrossRef]
19. Lu, H.; Koike, T.; Yang, K.; Hu, Z.; Xu, X.; Rasmy, M.; Kuria, D.; Tamagawa, K. Improving land surface soil moisture and energy flux simulations over the Tibetan plateau by the assimilation of the microwave remote sensing data and the GCM output into a land surface model. *Int. J. Appl. Earth Obs. Geoinf.* **2012**, *17*, 43–54. [CrossRef]
20. Liu, Q.; Shi, J.; Du, J.; Zhang, S. Soil moisture retrieval by remote sensing and multi-year trend analysis of the soil moisture in Tibetan Plateau. In Proceedings of the 2012 IEEE International Geoscience and Remote Sensing Symposium, Munich, Germany, 22–27 July 2012.
21. Zhao, T.J.; Zhang, L.X.; Shi, J.C.; Jiang, L.M. A physically based statistical methodology for surface soil moisture retrieval in the Tibet Plateau using microwave vegetation indices. *J. Geophys. Res.* **2011**, *116*, 5229. [CrossRef]
22. Zhao, L.; Yang, K.; Qin, J.; Chen, Y.; Tang, W.; Lu, H.; Yang, Z. The scale-dependence of SMOS soil moisture accuracy and its improvement through land data assimilation in the central Tibetan Plateau. *Remote Sens. Environ.* **2014**, *152*, 345–355. [CrossRef]
23. Yang, K.; Zhu, L.; Chen, Y.; Zhao, L.; Qin, J.; Lu, H.; Tang, W.; Han, M.; Ding, B.; Fang, N. Land surface model calibration through microwave data assimilation for improving soil moisture simulations. *J. Hydrol.* **2016**, *533*, 266–276. [CrossRef]
24. Wang, L.; Li, Z.; Ren, X. The effects of vegetation in soil moisture retrieval using microwave radiometer data. In Proceedings of the IGARSS 2004. 2004 IEEE International Geoscience and Remote Sensing Symposium, Anchorage, AK, USA, 20–24 September 2004.
25. Parkinson, C.L. Aqua: an Earth-Observing Satellite mission to examine water and other climate variables. *IEEE Trans. Geosci. Remote Sens.* **2003**, *41*, 173–183. [CrossRef]
26. Du, J.; Kimball, J.S.; Jones, L.A.; Kim, Y.; Glassy, J.; Watts, J.D. A global satellite environmental data record derived from AMSR-Eand AMSR2 microwave earth observations. *Earth Syst. Sci. Data Discuss.* **2017**, *9*, 791–808. [CrossRef]
27. Kolassa, J.; Gentine, P.; Prigent, C.; Aires, F.; Alemohammad, S.H. Soil moisture retrieval from AMSR-E and ASCAT microwave observation synergy. Part 2: Product evaluation. *Remote Sens. Environ.* **2017**, *195*, 202–217. [CrossRef]
28. Njoku, E.; Chan, T.; Crosson, W.; Limaye, A. Evaluation of the AMSR-E data calibration over land. *Holography* **2004**, *11*, 1–28.
29. Njoku, E.G.; Ashcroft, P.; Chan, T.K.; Li, L. Global survey and statistics of radio-frequency interference in AMSR-E land observations. *IEEE Trans. Geosci. Remote Sens.* **2005**, *43*, 938–947. [CrossRef]
30. Mo, T.; Choudhury, B.J.; Schmugge, T.J.; Wang, J.R.; Jackson, T.J. A model for microwave emission from vegetation-covered fields. *J. Geophys. Res. Ocean.* **1982**, *87*, 11229–11237. [CrossRef]

31. Lu, H.; Koike, T.; Fujii, H.; Ohta, T.; Tamagawa, K. Monitoring soil moisture change in North Africa with using satellite remote sensing and land data assimilaiton system. In Proceedings of the 2009 IEEE International Geoscience and Remote Sensing Symposium, Cape Town, South Africa, 12–17 July 2009.
32. Li, L.; Njoku, E.G.; Im, E.; Chang, P.S.; Germain, K.S. A preliminary survey of radio-frequency interference over the U.S. in Aqua AMSR-E data. *IEEE Trans. Geosci. Remote Sens.* **2004**, *42*, 380–390. [CrossRef]
33. Mladenova, I.; Lakshmi, V.; Jackson, T.J.; Walker, J.P.; Merlin, O.; de Jeu, R.A.M. Validation of AMSR-E soil moisture using L-band airborne radiometer data from National Airborne Field Experiment 2006. *Remote Sens. Environ.* **2011**, *115*, 2096–2103. [CrossRef]
34. Xie, Q.; Meng, Q.; Zhang, L.; Wang, C.; Sun, Y.; Sun, Z. A Soil Moisture Retrieval Method Based on Typical Polarization Decomposition Techniques for a Maize Field from Full-Polarization Radarsat-2 Data. *Remote Sens.-Basel.* **2017**, *9*, 168. [CrossRef]
35. Qi, Y.; Lu, L.; Jiang, L.; Tao, J.; Du, J.; Shi, J. *Tibetan Plateau Soil Moisture Products Intercomparison and the Field Observations*; American Geophysical Union Fall Meeting: Washington, DC, USA, 2002.
36. Xi, J.; Wen, J.; Tian, H.; Zhang, T. Applicability evaluation of AMSR-E remote sensing soil moisture products in Qinghai-Tibet plateau. *Trans. Chin. Soc. Agric. Eng.* **2014**, *30*, 194–202.
37. Zeng, J.; Zhen, L.; Quan, C.; Bi, H.; Ping, Z. A physically-based algorithm for surface soil moisture retrieval in the Tibet Plateau using passive microwave remote sensing. In Proceedings of the 2013 IEEE International Geoscience and Remote Sensing Symposium—IGARSS, Melbourne, Australia, 21–26 July 2013.

MDPI
St. Alban-Anlage 66
4052 Basel
Switzerland
Tel. +41 61 683 77 34
Fax +41 61 302 89 18
www.mdpi.com

Remote Sensing Editorial Office
E-mail: remotesensing@mdpi.com
www.mdpi.com/journal/remotesensing

www.ingramcontent.com/pod-product-compliance
Lightning Source LLC
Chambersburg PA
CBHW061543220926
51850CB00005B/238
* 9 7 8 3 0 3 6 5 2 9 5 6 1 *